W9-BSZ-093

MERCRUISER

Stern Drives
1964-91 REPAIR MANUAL
TYPE 1, MR, ALPHA AND BRAVO I&II

SELOC ®

Managing Partners	Dean F. Morgantini
	Barry L. Beck
Executive Editor	Kevin M. G. Maher, A.S.E.
Production Managers	Melinda Possinger
	Ronald Webb

Manufactured in USA
© 1995 Seloc Publishing, Inc.
104 Willowbrook Lane
West Chester, PA 19382
ISBN 13: 978-0-89330-005-0
ISBN 10: 0-89330-005-5
23rd Printing 1098765432

www.selocmarine.com

MARINE
TECHNICIAN
TRAINING

INDUSTRY SUPPORTED PROGRAMS
OUTBOARD, STERNDRIVE & PERSONAL WATERCRAFT

- *Dyno Testing • Boat & Trailer Rigging • Electrical & Fuel System Diagnostics*
- *Powerhead, Lower Unit & Drive Rebuilds • Powertrim & Tilt Rebuilds*
- *Instrument & Accessories Installation*

TRAIN IN SUNNY FLORIDA!

For information regarding housing, financial aid and employment opportunities in the marine industry, contact us today:

CALL TOLL FREE
1-800-528-7995

An Accredited Institution

SM

Name
Address
City   State   Zip
Phone

MARINE MECHANICS INSTITUTE
A Division of CTI

MEMBER
NMMA

9751 Delegates Drive • Orlando, Florida 32837
2844 W. Deer Valley Rd. • Phoenix, AZ 85027

FINANCIAL ASSISTANCE AVAILABLE FOR THOSE WHO QUALIFY!

ACKNOWLEDGMENTS

Nichols Publishing expresses sincere appreciation to Mercury Marine for their assistance in the production of this manual.

Nichols Publishing would like to express thanks to all of the fine companies who participate in the production of our books:
- Hand tools supplied by Craftsman are used during all phases of our vehicle teardown and photography.
- Many of the fine specialty tools used in our procedures were provided courtesy of Lisle Corporation.
- Lincoln Automotive Products (1 Lincoln Way, St. Louis, MO 63120) has provided their industrial shop equipment, including jacks (engine, transmission and floor), engine stands, fluid and lubrication tools, as well as shop presses.
- Rotary Lifts (1-800-640-5438 or www.Rotary-Lift.com), the largest automobile lift manufacturer in the world, offering the biggest variety of surface and in-ground lifts available, has fulfilled our shop's lift needs.
- Much of our shop's electronic testing equipment was supplied by Universal Enterprises Inc. (UEI).
- Safety-Kleen Systems Inc. has provided parts cleaning stations and assistance with environmentally sound disposal of residual wastes.
- United Gilsonite Laboratories (UGL), manufacturer of Drylock® concrete floor paint, has provided materials and expertise for the coating and protection of our shop floor.

FOREWORD

This is a comprehensive tune-up and repair manual for Mercruiser stern drives manufactured between 1964 and 1992. Competition, high-performance, and commercial units (including aftermarket equipment), are not covered. The book has been designed and written for the professional mechanic, the do-it-yourselfer, and the student developing his mechanical skills.

Professional Mechanics will find it to be an additional tool for use in their daily work on outboard units because of the many special techniques described.

Boating enthusiasts interested in performing their own work and in keeping their unit operating in the most efficient manner will find the step-by-step illustrated procedures used throughout the manual extremely valuable. In fact, many have said this book almost equals an experienced mechanic looking over their shoulder giving them advice.

Students and Instructors have found the chapters divided into practical areas of interest and work. Technical trade schools, from Florida to Michigan and west to California, as well as the U.S. Navy and Coast Guard, have adopted these manuals as a standard classroom text.

Troubleshooting sections have been included in many chapters to assist the individual performing the work in quickly and accurately isolating problems to a specific area without unnecessary expense and time-consuming work. As an added aid and one of the unique features of this book, many worn parts are illustrated to identify and clarify when an item should be replaced.

Illustrations and procedural steps are so closely related and identified with matching numbers that, in most cases, captions are not used. Exploded drawings show internal parts and their interrelationship with the major component.

TABLE OF CONTENTS

5 IGNITION

6 ELECTRICAL

7 REMOTE CONTROLS

10 STERN DRIVE

NOTE

PAGES 10-2 THRU 10-58 COVER
  ALL UNITS, EXCEPT BRAVO
PAGES 10-58 THRU 10-110 COVER
  BRAVO STERN DRIVE

APPENDIX (CONTINUED)

1
SAFETY

1-1 INTRODUCTION

Your boat probably represents a sizeable investment for you. In order to protect this investment and to receive the maximum amount of enjoyment from your boat it must be cared for properly while being used and when it is out of the water. Always store your boat with the bow higher than the stern and be sure to remove the transom drain plug and the inner hull drain plugs. If you use any type of cover to protect your boat, plastic, canvas, whatever, be sure to allow for some movement of air through the hull. Proper ventilation will assure evaporation of any condensation that may form due to changes in temperature and humidity.

Mercruiser stern drive unit tilted and secured in place with a safety bracket. The unit is now ready to be safely trailered to the water and a day of fun for its owner.

1-2 CLEANING, WAXING, AND POLISHING

Any boat should be washed with clear water after each use to remove surface dirt and any salt deposits from use in salt water. Regular rinsing will extend the time between waxing and polishing. It will also give you "pride of ownership", by having a sharp looking piece of equipment. Elbow grease, a mild detergent, and a brush will be required to remove stubborn dirt, oil, and other unsightly deposits.

Stay away from harsh abrasives or strong chemical cleaners. A white buffing compound can be used to restore the original gloss to a scratched, dull, or faded area. The finish of your boat should be thoroughly cleaned, buffed, and polished at least once each season. Take care when buffing or polishing with a marine cleaner not to overheat the surface you are working, because you will burn it.

1-3 CONTROLLING CORROSION

Since man first started out on the water, corrosion on his craft has been his enemy. The first form was merely rot in the wood and then it was rust, followed by other forms of destructive corrosion in the more modern materials. One defense against corrosion is to use similar metals throughout the boat. Even though this is difficult to do in designing a new boat, particularily the undersides, similar metals should be used whenever and wherever possible.

A second defense against corrosion is to insulate dissimilar metals. This can be done by using an exterior coating of Sea Skin or by insulating them with plastic or rubber gaskets.

Transom plate equipped with anodic bolt heads for corrosion prevention. These inexpensive heads perform the same function as zinc plates to protect more valuable parts.

Using Zinc

The proper amount of zinc attached to a boat is extremely important. The use of too much zinc can cause wood burning by placing the metals close together and they become "hot". On the other hand, using too small a zinc plate will cause more rapid deterioration of the the metal you are trying to protect. If in doubt, consider the fact that it is far better to replace the zincs than to replace planking or other expensive metal parts from having an excess of zinc.

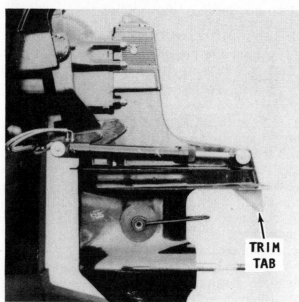

Profile view of a Mercruiser stern drive unit with the new trim tab properly installed.

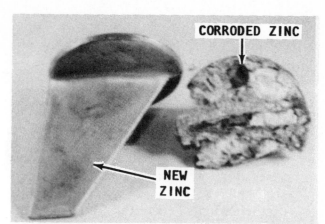

A new trim tab zinc, left, with a corroded zinc, right. An excellent example of the inexpensive zinc saving more costly parts of the engine.

When installing zinc plates, there are two routes available. One is to install many different zincs on all metal parts and thus run the risk of wood burning. Another route, is to use one large zinc on the transom of the boat and then connect this zinc to every underwater metal part through internal bonding. Mercruiser stern drive trim tabs are zinc. An additional zinc may be installed by using anodic bolt heads in the transom plate. With the additional zinc, the owner will feel more confident he is doing a little more to protect more expensive parts and his recreation investment.

1-4 PROPELLERS

As you know, the propeller is actually what moves the boat through the water. This is how it is done. The propeller operates in water in much the manner as a wood screw does in wood. The propeller "bites" into the water as it rotates. Water passes between the blades and out to the rear in the shape of a cone. The water in the shape of this cone pushing on the surrounding water is what propels the boat.

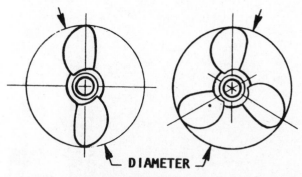

Diameter and pitch are the two basic dimensions of a propeller. The diameter is measured across the circumference of a circle scribed by the propeller blades, as shown.

Diameter and Pitch

Only two dimensions of the propeller are of real interest to the boat owner: the diameter and the pitch. These two dimensions are stamped on the propeller hub and always appear in the same order: the diameter first and then the pitch. For instance, the number 15-19 stamped on the hub, would mean the propeller had a diameter of 15 inches with a pitch of 19.

The diameter is the measured distance from the tip of one blade to the tip of the other as shown in the accompanying illustration.

The pitch of a propeller is the angle at which the blades are attached to the hub. This figure is expressed in inches of water travel for each revolution of the propeller. In our example of a 15-19 propeller, the propeller should travel 19 inches through the water each time it revolves. If the propeller action was perfect and there was no slippage, then the pitch multiplied by the propeller rpms would be the boat speed.

Most stern drive manufacturers equip their units with a standard propeller with a diameter and pitch they consider to be best suited to the engine and the boat. Such a propeller allows the engine to run as near to the rated rpm and horsepower (at full throttle) as possible for the boat design.

The blade area of the propeller determines its load-carrying capacity. A two-blade propeller is used for high-speed running under very light loads.

A four-blade propeller is installed in boats intended to operate at low speeds under very heavy loads such as tugs, barges, or large houseboats. The three-blade propeller is the happy medium covering the wide range between the high performance units and the load carrying workhorses.

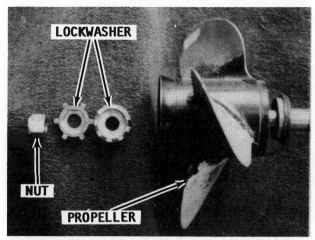

Standard attaching hardware for a propeller.

Propeller Selection

There is no standard propeller that will do the proper job in very many cases. The list of sizes and weights of boats is almost endless. This fact coupled with the many boat-engine combinations makes the propeller selection for a specific purpose a difficult job. In fact, in many cases the propeller is changed after a few test runs. Proper selection is aided through the use of charts set up for various engines and boats. These charts should be studied and understood when buying a propeller. However, bear in mind, the charts are based on average boats with average loads, therefore, it may be necessary to make a change in size or pitch, in order to obtain the desired results for the hull design or load condition.

Propellers are available with a wide range of pitch. Remember, a low pitch takes a smaller bite of the water than the high pitch propeller. This means the low pitch propeller will travel less distance through the water per revolution. The low pitch will require less horsepower and will allow the engine to run faster and more efficiently.

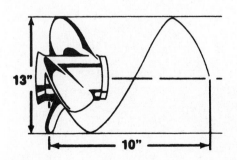
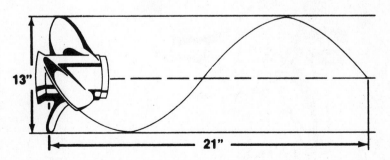

Diagram to explain the pitch dimension of a propeller. The pitch is the theoretical distance a propeller would travel through water if there were no friction.

It stands to reason, and it's true, that the high pitch propeller will require more horsepower, but will give faster boat speed if the engine is allowed to turn to its rated rpm.

If a higher-pitched propeller is installed on a boat, in an effort to get more speed, extra horsepower will be required. If the extra power is not available, the rpms will be reduced to a less efficient level and the actual boat speed will be less than if the lower-pitched propeller had been left installed.

All engine manufacturers design their units to operate with full throttle at, or slightly above, the rated rpm. If you run your engine at the rated rpm, you will increase spark plug life, receive better fuel economy, and obtain the best performance from your boat and engine. Therefore, take time to make the proper propeller selection for the rated rpm of your engine at full throttle with what you consider to be an average load. Your boat will then be correctly balanced between engine and propeller throughout the entire speed range.

A reliable tachometer must be used to measure engine speed at full throttle to ensure the engine will achieve full horsepower and operate efficiently and safely. To test for the correct propeller, make your run in a body of smooth water with the lower unit in forward gear at full throttle.

Observe the tachometer at full throttle. **NEVER** run the engine at a high rpm when a flush attachment is installed. If the reading is above the manufacturer's recommended operating range, you must try propellers of greater pitch, until you find the one that allows the engine to operate continually within the recommended full throttle range.

If the engine is unable to deliver top performance and you feel it is properly tuned, then the propeller may not be to blame. Operating conditions have a marked effect on performance. For instance, an engine will lose rpm when run in very cold water. It will also lose rpm when run in salt water as compared with fresh water. A hot, low-barometer day will also cause your engine to lose power.

Cavitation

Cavitation is the forming of voids in the water just ahead of the propeller blades. Marine propulsion designers are constantly fighting the battle against the formation of these voids due to excessive blade tip speed and engine wear. The voids may be filled with air or water vapor, or they may actually be a partial vacuum. Cavitation may be caused by installing a piece of equipment too close to the lower unit, such as the speedometer pickup, depth sounder, or bait tank pickup.

Vibration

Your propeller should be checked regularly to be sure all blades are in good

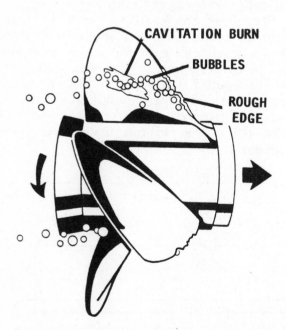

Cavitation air bubbles formed at the propeller. Manufacturers are constantly fighting this problem, as explained in the text.

Example of a damaged propeller. This unit should have been replaced long before this amount of damage was sustained.

condition. If any of the blades become bent or nicked, this condition will set up vibrations in the drive unit and the motor. If the vibration becomes very serious it will cause a loss of power, efficiency, and boat performance. If the vibration is allowed to continue over a period of time it can have a damaging effect on many of the operating parts.

Vibration in boats can never be completely eliminated, but it can be reduced by keeping all parts in good working condition and through proper maintenance and lubrication. Vibration can also be reduced in some cases by increasing the number of blades. For this reason, many racers use two-blade props and luxury cruisers have four- and five-blade props installed.

Shock Absorbers

The shock absorber in the propeller plays a very important role in protecting the shafting, gears, and engine against the shock of a blow, should the propeller strike an underwater object. The shock absorber allows the propeller to stop rotating at the instant of impact while the power train continues turning.

How much impact the propeller is able to withstand before causing the clutch hub to slip is calculated to be more than the force needed to propel the boat, but less than the amount that could damage any part of the power train. Under normal propulsion loads of moving the boat through the water, the hub will not slip. However, it will slip if the propeller strikes an object with a force that would be great enough to stop any part of the power train.

If the power train was to absorb an impact great enough to stop rotation, even for an instant, something would have to give and be damaged. If a propeller is subjected to repeated striking of underwater objects, it would eventually slip on its clutch hub under normal loads. If the propeller would start to slip, a new hub and shock absorber would have to be installed.

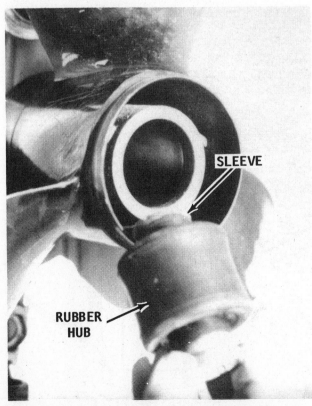

Rubber hub removed from a propeller. This hub was removed because the hub was slipping in the propeller.

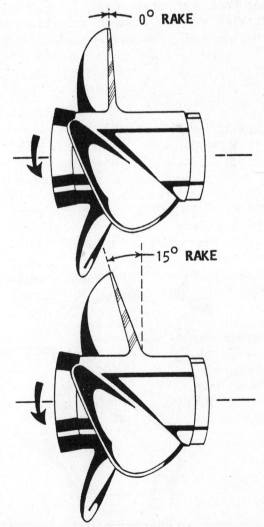

Illustration depicting the rake of a propeller, as explained in the text.

Propeller Rake

If a propeller blade is examined on a cut extending directly through the center of the hub, and if the blade is set vertical to the propeller hub, as shown in the accompanying illustration, the propeller is said to have a zero degree (0°) rake. As the blade slants back, the rake increases. Standard propellers have a rake angle from 0° to 15°.

A higher rake angle generally improves propeller performance in a cavitating or ventilating situation. On lighter, faster boats, higher rake often will increase performance by holding the bow of the boat higher.

Progressive Pitch

Progressive pitch is a blade design innovation that improves performance when forward and rotational speed is high and/or the propeller breaks the surface of the water.

Progressive pitch starts low at the leading edge and progressively increases to the trailing edge, as shown in the accompanying illustration. The average pitch over the entire blade is the number assigned to that propeller. In the illustration of the progressive pitch, the average pitch assigned to the propeller would be 21.

Cupping

If the propeller is cast with a edge curl inward on the trailing edge, the blade is said to have a cup. In most cases, cupped blades improve performance. The cup helps the blades to **"HOLD"** and not break loose, when operating in a cavitating or ventilating situation. This action permits the engine to be trimmed out further, or to be mounted higher on the transom. This is especially true on high-performance boats. Either of these

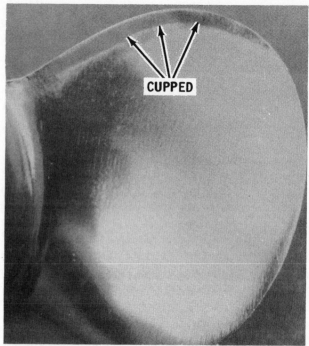

Propeller with a "cupped" leading edge. "Cupping" gives the propeller a better "hold" in the water.

two adjustments will usually add to higher speed.

The cup has the effect of adding to the blade pitch, as well as the rake. Cupping usually will reduce full-throttle engine speed about 150 to 300 rpm below the same pitch propeller without a cut to the blade. A propeller repair shop is able to increase or decrease the cup on the blades. This change, as explained, will alter engine rpm to meet specific operating demands. Cups are rapidly becoming standard on propellers.

In order for a cup to be the most effective, the cup should be completely concave (hollowed) and finished with a sharp corner. If the cup has any convex rounding, the effectiveness of the cup will be reduced.

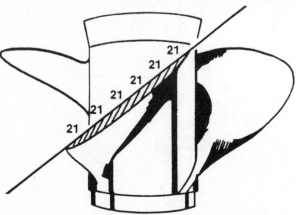

CONSTANT PITCH

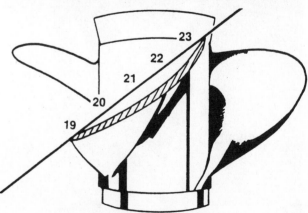

PROGRESSIVE PITCH

Comparison of a constant and progressive pitch propeller. Notice how the pitch of the progressive pitch propeller, right, changes to give the blade more thrust and therefore, the boat more speed.

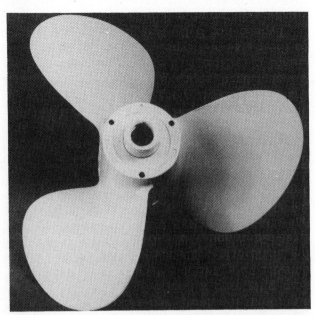

A rebuilt propeller ready for service.

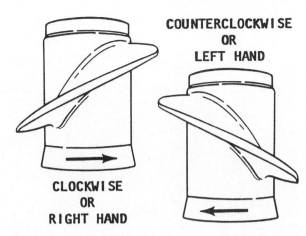

Right- and left-hand propellers showing how the angle of the blades is reversed. Right-hand propellers are by far the most popular.

Rotation

Propellers are manufactured as right-hand rotation (RH), and as left-hand rotation (LH). The standard propeller for outboards is RH rotation.

A right-hand propeller can easily be identified by observing it as shown in the accompanying illustration. Observe how the blade slants from the lower left toward the upper right. The left-hand propeller slants in the opposite direction, from upper left to lower right, as shown.

When the propeller is observed rotating from astern the boat, it will be rotating clockwise when the engine is in forward gear. The left-hand propeller will rotate counterclockwise.

Propeller Exhaust

To improve engine and boat performance, most Quicksilver propellers feature a hub design with a flared trailing edge or "Diffuser Ring". This feature assists exhaust gas flow, and provides a pressure barrier to help prevent exhaust gases from feeding back into the blades. This arrangement results in more quiet engine operation and the exhaust fumes are buried far behind the boat.

Mercruiser propeller exhaust hub. This arrangement of exhaust gases passing through the hub results in much quieter engine operation and the fumes are buried far behind the boat.

Taking on fuel at an automobile service station. Safety grounding is accomplished through the hose fixtures. The fuel tank should be kept full to prevent condensation.

A neglected boat and stern drive. Such corrosion and marine growth will be costly to the owner and greatly reduce his boating enjoyment through poor performance.

1-5 FUEL SYSTEM

With Built-in Fuel Tank

All parts of the fuel system should be selected and installed to provide maximum service and protection against leakage. Reinforced flexible sections should be installed in fuel lines where there is a lot of motion, such as at the engine connection. The flaring of copper tubing should be annealed after it is formed as a protection against hardening.

CAUTION: Compression fittings should **NOT** be used because they are so easily overtightened, which places them under a strain and subjects them to fatigue. Such conditions will cause the fitting to leak after it is connected a second time.

The capacity of the fuel filter must be large enough to handle the demands of the engine as specified by the engine manufacturer.

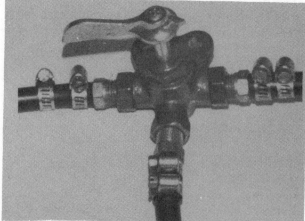

A three-position valve permits fuel to be drawn from either of two tanks or shut off completely. Such an arrangement prevents accidental siphoning of fuel from the tank.

All fittings and outlets must come out the top of the tank. An anti-siphon device should be installed close to the tank. This spring-loaded valve will automatically prevent fuel from being siphoned out of the tank if the line from the tank to the fuel pump should be damaged and begin to leak.

A manually-operated valve should be installed if anti-siphon protection is not provided. This valve should be installed in the fuel line as close to the gas tank as possible. Such a valve will maintain anti-siphon protection between the tank and the engine.

Fuel tanks should be mounted in dry, well ventilated places. Ideally, the fuel tanks should be installed above the cockpit floors, where any leakage will be quickly detected.

In order to obtain maximum circulation of air around fuel tanks, the tank should not come in contact with the boat hull except through the necessary supports. The supporting surfaces and hold-downs must fasten the tank firmly and they should be insulated from the tank surfaces. This insulation material should be non-abrasive and non-absorbent material. Fuel tanks installed in the forward portion of the boat should be

A one gallon minimum of fuel should be kept on board in an approved container.

especially well secured and protected because shock loads in this area can be as high as 20 to 25 g's ("g" equals force of gravity).

Engine Compartment Ventilation

All motorboats built after April 25, 1940 and before August 1, 1980, powered by a gasoline engine or by fuels having a flash-point of 110° F. or less **MUST** have the following, which is quoted from a Coast Guard publication dated 1984:

At least two ventilation ducts fitted with cowls or their equivalent for the purpose of properly and efficiently ventilating the bilges of every engine and fuel tank compartment. There shall be at least one exhaust duct installed so as to extend to the lower portion of the bilge and at least one intake duct installed so as to extend to a point at least midway to the bilge or at least below the level of the carburetor air intake.

All boats built after July 31, 1978 but prior to August 1, 1980, the requirement for ventilation of the fuel tank compartment can be omitted if there is no electrical source of ignition in the fuel tank compartment and if the fuel tank vents to the outside of the boat. After August 1, 1980, all boats with gasoline engines must be built with ventilation systems which comply with Coast Guard standards. The standard requires the engine compartment to be equipped with a blower. Now, if the blower and

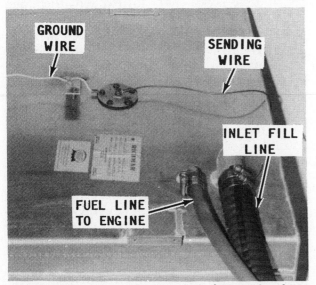

A three-position valve permits fuel to be drawn from either tank or to be shut off completely. Such an arrangement prevents accidental siphoning of fuel from the tank.

ventilation system do not meet Coast Guard regulations, the owner can recall the manufacturer to make it right. Owners can not recall the manufacturer for boats built prior to August 1, 1980. The operator is required to keep the system in proper operating condition.

In addition to the blower requirement for the engine compartment, a tag **MUST** be affixed to the control panel stating the blower must be operated for at least four minutes before an attempt is made to start the engine.

This tag requirement is made to the boat manufacturer and the boat owner is responsible for keeping it in place and in readable condition. The tag must be yellow and able to withstand salt spray.

Citation

An owner may receive a citation --ticket and fine -- if the ventilation system is not Coast Guard Approved, in good operating order, and the blower tag is not affixed to the control panel.

Flame Arrestors

A gasoline engine installed in a motorboat or motor vessel after April 25, 1940, except outboard motors, must have a Coast Guard Approved flame arrestor fitted to the carburetor. This requirement applies if the engine is enclosed or if the carburetor is below the gunwale of the boat.

Automotive Replacement Parts

When replacing fuel, electrical, and other parts, check to be sure they are marine type and Coast Guard Approved. Automotive parts are not made to the high standards of marine parts. The carburetors must **NOT** leak fuel; alternators, generators, and voltage regulators **MUST** be able to operate in a gas fume enclosed area without exploding; etc. Automotive parts could cause a fire endangering the crew and the craft. The part may look the same and even have a similar number, but if it is not **MARINE** it is not safe to use on the boat and will not pass Coast Guard inspection.

Taking On Fuel

The fuel tank of the boat should be kept full to prevent water from entering the system through condensation caused by temperature changes. Water droplets forming are one of the greatest enemies of the fuel system. By keeping the tank full, the air space in the tank is kept to an absolute minimum and there is no room for moisture to form. It is a good practice not to store fuel in the tank over an extended period, say for six months. Today, fuels contain ingredients that change into gums when stored for any length of time. These gums and varnish products will cause carburetor problems and poor spark plug performance. An additive (Sta-Bil) is available and can be used to prevent gums and varnish from forming.

Static Electricity

In very simple terms, static electricity is called frictional electricity. It is generated by two dissimilar materials moving over each other. One form is gasoline flowing through a pipe or into the air. Another form is when you brush your hair or walk across a synthetic carpet and then touch a metal object. All of these actions cause an electrical charge. In most cases, static electricity is generated during very dry weather conditions, but when you are filling the fuel tank on your boat it can happen at any time.

Fuel Tank Grounding

One area of protection against the build-up of static electricity is to have the fuel tank properly grounded (also known as bonding). A direct metal-to-metal contact from the fuel hose nozzle to the water in which the boat is floating. If the fill pipe is made of metal, and the fuel nozzle makes a good contact with the deck plate, then a good ground is made.

As an economy measure, some boats use rubber or plastic filler pipes because of compound bends in the pipe. Such a fill line does not give any kind of ground and if your boat has this type of installation and you do not want to replace the filler pipe with a metal one, then it is possible to connect the deck fitting to the tank with a copper wire. The wire should be 8 gauge or larger.

The fuel line from the tank to the engine should provide a continuous metal-to-metal contact for proper grounding. If any part of this line is plastic or other non-metallic material, then a copper wire must be connected to bridge the non-metal material. The power train provides a ground through the engine and drive shaft, to the propeller in the water.

Fiberglass fuel tanks pose problems of their own. One method of grounding is to run a copper wire around the tank from the fill pipe to the fuel line. However, such a wire does not ground the fuel in the tank. Manufacturers should imbed a wire in the fiberglass and it should be connected to the intake and the outlet fittings. This wire would avoid corrosion which could occur if a wire passed through the fuel. **CAUTION: It is not advisable to use a fiberglass fuel tank if a grounding wire was not installed.**

Anything you can feel as a "shock" is enough to set off an explosion. Did you know that under certain atmospheric conditions you can cause a static explosion yourself, particularly if you are wearing synthetic clothing. It is almost a certainty you could cause a static spark if you are **NOT** wearing insulated rubber-soled shoes.

As soon as the deck fitting is opened, fumes are released to the air. Therefore, to be safe you should ground yourself before opening the fill pipe deck fitting. One way to ground yourself is to dip your hand in the water overside to discharge the electricity in your body before opening the filler cap. Another method is to touch the engine block or any metal fitting on the dock which goes down into the water.

1-6 LOADING

In order to receive maximum enjoyment, with safety and performance, from your boat, take care not to exceed the load capacity given by the manufacturer. A plate attached to the hull indicates the U.S. Coast Guard capacity information in pounds for persons and gear. If the plate states the maximum person capacity to be 750 pounds and you assume each person to weigh an average of 150 lbs., then the boat could carry five people safely. If you add another 250 lbs. for motor and gear, and the maximum weight capacity for persons and gear is 1,000 lbs. or more, then the five persons and gear would be within the limit.

Try to load the boat evenly port and starboard. If you place more weight on one side than on the other, the boat will list to the heavy side and make steering difficult. You will also get better performance by placing heavy supplies aft of the center to keep the bow light for more efficient planning.

U.S. Coast Guard plate affixed to all new boats. When the blanks are filled in, the plate will indicate the Coast Guard's recommendations for persons, gear, and horsepower to ensure safe operation of the boat. These recommendations should not be exceeded, as explained in the text.

Clarification

Much confusion arises from the terms, certification, requirements, approval, regulations, etc. Perhaps the following may clarify a couple of these points.

1- The Coast Guard does not approve boats in the same manner as they "Approve" life jackets. The Coast Guard applies a formula to inform the public of what is safe for a particular craft.

2- If a boat has to meet a particular regulation, it must have a Coast Guard certification plate. The public has been led to believe this indicates approval of the Coast Guard. Not so.

3- The certification plate means a willingness of the manufacturer to meet the Coast Guard regulations for that particular craft. The manufacturer may recall a boat if it fails to meet the Coast Guard requirements.

4- The Coast Guard certification plate, see accompanying illustration, may or may not be metal. The plate is a regulation for the manufacturer. It is only a warning plate and the public does not have to adhere to the restrictions set forth on it. Again, the plate sets forth information as to the Coast Guard's opinion for safety on that particular boat.

5- Coast Guard Approved equipment is equipment which has been approved by the Commandant of the U.S. Coast Guard and has been determined to be in compliance with Coast Guard specifications and regulations relating to the materials, construction, and performance of such equipment.

1-7 HORSEPOWER

The maximum horsepower engine for each individual boat should not be increased by any great amount without checking requirements from the Coast Guard in your area. The Coast Guard determines horsepower requirements based on the length, beam, and depth of the hull. **TAKE CARE NOT** to exceed the maximum horsepower listed on the plate or the warranty and possibly the insurance on the boat may become void.

1-8 FLOTATION

If your boat is less than 20 ft. overall, a Coast Guard or BIA (Boating Industry of America) now changed to NMMA (National Marine Manufacturers Association) requirement is that the boat must have buoyant material built into the hull (usually foam) to keep it from sinking if it should become swamped. Coast Guard requirements are mandatory, but the NMMA is voluntary.

"Kept from sinking" is defined as the ability of the flotation material to keep the boat from sinking when filled with water and with passengers clinging to the hull. One restriction is that the total weight of the motor, passengers, and equipment aboard does not exceed the maximum load capacity listed on the plate.

Life Preservers —Personal Flotation Devices (PFDs)

The Coast Guard requires at least one Coast Guard approved life-saving device be carried on board all motorboats for each person on board. Devices approved are identified by a tag indicating Coast Guard approval. Such devices may be life preservers, buoyant vests, ring buoys, or buoyant cushions. Cushions used for seating are serviceable if air cannot be squeezed out of it. Once air is released when the cushion is squeezed, it is no longer fit as a flotation device. New foam cushions dipped in a rubberized material are almost indestructible.

Life preservers have been classified by the Coast Guard into five type categories. All PFDs presently acceptable on recreational

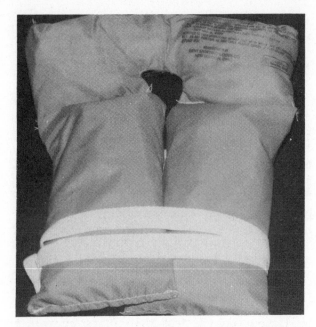

*Type I PFD Coast Guard approved life jacket. This type flotation device provides the greatest amount of buoyancy. **NEVER** use them for cushions or other purposes.*

boats fall into one of these five designations. All PFDs **MUST** be U.S. Coast Guard approved, in good and serviceable condition, and of an appropriate size for the persons who intend to wear them. Wearable PFDs **MUST** be readily accessible and throwable devices **MUST** be immediately available for use.

Type I PFD has the greatest required buoyancy and is designed to turn most **UNCONSCIOUS** persons in the water from a

A Type IV PFD cushion device intended to be thrown to a person in the water. If air can be squeezed out of the cushion, it is no longer fit for service as a PFD.

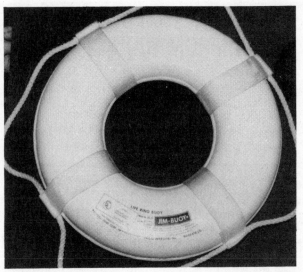

Type IV ring buoy also designed to be thrown to a person in the water. On ocean cruisers, this type device usually has a weighted pole with flag and light attached to the buoy.

face down position to a vertical or slightly backward position. The adult size device provides a minimum buoyancy of 22 pounds and the child size provides a minimum buoyancy of 11 pounds. The Type I PFD provides the greatest protection to its wearer and is most effective for all waters and conditions.

Type II PFD is designed to turn its wearer in a vertical or slightly backward position in the water. The turning action is not as pronounced as with a Type I. The device will not turn as many different type persons under the same conditions as the Type I. An adult size device provides a minimum buoyancy of $15\frac{1}{2}$ pounds, the medium child size provides a minimum of 11 pounds, and the infant and small child sizes provide a minimum buoyancy of 7 pounds.

Type III PFD is designed to permit the wearer to place himself (herself) in a vertical or slightly backward position. The Type III device has the same buoyancy as the Type II PFD but it has little or no turning ability. Many of the Type III PFD are designed to be particularly useful when water skiing, sailing, hunting, fishing, or engaging in other water sports. Several of this type will also provide increased hypothermia protection.

Type IV PFD is designed to be thrown to a person in the water and grasped and held by the user until rescued. It is **NOT** designed to be worn. The most common Type IV PFD is a ring buoy or a buoyant cushion.

Type V PFD is any PFD approved for restricted use.

Coast Guard regulations state, in general terms, that on all boats less than 16 ft. overall, one Type I, II, III, or IV device shall be carried on board for each person in the boat. On boats over 26 ft., one Type I, II, or III device shall be carried on board for each person in the boat **plus** one Type IV device.

It is an accepted fact that most boating people own life preservers, but too few actually wear them. There is little or no excuse for not wearing one because the modern comfortable designs available today do not subtract from an individual's boating pleasure. Make a life jacket available to your crew and advise each member to wear it. If you are a crew member ask your skipper to issue you one, especially when boating in rough weather, cold water, or when running at high speed. Naturally, a life jacket should be a must for non-swimmers any time they are out on the water in a boat.

1-9 EMERGENCY EQUIPMENT

Visual Distress Signals
The Regulation

Since January 1, 1981, Coast Guard Regulations require all recreation boats when used on coastal waters, which includes the Great Lakes, the territorial seas and those waters directly connected to the Great Lakes and the territorial seas, up to a point where the waters are less than two miles wide, and boats owned in the United States when operating on the high seas to be equipped with visual distress signals.

The only exceptions are during daytime (sunrise to sunset) for:

Recreational boats less than 16 ft. (5 meters) in length.

Boats participating in organized events such as races, regattas or marine parades.

Open sailboats not equipped with propulsion machinery and less than 26 ft. (8 meters) in length.

Manually propelled boats.

The above listed boats need to carry night signals when used on these waters at night.

Pyrotechnic visual distress signaling devices **MUST** be Coast Guard Approved, in serviceable condition and stowed to be read-

A sounding device should be mounted close to the helmsman for use in sounding an emergency alarm.

ily accessible. If they are marked with a date showing the serviceable life, this date must not have passed. Launchers, produced before Jan. 1, 1981, intended for use with approved signals are not required to be Coast Guard Approved.

USCG Approved pyrotechnic visual distress signals and associated devices include:

Pyrotechnic red flares, hand held or aerial.

Pyrotechnic orange smoke, hand held or floating.

Launchers for aerial red meteors or parachute flares.

Moisture protected flares should be carried on board for use as a distress signal in an emergency.

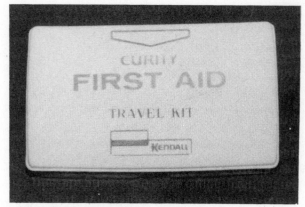

An adequately stocked first aid kit should be on board for the safety of crew and guests.

None-pyrotechnic visual distress signaling devices must carry the manufacturer's certification that they meet Coast Guard requirements. They must be in serviceable condition and stowed so as to be readily accessible.

This group includes:

Orange distress flag at least 3 x 3 feet with a black square and ball on an orange background.

Electric distress light -- not a flashlight but an approved electric distress light which **MUST** automatically flash the international **SOS** distress signal (. . . - - - . . .) four to six times each minute.

Types and Quantities

The following variety and combination of devices may be carried in order to meet the requirements.

1- Three hand-held red flares (day and night).

2- One electric distress light (night only).

3- One hand-held red flare and two parachute flares (day and night).

4- One hand-held orange smoke signal, two floating orange smoke signals (day) and one electric distress light (day and night).

If young children are frequently aboard your boat, careful selection and proper stowage of visual distress signals becomes especially important. If you elect to carry pyrotechnic devices, you should select those in tough packaging and not easy to ignite should the devices fall into the hands of children.

Coast Guard Approved pyrotechnic devices carry an expiration date. This date can **NOT** exceed 42 months from the date of manufacture and at such time the device can no longer be counted toward the minimum requirements.

SPECIAL WORDS

In some states the launchers for meteors and parachute flares may be considered a firearm. Therefore, check with your state authorities before acquiring such a launcher.

First Aid Kits

The first-aid kit is similar to an insurance policy or life jacket. You hope you don't have to use it but if needed, you want it there. It is only natural to overlook this essential item because, let's face it, who likes to think of unpleasantness when planning to have only a good time. However, the prudent skipper is prepared ahead of time, and is thus able to handle the emergency without a lot of fuss.

Good commercial first-aid kits are available such as the Johnson and Johnson "Marine First-Aid Kit." With a very modest expenditure, a well-stocked and adequate kit can be prepared at home.

Any kit should include instruments, supplies, and a set of instructions for their use. Instruments should be protected in a watertight case and should include: scissors, tweezers, tourniquet, thermometer, safety pins, eye-washing cup, and a hot water bottle.

The supplies in the kit should include: assorted bandages in addition to the various sizes of "band-aids", adhesive tape, absorbent cotton, applicators, petroleum jelly, antiseptic (liquid and ointment), local ointment, aspirin, eye ointment, antihistamine, ammonia inhalent, sea-sickness pills, antacid pills, and a laxative. You may want to consult your family physician about including antibiotics. Be sure your kit contains a first-aid manual because even though you have taken the Red Cross course, you may be the patient and have to rely on an untrained crew for care.

Fire Extinguishers

All fire extinguishers must bear Underwriters Laboratory (UL) "Marine Type" approved labels. With the UL certification, the extinguisher does not have to have a Coast Guard approval number. The Coast Guard classifies fire extinguishers according to their size and type.

Type B-I or B-II Designed for extinguishing flammable liquids. Required on all motorboats.

The Coast Guard considers a boat having one or more of the following conditions as a "boat of closed construction" subject to fire extinguisher regulations.

1- Inboard engine or engines.

2- Closed compartments under thwarts and seats wherein portable fuel tanks may be stored.

3- Double bottoms not sealed to the hull or which are not completely filled with flotation materials.

4- Closed living spaces.

5- Closed stowage compartments in which combustible or flammable material is stored.

6- Permanently installed fuel tanks.

Detailed classification of fire extinguishers is by agent and size:

B-I contains 1-1/4 gallons foam, 4 pounds carbon dioxide, 2 pounds dry chemical, and 2-1/2 pounds freon.

B-II contains 2-1/2 gallons foam, 15 pounds carbon dioxide, and 10 pounds dry chemical.

The class of motorboat dictates how many fire extinguishers are required on board. One B-II unit can be substituted for two B-I extinguishers. When the engine compartment of a motorboat is equipped with a fixed (built-in) extinguishing system, one less portable B-I unit is required.

Dry chemical fire extinguishers without gauges or indicating devices must be weighed and tagged every 6 months. If the gross weight of a carbon dioxide (CO_2) fire extinguisher is reduced by more than 10% of the net weight, the extinguisher is not acceptable and must be recharged.

READ labels on fire extinguishers. If the extinguisher is U.L. listed, it is approved for marine use.

DOUBLE the number of fire extinguishers recommended by the Coast Guard, because their requirements are a bare MINIMUM for safe operation. Your boat, family, and crew, must certainly be worth much more than "bare minimum".

1-10 COMPASS

Selection

The safety of the boat and her crew may depend on her compass. In many areas weather conditions can change so rapidly that within minutes a skipper may find himself "socked-in" by a fog bank, a rain squall,

A suitable fire extinguisher should be mounted close to the helmsman for emergency use.

or just poor visibilty. Under these conditions, he may have no other means of keeping to his desired course except with the compass. When crossing an open body of water, his compass may be the only means of making an accurate landfall.

During thick weather when you can neither see nor hear the expected aids to navigation, attempting to run out the time on a given course can disrupt the pleasure of the cruise. The skipper gains little comfort in a chain of soundings that does not match those given on the chart for the expected area. Any stranding, even for a short time, can be an unnerving experience.

Do not hesitate to spend a few extra dollars for a good reliable compass. If in doubt, seek advice from fellow boaters.

A pilot will not knowingly accept a cheap parachute, a good boater should not accept a bargain in lifejackets, fire extinguishers, or compass. Take the time and spend the few extra dollars to purchase a compass to fit your expected needs. Regardless of what the salesman may tell you, postpone buying until you have had the chance to check more than one make and model.

Lift each compass, tilt and turn it, simulating expected motions of the boat. The compass card should have a smooth and stable reaction.

The card of a good quality compass will come to rest without oscillations about the lubber's line. Reasonable movement in your hand, comparable to the rolling and pitching of the boat, should not materially affect the reading.

Installation

Proper installation of the compass does not happen by accident. Make a critical check of the proposed location to be sure compass placement will permit the helmsman to use it with comfort and accuracy.

First, the compass should be placed directly in front of the helmsman and in such a position that it can be viewed without body stress as he sits or stands in a posture of relaxed alertness. The compass should be in the helmsman's zone of comfort. If the compass is too far away, he may have to

The compass is a delicate instrument and deserves respect. It should be mounted securely and in position where it can be easily observed by the helmsman.

bend forward to watch it; too close and he must rear backward for relief.

Secondly, give some thought to comfort in heavy weather and poor visibilty conditions during the day and night. In some cases, the compass position may be partially determined by the location of the wheel, shift lever, and throttle handle.

Thirdly, inspect the compass site to be sure the instrument will be at least two feet from any engine indicators, bilge vapor detectors, magnetic instruments, or any steel or iron objects. If the compass cannot be placed at least two feet (six feet would be better) from one of these influences, then either the compass or the other object must be moved, if first order accuracy is to be expected.

Once the compass location appears to be satisfactory, give the compass a test before installation. Hidden influences may be concealed under the cabin top, forward of the cabin aft bulkhead, within the cockpit ceiling, or in a wood-covered stanchion.

Move the compass around in the area of the proposed location. Keep an eye on the card. A magnetic influence is the only thing that will make the card turn. You can quickly find any such influence with the compass. If the influence can not be moved away or replaced by one of non-magnetic material, test to determine whether it is merely magnetic, a small piece of iron or steel, or some magnetized steel. Bring the north pole of the compass near the object, then shift and bring the south pole near it. Both the north and south poles will be attracted if the compass is demagnetized. If the object attracts one pole and repels the other, then the compass is magnetized. If your compass needs to be demagnetized, take it to a shop equipped to do the job **PROPERLY.**

After you have moved the compass around in the proposed mounting area, hold it down or tape it in position. Test everything you feel might affect the compass and cause a deviation from a true reading. Rotate the wheel from hard over to hard over. Switch on and off all the lights, radios, radio direction finder, radio telephone, depth finder and the shipboard intercom, if one is installed. Sound the electric whistle, turn on the windshield wipers, start the engine (with water circulating through the engine), work the throttle, and move the gear shift lever.

If the boat has an auxiliary generator, start it.

If the card moves during any one of these tests, the compass should be relocated. Naturally, if something like the windshield wipers cause a slight deviation, it may be necessary for you to make a different deviation table to use only when certain pieces of equipment is operating. Bear in mind, following a course that is only off a degree or two for several hours can make considerable difference at the end, putting you on a reef, rock, or shoal.

Check to be sure the intended compass site is solid. Vibration will increase pivot wear.

Now, you are ready to mount the compass. To prevent an error on all courses, the line through the lubber line and the compass card pivot must be exactly parallel to the keel of the boat. You can establish the fore-and-aft line of the boat with a stout cord or string. Use care to transfer this line to the compass site. If necessary, shim the base of the compass until the stile-type lubber line (the one affixed to the case and not gimbaled) is vertical when the boat is on an even keel. Drill the holes and mount the compass.

Magnetic Items After Installation

Many times an owner will install an expensive stereo system in the cabin of his boat. It is not uncommon for the speakers to be mounted on the aft bulkhead up against the overhead (ceiling). In almost every case, this position places one of the speakers in very close proximity to the compass, mounted above the ceiling.

As we all know, a magnet is used in the operation of the speaker. Therefore, it is very likely that the speaker, mounted almost under the compass in the cabin will have a very pronounced affect on the compass accuracy.

Consider the following test and the accompanying photographs as prove of the statements made.

First, the compass was read as 190 degrees while the boat was secure in her slip.

Next a full can of diet Coke in an **aluminum** can was placed on one side and the compass read as 204 degrees, a good 14 degrees off.

Next, the full can was moved to the opposite side of the compass and again a

reading was observed. This time as 189 degrees, 11 degrees off from the original reading. (Can was probably not exactly opposite from the previous reading.)

Finally, the contents of the can were consumed, the can placed on both sides of the compass with **NO** affect on the compass reading.

Two very important conclusions can be drawn from these tests.

1- Something must have been in the contents of the can to affect the compass so drastically.

"Innocent" objects close to the compass, such as diet coke in an alluminum can, may cause serious problems and lead to disaster, as these three photos and the accompanying text prove.

2- Keep even "innocent" things clear of the compass to avoid any possible error in the boat's heading.

REMEMBER, a boat moving through the water at 10 knots on a compass error of just 5 degrees will be almost 1.5 miles off course in only **ONE** hour. At night, or in thick weather, this could very possibly put the boat on a reef, rock, or shoal, with disastrous results.

1-11   STEERING

USCG or BIA certification of a steering system means that all materials, equipment, and installation of the steering parts meet or exceed specific standards for strength, type, and maneuverability. Avoid sharp bends when routing the cable. Check to be sure the pulleys turn freely and all fittings are secure. See Chapter 7 for details.

1-12   ANCHORS

One of the most important pieces of equipment in the boat next to the power plant is the ground tackle carried. The engine makes the boat go and the anchor and its line are what hold it in place when the boat is not secured to a dock or on the beach.

The anchor must be of suitable size, type, and weight to give the skipper peace

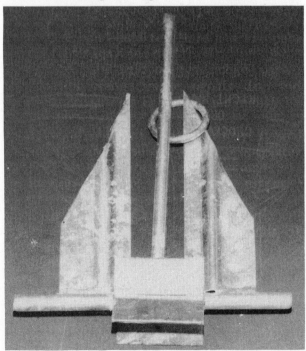

*The weight of the anchor **MUST** be adequate to secure the boat without dragging.*

of mind when his boat is at anchor. Under certain conditions, a second, smaller, lighter anchor may help to keep the boat in a favorable position during a non-emergency daytime situation.

In order for the anchor to hold properly, a piece of chain must be attached to the anchor and then the nylon anchor line attached to the chain. The amount of chain should equal or exceed the length of the boat. Such a piece of chain will ensure that the anchor stock will lay in an approximate horizontal position and permit the flutes to dig into the bottom and hold.

1-13 MISCELLANEOUS EQUIPMENT

In addition to the equipment you are legally required to carry in the boat and those previously mentioned, some extra items will add to your boating pleasure and safety. Practical suggestions would include: a bailing device (bucket, pump, etc.), boat hook, fenders, spare propeller, spare engine parts, tools, an auxiliary means of propulsion (paddle or oars), spare can of gasoline, flashlight, and extra warm clothing. The area of your boating activity, weather conditions, length of stay aboard your boat, and the specific purpose will all contribute to the kind and amount of stores you put aboard. When it comes to personal gear, heed the advice of veteran boaters who say, "Decide on how little you think you can get by with, then cut it in half."

Bilge Pumps

Automatic bilge pumps should be equipped with an overriding manual switch. They should also have an indicator in the operator's position to advise the helmsman when the pump is operating. Select a pump that will stabilize its temperature within the manufacturer's specified limits when it is operated continuously. The pump motor should be a sealed or arcless type, suitable for a marine atmosphere. Place the bilge pump inlets so excess bilge water can be removed at all normal boat trims. The intakes should be properly screened to prevent the pump from sucking up debris from the bilge. Intake tubing should be of a high quality and stiff enough to resist kinking and not collapse under maximum pump suction condition if the intake becomes blocked.

To test operation of the bilge pump, operate the pump switch. If the motor does not run, disconnect the leads to the motor. Connect a voltmeter to the leads and see if voltage is indicated. If voltage is not indicated, then the problem must be in a blown fuse, defective switch, or some other area of the electrical system.

If the meter indicates voltage is present at the leads, then remove, disassemble, and inspect the bilge pump. Clean it, reassemble, connect the leads, and operate the switch again. If the motor still fails to run, the pump must be replaced.

To test the bilge pump switch, first disconnect the leads from the pump and connect them to a test light or ohmmeter. Next, hold the switch firmly against the mounting location in order to make a good ground. Now, tilt the opposite end of the switch upward until it is activated as indicated by the test light coming on or the ohmmeter showing continuity. Finally, lower the switch slowly toward the mounting position until it is deactivated. Measure the distance between the point the switch was activated and the point it was deactivated. For proper service, the switch should deact-

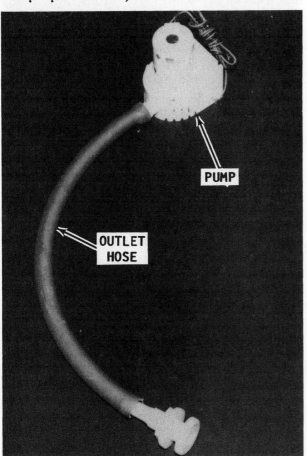

The bilge pump line must be cleaned frequently to ensure the entire bilge pump system will be able to perform properly in an emergency.

ivate between 1/2-inch and 1/4-inch from the planned mounting position. **CAUTION: The switch must never be mounted lower than the bilge pump pickup.**

1-14 BOATING ACCIDENT REPORTS

New federal and state regulations require an accident report to be filed with the nearest state boating authority within 48 hours if a person is lost, disappears, or is injured to the degree of needing medical treatment beyond first aid.

Accidents involving only property or equipment damage **MUST** be reported within 10 days if the damage is in excess of $200. Some states require reporting of accidents with property damage less than $200 or total boat loss.

A **$500 PENALTY** may be asessed for failure to submit the report.

WORD OF ADVICE

Take time to make a copy of the report to keep for your records or for the insurance company. Once the report is filed, the Coast Guard will not give out a copy, even to the person who filled the report.

The report must give details of the accident and include:

1- The date, time, and exact location of the occurrence.

2- The name of each person who died, was lost, or injured.

3- The number and name of the vessel.

4- The names and addresses of the owner and operator.

If the operator cannot file the report for any reason, each person on aboard **MUST** notify the authorities, or determine that the report has been filed.

1-15 NAVIGATION

Buoys

In the United States, a buoyage system is used as an assist to all boaters of all size craft to navigate our coastal waters and our navigable rivers in safety. When properly read and understood, these buoys and markers will permit the boater to cruise with comparative confidence that he will be able to avoid reefs, rocks, shoals, and other hazards.

In the spring of 1983, the Coast Guard began making modifications to U.S. aids to navigation in support of an agreement spon-

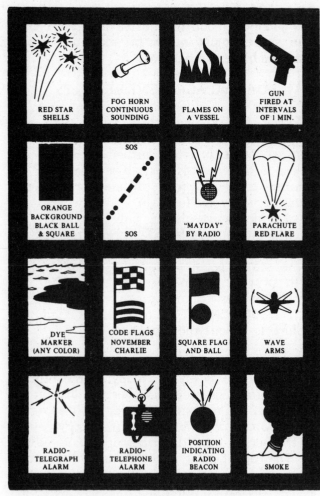

Internationally accepted distress signals.

sored by the International Associaiton of Lighthouse Authorities (IALA) and signed by representatives from most of the maritime nations of the world. The primary purpose of the modifications is to improve safety by making buoyage systems around the world more alike and less confusing.

The modifications shown in the accompanying illustrations should be completed by the end of 1989.

Lights

The following information regarding lights required on boats between sunset and sunrise or during restricted visibility is taken directly from a U.S. Coast Guard publication dated 1984.

The terms **"PORT"** and **"STARBOARD"** are used to refer to the left and right side of the boat, when looking forward. One easy way to remember this basic fundamental is to consider the words "port" and "left" both have four letters and go together.

Waterway Rules

On the water, certain basic safe-operating practices must be followed. You should learn and practice them, for to **know**, is to be able to handle your boat with confidence and safety. Knowledge of what to do, and not do, will add a great deal to the enjoyment you will receive from your boating investment.

Rules of the Road

The best advice possible and a Coast Guard requirement for boats over 39' 4" (12 meters) since 1981, is to obtain an official copy of the "Rules of the Road", which includes Inland Waterways, Western Rivers, and the Great Lakes for study and ready reference.

The following two paragraphs give a **VERY** brief condensed and abbreviated -- almost a synopsis of the rules and should not be considered in any way as covering the entire subject.

Powered boats must yield the right-of-way to all boats without motors, except when being overtaken. When meeting another boat head-on, keep to starboard, unless you are too far to port to make this practical. When overtaking another boat, the right-of-way belongs to the boat being overtaken. If your boat is being passed, you must maintain course and speed.

When two boats approach at an angle and there is danger of collision, the boat to port must give way to the boat to starboard. Always keep to starboard in a narrow channel or canal. Boats underway must stay clear of vessels fishing with nets, lines, or trawls. (Fishing boats are not allowed to fish in channels or to obstruct navigation.)

2
TUNING

2-1 INTRODUCTION

The efficiency, reliability, fuel economy and enjoyment you receive from the engine's performance are all directly dependent on having it tuned properly. The importance of performing service work in the sequence detailed in this chapter cannot be over-emphasized. Before making any adjustments, check the specifications in the Appendix. Do not rely on your memory.

Before beginning to tune any engine, check to be sure the engine cylinders have satisfactory compression. An engine with worn piston rings, burned valves, or a blown gasket cannot be made to perform properly no matter how much time and expense is spend on the tune-up. Poor compression must be corrected or the tune-up will not give the desired results.

2-2 TUNING FOR PERFORMANCE

First, check the battery to be sure it will deliver enough energy. The battery must not only crank the engine rapidly enough to draw in the proper amount of air/fuel mixture and compress it, but the battery must also have enough energy left to energize the ignition system for a hot spark to ignite the mixture.

Mechanical checks and service includes adjusting the drive belt, checking the compression, tightening the cylinder head and manifold bolts to meet torque specifications, and adjusting the valves.

Ignition system service includes replacing and adjusting the contact points on models equipped with breaker point distributors, checking the ignition advance, and adjusting the timing.

Fuel system service and carburetor adjustments are made after all ignition checks

and adjustments have been completed. It is very important to complete the ignition work before moving to the fuel system because the ignition adjustments have so much affect on the carburetion.

2-3 MECHANICAL TASKS

Drive Belt Adjustment

Check the drive belt at least once a year or during each tune-up. Replace the belt if there is any evidence of a crack or tear on the undersurface.

If it is necessary to install a new belt, be sure to use only the size and type recommended by the engine manufacturer.

Adjust the belt tension by first loosening the brace and pivot bolts, and then pivoting

FINGER PRESSURE

Adjust the drive belt until the belt will depress approximately 1/4-inch under finger pressure.

the alternator away from the engine until the belt deflection is 1/4 inch (6.35mm) when you exert a downward pressure against the belt. Make the belt deflection measurement midway between the circulating pump pulley and the alternator pulley. Tighten the alternator bracket and pivot bolts securely.

If the battery has a record of not holding a full charge, check the tension of the drive belt. The failure to hold a full charge may also be caused by a faulty alternator or a defective voltage regulator. The cause of a battery using too much water is usually attributed to the voltage regulator being set too high. In this case, replace the regulator. Refer to Chapter 6 for service details of the electrical system.

Battery Service

Check the battery, cables, and surrounding area for signs of corrosion. Look for loose or broken carriers; cracked or bulged cases; dirt and acid; electrolyte leakage; and low electrolyte level. Add distilled water to the cells to bring them up to the proper level. Keep the top of the battery clean and be sure the battery is securely fastened in position.

When cleaning the battery **BE SURE** the vent plugs are tight to prevent any of the cleaning solution from entering the cells. First, wash the battery with a diluted ammonia or soda solution to neutralize any acid, and then flush the solution off with clean water.

Keep the battery hold-down device tight enough to prevent any movement but not so tight that it puts the battery case under a strain.

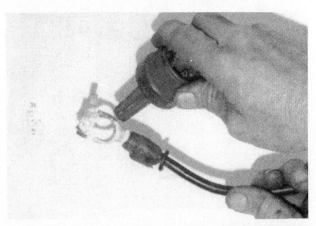

An inexpensive brush is the best tool for cleaning the inside diameter of battery cable connectors.

An explosive hydrogen gas is released from the cells when the caps are removed. This battery exploded when the gas ignited from smoking in the area with the caps removed, or possibly from a spark at the terminal post.

Tighten the battery cables on the battery posts to ensure a good contact. Clean the posts or cable terminals with a wire brush if they have become corroded. After the posts and terminals are clean, apply a thin coating of Multipurpose Lubricant to both as a preventative measure against corrosion forming.

Using a special tool to clean the battery terminals. After cleaning and the cables have been connected, coat the terminals and connectors with lubricant to prevent further corrosion.

Test the battery electrolyte at regular intervals.

Spark Plug Cleaning

Grasp the molded cap, then twist slightly and pull it loose from the plug. Do not pull on the wire, or the connection inside the cap may become separated or the boot may be damaged. Remove the spark plugs and be careful not to tilt the socket, to prevent cracking the insulator. Compare the spark plugs with the illustration on this page and the following page to determine how the engine has been running. Clean and gap the spark plugs. Use the specifications in the Appendix for the proper gap dimension.

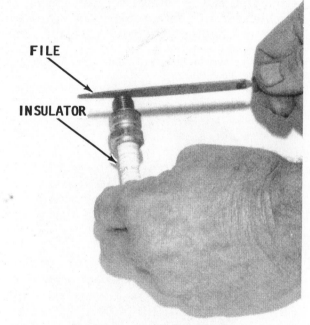

FILE

INSULATOR

Engine misfire and poor fuel economy can be caused by corroded spark plug electrodes. To regain performance, file and reset each spark plug, if they are the least bit corroded.

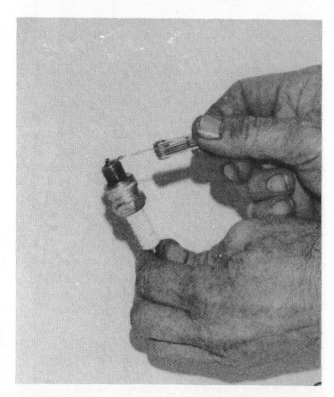

ALWAYS use a round gauge to measure spark plug gap. A flat gauge will not give an accurate reading. NEVER use a gasket on a spark plug with a tapered seat.

Spark plug fouled by oil. The engine may need an overhaul.

Excessive overheating, heavy load. Use a spark plug with a lower heat rating.

Powdery deposits have melted and shorted-out this spark plug.

Red, brown, or yellow deposits are by-products of combustion from fuel and lubricating oil.

Cylinder and Manifold Bolt Tightening

Tighten the cylinder head and manifold bolts in the sequence detailed by the diagrams in Chapter 3, Engine Service, and according to the torque specification given in the Appendix.

Compression Testing

Insert a compression gauge in each spark plug opening one-at-a-time; crank the engine and check the compression. A significant variation between cylinders is far more important than the actual reading of each one individually. A difference of over 20 psi, indicates a ring or valve problem. To determine which needs attention, insert a teaspoonful of oil into the spark plug opening of the low reading cylinder, and then crank the engine a few times to distribute the oil. Now, check the compression again to see if inserting the oil caused a change. If the reading went up, then the compression loss is due to worn rings. If the reading remained the same, the loss is due to a burned valve.

A compression gauge inserted into the spark plug opening for a compression check. A variation between cylinders is more important than the actual individual reading.

Valve Adjusting

Proper valve adjustment allows the hydraulic lifters to operate in the center of their designed travel. Valve adjusting is not a simple one-two operation. Therefore, this procedure is covered in detail in Chapter 3, Section 3-10, beginning on Page 3-46.

2-4 IGNITION SERVICE

All Models

It is not possible to do a good job of replacing the contact points or performing other service with the distributor in the engine. To remove the distributor, turn the crankshaft until the rotor points to the No. 1 cylinder position.

WORDS FROM EXPERIENCE

From this point on **DO NOT** attempt to crank or start the engine. If the position of the crankshaft is disturbed while the distributor is out of the engine, the engine must be timed again to synchronize the distributor with the crankshaft. This is a fairly lengthy and complicated procedure, so save yourself some work and leave the engine

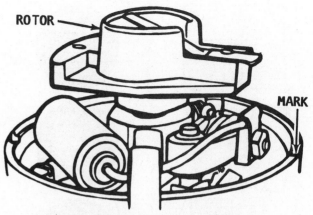

Mark the position of the rotor BEFORE removing the distributor from the engine.

alone while the distributor is removed. If the crankshaft is disturbed, then follow the instructions under the heading Engine Timing to retime the engine.

Distributor with Breaker Point Set

Most distributor caps have a "1" stamped on the cap. Remove the hold-down bolt. Remove the distributor slowly and you will notice the rotor turn. When the rotor stops turning, the distributor gear is free of the camshaft gear. At this point scribe a mark on the distributor housing in line with the edge of the rotor as an aid to installation.

At the time of installation, align the rotor on the mark you scribed on the housing, then lower the distributor slowly and you will see the rotor turn back until it is pointing to the No. 1 cylinder position.

Use a couple squirts of engine oil into the cylinder to determine if the compression loss is a burned valve or worn piston rings, as explained in the text.

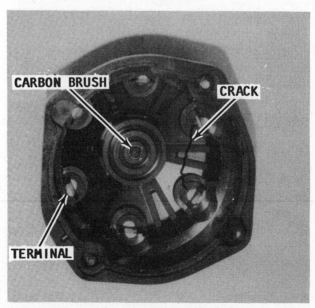

A crack in the distributor cap can cause hard starting and misfire, especially at high speed.

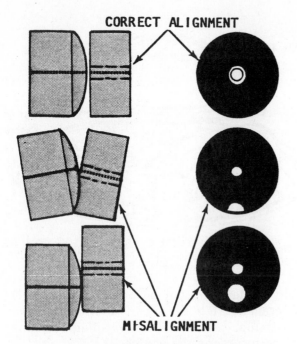

CORRECT ALIGNMENT

MISALIGNMENT

Breaker point alignment guide.

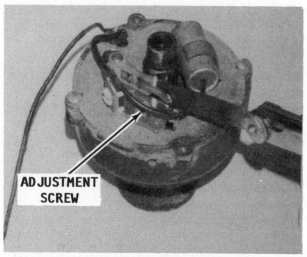

ADJUSTMENT SCREW

Adjusting the point gap.

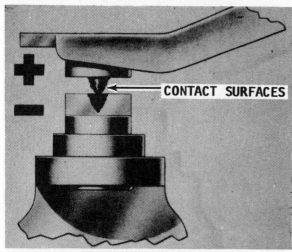

CONTACT SURFACES

A faulty condenser will cause abnormal contact point wear and loss of engine performance.

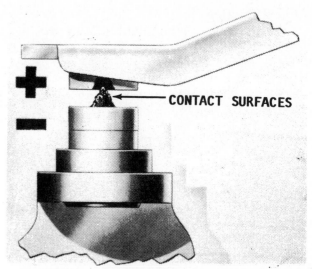

CONTACT SURFACES

Check the contact points for abnormal wear, burning, or pitting.

Always replace the contact points rather than filing them. It is seldom necessary to replace the condenser.

Adjust the point gap according to the specifications listed in the Appendix. Keep the feeler gauge blade clean, because the slightest amount of oil film will cause trouble when it oxidizes.

It is best to add 0.003" to the clearance specification when installing a new set of contact points to compensate for initial rubbing block wear. **ALWAYS** keep the contact point retaining screw **SNUG** during the adjustment to prevent the gap from changing when it is finally tightened.

After the proper gap adjustment has been made, apply a light layer of heavy grease to coat the distributor cam. Turn the distributor shaft in the normal direction of rotation so the lubricant is wiped off against the back of the rubbing block. The grease will remain on the rubbing block as a

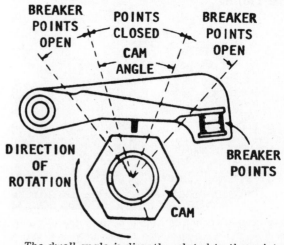

BREAKER POINTS OPEN   POINTS CLOSED   BREAKER POINTS OPEN

CAM ANGLE

DIRECTION OF ROTATION

BREAKER POINTS

CAM

The dwell angle is directly related to the point gap.

reservoir to supply lubricant as the block wears. Wipe off any excess lubricant. Leave only the grease stored on the rubbing block.

Replace the distributor in the engine, with the rotor pointing toward the mark you scribed on the housing. Tighten the hold-down bolt. The ignition timing will be adjusted after the engine is running, as described in the following paragraphs.

Distributor with Thunderbolt (Breakerless) Ignition

This system does not utilize breaker points or a mechanical advance mechanism. It was designed to be almost maintenance free without the requirement for periodic adjustment. There should be no reason to ever remove the distributor unless the engine was being overhauled or the entire distributor was to be replaced.

Most distributor caps have a "1" stamped on the cap. Remove the hold-down bolt. Remove the distributor slowly and you will notice the rotor turn. When the rotor stops turning, the distributor gear is free of the camshaft gear. At this point scribe a mark on the distributor housing in line with the edge of the rotor as an aid to installation.

At the time of installation, align the rotor on the mark you scribed on the housing, then lower the distributor slowly and you will see the rotor turn back until it is pointing to the No. 1 cylinder position.

Replace the distributor in the engine, with the rotor pointing toward the mark you scribed on the housing. Tighten the hold-down bolt. The ignition timing will be adjusted after the engine is running, as described in the following paragraphs.

Engine Timing

If the crankshaft was turned for any reason while the distributor was removed, the timing was lost and it will be necessary to retime the engine.

To time the engine, first remove the rocker arm cover. Next, rotate the crankshaft in the normal direction with a wrench on the harmonic balancer bolt until both valves for No. 1 cylinder are closed and the timing mark on the balancer is aligned with the "O" on the timing indicator. **NEVER** rotate the crankshaft in the opposite direction from the normal or the water pump in the stern drive will be damaged. Now, with both valves closed, and the timing mark

aligned with the timing indicator, the No. 1 cylinder is in firing position.

Align the rotor with the No. 1 cylinder wire terminal in the distributor cap, and then install the distributor in the block. If the distributor will not seat fully in the block, press down lightly on the housing while a partner turns the crankshaft slowly until the distributor tang snaps into the oil pump shaft slot and the distributor moves into its full seated position. Tighten the distributor hold-down bolt.

Ignition fine-tuning will be accomplished after the engine is running.

Wipe the distributor cap and the coil of any moisture to be sure it does not cause a leakage path.

Ignition Timing

Connect a tachometer to the engine. The tachometer input lead (usually Red or White) is connected to the primary **NEGATIVE** side of the ignition coil. The tachometer ground lead (usually Black) is connected to a suitable engine ground.

Connect a power timing light to the No. 1 spark plug. Do not puncture the wire or boot when you install the timing light because a puncture could start a voltage leak and lead to problems later.

Start the engine and adjust the speed and timing to the specifications listed in the Appendix.

CAUTION

Water must circulate through the stern drive, also to and from the engine, anytime the engine is run to prevent damage to the water pump located in the lower unit (original and Alpha drives); or on the front of the engine (Bravo drive). Just a few seconds without water will damage the water pump.

If a point adjustment is required, stop the engine, remove the cap, and make the adjustment. Remember, the dwell setting (gap) of the contact points affects the ignition timing; therefore, it is essential the dwell be set before adjusting the ignition timing.

Idle the engine at 600 rpm or less to be sure no centrifugal advance is taking place. Adjust the ignition timing by first loosening the distributor hold-down bolt, then rotate the distributor housing until the appropriate timing marks align. Tighten the hold-down bolt.

NO ADVANCE   FULL ADVANCE

Twist the rotor and check the automatic advance weights for good movement. The rotor must feel springy in one directon and solid in the opposite direction. Engine performance will suffer if the rotor turns sluggishly.

Checking Ignition Advance

Ignition timing is varied according to engine speed by means of a centrifugal advance unit. During tune-up, it is essential to check the operation of this unit. An accurate check can be made using a timing light with a timing gauge and an advance control knob. If you do not have this piece of test equipment, a rough check can be made to be sure the system is functioning properly.

With the ignition timing properly adjusted and the timing light connected to No. 1 spark plug wire, increase engine speed to the rpm given in the Centrifugal Timing Table in the Appendix. The ignition timing must advance to the range (in degrees) shown in the table. If the speed fails to advance to the required range, remove the distributor and check the advance mechanism under the breaker plate.

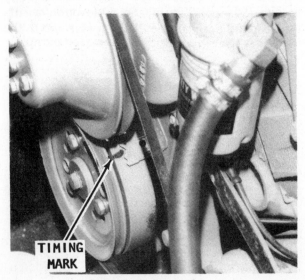

TIMING MARK

Typical timing marks.

DAMAGED PISTON

Damage to this piston was caused by abnormal combustion because the fuel exploded violently, causing the spark plug, piston, and valves to overheat. Proper adjustment of the spark advance or a change in fuel to a higher octane rating may correct the problem.

2-5 FUEL SYSTEM SERVICE

General Practices

Change the fuel filter in the base of the carburetor at least once a year. When the filter is changed, be sure to use a new gasket under the nut to prevent fuel from leaking out of the filter.

Change the water separator every year. Clean the flame arrestor after every 100 hours of operation.

Dual-Diaphragm Fuel Pump

As the name implies, the dual-diaphragm fuel pump has two diaphragms separated by a metal spacer and an attached sight gauge. Three important safety features are built into this type of pump. The pump will continue to operate on the second diaphragm if the main diaphragm fails. Gasoline can only leak into the space between

REPLACE FUEL PUMP IF GAS APPEARS IN THIS BOWL

SIGHT GLASS

Fuel in the sight glass means the diaphragms are leaking and the fuel pump must be replaced.

the two diaphragms and not out of the pump. If gasoline is detected in the sight gauge, it means the pump is defective.

If the dual-diaphragm becomes defective for any reason, it must be replaced. Do not attempt to repair this type of pump.

Carburetor Adjustment

Because the carburetor is required to accurately control and mix the air and fuel quantities entering the conbustion chamber, proper adjustments are critical to efficient engine operation. Dirt and gum in the passages restrict the flow of air or fuel causing a lean operating condition; hesitation on acceleration; and lack of power on demand.

The carburetor control linkage is subject to wear which will change the synchronization and fuel mixture. These changes will affect engine performance and fuel economy. Therefore, accurate fine carburetor adjustments can hardly been made or expected if the carburetor is not in satisfactory condition. If considerable difficulty is encountered in making the adjustments, the remedy may be to take time for a carburetor overhaul, see Chapter 4.

To make a preliminary adjustment, turn the idle mixture adjusting needles inward until they **BARELY** make contact with their seats, then back the needle out the specified number of turns.

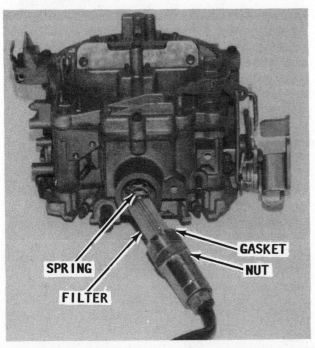

SPRING
FILTER
GASKET
NUT

The fuel filter in the base of the carburetor should be replaced every year.

NEVER turn the idle mixture screws **TIGHTLY** against their seats or they will be **DAMAGED.** Disconnect the throttle cable. Start the engine and run it at idle speed.

CAUTION

Water must circulate through the stern drive, also to and from the engine, anytime the engine is run to prevent damage to the water pump located in the lower unit (original and Alpha drives); or on the front of the engine (Bravo drive). Just a few seconds without water will damage the water pump.

Adjust each idle mixture needle to obtain the highest and steadiest manifold vacuum reading. If a vacuum gauge is not available, obtain the smoothest running, maximum idle speed by turning one of the idle adjusting needles in until engine speed begins to fall off, then back the needle off over the "high spot" until the engine rpm again drops off. Set the idle adjusting needle halfway between the two points for an acceptable idle mixture setting. Repeat this procedure with the other needle.

If these adjustments result in an increase in idle rpm, reset the idle speed adjusting screw to obtain the specified idle rpm and again adjust the idle mixture adjusting needles.

Shift the unit into forward gear and readjust the idle speed screw to obtain the recommended idle speed as given in the Appendix.

Stop the engine and install the throttle cable. Check to be sure the throttle valves are in the full open position when the remote-control is in the full forward position. On Models 120 to 165, with the throttle valves fully open, turn the wide-open throttle stop adjusting screw clockwise until the screw just touches the throttle lever. Tighten the set nut securely to prevent the adjustment screw from turning. Return the shift lever to the neutral gear and idle position.

The idle-stop screw should be against its stop.

Calibration for High Altitudes

Increased spark advance or carburetor recalibration have very little affect on high altitude performance. Tests have proven this statement to be true. However, a marked increase in performance can be obtained by changing propellers for high altitude operation.

Changing the prop should be the only modification considered to obtain the rated rpm. Any other recommendation should be considered a special case and should be referred to the factory branch or area distributor for specific jet sizes and spark timing settings.

Changing jet sizes and spark timing settings will: Cause engine failure if operated at lower elevations; result in increased fuel economy but will not have any significant affect on performance; and may cause added problems at much lower elevations.

2-6 ADJUSTING IDLE SPEED AND MIXTURE

Conncect a tachometer, then start the engine and allow it to warm to normal operating temperature.

CAUTION

Water must circulate through the stern drive, also to and from the engine, anytime the engine is run to prevent damage to the water pump located in the lower unit (original and Alpha drives); or on the front of the engine (Bravo drive). Just a few seconds without water will damage the water pump.

Adjust the idle speed screw until the engine is running at idle specification given in the Appendix. Turn the idle mixture adjusting needles and observe the tachometer reading until the highest rpm is obtained. **ALWAYS** turn the needles very **SLOWLY** because there is a time lag until the engine responds to the new mixture and stablizes.

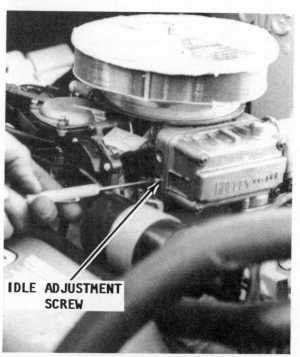

The idle mixture adjusting screw on early Ford V8 engines. A second adjusting screw is located on the opposite side of the carburetor.

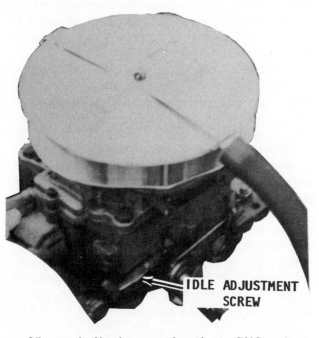

Idle speed adjusting screw location on GMC engines.

3
POWERPLANT

3-1 INTRODUCTION AND CHAPTER COVERAGE

This chapter is divided into twelve main sections. This first section is merely an introduction to explain how the chapter is divided to enable the reader to quickly move to the area needed to perform the work required.

The next section explains, in detail the workings of a four-cycle engine, followed by a troubleshooting section for the powerplant.

Service work is divided into ten logical sections. Most sections are complete in this chapter. When other tasks are involved, such as timing, the chapter reference is clearly indicated.

Many tasks require the engine to be removed from the boat. If engine removal is not necessary, the reader may still elect to remove the engine in order to perform the work more easily. Therefore, a separate section is presented, following troubleshooting, to remove and install the engine.

The remaining sections cover work on the unit in a logical sequence. To prevent unnecessary duplication of information, and to keep the size of this manual to a reasonable weight, many tasks necessary to return the unit to satisfactory operation are simply referenced by section number

Throughout the chapter, check the headings to be sure the proper procedures are being followed for the unit being serviced.

3-2 GENERAL PRINCIPLES

Engine specification charts are located in the Appendix. These charts can be used to determine the engine type, size, and specifications.

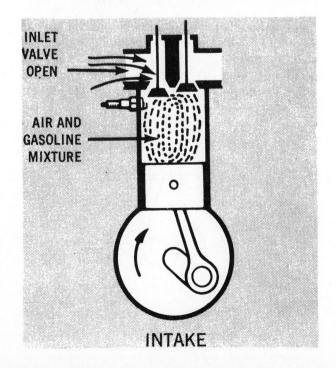

INLET VALVE OPEN

AIR AND GASOLINE MIXTURE

INTAKE

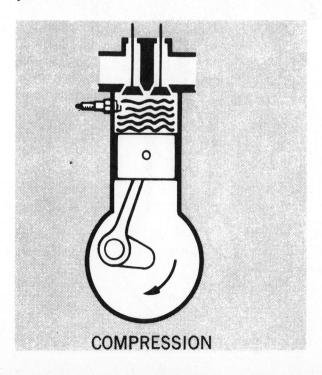

COMPRESSION

All engines used in power boats equipped with a Mercruiser Stern Drive, operate on the four stroke principle. During this cycle, the piston travels the length of its stroke four times. As the piston travels up and down, the crankshaft is rotated halfway (180 degrees). To complete one full cycle, the crankshaft rotates two complete turns; the camshaft, which controls the valves, is driven by the crankshaft at half crankshaft speed. Valve action, intake and exhaust, occurs once in each four-stroke cycle, and the piston acts as an air pump during the two remaining strokes.

INTAKE STROKE

The intake valve is opened as the piston moves down the cylinder, and this creates an area of pressure lower than the surrounding atmosphere. Atmospheric pressure will cause air to flow into this low-pressure area. By directing the air flow through the carburetor, a measured amount of vaporized fuel is added. When the piston reaches the bottom of the intake stroke, the cylinder is filled with air and vaporized fuel. The exhaust valve is closed during the intake stroke.

COMPRESSION STROKE

When the piston starts to move upward, the compression stroke begins. The intake valve closes, trapping the air-fuel mixture in the cylinder. The upward movement of the piston compresses the mixture to a fraction of its original volume; exact pressure depends principally on the compression ratio of the engine.

POWER STROKE

The power stroke is produced by igniting the compressed air/fuel mixture. When the spark plug arcs, the mixture ignites and burns very rapidly during the power stroke. The resulting high temperature expands the gases, creating very high pressure on top of the piston, which drives the piston down. This downward motion of the piston is transmitted through the connecting rod and is converted into rotary motion by the crankshaft. Both the intake and exhaust valves are closed during the power stroke.

EXHAUST STROKE

The exhaust valve opens just before the piston completes the power stroke. Pressure in the cylinder at this time causes the exhaust gases to rush into the exhaust manifold. The upward movement of the piston on its exhaust stroke expels most of the remaining gases.

As the piston pauses momentarily at the top of the exhaust stroke, the inertia of the exhausting gases tends to remove any re-

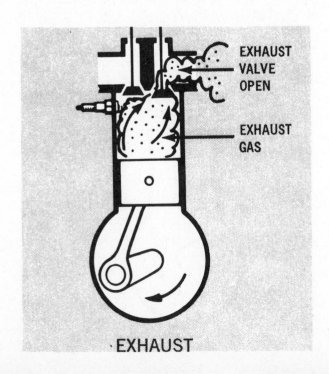

EXHAUST VALVE OPEN

EXHAUST GAS

EXHAUST

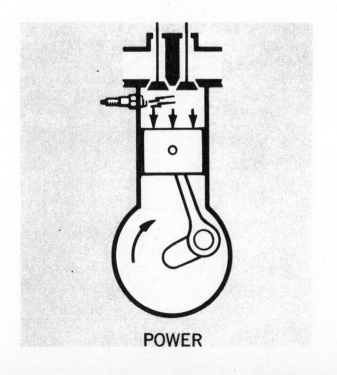

POWER

maining gas in the combustion chamber; however, a small amount always remains to dilute the incoming mixture. This unexpelled gas is captured between the piston and the cylinder head.

COMBUSTION

The power delivered from the piston at the crankshaft is the result of a pressure increase in the gas mixture above the piston. This pressure increase occurs as the mixture is heated, first by compression, and then (on the downward stroke) by burning. The burning fuel supplies heat, which raises the temperature and, at the same time raises the pressure. Approximately 75 percent of the mixture in the cylinder is composed of nitrogen gas, which does not burn but expands when heated by the burning of the combustible elements, and it is this expanding nitrogen which supplies most of the pressure on the piston.

The fuel and oxygen must burn smoothly within the combustion chamber to take full advantage of this heating effect. Maximum power would not be delivered to the piston if an explosion took place, because the entire force would be spent in one short hammer-like blow, occurring too fast for the piston to follow.

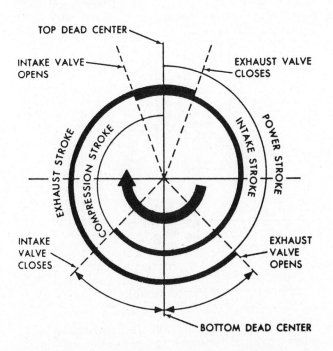

Diagram to illustrate the relationship between crankshaft rotation and valve timing.

Instead, burning takes place evenly as the flame moves across the combustion chamber. Burning must be completed by the time the piston is about half-way down so maximum pressure will be developed in the cylinder at the time the piston applies its greatest force to the crankshaft. This will be when the mechanical advantage of the connecting rod and crankshaft is at a maximum.

At the beginning of the power stroke (as the piston is driven down by the pressure), the volume above the piston increases, which would normally allow the pressure in the cylinder to drop. However, combustion is still in progress, and this continues to raise the temperature of the gases, expanding them and maintaining a continuous pressure on the piston as it travels downward. This action provides a smooth application of power throughout the effective part of the power stroke to make the most efficient use of the energy released by the burning fuel.

VALVE TIMING

On the power stroke, the exhaust valve opens before bottom dead center in order to get the exhaust gases started out of the combustion chamber under the remaining pressure (blowdown). At the end of the exhaust stroke, the intake valve opens before top dead center in order to start the air/fuel mixture moving into the combustion chamber. These processes are functions of the camshaft design and valve timing.

Valves always open and close at the same time in the cycle; the timing is not variable with speed and load as is ignition timing. There is, however, one particular speed for each given engine at which the air/fuel mixture will pack itself into the combustion chambers most effectively.

This is the speed at which the engine puts out its peak torque value. At low engine speeds, compression is somewhat supressed due to the slight reverse flow of gases through the valves just as they open or close when the mixture is not moving fast enough to take advantage of the time lag. At high speeds, the valve timing does not allow enough time during the valve opening and closing periods for effective packing of the air/fuel mixture into the cylinders.

ENGINE TYPES

The most popular engines used with Mercruiser Stern Drive units are:

Mercury Marine 4-cylinder 1976-84
Models 470, 485, and 488 -- 224 CID

Mercury Marine 4-cylinder since 1984
Models 165, 170, 180, 190,
and the 3.7 litre -- 224 CID

GMC 4-cylinder In-line -- 153 CID
and 181 CID

GMC 6-cylinder In-line -- 194, 230,
and 250 CID

GMC V6 -- 229 and 262 CID

GMC V8 -- 283, 292, 305, 307, 327, 350,
and 454 CID

Ford V8 -- 302 CID & 351 CID

The engine serial number is most important when ordering parts, corresponding with the factory, or when servicing the unit. The "year" is of little or no consequence. A "new" engine may not be installed the year it is produced.

VERY IMPORTANT WORDS

GMC or Ford automotive parts must **NEVER** be substituted for marine engine parts! Marine engines operate at maximum rpm for most of the time. Stock automotive parts will not hold up for prolonged periods of time under such conditions. Automotive parts may appear identical to marine parts, but be assured, marine parts are specially manufactured to meet Mercury Marine specifications. Most are super heavy duty units, often coated to protect against corrosive salt water atmosphere.

Marine electrical and ignition parts are extremely critical. Many times the unit must meet Coast Guard requirements for spark suppression. A spark could ignite an explosive atmosphere of gasoline vapors.

MERCURY MARINE
FOUR-CYLINDER IN-LINE ENGINE
1976-84

The powerplants on Models 470, 485, and 488 all have 224 CID. The block is manu-

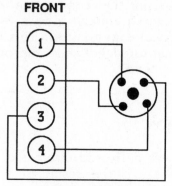

FRONT

**FIRING ORDER
1-3-4-2**

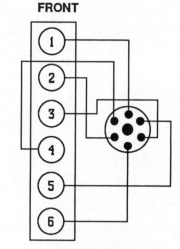

FRONT

**FIRING ORDER
1-5-3-6-2-4**

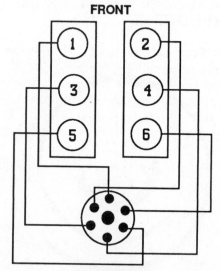

FRONT

**FIRING ORDER
1-6-5-4-3-2**

Diagrams to illustrate the firing order of a 4-cylinder (top), a 6-cylinder in-line (center), and a V6 engine (bottom). As the rotor on the upper end of the distributor shaft rotates, contact is made with the terminals to "fire" the cylinders as indicated.

factured by Mercury Marine. Internal moving parts and other components originate at the Ford Motor Company.

The crankshaft is supported by five main bearings. Three special features make this unit unique for small boat installation.

1- The engine is cooled by a closed-circuit fresh water system. After water is circulated through the block, it passes through a heat exchanger where heat is drawn off by outside water circulated through the exchanger and then discharged overboard.

2- The water pump is mounted on the front timing chain cover. The pump is driven by an extension on the camshaft at the same speed as the camshaft.

3- The alternator is mounted on the front end of the crankshaft and therefore, is not driven in the conventional manner by a pulley arrangement.

For many years, the standard carburetor was the Rochester 2GC. In the early 1980's a Mercarb replaced the 2GC. This new carburetor is almost identical to the 2GC. The major difference is that the 2GC has two idle adjustment needles whereas the MerCarb has only one, in the center of the carburetor base.

The firing order for these engines is 1-3-4-2.

MERCURY MARINE FOUR-CYLINDER IN-LINE ENGINE POWERPLANTS SINCE 1988

This four cylinder cast aluminium block with non-removable sleeves is also manufactured by Mercury Marine. The crankshaft is supported by five main bearings. The cylinder head is cast iron. Internal moving parts and other components are from the Ford Motor Company.

The 3.7 Litre -- Models 165 and 170 are equipped with a Mercarb carburetor. Models 180, 190 and the 3.7 Litre LX are equipped with a Rochester 4MV or 4MC Quadrajet carburetor.

The firing order for these engines is 1-3-4-2.

GMC IN-LINE ENGINES

For many years GMC in-line engines have been mated with Mercruiser stern drive units. The engines used are the 153, 181, 194, 230, and 250 CID. The 153 and 181 engines are four-cylinder powerplants. The 194, 230, and 250 CID engines have six cylinders. The four-cylinder engine has four main bearings, while the six-cylinder engine has seven. These engines have many interchangeable parts.

The firing order is 1-5-3-6-2-4 for 6-cylinder engine and 1-3-4-2 for the 4-cylinder engine.

GMC V6 ENGINES

The Mercruiser V6 engine is manufactured by GMC. The original engine had a bore of 3.736" and a stroke of 3.480", for 229 CID. In late 1983, the bore was increased to 4.00", but the stroke remained at 3.480" giving it a 262 CID. With a two-barrel carburetor, the engine is rated at 185hp. With a four-barrel, the horsepower is increased to 205hp.

From the forward end, the port bank of cylinders is numbered 1-3-5 and the starboard bank 2-4-6. The firing order begins with the port bank, 1-6-5-4-3-2.

The 229 CID V6 engine is considered an uneven firing engine because the crankshaft angle alternately changes from 132° to 108° through the firing cycle. To obtain this angle change, the cam in the distributor has alternating sharp and rounded lobes. Care must be exercised when installing the distributor, as discussed in Chapter 5.

The 185/205 262 CID V6 are conventional even firing engines.

These V6 engines have an unusual valve arrangement for the port bank of E-I-E-I-I-E, and for the starboard bank of E-I-I-E-I-E.

GMC V8 ENGINES

The first of the modern large-block engines was produced in 1965 with a 396 CID.

Since 1965, the call has been for increased power to move the boat faster through the water or pull more water skiers at greater speeds. Engineering answered the challenge. In addition to changes in carburetion and ignition, the engine has been progressively increased in size to 427, and 454 CID. These engines are called Mark IV to separate them from the small block engines. These 90° blocks have a center-to-center bore spacing of 4.84". The port bank cylinders are numbered 1-3-5-7 and the starboard bank 2-4-6-8. The firing order is 1-8-4-3-6-5-7-2.

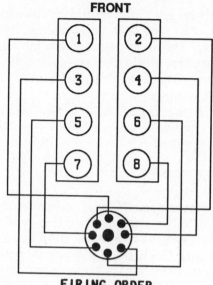

FRONT

FIRING ORDER
1-8-4-3-6-5-7-2

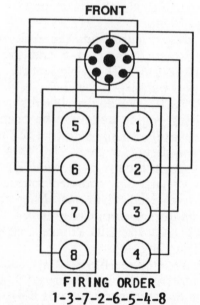

FRONT

FIRING ORDER
1-3-7-2-6-5-4-8

Firing order for a GMC V8 engine (top), and a Ford V8 engine (bottom).

The block and cylinder heads of production models are made of cast iron. An unusual feature of the heads is the angled pattern of the tilted valves to allow gas to flow smoother through intake and exhaust ports. This feature contributes to increased volumetric efficiency.

One of these larger engines may have been installed in your boat as original equipment or you may have had one installed as a replacement engine.

In order to identify the engines covered in a particular procedure, and to avoid repetition, the engines are classified as follows: The 283, 292, 300, 305, 307, 327, 350, and 400 CID engines are identified as small block V8's. The 5.0 Litre and 5.7 Litre engines fall into this category. The 427, and 454 CID engines are identified as Mark IV's or big block V8's. Typical illustrations and procedures are provided, except where specific details differ.

FORD V8 ENGINES

From about 1972 through 1978, some Mercruiser stern drive units were driven by Ford powerplants. The Ford engines used were the 302 and 351W CID. These engines were considered light-weight in design. The 302 CID engine has a bore of 4.00" and a stroke of 3.00". The 351 engine has the same 4.00" bore with the stroke increased to 3.50". Crankshaft rotation is **LEFT-HAND** when viewed from the **FLYWHEEL** end.

Starting from the forward end of the engine, the valve arrangement on the starboard bank is **I-E-I-E-I-E-I-E** and on the port bank, **E-I-E-I-E-I-E-I.** Hydraulic lifters are used for more quiet valve train operation.

Since about 1978, Mercruiser has not used Ford engines as factory installed equipment.

SERVICE SECTIONS

Service procedures for the various engines have been consolidated. In most cases the work is identical, regardless of the engine size or number of cylinders.

The following service sections are coded to assist you in referring to a particular section quickly when performing other work. You may find these section numbers referenced in other chapters.

The next three major sections of this chapter are divided into the following logical portions. First, troubleshooting specific engine problems; next, service procedures entailing removal and disassembling of the various components; and the third section involves cleaning and inspecting of the major parts -- pistons, rings, valves, etc.

Most major repairs require pulling the engine out of the boat. The job sequence covered for each engine type is:

1- Removing the engine for service.
2- Servicing the cylinder head.
3- Servicing the cylinders.
4- Servicing the camshaft, timing gear and chain.
5- Servicing the crankshaft.

Tuning the engine is covered in Chapter 2, Tuning.

3-3 TROUBLESHOOTING

The following troubleshooting procedures are common to all engines covered in this chapter. Reference should be made to this section before doing any engine work and it should be used in conjunction with the specific model instructions following these general procedures.

Troubleshooting must be a well thought-out procedure. Always attempt to proceed with the troubleshooting in an orderly manner. The "shot in the dark" approach will only result in wasted time, incorrect diagnosis, replacement of unnecessary parts, and frustration.

Obviously, if the instructions are to be of maximum benefit as a guide, they must be fully understood and followed in the proper sequence.

When an engine fails to start, the trouble must be localized to one of four general areas: cranking system; ignition system; fuel system; or compression. Each of these areas must be systematically inspected until the trouble is isolated. At that time, detailed tests of the system or area must be made to determine the part causing the problem.

TROUBLESHOOTING CHECK

When using this Troubleshooting Check, proceed sequentially through each test until the defect is uncovered. Then skip to the detailed testing procedure and check for that system. For example, if, when using the Troubleshooting Check procedure, the first two systems, cranking and ignition tests OK, but the third test shows there is trouble in the fuel system, then proceed to the detailed test under the Fuel System Troubleshooting Check in Chapter 4.

CRANKING SYSTEM TEST

1- Turn the key switch to the **START** position. The cranking motor should crank the engine at a normal rate of speed.

If the cranking motor cranks the engine slowly or fails to crank it at all, the trouble is in the cranking system --proceed directly to the Cranking System in Chapter 6 for detailed testing procedure to isolate the problem.

IGNITION SYSTEM TEST

2- Disconnect the wire from the center of the distributor cap and hold it about 1/4" from a good ground. An alternate method is to connect a standard spark tester (available at almost any automotive parts store), between the spark plug and the spark plug lead. Hold the spark plug lead with an insulated holder (to avoid receiving a "shock" when the current passes through). Crank the engine with the cranking motor. If there is **NO** spark, or if the spark is very weak, the trouble is in the ignition system. Proceed directly to the Ignition Troubleshooting in Chapter 5 for detailed testing procedures to uncover the problem.

3- Remove the flame arrestor. Look down into the throat of the carburetor and at the same time open and close the throttle several times. Observe if fuel is squirting out of the pump jets. The top of the carburetor has been removed in the accompanying illustration for photographic clarity.

Attempt to determine the age of the fuel in the fuel tank. Many fuels tend to "sour" in three to four months, especially if

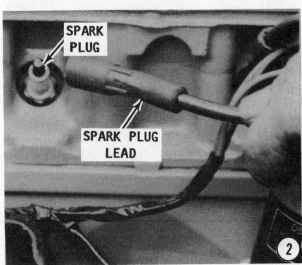

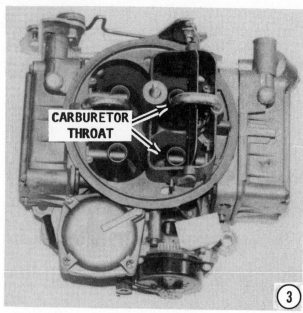

an additive such as Sta-Bil has not been added. Other fuels take longer to "sour". In no case should an attempt be made to start the engine if the fuel in the tank is more than 12 months old.

If no fuel is discharged from the pump jets, then the trouble is in the fuel system, proceed directly to the Fuel Troubleshooting in Chapter 4 to determine the problem.

COMPRESSION TEST

4- Good compression is the key to proper engine performance. An engine with worn piston rings, burned valves, worn guides, or blown gaskets cannot be made to

perform satisfactorily until the mechanical defects contributing to low compression are corrected. Generally, a compression gauge is used to determine the cranking pressure within each cylinder. However, today's large displacement engines generally have considerable valve overlap, and the resulting compression reading may be much lower than the manufacturer's specifications of about 150-170 psi (1034-1172kPa). It is entirely possible to obtain a reading as low as 120 psi (827kPa) on a modern engine which is in good mechanical condition. Such an engine is said to "exhale" at cranking speed, even though everything is perfectly normal at operating speeds.

To make a compression test, remove the spark plugs and lay them out in the order of removal. This is extremely important so you can "read" the firing end of each spark plug. After the spark plugs have been removed, install a compression gauge into one cylinder and have an assistant crank the engine.

GROUND the primary side of the coil to prevent damage to it. The throttle valve and choke **MUST** be in the **WIDE OPEN** position in order to obtain maximum readings. Crank

Use compressed air in a cylinder when the piston is at TDC, firing position and listen carefully. The source of leaking air will indicate the source of the problem. A hissing sound through the carburetor indicates a leaking intake valve.

the engine through several revolutions to obtain the highest reading on the compression gauge.

The significance in a compression test is the variation in pressure readings between the cylinders. As long as this variation is between 20-30 psi (138-207kPa) the engine is normal. If a greater variation exists, then the lower reading cylinder should be checked by making a cylinder leak test. A simple leak test can be made by first inserting a teaspoonful of oil into the spark plug opening of the lower reading cylinder, and then cranking the engine a few times to distribute the oil.

Check the compression again to see if inserting the oil caused a change. The oil helps make a temporary seal around the rings and increases the compression. If the reading went up, then the compression loss was probably due to worn rings. If the reading remained the same, the loss may be due to a burnt valve or worn valve guides.

TROUBLESHOOTING MECHANICAL ENGINE PROBLEMS

A more definite and scientific method of performing a leak test and determining the problem area, than the simple method described in the previous paragraph, is to use compressed air.

Install an air hose adaptor into the spark plug port. With the piston at top dead center firing position, apply 60-70 psi (418-483kPa) of air. On commercially built air adaptor units, a gauge indicates the percentage of leakage. Over 20% leakage, in most cases, is considered excessive.

Listening at the point from which the compressed air is escaping indicates the nature of the defect. Insert a short length of heater hose into the various areas being tested and listen at the other end. The hose helps to amplify the leakage noise. Air hissing from the exhaust manifold indicates a leaking exhaust valve, or worn guide. Air heard leaking from the carburetor air horn indicates a leaking intake valve or a worn intake guide. If you hear air hissing at the oil filler pipe, the rings are worn.

VACUUM GAUGE

A vacuum gauge is a relatively inexpensive piece of test equipment which can be very useful in isolating the mechanical problems of an internal combustion engine. As with the compression gauge, you cannot rely soley on the actual numerical reading of the vacuum gauge. Instead, relative readings and typical actions of the needle provide clues to some types of problems.

On late model powerplants, a vacuum plug is located on the upper surface of the intake manifold just aft of the carburetor. Simply remove the plug and install the vacuum gauge.

Normal idle vacuum in the intake manifold ranges from 15 to 22" Hg. On later model engines, lower and less steady intake manifold vacuum readings are becoming increasingly common because of the greater use of high-lift cams and the increase in the amount of valve overlap.

In addition to these factors, altitude affects a vacuum reading. At high elevations, a vacuum gauge will read about one inch lower for each 1,000 ft. (305 m) of elevation above sea level. Another outside influence on the vacuum gauge is a change in barometric pressure. Because of these factors, which are not determined by the condition of the engine, it is much more important for you to watch the action of the needle, than its actual reading. After you

With motor at idling speed floating motion right and left of vacuum pointer indicates carburetor too rich or too lean.

With motor at idling speed low reading of vacuum pointer indicates late timing or intake manifold air leak.

With motor at idling speed vacuum pointer should hold steady between 15 and 21.

With motor at idling speed dropping back of vacuum pointer indicates sticky valves.

have worked with a vacuum gauge just a few times, you will be able to recognize such problem areas as sticking valves, a tight valve lash adjustment, or a restriction in the exhaust system.

DYNAMOMETER TESTING

An inexpensive method of making a dynamometer-type of engine test, is to use a vacuum gauge and a set of shorting wires. With these two items, it is possible to isolate mechanical problems to one or more cylinders. When making this type of test, the spark plugs are shorted out one-at-a-time until the engine is running on one cylinder. Now, if a vacuum gauge is attached to the intake manifold, a reading can be obtained to compare the efficiency of each cylinder.

A low-reading cylinder can be the result of inefficiency in the ignition, fuel, or compression system.

To make this test, the ground clip of the shorting wires must be attached to a good ground.

CAUTION

Water must circulate through the stern drive, also to and from the engine, anytime the engine is run to prevent damage to the water pump located in the lower unit (original and Alpha drives); or on the front of the engine (Bravo drive). Just a few seconds without water will damage the water pump.

Start the engine.

Ignition system defects will cause engine misfire. If the spark plug electrodes are improperly gapped, excessively worn, or corroded, the engine will misfire during acceleration.

Have a helper advance the throttle. At the same time, short out the cylinders until the engine is running on one cylinder. It will be necessary to open the throttle wide in order to keep the engine running on one cylinder at a time.

Observe the vacuum gauge reading, and then move the shorting clip from one of the spark plugs to the spark plug of the cylinder just tested. Observe the vacuum gauge reading, and then test each of the other cylinders in turn by running the engine on that one spark plug. A weak cylinder, or a cylinder which is not firing, is easily determined, but the most important part of this test is the ability to compare the relative power (vacuum) of each firing stroke.

EXCESSIVE OIL CONSUMPTION

High oil consumption can usually be traced to one of four general areas:

1- A clogged positive crankcase ventilation system,

2- Piston rings not sealing.

3- Excessive valve stem-to-guide clearance.

4- Cracked intake manifold (the type which serves as a valve chamber cover).

High oil consumption complaints many times are the result of oil leaks rather than actual consumption of the engine. For this reason, before assuming an engine is burning oil, examine the exterior for evidence of oil leaks. In analyzing the problem, consideration must be given to the fact that oil has only three routes available to enter the combustion chambers. The oil may get past the piston rings, enter through the valve guides, or seep in through the intake manifold. Two very definite clues will indicate the engine is actually burning excessive oil. One is carbon deposits in the exhaust outlet and the other is oil-fouled spark plugs.

ENGINE NOISES

Engine noises can be generally classified as knocks, slaps, clicks, or squeaks. These noises are usually caused by loose bearings, sloppy pistons, worn gears, or other moving parts of the engine. Most common types of noises are either synchronized to engine speed or to one-half of engine speed. Noises that are timed to engine speed are sounds that are related to the crankshaft, rods, pistons, and pins. The cause of noises that seem to be one half of the engine speed are usually in the valve train. To determine whether the noise is timed to the engine speed or one-half the engine speed, operate the engine at a slow idle and observe whether the noise is synchronized with the flashes of a timing light.

A main bearing knock is usually identified by a dull thud which is noticeable under engine load. Attempting to move the boat under power with it tied to the dock will bring out a main bearing knock. If you pull the spark plug wire from one plug at a time and the noise disappears when a particular wire is removed, then the noise is probably coming from that cylinder. The cause may be either the rod bearing, piston pin, or the piston, which has quieted down because the load has been removed from the part. If a rod bearing is loose, the noise will be loudest when the engine is decelerating. Piston pin noise and piston pin slap are generally louder when a cold engine is first started.

Many times the source of an unusual sound may be isolated by using a stethoscope or other listening device. One such device is a long shank screwdriver. Allow your hand to extend over the end of the handle, press an ear to your hand, and then probe with the other end of the screwdriver

by touching it firmly against the block at each cylinder or noise area. Take care and use good judgement when using this method of attempting to detect the cause of a problem because the noise will travel through other metallic parts of the engine and could lead to a false interpretation of what you are hearing.

Carbon build up in the combustion chamber can cause interference with a piston. Fuel pumps can knock, belts can be noisy, and alternators can contribute to unusual sounds. Flywheels, water pumps, and loose manifolds can also cause noise problems.

3-4 ENGINE REMOVAL, INSTALLATION AND ALIGNMENT

In many cases the design of the boat and the method of engine installation will cause extra work in order to pull the powerplant. Engine covers and panels around the engine might have to be removed. The engine may have to move about 6" (15cm) forward in order to clear the driveshaft assembly. If there is not room to move the engine forward, then it will be necessary to remove the stern drive in order to clear the driveshaft assembly.

If the stern drive was not removed, the installation procedure may be continued without aligning the engine, because the engine mount height was undisturbed. However, if the vertical drive was removed,

Use a timing light to accurately adjust the ignition timing to the manufacturer's specifications. Performance and fuel consumption are both directly dependent on precise timing.

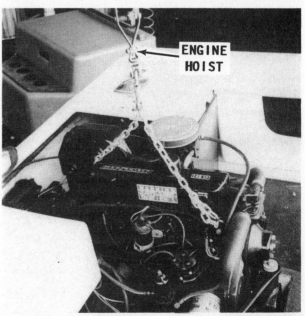

Lifting the engine with a bracket and assembled cables. If a chain is used, pass the chain through the lifting bracket and eye bolts, then secure the ends together with a bolt and nut.

engine alignment is essential to prolong drivetrain life.

Refer to Chapter 13 for detailed instructions to remove and install the powerplant and for engine alignment procedures.

3-5 OIL PUMP SERVICE
ALL MODELS

The oil pump seldom causes problems alone. However, the pressure relief valve may fail; the pickup screen may become clogged; or the gears or rotor may become worn creating excessive clearance; requiring service to the oil pump. This short section provides detailed procedures to service only the oil pump. Good shop practice would require an exhaustive check of many other areas, besides work on the oil pump, while the engine is out of the boat and the pan is removed.

OIL PUMP REMOVAL
ALL UNITS

Open the drain valves and drain the coolant from the block and exhaust manifold.

Remove the engine. See Chapter 13.

Remove the oil drain plug and drain the oil. Remove the oil pan bolts, and then separate the pan from the engine block.

The oil filter bypass valve is installed with the valve side facing toward the engine, as shown.

GOOD WORDS

If servicing a 153 or 181 CID engine, observe the oil pan gasket. Some engines come with a four piece oil pan gasket, while others have a one piece gasket. When purchasing a replacement gasket set, **TAKE TIME** to verify the new set contains the correct number of pieces.

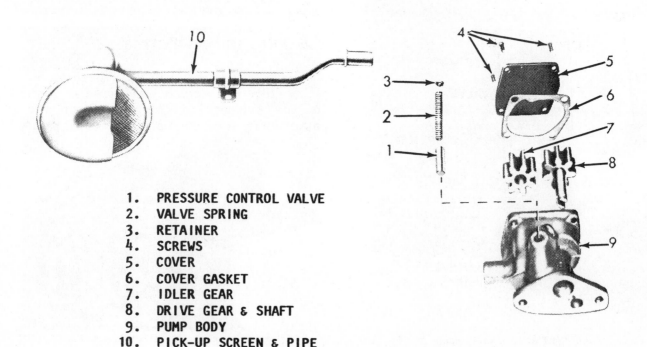

1. PRESSURE CONTROL VALVE
2. VALVE SPRING
3. RETAINER
4. SCREWS
5. COVER
6. COVER GASKET
7. IDLER GEAR
8. DRIVE GEAR & SHAFT
9. PUMP BODY
10. PICK-UP SCREEN & PIPE

Exploded view of the oil pump installed on GMC in-line engines.

Remove the oil pump pickup tube and screen assembly.

GMC Engines Only

The oil pump in a GMC engine consists of two gears and a pressure regulator valve, enclosed in a two-piece housing. The oil pump is driven by the distributor shaft, which is driven by a helical gear on the camshaft. A baffle is installed on the pickup screen to eliminate pressure loss due to sudden stops.

Make a mark with a marker or pencil across two indexing teeth of the oil pump to ensure they can be reinstalled in their original positions. **DO NOT** use an implement which will scratch the machined gear surface. Remove the pump to rear main bearing cap bolt, and then remove the pump and extension shaft.

Mercury Marine and Ford Engines Only

This oil pump consists of a single rotor and shaft, housed in a one-piece body; a relief valve; and an inlet tube and screen assembly. Oil pumps installed on early engines were cast iron. Newer pumps have a cast aluminium body. Both pumps are interchangeable **IF** matched with the correct pickup tube.

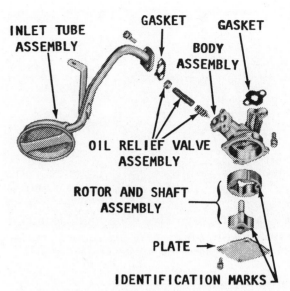

Exploded drawing showing principle parts of the oil pump on a Mercury Marine or Ford engine.

The oil pump is driven by the distributor shaft, which is driven by a helical gear on the camshaft.

A baffle is installed on the pickup screen to eliminate pressure loss due to sudden stops.

Remove the bolts securing the oil tube retaining brackets, the oil pump tube and screen assembly. Remove the oil pump attaching bolts and remove the oil pump, gasket, and the oil pump to distributor shaft.

Remove the oil pressure relief valve cap, spring, and valve.

CLEANING AND INSPECTING

Clean the pan with solvent. Clean the oil pickup tube and screen and examine them for any evidence of clogging. Clean the gasket surfaces of the block and the oil pan.

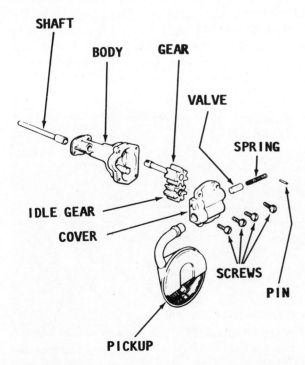

Arrangement of parts for the oil pump assembly used on a GMC engine.

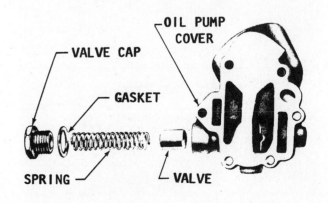

Arrangement of oil pump pressure valve parts on a GMC engine.

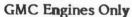

Proper arrangement of the side gaskets. The gaskets are held in place with a sealer on GMC engines.

GMC Engines Only

Wash the gears thoroughly and inspect them for wear and scores. If either gear is defective, they must be replaced as a **PAIR**.

Remove the oil pressure regulator valve cap, spring, and valve. The oil filter bypass valve and spring **MUST NOT** be removed because they are staked in place.

Wash the parts removed and check each one carefully. Inspect the regulator valve for wear and scores. Check the regulator valve spring to be sure it has not worn on its side or has collapsed. If in doubt about the condition of the spring, install a new one. Clean the screen staked in the cover.

Check the regulator valve in its bore in the cover. The clearance in the valve should only be a slip fit. If any side clearance can be felt, the valve or the cover should be replaced.

Inspect the filter bypass valve for nicks,

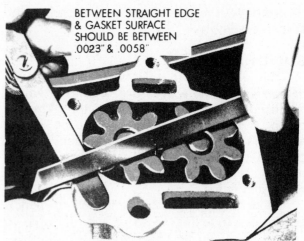

Using a straightedge and feeler gauge to check the rotor play on a GMC engine. The play should be 0.0023"-0.0058".

cracks, or warping. The valve should be flat with no nicks or scratches.

Mercury Marine and Ford Engines Only

Wash the rotor and the shaft thoroughly and inspect them for wear and scores. If either the rotor or the shaft is defective, they **MUST** be replaced as a **PAIR**.

Wash the pump parts carefully. Inspect the relief valve for wear or scores. Check the relief valve spring to be sure it is not worn on its side or has not collapsed. If in doubt about the condition of the spring, install a new one. Check the relief valve in its bore in the housing. The clearance for the valve should only be a slip fit. If any side clearance can be felt, the valve or the housing should be replaced. Clean the pick-up tube and screen.

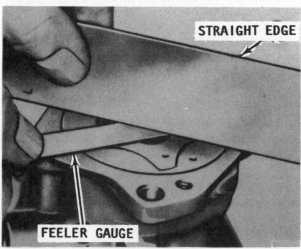

Using a straightedge and feeler gauge to check the rotor play on a Mercury Marine or Ford engine. The play should be 0.001"-0.005".

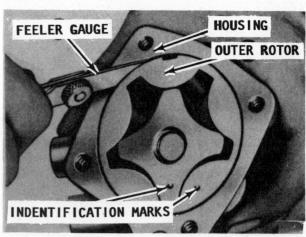

Using a feeler gauge to check the clearance between the outer rotor and the oil pump housing on a Mercury Marine or Ford engine. The identification marks must align, as shown.

OIL PUMP ASSEMBLING AND INSTALLATION

Assembling and installation of the oil pump is divided into two sets of procedures. The first set covers GMC engines and the second set Mercury Marine and Ford engines. Installation of the pan is the same for both GMC, Mercury Marine and Ford engines.

GMC Engines Only

Apply a generous amount of oil to the pressure regulator valve and spring. Install the lubricated valve and spring into the bore of the oil pump cover and then slide the retaining pin in place.

Install the gears and shaft onto the oil pump body aligning the two marked teeth together. Check the gear end clearance. Place a straightedge over the gears, and then measure the clearance between the straightedge and the gasket surface. The clearance should be 0.0023" (0.06mm) to 0.0058" (0.15mm).

If the gear end clearance is acceptable, remove the gears and pack the gear pocket full of petroleum jelly, **DO NOT** use chassis lube.

Arrangement of oil pump parts on a Mercury Marine and Ford engine. Identified parts are: Welsh plug (a), relief valve spring (b), relief valve plunger (c), pump housing (d), outer rotor (e), inner rotor (f), cover (g), lockwashers (h), and screws (i).

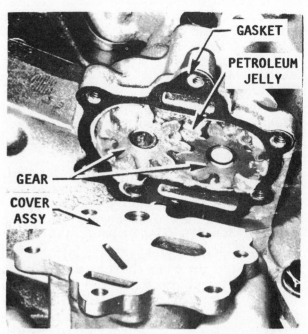

Serious consequences can result if the pump is not packed with petroleum jelly at time of assembling.

CAUTION

Unless the pocket is packed with petroleum jelly, it may not prime itself when the engine is started

Install the gears and shaft again, pushing the gears into the petroleum jelly. Place a new gasket in position, and then install the cover screws. Tighten the screws alternately and evenly to a torque value of 7 ft lb (10Nm).

If the oil pump pickup screen was removed, apply sealer to the mating surfaces

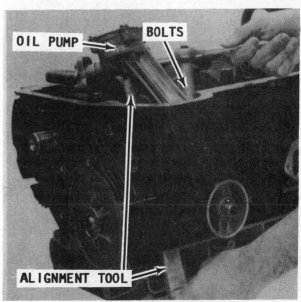

Using an alignment tool to properly position the oil pump before tightening the attaching bolts.

before the pipe is driven into position. An **AIR LEAK** could cause a **LOSS** of **OIL PRESSURE.**

Mercury Marine and Ford Engines Only

Prime the oil pump by filling either the inlet or the outlet port with engine oil. Rotate the pump shaft to distribute the oil inside the pump housing.

Position the oil pump to distributor shaft into the distributor socket. Remove the shaft and position the stop, if necessary.

Position a new gasket onto the pump housing. With the stop properly positioned, insert the pump and driveshaft into the oil pump. Install the pump shaft as an assembly. **NEVER** attempt to force the pump into position if it does not seat **EASILY.** The driveshaft hex may be misaligned with the distributor shaft. To align the hex shaft, rotate the oil pump to distributor shaft.

Tighten the oil pump attaching screws to a torque value of 10 ft lb (13Nm).

Install the pickup tube and screen assembly. Tighten the bolts securing the oil tube to the block to a torque value of 20 ft lb (27Nm).

The correct position for the oil pump screen is with the bottom edge parallel to the oil pan rails.

The oil pump rotor **MUST** be free to rotate after the attaching bolts have been installed. Tighten the attaching bolts to a torque value of 25 ft lb (34Nm).

Oil pump, pickup, and screen installed on a GMC engine.

PAN INSTALLATION ALL ENGINES

Install the rear seal in the rear main bearing cap. **BE SURE** the tabs on the seal are over the oil pan gasket. Use a sealer with enough body to act as a retainer.

ALWAYS install a new pan gasket set.

GOOD WORDS

If servicing a 153 or 181 CID engine, be sure to install the correct oil pan gasket. Some engines have a four piece gasket, while others have only a one piece gasket. Compare the new with the old to ensure the proper gasket is being installed.

Install the pan attaching bolts just fingertight. After all bolts have been started, tighten them to the following torque value working alternately and evenly from the center of the pan toward both ends.

M/M 4-cyl -	All -	10.8 ft lb 15Nm
GMC 4-cyl - & L6	Side 1/4" -	7 ft lb 9Nm
	End 5/16" -	10 ft lb 14Nm
GMC V6 -	Side 1/4" -	6.5 ft lb 9Nm
	End 5/16" -	14 ft lb 18Nm
GMC V8 -	Side 1/4" -	12 ft lb 10Nm
	End 5/16" -	5.5 ft lb 7Nm
Ford V8 -	Side 1/4" -	7 ft lb 9Nm
	End 5/16" -	10 ft lb 14Nm

Applying sealer to the ends of the oil pan gasket, as described in the text.

TAKE CARE to tighten the bolts evenly all the way around. Do not overtighten them as the bolts may be easily snapped off. Should this happen, it will be necessary to drill and tap an oversize hole, and install an oversize bolt.

Install the engine; see Chapter 13.

Fill the crankcase with the proper weight oil.

Close the water drain valves.

CAUTION

Water must circulate through the stern drive, also to and from the engine, anytime the engine is run to prevent damage to the water pump located in the lower unit (original and Alpha drives); or on the front of the engine (Bravo drive). Just a few seconds without water will damage the water pump.

Start the engine and check the completed work.

3-6 INTAKE MANIFOLD SERVICE

Intake manifolds seldom create any problems. The following short sections are included to provide procedures to remove the manifold in order to accomplish other work. Check each heading and find the procedures for the unit being serviced.

Mercury Marine Engines

Disconnect the negative battery cable and remove the spark plug leads. Remove the choke cover wire and the fuel pump vent tube from the carburetor. Remove the throttle cable if necessary. Do not lose the throttle return spring. Remove the inlet fuel line at the carburetor and plug the line to prevent loss of fuel and prevent contaminants from entering the fuel system. Remove the crankcase ventilation hose from the flame arestor or the rocker arm cover. Disconnect the water hoses from the manifold.

Remove the bolts securing the manifold to the cylinder head. Make note of any bolts which support retaining clamps for the wire harness.

Lift the intake manifold, with the carburetor attached, from the engine. Discard all gaskets and seals.

GMC In-line 4- and 6-Cylinder Engines

On GMC in-line engines, the intake manifold and the exhaust manifold are one assembly.

Remove the flame arrestor nut and the flame arrestor. Disconnect the throttle cable from the anchor block and actuating bracket. Disconnect the fuel and vacuum lines from the carburetor and the crankcase ventilation hose. Disconnect the fuel pump overflow and the choke heat tube.

Remove the carburetor. Remove the water hoses to the exhaust manifold to the thermostat housing. On Power Steering models, remove the alternator, cooler and

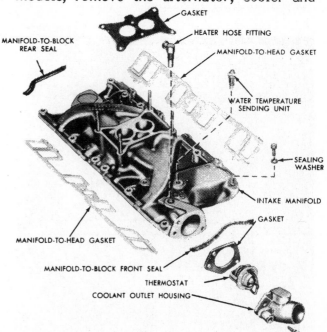

Exploded drawing showing arrangement of the intake manifold parts on a V8 engine.

Using a lifting bracket to remove the intake manifold from a V8 engine.

mounting plate. Remove the exhaust hose. Remove the manifold to head attaching bolts and clamps. Remove the manifold. The intake manifold and exhaust manifold are one unit.

GMC V6 and V8 Engines

Open all drain valves and drain the water from the block. Remove the flame arrestor from the carburetor. Disconnect the battery cables at the battery. Disconnect the hoses, fuel line, and throttle linkage at the carburetor or the fuel rail on engines equipped with EFI.

Remove the crankcase ventilation hose from the valve covers port and starboard and remove the hoses from the thermostat housing. Remove the fuel pump vent hose.

Disconnect the wires to the coil and temperature sending switch.

Remove the distributor cap and mark the distributor housing indicating the position of the rotor. Remove the distributor clamp and then pull the distributor free of the block. Move the distributor cap out of the way. Remove the coil and bracket.

On units equipped with Power Steering: Disconnect the engine wiring harness from the alternator. Remove the oil pressure sending unit.

Remove the bolts attaching the manifold to the head together with the solenoid bracket. Lift the intake manifold, with the carburetor attached, from the engine. Discard all gaskets and seals.

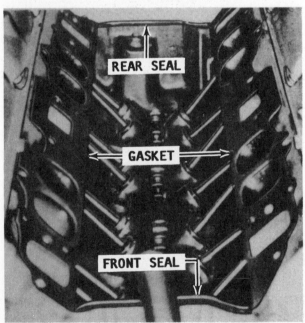

Placement of manifold gaskets and seals prior to installing the manifold.

Ford V8 Engines

Open the drain valves and drain the coolant from the block and exhaust manifold.

Remove the flame arrestor from the carburetor. Disconnect the battery cables at the battery. Disconnect the hoses, fuel line, and throttle linkage at the carburetor. Disconnect the wires to the coil and temperature sending switch.

Remove the distributor cap and mark the distributor housing indicating the position of the rotor. Remove the distributor clamp, and then pull the distributor free of the block. Move the distributor cap out of the way. Remove the coil and bracket.

Remove the bolts attaching the manifold to the head, and then lift the manifold, with the carburetor attached, from the engine. Discard all gaskets and seals.

CLEANING AND INSPECTING ALL UNITS

Clean the gasket surfaces of the intake manifold, the cylinder head or heads, and the block.

Check the old gaskets to determine if there is any history of an exhaust leak. Any evidence of exhaust leakage might indicate a crack in the head. Any sign of water in an intake manifold part would indicate either a crack in the manifold or a crack in the head.

INSTALLATION — INTAKE MANIFOLD

If a sealer is used on the manifold gasket, use the sealer sparingly. The sealer has a bad habit of "oozing" into the intake ports and restricts flow of the air/fuel mixture.

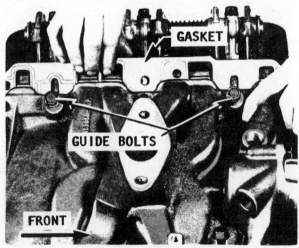

Installing the intake manifold.

Mercury Marine Engines

A **FACTORY GASKET** is used between the intake manifold and the cylinder head.

Apply a coat of Perfect Seal, Permatex, Form-A-Gasket, or equivalent, to both surfaces of the **NEW** gasket. Position the gasket against the cylinder head.

Align the manifold to the cylinder head matching all bolt holes. Hold the manifold in place with one hand and start the bolts with the other hand. Remember to position the retaining clamps supporting the wire harness in their original locations. After all bolts have been started, tighten the nuts alternately and evenly to a torque value of 25 ft lb (33Nm). Work from the center alternately toward both ends of the manifold.

Close the water drain valves.

Connect the water hoses to the manifold and connect the crankcase ventilation hose to the flame arrestor or the rocker arm cover. Connect the fuel line and the fuel line vent tube to the carburetor. Install the throttle cable and cable return spring. Install the choke cover wire.

Install the spark plug leads and connect the negative cable to the battery.

CAUTION

Water must circulate through the stern drive, also to and from the engine, anytime the engine is run to prevent damage to the water pump located in the lower unit (original and Alpha drives), or on the front of the engine (Bravo drive). Just a few seconds without water will damage the water pump.

Start the engine and check the completed work.

GMC In-line 4- and 6-Cylinder Engines

Apply a light coating of Perfect Seal, Permatex, Form-A-Gasket, or equivalent, to both sides of the manifold gasket. Place the gasket over the manifold end studs and carefully install the manifold. The Form-A-Gasket material will hold the gasket in proper place. Install the bolts and clamps with one hand while holding the manifold in place with the other.

Tighten all manifold bolts alternately and evenly to a torque value of 22 ft lb (30Nm) for 4-cylinder engines or 25 ft lb (33Nm) for 6-cylinder engines. Work from the center alternately towards both ends of the manifold.

Connect the exhaust hoses to the manifold. Replace the water hoses from the thermostat to exhaust manifold.

Install the carburetor, flame arrestor, the crankcase ventilation hose, fuel pump overflow hose, and the choke heat tube.

Connect the throttle cable. Install the flame arrestor. On Power Steering units: install the alternator, oil cooler, and mounting plate.

CAUTION

Water must circulate through the stern drive, also to and from the engine, anytime the engine is run to prevent damage to the water pump located in the lower unit (original and Alpha drives); or on the front of the engine (Bravo drive). Just a few seconds without water will damage the water pump.

Start the engine and check the completed work for leaks.

GMC V6 and V8 Engines

Place new rubber intake manifold seals in position at the front and rear rails of the cylinder block. **BE SURE** the pointed ends of the seal fit snugly against the block and head.

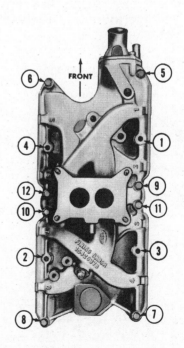

Tightening sequence for the intake manifold bolts.

SPECIAL WORDS

Some GMC engines do not have front and rear manifold seals. If this is the case, place a 3/16" (4.8mm) bead of GMC Silicone Rubber Sealer, or equivalent, on the front and rear ridges of the cylinder case. Extend the bead of silicone approximately 1/2" (12.7mm) up each cylinder head to retain and seal the side gaskets.

Apply a light coating of Perfect Seal, Permatex, Form-A-Gasket, or equivalent, to the area between the water passages on the head and the manifold. If one gasket is marked "Right Side Only", install this gasket onto the starboard cylinder head. The other gasket, portside, will be unmarked. If no such markings are found, the gaskets may be installed port or starboard. Most gaskets should be marked **"UP"** on one edge. Take time to ensure intake manifold gaskets are installed correctly.

If servicing a 305 CID engine equipped with a two barrel carburetor: Remove the metal insert on the starboard manifold gasket. This will provide clearance for the manifold heat pipe.

If servicing an engine equipped with an automatic choke, the intake manifold gasket **MUST** have an opening provided for the exhaust crossover port. If the gasket being used does not have this opening, the automatic choke will not recieve sufficient heat to open when the engine has warmed to operating temperature. Therefore, the choke will restrict the flow of air for longer than necessary and cause rough engine operation and use an excessive amount of fuel.

Install the intake manifold gaskets onto the heads. Carefully set the intake manifold and solenoid bracket in place and start the two guide bolts on each side.

Check to be sure everything is in order, and then start the remaining attaching bolts. Tighten the bolts alternately and evenly to a torque value of 35 ft lb (47Nm). Work from the center alternately towards both ends of the manifold.

Install the coil. Slide the distributor into place with the rotor pointing to the mark made prior to removing the distributor. Snap the distributor cap in place. If the crankshaft was rotated while the distributor was out of the block -- **BAD NEWS**. The engine **MUST** be timed. See Page 2-7 for detailed procedures to properly time the engine.

Units With Power Steering

Install the alternator with mounting bracket.

Install the oil pressure sending unit. Connect the crankcase ventilation hoses to the valve covers.

Connect the fuel pump vent hose.

Connect the battery cables at the battery; the hoses to the thermostat housing; the throttle linkage and fuel line at the carburetor or to the fuel rail on engines equipped with EFI; and the wires to the coil and temperature sending switch.

Close the water drain valves.

CAUTION

Water must circulate through the stern drive, also to and from the engine, anytime the engine is run to prevent damage to the water pump located in the lower unit (original and Alpha drives); or on the front of the engine (Bravo drive). Just a few seconds without water will damage the water pump.

Start the engine and check the completed work for leaks.

Ford V8 Engines
First, these words:

Sealer is used to install the intake manifolds on Ford V8 engines. This sealer will **SET UP** in **15 MINUTES**. Therefore, keep

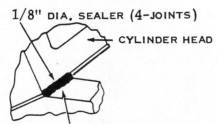

1/8" DIA. SEALER (4-JOINTS)

CYLINDER HEAD

SEAL MOUNTING SURFACE
OF CYLINDER BLOCK

INTAKE MANIFOLD GASKET

INTAKE
MANIFOLD
SEAL

1/16" DIA. SEALER
(4-SEAL ENDS)

Places to apply sealer to minimize manifold oil and water leaks.

the work moving along without delays for coffee, tea, trip to the head -- whatever.

Position new seals on the cylinder block and new gaskets on the cylinder heads, with the gaskets interlocked with the seal tabs. **BE SURE** the holes in the gaskets are aligned with the holes in the cylinder heads. Apply a bead of sealer to the outer end of each intake manifold seal for the full width of the seal. This will be done in four places. Remember, the sealer will set up in about 15 minutes.

Carefully lower the intake manifold into position on the cylinder block. After the manifold is in place, check to be sure the seals are in place by running your finger around the seal area. If the seals are not in place, lift the manifold and move the seals into their proper position. Start the intake manifold bolts. Tighten them alternately and evenly to a torque value of 25 ft lb (33Nm). Work from the center alternately towards both ends of the manifold.

Install the water pump bypass hose on the coolant outlet housing. Slide the clamp into position, and then tighten the clamp. Connect the water and exhaust hoses. Install the carburetor fuel inlet line.

Slide the distributor into place in the block with the rotor pointing to the mark made prior to removing the distributor. Install and tighten the hold down clamp. Install the distributor cap and return the spark plug wires back in the harness brackets on the valve rocker arm covers. Connect the spark plug wires to the spark plugs. Connect the high-tension lead and the coil wires. Connect the throttle cable.

If the crankshaft was rotated while the distributor was out of the block **BAD NEWS**. The engine **MUST** be retimed. See Page 2-7 for detailed procedures to properly time the engine.

Close the water drain valves.

CAUTION
Water must circulate through the stern drive, also to and from the engine, anytime the engine is run to prevent damage to the water pump located in the lower unit (original and Alpha drives); or on the front of the engine (Bravo drive). Just a few seconds without water will damage the water pump.

Start the engine and check the completed work.

After the engine temperature has stabilized, adjust the engine idle speed and the idle fuel mixture. See Chapter 4.

Tighten the intake manifold bolts again, alternately and evenly to a torque value of 25 ft lb (33Nm).

3-7 EXHAUST MANIFOLD SERVICE

Many engine failures and heating problems can be traced to the exhaust manifold and elbow systems. However, many engine tasks require the removal of the exhaust manifold in order to perform other work. The following short sections are included to provide procedures to remove the manifold.

Check each heading and find the procedures for the unit being serviced.

See Chapter 9, Cooling, for detailed procedures to clean and inspect the manifold and/or elbows.

Mercury Marine Engines
Open all drain valves and drain the water from the block. Disconnect the lead at the temperature sending unit located on the thermostat housing. Remove all water hoses from the manifold.

Disconnect the shift cables from the shift plate assembly. Make an effort not to alter the length of the cables.

It may be necessary to drain the engine oil and remove the dipstick oil tube in order to remove the manifold.

Note the location of the wiring harness clamps before disconnecting the engine harness shift cutout switch leads from the terminal block.

Remove the attaching hardware and lift the manifold from the cylinder head. Discard all gaskets and seals. Remove the closed cooling reservoir from the thermostat housing and then remove the thermostat housing from the manifold. Remove the end cap from the exhaust elbow and finally remove the elbow from the manifold. Remove and discard all gaskets and seals.

GMC In-line 4- and 6-Cylinder Engines
On GMC in-line engines, the intake manifold and exhaust manifold are one assembly.

Open all drain valves and drain the coolant from the block and exhaust manifold. See Chapter 8 for detailed location of the drain valves.

Remove the flame arrestor nut and the flame arrestor. Disconnect the throttle cable from the anchor block and actuating bracket.

Disconnect the fuel and vacuum lines from the carburetor and the crankcase vent-ilation hose. Disconnect the fuel pump overflow and the choke heat tube.

Remove the carburetor. Remove the water hoses to the exhaust manifold to the thermostat housing. On Power Steering models, remove the alternator, cooler and mounting plate. Remove the exhaust hose. Remove the manifold to head attaching bolts and clamps. Remove the manifold. The intake manifold and exhaust manifold are one unit.

GMC V6 and V8 Engines

Remove all water hoses and exhaust hoses from the elbow to the side of the exhaust housing.

Remove and plug the fuel lines. Remove the fuel filter assembly.

Disconnect the instrument harness plug from the main engine harness. Remove the shift plate assembly from the exhaust elbow on the starboard manifold. Disconnect both shift cables.

Remove the circuit breaker solenoid and retaining bracket from the starboard manifold.

If servicing an engine equipped with electronic ignition: Remove the ignition amplifier from the port exhaust elbow.

Remove the nuts from the manifold studs, and then lift the manifold free of the cylinder head.

For other service on the exhaust manifold, see Chapter 9, Cooling.

Exhaust manifold cleaned and ready for installation on a V8 engine.

Ford V8 Engines

Open all drain valves and drain the coolant from the block and exhaust manifolds.

Remove the water hoses, and then take off the exhaust hoses from the exhaust housing to the block.

Studs are located along the side of each head to support the exhaust manifold on that side. Nuts on these studs secure each manfold to the head. Remove the nuts from the studs, and then lift the manifold free.

Repeat the procedure for the other manifold.

For other service on the exhaust manifold, see Chapter 9, Cooling.

INSTALLATION -- EXHAUST MANIFOLD

Mercury Marine Engines

Apply a light coating of Permatex, Form-A-Gasket, or equivalent, to both sides of all new gaskets just prior in installation.

If the new elbow replacement gasket has a single long slot, the gasket is installed with the slot facing **FORWARD**. Tighten the bolts to a torque value of 30 ft lb (41Nm).

Apply Perfect Seal to the threads of the temperature sender and install the sender into the thermostat housing. Install the thermostat, thermostat housing together with a new gasket to the manifold. Tighten the securing bolts to a torque value of 20 ft lb (27Nm). Connect the temperature sender lead to the sender. Attach the closed cooling reservoir to the thermostat housing.

Place the gasket over the manifold end studs and carefully install the manifold. The Form-A-Gasket material will hold the gasket in proper place. Install the bolts and clamps with one hand while holding the manifold in place with the other. Tighten all manifold bolts alternately and evenly to a torque value of 25 ft lb (34Nm). Work from the center alternately towards both ends of the manifold.

If the dipstick tube was removed, install the tube and clamp to the manifold bolt in its original location.

Connect the exhaust hoses and manifold bellows, tighten all clamps securely.

Install the wiring harness clamps and the attach the engine harness shift cutout switch leads to the terminal block. Connect the shift cables. If the length of the cables was not altered, no adjustment is necessary.

Fill the closed cooling system.

CAUTION

Water must circulate through the stern drive, also to and from the engine, anytime the engine is run to prevent damage to the water pump located in the lower unit (original and Alpha drives); or on the front of the engine (Bravo drive). Just a few seconds without water will damage the water pump.

Start the engine and check the completed work.

GMC In-line 4- and 6-Cylinder Engines

On GMC in-line engines the intake and exhaust manifold are one assembly.

Apply a light coating of Permatex, Form-A-Gasket, or equivalent, to both sides of the manifold gasket. Place the gasket over the manifold end studs and carefully install the manifold. The Form-A-Gasket material will hold the gasket in proper place. Install the bolts and clamps with one hand while holding the manifold in place with the other. Tighten all manifold bolts alternately and evenly to a torque value of 22 ft lb (30Nm). Work from the center alternately towards both ends of the manifold.

Connect the exhaust hoses to the manifold. Replace the water hoses from the thermostat to exhaust manifold.

Install the carburetor, flame arrestor, the crankcase ventilation hose, fuel pump overflow hose, and the choke heat tube.

Connect the throttle cable. Install the flame arrestor. On Power Steering units: Install the alternator, oil cooler, and mounting plate.

CAUTION

Water must circulate through the stern drive, also to and from the engine, anytime the engine is run to prevent damage to the water pump located in the lower unit (original and Alpha drives); or on the front of the engine (Bravo drive). Just a few seconds without water will damage the water pump.

Start the engine and check the completed work.

GMC V6 and V8 Engines
First, these words:

The manufacturer does not always install a gasket between the exhaust manifold and the block. However, if any evidence indicates an exhaust leak, a new gasket should be installed. A new gasket should be available at the nearest automotive parts dealer.

If a new gasket is not available, a substitute would be to coat the surfaces of the manifolds and the matching surfaces on the block with Permatex Form-A-Gasket material or the equivalent.

Position the manifolds to the cylinder head and start the nuts onto the studs. Tighten all manifold nuts alternately and evenly to a torque value of 20 ft lb (27Nm) for V6 engines or 25 ft lb (34Nm) for V8 engines. Work from the center alternately towards both ends of the manifold.

Install the water hoses and exhaust hoses to the manifold. Close the water drain valves.

If servicing an engine equipped with electronic ignition: Install the ignition amplifier to the port exhaust elbow.

Install the circuit breaker slave solenoid and bracket onto the starboard manifold. Tighten the bolts securely.

Connect the fuel line to the carburetor or the fuel rail if working on an engine equipped with EFI.

Install the shift plate assembly onto the port manifold and connect the shift cables. If the length of the cables was not altered, no adjustment will be necessary.

Connect the instrument harness to the main engine harness plug.

CAUTION

Water must circulate through the stern drive, also to and from the engine, anytime the engine is run to prevent damage to the water pump located in the lower unit (original and Alpha drives); or on the front of the engine (Bravo drive). Just a few seconds without water will damage the water pump.

Start the engine and check the completed work.

Ford V8 Engines

FACTORY GASKETS are used between the exhaust manifolds and the engine block.

Position a **NEW** gasket in place on the manifold studs on one side.

Place the manifold to the cylinder head on the studs. Hold the manifold in place with one hand and start the nuts onto the studs with the other hand. After all nuts have been started, tighten the nuts alterna-

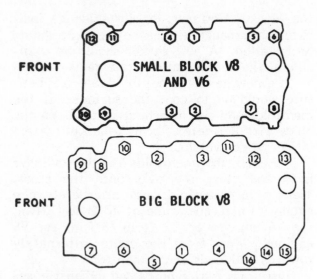

Tightening sequence for the intake manifold bolts on GMC small block V8 and V6 engines (top), and GMC V8 big block (bottom).

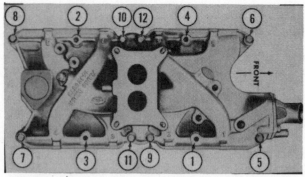

Tightening sequence for the intake manifold bolts on a Ford V8 engine.

tely and evenly to a torque value of 25 ft lb (33Nm). Work from the center alternately toward both ends of the manifold.

Repeat the procedure for the manifold on the other side.

Close the water drain valves.

CAUTION

Water must circulate through the stern drive, also to and from the engine, anytime the engine is run to prevent damage to the water pump located in the lower unit (original and Alpha drives); or on the front of the engine (Bravo drive). Just a few seconds without water will damage the water pump.

Start the engine and check the completed work.

3-8 CYLINDER HEAD SERVICE

Service work performed on the Mercury Marine 4-cylinder in-line head, GMC 4-cylinder in-line head, GMC 6-cylinder in-line head and the heads from a GMC or Ford V6 or V8 engine are almost identical except for the number of valves, etc. Naturally the "V" engines have two heads instead of one, but the service work is still valid.

THEREFORE, the service work on the head is grouped together for all units. Any differences will be clearly indicated.

SPECIAL WORDS

The cylinder head and valve mechanism are the most important areas affecting the power, performance, and economy of an engine. Time and much care are required when reconditioning the cylinder head and valves to maintain the correct valve stem-to-guide clearance; to grind the valves properly; and to make the correct valve adjustment.

The procedures in this section provide removing, disassembling, cleaning and inspecting, assembling, and installation of the cylinder head. Work which may be accomplished while the head is removed is also included.

REMOVAL — CYLINDER HEAD

Remove the flame arrestor. Disconnect the crankcase ventilation hoses at the rocker arm cover/s. Disconnect all wires from the rocker arm cover clips. Remove the rocker arm cover/s. Remove the intake and exhaust manifolds.

Open all drain valves and drain water from the block. Disconnect the spark plug wires. Remove the spark plugs and take care not to tilt the spark plug socket to prevent cracking the insulator.

Keep the spark plugs in order for evaluation of how each individual cylinder has been performing.

Remove the alternator mounting bracket and brace attaching bolts. Move the alternator out of the way.

Remove the screws attaching the rocker arm cover to the cylinder head.

Remove the rocker arm covers and gaskets.

GMC Engines Only

Remove the rocker arm nuts, rocker arm balls, and rocker arms.

Mercury Marine and Ford Engines Only

Remove the valve rocker arm bolt, oil

deflector (if equipped), fulcrum seat, and the rocker arm. An oil baffle is sometimes mounted under the rear bolts on the rocker arm-and-shaft assembly.

STOP

Take time and care to devise a system to keep all parts in order (in a rack or similar object) to **ENSURE** each part will be installed back into the same position from which it was removed. Such a system will also assist in keeping the various parts ready for installation after they have been cleaned and inspected.

Withdraw each push rod and keep them in order to ensure they will be installed back into the same opening from which they were removed.

Loosen, and then remove the cylinder head bolts. Lift the head/s free of the block.

DISCARD the gasket/s.

HEAD DISASSEMBLING

Use a tool to compress the valve springs and remove the keepers. Release the compressor, and then take off the spring caps, spring shields, springs, spring dampers, oil seals, and valve spring shims.

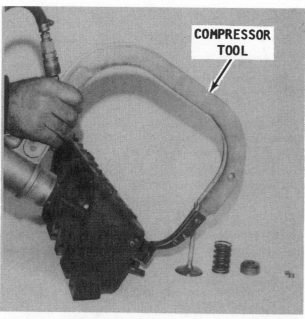

Valve locks are removed by first compressing the spring, and then taking off the keepers, rotator, spring, and oil seal. Push the valve out and ALWAYS keep it in sequence to ENSURE installation back in the same location from which it was removed.

GOOD WORDS

Pay attention and record the number of shims under each valve spring. This will be **MOST** important during assembling.

Remove the valves from the cylinder head and place them in a rack in proper sequence to **ENSURE** they will be installed back into the original position from which they were removed.

CLEANING AND INSPECTING ALL HEADS

Clean all carbon from the combustion chambers and valve ports. Use lacquer thinner to cut the gum on the valve guides. This gum causes sticky valves.

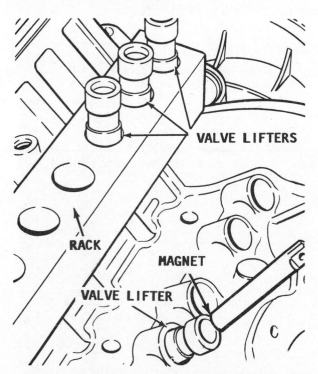

Removing the valve lifter using a magnet and a device for keeping the lifters in sequence. The rack will help ENSURE each lifter will be installed in the same location from which it was removed.

Take time to set-up a system for keeping the push rods in order as they are removed. This is the ONLY way to ensure each rod will be installed back in the same location from which it was removed. Note the rack for this purpose at the bottom of this illustration.

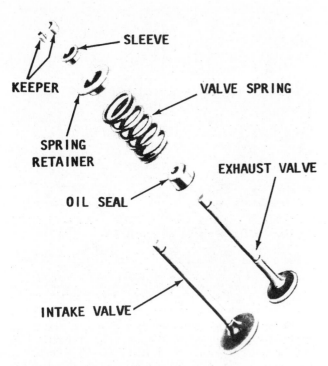

Arrangement of valve mechanism parts.

Clean all carbon and sludge from the push rods, rocker arms, and push rod guides. Check each push rod for straightness by rolling it across a flat surface and observing any indication of the rod rising off the surface. If any portion of the rod rises more than 0.030" (.74mm), the rod **MUST** be replaced. Push rods **CANNOT** be straightened.

Clean all carbon deposits from the head gasket mating surface. **TAKE CARE** not to damage the gasket surface.

Check the flatness of the gasket surface with a straightedge and a feeler gauge. Surface irregularity must not exceed 0.003"

Using a wire brush and electric drill to clean carbon from the cylinder head.

(0.08mm) in any 6" (15cm) space, and the total must not exceed 0.007" (0.18mm) for the entire length of the head. If necessary, the cylinder head gasket surface may be machined at a shop equipped for such work.

CAUTION: Do not remove more than 0.010" (0.18mm) of stock from the head surface.

Clean the inside of the valve guides with a wire brush and lacquer thinner to remove all gum and carbon deposits. Any gum or deposits could prevent the valve from closing properly.

Inspect the cylinder head for cracks in the exhaust ports, combustion chambers, or external cracks to the water chamber. Inspect the valves for burned heads, cracked faces, or damaged stems. Bear in mind, excessive valve stem to bore clearance will increase oil consumption and may cause valve breakage. Insufficient clearance will result in noisy and sticky valves and also disturb engine operating smoothness.

ROCKER ARM STUD REPLACEMENT

GMC Engines Only

Replace any rocker arm stud with damaged threads with standard studs. If the studs are loose in the head, oversize studs are available in 0.003" (0.08mm) or 0.013" (0.33mm) oversize. These oversize studs

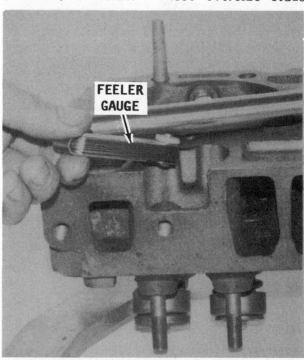

*The cylinder head gasket surface must be checked for any uneven condition. Surface irregularities **MUST NOT** exceed 0.003" in any six-inch space.*

can be installed after reaming the holes with Tool J-7515 for 0.003" (0.08mm) oversize, and Tool J-6036 for 0.013" (0.33mm) oversize as follows:

First, remove the old stud by placing Tool J-5802 over the stud.

Next, install the nut and flat washer and remove the stud by turning the nut.

Now, ream the hole for an oversize stud using the proper size tool just listed. After the hole has been properly reamed, coat the press-fit area of the stud with hypoid axle lubricant.

Finally, install a new stud using Tool J-6880. This tool **MUST** bottom on the head.

Mercury Marine and Ford Engines Only

Broken or damaged rocker arm studs can be replaced with standard studs. Loose studs in the head may be replaced with 0.006" (0.15mm), 0.010" (0.25mm), or 0.015" (0.38mm) oversize studs.

MANY GOOD WORDS

If it is necessary to remove a rocker arm stud, tool kit T62F-6A527B is available from a marine or automotive parts dealer. This kit includes: a stud remover, a 0.006" (0.15mm) over size reamer, and a 0.015" (0.38mm) oversize reamer. For 0.010"

(0.25mm) oversize studs, use reamer T66P-6A527B. To press in replacement studs, use stud replacer T69P-6049-D.

Standard and oversize studs can be identified by measuring the stud diameter 1-1/8" (28.6mm) or less from the pilot end of the stud. Stud diameters are:

Standard	0.3714-0.3721" 9.43-9.45mm
0.006" Oversize 0.15mm	0.3774-0.3781" 9.58-9.60mm
0.010" oversize 0.25mm	0.3814-0.3821" 9.68-9.70mm
0.015" oversize 0.38mm	0.3864-0.3871" 9.81-9.83mm

Any time a standard size rocker arm stud is replaced with a 0.010" (0.25mm) or 0.015" (0.38mm) oversize stud, **ALWAYS USE** the 0.006" (0.15mm) oversize reamer **BEFORE** finish-reaming with the 0.010" (.25mm) or 0.015" (0.38mm) oversize reamer.

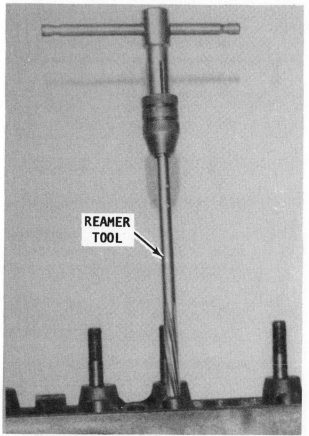

To restore production clearances, worn valve guides can be reamed oversize, and then valves with oversize stems installed.

A pressed in valve rocker arm stud can be removed using tool J-5802, or equivalent.

To remove a broken stud, position the sleeve of the rocker arm stud remover, Tool T62F-6A527-B, over the stud with the bearing end down. Thread the puller into the sleeve and over the stud until it is fully bottomed. Hold the sleeve with a wrench, and at the same time, rotate the puller clockwise and remove the damaged stud.

If a loose rocker arm stud is being replaced, ream the stud bore using the proper reamer, or reamers in sequence, for the required oversize stud. **TAKE EXTRA CARE** to be sure metal particles do not enter the valve area.

Coat the end of the new stud with Lubriplate, or equivalent. Align the stud and installer T69P-6049-D with the stud bore. Next, tap the sliding driver until it bottoms. After the installer makes contact with the stud boss, the stud is installed to its correct height.

VALVE MECHANISM SERVICE

Clean the valves, springs, spring retainers, keepers, and sleeves in solvent, and then blow the parts dry with compressed air.

Inspect the valve face and head for pits, grooves, and scores. Inspect the stem for wear and the end for grooves. The face must be trued on a valve grinding machine, which will remove minor pits and grooves.

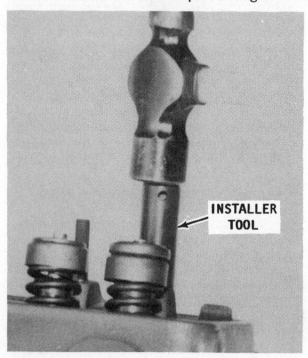

*Before driving the new rocker arm stud into place with an installer tool, **ALWAYS** coat the parts with hypoid gear lubricant.*

Valves with serious defects, or those having heads with a knife edge, **MUST** be replaced.

VALVE GUIDE

Measure the valve stem clearance with a dial indicator clamped on one side of the cylinder head. Locate the indicator so movement of the valve stem will cause a direct movement of the indicator stem. The indicator stem must contact the side of the valve stem just above the valve guide. With a new valve and the head dropped about 1/16" (1.6mm) off the valve seat, move the valve stem from side-to-side, with a light pressure to obtain a clearance reading. Refer to the following table for the correct valve stem-to-guide clearance for different valve stem diameters.

STEM DIAMETER	INTAKE VALVE	EXHAUST VALVE
5/16" .31mm	.0015-.0035" .038-.088mm	.0025-.0035" .063-.088mm
11/32" .34mm	.0015-.0035" .038-.088mm	.0025-.004" .063-.102mm
3/8" .37mm	.0015-.0035" .038-.088mm	.0025-.004" .063-.102mm

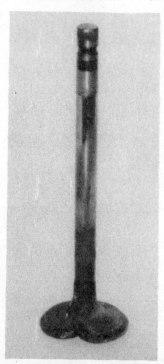

*A severely burned exhaust valve face. The valve was sticking in the guide as evidenced by the gum on the neck of the stem. **TAKE TIME** to clean the valve guide thoroughly.*

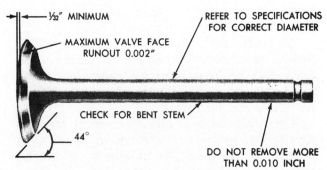

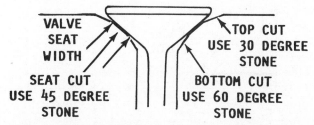

Critical valve tolerances.

Line drawing to depict the valve making proper contact with the valve seat. Terminology and the stones used are explained in the text.

Valve stem guides are not replaceable. However, valves with oversize stems are available for intake and exhaust valves. Reamers are also available to ream the bores for new valves.

VALVE SEATS

Remove the carbon from the combustion chambers. Use care to avoid scratching the head or the valve seats. Clean all carbon and gum deposits from the valve guide bores. Clean the valves. Inspect the valve faces and seats for pits, burnt spots, or evidence of poor seating.

"True" the valve seats to a 45° angle. Cutting a valve seat results in lowering the

valve spring pressure and increases the width of the seat. The nominal width of the valve seat is 1/16" (1.58mm). If a valve seat is over 5/64" (1.98mm) wide after truing, it should be narrowed to the specified width by using 30° to 60° stones.

If the valve and seat are refinished to the extent the valve stem is raised more than 0.050" (1.27mm) above its normal height, the hydraulic valve lifter will not operate properly. If this condition exists, the worn part **MUST** be replaced. The normal height of the valve stem retainer above the valve spring seat surface of the head is 1.925" (48.9mm).

SPECIAL WORDS

Lapping valve seats is now considered by most engine rebuilders as an obsolete procedure. The following fact has been proven: A valve will seat **ITSELF** in the first 30 seconds of engine operation regardless if the seat has or has not been lapped.

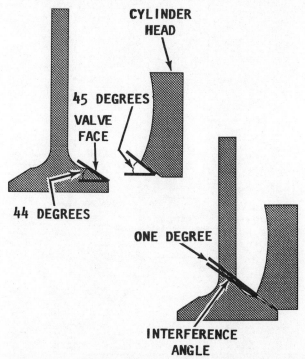

The angle of the cylinder head seat is ground to 45° using a 45° stone. The angle on the valve face is ground to 44° on a valve grinding machine. The 1° difference between the two angles is called the interference angle and provides a positive seal to the combustion chamber.

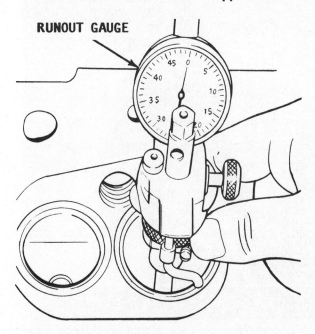

After the seat has been reground, check to be sure it is concentric with the guide by using a dial gauge. The run-out should not exceed 0.002".

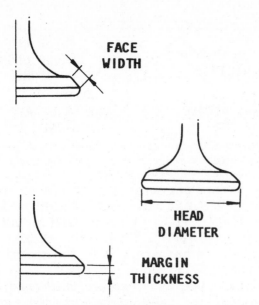

Common valve terminology. Measurements are made with a pair of calipers and must meet the limits listed for maximum performance.

VALVES

Grind the 45° valve face to 44° angle for a 1° interference angle. Remove only enough stock to correct runout or to dress off the pits and grooves. If the edge of the valve head is less than 1/32" (0.79mm) after

Using a professional type grinding machine to resurface a valve tip. The same machine can also be used to resurface the valve face -- the long bevelled edge. A constant liberal flow of cutting fluid is essential.

grinding, replace the valve as it will run too hot during engine operation.

Measure the valve seat widths, which should be as specified in the Appendix. The seats can be narrowed by removing stock from the top and bottom edges by using a 30° stone and 60° stone.

The finished seat should contact the approximate center of the valve face. To determine the position of the seat on the valve face. Coat the seat with Prussian blue, or equivalent, and then rotate the valve in place with a light pressure. The blue pattern on the valve face will show the position of the seat.

VALVE SPRINGS

Check the valve springs for the correct tension against the Specifications in the Appendix. A quick check can be made by laying all the springs on a flat surface and comparing the heights which **MUST** be even. Also, the ends **MUST** be **SQUARE** or the spring will tend to cock the valve stem. Weak valve springs cause poor engine performance; therefore, replace a weak spring

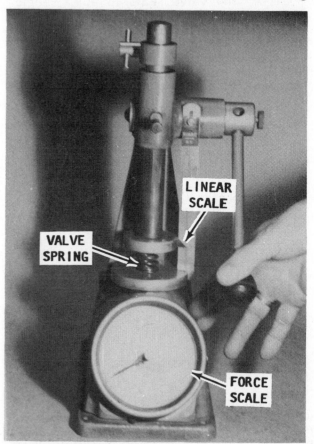

Using a valve spring tester to determine the force required to compress a valve spring to a definite length.

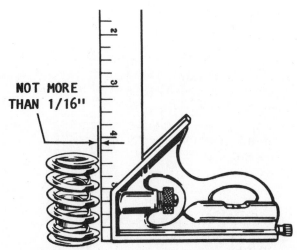

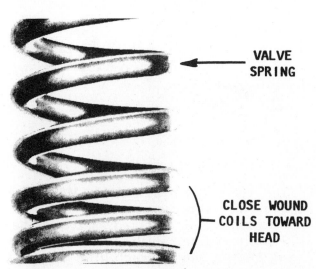

A valve spring should not be out by more than 1/16" when it is rotated against a square on a flat surface, as shown. If it is not square within 1/16", the spring should be replaced.

or if the spring is out of square by more than 1/16" (1.58mm).

HYDRAULIC LIFTERS

Dirt, deposits of gum, and air bubbles in the lubricating oil can cause the hydraulic

Enlarged view of a coil spring to show the differences in the coil windings at the ends. Install the spring with the close wound coil end towards the head.

lifters to wear enough to cause failure. The dirt and gum can keep a check valve from seating, which in turn will cause the oil to return to the reservoir during the time the push rod is being lifted. Excessive movement of the lifter parts causes wear, which soon destroys the lifters' effectiveness.

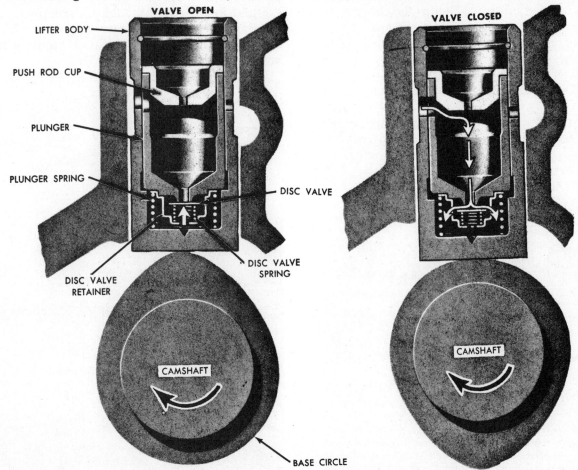

Relationship of the cam lobe and the valve lifter in the valve open position (left) and in the valve closed position (right).

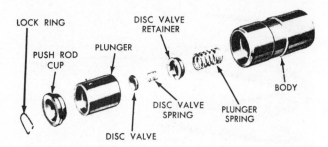

Arrangement of hydraulic lifter parts.

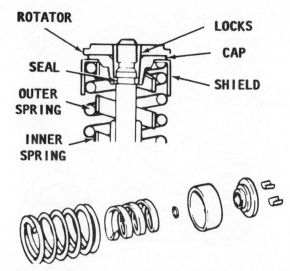

Arrangement of valve mechanism parts before and after installation.

The valve lifter assemblies **MUST** be kept in the proper sequence in order to ensure each will be installed back into the same position from which it was removed. Clean, inspect, and test each lifter separately so as not to intermix the internal parts. If only one part of a lifter is defective, replace the entire assembly.

To test a cleaned lifter, assemble the parts dry, and then quickly depress the plunger with your finger. The trapped air should partially return the plunger if the lifter is operating properly. If the lifter is worn, or if the check valve is not seating, the plunger will not return.

Devise a system to keep the lifters in order and soak them in 30W engine oil for a few hours prior to installation, preferably overnight.

Apply Lubriplate to the valve tips before installation. Install each valve into the same port from which it was removed. The spring end with the closed coil **MUST** be against the cylinder head. Place the spring and rotator on the exhaust valves, and then compress the spring with the valve compressor and install the oil seal and keepers. Pay special attention to ensure the seal is **FLAT** and not twisted in the valve stem groove and the keepers seat properly.

CYLINDER HEAD ASSEMBLING

Coat the valves, valve stems, and valve guides with a liberal coating of engine oil.

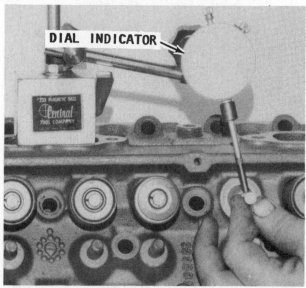

Checking the valve stem clearance with a new valve and dial indicator. If the clearance exceeds 0.0037" to 0.0052", the guide must be reamed oversize and a valve with an oversize stem installed to obtain the proper clearance.

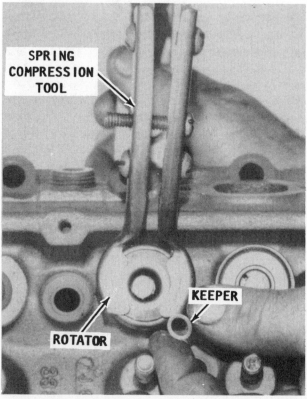

Installing the valve into the cylinder head, as explained in the text.

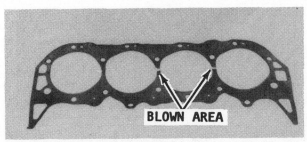

Compression loss may be caused by a blown head gasket between cylinders, as indicated.

Measuring the height of the installed valve spring. Compare this measurement with the Specifications in the Appendix.

Assemble the remaining valves, valve springs, rotators, spring caps, oil seals, and keepers in the cylinder head in the same manner as the first.

Measure the assembled height of the valve spring from the surface of the cylinder head spring pad to the underside of the rotator.

If the assembled height is greater than the amount listed in the Specifications in the Appendix, it will be necessary to install enough spacers between the cylinder head spring pad and the valve spring until the assembled height is within specifications. **NEVER** install spacers unless it is necessary because the spacers will overstress the valve spring and exert an extra load on the camshaft lobe. This extra strain could lead to a worn lobe or possibly cause the spring to break.

CYLINDER HEAD INSTALLATION

The gasket surfaces on both the head and the block must be clean of any foreign material and free of nicks or heavy scratches. Bolt threads in the block and threads on the cylinder head bolts **MUST** be clean. Dirt on the threads will affect the bolt torque value.

ALWAYS use a marine head gasket because the boat may be used in salt water. Marine head gaskets use a special copper or stainless steel coating to resist corrosion. Automotive head gaskets are manufactured from steel which corrodes rapidly when exposed to salt water. Should this occur,

water could enter the cylinder and cause considerable damage.

If the head gasket has the word **FRONT** stamped onto one edge, this edge **MUST** face **FORWARD** and the stamped side must face **DOWN**.

Place a new cylinder head gasket over the cylinder dowels on the block. Dowels in the block will hold the gaskets in place. **USE CARE** when handling the gaskets to prevent kinking or damaging the surfaces. **DO NOT** use any sealing compound on head gaskets, because they are coated with a special lacquer to provide a good seal, once the parts have warmed to operating temperature.

Carefully guide the cylinder head into place over the dowel pins and gasket. Coat the threads of the cylinder head bolts with sealing compound and install them finger-tight. Tighten the cylinder head bolts alternately and evenly in three sequences to a torque value listed in the following table.

Installing the head on an in-line pushrod engine. A new head gasket should ALWAYS be used whenever the head is removed for any reason.

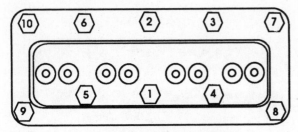

GMC 4 CYLINDER (L6 IS SIMILAR)

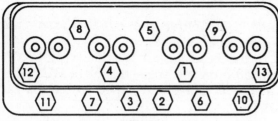

GMC V6

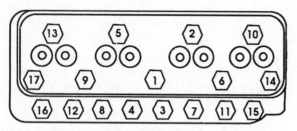

GMC SMALL BLOCK V8

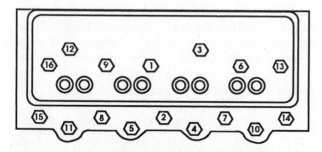

GMC BIG BLOCK V8

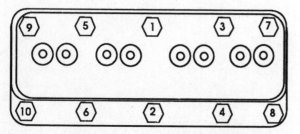

MERCURY MARINE 4 CYLINDER AND FORD V8

Cylinder head bolt tightening sequence for the engines indicated.

Begin in the center and work alternately toward both ends of the head. Tighten to 1/3 the torque value on the first sequence; to 2/3 the torque value on the second sequence; and to the full torque value on the third and final sequence.

M/M 4-Cylinder	130 ft lb (176Nm)
GMC 4-Cylinder	100 ft lb (136Nm)
GMC L6	95 ft lb (128Nm)
GMC V6	70 ft lb (100Nm)
GMC V8 Sm. Block	70 ft lb (95Nm)
GMC V8 Big Block	80 ft lb (109Nm)
Ford V8	70 ft lb (95Nm)

Install the exhaust manifolds, see Section 3-7, this chapter.

Apply Lubriplate, or equivalent, to both ends of the push rods. Install the push rods in the same position from which they were removed. Make sure each pushrod seats inside the lifter socket. On some engines, the push rod is hardened on one end. This end is identified by strips of color. The push rod **MUST** be installed with the hardened end upward to make contact with the rocker arm.

Apply a coating of Lubriplate, or equivalent, to the valve stem tips. Install the rocker arms and adjust the valve lash, see Section 3-10, this chapter.

SPECIAL WORDS
MERCURY MARINE AND FORD
ROCKER ARM ATTACHING BOLTS

The rocker arm attaching bolts **MUST** be tightened to the specified torque value while the valve is in the closed position. If the bolts are tightened while the valve is open, a pushrod may be bent the first time the engine is started.

Rotate the engine until the number one piston is at TDC (top dead center) on the compression stroke. The following bolts may be tightened to a torque value of 20 ft lb (27Nm) with the engine in this position:

No. 1 Cylinder - Intake & Exhaust
No. 2 Cylinder - Intake
No. 3 Cylinder - Exhaust
No. 4 Cylinder - Intake

Rotate the crankshaft through 360° and tighten the following bolts to the same torque value.

No. 2 Cylinder - Exhaust
No. 3 Cylinder - Intake
No. 4 Cylinder - Exhaust

Connect the exhaust manifolds at the exhaust housing, see Section 3-7, this chapter.

Install the alternator attaching bracket and the alternator. Adjust the drive belt tension.

Clean the valve rocker arm covers. Place the valve rocker arm cover gaskets in each cover. Always use **NEW** gaskets. Check to be sure the tabs of the gasket engage the notches in the cover. Install the valve rocker arm covers, and then tighten the securing bolts in two stages. First, tighten the bolts to a torque value of 3 to 5 ft lbs. The gasket will have a tendancy to "settle". Therefore, after a short time --say five to ten minutes, tighten them again to the same torque value.

NOW, THESE WORDS

Sealer is used to install the intake manifolds on Ford V8 units. This sealer will **SET UP** in **15 MINUTES**. Therefore, keep the work moving along without delays for coffee, tea, trip to the head -- whatever.

Clean the mating surfaces of the intake manifold, the cylinder heads, and the cylinder block with solvent. Apply a bead of silicone rubber sealer onto the cylinder head at four places.

Place new seals onto the cylinder block and new gaskets on the cylinder heads, with the gaskets interlocked with the seal tabs. Check to be sure the holes in the gaskets are properly aligned with the holes in the cylinder heads.

Install the intake manifold, see Section 3-6, this chapter.

Install the coil, spark plugs, and high-tension leads. Connect the upper water hose and the engine ground strap. Connect the temperature sending unit wires and install the fuel and vacuum lines in the clip at the water outlet.

Close the water drain valves.

CAUTION

Water must circulate through the stern drive, also to and from the engine, anytime the engine is run to prevent damage to the water pump located in the lower unit (original and Alpha drives); or on the front of the engine (Bravo drive). Just a few seconds without water will damage the water pump.

Start the engine and check the completed work for leaks and performance.

3-9 PISTON, RING, AND ROD SERVICE

This section provides detailed procedures for removing, disassembling, cleaning, inspecting, assembling, and installing the complete piston/rod assembly. All parts **MUST** be kept together because if the old pistons are serviceable, they **MUST** be installed onto the rods and back into the same cylinder bore from which they were removed.

Pistons are pistons, rods are rods, and rings are rings. Therefore, the procedures apply to all engines. However, any differences, should they occur, such as special markings, will be clearly identified.

REMOVAL -- PISTON/ROD ASSEMBLY

Open all valves and drain water from the block.

Remove the engine, see Chapter 13.
Remove the oil pan, see Section 3-5.
Remove the oil pump, see Section 3-5.
Remove the head/s, see Section 3-8.

After the oil pan, oil pump, and cylinder head/s have been removed, use a ridge reamer to remove the ridge and deposits from the upper end of the cylinder bore. Before the ridge and/or deposits are removed, turn the crankshaft until the piston is at the bottom of its stroke and place a cloth on top of the piston to collect the cuttings. After the ridge is removed, turn the crankshaft until the piston is at the top of its stroke to remove the cloth and cuttings.

A ridge almost always forms on the upper edge of the cylinder wall because this area receives the least wear from the piston. A ridge remover (shown below), is used to cutaway the ridge.

Repeat the procedure until the ridge has been removed from the upper end of each cylinder bore.

Remove the connecting rod cap nuts and the cap from the first piston. Slide a short piece of hose, plastic, or rubber, onto each rod cap stud, to provide protection against scratching the cylinder wall with the bare studs as the piston assembly is removed.

Push the piston assembly out the top of the cylinder block. Remove the protective pieces of hose material from the rod cap studs, and then temporarily secure the cap to the rod with the two nuts. This will **GUARANTEE** the cap stays with the rod.

As each piston assembly is removed, immediately identify the piston, connecting rod, and cap with a quick drying paint or silver pencil to **ENSURE** each part will be replaced in the exact position from which it was removed.

CRITICAL WORDS

Never scratch, engrave, emboss, or notch a connecting rod. Such action will cause a stress riser -- a weak point -- in the molecular structure of the material. A weak spot could cause rod failure.

GMC and Ford use a different numbering system for the cylinders in "V" engine blocks. The accompanying illustration will help to properly identify the cylinders of the unit being serviced.

Repeat the removal procedure until all piston assemblies have been removed, identified, and partially reassembled to keep the parts matched. It will be necessary to turn the crankshaft slightly to remove the rod

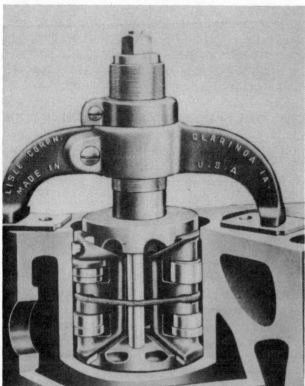

*A ridge remover **MUST** be used to cut the ridge from the top of the cylinder walls. A stop under the blade prevents cutting into the wall too deeply. **NEVER** cut more than 1/32" below the bottom of the ridge.*

*The rod bolt threads must **ALWAYS** be covered with a piece of rubber hose to prevent damage to the bearing surface by the rod threads scraping as the piston assembly is removed.*

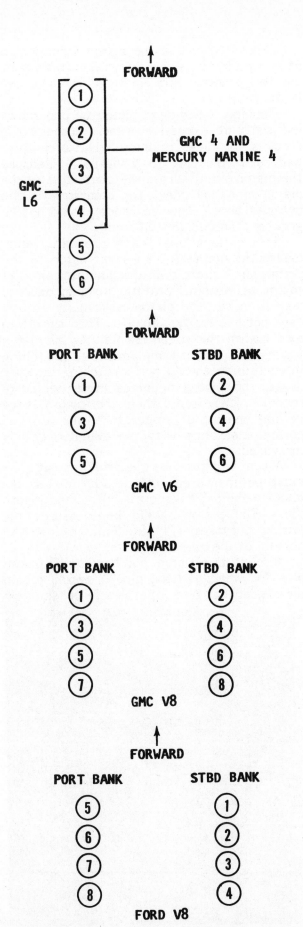

FORWARD

① ②
③ GMC 4 AND
④ MERCURY MARINE 4

GMC
L6
⑤
⑥

FORWARD

PORT BANK   STBD BANK

① ②
③ ④
⑤ ⑥

GMC V6

FORWARD

PORT BANK   STBD BANK

① ②
③ ④
⑤ ⑥
⑦ ⑧

GMC V8

FORWARD

PORT BANK   STBD BANK

⑤ ①
⑥ ②
⑦ ③
⑧ ④

FORD V8

Cylinder numbering system for the engines covered in this manual.

caps and to push the piston assembly clear of the cylinder.

CLEANING AND INSPECTING

All cleaning and inspection work should be done in an organized and thorough manner. A complete and proficient job in this area will justify the hours of work and the cost of new parts involved in the engine reconditioning task.

An oversight of letting a defective part be reinstalled, or assembling parts which have not been properly cleaned will nullify much of the other work and may restrict performance of the engine after the job is completed.

Check the rod bolts and nuts for defects in the threads. Inspect the inside of the rod bearing bore for evidence of galling, which indicates the insert is loose enough to move around. Check the mating surfaces of the rod and rod cap for any sign of damage caused by the rod bolts having been over-tightened.

If the engine has over 750 hours of service, or if the piston and rod assembly has been removed, it is considered good shop practice to replace the piston pins. Loose piston pins, coupled with tight piston assemblies such as new piston rings, will cause piston pin noises. These noises may disappear as the engine loosens, but strange noises are difficult to explain following an engine overhaul.

Most mechanics have the piston pin work done by a machine shop with the necessary equipment and trained people to perform a precision job. If done in the machine shop, the connecting rods will be aligned resulting in the pistons and rings running true with the cylinder walls.

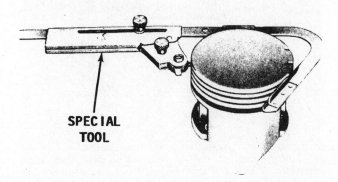

SPECIAL TOOL

*The ring grooves must be thoroughly cleaned to permit the new rings to seat properly. **TAKE CARE** not to nick the sealing surfaces.*

Pre-ignition caused this piston crown to melt.

Wash the connecting rods in cleaning solvent and dry with compressed air. Check for twisted or bent rods and inspect for nicks or cracks. Replace any damaged connecting rods.

Remove the compression rings with an expander. Remove the oil ring by removing the two rails and the spacer-expander, which are separate pieces in each piston's third groove.

Use a cleaning solvent to remove the varnish from the piston skirts and pins. **NEVER** use a wire brush on any part of the piston. Clean the ring grooves with a groove cleaner and make sure the oil ring holes and slots are clean. Inspect the pistons for cracked ring lands, skirts, or pin bosses; wavy or worn ring lands; scuffed or damaged skirts; and eroded areas at the top of the pistons.

Check the top of the piston for any sign of "piston burn" caused by pre-ignition, poor quality gasoline, improper timing, or water in the gas.

Replace a damaged piston or one showing signs of excessive wear. Inspect the grooves for nicks or burrs. Nicks or burrs will cause the ring to "hang-up". Measure the piston skirt (across the centerline of the piston pin), and check the clearance in the cylinder bore. The clearance should not be greater than 0.0025" (0.06mm).

The pistons are "cam ground", which means the diameter at a right angle to the piston pin is more than the diameter parallel to the piston pin. Stating this fact another way is to say the piston is actually elliptical, not perfectly round. This condition exists when the piston is cold. As the piston warms to operating temperature, stretching in certain areas and non stretching in other areas, the piston becomes more perfectly round. Therefore, when the piston is at normal operating temperature, the area of normal clearance with the cylinder wall is increased.

When a piston is checked for size, it must be measured with a micrometer applied to the skirt at points 90° to the piston pin. The piston should be measured for fitting purposes 1/4" (6.35mm) below the bottom of the oil ring groove.

Inspect the piston pin bores and piston pins for wear. Piston pin bores and piston pins must be free of varnish or scuffing

The connecting rod of this piston bent, and then snapped, due to the tremendous forces exerted on the rod when the engine "ranaway". A "runaway" engine may be caused by operating the unit at high rpm using a flush attachment on the lower unit and no load on the propeller.

The rings on this piston became stuck due to lack of adequate lubrication, incorrect timing, or overheating.

It is believed, this crown seized with the cylinder wall when the unit was operated at high rpm and the timing was not adjusted properly. At the same instant, the rod pulled the lower part of the piston downward severing it from the crown.

when being measured with a dial bore gauge or an inside micrometer. If the clearance is greater than 0.0001" (0.025mm), the piston and/or the piston pin should be replaced.

Inspect the bearing surfaces of the piston pins. Use a micrometer to check for wear by measuring the worn and unworn surfaces. Rough or worn pins MUST be replaced. Test the fit of each pin with its piston boss. If the boss is worn out-of-round

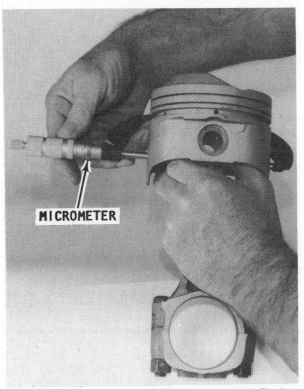

Measuring the diameter of a "cam ground" piston using a micrometer, as explained in the text.

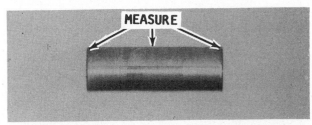

Three measurements with a micrometer should be made to determine piston pin wear: One on the center section; the second and third, at both ends.

or is oversize, the piston and pin assembly MUST be replaced. NEVER attempt to use an oversize pin because the pin is a press fit in the connecting rod. An easy finger push fit at 70°F temperature should insert the pin into the piston. Such a fit should allow 0.0003" (0.07mm) to 0.0005" (0.13mm) clearance.

PISTON DISASSEMBLING

Place the piston assembly on a support and position the assembly in an arbor press, as shown. Press the pin out of the connecting rod. Remove the assembly from the press. Lift the piston pin from the support, and remove the tool from the piston and rod.

PISTON ASSEMBLING

Apply a liberal amount of oil in the piston holes and the connecting rod holes to aid the press fit of the pin. Place the connecting rod in its respective piston with the flange or heavy side of the rod at the bearing end toward the FRONT of the piston. The front of the piston is identified by

After a piston pin has been coated with light engine oil, it should support its own weight in either pin boss at room temperature.

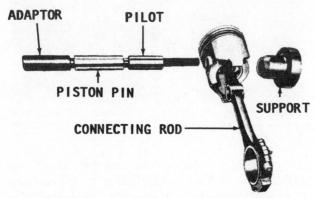

Arrangement of special tools required to properly install a piston pin through the piston.

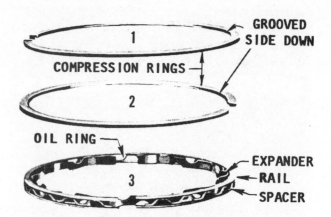

Proper arrangement of piston rings.

the cast depression in the top of the piston head.

Install the piston pin on the installer. Place the pilot spring and the pilot in the support. Install the piston and rod on the support, with the pilot indexing through the piston and rod. Place the support on the arbor press. Start the pin into the piston, and press on the installer until the pin pilot bottoms. Remove the installer and support assembly from the piston and connecting rod assembly. Check the piston pin to be sure there is free movement in the piston bores.

Ford V8 Engine

Insert the connecting rod in the piston with the numbered side of the rod facing outward and the arrow on the piston facing forward when installed in the block, as shown in the accompanying illustration. Press the pin into place, and then keep the assembly in order to **ENSURE** it will be

installed in the correct bank of the block -- port or starboard.

RING INSTALLATION
ALL ENGINES

Good shop practice dictates a new set of rings be installed whenever the piston is out of the block. Order the ring set according to the amount of cylinder wall wear.

If the wear is less than 0.005" (0.13mm), a standard set of piston rings can be used. If the cylinder wall wear is between 0.006" (0.15mm), and 0.012" (0.30mm), a set of piston rings with a special oil ring and expanders **MUST** be used to keep the engine from pumping oil into the cylinder. If the

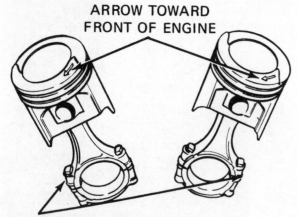

View looking **AFT** *to indicate piston and rod location for both banks of a Ford V8 engine.*

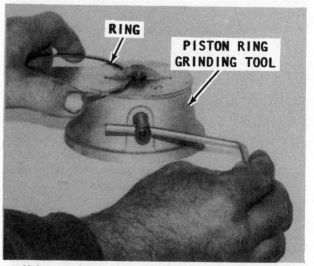

Using a piston ring grinding tool to obtain proper ring end gap clearance. Remove only a small amount at a time, and then check the gap in the cylinder bore as shown in the adjacent illustration on the next page. If too much material is removed, the end gap will be excessive resulting in some compression loss. Even a very small amount of compression will seriously affect performance of the powerplant. Therefore, the end gap grinding work must move ahead **SLOWLY** *with frequent check being made in the cylinder.*

cylinder bore is worn over 0.012" (0.30mm) the cylinder should be reconditioned by boring or honing in order to straighten the cylinder walls. This operation is necessary to enable the new piston rings to make a proper seal.

Before installing a set of piston rings, the end gaps and the side clearance between the piston ring groove and the ring must be checked. The correct side clearance should be 0.002" (0.05mm) to 0.004" (0.10mm), with a wear limit of 0.006" (0.15mm). The side clearance of a new ring is not excessive unless the ring groove is worn.

A burr in the soft piston metal may cause the ring to bind. To check for burrs, rotate the back of each piston ring around the groove and be sure it does not bind anywhere around the piston.

The end gap of each ring must be checked to be sure the ring is not too long and there is enough gap where the two ends come together to handle expansion of the ring when it becomes hot. The two ends must never come completely together during operation of the piston. The oil rings are flexible and do not need to be checked for end gap.

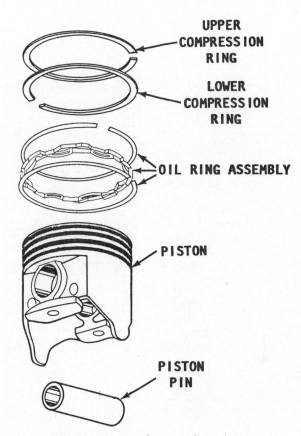

Ring arrangement for a modern piston.

Checking End Gap

To check the end gap of compression rings: first, place each ring in the cylinder into which it will be used. Next, square the ring in the bore using the upper end of a piston. Now, measure the end gap between the ends of the ring. The standard clear-

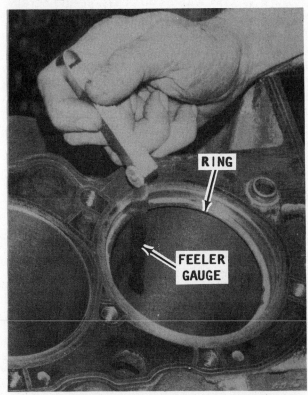

*Measuring piston ring end gap with a feeler gauge. The end gap for both top rings must be 0.0025" — 0.0040". **ALWAYS** use an inverted piston to push the ring squarely into the cylinder in order to obtain an accurate measurement.*

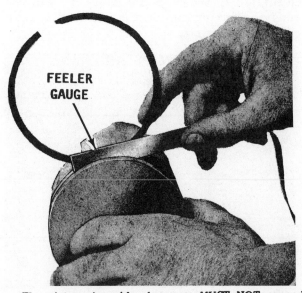

*The piston ring side clearance **MUST NOT** exceed 0.004".*

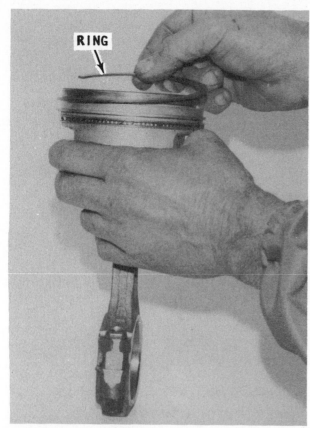

Installing a new ring into the ring groove. Notice how the ring is bent slightly upward. After the ring is started, rotate the ring around and over the top of the piston. The ring will feed into the groove and the end will finally snap into place.

ance is 0.010" (0.25mm) to 0.020" (0.50mm), except for the steel rails of the oil rings. The gap for these rings is 0.015" (0.38mm) to 0.030" (0.76mm). The exception to all of these measurements would be as listed in the specifications for a particular group of engines. If the end gap is too small, the ends of the ring can be filed to increase the gap measurement.

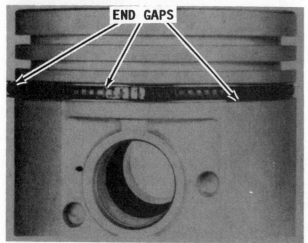

Proper method of spacing the rail gaps and the spacer gap on a compound oil ring.

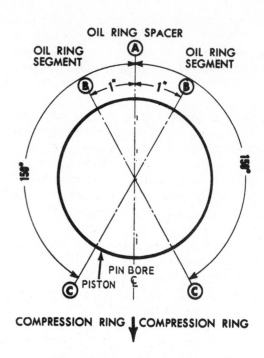

FRONT OF ENGINE

Diagram to indicate piston ring spacing.

Ring Installation

When installing the rings onto the piston, be sure to check the compression and scraper rings for the proper method of installation. Some rings have the word **TOP** stamped on the side. This side **MUST** face **UP**, toward the top of the piston. A compression ring with a groove in its outer face must be installed with this groove facing **DOWN**. If the groove is cut into the rear face of the ring, the groove **MUST** face **UP** when installed.

If a single chrome plated compression ring is used, it **MUST** be installed in the **TOP** groove. All compression rings are marked with a dimple, a letter "T", a letter "O", or the word **TOP** to identify the side of the ring which must be assembled toward the top of the piston.

Install the oil ring spacer in the oil ring groove and align the ring gap with the piston pin hole. Hold the spacer ends butted, and then install a steel rail on the top side of the spacer. Position the gap at least 1" (25cm) to the left of the spacer gap, and then install the second rail on the lower side of the spacer. Position the gap at least 1" (25cm) to the right of the spacer gap. Flex the oil ring assembly in its groove to make sure the ring is free and does not bind in the groove at any point.

If a steel spacer is used along with a top compression ring in order to compensate for machine work on the groove, the steel spacer **MUST** be installed **ABOVE** the cast iron ring.

After the rings are installed onto the piston, the clearance in the grooves needs to be checked with a feeler gauge. The recommended clearance between the ring and the upper "land" (upper surface of the groove), is 0.006" (0.15mm). Ring wear forms a step at the inner portion of the upper land. If the piston grooves have worn to the extent to cause high steps on the upper land, the step will interfere with the operation of new rings and the ring clearance will be too much. Therefore, if steps exist in any of the upper lands, replace the piston.

New compression rings on new pistons should have a side clearance of 0.003" (0.07mm) to 0.005" (0.13mm). The oil ring should have a side clearance of 0.0035" (0.09mm) to 0.0095" (0.24mm).

CYLINDER BLOCK
ALL UNITS

Check the cylinder block for cracks in the cylinder walls, water jacket, and the main bearing webs.

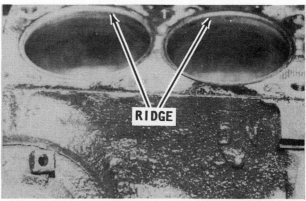

*Naturally the top ring does not travel to the top of the cylinder bore. Therefore, a ridge almost always forms at the top of the cylinder above the top ring travel. All measurements **MUST** be made below this ridge area or after the ridge has been removed. After the ridge has been removed, this cylinder bore will be resurfaced with a fine hone to remove the glaze and provide piston ring seating.*

Inspect the cylinder walls for scoring, roughness, or ridges which indicate excessive wear. Check the cylinder bores to taper, out-of-round, or excessive ridge wear at the top of the ring travel. This should be done with a dial indicator.

Set the gauge so the thrust pin is forced in about 1/4" (6.35mm) to enter the gauge in the cylinder bore. Center the gauge in

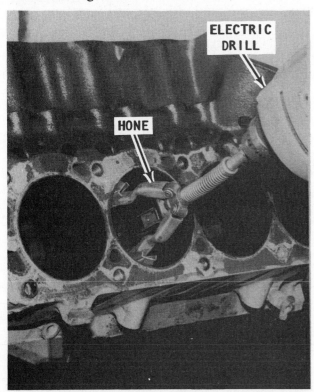

*Refinishing the cylinder wall using an electric drill and a hone. **ALWAYS** keep the tool moving in long even strokes over the entire depth of the cylinder.*

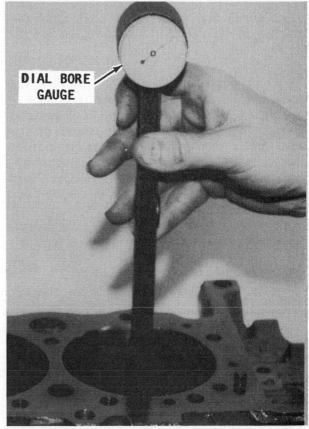

Cylinder wall taper and wear measured with a bore gauge indicator.

the cylinder and turn the dial to "0". Slowly move the gauge up and down the cylinder to determine the amount of taper. Move the gauge around in the cylinder to determine if it is out-of-round.

A cylinder bore which is tapered 0.005" (0.13mm) or more, or is out-of-round 0.003" (0.07mm) or more, must be reconditioned. If the cylinder bores are not worn excessively, use a 220-grit stone to remove the wall glaze to enable the new rings to seat quickly. To prevent excessive engine wear, **ALWAYS** use a solution of soap and hot water to remove all traces of abrasives.

PISTON AND ROD ASSEMBLY
INSTALLATION
ALL ENGINES

Install each piston in its respective bore. If old pistons are being installed for further service, **ENSURE** each piston is installed back into the same bore from which it was removed.

V6 and V8 Engines Only

The piston and rod assembly must be mated as shown for right and left bank (port and starboard) cylinders. The notch in the piston **MUST** face forward for the right bank (starboard) cylinders, and **MUST** face aft for the left bank (port) cylinders.

Assemble the piston and rod on the spring loaded guide pin. Lubricate the piston pin to avoid damage when it is pressed through the connecting rod. Install the drive pin in the upper end of the piston pin. Press on the drive pin until the piston pin bottoms. Remove the piston and rod assembly from the press.

Rotate the piston on the pin to be sure the pin was not damaged during the pressing operation.

Apply a liberal amount of light engine oil to the piston rings, pistons, connecting rod bearings, crankshaft journals and cylinder bores. Position the crankpin straight up and down. Remove the connecting rod cap. Seat the bearing upper shell in the rod. Slide connecting rod protectors — short pieces of rubber or plastic hose over the rod cap studs. These protectors will hold the upper bearing shell in place and prevent damage to the cylinder wall or to the crankpin during installation.

Rotate the oil ring rails until the gaps are toward the center of the engine. Rotate the compression rings until the gaps are **NOT** in line with each other and **NOT** in line with the gaps in the oil ring rails. Make sure the ends of the oil ring spacer-expander are not lapped over, but just butt together.

Compress the rings with a ring compressor.

SPECIAL WORDS FOR
MERCURY MARINE AND FORD ENGINES

A- Be sure the arrow on the top of the piston is facing **FORWARD**, toward the front of the engine. Pistons with no markings can be installed in either direction.

B- The numbers on the connecting rods and the bearing caps must be on the **OUTBOARD** side when the rod is installed in the cylinder bore.

SPECIAL WORDS FOR
ALL GMC ENGINES

A- The side of the piston with the depression cast into the top **MUST** face **FORWARD**, toward the front of the block.

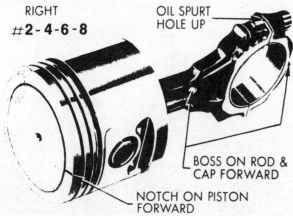

*Arrangement of piston and rod assembly parts for the **PORT** bank of a GMC V6 or V8 engine.*

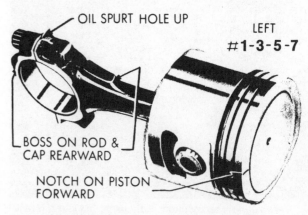

*Arrangement of piston and rod assembly parts for the **STARBOARD** bank of a GMC V6 or V8 engine.*

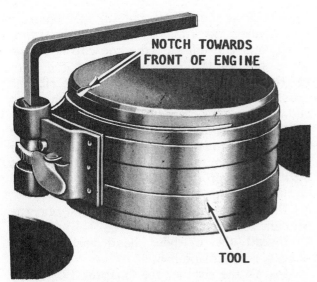

A piston ring compression tool must *ALWAYS* be used when installing a piston into the cylinder. If such a tool is not used, a ring may be distorted and break during piston installation.

B- The connecting rod bearing tang slot **MUST** face the side **OPPOSITE** to the camshaft.

GOOD WORDS FOR ALL MODELS

Take care to guide the connecting rods during the piston installation to prevent damaging the crankshaft journals.

Installation of a well-lubricated piston and ring assembly into its proper cylinder. *NEVER* "hammer" on the piston crown because a ring may have popped out of its groove and be broken. Use a gentle "tapping" action with the end of a wooden handle, as shown.

Use the wooden end of a hammer handle to push the piston down into the cylinder. **NEVER** hammer on a piston in an attempt to get it into a cylinder because a ring could be snapped during installation. Install the cap with a new lower bearing shell and temporarily hand tighten the connecting rod bolts. The main bearing clearance must be first checked before the caps are permanently installed and the rod bolts tightened to specification.

Install the other piston assemblies in a similar manner paying particular attention to install each used piston assembly into the same cylinder from which it was removed.

Clean the journal thoroughly of all traces of oil, and then place a piece of Plastigage on the bearing surface, the full width of the cap. Install the cap and torque the retaining bolts according to the Specifications listed in the following table. **DO NOT** turn the crankshaft with the plastigage in place or it will distort and the reading will have no value.

INSTALLING PLASTIGAGE

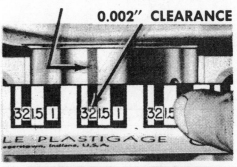

MEASURING PLASTIGAGE

After the connecting rod cap has been properly tightened, and then removed, the flattened Plastigage can be compared with the scale on the side of the package. In this manner, the amount of clearance can accurately be determined.

M/M 4-cyl	40 ft lb
	54Nm
GMC 4-cyl	40 ft lb
	54Nm
GMC L6 -	35 ft lb
	47Nm
GMC V6 -	45 ft lb
	61Nm
GMC V8 -	45 ft lb
	61Nm
Ford V8 -	45 ft lb
	61Nm

Remove the bearing cap. Use the scale on the package to determine the clearance. The bearing journal is tapered and so is the Plastigage. Measure the Plastigage at the widest point and also at the narrowest point. These measurements will give you the minimum and maximum clearances. If the clearance exceeds 0.0025" (0.06mm), a new insert should be installed. If a new insert is installed, again make another Plastigage measurement. If the new clearance still does not give the proper clearance according to the Specifications, then an undersized insert should be used.

Guide the connecting rod bearing into place on the crankshaft journal.

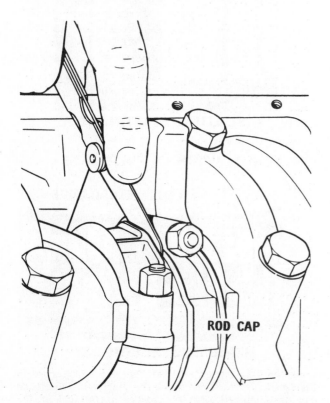

Using a feeler gauge to check end clearance between adjacent rod caps on the same crankshaft throw.

Check to be **SURE** the oil spit holes in the connecting rods are facing the camshaft, and the edge of the rod cap is on the same side as the conical boss on the connecting rod web. These marks (the rib and boss), will be toward the other connecting rod on the same crankpin.

Check the end clearance between the connecting rods on each crankpin. The clearance should be 0.005" (0.13mm) to 0.012" (0.30mm).

Install the oil pump, see Section 3-5.

Install the oil pan gaskets, seals, and the oil pan, see Section 3-5.

Install the cylinder head gasket/s and head/s, see Section 3-8.

Install the engine, see Chapter 13.

Connect all water hoses and wiring. Fill the crankcase with the proper amount and grade oil.

CAUTION

Water must circulate through the stern drive, also to and from the engine, anytime the engine is run to prevent damage to the water pump located in the lower unit (original and Alpha drives); or on the front of the engine (Bravo drive). Just a few seconds without water will damage the water pump.

Start the engine and check the completed work for leaks and performance.

CRITICAL WORDS

Operate the engine **ONLY** at reduced speed for the first hour.

3-10 VALVE LASH ADJUSTMENTS

Different procedures are to be followed when making the valve lash adjustment according to the engine manufacturer and the size engine being serviced. Therefore, move through this section until the heading for the particular model engine being serviced is located, and then follow only those procedures.

CHECKING HYDRAULIC LIFTER COLLAPSED CLEARANCE MERCURY MARINE ENGINE

Anytime the valves and the seats have been ground, or if you detect noise in the valve train which is not due to a collapsed lifter, the valve clearance **MUST** be checked. The valve clearance is **NOT** adjustable.

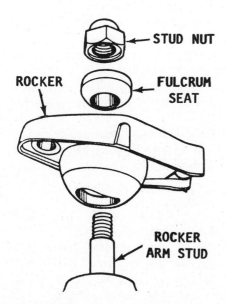

Arrangement of rocker arm parts for a Mercury Marine or Ford engine. Adjustment of this type valve train requires changing the length of the pushrods.

However, if the clearance is incorrect the pushrod can be replaced with a shorter or longer one to bring the clearance to specification.

To make the actual valve lash clearance check, first apply pressure on the valve lifter with Tool T711P-6513-A to bleed down the lifter plunger until it is fully collapsed, and then check the clearance with a feeler gauge. Insert the gauge bet-

ween the end of the valve stem and the face of the rocker arm. The clearance should be 0.110" (1.905mm), to 0.211" (4.445mm). **BE SURE** the feeler gauge is no wider than 3/8" (9.52mm). If the clearance is not within the prescribed limits, replace the push rod with a longer or shorter one as required.

The collapsed lifter clearance is checked by first turning the crankshaft to one of two positions, and then checking the valve clearances at each position as follows:

Position A

With the No. 1 piston at TDC, at the end of the compression stroke, check the following valves:

No. 1 Intake
No. 1 Exhaust
No. 2 Intake
No. 3 Exhaust
No. 4 Intake

Position B

Rotate the crankshaft through **EXACTLY** 360^{0} and then check the following valves:

No. 2 Exhaust
No. 3 Intake
No. 4 Exhaust

If the clearance is found to be **LESS** than 0.110" (1.905mm), replace the original pushrod with a **SHORTER** pushrod.

If the clearance is found to be **MORE** than 0.210" (4.445mm), replace the original pushrod with a **LONGER** pushrod.

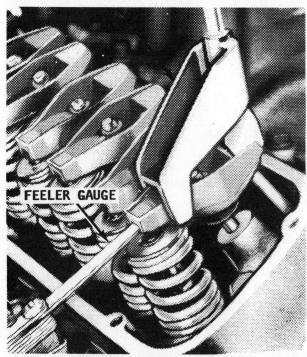

Using a feeler gauge to check the collapsed hydraulic lifter lash on a 4-cylinder Mercury Marine engine. To keep the clearance 0.110" to 0.211", shorter or longer push rods may be installed.

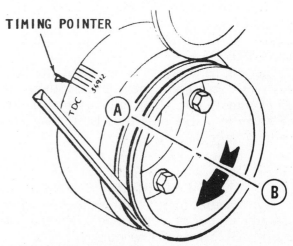

Crankshaft position for checking the valve lash of a 4-cylinder Mercury Marine engine. The No. 1 piston is at TDC at the end of the compression stroke for the "A" position. A chalk mark is then made 180^{0} opposite for point "B".

Pushrod Length	Stripe Color
8.505-8.535" 216.0-216.8mm	No stripe
8.565-8.595" 217.5-218.3mm	Black
8.596- 8.625" 218.3-219.1mm	Red
8.655-8.685" 219.8-220.6mm	Blue
8.625-8.655" 219.1-219.8mm	No stripe
8.685-8.715" 220.6-221.4mm	Yellow

Install the rocker arm covers. **ALWAYS** use new gaskets. Check to be sure the cover hole reinforcements are in place. Tighten the screws to a torque value of 7 ft lb (10Nm).

Close the water drain valves.

CAUTION

Water must circulate through the stern drive, also to and from the engine, anytime the engine is run to prevent damage to the water pump located in the lower unit (original and Alpha drives); or on the front of the engine (Bravo drive). Just a few seconds without water will damage the water pump.

Start the engine and check the completed work.

GOOD WORDS

The engine may run rough until the lash adjusters stabilize themselves, then it should smooth out. If it fails to smooth out, one or more of the push rods is sticking and needs to be replaced.

GMC 4- AND 6-CYL. IN-LINE ENGINES

The manufacturer recommends the valve lash adjustment be performed when the engine is **COLD** and **NOT** running.

The following reason is given for this recommendation: An attempt to turn the valve adjusting nut while the engine is running may not allow the lifters to bleed

down. This condition could result in valve train failure and may result in a bent rod.

However, procedures are presented in this section for adjustment with and without the engine operating. The "engine running method" is acceptable by the manufacturer.

Engine Cold and Not Operating

The valve lash is adjusted when the lifter is on the base of the cam.

STAGE ONE

Disconnect the spark plug wires, and then remove the spark plugs. Remove the distributor cap and observe the points.

Crank the engine, preferable with a remote starter switch, or have an assistant stationed at the key switch. Continue cranking for very brief moments until the rotor aligns with number one cylinder terminal and the points are wide open.

In this position, the number one piston is at TDC (top dead center) on the compression stroke. The following valves may be adjusted with the engine in this position:

4- and 6-Cylinder in-line
No. 1 Cylinder - Intake & Exhaust
No. 2 Cylinder - Intake
No. 3 Cylinder - Exhaust
No. 4 Cylinder - Intake

6-Cylinder
No. 5 Cylinder - Exhaust
No. 6 Cylinder - No Adj.

Adjust the nut until all lash is removed. This position may be determined by checking push rod side play. Turn the push rod and at the same time adjust the nut very **SLOWLY**. When no movement is felt on the push rod, turn the nut one full turn down to place the lifter plunger in the center of its travel.

STAGE TWO

Crank the engine, preferable with a remote starter switch, or have an assistant stationed at the key switch. Continue cranking for very brief moments until the rotor aligns with number four cylinder terminal if working on a four cylinder engine or number six cylinder terminal if working on a six cylinder engine. In this position the points are wide open.

The number four or six piston is at TDC (top dead center) on the compression stroke.

The following valves may be adjusted with the engine in this position:

    4- and 6-Cylinder In-line
    No. 2 Cylinder - Exhaust
    No. 3 Cylinder - Intake
    No. 4 Cylinder - Exhaust

    6-Cylinder
    No. 5 Cylinder - Intake
    No. 6 Cylinder - Intake & Exhaust

Adjust the nut until all lash is removed. This position may be determined by checking push rod side play. Turn the push rod and at the same time adjust the nut very **SLOWLY**. When no movement is felt on the push rod, turn the nut one full turn down to place the lifter plunger in the center of its travel.

All Models
Install the spark plugs and connect the spark plug wires.

Engine Warm and Operating
Loosen the adjusting nut until the valve clatters, indicating a loose adjustment. Next, rotate the push rod with one hand and at the same time tighten the adjusting nut with the other hand. Continue tightening very **SLOWLY** until all noise just disappears and a slight resistance may be felt as the

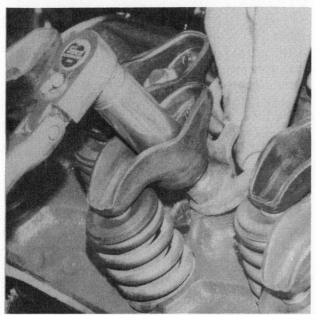

Rotate the push rod at the same time the rocker arm nut is being tightened to determine the Zero lash position, and then tighten the adjusting nut one additional full turn. The proper valve lash adjustment is then established.

push rod is rotated. At this point, **SLOWLY** turn the adjusting nut 1/4 turn more, and then wait through several engine revolutions for the lifter to bleed down before making any further adjustment.

DO NOT attempt to turn the adjusting nut a full turn while the engine is operating. Adjustment in this manner will not allow the lifters to bleed down. Failure of the lifter to bleed down could result in valve train damage, most likely a bent push rod or rods.

Continue making this adjustment 1/4 turn at-a-time until the nut is one complete turn down from the "zero lash" point. Repeat the sequence until all valves have been properly adjusted.

Install the rocker arm cover using a **NEW** gasket. Tighten the screws to a torque value of 4 ft lb (5Nm).

CAUTION
Water must circulate through the stern drive, also to and from the engine, anytime the engine is run to prevent damage to the water pump located in the lower unit (original and Alpha drives); or on the front of the engine (Bravo drive). Just a few seconds without water will damage the water pump.

Start the engine and check the completed work for performance and leaks.

GOOD WORDS
The engine may run rough until the lifters stabilize themselves, then it should smooth out. If it fails to smooth out, one or more of the push rods is sticking and needs to be replaced.

VALVE LASH ADJUSTMENT GMC V6 AND V8 ENGINES

Valve lash adjustment is divided into two sections. One section for certain numbered exhaust and intake valves and the other section for the remaining valves.

GMC V6 Only
Exhaust Valves 1, 5, and 6
Intake Valves 1, 2, and 3

GMC V8 Only
Exhaust Valves 1, 3, 4, and 8
Intake Valves 1, 2, 5, and 7

Crank the engine **SLOWLY** until two conditions exist: (1) the No. 1 piston is in the firing position, and (2) the mark on the harmonic balancer is aligned with the center or "O" mark on the timing tab fastened to the timing chain cover.

These two conditions may be determined by placing your fingers on the No.1 cylinder valves as the mark on the balancer comes near the "O" mark. If the valves are not moving, the engine is in the No. 1 firing position. If the valves move as the mark comes up to the timing tab, the engine is in the No. 6 firing position. This means the engine must be turned over one more time to reach the No. 1 position.

The actual valve lash adjustment is made by first, backing off the adjusting nut (rocker arm stud nut) until there is a small amount of play in the push rod. Next, tighten the adjusting nut to barely remove the clearance between the push rod and rocker arm. This position may be determined by rotating the push rod with one hand at the same time you tighten the nut with the other, as shown in the accompanying illustration.

At the point when you cannot rotate the push rod easily, the clearance has been eliminated. Now, tighten the adjusting nut 1 turn more to place the hydraulic lifter plunger in the center of its travel. No further adjustment is required.

GMC V6 Only
Exhaust Valves 2, 3, and 4
Intake Valves 4, 5, and 6

GMC V8 Only
Exhaust Valves 2, 5, 6, and 7
Intake Valves 3, 4, 6, and 8

Crank the engine **SLOWLY** one revolution from its position in the previous procedure until the No. 6 piston is in the firing position and the mark on the harmonic balancer is again aligned with the center or "O" mark on the timing tab fastened to the timing chain cover.

Repeat the sequence given in the previous procedure until the valves listed have been adjusted.

Continue in the same manner for the same numbered valves on the other head.

For those who prefer to adjust the valve lash while the engine is running, the preferred method is to find the "zero lash" described in the first procedure of this section, and then to slowly turn the adjusting

nut 1/4 turn and wait several engine revolutions for the lifter to bleed down before making a further adjustment. **DO NOT** attempt to turn the adjusting nut one full turn while the engine is operating. Adjustment in this manner will not allow the lifters to bleed down which could result in valve train damage, most likely a bent push rod or rods.

Continue making this adjustment 1/4 turn at-a-time until the nut is one complete turn down from "zero lash" point. Repeat the sequence until all of the valves have been properly adjusted.

Install the rocker arm covers. **ALWAYS** use new gaskets. Check to be sure the cover hole reinforcements are in place. Tighten the screws to a torque value of 5 ft lb (6Nm).

Close the water drain valves.

CAUTION
Water must circulate through the stern drive, also to and from the engine, anytime the engine is run to prevent damage to the water pump located in the lower unit (original and Alpha drives); or on the front of the engine (Bravo drive). Just a few seconds without water will damage the water pump.

Start the engine and check the completed work.

GOOD WORDS
The engine may run rough until the lash adjusters stabilize themselves, then it should smooth out. If it fails to smooth out, one or more of the push rods is sticking and needs to be replaced.

CHECKING HYDRAULIC LIFTER COLLAPSED CLEARANCE FORD V8 ONLY

Anytime the valves and the seats have been ground, or if you detect noise in the valve train which is not due to a collapsed lifter, the valve clearance **MUST** be checked. The valve clearance is **NOT** adjustable. However, if the clearance is incorrect the pushrod can be replaced with a shorter or longer pushrod to bring the clearance to specification.

Check the accompanying diagram on this page and notice that three crankshaft positions are designated by the letters **A, B,** and **C.**

The collapsed lifter clearance is checked by first turning the crankshaft to each of the three positions indicated in the diagram, and then checking the valve clearances at each position as follows:

Position A

With the No. 1 piston at TDC, at the end of the compression stroke, check the following valves:

No. 1 intake, No. 1 exhaust
No. 4 intake, No. 3 exhaust
No. 8 intake, No. 7 exhaust

Position B

Turn the crankshaft to Position B, and then check the following valves:

No. 3 intake, No. 2 exhaust
No. 7 intake, No. 6 exhaust

Position C

Turn the crankshaft to Position C, and then check the following valves:

No. 2 intake, No. 4 exhaust
No. 5 intake, No. 5 exhaust
No. 6 intake, No. 8 exhaust

To make the actual valve lash clearance check, first apply pressure on the valve lifter with Tool T711P-6513-A to bleed down the lifter plunger until it is fully collapsed, and then check the clearance with a feeler gauge. The clearance should be 0.083" (2.11mm) to 0.183" (4.65mm). **BE SURE** the feeler gauge is no wider than 3/8" (9.52mm). If the clearance is not within the prescribed limits, replace the push rod with a longer or shorter one as required.

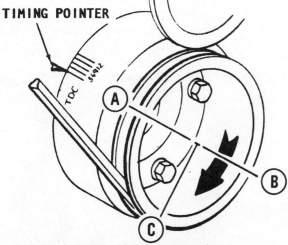

Crankshaft position for checking the valve lash on a Ford V8 engine. The No. 1 piston is at TDC at the end of the compression stroke for the "A" position. A chalk mark is then made 90° apart for points "B" and "C", as indicated.

Install the rocker arm covers. **ALWAYS** use new gaskets. Check to be sure the cover hole reinforcements are in place. Tighten the screws to a torque value of 4 ft lb (5Nm).

Close the water drain valves.

CAUTION

Water must circulate through the stern drive, also to and from the engine, anytime the engine is run to prevent damage to the water pump located in the lower unit (original and Alpha drives); or on the front of the engine (Bravo drive). Just a few seconds without water will damage the water pump.

Start the engine and check the completed work.

GOOD WORDS

The engine may run rough until the lash adjusters stabilize themselves, then it should smooth out. If it fails to smooth out, one or more of the push rods is sticking and needs to be replaced.

3-11 SERVICING TIMING CHAIN, GEAR AND CAMSHAFT

This section provides detailed procedures to service the timing chain, timing gear and the camshaft in that order. The timing

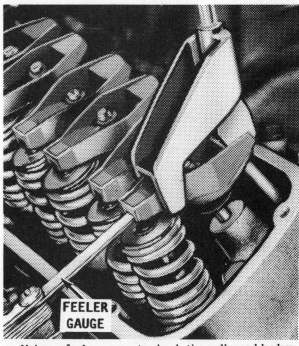

Using a feeler gauge to check the collapsed hydraulic lifter lash on a Ford V8 engine. To keep the clearance 0.090" to 0.190", shorter or longer push rods may be installed.

chain is the most likely of the three to require service, because of stretching. Therefore this task is presented first, followed by the timing gear and finally the camshaft.

SPECIAL WORDS

Two methods are used to drive the camshaft on Mercruiser engines covered in this manual.

1- Meshing gears -- one gear on the end of the camshaft and the other on the end of the crankshaft.

2- Timing chain -- indexed over metal or nylon sprockets -- one mounted on the end of the camshaft and the other mounted on the end of the crankshaft.

TIMING GEAR SERVICE
GMC 4- AND 6-CYLINDER IN-LINE ENGINES

The engine does not have to be removed from the boat if the camshaft merely needs servicing. However, if the camshaft bearings require replacement, then considerable work must be done including the removal of the engine as described in Chapter 13.

SPECIAL WORDS

Removal and replacement of the cam shaft bearings should be performed by qualified mechanics in a shop equipped to handle such work.

Timing Gear Cover Removal

Open the drain valves and drain the coolant from the block and exhaust manifold. Loosen the alternator bracket to provide slack, and then remove the belt.

Remove the valve cover and gasket. Loosen the valve rocker arm nuts until the pivot rocker arms clear the push rods. Note the position of the distributor rotor. Remove the distributor.

Remove the ignition coil and side cover gasket. **Take Time** to set up a system to keep the push rods and valve lifters in order to ensure each will be installed back into the exact location from which it was removed.

Withdraw each push rod and valve lifter in order.

If servicing an engine with a crankshaft pulley, remove the drive belt and remove the pulley from the pulley hub. Obtain

Using a puller to remove the harmonic balancer on a GMC engine.

puller tool J6978-E. Install the puller to the pulley hub. Some engines use two 3/8-24 x 2" and one 5/16-24 x 2", others use three 3/8-24 x 2" bolts. Remove the pulley hub.

If servicing an engine with a harmonic balancer (torsional damper), obtain the same puller tool J6978E.

GOOD WORDS

The harmonic balancer is the outer ring with drilled holes. The torsional damper is the center ribbed rubber hub. The balancer and the damper are a **ONE PIECE** assembly and **MUST** always be removed together. **NEVER** use a puller which attaches to the outer circumference of the balancer ring. This type puller will pull the balancer ring off the rubber damper and there is **NO WAY** they can be reassembled.

Install the tool to the inner bolt holes of the harmonic balancer. Rotate the center tool bolt to release both the damper and balancer from the crankshaft.

Remove the two screws through the oil pan into the timing gear cover. Pull the timing gear cover forward just a bit.

Use a sharp knife and cut the oil pan front seal flush with the cylinder block at both sides of the cover.

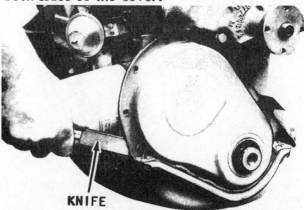

KNIFE

Before removing the timing cover, the ends of the oil pan gaskets MUST be cut with a knife.

Remove the front cover and attached portion of the oil pan front seal. Remove the front cover gasket. Clean all gasket material from the cover and block mating surfaces.

Pry the oil seal from the timing gear cover with a large screwdriver. Remove the oil nozzle from the block using a pair of needle nose pliers.

Camshaft Removal

Rotate the camshaft gear until the holes in the gear are aligned with the thrust plate screws. Remove the two screws. **CARE-FULLY** withdraw the camshaft gear and camshaft by pulling the gear straight forward and the shaft out of the block.

Check the gear and thrust plate end play. This clearance should be 0.001" to 0.005" (.025mm). If the decision is made to replace the camshaft gear, or the thrust plate, the gear must be pressed from the shaft. Gear removal from the camshaft requires the use of camshaft gear removal support sleave. Place the camshaft through the gear removal tool. Place the end of the removal tool onto the table of an arbor press, and then press the shaft free of the gear.

The thrust plate **MUST** be positioned to prevent the woodruff key in the shaft from damaging the shaft when the shaft is pressed out of the gear. Also, be sure to support the end of the gear or the gear will be seriously **DAMAGED.**

To assemble the camshaft parts, first firmly support the camshaft at the back of the front journal in an arbor press. Next, place the gear spacer ring and the thrust plate over the end of the shaft, and install the woodruff key in the shaft keyway. Install the camshaft gear and press it onto the shaft until it bottoms against the gear spac-

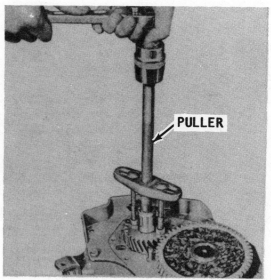
A puller **MUST** be used to remove the crankshaft gear.

er ring. Measure the end play of the thrust plate. This clearance should be 0.001" to 0.005" (.025mm). Camshaft cleaning and inspecting begins on Page 3-65.

Camshaft Installation

Coat the cam lobes with GMC Super Engine Oil Supplement (GMC P/N 1051858) or equivalent, and then add the remainder in the can to the crankcase oil. Insert the camshaft assembly most of the way into the engine block. Take care to move the shaft straight into the block and not to damage the bearings or cams.

Rotate the crankshaft and camshaft until the valve timing marks on the gear teeth align, then push the camshaft into final position. Install the camshaft thrust plate screws and tighten them to a torque value of 6 ft lb (8Nm).

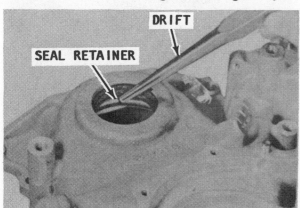
Using a drift to remove the front oil seal.

Removing the two thrust plate screws before the camshaft gear and camshaft are withdrawn straight forward out of the block.

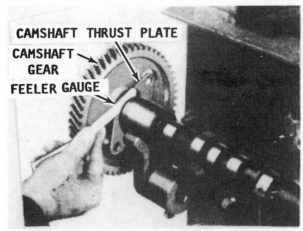

Using a feeler gauge to check camshaft gear end play.

Check the camshaft and crankshaft gear runout with a dial indicator. The camshaft gear runout should not exceed 0.004" (.102mm) and the crankshaft gear runout should not exceed 0.003" (0.076mm).

If the gear runout is more than the 0.003" (0.076mm), then either the shaft or the gear will have to be replaced.

Check the backlash between the timing gear teeth with a narrow feeler gauge or dial indicator. The backlash should not be less than 0.004" (0.102mm) nor more than 0.006" (0.152mm).

Install the seal on the front cover, pressing the tips into the holes provided in the cover. Coat the gasket with sealer and place it in position on the cover. Apply a 1/8" (0.125mm) bead of RTV (Part No. 105-1435) Silicone Rubber Seal to the joint at the oil pan and cylinder block as shown.

Install the centering tool J-23042 into the crankcase front cover seal. A centering tool **MUST** be used to align the crankcase front cover, to ensure the harmonic balancer installation will not damage the seal. The centering tool will also position the seal **EVENLY** around the balancer.

Install, and partially tighten, the two oil pan to front cover screws. Install the front cover-to-block attaching screws. Remove the centering tool. Tighten all the cover attaching screws alternately and evenly to a torque value of 6 ft lb (8Nm).

To install the harmonic balancer (torsional damper), assembly: Coat the oil seal contact area on the hub with engine oil. Position the assembly over the crankshaft and key, then start the center hub into position with a mallet. Drive the hub onto the crankshaft using tool J-5590 and a mallet until the hub bottoms against the crankshaft gear. The crankshaft extends slightly through the hub, therefore a hollow tool **MUST** be used to drive the hub completely to the bottom position. Install the pulley (if used) onto the hub. On some engines there are two 3/8" holes and one 5/16" hole, on others there are three 3/8-24 x 2" bolts. All holes **MUST** be matched on the hub in order to properly position the timing mark.

Install the valve lifters and push rods into the same position from which they were removed. Install the rocker arms onto the push rods and tighten the rocker arm nut **ONLY** enough to hold the push rod until the valves are adjusted. Replace the side cover

Using a dial indicator gauge to check camshaft and crankshaft run out.

Dial indicator gauge setup to check timing gear backlash.

using a new gasket. Install the ignition coil. Install the distributor rotor and cap.

Adjust the valves, see Section 3-10.

Install the drive belt, and then test the tension according to the procedure listed in Chapter 2, Section 2-3.

CAUTION

Water must circulate through the stern drive, also to and from the engine, anytime the engine is run to prevent damage to the water pump located in the lower unit (original and Alpha drives); or on the front of the engine (Bravo drive). Just a few seconds without water will damage the water pump.

Start the engine and check the completed work. Operate the unit at reduced rpm for the first hour.

SPECIAL WORDS

A new camshaft is "broken in" during the first 15 minutes of engine operation. Valve train loads are greatest at **LOW** rpm. Never operate the engine with a new camshaft at less than 1500 rpm for the first 15 minutes. Operation at less than 1500 rpm will cause considerable and unnecessary wear on the valve train parts.

TIMING CHAIN SERVICE
GMC V6 AND V8 ENGINES

Open the drain valves and drain the coolant from the block and exhaust manifold.

BEFORE *the camshaft is installed, check to be sure the oil hole in the block is clear.*

Remove the intake manifold, see Section 3-6.

Pull the spark plug wires from the brackets on the rocker arm covers. Remove the spark plug wires from the spark plugs.

Remove the rocker arm covers. Loosen the rocker arm nuts, and then turn the rocker arms aside.

Set up some kind of system to hold the push rods, and then remove each one in order so they will be installed back in the same position from which they were removed.

TAKE CARE not to get any dirt into the engine particularly into the valve lifters. Use cloths and compressed air to clean the cylinder heads and adjacent parts.

Remove the valve lifters and keep them in order. Each **MUST** be installed into the same lifter bore from which it was removed.

Remove the fuel pump and fuel pump push rod.

CAUTION

Be sure to plug the fuel line to prevent fuel from siphoning out of the tank.

Remove the drive belt and pulley. Remove the water pump.

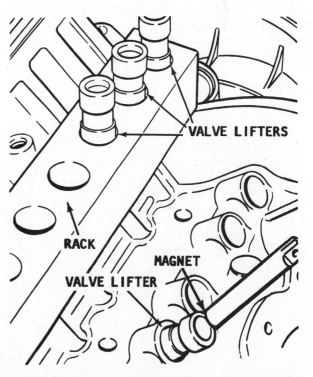

Devise a system to keep the valve lifters in order to ***ENSURE*** *each will be installed back in the same position from which it was removed.*

GOOD WORDS

The harmonic balancer is the outer ring with drilled holes. The torsional damper is the center ribbed rubber hub. The balancer and the damper are a **ONE PIECE** assembly and **MUST** always be removed together. **NEVER** use a puller which attaches to the outer circumference of the balancer ring. This type puller will pull the balancer ring off the rubber damper and there is **NO WAY** they can be reassembled.

Install tool J-23523-03 to the inner bolt holes of the harmonic balancer. Rotate the center tool bolt to release both the damper and balancer from the crankshaft.

Remove the front cover to block attaching bolts, and then pull the cover forward a little to allow room to cut the oil pan front seal with a sharp knife. Cut the seal, and then remove the front cover. Discard the gaskets.

Rotate the crankshaft until the timing mark on the camshaft sprocket is aligned with the timing mark on the crankshaft sprocket, as shown.

Timing Chain and Camshaft Removal

Remove the three camshaft sprocket to camshaft bolts. The camshaft sprocket is a light press fit onto the camshaft. Tap the sprocket lightly on the lower edge of the camshaft to dislodge it.

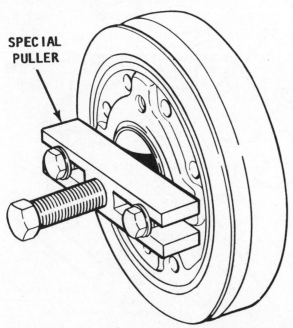

SPECIAL PULLER

A special tool MUST be used to remove the harmonic balancer. A standard wheel puller will not do the job.

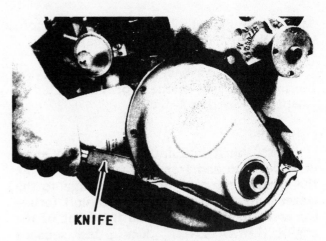

KNIFE

The ends of the oil pan gaskets must be cut prior to removing the timing cover.

Remove the camshaft sprocket and the timing chain together. If the crankshaft sprocket is to be replaced, remove the sprocket using Tool J-5825 on a small block engine or J-1619 on a big block engine.

Install two 5/16-18x5" bolts in the camshaft sprocket and carefully pull the camshaft straight out of the block. Use **UTMOST CARE** not to damage the bearing surfaces in any way.

Camshaft cleaning and inspecting begins on Page 3-65.

Camshaft Installation

Lubricate the camshaft journals with engine oil and the camshaft lobes with "Moly-Kote" or equivalent. Add the remainder in the can to the crankcase oil.

Install the two 5/16-18x4" bolts in the camshaft bolt holes. Insert the camshaft assembly into the engine block. **TAKE CARE** to move the shaft straight into the block and not to damage the bearings or cams.

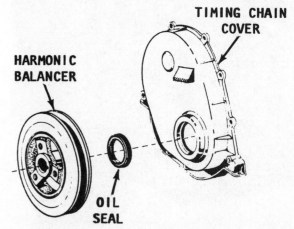

TIMING CHAIN COVER

HARMONIC BALANCER

OIL SEAL

The position of the harmonic balancer in front of the timing chain cover oil seal, makes it difficult to detect a leaking oil seal.

Removing the timing chain, camshaft sprocket, and crankshaft sprocket together as an assembly.

Timing mark on the camshaft sprocket (top), and the mark on the crankshaft sprocket (bottom), properly aligned through the centers of the camshaft and the crankshaft.

Timing Chain Installation

Position the key in the keyway of the crankshaft. Align the keyway of the crankshaft sprocket with the key, and then install the sprocket, using the bolt and nut from Tool J-23523. Engage the timing chain onto the camshaft sprocket.

Hold the sprocket in the vertical position, with the chain hanging down, and then align the timing mark on the camshaft with the timing mark on the crankshaft sprocket. Now, align the dowel in the camshaft with

the dowel hole in the camshaft sprocket, and then slide the sprocket onto the camshaft. **NEVER** try to drive the sprocket onto the camshaft because the **WELSH PLUG** at the rear of the engine will be dislodged.

Take up on the three mounting bolts to draw the camshaft sprocket onto the shaft. Check to be sure the dowel in the shaft indexes with the hole in the sprocket. Tighten the bolts alternately and evenly to a torque value of 20 ft lb (27Nm). Coat the timing chain with engine oil.

Installing the distributor drive gear after the timing chain is in place.

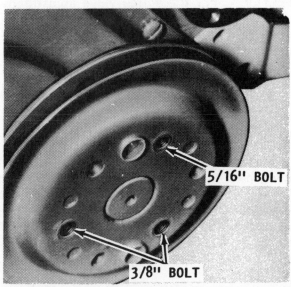

TAKE CARE to align the one 5/16" and the two 3/8" holes in the crankshaft pulley with the proper holes in the harmonic balancer.

Timing Chain Cover Installation

Coat the cover gasket with sealer and place it in position over the dowel pins in the cylinder block. Coat the timing cover seal lip with engine oil and place the cover in position over the dowel pins.

Install the attaching bolts and tighten them to a torque value of 7ft lb (9Nm).

Install the water pump, pulley, and the drive belt.

Coat the timing chain cover seal area on the harmonic balancer hub with engine oil. Align the keyway and start the balancer onto the crankshaft. Seat the harmonic balancer onto the crankshaft using a large washer and a 7/16-18x4" bolt.

Remove the bolt and washer. Install the belt pulley and tighten the screws to a torque value of 7ft lb (9Nm).

Thread a 7/16-20x2" bolt and thick washer into the crankshaft and tighten it to a torque value of 60 ft lb (81Nm).

Install the drive belt and adjust the tension according to the procedure given in Chapter 2, Section 2-3.

Install the fuel pump and fuel pump push rod. Install the valve lifters and push rods in the **SAME POSITION** from which they were removed. Position the rocker arm over the push rods and start the rocker arm nuts. Tighten the rocker arm nuts slowly and at the same time rotate the push rod. **STOP** tightening the nut when the push rod turns with difficulty, and move on to the next valve. Continue in a similar manner until all the nuts have been tightened as just described. For the actual valve adjustment, see Section 3-10.

Install the valve covers with **NEW** gaskets.

Install the intake manifold, see Section 3-6, this chapter.

Close the water drain valves.

CAUTION

Water must circulate through the stern drive, also to and from the engine, anytime the engine is run to prevent damage to the water pump located in the lower unit (original and Alpha drives); or on the front of the engine (Bravo drive). Just a few seconds without water will damage the water pump.

Start the engine and check the completed work.

SPECIAL WORDS

A new camshaft is "broken in" during the first 15 minutes of engine operation. Valve train loads are greatest at **LOW** rpm. Never operate the engine with a new camshaft at less than 1500 rpm for the first 15 minutes. Operation at less than 1500 rpm will cause considerable and unnecessary wear on the valve train parts.

TIMING CHAIN
AND ALTERNATOR SERVICE
EARLY MERCURY MARINE ENGINES

The alternator is mounted on the front end of the crankshaft and therefore, it is not driven in the conventional manner by a pulley arrangement.

The following paragraphs give detailed procedures for removing, disassembling, assembling, and installing the alternator in addition to the steps required to service the timing chain.

Alternator Removal

In order to remove the alternator, the front engine mount must be removed. Therefore, the engine must be raised slightly with a lifting device before the front mount can be removed.

Run a length of chain through the holes in the two lifting brackets, and then fasten the ends together with a bolt and nut. Attach the lifting device in the center of the chain, and then tie the chains together to prevent the lifting device from riding down the chain as a strain is taken.

To remove the alternator rotor, the engine must first be raised slightly, the rotor center bolt and washer removed, and a special tool used, as described in the text.

Remove the front engine mount. Drain the water from the closed-circuit cooling system. Drain the oil from the crankcase.

Remove the bolt from the center of the alternator rotor. Install Tool J-6978-04 and remove the rotor.

Disconnect the alternator stator wires from the rectifier, and then remove the stator wiring harness from the clamp. Remove the five stator attaching screws, and then remove the stator.

Remove the water pump cover and the pump impeller.

CAUTION: Impeller is secured with a left-hand nut.

Remove the remaining timing chain cover attaching bolts, and then remove the cover.

Measuring Timing Chain Deflection

Remove the spark plugs as an aid to rotating the crankshaft. Rotate the crankshaft **CLOCKWISE** (as viewed from the front of the engine) to take up all of the slack on the left side of the timing chain. Make a reference mark on the block, and then force the chain inward and measure the distance from your mark to the chain.

Now, rotate the crankshaft **COUNTERCLOCKWISE** to take up the slack on the right side of the chain. Force the chain outward, and again measure the distance from your mark on the block to the chain. The difference between the two measure-

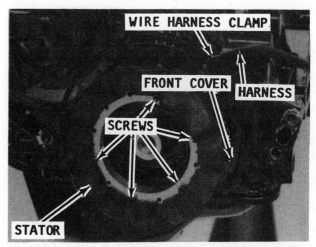

Stator assembly installed with major parts identified.

ments is the chain deflection. If the chain deflection is more than one inch, replace the timing chain and inspect the sprockets and tensioner for abnormal wear.

Timing Chain Removal

Rotate the crankshaft until the timing marks on the camshaft sprocket and the crankshaft sprocket are aligned through the center of the sprocket hubs. Remove the chain tensioner and the camshaft sprocket fasteners.

Work the camshaft sprocket and the crankshaft sprocket off together by prying first one and then the other forward until the camshaft sprocket is free. Remove the timing chain and then finish removing the crankshaft sprocket.

Puller installed to remove the alternator rotor.

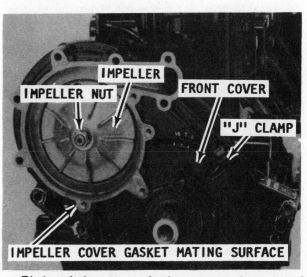

Timing chain cover and other parts to be removed before the timing chain and sprockets can be removed.

Timing chain tensioner with normal wear pattern at the bottom and excessive wear at the top. When the wear pattern moves into the area indicated, the tensioner should be replaced.

CLEANING AND INSPECTING

Clean all of the parts removed with solvent and blow them dry with compressed air. Inspect the timing chain for worn or broken links. Inspect both sprockets for cracks, scores, nicks, and worn or damaged teeth. Inspect the chain tensioner for signs of wear or damage. A tensioner worn beyond normal groove wear **MUST** be replaced. Place the tensioner on a flat surface and measure the distance from the surface to the top of the tensioner at the highest point. This measurement must be 1-1/4" or more.

If the chain deflection is within specifications and the tensioner does not touch the chain the tensioner **MUST** be replaced.

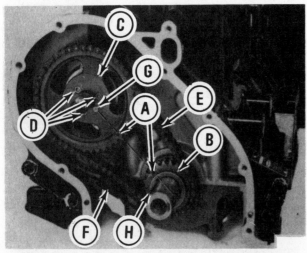

Timing chain cover removed with major parts identified as: Timing marks (a), crankshaft sprocket (b), camshaft sprocket (c), sprocket attaching screws (d), timing chain (e), chain tensioner (f), sprocket locating pin (g), and key (h).

Thoroughly clean the gasket surface of the timing chain cover and the cylinder block. Clean the parts of the cover with solvent and blow them dry with compressed air.

ALWAYS install a new oil seal. **NEVER** attempt to use the old seal because you will most likely get an oil leak at this point.

Timing Chain Installation

Position the key in the keyway of the crankshaft. Rotate the camshaft until the camshaft sprocket locating pin is aligned with the key on the crankshaft with an imaginary line running through the centers of the shafts.

Lay out the camshaft sprocket and the crankshaft sprocket with the timing marks aligned opposite each other, as shown. Now, engage the chain onto both sprockets, then slide both sprockets onto their shafts with the timing marks still aligned through the center of the shafts and the keyway of the crankshaft sprocket aligned with the key. Continue pushing both sprockets at the **SAME TIME** until they are fully seated, with the camshaft sprocket indexed over the locating pin and the crankshaft sprocket indexed with the key. **NEVER** try to drive the sprocket onto the camshaft because the **WELSH PLUG** at the rear of the engine will be dislodged.

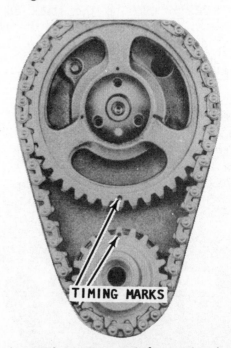

Timing mark on the camshaft sprocket (top), and the mark on the crankshaft sprocket (bottom), properly aligned through the centers of the camshaft and the crankshaft.

Install the camshaft attaching screws and tighten them to a torque value of 13 ft lb (15Nm). Install the chain tensioner, and then apply a liberal coating of engine oil to the chain, chain tensioner, and both sprockets.

Timing Chain Cover
Installation

Lay down a bead of Permatex, Form-A-Gasket, or equivalent, to the outside diameter of the new water pump shaft oil seals and to the timing cover oil seal. **TAKE CARE** not to get any sealer on the lip of the seals. **BE SURE** to support the underside of the cover during installation of new oil seals. **PAY ATTENTION** that both water pump seal lips face the water side or front of the engine.

Press the first water pump seal in until it bottoms-out. Press the second seal in until it is even with the cover surface. Fill the space between the water pump seals with Multi-purpose Lubricant.

Install the timing cover crankshaft seal. **BE SURE** the seal lip faces toward the engine block. Press the seal into place until it bottoms-out. Coat the seal lip, the crankshaft end, and the water pump shaft with Multi-purpose Lubricant.

Place a **NEW** timing chain cover gasket in position with the hole in the gasket indexed over the dowel pin on the block. Apply a coating of Permatex, Form-A-Gasket, or equivalent, to the threads of the cover attaching bolts and then install the bolts. Tighten the bolts to a torque value of 15 ft lb (20Nm). Install the J-clamp.

Install the water pump impeller with the flat washer on the larger diameter of the shaft and against the impeller. Install the cupped washer with the concave side toward the impeller. Install the **LEFT-HAND** nut

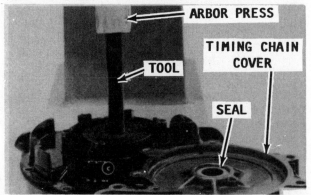

Set-up in an arbor press to install new seals into the timing chain cover.

and tighten it to a torque value of 40 ft lb (55Nm).

Install the water pump cover with a **NEW** gasket. Apply a coating of Permatex, Form-A-Gasket, or equivalent, to the attaching bolt threads and then install and tighten them to a torque value of 15 ft lb (21Nm).

Alternator Installation

Place the stator on the timing chain cover with the output wires toward the recess cast in the cover. Coat the attaching screw threads with Permatex, Form-A-Gasket, or equivalent, and then install the screws and tighten them to a torque value of 4 ft lb (5Nm). Install the wire harness in the clamp and connect the stator wires to the rectifier.

Coat the rotor surface, the area where the oil seal rides, and the lip of the oil seal with Multi-purpose Lubricant. Lubricate the end of the crankshaft and also the inside diameter of the rotor assembly with anti-scoring, extreme-pressure lubricant. Verify the rotor locking key is properly seated in its groove on the crankshaft.

Start the rotor onto the crankshaft, with the slot in the rotor aligned with the key in the crankshaft. Install Tool J-21058-8 onto the end of the crankshaft. Keep the center shaft of the tool from turning and at the same time turn the large nut to force the rotor into position. Continue turning the nut until the rotor bottoms out against the cover.

Install the rotor attaching bolt and washer and tighten the bolt to a torque value of 50 ftlb (70Nm).

Close the drain valves, and then fill the closed-circuit cooling system with water.

Fill the crankcase with the proper weight oil.

Start the engine, run it at idle speed, and check for leaks.

CAUTION

Water must circulate through the stern drive, also to and from the engine, anytime the engine is run to prevent damage to the water pump located in the lower unit (original and Alpha drives); or on the front of the engine (Bravo drive). Just a few seconds without water will damage the water pump.

TIMING CHAIN COVER REMOVAL
MERCURY MARINE AND FORD ENGINES

Open the drain valves and drain the coolant from the block and exhaust manifold.

Remove the drive belt. Remove all accessory brackets attached to the water pump. Remove the water pump. Remove the water pump pulley. Disconnect the water hoses from the water pump. Remove the bolts attaching the pump to the timing chain cover. Remove the intake manifold.

Drain the crankcase. Remove the crankshaft pulley from the crankshaft vibration damper (V8). Remove the bolt and washer attaching the damper to the crankshaft. Install a puller onto the damper, and then remove the damper. Be sure to use a puller which attaches to the front face of the damper, not the outer circumference to avoid damage to the damper.

Disconnect the outlet line from the fuel pump. Remove the bolts attaching the fuel pump, and then move the fuel pump to one side out of the way with the inlet line still connected.

MERCURY MARINE SPECIAL WORDS

Remove the engine water pump impeller cover and the impeller.

V8 CRITICAL WORDS

Two pan bolts pass up through the pan and into the timing chain cover. These two bolts, one on each side, **MUST** be removed before the cover will come free.

All Engines

Use a thin bladed knife to cut the oil pan gasket flush with the block face **BEFORE**

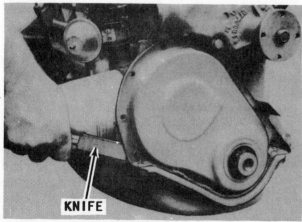

Using a knife to cut the ends of a new oil seal off flush with the cap faces after the timing cover bolts have been started, but **BEFORE** *they are tightened.*

separating the cover from the block. Cutting the pan gasket at this time means the pan does not have to be removed in order to replace the front part of the pan gasket when installing the timing chain cover. Remove the timing chain cover and the water pump together as an assembly. Discard the cover gasket.

Timing Chain Removal

Remove the crankshaft front oil slinger.

Reinstall the damper attaching bolt into the crankshaft. Remove the spark plugs from both banks. With a socket wrench on the damper bolt head, rotate the crankshaft clockwise until the timing mark on the camshaft sprocket aligns with the timing mark on the crankshaft sprocket, as shown in the accompanying illustration.

Remove the camshaft sprocket cap screw, washers, the two piece fuel pump eccentric, and front oil slinger. Slide both sprockets forward and clear of the shafts with the chain still engaged.

Camshaft Removal

Remove the camshaft thrust plate attaching bolts. Carefully pull the camshaft straight out the front of the block. Use the **UTMOST CARE** not to damage the bearing surfaces in any way.

Camshaft cleaning and inspecting begins on Page 3-65.

Camshaft Installation

Cover the camshaft journals with a coating of heavy engine oil and apply Lubriplate or equivalent to the camshaft lobes. If a new camshaft is being installed, coat the cam lobes with engine oil supplement, and then add the rest of the contents to the crankcase oil. **CAREFULLY** move the camshaft through the bearings into position. Install the camshaft thrust plate with the groove towards the cylinder block.

Check the camshaft end play by first pushing the camshaft aft as far as possible. Next, install a dial indicator in such a manner to allow the indicator point to be on the camshaft sprocket attaching screw. Set the dial indicator at zero.

Now, place a large screwdriver between the camshaft sprocket and the block. Finally, pull the camshaft forward and release it. Compare the dial indicator reading with the Specifications in the Appendix. If the end play is excessive, replace the thrust plate. Remove the dial indicator.

Lubricate the lifters and bores with heavy engine oil. Install the valve lifters **PAYING ATTENTION** each one is replaced in the exact position from which it was removed.

Apply Lubriplate or equivalent to each end of the push rods and install the rods into the same guides from which they were removed. Coat the valve stem tips with Lubriplate, and then the rocker arms and fulcrum seats with heavy engine oil. Position the rocker arms over the push rods.

Timing Chain Installation

Position the key in the crankshaft keyway. Lay out the camshaft sprocket and the crankshaft sprocket with the timing marks aligned opposite each other, as shown. Now, engage the chain onto both sprockets, then slide both sprockets onto their shafts with the timing marks still aligned through the center of the shafts and the keyway of the crankshaft sprocket aligned with the key. Continue pushing both sprockets at the same time until they are fully seated. **NEVER** try to drive the sprocket onto the camshaft because the **WELSH PLUG** at the rear of the engine will be dislodged.

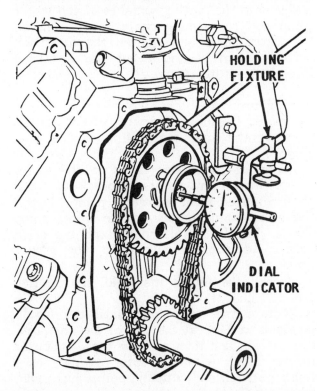

Set-up to check camshaft end play with a dial indicator.

Install the fuel pump eccentric, washers, and camshaft sprocket cap screw. Tighten the cap screw to a torque value of 13 ft lb (17Nm). Install the crankshaft front oil slinger.

Timing Chain Cover Installation

Clean the gasket surfaces of the timing chain cover, the oil pan, and the engine block. Install a new oil seal in the timing chain cover. **ALWAYS** use a new oil seal. Attempting to use the old seal is just asking for an oil leak and one will surely develop here.

Apply a coating of engine oil to the timing chain. Lay down a bead of sealer onto the gasket surface of the oil pan. Cut, and then position the required section of a new gasket onto the oil pan. Lay down a second bead of sealer on top of the pan gasket section. Now, coat the gasket sur-

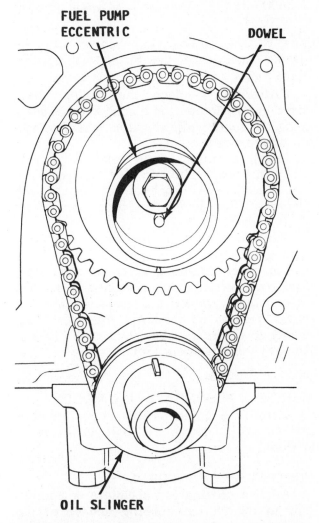

Illustration to depict the fuel pump eccentric installed on the camshaft sprocket and the oil slinger installed on the crankshaft sprocket.

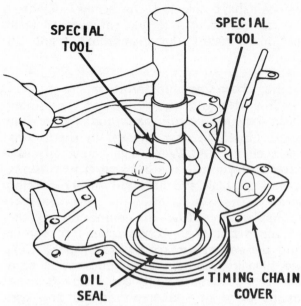

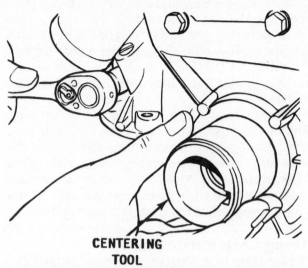

OIL SEAL

TIMING CHAIN COVER

SPECIAL TOOL   SPECIAL TOOL

*Using special tools to drive a new oil seal into the timing chain cover of a Ford V8 engine. A new seal should **ALWAYS** be installed or an oil leak at this point will surely develop.*

CENTERING TOOL

*A special centering tool **MUST** be used when installing the timing chain cover to **ENSURE** the seal makes even contact around the crankshaft. An improperly installed seal wil cause an oil leak.*

face of the timing chain cover and the mating surface of the block with sealer. Place a new gasket in position on the block.

CAREFULLY place the timing chain cover in position on the block, without damaging the seal or allowing the gasket to slip out of alignment.

Install a timing chain cover alignment tool. This tool is **NECESSARY** to ensure the seal makes even contact all around the crankshaft to prevent the possibility of an oil leak at this point.

Push downward on the cover, and while holding the cover in this position, install the pan to cover attaching bolts, on each side. Coat the cover to block attaching bolts with oil-resistant sealer, and then install the bolts. Tighten all the bolts alternately and evenly to a torque value of 15 ft lb (20Nm). Remove the alignment tool.

MERCURY MARINE SPECIAL WORDS

Install the engine water pump impeller and cover. Tighten the bolt to 15 ft lb (20Nm).

FORD V8 SPECIAL WORDS

Apply a coating of Lubriplate or equivalent to the oil seal rubbing surface of the inner hub of the vibration damper, to prevent damage to the seal. Cover the crankshaft with a coating of engine oil. Align the

crankshaft vibration damper keyway with the key on the crankshaft, and then slide the damper onto the crankshaft. Install the cap screw and washer. Tighten the cap screw to a torque value of 85 ft lb (115Nm).

Remove any old gasket material from the water pump and mating surface of the timing chain cover. Coat both sides of a **NEW** gasket with sealer. Lay a bead of sealer onto the mating surfaces of the water pump and the timing chain cover. Install the water pump and tighten the attaching bolts alternately and evenly to a torque value of 12 ft lb (16Nm).

All Engines

Install the fuel pump with a **NEW** gasket. Connect the fuel outlet line to the fuel pump.

Connect the water hoses to the water pump. Install any accessory brackets which were removed from the water pump. Install the water pump pulley on the water pump shaft. Install the alternator and the drive belt. Adjust the drive belts for proper tension.

Install the intake manifold, see Section 3-6, this chapter.

Rotate the crankshaft until the No. 1 piston is at TDC at the end of the compression stroke, and then position the distributor in the block with the rotor at the No. 1 firing position and with the points just open. Install the hold down clamp.

Adjust the valve clearance, see Section 3-10, this chapter.

Install the valve covers with **NEW** gaskets and tighten the attaching screws to a torque value of 7 ft lb (10Nm) for Mercury Marine engines or 4 ft lb (5Nm) for Ford V8 engines.

Fill the crankcase with the proper weight oil.

Close the water drain valves.

CAUTION

Water must circulate through the stern drive, also to and from the engine, anytime the engine is run to prevent damage to the water pump located in the lower unit (original and Alpha drives); or on the front of the engine (Bravo drive). Just a few seconds without water will damage the water pump.

Start the engine and check the completed work.

SPECIAL WORDS

A new camshaft is "broken in" during the first 15 minutes of engine operation. Valve train loads are greatest at **LOW** rpm. Do **NOT** operate an engine with a new camshaft at less than 1500 rpm for the first 15 minutes. Operation at less than 1500 rpm will cause considerable and unnecessary wear on the valve train parts.

CAMSHAFT BEARING SERVICE
ALL UNITS

Removal and replacement of the camshaft bearings should be performed by qualified mechanics in a shop equipped to handle such work. However, in most cases the block bores for the bearings can be bored to a larger size and oversize bearings installed. **BE SURE** to check with your local marine dealer for available oversize bearings.

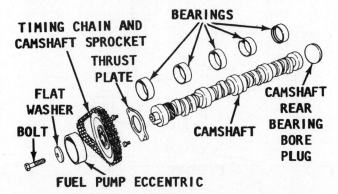

Exploded drawing of a typical camshaft and associated parts.

CLEANING AND INSPECTING

Clean the camshaft with solvent and wipe the journals dry with a lint-free cloth. **ALWAYS** handle the shaft **CAREFULLY** to avoid damaging the highly finished journal surfaces. Blow out all oil passages with compressed air.

Clean the gasket surfaces on the block and crankcase front cover. Check the diameter of the camshaft bearings with a micrometer for out-of-round condition, taper, and wear. If the journals are out-of-round more than 0.001" (0.025mm), the camshaft should be replaced.

Inspect the camshaft lobes for scoring and signs of abnormal wear. Normal lobe wear may result in pitting in the general area of the lobe toe. This pitting is not detrimental to the operation of the camshaft; therefore the camshaft need not be replaced unless the lobe lift loss has exceeded 0.005" (0.127mm). The camshaft lobe lift can be checked with a micrometer and the results compared with the Specifications given in the Appendix.

Check the camshaft for alignment. This is best done using **"V"** blocks and a dial indicator. The dial indicator will indicate the exact amount the camshaft is out of true. If it is out more than 0.002" (0.051mm) dial indicator reading, the camshaft should be replaced. When using the dial indicator in this matter the high reading indicates the high point of the shaft.

Remove any light scuffs, scores, or nicks from the camshaft machined surfaces with a smooth oil stone.

Check the camshaft bearings. If any one needs to be replaced, replace **ALL** never just one. Removal and replacement of the camshaft bearings should be performed by qualified mechanics in a shop equipped to handle such work.

Check the distributor drive gear for broken or chipped teeth and replace the camshaft if the gear is damaged.

Always handle a camshaft with UTMOST care to prevent damage to the lobes or the bearing surfaces. Would you believe, this camshaft was accidently dropped and snapped in half. No, the authors did not do it.

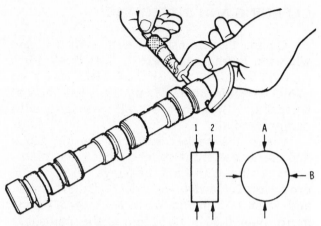

Camshaft bearing journal taper and out of round (if any), must be accurately determined using a micrometer. The difference between measurements indicated as No. 1 and No. 2 is the bearing journal taper. The difference between "A" and "B" is the bearing journal out of round. Compare the values obtained with the specifications in the Appendix.

**DIMENSION "A" MINUS
DIMENSION "B" EQUALS
CAMSHAFT LOBE LIFT**

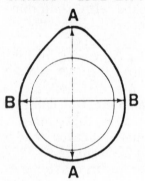

Camshaft lobe lift may be measured with a micrometer, if the camshaft has been removed. A dial indicator gauge and holding fixture is required if the camshaft remains in place in the block.

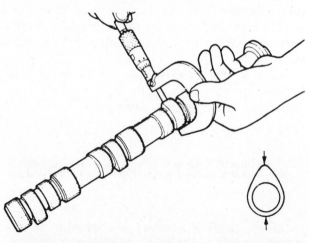

Using a micrometer to measure camshaft lobe wear and lobe lift, as indicated in the illustration above.

CAMSHAFT LIFT MEASUREMENT
CAMSHAFT INSTALLED
GMC ONLY

If the engine is to perform properly, the valves must open and close at a predetermined precise moment for maximum efficiency.

On the **POWER** stroke, the exhaust valve must open before bottom dead center in order to permit the exhaust gases to leave the combustion chamber under the remaining pressure (blowdown).

On the **EXHAUST** stroke, the intake valve must open just before top dead center in order to permit the air/fuel mixture to enter the combustion chamber. The movement of the valves are functions of the camshaft design and the valve timing. Therefore, excessive wear of any camshaft part will affect engine performance.

If improper valve operation is indicated, measure the lift of each push rod in consecutive order and record the readings. Remove the valve mechanism. Position a dial indicator with ball socket adapter (Tool J-8520) on the push rod. Rotate the crankshaft slowly in the operating direction until the lifter is on the heel of the cam lobe. At this point, the push rod will be in its lowest position.

Set the dial indicator on zero, then rotate the crankshaft slowly, until the push rod is in the fully raised position.

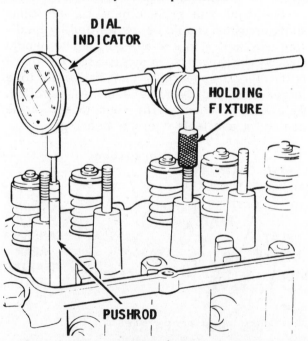

Set-up using a holding fixture to secure a dial indicator gauge over a pushrod to check cam lift.

The distributor primary lead **MUST** be disconnected from the negative post on the coil and the ignition switch **MUST** be in the **ON** position, or the grounding circuit in the ignition switch will be **DAMAGED.**

Compare the total lift recorded from the dial indicator with the specifications listed in the Appendix.

Continue to rotate the camshaft until the dial indicator reads zero. This point will be a check on the accuracy of the original indicator reading.

If the camshaft readings for all lobes are within specifications, remove the dial indicator assembly.

Install the valve mechanism, see Section 3-8, this chapter.

Adjust the valve mechanism, see Section 3-10, this chapter.

CAUTION

Water must circulate through the stern drive, also to and from the engine, anytime the engine is run to prevent damage to the water pump located in the lower unit (original and Alpha drives); or on the front of the engine (Bravo drive). Just a few seconds without water will damage the water pump.

Start the engine and check the completed work.

CAMSHAFT LIFT MEASUREMENT
CAMSHAFT INSTALLED
MERCURY MARINE AND FORD ENGINES

The camshaft lobe lift measurement is made to determine if the camshaft lobe is worn and the camshaft and the lifter operating on the worn lobe must be replaced.

Remove the valve rocker arm covers. Remove the rocker arm stud nut, fulcrum seat, and rocker arm. Use the adaptor for the ball end push rods. Remove the pedestal mounted rocker arms. **MAKE SURE** the push rod is in the valve lifter socket. Install a dial indicator in such a manner as to have the ball socket adaptor of the indicator on the end of the push rod and in the same plane as the push rod movement.

Connect an auxiliary starter switch in the starting circuit. Crank the engine with the ignition switch **OFF.** Rotate the crankshaft in small amounts until the lifter is on the base circle of the camshaft lobe. At this point, the push rod will be in its lowest position.

Set the dial indicator at **ZERO.** Rotate the crankshaft slowly until the push rod is in the fully raised position, as determined when the dial indicator is at the highest reading. Now, compare the total lift recorded on the indicator with the Specifications in the Appendix.

To double check the accuracy of the reading, continue to rotate the crankshaft until the dial indicator is back to zero. If the lift on any lobe is below specified wear limits, the camshaft and the valve lifter(s) operating on the worn lobe(s) **MUST** be replaced.

Remove the dial indicator and the auxiliary starter switch. Install the rocker arms. After the rocker arms have been installed **DO NOT** rotate the crankshaft until the hydraulic valve lifters have had enough time to bleed down, or serious damage may be caused to the valve.

Install the rocker arm covers.

CAUTION

Water must circulate through the stern drive, also to and from the engine, anytime the engine is run to prevent damage to the water pump located in the lower unit (original and Alpha drives); or on the front of the engine (Bravo drive). Just a few seconds without water will damage the water pump.

Start the engine and check the completed work.

3-12 CRANKSHAFT CONNECTING ROD BEARINGS AND MAIN BEARINGS

REMOVAL
GMC 4- AND 6-CYLINDER IN-LINE ENGINES

Remove the engine, see Chapter 13.

Remove the oil pan, see Section 3-5.

Remove the oil pump, see Section 3-5.

Remove the flywheel and coupler from the crankshaft.

Remove the main bearing caps and the connecting rod caps. **TAKE TIME** to mark each bearing cap to **ENSURE** it will be installed back onto the same rod from which it was removed.

Lift the crankshaft out of the block. **ALWAYS** handle the shaft carefully to avoid damaging the highly finished journal surfaces.

Carefully push the pistons up out of the way to the top of the cylinders. Take care not to push past any piston rings.

CLEANING AND INSPECTING

Clean the crankshaft with solvent and wipe the journals dry with a lint-free cloth. Blow out all oil passages with compressed air. Oil passageways lead from the rod journal to the main bearing journal. **BE CAREFUL** not to blow the dirt into the main bearing journal bore.

Measure the diameter of each journal at four places to determine the out of round taper, and wear. The out of round limit is 0.001" (0.025mm); the taper must not exceed 0.001" (0.025mm); and the wear limit is 0.0035" (0.0889mm). If any one of the limits is exceeded, the crankshaft **MUST** be reground to an undersize, and undersized bearing inserts must be installed. Check the Specifications listed in the Appendix.

Support the crankshaft on **"V"** blocks at the front and rear main bearing journals. Use a dial indicator to check the runout at the front center and rear center journals. Total indicator readings at each journal should not exceed .002" (0.051mm).

While checking the runout at each journal, **TAKE TIME** to note the relation of the "high" spot (or maximum eccentricity) on each journal to the others. "High" spots on all journals should come at the same angular location. If the "high" spots do not appear to be in the same angular locations, the crankshaft has a "crook" or "dogleg", making it unsatisfactory for service.

Remove the old bearing shells and the rear oil seal.

Crankshaft bearings are of the precision fit type, which do not require reaming to

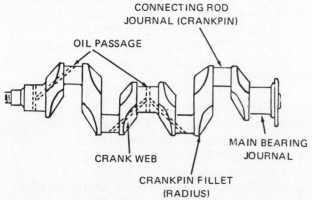

Common crankshaft terminology used in this book and in shops throughout the world.

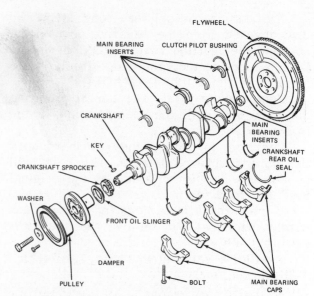

Crankshaft and associated parts for a GMC 4-cylinder engine. The 6-cylinder in-line crankshaft is an extended version of the crankshaft shown.

size. Shims are not provided for adjustments since worn bearings are readily replaced with new bearings of the proper size. Bearings for service replacement are furnished in standard size and undersizes. **NEVER** file a bearing cap to adjust for wear in old bearings.

INSTALLATION

Install new main and connecting rod bearing shells. The upper half of the main bearing shell has a hole. This half of the shell **MUST** be inserted between the crankshaft and the block. **CAREFULLY** place the crankshaft into the bearing halves on the block.

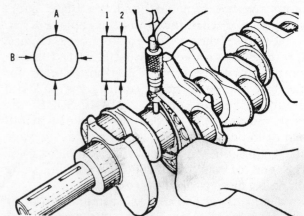

Crankshaft bearing journal taper and out of round (if any), must be accurately determined using a micrometer. The difference between measurements indicated as No. 1 and No. 2 is the bearing journal taper. The difference between "A" and "B" is the bearing journal out of round. Compare the values obtained with the specifications in the Appendix.

Clean the journal thoroughly of all traces of oil, and then place a piece of Plastigage on the bearing surface, the full width of the cap. Install the cap and tighten the retaining bolts alternately and evenly to the torque value of 65 ft lb (88Nm) for the main bearing caps and 33 ft lb (45Nm) for 11/32" rod bolts or 40 ft lb (54Nm) for 3/8" rod bolts.

SPECIAL WORDS

DO NOT allow the crankshaft to rotate with the Plastigage in place or the Plastigage will be distorted and the reading will have no value.

Remove the bearing cap and measure the flattened Plastigage at its widest point using the scale printed on the Plastigage envelope. The number within the graduation which most closely corresponds to the width of the Plastigage indicates the bearing clearance in thousandths of an inch. The bearing journal is tapered and so is the Plastigage. Measure the Plastigage at the widest point and also at the narrowest point. These measurements will give the minimum and maximum clearances. The new connecting rod bearing shell insert clearance should be 0.0007-0.0027" (0.0178-0.0686mm). The new main bearing shell insert clearance should be 0.0003-0.0029" (0.008-0.074mm).

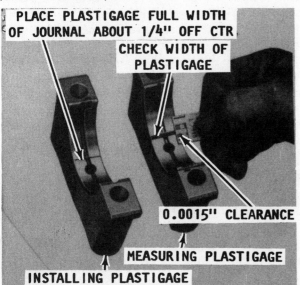

After the crankshaft main bearing caps have been tightened to the full torque value, and then removed, the widest part of the flattened Plastigage can be compared with the scale on the side of the package. In this manner, the amount of bearing clearance can be fairly accurately determined. The crankshaft must NOT be allowed to rotate, even the slightest amount, or the Plastigage will be distorted and the measurement will have no value.

If either shell insert clearance exceeds specifications, a new insert should be installed. After the new insert has been installed, make another Plastigage measurement. If the new clearance still does not give the proper clearance according to the specifications, then an undersized insert should be used.

Lubricate the rear bearing seal with oil. **DO NOT** allow any oil to get on the parting surface. Gradually push the seal with a hammer handle until the seal is rolled into place. To replace the upper half of the seal, use a small hammer and brass punch. Tap one end of the seal into the block groove, and then push the seal until it protrudes out the other side.

Install the thrust bearing cap and tighten the bolts just fingertight. Pry the crankshaft forward against the thrust surface of the upper half of the bearing. Hold the crankshaft forward and pry the thrust bearing cap to the rear. This action will align the thrust surfaces of both halves of the bearing. Retain the forward pressure on the crankshaft, and tighten the cap bolts alternately and evenly to the torque value given in the previous paragraphs.

Install the rod caps, see Section 3-9.

Install the oil pump, see Section 3-5.

Install the oil pan, see Section 3-5.

Install the flywheel and coupler. The flywheel bolts are tightened to a torque value of 21 ft lb (28Nm).

Install the engine, see Chapter 13.

CAUTION

Water must circulate through the stern drive, also to and from the engine, anytime the engine is run to prevent damage to the water pump located in the lower unit (original and Alpha drives); or on the front of the engine (Bravo drive). Just a few seconds without water will damage the water pump.

Start the engine and check the completed work.

CRANKSHAFT, CONNECTING ROD, AND MAIN BEARING SERVICE GMC V6 AND V8 ENGINES

Service work affecting the crankshaft or the crankshaft bearings may also affect the connecting rod bearings. Therefore, it is highly recommended that the rod bearings be carefully inspected as they are removed.

If the crankpins are worn to the extent the crankshaft should be replaced or reground, then this work should be done in a qualified shop. Attempting to save time and money by merely replacing the crankshaft bearings will not give satisfactory engine performance.

CRANKSHAFT REMOVAL

Remove the engine, see Chapter 13.
Remove the oil pan, see Section 3-5.
Remove the oil pump, see Section 3-5.
Remove the flywheel and coupler.
Remove the main bearing caps and connecting rod caps. Push the pistons to the top of the cylinders. Lift the crankshaft out of the cylinder block. V8 engines have 5 main bearings, V6 have 4 main bearings.

Check the Specifications in the Appendix for the main journal diameter, not all journals on the same crankshaft have the same diameter. Check these dimensions with a micrometer for out of round, taper, or undersize. If the journals exceed 0.001" (0.0254mm) out of round or taper, the crankshaft should be replaced or reconditioned to an undersize figure to permit installation of undersize precision type bearings.

CLEANING AND INSPECTING

Clean the crankshaft with solvent and wipe the journals dry with a lint-free cloth. **ALWAYS** handle the shaft carefully to avoid damaging the highly finished journal surfaces. Blow out all of the oil passages with compressed air. Oil passageways lead from the rod to the main bearing journal. **TAKE CARE** not to blow dirt into the main bearing journal bore.

Measure the diameter of each journal at four places to determine the out of round taper and wear. The out of round limit is 0.001" (0.025mm); the taper must not exceed 0.001" (0.025mm); and the wear limit is 0.0035" (0.0889mm). If any one of the limits is exceeded, the crankshaft **MUST** be reground to an undersize, and undersized bearing inserts must be installed. Check the Specifications listed in the Appendix.

While checking the runout at each journal take time to note the relation of the "high" spot (or maximum eccentricity) on each journal to the others. "High" spots on all journals should come at the same angular location. If the "high" spots do not appear to be in the same angular locations, the crankshaft has a "crook" or "dogleg", making it unsatisfactory for service.

BEFORE INSTALLATION

TAKE TIME to read the good words in the next few paragraphs before starting any crankshaft or bearing insert work. Many of the facts given are probably already known, others may be new, and all of the information will assist in completing the work in the shortest time and with assurance of satisfactory engine performance after completion.

Main bearings and connecting rod bearings are of the precision insert type and do not utilize shims for adjustment. If clearances are found to be excessive, a new bearing, both upper and lower halves, must

SCRATCHES   DIRT IMBEDDED INTO BEARING MATERIAL
SCRATCHED BY DIRT

OVERLAY WIPED OUT
LACK OF OIL

BRIGHT (POLISHED) SECTIONS
IMPROPER SEATING

OVERLAY GONE FROM ENTIRE SURFACE
TAPERED JOURNAL

RADIUS RIDE
RADIUS RIDE

CRATERS OR POCKETS
FATIGUE FAILURE

Examples of various wear patterns on bearing halves, including possible reasons for the condition.

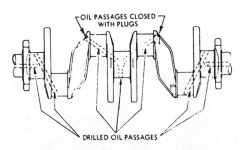

The oil passage holes MUST be cleaned to ensure lubrication for connecting rod and main bearings.

be installed. Main bearings are available in standard size and as 0.001", 0.002", 0.009", and 0.020" undersize. Connecting rod bearings are available in standard size and 0.001" undersize for use with a standard size crankshaft.

In order to obtain close tolerances in production, selective fitting of both rod and main bearing inserts is necessary. For this reason you may find one half of a standard insert with one half of a .0001" undersize insert which will decrease the clearance 0.0005" from using a full standard bearing.

If a production crankshaft cannot be precision fitted by this method, it is then ground 0.009" undersize only on the main journals. A 0.009" undersize bearing and a 0.010" undersize bearing may be used for precision fitting in the same manner as just described. Any engine fitted with a 0.009" undersize crankshaft will be identified by the following markings: ".009" will be stamped on the crankshaft counterweight forward of the center main bearing journal. A figure "9" will be stamped on the block at the left front oil pan rail.

If, for any reason, the main bearing caps are replaced, it may be necessary to shim the bearing. Laminated shims for each cap are available for service. The amount of shimming required will depend on the bearing clearance.

In general, except for the No. 1 main bearing, the lower half of the bearing will show more wear and fatigue than the upper half. If the determination is made from inspection, that the lower half is suitable for use, then it is safe to assume the upper half is also satisfactory. **HOWEVER,** if the lower half shows evidence of wear or damage, both the upper and lower halves **MUST** be replaced. **NEVER** replace one half of the bearing without replacing the other half.

A crankshaft main bearing consists of two halves. These halves are **NOT** alike and

are **NOT** interchangeable in the cap and crankcase. The upper (crankcase) half of the bearing is grooved to supply oil to the connecting rod bearings while the lower (bearing cap) half of the shell is not grooved. The two bearing halves **MUST** not be interchanged. Some V6 engines have a thrust bearing in place of a No. 2 standard bearing. All crankshaft bearings, except the No. 2 thrust bearing (on some V6 engines only) and the rear main bearing are identical.

The thrust bearing is longer and flanged to take end thrust. When the shells are placed in the crankcase and bearing cap, the ends extend slightly beyond the parting surfaces. The reason for them to extend slightly in this manner is to ensure positive seating and to prevent turning when the cap bolts are tightened. The ends of the shells must **NEVER** be filed flush with the parting surface of the crankcase or bearing cap.

CHECKING CONNECTING ROD AND MAIN BEARING CLEARANCE

Plastigage is soluble in oil, therefore, clean the crankshaft journal thoroughly to

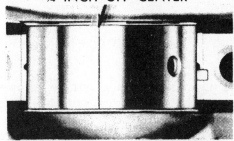

INSTALLING PLASTIGAGE

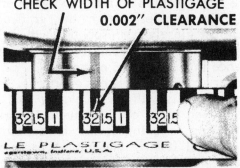

MEASURING PLASTIGAGE

After the connecting rod cap has been tightened to the required full torque value, and then removed, the flattened Plastigage can be compared with the scale on the side of the package. In this manner, the amount of clearance can be fairly accurately determined.

remove any trace of oil. With engine held in the bottom-up position, rotate the crankshaft until the oil hole is down to avoid getting oil on the Plastigage. Whenever the crankshaft is rotated with the rear bearing cap removed, hold the oil seal to prevent it from rotating out of position in the crankcase. Place paper shims in the upper half of adjacent bearings and tighten the cap bolts to take the weight off the bearing shell being checked.

Place a piece of Plastigage lengthwise along the bottom center of the lower bearing shell, and then install the cap with the shell and tighten the bolt nuts alternately and evenly to the following torque values:

Connecting Rod Bolts

GMC V6	45 ft lbs
	(61Nm)
GMC V8	
Small Block	45 ft lbs
	(61Nm)
Big Block	50 ft lbs
	(68Nm)

Main Bearing Bolts

GMC V6	85 ft lbs
	(115Nm)
GMC V8	
Small Block	80 ft lbs
	(108Nm)
Big Block	110 ft lbs
	(150Nm)

SPECIAL WORDS

The crankshaft must not be allowed to rotate or the Plastigage will be distorted and the measurement will have no value.

Remove the bearing cap and measure the flattened Plastigage at its widest point using the scale printed on the Plastigage envelope. The number within the graduation which most closely corresponds to the width of the Plastigage indicates the bearing clearance in thousandths of an inch.

Under normal service conditions, main bearing journals will wear evenly and are seldom found to be out of round. However, if a bearing is being fitted to an out of round journal (0.001" maximum) be sure to fit the bearing to the maximum diameter of the journal.

If the bearing is fitted to the minimum diameter and the journal is out of round by 0.001", interference between the bearing and journal will result in bearing failure.

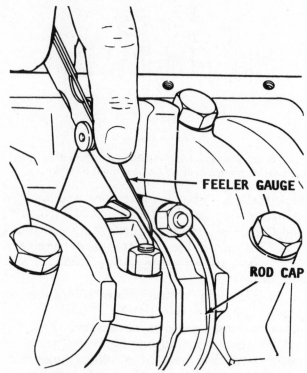

A feeler gauge can be used to accurately determine connecting rod side clearance. Acceptable clearance is 0.010"-0.020".

If the flattened Plastigage tapers toward the middle or ends, there is a difference in clearance indicating taper, low spot, or other irregularity of the bearing or journal. **BE SURE** to measure the journal with a micrometer if the flattened Plastigage indicates more than 0.001" difference.

If the bearing clearance is within specifications, the bearing insert is satisfactory. If the clearance is not within specifications, replace the insert. **ALWAYS** replace both the upper and lower insert together as a unit. **NEVER** just one half.

If a new bearing is being installed in the cap, and the clearance is less than 0.001", inspect the cap for burrs or nicks. If none are found, then install shims on the bearing shoulders as required.

A standard 0.001" undersize bearing may give the proper clearance. If the undersize bearing does not give proper clearance, the crankshaft journal will have to be reground for use with the next undersize bearing.

Continue the procedure until all the bearings have been checked. Rotate the crankshaft to be sure there is an even drag for the complete turn without binding in any one spot.

Measure the crankshaft end play by first forcing the crankshaft to the extreme front position. Next measure the clearance at the

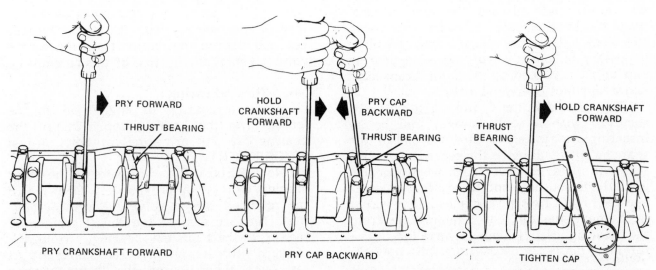

*The thrust bearing **MUST** be aligned properly before it is tightened. Alignment is accomplished by first prying the crankshaft forward, then prying the main bearing cap backward, and finally tightening the bolts to the torque values given in the text.*

front end of the rear main bearing with a feeler gauge and compare the results with the Specifications given in the Appendix. Measure the connecting rod side clearance between connecting rod caps, 0.008-0.014" (0.203 to 0.356mm) is acceptable.

Install a new rear main bearing oil seal in the cylinder block and main bearing cap, with the **ARROWS** pointing toward the **FRONT** of the engine.

Tighten the main bearing cap bolts alternately and evenly, a little at a time, to the torque value given on the previous page.

Apply a coat of Loctite to the threads of the flywheel bolts.

Align the flywheel bolt holes with the coupler holes. They will only align **ONE** way. Install the flywheel to the coupler. Tighten the attaching bolts to a torque value of 40 ft lb (54Nm) for V6 engines, 60 ft lb (81Nm) for small block V8 engines and 65 ft lb (88Nm) for big block V8 engines.

Install the oil pump, see Section 3-5.
Install the oil pan, see Section 3-5.
Install the engine, see Chapter 13.
Fill the crankcase with the proper weight oil.
Close the water drain valves.

CAUTION

Water must circulate through the stern drive, also to and from the engine, anytime the engine is run to prevent damage to the water pump located in the lower unit (original and Alpha drives); or on the front of the engine (Bravo drive). Just a few seconds without water will damage the water pump.

Start the engine and check the completed work. Operate the unit at reduced rpm for the first hour.

SERVICING MAIN BEARINGS AND/OR SEALS WITHOUT REMOVAL OF THE CRANKSHAFT MERCURY MARINE AND FORD ENGINES

REMOVAL

Remove the engine, see Chapter 13.
Remove the oil pan, see Section 3-5.
Remove the oil pump, see Section 3-5.

Remove the main bearing cap to which new bearings are to be installed. Work on **ONE** bearing at a time.

Insert the upper bearing removal tool (Tool 6331) in the oil hole in the crankshaft. If the tool is not available, a cotter pin may be bent as required to do the job. Rotate the crankshaft clockwise as viewed from the front of the engine to roll the upper bearing out of the block. If difficulty is encountered removing the bearing, it may be necessary to remove the crankshaft, see the first part of this Section.

Clean the crankshaft journal with solvent and wipe it dry with a lint-free cloth. Inspect the journals and thrust faces (thrust bearing) for nicks, burrs, or bearing particles that would cause bearing wear.

If the rear main bearing is being replaced, remove and discard the rear oil seal from the bearing cap. In order to remove the block half of the rear oil seal, it will be necessary to loosen all of the main bearing cap bolts, and then to raise the crankshaft just a hair (not over 1/32 inch).

The oil seal in the cylinder block may be removed with a seal removal tool or by inserting a metal screw into one end of the seal and then pulling on the screw to remove the seal. **USE EXTRA CARE** when working around the journals not to scratch or damage the highly finished surfaces in any way.

Remove the oil seal retaining pin from the bearing cap, if one is installed. The retaining pin is not used with a split-lip seal.

INSTALLATION

Upper No. 1, 2, and 4 Main Bearings

Place the plain end of the bearing over the shaft on the locking tang side of the block. Partially install the bearing to permit Tool 6331 to be inserted into the oil hole in the crankshaft. Insert the tool, and then rotate the crankshaft counterclockwise (when viewed from the front of the engine)

Removing the rear main seal without disturbing the crankshaft.

until the bearing seats in position. Remove the tool. Apply a light coating of oil to the journal and bearing cap. Install the bearing cap and tighten the cap bolts to the torque value given at the bottom of this column.

No. 3 Thrust Bearing

Apply a coating of engine oil to the journal and the bearing cap. Install the bearing cap, but bring the cap bolts up just **FINGERTIGHT**.

Now, **RELEASE** the pressure on the other bearing caps to approximately **ONE-HALF** the required torque value. Pry the crankshaft forward against the thrust surface of the upper half of the bearing. Hold the forward pressure on the crankshaft and at the same time tighten the thrust bearing cap bolts to the specified torque value.

Installation of the **REAR MAIN BEARING**: First, clean the oil seal groove with a brush and lacquer thinner. If the bearing cap is equipped with a retaining pin, remove the pin. The pin is not used with a split lip seal.

Next, install the lower seal into the rear main bearing cap with the **UNDERCUT SIDE** of the seal toward the **FRONT** of the engine. Permit the seal to extend about 3/8 inch above the parting surface in order for it to mate with the upper seal when the cap is installed.

Dip both halves of the split lip type seal in engine oil. Install the upper seal into its groove in the cylinder block with the **UNDERCUT SIDE** of the seal toward the **FRONT** of the engine. This is accomplished by rotating it on the seal journal of the crankshaft until about 3/8 inch extends below the parting surface. Check to be sure **NO RUBBER** has been shaved from the outside diameter of the seal by the bottom edge of the groove. **DO NOT** allow oil to get on the sealer area.

Tighten the cap bolts to the specified torque value.

Mercury Marine	55 ft lb (75Nm)
Ford V8 - (302 CID)	100 ft lb (135Nm)
Ford V8 - (351 CID)	65 ft lb (85Nm)

Install the timing chain and sprockets, the cylinder front cover, and the crankshaft pulley and adapter, see Section 3-11.

Coat the threads of the flywheel attaching bolts with oil resistant sealer. Position the flywheel on the crankshaft flange. Install and tighten the bolts to the following torque value:

Mercury Marine   30 ft lb (41Nm)
Ford V8 (All)   80 ft lb (110Nm)

Install the oil pump, see Section 3-5. **BE SURE** to prime the pump.
Install the oil pan, see Section 3-5.
Install the engine, see Chapter 13.
Fill the crankcase with the proper weight oil.

CAUTION

Water must circulate through the stern drive, also to and from the engine, anytime the engine is run to prevent damage to the water pump located in the lower unit (original and Alpha drives); or on the front of the engine (Bravo drive). Just a few seconds without water will damage the water pump.

Start the engine and check the completed work. Operate the unit at reduced rpm for the first hour.

CRANKSHAFT, CONNECTING ROD BEARINGS, AND MAIN BEARING SERVICE MERCURY MARINE AND FORD ENGINES

Remove the engine, see Chapter 13.
Remove the oil pan, see Section 3-5.
Remove the oil pump, see Section 3-5.
Remove the flywheel and coupler.

Remove each rod bearing cap; clean and inspect the lower bearing shell; and identify the cap plainly to ensure it will be installed onto the same rod from which it was removed. Inspect each rod bearing journal surface. If the journal surface is scored or ridged, the crankshaft must be replaced or reground to ensure satisfactory service with the new bearings. A slight roughness may be polished out with fine grit polishing cloth thoroughly wetted with engine oil. Burrs may be honed off with a fine stone.

Remove and inspect the main bearing caps and journals for signs of excessive wear. Identify each bearing cap to ensure it will be replaced in the same position from which it was removed. **CAREFULLY** remove the crankshaft from the block and support it on "V" blocks at the No. 1 and No. 4 main bearing journals.

CLEANING AND INSPECTING

Clean the crankshaft with solvent and wipe the journals dry with a lint-free cloth. Inspect the main and connecting rod journals for cracks, scratches, grooves, or scores. Inspect the crankshaft oil seal surface for nicks, sharp edges or burrs which might damage the oil seal during installation or might cause premature seal wear. **ALWAYS** handle the crankshaft carefully to avoid damaging the highly finished journal surfaces. Blow out all oil passages with compressed air. The oil passageway leads from the rod to the main bearing journal. **TAKE CARE** not to blow dirt into the main bearing journal bore.

Measure the diameter of each journal at four places to determine the out of round, taper, and wear. The out of round limit is 0.001"; the taper must not exceed 0.001"; and the wear limit is 0.0035". If any one of these limits is exceeded, the crankshaft must be reground to an undersize, and undersized bearing inserts must be installed.

Dress minor scores with an oil stone. If the journals are severely marred or exceed the wear limit, they should be refinished to size for the next undersize bearing.

Refinish the journals to give the proper clearance with the next undersize bearing. If the journal will not clean up to the maximum undersize bearing available, replace the crankshaft.

ALWAYS reproduce the same journal shoulder radius that existed originally. A radius too small will result in fatigue failure of the crankshaft. A radius too large will

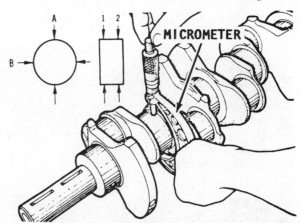

Crankshaft bearing journal taper and out of round (if any), must be accurately determined using a micrometer. The difference between measurements indicated as No. 1 and No. 2 is the bearing journal taper. The difference between "A" and "B" is the bearing journal out of round.

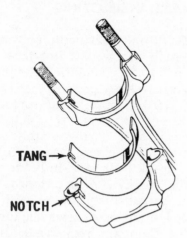

Proper installation of the rod cap bearing insert requires the tang to be "rolled" into the notch on the rod cap.

result in bearing failure due to radius ride of the bearing.

After the journals have been refinished, bevel (chamfer) the oil holes, and then polish the journal with fine grit, No. 320, polishing cloth and engine oil. Crocus cloth may also be used as a polishing agent.

CRANKSHAFT INSTALLATION

Remove the rear journal oil seal from the block. Remove the rear main bearing cap. Remove the main bearing inserts from the block and from the bearing caps.

Remove the connecting rod bearing inserts from the connecting rods and caps.

If the crankshaft main bearing journals have been refinished to a definite undersize, install the correct undersize bearings. **BE SURE** the bearing inserts and the bearing bores are clean. A very small amount of

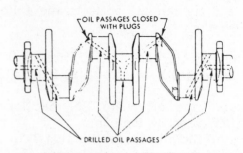

The oil passage holes MUST be cleaned to ensure lubrication for connecting rod and main bearings.

foreign material under the inserts could distort the bearing and cause a failure.

Place the upper main bearing inserts in position in the bores. **CHECK TO BE SURE** the tang indexes in the slot. Install the lower main bearing inserts in the bearing caps.

Clean the rear journal oil seal groove and the mating surfaces of the block and rear main bearing cap. Remove the oil seal retainer pin from the rear main bearing seal groove if one is installed. The retainer pin is not used with a split lip seal. Dip the lip seal halves in clean engine oil. Carefully install the seal halves in the block and in the rear main bearing cap with the **UNDERCUT** sides of the seal toward the **FRONT** of the engine and with approximately 3/8 inch protruding above the partial surface.

CAREFULLY lower the crankshaft into place. Be careful not to damage the highly finished journal surfaces.

Plastigage is soluble in oil, therefore, clean the crankshaft journal thoroughly of all traces of oil, and then turn the crankshaft until the oil hole is down away from the cap to prevent getting any oil on the

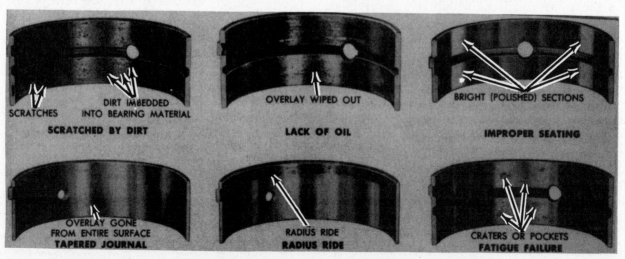

Examples of various wear patterns on bearing halves, including possible reasons for the condition.

PLACE PLASTIGAGE FULL WIDTH
OF JOURNAL ABOUT
¼ INCH OFF CENTER

INSTALLING PLASTIGAGE

CHECK WIDTH OF PLASTIGAGE

0.002" CLEARANCE

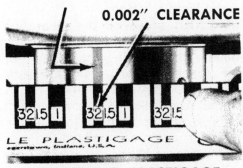

MEASURING PLASTIGAGE

After the connecting rod cap has been tightened to the required full torque value, and then removed, the flattened Plastigage can be compared with the scale on the side of the package. In this manner, the amount of clearance can be fairly accurately determined.

Plastigage. Whichever direction the crankshaft is turned with the rear bearing cap removed, hold the oil seal to prevent it from rotating out of position in the crankcase.

Place a piece of Plastigage on the bearing surface across the full width of the bearing cap and about 1/4 inch off center. Install the bearing cap, and then tighten the bolts to the torque value listed in the table on Page 3-74.

DO NOT turn the crankshaft while the Plastigage is in place.

Remove the bearing cap and measure the flattened Plastigage at its widest point using the scale printed on the Plastigage envelope. The number within the graduation which most closely corresponds to the width of the Plastigage indicates the bearing clearance in thousands of an inch. The widest point of the Plastigage is the minimum clearance and the narrowest point is the maximum clearance. The difference

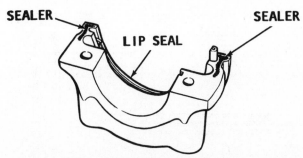

Apply sealer to the two areas shown on the main bearing cap and the matching area of the block to minimize the possibility of an oil leak.

between the readings indicates the taper of the journals. Compare your measurements with the Specifications given in the Appendix.

If the clearance exceeds the limits given, try 0.001" or 0.002" undersize bearings in combination with the standard bearings. Bearing clearance must be within the specified limits. If 0.002" undersize main bearings are used on more than one journal, **BE SURE** they are all installed in the cylinder block side of the bearing. If standard and 0.002" undersize bearings do not bring the clearance within the specified limits, refinish the crankshaft journal, then install undersize bearings.

After the bearing has been fitted, apply a light coat of engine oil to the journal and the bearing. Install the cap and tighten the bolts to the specified torque value given in the table on Page 3-74.

Repeat the procedure for the rear main bearing cap, and then for the remaining

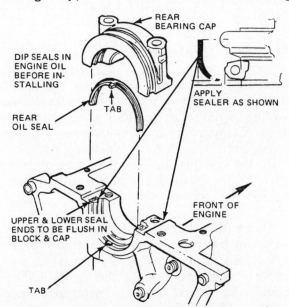

Details of installing the crankshaft rear oil seal when the crankshaft has been removed.

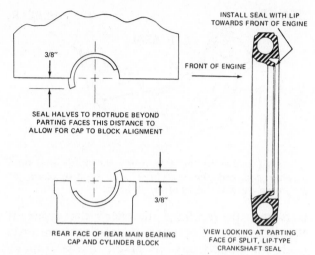

INSTALL SEAL WITH LIP
TOWARDS FRONT OF ENGINE

3/8"

FRONT OF ENGINE

SEAL HALVES TO PROTRUDE BEYOND
PARTING FACES THIS DISTANCE TO
ALLOW FOR CAP TO BLOCK ALIGNMENT

3/8"

REAR FACE OF REAR MAIN BEARING
CAP AND CYLINDER BLOCK

VIEW LOOKING AT PARTING
FACE OF SPLIT, LIP-TYPE
CRANKSHAFT SEAL

Cross-section drawing to depict installation of the rear oil seal in support of the instructions given in the text.

bearings except for No. 3 bearing, which is the thrust bearing. **CHECK TO BE SURE** each bearing cap has been installed in its original position and all of the cap bolts have been tightened to the required torque value.

Install the thrust, No. 3, bearing cap with the bolts fingertight. Pry the crankshaft forward with a large screwdriver against the thrust surface of the upper bearing half. Hold the crankshaft forward and at the same time, pry the thrust bearing cap to the rear. These movements will align the thrust surfaces of both halves of the bearings. Retain the forward pressure on the crankshaft, and tighten the cap bolts to the required torque value.

Force the crankshaft toward the rear of the engine. Install a dial indicator in such a position that the contact point rests against the crankshaft flange and the indicator axis is parallel to the crankshaft axis. Set the indicator to zero, and then push the crankshaft forward and note the reading. The reading will indicate the crankshaft end play. Compare your measurement with the limit given in the Specifications.

If the end play is less than the minimum limit, inspect both faces of the thrust bearing for scratches, burrs, nicks, or possible dirt.

If the thrust bearing faces are clean and not damaged, the problem of not enough end play is due to not having the thrust bearing properly aligned. Loosen the cap bolts and repeat the thrust bearing installation procedure. Check the end play again with the dial indicator.

After the main bearings have been properly installed and tightened to the required torque value, install the connecting rod caps.

Rotate the crankshaft until the throw for the piston being installed is at the bottom of its stroke. Now, push the piston all the way down until the connecting rod bearing seats on the crankshaft journal. Install the connecting rod cap. Tighten the cap nuts to the torque value given on Page 3-74.

After all of the piston assemblies have been installed and the connecting rod cap nuts tightened to the required torque value, check the side clearance between the connecting rods on each crankshaft journal. Compare your measurement with the limits given in the Specifications.

TAKE CARE to check the clearance of each rod bearing with Plastigage in the same manner as checking the main bearing clearance described earlier in this section. Compare your results with the Specifications in the Appendix. Oil the connecting rod bearings and crankshaft journals, then install the caps again and tighten the nuts to the required torque value.

Install the timing chain and sprockets, the cylinder front cover, and the crankshaft pulley and adapter, see Section 3-11.

Coat the threads of the flywheel attaching bolts with oil resistant sealer. Position the flywheel on the crankshaft flange. Install and tighten the bolts to the following torque value:

Mercury Marine   30 ft lb (41Nm)
Ford V8   80 ft lb (110Nm)

Install the oil pan, see Section 3-5.
Install the oil filter and the fuel pump.
Install the engine, see Chapter 13.
Fill the crankcase with the proper weight oil.

CAUTION

Water must circulate through the stern drive, also to and from the engine, anytime the engine is run to prevent damage to the water pump located in the lower unit (original and Alpha drives); or on the front of the engine (Bravo drive). Just a few seconds without water will damage the water pump.

Start the engine and check the completed work. Operate the unit at reduced rpm for the first hour.

4
FUEL

4-1 INTRODUCTION AND CHAPTER COVERAGE

The fuel system includes the fuel tank, fuel pump, fuel filters, water separating filter, carburetor, connecting lines, and the parts associated with these items. On models equipped with Electronic Fuel Injection (EFI), the carburetor is replaced with two fuel rails each supporting four injectors, a fuel regulator, an electric fuel pump, several electronic sensors, and a "black box" - Electronic Control Unit (ECU).

Regular maintenance of the carbureted system, as an aid to satisfactory performance from the engine, is limited to changing the fuel filter, water separating filter, and cleaning the flame arrester at regular intervals.

If a sudden increase in gas consumption is noticed, or if the engine does not perform properly, and if the flame arrester has been checked and found to be clean, then an overhaul, including boil-out, of the carburetor, or replacement of the fuel pump may be in order.

Regular maintenance of the EFI system is limited to changing the fuel filter and the water separating filter.

If a sudden increase in gas consumption is noticed on an engine equipped with EFI, the troubleshooting procedures are slightly more complicated as the entire system is electronically controlled.

COVERAGE

The following sections in this chapter, 4-2 thru 4-5, mainly apply to carbureted models, because from 1964 to the present day 99% of all units in the field supported carbureted systems.

The authors foresee in the years to come a great increase in the numbers of units supporting EFI systems. A full desciption of the operational principles of this system plus what minimal repair and troubleshoot-ing procedures exist are presented in Section 4-13.

4-2 TROUBLESHOOTING FUEL SYSTEM

The following paragraphs provide an orderly sequence of tests to pinpoint problems in the system. It is very rare for the carburetor by itself to cause failure of the engine to start.

Many times fuel system troubles are caused by a plugged fuel filter, a defective fuel pump, or by a leak in the line from the fuel tank to the fuel pump. Would you believe, a good majority of starting troubles which are traced to the fuel system are the result of an empty fuel tank and to aged fuel.

Fuel will begin to sour in three to four months and will cause engine starting problems. A fuel additive such as Sta-Bil may be used to prevent gum from forming during storage or prolonged idle periods.

If the automatic choke should stick in the open position, the engine would have trouble starting. This condition can be quickly corrected by rapid movement of the accelerator which will discharge fuel into the intake manifold and the engine will start.

If the automatic choke should stick in the closed position, the engine will flood making it very difficult to start. To correct this condition, move the fast idle or warm up lever to the wide-open position as the engine is cranked. This action will activate the unloader linkage to open the choke valve and help the flooded engine to start.

When the engine is hot, the fuel system can cause starting problems. After a hot engine is shut down, the temperature inside the fuel bowl may rise to 200°F and cause the fuel to actually boil. All carburetors are vented to allow this pressure to escape to the atmosphere. However, some of the fuel may percolate over the high-speed nozzle and overflow into the intake manifold.

In order for this raw fuel to vaporize enough to burn, considerable air must be added to lean out the mixture. Therefore, the only remedy is to open the throttle as wide as possible and to crank the engine until enough air is drawn in to provide the proper mixture for the engine to start. **NEVER** move the throttle lever back-and-forth in an attempt to start a hot engine. This action will only compound the problem by adding more fuel to an already too-rich mixture.

If the needle valve and seat assembly is leaking, an excessive amount of fuel may enter the intake manifold in the following manner: After the engine is shut down, the pressure left in the fuel line will force fuel past the leaking needle valve. This extra fuel will raise the level in the fuel bowl and cause fuel to overflow into the intake manifold.

A continuous overflow of fuel into the intake manifold may be due to a sticking or defective flot which would cause an extra high level of fuel in the bowl and overflow into the intake manifold.

EXCESSIVE FUEL CONSUMPTION

Excessive fuel consumption can be the result of any one of three conditions, or a combination of all three.

1- Inefficient engine operation.

2- Faulty condition of the hull, including excessive marine growth.

3- Poor boating habits of the operator.

If the fuel consumption suddenly increases over what could be considered normal, then the cause can probably be attributed to the engine or boat and not the operator.

Marine growth on the hull can have a very marked effect on boat performance. This is why sail boats always try to have a haul out as close to race time as possible. While you are checking the bottom take note of the propeller condition. A bent blade or other damage will definitely cause poor boat performance.

If the hull and propeller are in good shape, then check the fuel system for a leak. Check the line between the fuel pump and the carburetor while the engine is running and the line between the fuel tank and the pump when the engine is not running. A leak between the tank and the pump many times will not appear when the engine is operating because the suction created by the pump sucking fuel will not allow the fuel to leak. Once the engine is turned off and the suction no longer exists, fuel may begin to leak.

If a minor tune-up has been performed and the spark plugs, points, and timing are properly adjusted, then the problem most likely is in the carburetor and an overhaul is in order. Check the power valve and the needle valve and seat for leaking. Use extra care when making any adjustments affecting the fuel consumption, such as the float level, automatic choke, vacuum-control, and the power valve. Any time the automatic choke is checked, **BE SURE** the heat tube is open and the vacuum system is operating properly.

The flame arrester should be cleaned at regular intervals.

ENGINE SURGE

If the engine operates as if the load on the boat were being constantly increased and decreased even though you are attempting to maintain a constant engine speed, the problem can most likely be attributed to the fuel pump.

ROUGH ENGINE IDLE

If an engine does not idle smoothly, the most reasonable approach to the problem is to perform a tune-up to eliminate such areas as: defective points, faulty spark plugs; and the timing out of adjustment.

Other problems that can prevent an engine from running smoothly include: An air leak in the intake manifold; uneven compression between the cylinders; and sticky valves.

Of course any problem in the carburetor affecting the air/fuel mixture will also prevent the engine from operating smoothly at idle speed. These problems usually include: Too high a fuel level in the bowl; a heavy float; leaking needle valve and seat; a dirty flame arrestor; defective automatic choke; and improper adjustments for idle mixture or idle speed.

4-3 FUEL PUMPS AND FILTERS

The next few paragraphs briefly describe operation of the fuel pump. This description is followed by detailed procedures for testing the pressure; testing the volume; removing; and installing the fuel pump.

SINGLE-DIAPHRAGM FUEL PUMP

The fuel pump sucks gasoline from the fuel tank and delivers it to the carburetor in sufficient quantities, under pressure, to satisfy engine demands under all operating conditions.

The pump is operated by a two-part rocker arm. The outer part rides on a eccentric on the camshaft and is held in constant contact with the camshaft by a strong return spring. The inner part is connected to the fuel pump diaphragm by a short connecting rod. As the camshaft rotates, the rocker arm moves up and down. As the outer part of the rocker arm moves downward, the inner part moves upward, pulling the fuel diaphragm upward. This upward movement compresses the diaphragm spring and creates a vacuum in the fuel chamber below the diaphragm. The vacuum causes the outlet valve to close and permits fuel from the gas tank to enter the chamber by way of the fuel filter and the inlet valve.

Now, as the eccentric on the camshaft allows the outer part of the rocker arm to move upward, the inner part moves downward, releasing its hold on the connecting rod. The compressed diaphragm spring then exerts pressure on the diaphragm, which closes the inlet valve and forces fuel out through the outlet valve to the carburetor.

Because the fuel pump diaphragm is moved downward only by the diaphragm spring, the pump delivers fuel to the carburetor only when the pressure in the outlet line is less than the pressure exert ed by the diaphragm spring. This lower pressure condition exists when the carburetor float needle valve is unseated and the fuel passages from the pump into the carburetor float chamber are open.

When the needle valve is closed and held in place by the pressure of the fuel on the flaot, the pump builds up pressure in the fuel chamber until it overcomes the pressure of the diaphragm spring. This pressure almost stops movement of the diaphragm until more fuel is needed in the carburetor float bowl.

From this description and the accompanying illustration, you can appreciate why the condition of the fuel pump diaphragm and the carburetor float must be in good condtion at all times for proper engine performance.

DUAL-DIAPHRAGM FUEL PUMP

Some fuel pumps have two diaphragms and a sight bowl attached on the outside of the pump. The diaphragms are separated by a metal spacer. This type of fuel pump has four important safety features:

1- If the primary diaphragm fails, the pump continues to function with the second diaphragm.

2- Fuel cannot leak outward from the pump. The only possible place it can leak to is into the space between the diaphragms.

3- Fuel observed in the sight bowl indicates a faulty pump.

4- The possibility of both diaphragms failing at the same time is extremely remote because they are made of different materials and are shaped differently.

No maintenance is required or possible on the dual-diaphragm pump. If fuel is detected in the sight bowl replace the pump.

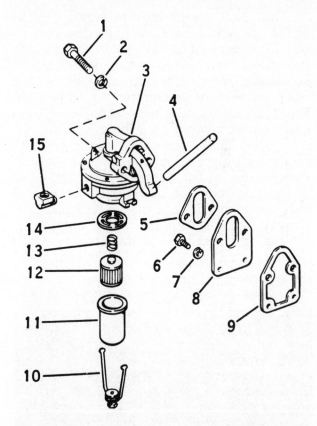

Exploded view of a Carter fuel pump used on a late model Chevrolet engine. A fuel pump repair kit is available which consists of all moving or wearing parts except the rocker arm. If the rocker arm or a casting is damaged, it is advisable to replace the pump rather than attempting to repair it. (1) screw, (2) lockwasher, (3) fuel pump assembly, (4) push rod, (5) gasket, (6) screw, (7) lockwasher, (8) plate, (9) gasket, (10) yoke, (11) bowl, (12) filter element, (13) spring, (14) gasket, (15) fitting.

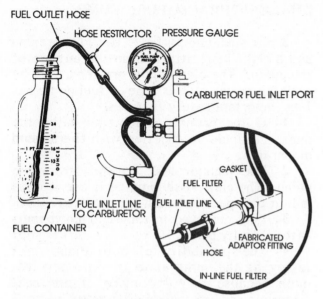

Arrangement of pressure gauge and container to test fuel pump pressure and volume.

SERVICE PROCEDURES

Most fuel pumps on late-model engines are of the sealed-type and cannot be repaired. The cost of a new pump is nominal, therefore, even if the pump is not sealed, it is usually more practical to replace the pump instead of attempting to repair it. For your safety, the new pump must be a Coast Guard approved marine-type unit.

Proper fuel pressure and volume are both necessary for proper engine performance. If the pressure is not adequate, the fuel level in the float bowl of the carburetor will be low and result in a lean mixture and fuel starvation at high speeds. If the pressure is too high the fuel level in the float bowl will rise and result in a rich mixture and flooding.

The fuel line at the carburetor is disconnected to perform the fuel pump test.

If the volume is not adequate, the engine will be starved at high speeds.

Service instructions consist of checking the output pressure of the pump and the volume delivered to the carburetor.

FUEL PUMP PRESSURE TEST

Remove the flame arrester. Disconnect the fuel line at the carburetor. **TAKE CARE** not to spill fuel on a hot engine because it may **IGNITE**.

ALWAYS have a fire extinguisher handy when working on any part of the fuel system. **REMEMBER**, a very small amount of fuel vapor in the bilge, has the potential explosive power of one stick of **DYNAMITE**.

Connect a pressure gauge, restrictor, container, and flexible hose between the fuel filter and the carburetor. Start the engine. With the engine idling, vent the outlet hose into the container by opening the hose restrictor momentarily. Close the hose restrictor and allow the pressure to stabilize. The pressure reading should be between 3.5 and 5.5 psi.

FUEL PUMP VOLUME TEST USING A GAUGE

If the fuel pump pressure is within the 3.5 to 5.5 psi range, test the volume by opening the hose restrictor with the engine idling and collect the fuel discharged into the graduated container. The fuel pump should discharge a pint of fuel in 30 seconds for an L6 or V6 engine and in 20 seconds for a V8 engine.

FUEL PUMP VOLUME TEST WITHOUT A GAUGE

CAUTION: Gasoline will be flowing in the engine compartment during this test. Therefore, guard against fire by grounding the high-tension wire to prevent it from sparking.

The high tension wire between the coil and the distributor can be grounded by either pulling it out of the distributor cap and grounding it, or by connecting a jumper wire from the primary (distributor) side of the ignition coil to a good ground.

Disconnect the fuel line at the carburetor. Place a suitable container over the end of the fuel line to catch the fuel dis-

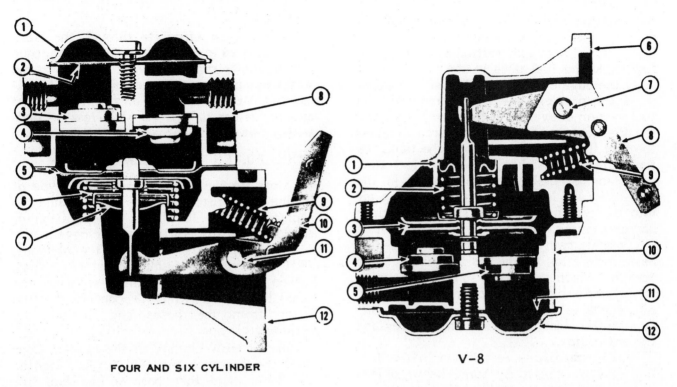

FOUR AND SIX CYLINDER

1	Pulsator Cover	7	Oil Seat
2	Pulsator Diaphragm	8	Fuel Cover
3	Outlet Valve	9	Rocker Arm Return Spring
4	Inlet Valve	10	Rocker Arm and Lever Assembly
5	Diaphragm Assembly	11	Pivot Pin
6	Diaphragm Spring	12	Pump Body

V-8

1	Oil Seal and Retainer	7	Pivot Pin
2	Diaphragm Spring	8	Rocker Arm and Lever Assembly
3	Diaphram Assembly	9	Rocker Arm Return Spring
4	Inlet Valve	10	Fuel Cover
5	Outlet Valve	11	Pulsator Diaphragm
6	Pump Body	12	Pulsator Cover

Cutaway views of two fuel pumps.

charged, and then crank the engine. If the fuel pump is operating properly, a healthy stream of fuel should pulse out of the line.

If the engine does not start even though there is adequate fuel flow from the fuel line, the fuel filter in the carburetor inlet may be plugged or the fuel inlet needle valve and the seat may be gummed together and prevent adequate fuel flow.

Continue cranking the engine and catching the fuel for about 15 pulses to determine if the amount of fuel decreases with each pulse or maintains a constant amount. A decrease in the discharge indicates a restriction in the line. If the fuel line is plugged, the fuel stream may stop. If there is fuel in the fuel tank but no fuel flows out of the fuel line while the engine is being cranked, the problem may be in one of three areas:

1- The line from the fuel pump to the carburetor may be plugged as already mentioned.

2- The fuel pump may be defective.

3- The line from the fuel tank to the fuel pump may be plugged or the line may be leaking air.

The following test explores these possibilities.

FUEL LINE TEST

The fuel line from the tank to the fuel pump can be quickly tested by disconnecting

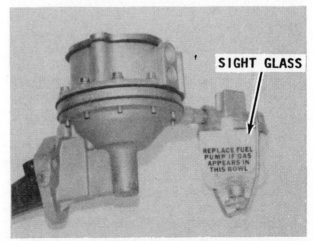

SIGHT GLASS

REPLACE FUEL PUMP IF GAS APPEARS IN THIS BOWL

Fuel in the sight glass indicates a ruptured diaphragm and the fuel pump must be replaced.

the existing fuel line at the fuel pump and connect a six-gallon portable tank and fuel line. This simple substitution eliminates the fuel tank and fuel lines in the boat. Now, start the engine and check the performance.

If the problem has been corrected, the fuel system between the fuel pump inlet and the fuel tank is at fault. This area includes the fuel line, the fuel pickup in the tank, the fuel filter, anti-siphon valve, the fuel tank vent, and excessive foreign matter in the fuel tank, and loose fuel fittings sucking air into the system. Improper size fuel fittings can also restrict fuel flow.

Possible cause of fuel line problems may be deterioration of the inside lining of the fuel line which may cause some of the lining to develop a blockage similar to the action of a check valve. Therefore, if the fuel line appears the least bit questionable, replace the entire line.

Another possible restriction in the fuel line may be caused by some heavy object lying on the line -- a tackle box, etc.

FUEL PUMP REMOVAL

Disconnect the fuel inlet and outlet lines from the fuel pump. **ALWAYS USE** two wrenches when disconnecting or connecting the outlet line fitting to avoid damaging the fuel pump.

Be sure to plug the fuel line to prevent fuel from siphoning out of the fuel tank.

Remove the fuel pump mounting bolts, and then the pump and gasket. If you are working on a V6 or V8 engine and the push rod is to be removed, remove the pipe plug, push rod, and the fuel pump adaptor.

FUEL PUMP INSTALLATION

If working on a V6 or V8 engine, install the fuel pump push rod and pipe fitting or the adaptor. Lay down a bead of Permatex, Form-A-Gasket, or equivalent, on the gasket and pipe fitting.

Install the fuel pump and a **NEW** gasket. Use sealer on the mounting bolt threads. Tighten the bolts securely.

On a V6 or V8 engine, a pair of mechanical fingers can be used to hold the fuel pump push rod up while installing the pump.

Connect the fuel lines to the pump.

Start the engine and check for leaks.

CAUTION
Water must circulate through the stern drive, also to and from the engine, anytime the engine is run to prevent damage to the water pump located in the lower unit (original and Alpha drives); or on the front of the engine (Bravo drive). Just a few seconds without water will damage the water pump.

FUEL FILTER REPLACEMENT

Most marine engines have some type of in-line fuel filter. This filter should be replaced every 100 hours of operation or sooner is you suspect it may be clogged.

To replace the in-line filter element, first remove the flame arrestor. Next, loosen the retaining clamps securing the hoses to the fuel filter. Disconnect the fuel filter from the hoses and discard the retaining clamps.

Install **NEW** clamps on the hoses. Connect the hoses to the new filter. Tighten the filter, and then position the fuel line hose clamps in place and crimp them securely. Start the engine and check for leaks.

CAUTION
Water must circulate through the stern drive, also to and from the engine, anytime the engine is run to prevent damage to the water pump located in the lower unit (original and Alpha drives); or on the front of the engine (Bravo drive). Just a few seconds without water will damage the water pump.

FLAME ARRESTER CLEANING

The flame arrester should be removed and cleaned every 50 hours. It is not

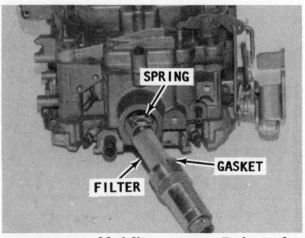

Arrangement of fuel filter parts on a Rochester four barrel carburetor.

necessary to replace the arrestor unless it is damaged and will not seat properly on the carburetor.

Remove and clean the arrestor with solvent and blow it dry with compressed air. Replace it on the carburetor.

WATER SEPARATING FUEL FILTER

Most new engines have a factory installed water separating fuel filter. This type filter is also available as an accessory for all other engines.

A water separating filter, as its name suggests, removes water and other fuel system contaminants before they reach the carburetor or EFI system and helps minimise potential problems. The presence of water in the fuel will alter the proportion of air/fuel mixture to the "lean" side, resulting in a higher operating temperature and possible damage to pistons if not corrected.

The filter consists of a mounting plate and disposable cannister filter. The manufacturer recommends the disposable cannister be changed at least one a year.

The filter is installed between the fuel tank and the mechanical fuel pump on carbureted engines or the electric fuel pump on fuel injected engines.

Notice there are two inlet fittings and two outlet fittings on the mounting plate. There are many ways to connect the fuel line incorrectly, but only **ONE** way to connect the lines correctly!

Refer to the accompanying illustration:

If **"C"** is the inlet fitting, use **"A"** as the outlet fitting.

If **"B"** is the inlet fitting, use **"D"** as the outlet fitting.

If the water separating fuel filter is installed, the fuel inlet fitting, either **"C"** or **"B"** will already be selected. If an installation kit has been purchased, then simply follow the instructions supplied with the kit for the engine being serviced.

Service

Remove the cannister filter from the mounting plate by rotating the cannister counterclockwise. An oil filter wrench may be necessary to break the filter free. Keep the filter upright to avoid spilling fuel. Properly dispose of the fuel and fuel saturated cannister.

Observe the top of the cannister. Two sealing rings **MUST** also be removed and discarded. If no rings or just one ring came

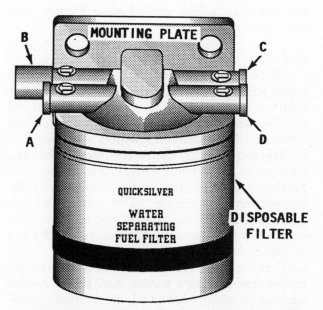

Most newer models come equipped with a water separating fuel filter. On older models, such a filter should be installed in the fuel line at the first opportunity. Such a kit is not expensive, and contains a disposable cannister filter similar to a standard oil filter element. Observe the four fuel fittings (A, B, C, and D as indicated), on the filter mounting bracket. Instructions are supplied with the filter for correct installation.

away with the cannister, check the mounting plate sealing surface to locate the old rings. All old sealing rings **MUST** be discarded. If an old sealing ring remained in place after the a new cannister filter is installed - with two new sealing rings, the filter will most definately leak fuel.

An old cannister filter **CANNOT** be cleaned and reinstalled. It must be discarded.

Coat the two sealing rings of a new cannister filter with clean engine oil. Install the filter onto the mounting plate and tighten securely **BY HAND**.
DO NOT use an oil filter wrench to tighten the cannister. This serves no purpose and will make the cannister will be very difficult to remove.

4-4 MODERN FUELS

The manufacturer of the units covered in this manual recommends the powerplants be operated using either regular unleaded or regular leaded gasoline with a minimum octane rating of 84 or higher.

Question

What is octane rating and why are we concerned with having a "minimum" rating for use in gasoline engine, specifically outboard units?

Answer

An octane number assigned to specific gasoline is the unit measure of the gaoline's ability to resist detonation. Stated in another term, the octane number is the unit measure of a gasoline's antiknock quality.

Detonation, in simple terms, is the uncontrolled "explosion" of any remaining air-fuel mixture in the combustion chamber after normal "burning" of the fuel has occurred. Detonation can be due to excessive temperature and certain pressure conditions in the chamber.

Detonation is "bad news" because it creates shock pressure waves, and therefore an audible "knock" instead of the uniform combustion and expansion of the air/fuel mixture which is desirable. Detonation can result in loss of power, excessive temperature in certain areas of the combustion chamber, and in worse case, actual damage to the powerhead.

If a particular gasoline is not of sufficient high octane as recommended by the manufacturer, the fuel's ability to resist detonation can be raised by blending a higher octane gasoline or by using an additive. One brand of additive on the American market is No. 104 Octane Booster available at most auto parts houses, including NAPA outlets. This type additive can also be used to help prevent detonation (knock) in unleaded gasoline.

Unknown to the general public, many refineries are adding alcohol in an effort to hold the octane rating.

Alcohol in gasoline can have a deteriorating effect on certain fuel system parts.

A lead substitute additive can help prevent detonation when unleaded gasoline is used.

Seals can swell, pump check valves can swell, diaphragms distort, and other rubber or neoprene composition parts in the fuel system can be affected.

Since about 1981, all manufacturers have made every effort to use materials that will resist the alcohol being added to fuels.

Fuels containing alcohol will slowly absorb moisture from the air. Once the moisture content in the fuel exceeds about 1%, it will separate from the fuel taking the alcohol with it. This water/alcohol mixture will settle to the bottom of the fuel tank.

In the United States, the Environmental Protection Agency (EPA) slated a proposed national phase-out of leaded fuel, "Regular" gasoline, by 1988. Fortunately, this has not happened. Tetraethyl Lead in gasoline boosts the octane rating (energy). Therefore, if the lead is removed, it must be replaced with another agent.

A lead substitute additive helps cushion the inpact between the valve and the seat. The valve seats on modern cylinder heads are induction hardened to protect them from the harmful affects of unleaded fuel.

"SOUR" FUEL

Under average conditions (temperate climates), fuel will begin to breakdown in about four months. A gummy substance forms in the bottom of the fuel tank and in other areas. The filter screen between the tank and the carburetor and small passages in the carburetor will become clogged. The gasoline will begin to give off an odor similar to rotten eggs. Such a condition can cause the owner much frustration, time in cleaning components, and the expense of replacement or overhaul parts for the carburetor.

Even with the high price of fuel, removing gasoline that has been standing unused over a long period of time is still the easiest and least expensive preventative maintenance possible. In most cases, this old gas can be used without harmful effects in an automobile using regular gasoline.

The gasoline preservative additive Quicksilver Gasoline Stabilizer and Conditioner, shown below, will keep the fuel "fresh" for up to twelve months. If this particular product is not available in your area, other similar additives are produced under various trade names.

4-5 CARBURETORS
GENERAL INFORMATION

In the simplest terms, a carburetor is merely a metering device which mixes the proper amount of fuel and air for delivery to the cylinders under all operating conditions.

When the engine is idling, the mixture is roughly 10 parts air to 1 part fuel. At high speed or under heavy load, the mixture is about 12 parts air to 1 part fuel.

The fuel is held in reserve in the float chamber of the carburetor. A float valve in this chamber admits fuel from the fuel pump to replace the fuel leaving the chamber and burned by the engine. Metering jets extend from the fuel chamber into the carburetor throat.

The downward movement of a piston creates a suction that draws air into the carburetor throat. There is a restriction in the throat called a venturi. This venturi reduces air pressure at this point by increasing the air velocity.

The difference between the air pressure in the float chamber and the pressure in the carburetor throat causes the fuel to be forced through the metering jets and into the air stream. This mixture of fuel and air is then burned in the engine cylinders.

From this description, you can appreciate why the jets must be clean, free of gum, and adjusted properly for satisfactory engine performance.

The low-speed jet has an adjustable needle to compensate for changing atmospheric conditions. The high-speed jet is a fixed orifice.

The volume of the air/fuel mixture drawn into the engine to regulate engine speed, is controlled by a throttle valve. An extra amount of fuel is required to start a cold engine. This extra fuel is delivered by a choke valve installed ahead of the metering jets and the venturi or venturis as the case may be. After the engine starts and warms to operating temperature, the choke is gradually opened to restore the normal air/fuel mixture.

The throat of a carburetor is usually referred to as a "barrel". Single, double, or four barrel carburetors have a metering jet, needle valve, throttle, and choke plate for each barrel or pair of barrels.

4-6 ROCHESTER BC CARBURETOR

This single-barrel carburetor was used on early model four- and six-cylinder engines and was replaced by the Rochester 2GC model.

A couple features of this carburetor include:

A concentric-type float bowl allowing fuel to surround the main carburetor bore and venturi. This design, plus the centrally located main fuel discharge nozzle, prevents fuel spill-over during sharp turns, quick starts, and sudden stops.

This assembly contains the main metering parts of the carburetor and is easily removed for inspection and service. It is attached to the air horn and suspended in the fuel in the float chamber. This arrangement insulates the main assembly from heat that may be transmitted from the engine directly to the bottom of the float bowl. This design helps maintain more accurate fuel metering because less fuel vapors enter the main metering parts of the assembly when the engine is hot.

REMOVAL

1- Remove the choke cover attaching screws and retainers. Remove the choke cover and thermostatic coil assembly from the choke housing. Remove the choke cover

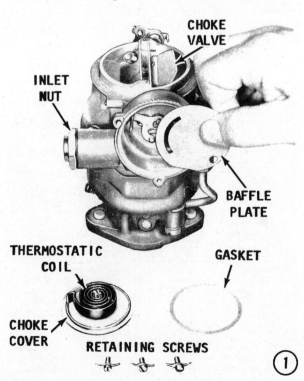

CHOKE VALVE

INLET NUT

BAFFLE PLATE

THERMOSTATIC COIL

GASKET

CHOKE COVER

RETAINING SCREWS

①

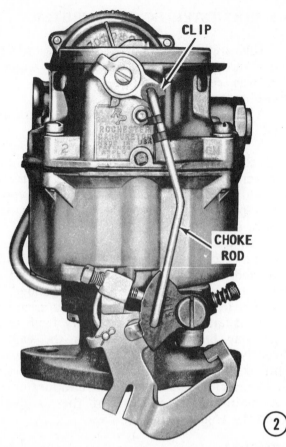

②

gasket and baffle plate. Remove the fuel filter inlet nut, gaskets, fuel filter, and spring. The choke valve and shaft **SHOULD NOT** be removed unless the shaft is binding or the valve is damaged. To repair this type of damage, file the choke valve screw staked ends level with the choke shaft. Remove the choke valve plate retaining screws from the shaft. Lift out the plate, and then slide the shaft out of the air horn. The shaft is

removed by first rotating the shaft to remove the piston and pin from the choke housing. Remove the choke housing from the bowl cover by removing the attaching screws and gasket.

2- Remove the choke rod retaining spring clip and rod from the choke shaft lever. Unscrew the vacuum tube connector nut from the choke housing. Remove the bowl cover screws, and then the re turn spring bracket. Lift the cover **STRAIGHT UP** to prevent damage to the floats.

3- With the cover up-ended, remove the float hinge pin, the floats, and the float needle. Remove the float seat and gasket. Remove the main well support, vacuum power piston, and spring, Separate the gasket from the cover. Remove the main metering jet and power valve retainer. Take out the spring and ball from the main well support.

4- Use a pair of needle nose pliers to remove the pump discharge guide. Hold the accelerating pump plunger all the way down, and at the same time remove the hairpin retainers and the pump link from the throttle lever and pump arm. Remove the pump assembly from the bowl. Emerse the plunger in gasoline or kerosene to prevent the leather from drying. Remove the pump return spring from the pump well. Turn the bowl upside down and carefully remove the pump discharge spring and ball. While the bowl is still upside down, remove the throttle body attaching screws and gasket. Remove the idle mixture needle adjusting screw and spring from the throttle body.

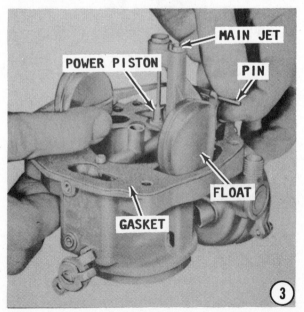

③

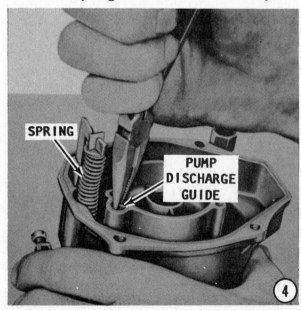

④

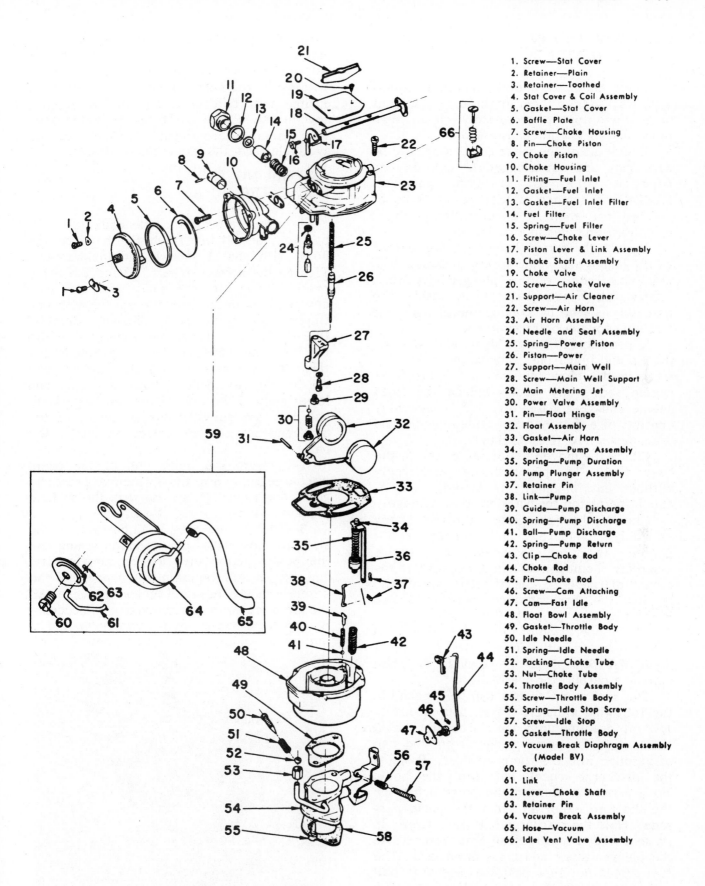

1. Screw—Stat Cover
2. Retainer—Plain
3. Retainer—Toothed
4. Stat Cover & Coil Assembly
5. Gasket—Stat Cover
6. Baffle Plate
7. Screw—Choke Housing
8. Pin—Choke Piston
9. Choke Piston
10. Choke Housing
11. Fitting—Fuel Inlet
12. Gasket—Fuel Inlet
13. Gasket—Fuel Inlet Filter
14. Fuel Filter
15. Spring—Fuel Filter
16. Screw—Choke Lever
17. Piston Lever & Link Assembly
18. Choke Shaft Assembly
19. Choke Valve
20. Screw—Choke Valve
21. Support—Air Cleaner
22. Screw—Air Horn
23. Air Horn Assembly
24. Needle and Seat Assembly
25. Spring—Power Piston
26. Piston—Power
27. Support—Main Well
28. Screw—Main Well Support
29. Main Metering Jet
30. Power Valve Assembly
31. Pin—Float Hinge
32. Float Assembly
33. Gasket—Air Horn
34. Retainer—Pump Assembly
35. Spring—Pump Duration
36. Pump Plunger Assembly
37. Retainer Pin
38. Link—Pump
39. Guide—Pump Discharge
40. Spring—Pump Discharge
41. Ball—Pump Discharge
42. Spring—Pump Return
43. Clip—Choke Rod
44. Choke Rod
45. Pin—Choke Rod
46. Screw—Cam Attaching
47. Cam—Fast Idle
48. Float Bowl Assembly
49. Gasket—Throttle Body
50. Idle Needle
51. Spring—Idle Needle
52. Packing—Choke Tube
53. Nut—Choke Tube
54. Throttle Body Assembly
55. Screw—Throttle Body
56. Spring—Idle Stop Screw
57. Screw—Idle Stop
58. Gasket—Throttle Body
59. Vacuum Break Diaphragm Assembly
    (Model BV)
60. Screw
61. Link
62. Lever—Choke Shaft
63. Retainer Pin
64. Vacuum Break Assembly
65. Hose—Vacuum
66. Idle Vent Valve Assembly

Exploded view of the Rochester B, BC, and BV carburetor with principle parts identified.

CLEANING AND INSPECTING THE ROCHESTER BC

Place all of the metal parts in a screen-type tray and dip them in carburetor solvent until they appear completely clean, then blow them dry with compressed air.

Check all of the parts and passages to be sure they are not clogged or contain any deposits. Blow out all of the passages. **NEVER** use a piece of wire or any type of pointed instrument to clean drilled passages or calibrated holes in a carburetor.

Inspect the pump plunger. If the leather or its garter expanding spring is damaged in any way, replace the plunger assembly. Verify the bypass ball check inside the assembly is free by shaking the plunger and listening for the ball movement.

Inspect the floats for dents and wear on the lip and hinge pin. Check the cover for wear in the hinge pin holes. If the float needle shows any wear, replace the float needle-and-seat assembly. This assembly consists of a matched and tested needle and seat, plus a new fiber washer.

Check the movement of the piston in the cover bore. The piston should move freely without any binding. If binding is felt, check the piston for burrs or other damage.

Check the complete throttle body assembly. If there is any evidence of abnormal wear or looseness, the entire assembly should be replaced, because of the close tolerance of the throttle valve and because the spark advance ports are drilled in relation to a properly fitted valve.

Check operation of the choke valve when it is assembled in the cover.

ALWAYS replace the fuel filter when making a carburetor overhaul.

The fuel pump system can be checked in the following manner: First, pour about 1-1/2" of gasoline into the carburetor bowl. Next, slide the pump plunger from the can of gasoline into the pump cylinder. Position the discharge check ball into the body. Now, raise the plunger and press lightly on the shaft to expel air from the pump passage. Hold the discharge ball down firmly in its seat with a small, clean brass rod. Raise the plunger again and press downward. The fuel should not flow past the discharge ball or back through the inlet ball in the pump assembly. If the pump plunger depresses easily, it could mean that either dirt is present or that the check balls are damaged.

Clean the passage and repeat the test. If leakage is still indicated, replace the check ball or the pump plunger assembly.

Most of the parts that should be replaced during a carburetor overhaul, including a new matched fuel inlet needle valve-and-seat assembly, are found in a carburetor kit available from the local marine dealer.

ASSEMBLING THE ROCHESTER BC

1- Place a **NEW** throttle body gasket in position, and then attach the bowl to the body with the 2 screws and lockwashers. Tighten the screws evenly. Place the pump return spring into the pump well, and then center it by depressing the spring with one finger. Install the pump plunger assembly. **USE CARE** and insert the leather part into the bowl and connect the pump link to the throttle lever and pump arm. Now, install the hairpin retainers at both the upper and lower ends of the link. Drop the large steel ball into the pump discharge cavity of the bowl and position the spring on top of the ball.

2- Insert the index end of the pump discharge guide into the spring and press the guide down until it is flush with the surface of the bowl.

3- Place a new gasket in position on the air horn. Insert the power piston spring and the power piston into the air horn cavity, and then attach the main well support to the cover with a screw and lockwasher. Install the main metering jet and tighten it securely. Hold the power piston stem down, and at the same time, install the power ball,

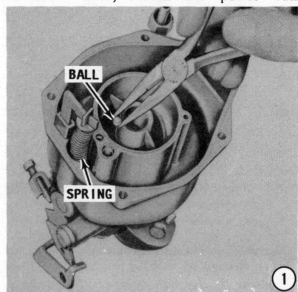

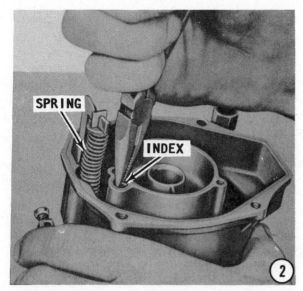

2

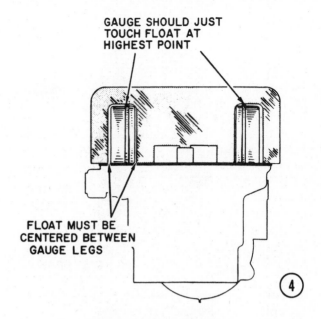

GAUGE SHOULD JUST
TOUCH FLOAT AT
HIGHEST POINT

FLOAT MUST BE
CENTERED BETWEEN
GAUGE LEGS

4

spring, and plug. Tighten the plug securely. Install a new fiber washer in the float needle seat well, and then install the seat and float needle. Attach the float and hinge pin with the float tang **FACING** the **COVER**. At this point, the float level and the float drop adjustments **MUST** be made.

4- Float Level Check: Place the proper gauge over the floats with the gasket in position. The floats should be equally centered in the float gauge cutouts and they should just touch at their highest point.

5- Float Level Adjustment: Bend the float arm until the distance from the air horn gasket to the top of each float is 1-9/32".

6- Float Drop Check: Measure the distance from the bottom of the float to the air horn gasket, as shown. Bend the float tang until this distance is 1-3/4".

7- CAREFULLY place the assembled air

① INVERT AIR HORN WITH
   GASKET IN PLACE

② GAUGE FROM GASKET SURFACE
   TO TOP OF EACH FLOAT

④ VISUALLY CHECK
   FLOAT ALIGNMENT

③ BEND TANG HERE
   TO ADJUST

5

① AIR HORN
   RIGHT SIDE UP
   TO ALLOW
   FLOATS TO
   HANG FREE
   (GASKET IN
   PLACE)

③ BEND FLOAT
   TANG TO ADJUST
   FOR PROPER
   SETTING

② MEASURE FROM GASKET SURFACE
   TO BOTTOM OF EACH FLOAT

6

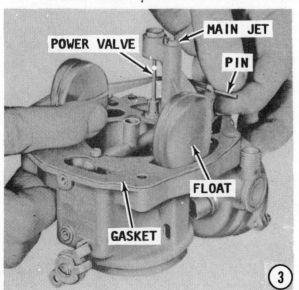

POWER VALVE   MAIN JET

PIN

FLOAT

GASKET

3

horn onto the main body. Install and tighten the retaining screws. Tighten the 1/2" brass fitting on the choke suction tube. Assemble the choke piston to the shaft and slide the assembly into the choke housing bore. Rotate the choke shaft counterclockwise until the piston enters its cylinder. Install the choke valve with the letters **RP** facing **UP-WARD**. Be sure to center the valve before tightening the two screws. Check to be sure the valve operates freely. Install the choke

SPRING

INDEX

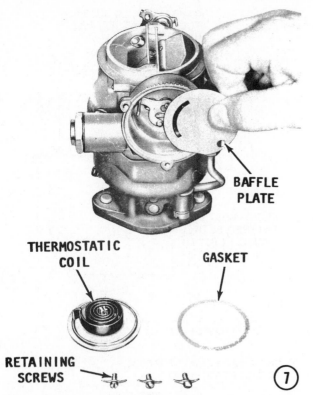

BAFFLE
PLATE

THERMOSTATIC
COIL

GASKET

RETAINING
SCREWS

⑦

baffle plate, the choke housing gasket, and the thermostatic coil cover. Rotate the cover clockwise until the index marks on the cover and the housing are aligned. Check the specifications in the Appendix. Tighten the three screws and retainers securely. The choke valve should touch the bore of the air horn lightly at room temperature.

8- Install the idle mixture adjusting needle. Turn it in **GENTLY** until it seats,

then back it out 2 turns. This position will give a rough adjustment at this time. Place the choke counterweight on the end of the choke shaft, with the **TANG** facing the choke **HOUSING**. Install the spacing washer and trip lever in such a position that the tang of the trip lever will be on top of the counterweight tang when the choke is fully open. Attach one end of the choke rod to the counterweight and the other end of the rod to the fast-idle cam. The dog leg of the rod **MUST FACE** the idle adjusting needle. Install the pin spring and the end clip which will secure the choke rod.

ROCHESTER BC ADJUSTMENTS

9- Choke Rod Adjustment: Check to be sure the idle screw contacts the fast-idle cam when the choke valve is completely open or completely closed. If the screw drops off the cam in either choke position, bend the choke rod until the cam is positioned correctly.

10- Unloader Adjustment: With the throttle in the wide-open position, use Tool J-9580, or measure 0.230" to 0.270", to verify the small end slides freely between the lower end of the choke valve and the bore of the carburetor. Bend the tang on the throttle lever as required to obtain the necessary clearance.

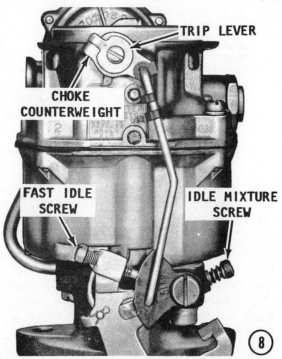

TRIP LEVER

CHOKE
COUNTERWEIGHT

FAST IDLE
SCREW

IDLE MIXTURE
SCREW

⑧

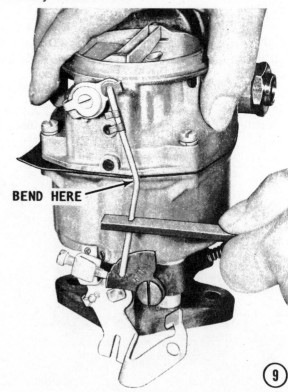

BEND HERE

⑨

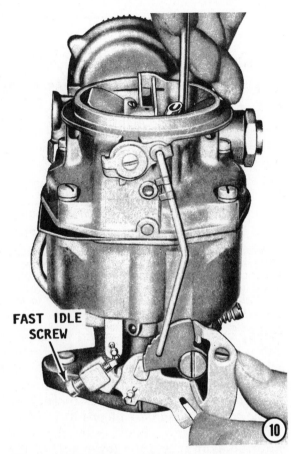

FAST IDLE
SCREW

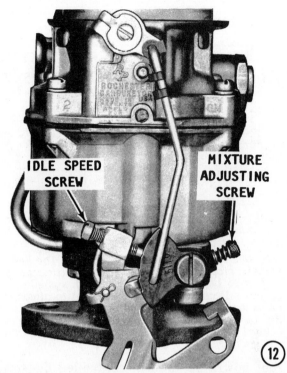

IDLE SPEED
SCREW

MIXTURE
ADJUSTING
SCREW

11- Automatic Choke Adjustment:
Loosen the three screws that secure the choke cover, and then turn the cover until the scribe line on the cover aligns with the index mark on the choke housing. Tighten the cover screws.

12- Idle Mixture and **Speed Adjustment:**
Start the engine and allow it to warm to operating temperature, until the choke is wide open.

CAUTION

Water must circulate through the stern drive, also to and from the engine, anytime the engine is run to prevent damage to the

water pump located in the lower unit (original and Alpha drives); or on the front of the engine (Bravo drive). Just a few seconds without water will damage the water pump.

Adjust the idle speed screw until the engine speed is 500 rpm, then adjust the mixture adjusting screw until the highest steady idle speed is obtained. A clean flame arrestor **MUST** be installed when these adjustments are made.

4-7 CARTER RBS CARBURETOR

DESCRIPTION AND OPERATION

The following paragraphs will give you a general description of the Carter RBS carburetor including an explanation of how the various systems operate and their influence on engine performance.

The RBS carburetor is built with a single light-weight aluminum casting with a pressed-steel fuel bowl. Most of the calibration points are located in the central casting making it easy to service and adjust because the adjustment points are readily accessible. The fuel pick-ups are located near the centerline of the carburetor bore to gain the benefits of a concentric float bowl carburetor, but they are located so engine heat being radiated through the carburetor bore and casting are not easily conducted to the

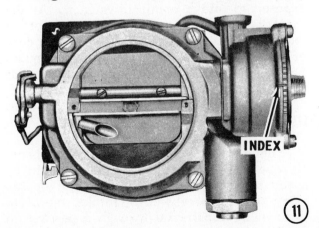

INDEX

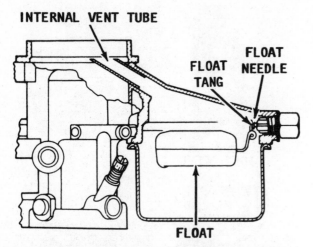

Details of the Carter RBS carburetor. The fuel level must be properly adjusted to allow the correct amount of fuel to be delivered to the other systems at all speeds. Because the bowl is vented to atmospheric pressure, the pressure difference between the area on the top of the fuel in the bowl and the pressure in the venturi area forces the fuel through the various systems.

fuel bowl. A vacuum-controlled diaphragm step-up rod assembly provides instant response to varying engine demands.

FLOAT SYSTEM

The float system maintains an adequate supply of fuel at the proper level in the fuel bowl for use by all of the other systems. The float assembly is compact for rigidity, and assures little or no change in the fuel level setting due to heat or vibration.

The float is made of a cellular nitro rubber material which is impervious to denting under normal handling conditions and it is not susceptible to punctures. These features assure a constant bouyancy factor and long service life. The single float extends around the metering portion of the casting to produce the effect of having twin floats.

The needle seat is resilient and has the unique ability to "digest" small foreign particles in the fuel, thereby minimizing leakage of fuel or flooding under extreme conditions. This resilient seal reduces wear and extends the life of the fuel intake needle. The fuel bowl is vented to the inside of the air horn to provide a balance between air horn pressure and fuel bowl pressure.

IDLE SYSTEM

Fuel for idling and early part-throttle operation is metered through the idle, low-

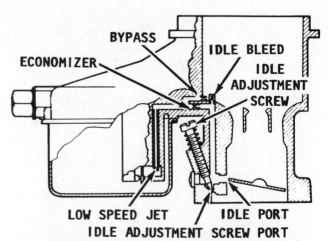

Details of the idle system. The air/fuel mixture at idle is adjusted with the idle adjustment screw. This system furnishes the proper air/fuel mixture at low speed when the throttle valve is almost closed and the high speed system is ineffective, due to the low velocity of air through the venturi.

speed, system. The low-speed jet is pressed into place within a passage in the casting to prevent damage. It cannot be removed, but it can be cleaned by blowing compressed air through the step-up rod jet or through either of the idle passaage air bleeds located in the carburetor bore. Fuel from the bowl is metered as it enters the lower end of the jet and flows up through the tube where air which has been metered through the by-pass mixes with the fuel.

Both fuel and air then pass through the economizer and on to the idle bleed where more air, which has been metered, is introduced. This air/fuel mixture is discharged into the intake manifold through the idle port and the idle adjustment screw port.

The idle adjustment screw controls the amount of air/fuel mixture admitted to the intake manifold. Backing the screw out increases the amount of mixture; turning the screw in decreases the mixture. At idle speed, only a small amount of idle port is exposed to intake manifold vacuum. As the throttle is opened, more of the idle port is exposed to allow an increase in the amount of air/fuel mixture admitted to the engine. The idle jet, economizer, and both air bleeds are calibrated and pressed into place.

HIGH-SPEED SYSTEM

Fuel for part throttle and for full-throttle operation is supplied through the high-speed system. During part-throttle operation, the relatively high intake manifold vacuum is transferred through a passage in

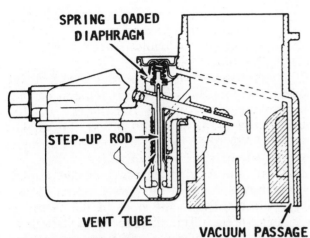

Details of the Carter RBS high speed system. When the throttle is opened, this system supplies the proper air/fuel mixture. The air velocity passing through the venturi sucks fuel out of the high speed nozzle. If the throttle is partially opened, vacuum transferred to the diaphragm of the step rod, lifts the rod. The larger part of the rod enters the jet and restricts fuel flow to provide an economical air/fuel mixture. As the throttle is opened more, intake manifold vacuum drops, and the step up rod is lowered into the jet to provide the richer mixture required by the additional power demand.

the manifold casting up to the upper surface of the spring-loaded diaphragm to which the step-up rod is attached. The manifold vacuum, exerting an opposing force to the calibrated spring, provides an economical air-fuel mixture to the engine under all conditions, except when full power is required. When the diaphragm is up, vacuum is high, the larger diameter (lower end) of the step-up rod is in the jet to provide the economy mixture.

When the throttle is opened, the manifold vacuum decreases. Once the difference in pressure applied to the two sides of the diaphragm is not great enough to offset the downward pressure of the calibrated spring, the diaphragm moves downward and the step-up rod, attached to the diaphragm, is lowered in the jet. The smaller diameter of the rod permits the metered increase in fuel flow to satisfy additional power demands. Additional fuel for these power requirements is fed through the constant-feed bushing as well as through the step-up rod jet.

Under an acceleration condition, the same action takes place to enrich the fuel mixture for the additional power needs. However, the step-up rod is raised in the jet just as soon as terminal acceleration is reached when engine load, and manifold vacuum, indicates the need for a less rich fuel mixture.

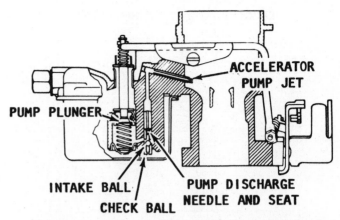

Details of the Carter RBS acceleration system. On demand during acceleration, the pump plunger is activated to force fuel past the discharge needle and out through the metered pump jet, as shown.

A vent tube is pressed into place within the high-speed well around the step-up rod. This tube has calibrated side holes. A metered amount of air is admitted from the bore of the carburetor to the ring around the vent tube. This air passes through the side holes in the vent tube to mix with the fuel before it flows through the nozzle into the air stream. This air/fuel mixture is known as an emulsion and it permits the fuel to vaporize immediately as it emerges from the tip of the nozzle and to assure equal distribution of the fuel to each cylinder. This action is further benefited by the additional air bleed located in the center of the nozzle.

During idle operation, or with the engine shut off with hot fuel on an extremely hot day, fuel sometimes boils in the fuel bowl and the various passages of the carburetor. When these vapor bubbles in the high-speed passageway force fuel out the nozzle, the carburetor is said to be percolating. The design of the RBS carburetor prevents this action as the fuel level in the high-speed well is far below the nozzle cross-over passage and only vapors can be emitted through the nozzle, not raw fuel.

ACCELERATION SYSTEM

The accelerating pump system supplies the necessary amount of measured fuel to ensure smooth engine performance during low speed acceleration. As the throttle is closed, the pump plunger is raised in the pump cylinder and the fuel from the bowl flows into the cylinder through the intake ball check. This ball check is located next to the pump cylinder. Air is prevented from

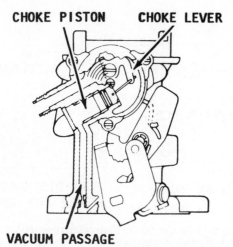

CHOKE PISTON   CHOKE LEVER

VACUUM PASSAGE

Details of the Carter RBS choke system. This system provides the necessary air/fuel mixture for starting a cold engine. The flow of air through the carburetor is restricted because of the tension on a thermostatic spring holding the choke valve closed. After the engine starts, intake manifold vacuum, applied to the underside of the choke piston helps to open the choke valve to admit more air. On a dual carburetor installation, the choke system only operates on the forward carburetor.

entering the cylinder due to the sealing action of the pump discharge needle being on its seat during the intake stroke. When the throttle is opened, mechanical action, transmitted through the pump arm, pushes the plunger downward in the cylinder. Fuel is forced past the discharge needle and out through the metered pump jet and into the air stream. During the discharge stroke, the intake ball is on its seat to prevent fuel from flowing back into the bowl. The spring on the connector link and the size of the pump jet provide a pump discharge of the required amount and for the proper length of time.

CHOKE SYSTEM

The Carter Climatic Control choke on the RBS carburetor provides the correct mixture necessary for quick cold engine starting and proper warm-up performance. On a dual carburetor installation, the choke system operates only on the forward carburetor. The thermostatic coil housing assembly on the rear carburetor is set to the index mark plus 1-1/4 turns clockwise. At this setting, the choke valve in the rear carburetor is always in the open position.

When the engine is cold, tension of the thermostatic coil spring holds the choke valve closed. As the engine is cranked, air pressure against the offset choke valve causes the choke valve to open slightly against the spring tension of the thermostatic coil. Intake manifold vacuum applied to the underside of the choke piston also tends to pull the choke valve open.

When the engine starts, the choke valve assumes a position where the tension of the thermostatic coil spring is balanced by the pull of the manifold vacuum on the choke piston **AND** the force of the air stream against the offset choke valve.

After the engine starts and the choke piston moves down in the choke piston cylinder, slots located in the sides of the cylinder are uncovered. The uncovered slots allow intake manifold vacuum to draw warm air, heated by the exhaust manifold, through the thermostatic coil housing assembly. This warm air causes the thermostatic coil spring to lose its tension gradually until the choke valve is in the wide-open position.

If the engine should be accelerated during the warm-up period, the drop in manifold vacuum allows the thermostatic coil spring to close the choke valve for just a moment which provides the necessary richer mixture.

If the engine becomes flooded during the starting period, the choke valve can be opened manually to clean out the excessive fuel in the intake manifold. This is accomplished by moving the throttle to the wide-open position while the engine is being cranked. Opening the throttle causes a projection on the throttle lever to contact a cam, which in turn, opens the choke valve to a predetermined position. This is known as the unloader.

DISASSEMBLING
THE CARTER RBS

1- Remove the pump rod retainer from the top of the pump plunger shaft. Remove

PUMP ARM LINK

PUMP PLUNGER SHAFT

the pump arm connector link nut from the end of the connector link, and then remove the spring and retainer from the connector link by pushing the pump plunger downward. **LEAVE** the connector link in place. It cannot be removed unless the throttle shaft and lever assembly is removed. Remove the pump arm retaining screw, retainer, pump arm, upper pump spring, bushing, and washer.

2- Remove the thermostatic coil housing, retainers, and gasket. If two carburetors are used, **MARK** the thermostatic coil assembly to identify the front and rear carburetor housing. When two carburetors are used, the rear thermostatic coil assembly, which is under extreme tension to maintain the choke in the open position, becomes stretched and loses its ability to properly operate the choke.

Remove the choke piston lever screw and the choke piston lever. Slide the choke piston out of the cylinder and disconnect the choke piston link from the choke lever.

3- Under normal conditions, the choke valve and shaft do not need to be removed. However, if the choke valve and shaft are worn remove the cam retainer, lift the cam collar, and disengage it from the connector rod. Use a screwdriver and spread the fork end of the choke lever. This lever is located on the choke shaft between the choke housing and the body casting. Now, slide the choke lever from the shaft and disengage it from the connector rod.

4- File off the staked ends of the choke valve attaching screws, and then remove the choke valve. Remove the idle adjustment screw and spring. Screw in the throttle adjustment screw until the throttle valve

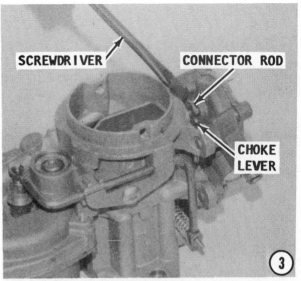

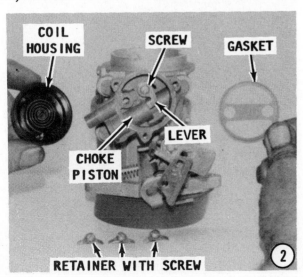

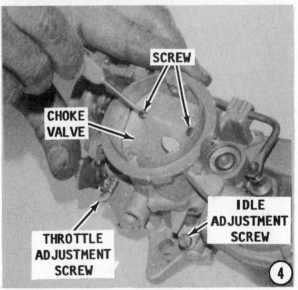

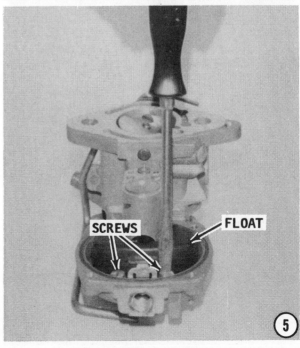

HAMMER

PLUNGER SHAFT

⑥

seats in the bore of the carburetor. Remove the choke valve and slide the choke shaft out of the casting.

5- Turn the carburetor upside down and remove the float bowl attaching screws, float bowl, and the bowl ring gasket. Remove the float pin attaching screws and the float. Remove the pin from the float. Remove the needle and seat assembly.

6- Turn the carburetor rightside up and place the float bowl under the lower portion

of the pump cylinder to catch parts. Press the pump plunger down until it bottoms. Hold the plunger down and at the same time tap the upper end of the plunger shaft with a light hammer and then remove the pump cylinder retainer, which is pressed on, and the spring.

7- Use a screwdriver and pry the washer out of the diaphragm cover cap. Pierce the diaphragm cover cap by driving a screwdriver blade or a punch into the cover. Remove the pump plunger and spring. Pry the cover out of the casting using the bowl cover vent boss as a fulcrum. Remove the diaphragm retainer, spring, and step-up rod-and-diaphragm assembly. Remove the intake check ball using a suitable tool **ONLY** if there have been acceleration problems. **ALWAYS** pry upward and **AVOID** any excessive side thrust in the seat to prevent **DAMAGE** to the hole in the casting. **BEFORE** removing the pump intake seat, be certain new parts are available because the old parts **CANNOT** be used. Turn the carburetor upright and remove the pump intake ball. **NEVER** attempt to remove the pump discharge needle if a new needle is not available.

8- If the throttle shaft needs to be replaced, remove the throttle valve attaching screws by first filing off the staked ends. Remove the throttle valve, and then slide the throttle shaft from the casting. **PAY ATTENTION** to the location of the connector link in the throttle lever as an aid during assembling.

SCREWDRIVER

DIAPHRAGM COVER CAP

STEP-UP ROD

⑦

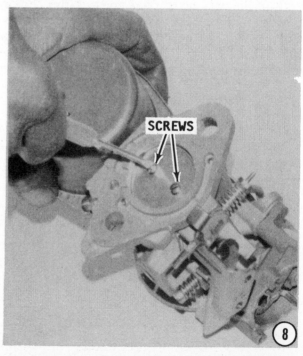

SCREWS

⑧

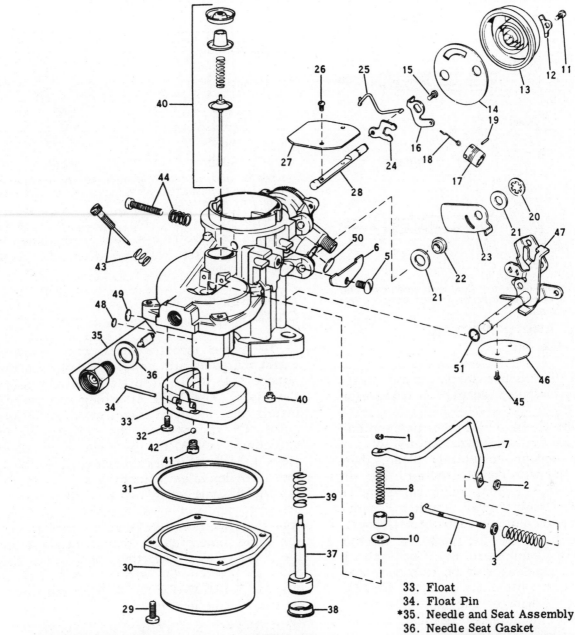

* 1. Pump Rod Retainer
  2. Connector Link Nut
  3. Spring and Retainer
  4. Connector Link
  5. Pump Arm Retainer Screw
  6. Pump Arm Retainer
  7. Pump Arm
* 8. Upper Pump Spring
  9. Bushing
  10. Washer
  11. Coil Housing Screw
  12. Coil Housing Retainer
  13. Thermostatic Coil and Housing
*14. Coil Housing Gasket
  15. Lever Screw
  16. Choke Piston Lever

17. Choke Piston
18. Choke Piston Link
19. Piston Pin
20. Cam Retainer
21. Washers
22. Cam Collar
23. Cam
24. Choke Lever
25. Connector Rod
26. Choke Valve Screw
27. Choke Valve
28. Choke Shaft
29. Bowl Attaching Screws
30. Float Bowl
*31. Bowl Ring Gasket
32. Float Pin Screw

33. Float
34. Float Pin
*35. Needle and Seat Assembly
36. Needle Seat Gasket
*37. Pump Plunger Assembly
*38. Pump Cylinder Retainer
39. Lower Pump Spring
*40. Step-up Rod and Diaphragm Assembly
*41. Intake Ball Check
*42. Intake Check Ball
43. Idle Adjustment Screw and Spring
44. Throttle Adjusting Screw and Spring
45. Throttle Valve Screw
46. Throttle Valve
47. Throttle Shaft
48. Fuel Well Welsh Plug
49. Idle Port Plug
50. Choke Piston Welsh Plug
51. "O"-Ring
*   Repair Kit Parts

Exploded view of the Carter RBS carburetor with principle parts identified.

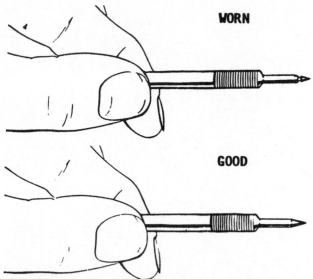

WORN

GOOD

Carburetor idle mixture adjustment needles. The top needle is worn and unfit for further service. The bottom needle is new.

CLEANING AND INSPECTING THE CARTER RBS

NEVER dip rubber parts, plastic parts, diaphragms, or pump plungers in carburetor cleaner.

Place all of the metal parts in a screen-type tray and dip them in carburetor solvent until they appear completely clean, then blow them dry with compressed air. The air horn has a plastic vent valve guide and a cranking enrichment valve. Both will withstand normal cleaning in carburetor cleaner.

Blow out all of the passages in the castings with compressed air. Check all of the parts and passages to be sure they are not clogged or contain any deposits. **NEVER** use a piece of wire or any type of pointed instrument to clean drilled passages or calibrated holes in a carburetor.

Inspect the idle mixture needle for any type of damage. Carefully check the float needle and seat assembly for wear. Good shop practice is to always install a new factory matched set to avoid leaks.

Inspect the holes in the levers for wear, especially for an out-of-round condition. Replace any levers or rods if they are worn. Check the throttle and choke levers and the valve plates for binding and damage. Inspect the fast idle cam for wear or damage.

Check the springs to be sure they have not become distorted or have lost their tension. If in doubt, compare them with a new one, if possible.

Inspect the sealing surfaces of the casting for damage.

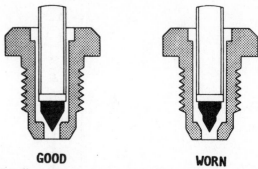

GOOD   WORN

Needle and seat arrangement on the carburetor covered in this section, showing a worn and new needle for comparison.

Most of the parts that should be replaced during a carburetor overhaul are included in an overhaul kit available from the local marine dealer.

ASSEMBLING THE CARTER RBS

1- If the throttle shaft was removed, insert the pump arm connector link into the outside hole of the throttle lever and insert the shaft. Use **NEW** attaching screws and install the throttle valve with the trade mark **"C"** stamped on the throttle valve extending **TOWARD** the idle port when viewed from the manifold flange side. With the throttle valve attaching screws loose, tap the throttle valve lightly with a screwdriver to center the valve in the bore. Use your finger to hold the valve in place and at the same time tighten the screws. After the screws have been tightened, stake the ends to prevent the screws from backing out, and **TAKE CARE** not to bend the throttle shaft in the process.

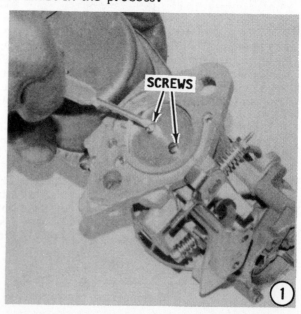

SCREWS

①

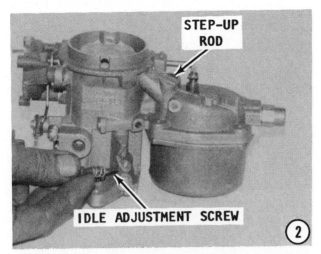

STEP-UP ROD

IDLE ADJUSTMENT SCREW

②

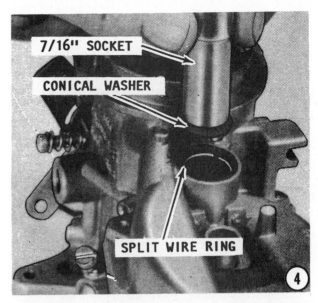

7/16" SOCKET

CONICAL WASHER

SPLIT WIRE RING

④

2- Install the idle adjustment screw and spring. Seat the screw **LIGHTLY**, and then back it out about 1-1/2 turns as a rough adjustment at this time. If the pump intake seat and ball were removed, install a **NEW** ball in the passage, and then install a new seat by driving it in place with a brass drift. Set the carburetor in the upright position, and place the step-up rod and diaphragm assembly in place. Turn the carburetor upside down and use the upper stem of the diaphragm to guide the step-up rod through the pressed-in metering jet. Place the brim of the hat-shaped retainer, without the spring, on top of the diaphragm, and then use it as a tool to press the diaphragm firmly against the gasket ledge in the casting. Remove the hat-shaped retainer and check the diaphragm assembly to be sure it is installed evenly.

3- Insert the step-up rod spring into the cup-shaped plate on top of the diaphragm. Place the hat-shaped retainer with the brim **DOWN** on top of the spring. Place a new diaphragm cover cap in position. Now, use a 5/8" socket as a seating tool, and **LIGHTLY** tap the socket until the cover cap seats in the housing.

4- Place the split-wire ring inside the diaphragm over the cap. Place the conical washer with the **SMALL** end of the cone pointing **UP** over the split-wire ring. Now,

SOCKET

③

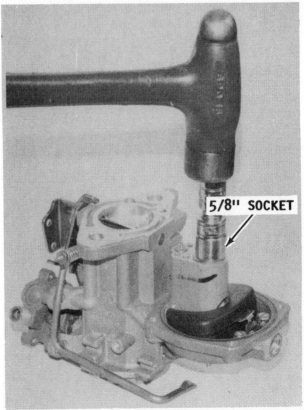

5/8" SOCKET

Installing the diaphragm cover onto the step-up rod.

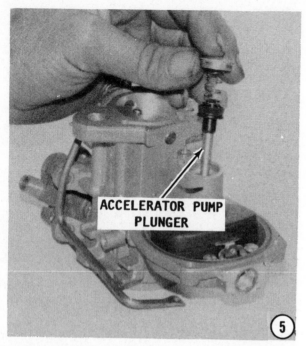

place a 7/16" socket over the inside diameter of the conical washer and **LIGHTLY** tap it into place until the washer is flat. **NEVER** use a smaller tool when installing the conical washer because it may enter the cover cap or strike the center portion of the cover. **DO NOT** drive the washer beyond the flat position.

5- Invert the carburetor and install a **NEW** accelerator pump plunger and spring. Install and seat a **NEW** pump retainer.

6- Install a **NEW** needle and seat assembly in the casting. **ALWAYS** use a new gasket as a precaution against leaks.

7- Insert the float pin in the float bracket and slide the float into place.

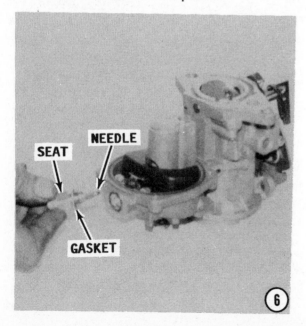

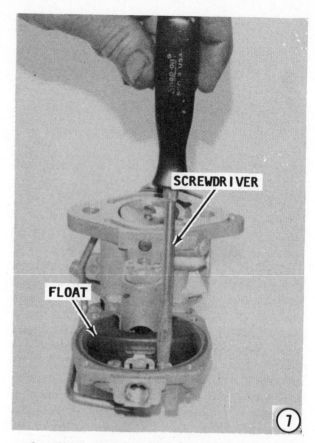

8- **Float Level Measurement:** Turn the carburetor upside down. Now, with only the weight of the float seating the needle, measure the vertical distance from the machined casting to the small "bump" at the outer end of the float. Measure the distance at both ends of the float. This distance should

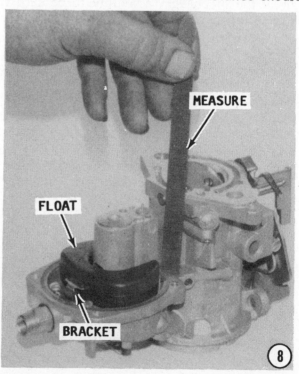

be 15/32". **TAKE EXTRA CARE** to ensure the needle is not pressed into the seat, because this carburetor has a resilient seat and pressing the needle would cause the seat material to take a "temporary set" which would result in an incorrect fuel level when the "set" is relieved. Adjustment: Hold the lip end of the float bracket with a pair of needle-nose pliers and bend the bracket at its narrowest point. Install the fuel bowl and a **NEW** bowl ring gasket using the attaching screws.

9- If the choke was completely disassembled, install the choke shaft in the casting. Attach the choke valve to the shaft with **NEW** screws. **DO NOT TIGHTEN** the screws at this time. Center the choke valve in the bore by tapping it lightly with a screwdriver. Hold the choke valve in place with your finger and at the same time tighten the screws. After the screws are tight, stake the ends to prevent the screws from backing out.

10- Attach the connector rod to the choke lever, and then install the lever on the choke shaft. Secure it in place by crimping the forked ends of the lever with pliers. Attach the cam to the lower end of the connector rod and attach the cam, collar, and retainer to the pivot shaft.

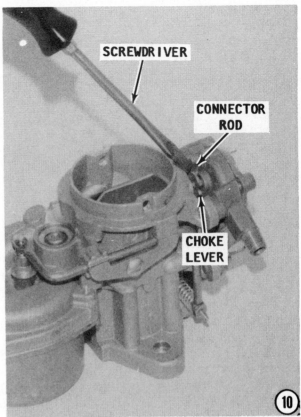

11- Assemble the choke piston and link using the choke piston pin. Attach the piston and link assembly to the choke lever, then slide the piston into the cylinder. Position the choke lever on the end of the choke shaft and attach it to the shaft using the lever screw. Check to be sure the choke linkage moves freely without binding.

12- Place the coil housing gasket in place, and then install the coil housing with the retainers and **JUST START** the screws. Check to be sure the hook engages the lever. If the choke being adjusted is on the forward carburetor of a dual installation,

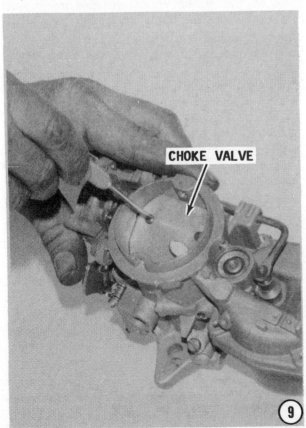

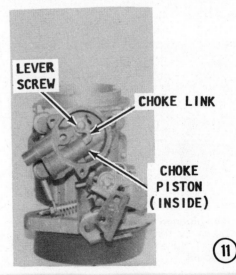

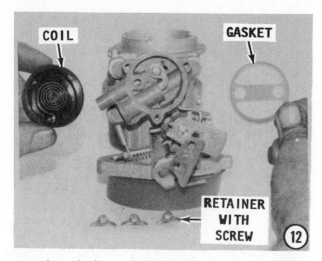

set the choke on the index mark, and then tighten the cover screws. The choke valve should now be spring-loaded in the closed position. If the choke being adjusted is on the rear carburetor, set the choke on the index mark, then turn it 1-1/4 turns clockwise. Now, tighten the cover screws. The choke valve should now be lightly spring-loaded. **REMEMBER**, the thermostatic coil assemblies **MUST** be installed back in their original location because the rear coil assembly is under tension to maintain the choke in the open position and this causes the coil to lose its ability to operate a choke properly. Place the carburetor in the upright position. Install the washer, bushing, and spring on the accelerator plunger shaft. Place the end of the pump arm over the plunger shaft and secure it with a retainer. Install the retainer washer on the throttle shaft connector link. Install the spring on the connector link through the lower hole of the pump arm and attach it with a nut.

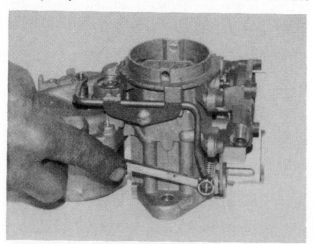

The plunger must start its downward movement as soon as the throttle starts to open. If the plunger fails to move, make the adjustment with a box-end wrench, as shown.

Align the pump arm so it fits in the groove in the carburetor casting and secure it with a retainer and screw. Check to be sure the pump arm moves freely.

BENCH ADJUSTMENTS FOR THE CARTER RBS

Pump

A pump adjustment **MUST** be made each time the carburetor is disassembled, and **MUST** be made **BEFORE** the unloader adjustment.

Back out the throttle adjusting screw and hold the choke valve wide open so the throttle valve seats in the bore of the carburetor.

With the throttle valve tightly closed, adjust the connector rod nut until there is no lag between the downward movement of the pump plunger and the opening of the throttle valve.

The plunger **MUST START** its downward movement as soon as the throttle valve starts to open.

Unloader, Forward Carburetor Only

Hold the throttle valve in the wide open position, and adjust the tang on the throttle lever until the clearance between the upper edge of the choke valve and the inner wall of the air horn is 7/64".

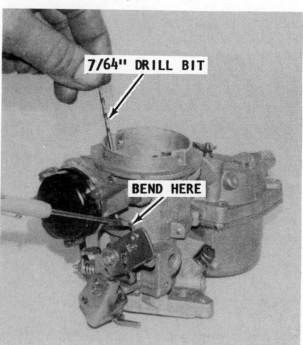

Adjusting the unloader at the forward carburetor using a 7/64" drill. The screwdriver points to the tab which is bent with a pair of pliers to make the adjustment.

SYNCHRONIZING DUAL CARBURETORS

Turn the idle mixture screws on both carburetors in until they seat, then back them out 1-1/2 turns. Disconnect the forward ball joint link. Back out both idle stop screws fully until both throttles are fully closed in the bores. Next, adjust the forward ball joint link until both throttles begin to open at the same time.

Operate the throttle by pushing on the throttle connector pin where the remote control attaches. Adjust the rear idle stop screw until it contacts the throttle arm, and then turn it in 1-1/2 turns to open the throttle slightly.

Start the engine.

CAUTION

Water must circulate through the stern drive, also to and from the engine, anytime the engine is run to prevent damage to the water pump located in the lower unit (original and Alpha drives); or on the front of the engine (Bravo drive). Just a few seconds without water will damage the water pump.

Adjust the speed to between 500 and 600 rpm in gear, and using only the rear idle stop screw to make the adjustment.

Adjust both idle mixture screws until the engine runs smoothly. Reset the idle rpm, if necessary, but use only the rear idle stop screw.

Finally, bring the forward idle stop screw up until it just touches the throttle arm.

The Rochester 2GC has two idle adjustment needles. The Mercarb has only one in the center.

4-8 ROCHESTER MODEL 2GC MERCARB -- SINCE 1984

The Rochester two-barrel carburetor is available in two different models: 2GC and 2GV. The 2GC and MerCarb are almost identical. The MerCarb has one idle adjustment needle and the 2GC has two. This is basically the only difference between the two carburetors.

The Model 2GV has an automatic choke with a thermostatic coil installed on the engine manifold.

The MerCarb and 2GC have the automatic choke as a part of the carburetor.

DISASSEMBLING

1- Remove the retaining clips at both ends of the accelerator pump link rod. On newer models only, only one clip is used at the lower end of the rod. The rod may be pivoted until the ear on the rod aligns with the slot in the upper bracket and then withdrawn. The choke rod may now be removed. Remove the fast-idle cam attaching screw, then remove the fast-idle

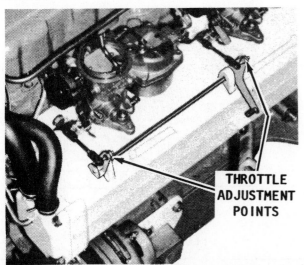

Throttle adjustment points to synchronize dual carburetors on an in-line engine.

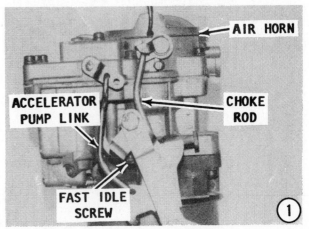

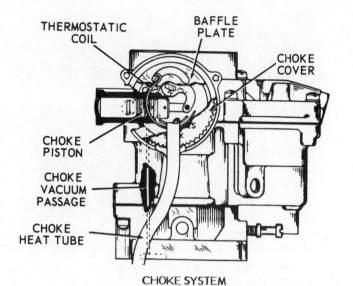

CHOKE SYSTEM

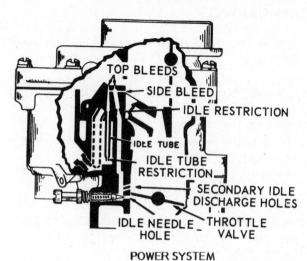

POWER SYSTEM

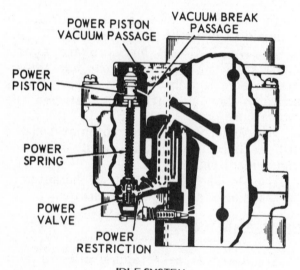

IDLE SYSTEM

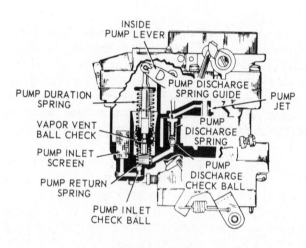

POWER SYSTEM

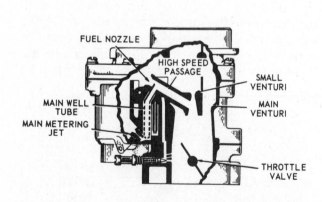

MAIN METERING SYSTEM

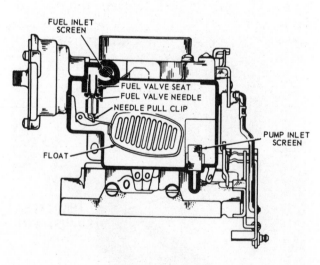

FLOAT SYSTEM

Details of the various systems of the Rochester 2GC carburetor, for study purposes.

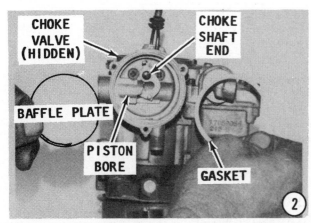

CHOKE VALVE (HIDDEN) — CHOKE SHAFT END — BAFFLE PLATE — PISTON BORE — GASKET — 2

linkage as an assembly. Remove the air horn attaching screws and lift the air horn straight up from the body.

2- Disassemble the butterfly choke valve by first removing the two choke valve retaining screws. However, under normal conditions, the choke shaft is not removed. You may have to file the staked ends of the screws in order to remove them. Lift out the choke valve.

Remove the plastic choke cover with the thermostatic coil attached, then remove the gasket and the baffle plate. Now, rotate the choke shaft counterclockwise to guide the piston from its bore. Remove the shaft and piston assembly.

3- Turn the air horn upside down and remove the float hinge pin. Lift off the float, and remove the float needle. Depress the power piston shaft which will allow the spring to snap and in turn will force the piston from the casting and allow the power piston to be removed. Loosed the screw inside the casting on the pump lever shaft and remove the accelerator pump plunger and the pump lever. Remove the two choke housing-to-air horn attaching screws and lift off the choke housing. Discard the gasket.

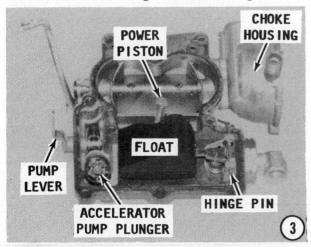

POWER PISTON — CHOKE HOUSING — FLOAT — PUMP LEVER — ACCELERATOR PUMP PLUNGER — HINGE PIN — 3

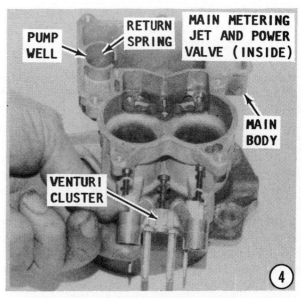

PUMP WELL — RETURN SPRING — MAIN METERING JET AND POWER VALVE (INSIDE) — MAIN BODY — VENTURI CLUSTER — 4

4- Remove the pump plunger return spring, main metering jet and power valve from the main body. Some models have an aluminum inlet ball in the bottom of the pump well. This ball will fall out when the carburetor is turned over. Remove the venturi cluster by removing the attaching screws.

5- Use a pair of needle-nosed pliers to remove the discharge ball spring T-shaped retainer. Now, take out the pump discharge spring and the steel discharge ball. Turn the carburetor over and remove the three throttle body-to-bowl attaching screws. Lift off the throttle body. **DO NOT** disassemble the throttle body. Replacement parts are **NOT** available because of the close relationships between the throttle plates and the idle ports.

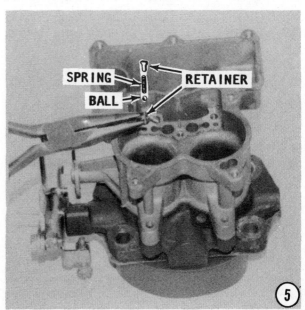

SPRING — BALL — RETAINER — 5

CLEANING AND INSPECTING THE ROCHESTER 2GC AND MERCARB

NEVER dip rubber parts, plastic parts, diaphragms, or pump plungers in carburetor cleaner. These parts should be cleaned **ONLY** in solvent, and then blown dry with compressed air.

Place all of the metal parts in a screen-type tray and dip them in carburetor cleaner until they appear completely clean, then blow them dry with compressed air.

Blow out all of the passages in the castings with compressed air. Check all of the parts and passages to be sure they are not clogged or contain any deposits. **NEVER** use a piece of wire or any type of pointed instrument to clean drilled passages or calibrated holes in a carburetor.

Move the throttle shaft back-and-forth to check for wear. If the shaft appears to be too loose, replace the complete throttle body because individual replacement parts are not available.

Inspect the main body, air horn, and venturi cluster gasket surfaces for cracks and burrs which might cause a leak.

Shake the float to determine is there is any liquid inside, and if there is, replace the float. Check the float arm needle contacting surface and replace the float if this surface has a groove worn in it.

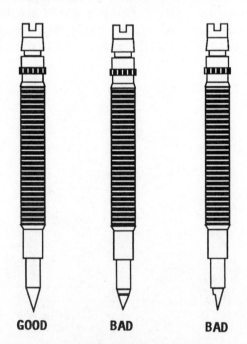

GOOD   BAD   BAD

Three carburetor idle adjustment needles lined-up for comparison. The far left needle is new, the other two are worn and unfit for further service.

Inspect the tapered section of the idle adjusting needles and replace any that have developed a groove.

Most of the parts that should be replaced during a carburetor overhaul are included in an overhaul kit available from your local marine dealer. This kit will also contain a matched fuel inlet needle and seat. This combination should be replaced each time the carburetor is disassembled as a precaution any leakage.

ASSEMBLING ROCHESTER 2GC AND MERCARB

1- Install the idle mixture adjusting needles and springs into the throttle body **FINGERTIGHT**. Now, back the screws out 1-1/2 turns as a rough adjustment at this time. Use a **NEW** gasket and assemble the throttle body onto the bowl. A **NON-VENT** gasket **MUST** be used because during hot-engine operation, the fuel in the carburetor tends to "percolate" due to engine heat. If a vented, automotive-type gasket is used, these fuel vapors will be vented directly into the atmosphere. A genuine replacement gasket will prevent these fuel vapors from venting. Insert the pump discharge check steel ball, spring, and T-shaped retainer into the top of the main body. Use a **NEW** gasket and install the venturi cluster. **BE SURE** the undercut screw has a gasket and is placed in the center hole. Install the main metering jets and the power valve, with a **NEW** gasket. Install the pump return spring in the pump well and the pump inlet screen in the bottom of the bowl.

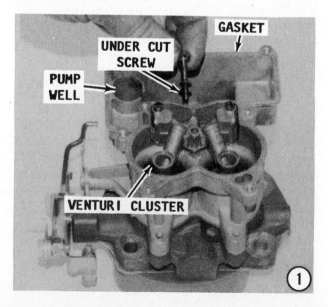

1. Idle Stop Lever Screw
2. Air Horn Gasket
3. Long Air Horn Screw
4. Short Air Horn Screw
5. Air Horn Lockwasher
6. Pump Lever
7. Pump Screw
8. Pump Clip
9. Pull Clip
10. Choke Gasket
11. Choke Valve
12. Pump Discharge Ball
13. Pump Discharge Spring
14. Pump Spring Guide
15. Venturi Gasket
16. Outer Venturi Screw
17. Center Venturi Screw
18. Outer Venturi Lockwasher
19. Center Venturi Gasket
20. Main Metering Jet
21. Power Valve Gasket
22. Choke Rod
23. Idle Speed Stop Lever
24. Throttle Body Gasket
25. Throttle Body Screw
26. Throttle Body Lockwasher
27. Pump Rod
28. Pump Rod Clip
29. Idle Stop Screw
30. Idle Adjusting Needle
31. Idle Needle Spring
32. Pump Shaft and Lever
    Assembly
33. Pump Assembly
34. Air Horn Assembly
35. Power Piston Assembly
36. Float Assembly
37. Needle and Seat Assembly
38. Needle Seat Gasket
39. Choke Housing Assembly
40. Thermostat Cover
41. Thermostat Cover Gasket
42. Choke Lever and Link
    Assembly
43. Choke Shaft Assembly
44. Choke Lever and Collar
    Assembly
45. Float Bowl Assembly
46. Venturi Cluster Assembly
47. Power Valve Assembly
48. Throttle Body Assembly
49. Float Hinge Pin
50. Choke Housing Screw
51. Baffle Plate
52. Thermostat Cover Retainer
53. Thermostat Cover Screw
54. Choke Lever Screw
55. Choke Valve Screw
56. Choke Piston
57. Choke Piston Pin
58. Lead Ball Plug
59. Expansion Plug
60. Pump Return Spring
61. Fuel Line Fitting

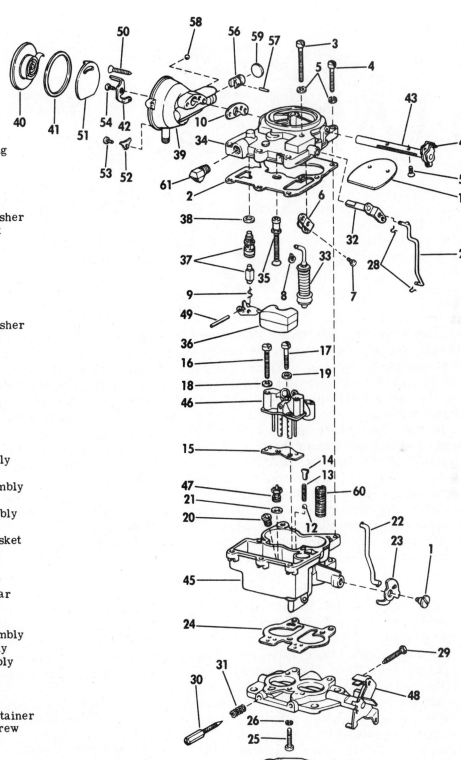

NOTE: The 2GC has two
Idle Adjustment Needles (30).
The MERCARB has only one
in the center.

Exploded view of the Rochester 2G and Mercarb carburetor with principle parts identified. The
Rochester 2G has two idle adjustment needles (30). The Mercarb has only one in the center.

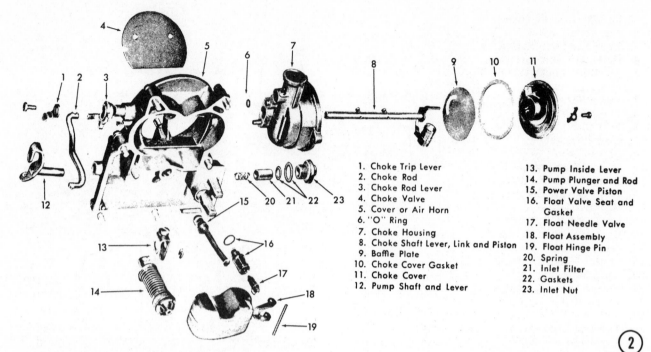

1. Choke Trip Lever
2. Choke Rod
3. Choke Rod Lever
4. Choke Valve
5. Cover or Air Horn
6. "O" Ring
7. Choke Housing
8. Choke Shaft Lever, Link and Piston
9. Baffle Plate
10. Choke Cover Gasket
11. Choke Cover
12. Pump Shaft and Lever

13. Pump Inside Lever
14. Pump Plunger and Rod
15. Power Valve Piston
16. Float Valve Seat and Gasket
17. Float Needle Valve
18. Float Assembly
19. Float Hinge Pin
20. Spring
21. Inlet Filter
22. Gaskets
23. Inlet Nut

2

2- The following referenced part numbers in this step will be found on the accompanying exploded drawing. Place a new gasket in position and install the choke housing (7). Secure the housing with the two attaching screws. Assemble the choke piston to the choke shaft and link (8). The piston pin and flat section on the side of the choke piston **MUST** face **OUTWARD** toward the air horn. Push the choke shaft into the air horn, and rotate the shaft until the piston enters the housing bore. Place the choke valve (4) on the choke shaft with the letters **RP** facing up. Install the two choke valve retaining screws just **FINGERTIGHT.** Place the choke rod lever (3) and the trip lever (1) on the end of the choke shaft. Center the choke valve to obtain 0.020" clearance between the choke lever and the air horn casting. Now, tighten and then stake the ends of the choke valve retaining screws to prevent them from backing out. Install the baffle plate (9) gasket (10), and thermostatic cover (11). Rotate the cover until the index marks align. Fine adjustments are made in Step 10. Install and tighten the three cover retainers and retaining screws.

Install the outer pump lever (12) in the air horn. Assemble the inner pump arm (13) and tighten the screw. Attach the pump plunger (14) to the inner arm (13), with the pump shaft pointing **INWARD.** Install the horseshoe retainer. Position the screen on the float needle seat (16) and screw the

assembly into the air horn. Install the power piston (15) into the vacuum cavity. Check to be sure the piston moves **FREELY.** Stake the retainer lightly to hold it in place. Install the air horn gasket and attach the needle (17) to the float (18). **CAREFULLY** insert the needle into the float needle seat while guiding the float between the bosses. Insert the hinge pin (19) to finish the air horn assembling.

Two float adjustments **MUST** be made at this time: The float level, and the float drop.

3- Float Level Adjustment on models with a **VERTICAL** seam in the float: Measure the distance from the top of the float to the air horn, with the gasket in place. Compare your measurement with the Specificaitons in the Appendix. **CAREFULLY** bend the float arm at the rear of the float as shown in the accompanying illustration, until the required measurement is reached.

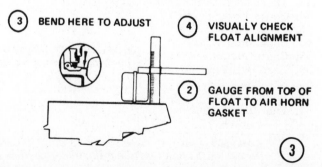

(1) INVERT AIR HORN WITH GASKET IN PLACE.

(3) BEND HERE TO ADJUST

(4) VISUALLY CHECK FLOAT ALIGNMENT

(2) GAUGE FROM TOP OF FLOAT TO AIR HORN GASKET

3

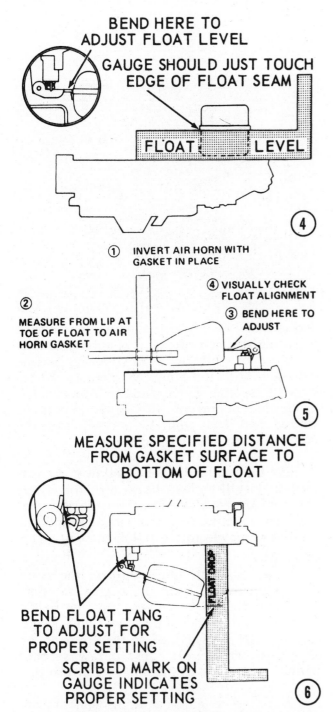

BEND HERE TO
ADJUST FLOAT LEVEL

GAUGE SHOULD JUST TOUCH
EDGE OF FLOAT SEAM

FLOAT   LEVEL

① INVERT AIR HORN WITH
GASKET IN PLACE

④ VISUALLY CHECK
FLOAT ALIGNMENT

② MEASURE FROM LIP AT
TOE OF FLOAT TO AIR
HORN GASKET

③ BEND HERE TO
ADJUST

④

⑤

MEASURE SPECIFIED DISTANCE
FROM GASKET SURFACE TO
BOTTOM OF FLOAT

BEND FLOAT TANG
TO ADJUST FOR
PROPER SETTING

SCRIBED MARK ON
GAUGE INDICATES
PROPER SETTING

⑥

4- **Float Level Adjustment** on models with a **HORIZONTAL** seam in the float: Measure the distance from the lower edge of the seam to the air horn, with the gasket in place. Compare your measurement with the Specifications. **CAREFULLY** bend the float arm at the rear of the float as shown in the accompanying illustration.

5- **Float Level Adjustment** on models with a **NITROPHYL-TYPE** (hollow) float: Measure the distance from the lip at the toe of the float to the air horn, with the gasket

in place. Compare your measurement with the Specifications. **CAREFULLY** bend the float arm at the rear of the float, as shown in the accompanying illustration, until the required measurement is reached.

6- **Float Drop Adjustment:** Turn the air horn right side up to allow the float to move to the wide-open position. Now, measure the distance from the air horn gasket to the bottom of the float. Compare your measurement with those given in the Specifications. **CAREFULLY** bend the float tang until the required measurement is reached, as shown in the accompanying illustration.

7- Replace the assembled air horn onto the bowl and guide the accelerator pump plunger into its well. Install and tighten the cover screws. Install the idle needle into the spring, then into the throttle body. Install the accelerator link, the choke rod, the fast-idle cam, and screw.

BENCH ADJUSTMENTS FOR THE ROCHESTER 2GC AND MERCARB

8- Back out the idle stop screw and completely close the throttle valves in their bores. Place a pump gauge across the top of the carburetor air horn ring, as shown, with the 1-5/32" leg of the gauge pointing downward towards the top of the pump rod. The lower edge of the gauge leg should just touch the top of the pump rod. **CAREFULLY** bend the pump rod, as required to obtain the proper setting.

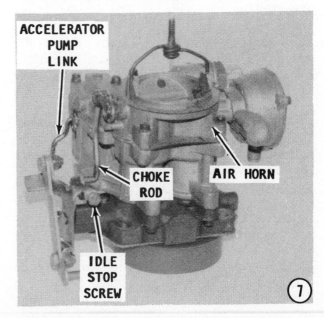

ACCELERATOR PUMP LINK
CHOKE ROD
AIR HORN
IDLE STOP SCREW
⑦

PLACE GAUGE ACROSS TOP
OF AIR HORN RING

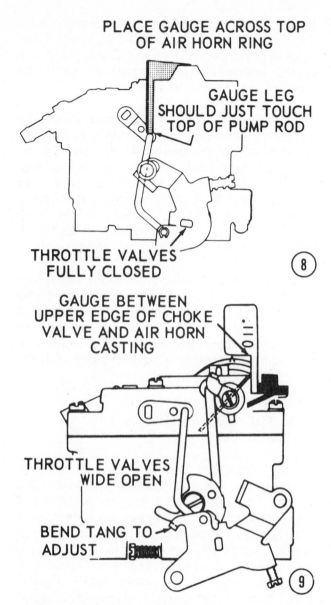

GAUGE LEG
SHOULD JUST TOUCH
TOP OF PUMP ROD

THROTTLE VALVES
FULLY CLOSED

⑧

GAUGE BETWEEN
UPPER EDGE OF CHOKE
VALVE AND AIR HORN
CASTING

THROTTLE VALVES
WIDE OPEN

BEND TANG TO
ADJUST

⑨

INDEX

⑩

9- Unloader Adjustment: Open the throttle valves wide. The choke valve should open only enough to allow the specified gauge between the upper end of the valve and the inner air horn wall. **CAREFULLY** bend the tang on the throttle lever until the proper adjustment is reached.

10- Automatic Choke Adjustment: Loosen the thermostat cover attaching screws and rotate the cover until the mark on the cover is aligned with the index line on the housing. Tighten the screws. Do not use any setting except the standard one, unless the engine is usually operated on special blends of fuel which do not give satisfactory warm-up performance with the standard setting. A lean setting may be required with high octane fuel because a standard thermostat setting would produce

too much loading of the engine during warm-up. A rich setting should be used only if a lot of spitting occurs during engine warm-up with the standard setting. Whenever the setting is changed for either richer or leaner operation, the cover should be moved just one point at-a-time, and the results tested with the engine cold.

11- Idle Speed and Mixture Adjustments: Install a clean flame arrestor. Start the engine and allow it to run until it has warmed to operating temperature and the choke has moved to the full open position.

CAUTION

Water must circulate through the stern drive, also to and from the engine, anytime the engine is run to prevent damage to the water pump located in the lower unit (original and Alpha drives); or on the front of the engine (Bravo drive). Just a few seconds without water will damage the water pump.

Stop the engine and disconnect the throttle cable from the throttle lever. Now, turn the idle mixture adjusting screws in until they just barely make contact with their seats, then back out 1-1/4 turns as a rough adjustment. **TAKE CARE** not to turn the screws in tightly against the seats or the needles and seats will be **DAMAGED.**

With water running through the engine and outdrive, start the engine again and shift the drive unit into forward gear and run at idle rpm. Adjust the idle mixture needle for the highest and steadiest manifold vacuum reading. If a vacuum gauge is not available, obtain the smoothest running, maximum idle speed by first turning the idle adjusting needle in until the engine rpm begins to drop slightly. From this point, back the needle out over the "high spot" until the engine rpm again begins to drop. Now, set the idle adjusting needle halfway between the two points for a proper idle mixture. If these adjustments result in an increase in idle rpm, reset the idle speed adjusting screw to obtain the specified idle rpm and again adjust the idle mixture adjusting needle. Shift the drive unit into neutral.

Stop the engine and install the throttle cable. Check to be sure the throttle valves are fully open when the remote control is in the full forward position. **ALWAYS** have a helper turn the propeller when shifting without the engine running in order to engage the shift dog.

The engine **MUST NOT BE RUNNING** to make the following test. With the throttle valves fully open, turn the wide-open throttle stop adjusting screw clockwise until the screw just makes contact with the throttle lever. Tighten the set nut securely to prevent the adjustment screw from turning. Now, return the shift control lever to the neutral position, idle, and check to see if the idle stop screw is against the stop. Shift into forward gear. Readjust the idle speed screw until the recommended idle rpm is reached.

4-9 ROCHESTER 4MV AND 4MC CARBURETOR

DESCRIPTION

The Rochester 4MV and 4MC carburetors are Quadrajet units. Both carburetors have the same bore and are identical except for the type of choke installed. The following service procedures are valid for both choke types. The air/fuel mixture is controlled with a secondary-side air valve and tapered metering rods.

SPECIAL WORDS ON "TORX" SCREWS AND BOLTS

Carburetors installed on newer units use "torx" screw or bolt heads instead of the traditional slotted or Phillips type head. A "torx" head has an indented star shape and needs a special torx screwdriver. These special screwdrivers can be found at most automotive parts stores. The manufacturer has used three different size torx heads on this carburetor in recent years: #9, #20, and #25.

The Quadrajet carburetor has two stages. The primary (fuel inlet) side has small 1-3/8" bores with a triple venturi set-up equipped with plain-tube nozzles. The carburetor operates much the same as other carburetors using the venturi principle. The triple venturi, plus the small primary bores, makes for a more stable and finer fuel control during idle and partial throttle operation. When the throttle is partially open, the fuel metering is accomplished with tapered metering rods, positioned by a vacuum-responsive piston and operating in specially designed jets.

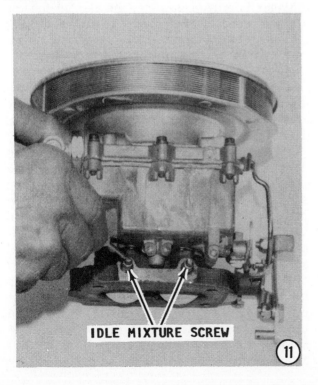

IDLE MIXTURE SCREW

⑪

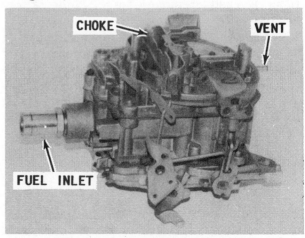

CHOKE   VENT

FUEL INLET

The Rochester 4bbl carburetor.

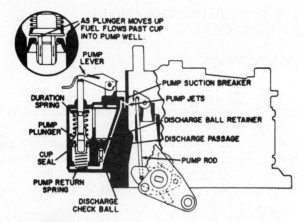

Cross-section illustrating the accelerating pump.

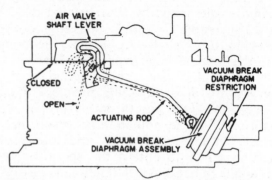

Illustration to indicate the air-valve dashpot operation.

The secondary side has two large, 2-1/4", bores. These large bores, when added to the primary side bores, provide enough air capacity to meet most engine requirements. The air valve is used in the secondary side for metering control and backs-up the primary bores to meet air and fuel demands of the engine.

The secondary air valve operates the tapered metering rods. These rods move in orifice plates and thus control fuel flow from the secondary nozzles in direct relation to the air flowing through the secondary bores.

The float bowl is designed to avoid problems of fuel spillage during sharp turns of the boat which could result in engine cutout and delayed fuel flow. The bowl reservoir is smaller than most four-barrel carburetors to reduce fuel evaporation during hot engine shut-down.

The float system has one pontoon float and fuel valve which makes servicing much easier than on some other model carburetors. A fuel filter is located in the float bowl ahead of the float needle valve. This filter is easily removed for cleaning or replacement.

The throttle body is made of aluminum as part of a weight-reduction program and also to improve heat transfer away from the fuel bowl and prevent the fuel from "percolating" during hot engine shut-down. A heat insulator gasket is used between the throttle body and bowl to help prevent "percolating".

ACCELERATING SYSTEM

When the throttle is opened suddenly during acceleration, the air flow and manifold vacuum change almost at the same time. The fuel, which is heavier, has a tendency to lag behind. This condition causes a lean mixture for just a moment. It is at this time, the accelerator pump comes into play by providing the extra fuel necessary to maintain the proper mixture.

The accelerator pump system is located in the primary side of the carburetor. The system consists of spring-loaded pump plunger and a return spring. The plunger is operated by a pump lever on the air horn which is connected to the throttle lever by a pump rod.

When the pump plunger moves upward in the pump well during throttle closing, fuel from the float bowl enters the pump well through a slot in the top of the pump well. This fuel flows past the synthetic pump cup seal into the bottom of the pump well. The pump cup floats and moves up and down on the pump plunger head. When the pump plunger is moved upward, the flat on the top of the cup unseats from the flat on the plunger head and allows fuel to move through the inside of the cup into the bottom of the pump well. This action also vents any vapors which may be in the bottom of the pump well and allows a solid charge of fuel to be maintained in the fuel well beneath the plunger head.

When the primary throttle valves are opened, the connecting linkage forces the pump plunger downward. The pump cup seats at once, and fuel is forced through the pump discharge check ball and passes on through the passage to the pump jets located in the air horn. From these jets, the fuel sprays into the venturi area of each primary bore. As mentioned earlier, the pump plunger is spring-loaded. The upper tine portion of the spring is balanced with the bottom pump return spring to permit a smooth sustained charge of fuel to be delivered during the acceleration period.

The pump discharge check ball seats in the pump discharge passage during upward motion of the pump plunger to prevent air from being drawn into the passage. Without this arrangement, there would be a momentary lag during acceleration.

A vacuum exists at the jets during high-speed operation. A cavity just beyond the pump jets is vented to the top of the air horn, outside the carburetor bores. This cavity serves as a suction breaker. Therefore, when the pump is not in operation, fuel cannot be pulled out of the pump jets into the venturi area, but ensures a full pump discharge when needed and still prevents any fuel spill over from the pump discharge passage.

CHOKE SYSTEM

The choke valve is located in the primary side of the carburetor and provides the correct air/fuel mixture for quick cold-engine starting and until the engine reaches its operating temperature. The air valve is locked closed until the engine is completely warmed and the choke valve is wide open.

The principle parts of the choke system are a choke valve located in the primary air horn bore, a vacuum diaphragm unit, fast-idle cam, connecting linkage, an air valve or secondary throttle valve lockout lever, and a thermostatic coil.

While the engine is being cranked, the tension of the thermostatic coil holds the choke valve closed. The closed choke valve restricts air flow through the carburetor to provide a richer mixture for starting.

After the engine starts, manifold vacuum applied to the vacuum-break unit opens the choke valve the proper amount for the engine to run without loading or stalling.

Other conditions exist after the engine starts: The cold-enrichment feed holes are no longer in a low-pressure area so they cease to feed fuel. These holes are then used as secondary main well air bleeds. The fast-idle cam follower lever on the end of the primary throttle shaft drops from the highest step on the fast-idle cam to a lower step when the throttle is opened. The lever in this position gives the engine enough fast idle and a correct fuel mixture for smooth operation until full operating temperature is reached. Once the engine has warmed-up, the thermostatic coil heats and eases its tension and allows the choke valve to open

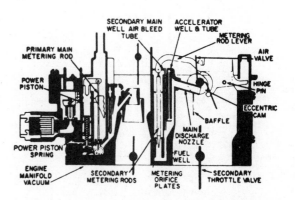

Cross-section illustrating the power system.

farther due to the intake air pushing on the off-set choke valve. The choke valve continues to open until the thermostatic coil is completely relaxed and the choke is fully open.

POWER SYSTEM

The power system provides extra fuel to meet power demands during heavy engine loads and during high-speed operation. This richer mixture is supplied through the main metering systems in the primary and secondary sides of the carburetor.

The power system enriches the fuel mixture in the two primary bores. This system is made up of a vacuum-operated power piston and a spring located in a cylinder connected by a passage to the intake manifold vacuum. The spring under the power piston operates against manifold vacuum and pushes the power piston upward.

During partial throttle and cruising ranges, the manifold vacuum is enough to hold the power piston down against spring tension so the larger diameter of the metering rod tip is held in the main metering jet orifice.

As engine load is increased until more fuel and a richer mixture is necessary, the power piston spring overcomes the vacuum pull on the power piston, and the tapered tips of the metering rods move upward in the main metering jet orifices. The smaller diameter of the metering rod tip permits more fuel to pass through the main metering jet and enriches the fuel mixture to meet the added power demands. When engine load decreases, the manifold vacuum rises and extra fuel and a richer mixture is no longer required. The higher vacuum pulls the power piston down against spring tension, and this action moves the larger diameter of the metering rod into the metering

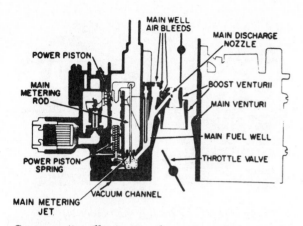

Cross-section illustrating the main metering system.

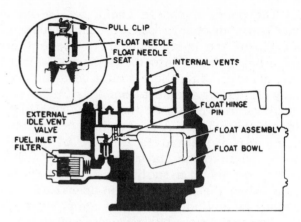

Cross-section illustrating the float system.

jet orifice, returning the fuel mixture to normal.

The primary side of the carburetor supplies enough air and fuel for low-speed operation. More air and fuel are required at higher speeds to meet engine demands and it is the secondary side of the carburetor that meets these requirements.

The secondary side of the 4bbl has a separate and independent metering system. This system consists of two large throttle valves connected by a shaft and linkage to the primary throttle shaft. Fuel metering is controlled by spring-loaded air valves, metering orifice plates, secondary metering rods, main fuel wells with bleed tubes, fuel-discharge nozzles, accelerating wells, and tubes.

A lever on the primary throttle shaft, through a connecting link to the secondary throttle shaft, begins to open the secondary throttle valves when the engine reaches a point where the primary bores cannot deliver the quantity of air and fuel demanded by the engine. As the secondary throttle valves are opened, engine manifold vacuum (reduced pressure) is applied directly beneath the air valves. Atmospheric pressure on top of the air valves overcomes spring tension and forces them open, allowing metered air to pass through the secondary bores of the carburetor.

The secondary main discharge nozzles are located above the secondary throttle valves and just below the center of the air valves. There is one nozzle for each bore.

Because the valves are located in a low-pressure area, they feed fuel in the following manner:

When the secondary throttle valves are opened, atmospheric pressure opens the air valves. This action rotates a plastic cam

attached to the center of the air-valve shaft. The cam movement lifts the secondary metering rods out of the secondary orifice plates through the metering rod lever. As the throttle valves are opened still farther and engine speed continues to increase, air flow through the secondary side increases and opens the air valve more. The opening valves lift the secondary metering rods farther out of the orifice plates. The metering rods are tapered. This design allows fuel flow through the secondary metering orifice plates in direct proportion to air flow through the secondary bores. This system allows correct air/fuel mixtures through the secondary bores to be controlled by the depth of the metering rods in the orifice plates.

The actual depth of the metering rods in the orifice plates is factory adjusted in relation to air valve position and to meet air/fuel requirements for each engine. If an adjustment should become necessary due to replacement of parts, a service setting is possible.

METERING SYSTEM

The main metering system supplies fuel to the engine from off-idle to wide-open throttle. The primary bores supply fuel and air during this range through plain-tube nozzles and the venturi principle. The main metering system starts to operate when air flow increases through the venturi system and more fuel is required to supply the correct air/fuel mixture to the engine. Fuel from the idle system gradually diminishes as the lower pressures are now in the venturi area.

The main metering system consists of the main metering jets, vacuum-operated

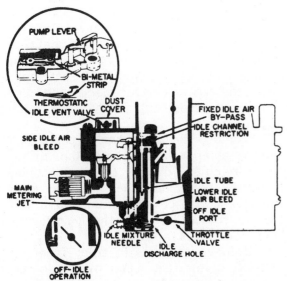

Cross-section illustrating the idle system.

metering rods, main fuel wells, main well air bleeds, fuel discharge nozzles, and triple venturis.

When the primary throttle valves open beyond the off-idle range and allow more air to enter the engine intake manifold, air velocity increases in the carburetor venturi. The increased velocity causes a drop in pressure in the large venturi, which increases many times in the boost venturi. Because the low pressure (vacuum) is now in the smallest boost venturi, fuel flows from the main discharge nozzle in the following manner:

Fuel from the float bowl flows through the main metering jets into the main fuel wells. It passes upward in the main well and is fed with air by an air bleed located at the top of the well. The fuel is further fed air through calibrated air bleeds located near the top of the well in the carburetor bores. The fuel mixture then passes from the main well through the main discharge nozzles into the boost venturis where the fuel mixture then combines with the air entering the engine through the carburetor bores. It then passes as a combustible mixture through the intake manifold and on into the engine cylinders. The main metering system is calibrated by tapered and stepped metering rods operating in the main metering jets and also through the main well air bleeds.

FLOAT SYSTEM

The float bowl is located between the primary bores and adjacent to the secondary bores. This position assures an adequate fuel supply to all carburetor bores and does

much to maintain excellent engine performance when the bow of the boat is high or during high-speed tight turns. The float pontoon is solid and made of a light plastic material. The combination of these two features gives added buoyancy to the float and allows the use of a single float to maintain a constant fuel level.

The parts of the float system include: The float bowl, a single pontoon float, float hinge pin-and-retainer combination, float valve and seat, and a slot valve pull clip. A plastic filler block is located in the top of the float chamber over the float valve to prevent fuel slosh into this area.

IDLE SYSTEM

The idle system is located on the fuel inlet (primary) side of the carburetor to supply the correct air/fuel mixture during idle and off-idle operation. The idle system is used during this period because air flow through the carburetor venturi is not great enough to obtain good metering from the main discharge nozzles.

DISASSEMBLING THE ROCHESTER 4MV AND 4MC

1- Remove the retaining clip from the vacuum-break link at the vacuum-break diaphragm. Disconnect the vacuum-break link from the vacuum-break assembly. Gently push apart the retaining ears of the bracket to release the vacuum-break cannister.

Place the carburetor on the work bench in the upright position. If servicing an older carburetor, remove the idle vent valve attaching screw, and then remove the idle vent valve assembly. Remove the clips

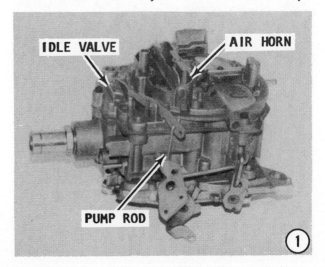

SECONDARY METERING ROD

2

from the upper end of the choke rod, disconnect the choke rod from the upper choke shaft lever, and then remove the choke rod from the bowl. Detach the spring clip from the upper end of the pump rod, and then disconnect the pump rod from the pump lever. Remove the nine air horn-to-bowl attaching screws. Two screws are located next to the primary venturi. Lift straight up on the air horn and remove it. **TAKE CARE** not to bend the accelerating well and air bleed tubes sticking out from the air horn.

2- Hold the air valve wide open and then remove the secondary metering rods by tilting and sliding the rods from the holes in the hanger. Remove the dashpot piston from the air-valve link by rotating the bend through the hole, and then remove the dashpot from the air horn by rotating the bend

through the air horn. Further disassembly of the air horn is not necessary. **DO NOT** remove the air valves, air valve shaft, and secondary metering rod hangers because they are calibrated. **DO NOT** attempt to remove the high-speed air bleeds and accelerating well tubes because they are pressed into position.

3- Remove the accelerating pump piston from the pump well. Release the air horn gasket from the dowels on the secondary side of the bowl, and then pry the gasket from around the power piston and primary metering rods. Remove the pump return spring from the pump well.

4- Remove the plastic filler from over the float valve. Use a pair of needle-nosed pliers and pull straight up on the metering rod hanger directly over the power piston and remove the power piston and the primary metering rods. Disconnect the tension spring from the top of each rod and then rotate the rod and remove the metering rods from the power piston. Pull up just a bit on the float assembly hinge pin until the pin can be removed by sliding it toward the pump well. Disengage the needle valve pull clip by sliding the float assembly toward the front of the bowl. **TAKE EXTRA CARE** not to distort the pull clip.

5- Remove the two screws from the float needle retainer, and then lift out the retainer and needle assembly. **NEVER** attempt to remove the needle seat because it is staked and tested at the factory. If the float assembly is damaged, replace the assembly. Remove both primary metering rod

ACCELERATING PUMP PISTON

3

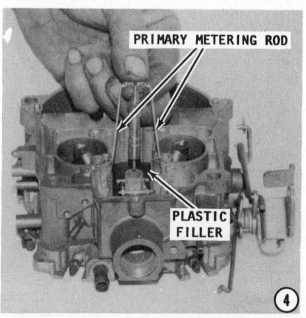

PRIMARY METERING ROD

PLASTIC FILLER

4

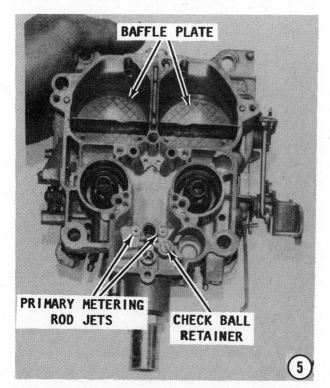

jets. Remove the pump discharge check ball retainer and the check ball. Remove the baffle plates from the secondary side of the bowl. Disconnect the vacuum hose from the tube connection on the bowl and from the vacuum break assembly. Remove the retaining screw, and then lift the assembly from the float bowl.

6- Remove the two screws from the hot-idle compensator cover. Lift the hot-idle compensator and O-ring from the float bowl. Remove the fuel inlet filter retaining nut, gasket, filter, and spring. Remove the throttle body by taking out the throttle body-to-bowl attaching screws, and then lift off the insulator gasket. Remove the idle

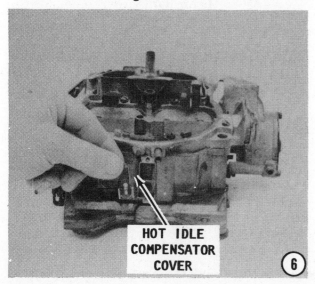

mixture screws and springs. **TAKE CARE** not to damage the secondary throttle valves.

CLEANING AND INSPECTING ROCHESTER 4MV AND 4MC

NEVER dip rubber parts, plastic parts, diaphragms, pump plungers or the vacuum-break assembly in carburetor cleaner. Place all of the metal parts in a screen-type tray and dip them in carburetor solvent until they appear completely clean, then blow them dry with compressed air.

Blow out all of the passages in the castings with compressed air. Check all of the parts and passages to be sure they are not clogged or contain any deposits. **NEVER** use a piece of wire or any type of pointed instrument to clean drilled passages or calibrated holes in a carburetor.

Inspect the idle mixture needles for damage. Check the float needle and diaphragm for wear. Inspect the upper and lower surfaces of the carburetor castings for damage. Inspect the holes in the levers for being out-of-round. Check the fast-idle cam for wear or damage. Check the air valve for binding. If the air valve is damaged, the complete air horn assembly must be replaced.

Most of the parts that should be replaced during a carburetor overhaul, including the latest updated parts, are found in a carburetor kit available at your local marine dealer.

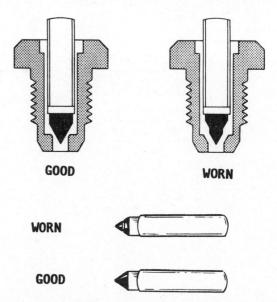

Needle and seat arrangement on the carburetor covered in this section, showing a worn and a new needle for comparison.

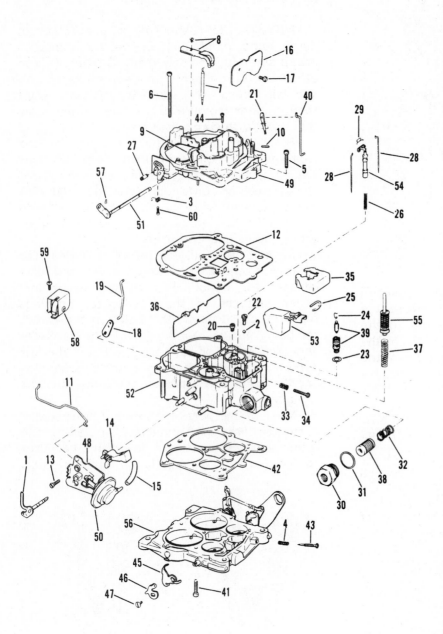

1. Vacuum break lever and shaft
2. Pump Discharge Ball
3. Secondary Air Valve Adjustment Spring
4. Idle Needle Spring
5. Air Horn Screw
6. Air Horn Screw
7. Secondary Metering Rod
8. Secondary Metering Rod Holder
9. Air Valve Lockout Lever
10. Roll Pin
11. Vacuum Break Rod
12. Air Horn Gasket
13. Control Attaching Screw
14. Fast Idle Cam
15. Vacuum Hose
16. Choke Valve
17. Choke Valve Screw
18. Intermediate Choke Lever
19. Choke Rod
20. Primary Jet
21. Pump Actuating Lever
22. Pump Discharge Ball Retainer
23. Needle Seat Gasket
24. Float Needle Pull Clip
25. Float Assembly Hinge Pin
26. Power Piston Spring
27. Secondary Air Valve Adjustment Shaft
28. Primary Metering Rod
29. Primary Metering Rod Spring
30. Fuel Inlet Filter Nut
31. Filter Nut Gasket
32. Fuel Filter Relief Spring
33. Idle Stop Screw Spring
34. Idle Screw
35. Float Bowl Insert
36. Secondary Baffle Plate
37. Pump Return Spring
38. Fuel Inlet Filter
39. Needle and Seat Assembly
40. Pump Rod
41. Throttle Body Screw
42. Throttle Body Gasket
43. Idle Needle
44. Air Horn Screw
45. Cam Lever
46. Fast Idle Lever
47. Idle Lever Screw
48. Vacuum Break Control Bracket
49. Air Horn Assembly
50. Vacuum Break Control Assembly
51. Choke Shaft & Lever Assembly
52. Float Bowl Assy
53. Float Assembly
54. Power Piston Assembly
55. Pump Assembly
56. Throttle Body Assembly
57. Choke Rod Clip
58. Choke Thermostat Assembly
59. Choke Screw
60. Set Screw

Exploded view of the Rochester 4MV and 4MC carburetor installed on GMC engines. Principle parts are identified.

In addition to the parts, most kits include the latest specifications which are so important when making bench adjustments.

ASSEMBLING
ROCHESTER 4MV AND 4MC

1- Turn the idle mixture adjusting screws in until they are barely seated, and then back them out two turns as a rough adjustment at this time. **NEVER** turn the adjusting screws down tight into their seats or they will be damaged. Install the pump rod in the lower hole of the throttle lever by rotating the rod. Place a new insulator gasket on the bowl with the holes in the gasket indexed over the two dowels. Install, and then tighten the throttle body-to-bowl screws evenly. Install the fuel inlet filter spring, filter, new gasket, and inlet nut. Tighten the nut. Position a **NEW** hot-idle compensator O-ring seal in the recess in the bowl, and then install the compensator.

2- Install the U-bend end of the vacuum break rod in the diaphragm link, with the end toward the bracket, and then slide the grooved end of the rod into the hole of the actuating lever. Install the spring clip to retain the rod in the vacuum-break rod. Install the fast-idle cam on the choke housing assembly. Check to be sure the fast-idle cam actuating pin on the middle choke shaft is located in the cutout area of the fast-idle cam. Connect the choke rod to the plain end of the choke rod actuating lever, and then hold the choke rod with the grooved end pointing inward and position the choke rod actuating lever in the well of the float

bowl. Install the choke assembly, with the shaft engaged with the hole in the actuating lever. Install and tighten the retaining screw. Remove the choke rod from the lever. This rod will be installed later. Install and connect the vacuum hose.

3- Install the baffle plates in the secondary side of the bowl with the notches facing up. Install the primary main metering jets. Install a **NEW** float needle seat. Install the pump discharge ball check and retainer in the passage next to the pump well.

4- Install the pull clip on the needle with the open end toward the front of the bowl. Install the float by sliding the float lever under the pull clip from the front to the back. Hold the float assembly by the toe and with the float lever in the pull clip,

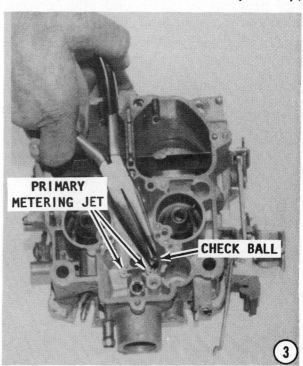

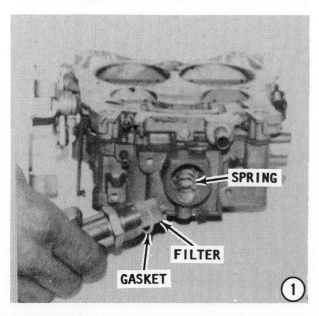

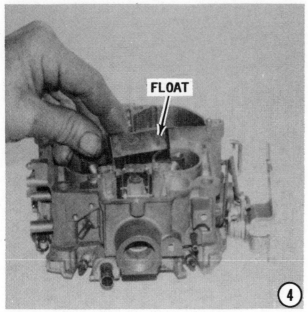

install the retaining pin from the pump well side. **TAKE CARE** not to distort the pull clip.

Float Level Adjustment

5- Measure the distance from the top of the float bowl gasket surface, with the gasket removed, to the top of the float at the toe end. **CHECK TO BE SURE** the retaining pin is held firmly in place and the tang of the float is seated on the float needle when making the measurement. Check your measurement with the Specifications in the Appendix. **CAREFULLY** bend the float up or down until the correct measurement is reached.

6- Install the power piston spring in the power piston well. Install the primary metering jets, if they were removed during disassembly. Be sure the tension spring is connected to the top of each metering rod. Install the power-piston assembly in the well with the metering jets. A sleeve around the

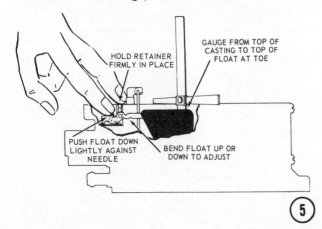

piston holds the piston in place during assembly.

7- Install the plastic filler over the float needle. Press it down firmly until it is seated.

Place the pump return spring in the pump well. Install the air horn gasket around the primary metering rods and piston. Install the gasket on the secondary side of the bowl with the holes in the gasket indexed over the two dowels. Install the accelerating pump plunger in the pump well.

8- Install the secondary metering rods.

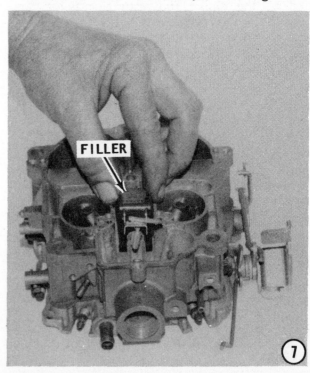

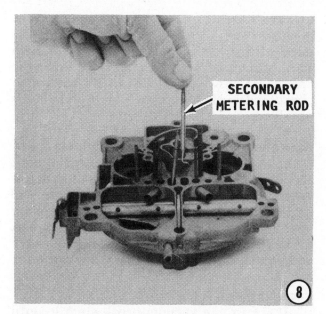

SECONDARY
METERING ROD

⑧

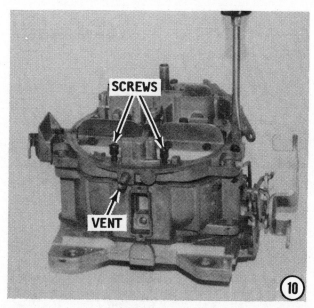

SCREWS

VENT

⑩

Hold the air valve wide open and check to be sure the rods are positioned with their upper ends through the hanger holes and pointing toward each other.

9- Slowly position the air horn assembly on the bowl and **CAREFULLY** insert the secondary metering rods, the high-speed air bleeds, and the accelerating well tubes through the holes of the air horn gasket. **NEVER** force the air horn assembly onto the float bowl. Such action may distort the secondary metering plates. If the air horn assembly moves slightly sideways it will center the metering rods in the metering plates. Install the attaching screws as

follows: The four long air horn screws around the secondary side; two short screws in the center section; one short screw above the fuel inlet; and the two countersunk screws in the primary venturi area. Tighten the screws evenly and in the sequence as shown in the accompanying illustration.

10- Install the idle vent actuating rod in the pump lever. Connect the pump rod to the inner hole of the pump lever and secure it with a spring clip. Connect the choke rod in the lower choke lever and secure it with a spring clip. Install the idle vent valve with the actuating rod engaged, and then tighten the attaching screws.

Place the vacuum-break cannister between the retaining ears on the bracket and gently squeeze the ears together around the cannister. Connect the vacuum-break link to the vacuum-break assembly and secure it in place with the retaining clip.

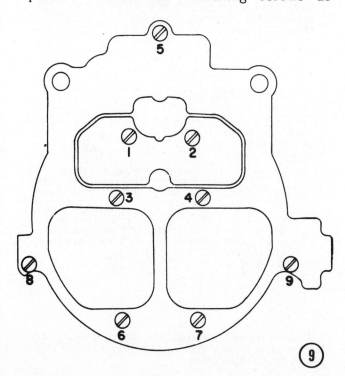

5

1   2

3   4

8   9

6   7

⑨

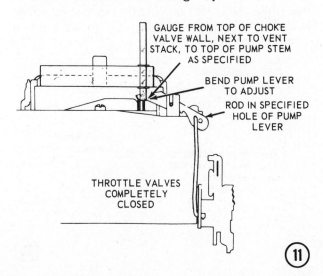

GAUGE FROM TOP OF CHOKE
VALVE WALL, NEXT TO VENT
STACK, TO TOP OF PUMP STEM
AS SPECIFIED

BEND PUMP LEVER
TO ADJUST

ROD IN SPECIFIED
HOLE OF PUMP
LEVER

THROTTLE VALVES
COMPLETELY
CLOSED

⑪

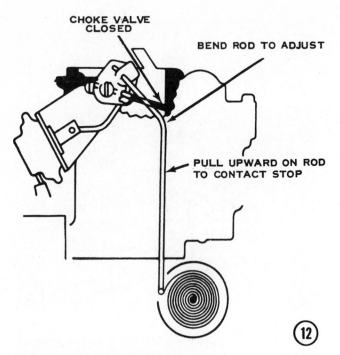

CHOKE VALVE CLOSED

BEND ROD TO ADJUST

PULL UPWARD ON ROD TO CONTACT STOP

⑫

BENCH ADJUSTMENTS ROCHESTER 4MV AND 4MC

Pump Adjustment

11- Disconnect the secondary actuating link. Measure the distance from the top of the choke valve wall, next to the vent stack, to the top of the pump stem, with the throttle valves completely closed and the pump rod in the inner hole in the lever. Compare your measurement with the Specifications in the Appendix. **CARE-FULLY** bend the pump lever until the specified dimension is obtained. Connect the actuating link.

Choke Rod Adjustment

12- First, place the cam follower on the second step of the fast-idle cam and against the high step. Next, rotate the choke valve toward the closed position by pushing down

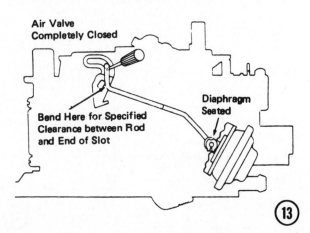

Air Valve Completely Closed

Bend Here for Specified Clearance between Rod and End of Slot

Diaphragm Seated

⑬

LIGHTLY on the vacuum-break lever. Now, measure the distance between the lower edge of the choke valve and the air horn wall. Compare your measurement with the Specifications. **CAREFULLY** bend the choke rod until the proper measurement is obtained.

Air Valve Dashpot Adjustment

13- Push the vacuum-break stem in until the diaphragm is seated. Measure the distance between the dashpot rod and the end of the slot in the air valve lever. Compare your measurement with the Specifications. **CAREFULLY** bend the rod at the air valve end until the proper adjustment is reached.

Vacuum Break Adjustment

14- Push the vacuum-break stem in until the diaphragm is seated. At the same time, hold the choke valve toward the closed position and measure the distance between the lower edge of the choke valve and the air horn wall. Compare your measurement with the Specifications. **CAREFULLY** bend the vacuum break tang until the proper adjustment is reached.

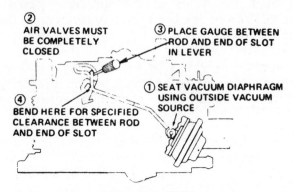

② AIR VALVES MUST BE COMPLETELY CLOSED

③ PLACE GAUGE BETWEEN ROD AND END OF SLOT IN LEVER

① SEAT VACUUM DIAPHRAGM USING OUTSIDE VACUUM SOURCE

④ BEND HERE FOR SPECIFIED CLEARANCE BETWEEN ROD AND END OF SLOT

THROUGH 1971

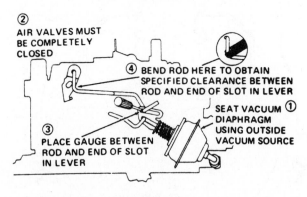

② AIR VALVES MUST BE COMPLETELY CLOSED

④ BEND ROD HERE TO OBTAIN SPECIFIED CLEARANCE BETWEEN ROD AND END OF SLOT IN LEVER

① SEAT VACUUM DIAPHRAGM USING OUTSIDE VACUUM SOURCE

③ PLACE GAUGE BETWEEN ROD AND END OF SLOT IN LEVER

SINCE 1972

Depending on the model, the air-valve dashpot on later models is adjusted by inserting a gauge into a slot in the dashpot lever.

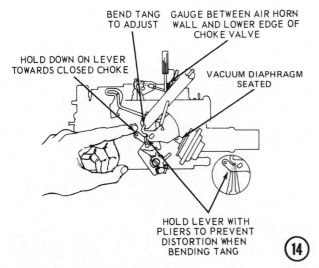

HOLD DOWN ON LEVER TOWARDS CLOSED CHOKE

BEND TANG TO ADJUST

GAUGE BETWEEN AIR HORN WALL AND LOWER EDGE OF CHOKE VALVE

VACUUM DIAPHRAGM SEATED

HOLD LEVER WITH PLIERS TO PREVENT DISTORTION WHEN BENDING TANG

(14)

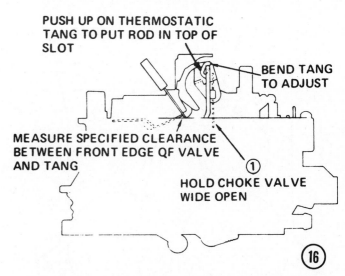

PUSH UP ON THERMOSTATIC TANG TO PUT ROD IN TOP OF SLOT

BEND TANG TO ADJUST

MEASURE SPECIFIED CLEARANCE BETWEEN FRONT EDGE OF VALVE AND TANG

①

HOLD CHOKE VALVE WIDE OPEN

(16)

Unloader Adjustment

15- Secure a rubber band on the vacuum-break to hold the choke valve in the closed position. Move the primary throttle valves to the wide-open position and at the same time measure the distance between the edge of the choke valve and the air horn wall. Compare your measurement with the Specifications. **CAREFULLY** bend the fast-idle lever tang until the proper adjustment is reached.

Air Valve Lockout Adjustment

16- With the choke valve wide open, force the thermostatic spring tang to move the choke rod to the top of the slot in the choke lever. Now, move the air valve toward the open position. Measure the distance between the lockout tang and the front edge of the air valve, as shown in the accompanying illustration. **CAREFULLY** bend the upper end of the air valve lockout lever tang until the required adjustment is reached. Finally, open the choke valve to its wide-open position by applying force to the underside of the choke valve. **BE SURE**

the choke rod is in the bottom of the slot in the choke lever, the air valve lockout tang holds the air valve closed.

Secondary Opening Adjustment

17- Open the primary throttle valves until the actuating link contacts the tang on the secondary lever, and then measure the distance between the link and the tang on the secondary lever, which should be 0.070". **CAREFULLY** bend the tang on the secondary lever until the proper adjustment is reached.

Secondary Closing Adjustment

18- After the idle speed has been adjusted and with the tang on the throttle lever against the actuating lever, measure the distance between the actuating link and the front of the slot in the secondary lever. Compare your measurement with the Specifications. **CAREFULLY** bend the tang on the throttle lever until the proper adjustment is reached.

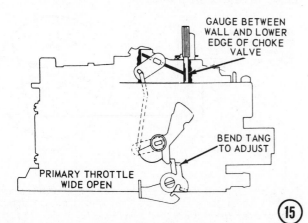

GAUGE BETWEEN WALL AND LOWER EDGE OF CHOKE VALVE

BEND TANG TO ADJUST

PRIMARY THROTTLE WIDE OPEN

(15)

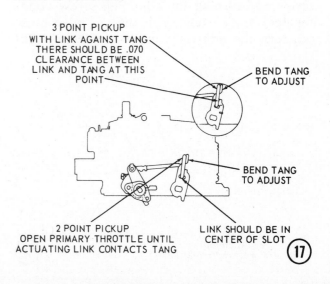

3 POINT PICKUP WITH LINK AGAINST TANG THERE SHOULD BE .070 CLEARANCE BETWEEN LINK AND TANG AT THIS POINT

BEND TANG TO ADJUST

BEND TANG TO ADJUST

2 POINT PICKUP OPEN PRIMARY THROTTLE UNTIL ACTUATING LINK CONTACTS TANG

LINK SHOULD BE IN CENTER OF SLOT

(17)

UNSEALED CANISTER PURGE PASSAGE

SEALED CANISTER PURGE PASSAGE

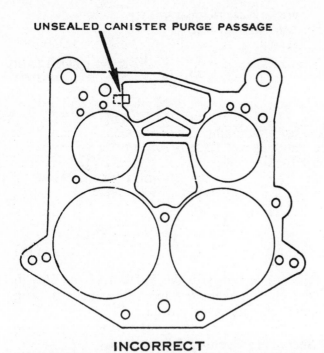

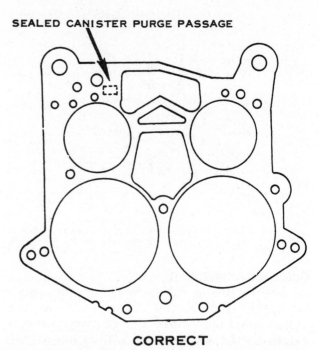

INCORRECT

CORRECT

The addition of a charcoal canister purge port on models since 1970 has resulted in a change in the gasket used between the float bowl and the throttle body. A vacuum leak will result if the wrong gasket is used because air will bypass the primary throttle valves through the canister purge passageway (left) and the engine will not idle smoothly.

Secondary Metering Rod Adjusment

19- Measure the distance from the top of the metering rod to the top of the air horn casting and also to the flame arrestor stud hole. Compare your measurement with the Specifications. **CAREFULLY** bend the metering rod hanger at the point shown in the accompanying illustration until the proper adjustment for both metering rods is reached. Both metering rods must be adjusted to the **SAME DIMENSION.**

Air Valve Spring Adjustment

20- First, remove all of the tension on the spring by loosening the Allen-head lockscrew and turning the adjusting screw counterclockwise. Now, with the air valve closed, turn the adjusting screw clockwise until

the torsion spring just contacts the pin on the shaft, and then 3/8 turn more. Hold the adjusting screw in this position, and tighten the lockscrew.

Choke Coil Rod Adjustment

21- Close the choke valve so the choke rod is at the bottom of the choke lever slot. Now, pull the choke coil rod up to the end of its travel. The top of the hole must be even with the bottom of the rod. **CAREFULLY** bend the choke coil rod until the proper adjustment is reached.

Idle Speed and Mixture Adjusment

22- Start the engine and allow it to warm to operating temperature with the flame arrestor in place.

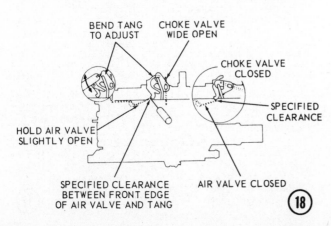

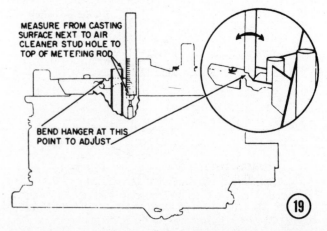

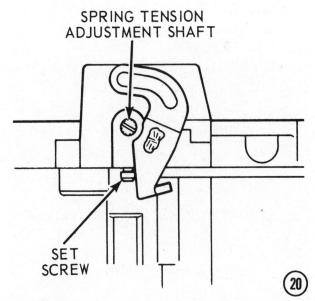

SPRING TENSION
ADJUSTMENT SHAFT

SET
SCREW

(20)

CAUTION

Water must circulate through the stern drive, also to and from the engine, anytime the engine is run to prevent damage to the water pump located in the lower unit (original and Alpha drives); or on the front of the engine (Bravo drive). Just a few seconds without water will damage the water pump.

Shut the engine down and disconnect the throttle cable. Turn the idle mixture adjusting screws in until they barely touch their seats, and then back them out one full turn as a rough adjustment at this time. **NEVER** turn the adjusting screws in hard or the needle and seat will be **DAMAGED**. Start the engine and run it at idle speed. Adjust the idle mixture needle for the highest and steadiest manifold vacuum reading. If you do not have a vacuum gauge, obtain

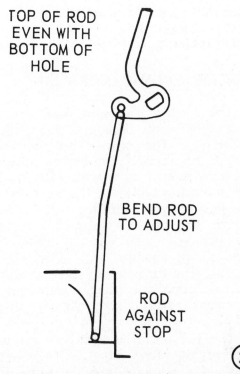

TOP OF ROD
EVEN WITH
BOTTOM OF
HOLE

BEND ROD
TO ADJUST

ROD
AGAINST
STOP

(21)

the smoothest running, maximum idle speed by turning the idle adjusting needle in until the engine rpm begins to drop off, then back the needle off over the "high spot" until the engine rpm again begins to drop off. Set the idle adjusting needle halfway between the two points as an idle speed setting. Repeat the procedure with the other needle. If the adjustments result in an increase in idle rpm, reset the idle speed adjusting screw to obtain between 550 and 600 rpm. Again adjust the idle mixture adjusting needles. Stop the engine and install the throttle cable. Check to be sure the throttle valves are fully open when the remote control is in the full forward position. Shift the unit into forward gear and readjust the idle speed screw until the rpm is between 550 and 600 rpm.

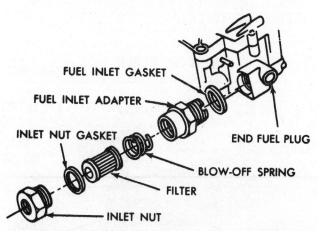

FUEL INLET GASKET

FUEL INLET ADAPTER

INLET NUT GASKET

END FUEL PLUG

BLOW-OFF SPRING

FILTER

INLET NUT

Arrangement of fuel filter parts including a self-tapping fuel filter adapter which may be installed if the threads in the inlet hole of the carburetor are damaged. Use of this type adapter will return an otherwise unfit carburetor to service.

IDLE SCREW

(22)

4-10 HOLLEY CARBURETORS
TWO BARREL - MODEL 2300C
FOUR BARREL - MODEL 4150, 4160 & 4011

DESCRIPTION AND OPERATION

The Holley two-barrel and four-barrel carburetors are designed and built very similarly. The four-barrel unit can be considered as dual two-barrel carburetors, mounted side-by-side, with each having its own metering bowl and float system. Each side has its own metering body. The two primary bores have a single choke valve, and the primary side has a power valve, accelerating pump, and adjustable idle system. The throttles on the secondary side are controlled by a vacuum diaphragm. The secondary metering body has only fixed idle and high-speed metering systems.

The two-barrel is basically the primary side of the four-barrel. Therefore, the overhaul procedures are the same for the two-barrel model as those provided in the following section for the four-barrel.

The four-barrel carburetors have a dual, high-speed system made up of primary and secondary circuits. The primary circuits are composed of the float, idle, acceleration, main metering, power, and choke circuits. The secondary circuit becomes operational when the engine demands extra power. Each circuit will be described under separate headings.

The illustrations accompanying this section are keyed by number to the paragraphs which describe the parts being shown.

Holley four-barrel carburetor.

FLOAT CIRCUIT

1- Fuel from the fuel pump line enters the fuel bowl through the inlet fitting, then passes through a filter, and finally into the fuel bowl through the needle and seat. The amount of fuel entering the fuel bowl is controlled by the fuel pump pressure, the size of the hole in the needle seat, and by the distance the needle is permitted to rise out of the seat as determined by the float drop. Therefore, the fuel level in the bowl is determined by the float level setting. As the fuel level drops, the float lowers, which allows more fuel to enter the bowl. When the level rises to the setting level, the float pushes the needle into the seat and thus shuts off the flow of fuel or at least restricts the amount entering the bowl. In this manner, the float rising and lowering allows only enough fuel to enter the bowl to replace the fuel used.

Holley two-barrel carburetor.

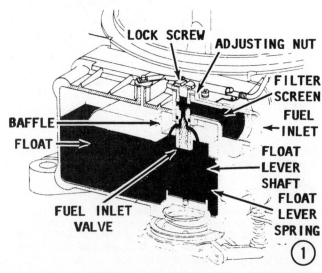

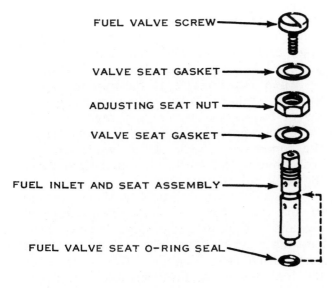

FUEL VALVE SCREW

VALVE SEAT GASKET

ADJUSTING SEAT NUT

VALVE SEAT GASKET

FUEL INLET AND SEAT ASSEMBLY

FUEL VALVE SEAT O-RING SEAL

ADJUSTABLE

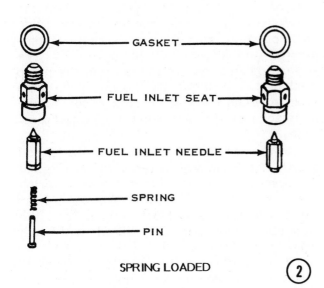

GASKET

FUEL INLET SEAT

FUEL INLET NEEDLE

SPRING

PIN

SPRING LOADED   ②

2- Holley carburetors are equipped with one of three common types of needle and seat assemblies, as shown in the accompanying illustration: (1) spring loaded; (2) solid needle; and (3) externally adjustable.

To prevent dirt in the fuel from flooding, most of the needles have a special soft plastic tip. A spring under the float stabilizes and maintains a normal fuel level in most models. The proper fuel level is critical on any carburetor because all of the basic settings and calibrations of the other systems are based on the fuel level in the bowl. The float system is equipped with a vent valve in order to pressurize the fuel and to create a pressure differential. The reduced pressure in the venturi creates a pressure differential forcing the fuel to flow out of the fuel bowl to the discharge nozzle.

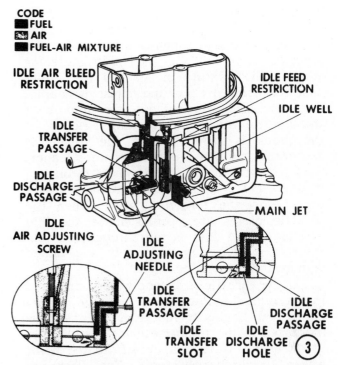

CODE
■ FUEL
▨ AIR
■ FUEL-AIR MIXTURE

IDLE AIR BLEED RESTRICTION

IDLE FEED RESTRICTION

IDLE WELL

IDLE TRANSFER PASSAGE

IDLE DISCHARGE PASSAGE

MAIN JET

IDLE AIR ADJUSTING SCREW

IDLE ADJUSTING NEEDLE

IDLE TRANSFER PASSAGE

IDLE DISCHARGE PASSAGE

IDLE TRANSFER SLOT

IDLE DISCHARGE HOLE   ③

3- Fuel passes through the main metering jet to the vertical passageway. Near the top of this passageway is an idle feed restriction. This restriction performs the same function as an idle tube in other carburetors by metering the fuel for low-speed operation. From the restriction, the fuel moves horizontally across the carburetor to another vertical passageway, and then down the second passageway to the idle mixture screw port.

An idle air bleed is located in the low-speed fuel passageway, above the idle mixture adjustment screw. The idle air mixes with the fuel to form an emulsion that is easier to atomize when it is discharged. This air-fuel mixture moves much faster than solid fuel. The idle air bleed prevents fuel from siphoning from the fuel bowl when the engine is not running.

4- The fuel branches into two passageways at the idle mixture adjusting screw. One branch is controlled by the idle mixture needle. This branch exits below the throttle

IDLE PORT PASSAGE

TRANSFER PORT PASSAGE   ④

valve and supplies the fuel for the hot curb idle. The other branch exits above the throttle valve and supplies the fuel for the transfer from idle to the main metering stage. The mixture of air and fuel flows past the pointed tip of the adjusting needle. If the needle is backed out, the volume of the mixture is increased making it richer. If the needle is turned inward, the volume is decreased and the mixture becomes leaner. Fuel from the needle then moves through a passage and is discharged into the throttle bore below the throttle plate. When the throttle plates are opened above the idle position, more fuel is fed through the idle transfer passageway to supply the added demands of the engine.

If the throttle is opened wider, air speed in the venturi increases and the main metering system begins to function. As the flow increases in the main metering system, the idle transfer system tapers off to the point where its discharge stops. The discharge from the idle transfer system is stopped because of the loss of manifold vacuum due to the opening of the throttle valve and the loss of air velocity between the edge of the throttle valve and the transfer port.

HOLLEY ACCELERATION SYSTEM

5- During acceleration, the air flow through the carburetor responds almost immediately to the increased throttle opening. Since fuel is heavier than air, it has a slower response. The accelerator pump system mechanically supplies fuel until the other fuel metering systems can once again supply the proper mixture. The diaphragm-type pump is located in the bottom of the primary fuel bowl. This location assures a more solid charge of fuel (fewer bubbles).

When the throttle is opened, the pump linkage, actuated by a cam on the throttle lever, forces the pump diaphragm up. As the diaphragm moves up, the pressure forces the pump inlet check ball or valve onto its seat, thereby preventing the fuel from flowing back into the fuel bowl.

The fuel passes through a short passage in the fuel bowl into the long diagonal passage in the metering body. It next flows into the main body passage and then into the pump discharge chamber. The pressure of the fuel causes the discharge valve to raise, and fuel is then discharged into the venturi.

The pump override spring is an important part of the accelerator system. When the accelerator is moved rapidly to the wide open position, the override spring is compressed and allows full pump travel. The spring applies pressure to maintain the pump discharge. Without the spring, the pump linkage would be bent or broken due to the resistance of the fuel which is not compressible.

As the throttle moves toward the closed position, the linkage returns to its original position and the diaphragm return spring forces the diaphragm down. The pump inlet check valve is moved off its seat and the diaphragm chamber is refilled with fuel from the fuel bowl.

The accelerator pump delivery rate is controlled by the pump cam, linkage, the override spring, and the size of the discharge holes.

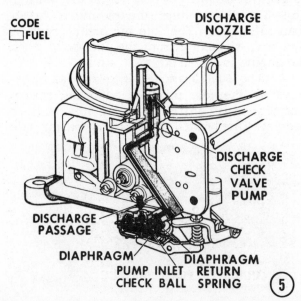

CODE
☐ FUEL

DISCHARGE NOZZLE

DISCHARGE CHECK VALVE PUMP

DISCHARGE PASSAGE

DIAPHRAGM
PUMP INLET CHECK BALL

DIAPHRAGM RETURN SPRING

5

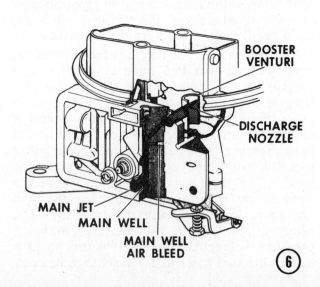

BOOSTER VENTURI

DISCHARGE NOZZLE

MAIN JET
MAIN WELL

MAIN WELL AIR BLEED

6

HOLLEY MAIN METERING

6- At higher speeds the vacuum is increased at the main discharge nozzle in the center of the booster venturi.

This vacuum or pressure differential causes fuel to flow through the main metering jet into the main well. The fuel moves up the main well past one or more air bleed holes from the main airwell. These air bleed holes are supplied with the filtered air from the "high speed" air bleeds in the air horn. The mixture of fuel and air moves up the main well and through a channel to the main discharge nozzle in the booster venturi.

POWER ENRICHMENT SYSTEM

7- During high speed or heavy load operation, when manifold vacuum is low, the power system provides added fuel for power operation. A vacuum passage in the throttle body transmits vacuum to the power valve vacuum chamber in the main body. All of the power valves used in this series of carburetors are actuated by a vacuum diaphragm. Manifold vacuum is applied to the vacuum side of the diaphragm to hold it closed at idle and normal moderate load conditions.

When manifold vacuum drops below the power valve's calibration, the power valve spring opens the valve to admit additional fuel. This fuel is metered by the power valve channel restrictions into the main well

and is added to the fuel flowing from the main metering jets.

HOLLEY ELECTRIC CHOKE

8- The choke system has a bi-metal spring. When this spring cools, it closes the choke. An electric heating coil is installed in the choke spring cover. This heating element substitutes for the normal heat tube of other carburetors. The heating element is on whenever the ignition switch is turned on and provides the heat necessary to allow the bi-metal spring to open the choke. The time lapse between switch turn on until the choke opens is carefully calculated. For this reason, it is important that accessories such as bilge pumps, bilge blowers, etc., not be connected to the ignition circuit which would make it necessary to turn the ignition on for a period of time before starting the engine. Such an arrangement would open the choke before the engine starts and make starting difficult.

SECONDARY SYSTEMS

VACUUM SECONDARY OPERATION

9- At lower speeds the secondary throttle valves remain closed, allowing the engine to maintain proper air-fuel velocities and distribution for lower speed, light load operation. When engine demand increases to a point where additional breathing capacity is needed, the vacuum controlled secondary throttle valves begin to open automatically.

Vacuum from one of the primary venturi

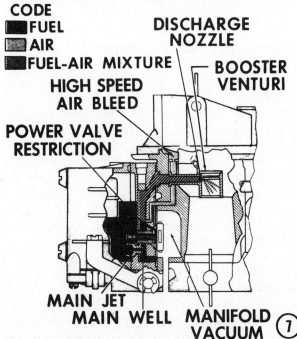

CODE
■ FUEL
▨ AIR
▨ FUEL-AIR MIXTURE
DISCHARGE NOZZLE
BOOSTER VENTURI
HIGH SPEED AIR BLEED
POWER VALVE RESTRICTION
MAIN JET
MAIN WELL
MANIFOLD VACUUM ⑦

SECONDARY BOWL   PRIMARY BOWL
CHOKE ⑧

and one of the secondary venturi is channeled to the top of the secondary diaphragm. The bottom of the diaphragm is open to atmospheric pressure. At higher speeds and higher primary venturi vacuum, the diaphragm, operating through a rod and secondary throttle lever, will start to open the secondary throttle valves. This action will start to compress the secondary diaphragm spring. As the secondary throttle valves open further, a vacuum signal is created in the secondary venturi. This additional vacuum assists in opening the secondary throttle valves to the maximum designed opening. The secondary opening rate is controlled by the diaphragm spring and the size of the vacuum restrictions in the venturi.

When the engine speed is reduced, venturi vacuum decreases and the diaphragm spring starts to push the diaphragm down to start closing the secondaries. Closing the primary throttle valves moves the secondary throttle connecting link.

Most production model carburetors have a ball check and bypass bleed installed in the diaphragm passage. The ball permits a smooth, even opening of the secondaries, but lifts off the inlet bleed to cause rapid closing of the secondaries when the primary throttle valves are closed.

SECONDARY FUEL METERING
SECONDARY FLOAT CIRCUIT

10- The secondary system has a separate fuel bowl. Fuel is usually supplied to the secondary bowl by a transfer tube from the primary fuel inlet fitting.

The secondary fuel bowl is equipped with a fuel inlet valve and float assembly similar to the primary side. The specified fuel level

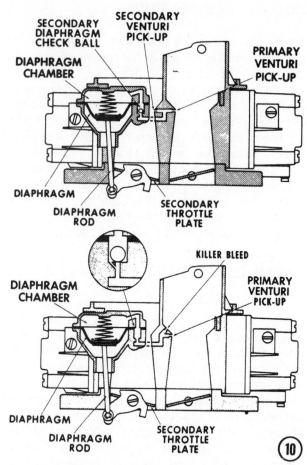

on the secondary side is usually slightly lower than the primary side.

Many Holley carburetors use a fuel balance tube between the primary and secondary fuel bowls. This tube is made of brass and connects the bowls just above the fuel level to prevent flooding of the secondary fuel bowl.

SECONDARY IDLE CIRCUIT

11- The secondary side of the carburetor is seldom used. Therefore, the fuel in the secondary fuel bowl could become stale and develop gum and varnish. To avoid this problem, a fixed idle is designed into the secondary side. Anytime the engine is running, some fuel is used from the secondary bowl and fresh fuel flows in to replace it. The secondary idle mixture is not adjustable. The transfer operates in the same manner as on the primary side.

SECONDARY MAIN
METERING CIRCUIT

12- As the secondary throttle valves are opened further, air velocity increases in the

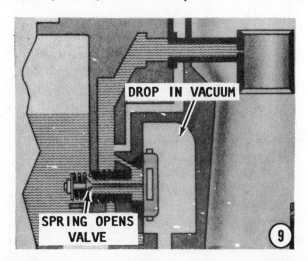

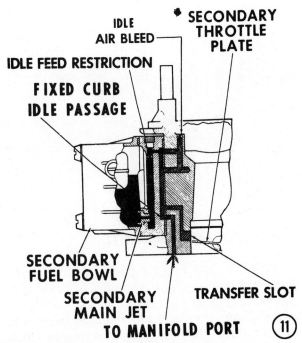

IDLE AIR BLEED
SECONDARY THROTTLE PLATE
IDLE FEED RESTRICTION
FIXED CURB IDLE PASSAGE
SECONDARY FUEL BOWL
SECONDARY MAIN JET
TRANSFER SLOT
TO MANIFOLD PORT (11)

FLOAT
FUEL TRANSFER TUBE
(1)

boost venturi. This action allows fuel to flow through the main metering system.

Fuel from the secondary fuel bowl enters the lower main metering holes in the metering body and then moves into the main well passageway. Air in the main well air bleed mixes with fuel that is still liquid at this point. The air-fuel mixture then moves horizontally through the discharge nozzle and into the boost venturi. The secondary circuit does not have a power circuit.

4-11 SERVICING
TWO BARREL — MODEL 2300C AND
FOUR BARREL — MODEL 4150 AND 4160
(Model 4011 Covered in Next Section 4-12

DISASSEMBLING

1- Remove the four primary fuel bowl retaining screws. Slide the bowl straight

off. Discard the gaskets. Pull off the fuel transfer tube and discard the O-ring seals.

2- Remove the secondary fuel bowl assembly. Detach the secondary metering body. TAKE CARE and remove the secondary metering body, plate, and gaskets. Slide the balance tube, washers, and O-rings out of the main body from either end.

TAKE NOTE of the position of the fast-idle cam to the cam lever for proper assembly. Disconnect the link from the fast-idle cam lever and cam, and then slide the lever and cam off of the stub shaft, and at the same time, disengage the choke rod from the cam lever. Disconnect the secondary vacuum diaphragm assembly stem from the secondary stop lever, and then remove the assembly. Discard the gasket. Remove the choke plate retaining screws by first filing the backs of the screws to remove the stake marks. Remove the choke plate. Remove the choke shaft.

3- Remove the accelerator pump discharge nozzle by first removing the retaining screw. Discard both gaskets. Turn the carburetor over and catch the accelerating pump discharge needle as it falls out.

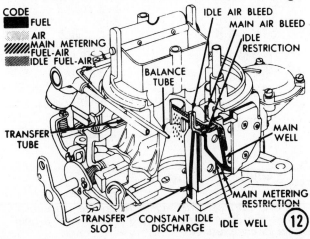

CODE
FUEL
AIR
MAIN METERING
FUEL-AIR
IDLE FUEL-AIR

IDLE AIR BLEED
MAIN AIR BLEED
IDLE RESTRICTION
BALANCE TUBE
TRANSFER TUBE
MAIN WELL
TRANSFER SLOT
CONSTANT IDLE DISCHARGE
IDLE WELL
MAIN METERING RESTRICTION
(12)

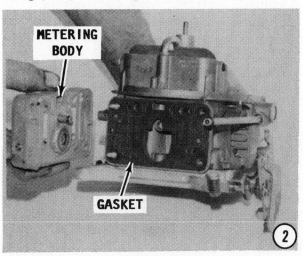

METERING BODY
GASKET
(2)

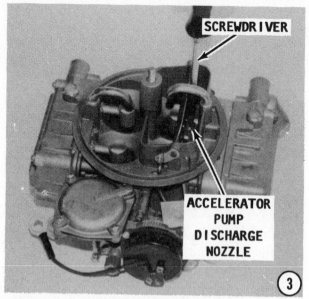

Remove the six screws securing the throttle body and main body together and then separate the parts. Discard the gasket.

4- Remove the float retainer E-clip, and then slide out the float and spring. Remove the fuel inlet needle, and then take out the float baffle. Remove the fuel inlet needle seat and gasket. Discard the gasket. Remove the fuel inlet fitting and discard the gasket. Remove the bowl vent valve assembly and accelerator pump cover on the primary fuel bowl. Remove the diaphragm and spring.

5- The power valve assembly should be replaced each time the carburetor is overhauled. This assembly should be removed from the primary metering body using Tool No. 3747. The metering jets can, but do not have to be removed because they can be cleaned with compressed air while they are still in place in the metering body. Remove the idle adjusting needles and gaskets. The secondary metering body does not have to be disassembled because the metering restrictions can be adequately cleaned with compressed air.

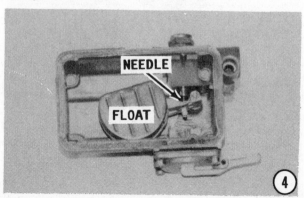

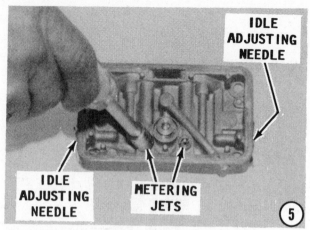

CLEANING AND INSPECTING

NEVER dip rubber parts, plastic parts, diaphragms, or pump plungers in carburetor cleaner. Place all of the metal parts in a screen-type tray and dip them in carburetor solvent until they appear completely clean, then blow them dry with compressed air.

Blow out all passages in the castings with compressed air. Check all of the parts and passages to be sure they are not clogged or contain any deposits. **NEVER** use a piece of wire or any type of pointed instrument to clean drilled passages or calibrated holes in a carburetor.

Inspect the choke and throttle shafts for excessive wear and replace the complete carburetor if the throttle shaft is worn.

Check the floats for leaking by shaking them and listening for fluid movement inside. If a float contains fluid, it must be replaced. Inspect the arm needle contact surface and the float shaft and replace them if either has any grooves worn in it.

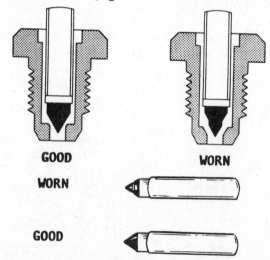

Needle and seat arrangement on the carburetor covered in this section, showing a worn and a new needle for comparison.

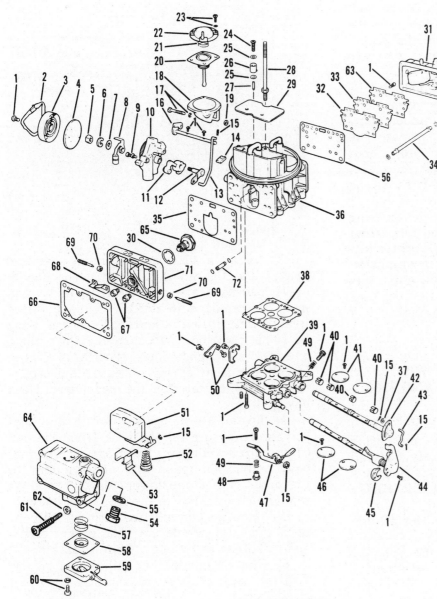

1. Screw
2. Choke Thermostat Housing Clamp
3. Choke Thermostat Housing and Spring
4. Choke Thermostat Housing Gasket
5. Nut
6. Lockwasher
7. Spacer
8. Choke Thermostat Lever Link and Piston Assembly
9. Screw and Washer
10. Choke Housing
11. Fast Idle Cam Assembly
12. Choke Housing Shaft and Lever
13. Choke Rod
14. Choke Rod Seal
15. Retainer

16. Choke Shaft
17. Screw and Washer
18. Secondary Housing
19. Gasket
20. Diaphragm Assembly
21. Diaphragm Spring
22. Cover
23. Screw and Washer
24. Screw
25. Gasket
26. Accelerating Pump Discharge Nozzle
27. Accelerating Pump Discharge Needle
28. Flame Arrestor Anchor Screw
29. Choke Plate
30. Power Valve Gasket
31. Secondary Fuel Bowl
32. Secondary Plate

33. Secondary Plate
34. Fuel Line
35. Primary Meter Block Gasket
36. Main Body
37. Washer
38. Throttle Body-to-main Body Gasket
39. Throttle Body
40. Shaft Bushings
41. Secondary Throttle Plates
42. Secondary Throttle Shaft
43. Throttle Connecting Rod
44. Primary Throttle Shaft Assembly
45. Accelerating Pump Cam
46. Primary Throttle Plates
47. Accelerating Pump Operating Lever
48. Sleeve Nut
49. Spring
50. Diaphragm Lever Assembly
51. Float
52. Float Spring
53. Baffle Plate
54. Fuel Inlet Fitting
55. Gasket
56. Gasket
57. Diaphragm Spring
58. Diaphragm Assembly
59. Accelerating Pump Cover
60. Retaining Screw and Lockwasher
61. Screw
62. Gasket
63. Secondary Metering Block
64. Primary Fuel Bowl
65. Power Valve
66. Primary Fuel Bowl Gasket
67. Main Jets
68. Baffle Plate
69. Idle Adjusting Needle
70. Seal
71. Primary Metering Block
72. Tube and O-Rings

Exploded view of the Holley four-barrel carburetor with major parts identified.

ALWAYS replace the needle valve-and-seat assemblies. These parts receive the most wear in the carburetor and the proper fuel level cannot be maintained if they are worn.

Check the choke vacuum diaphragm by first depressing the diaphragm stem, and then placing a finger over the vacuum fitting to seal the opening. Now, release the diaphragm stem. If the stem moves out more than 1/16" in ten seconds, the diaphragm has an internal leak and it should be replaced.

Most parts requiring replacement during a carburetor overhaul, including the latest updated parts, are found in a carburetor kit. These kits are available at your local marine dealer. In addition to the parts, most kits include the latest specifications, which are so important when making adjustments.

ASSEMBLING
TWO BARREL — MODEL 2300C AND
FOUR BARREL — MODEL 4150 AND 4160

1- Install a new power valve and gasket in the primary main metering body using Tool C-3747 to tighten it securely. Install the idle mixture adjusting needles and new gaskets. Tighten the needles FINGERTIGHT, then back each one out one full turn as a rough adjustment at this time. Install the main metering jets.

2- Assemble the parts to the primary fuel bowl by first placing the accelerator pump spring in position. Next place the diaphragm in position with the contact button facing the PUMP LEVER in the cover, as shown in the accompanying illustration. Place the cover over the diaphragm with the pump lever FACING the FUEL INLET side, as shown. Start the attaching screws, then depress the pump lever to center the diaphragm, then tighten the screws. Install a

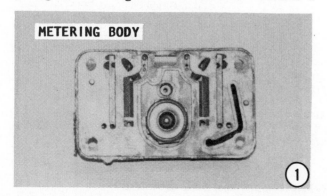

new gasket over the fuel inlet fitting, then the fitting and tighten it securely.

3- Install a NEW needle valve seat and gasket and tighten it securely. The needle valve and seat are a MATCHED SET and must be replaced in PAIRS. Slide the fuel inlet needle into its seat. Place the float baffle in position, then slide the float hinge over the pivot and secure it with an E-clip. Install the float spring. Repeat the procedures in this step for the secondary float bowl.

4- Invert (turn upside down) the fuel bowl and measure the distance from one end of the lower edge of the float to the fuel bowl body. Compare this measurement to the other end of the lower edge. The float should hang parallel to the fuel bowl body.

5- Bend the curved float arm which contacts the inlet needle until the proper alignment of float to fuel bowl body is obtained.

6- Place a new gasket in position on the throttle body. Lower the throttle body and at the same time, align the roll pin guides with the openings in the throttle body. The primary bores of the throttle body MUST be on the SAME side as the primary venturi. Hold the parts together, turn them over and

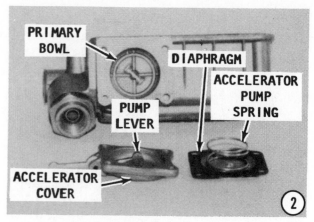

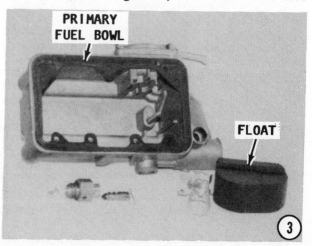

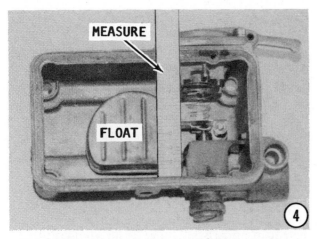

install and tighten them securely. Push an
O-ring seal into each recess, then install a
flat washer at each end. Install a **NEW**
gasket, then the plate, another **NEW** gasket,
and the secondary metering body with the
restrictions at the **BOTTOM**. Secure it all
together with the attaching screws.

7- Slide the assembled secondary
metering bowl and a **NEW** gasket over the
fuel transfer tube. Install and tighten the
attaching screws. Slide a new O-ring seal
on each end of the fuel transfer tube, and
then press the tube into the opening in the
secondary bowl. Place a **NEW** gasket over
the primary metering body pin, then slide
the primary metering body over the fuel
transfer tube and into place on the throttle
body. Slide a **NEW** gasket over the other
side of the primary metering body, and then
install the primary fuel bowl over the fuel
transfer tube and down against the primary
metering body. Slide **NEW** gaskets over the
long fuel bowl mounting screws, then install
and tighten them securely. **NEVER** use an
old gasket because they will always leak
following a second installation.

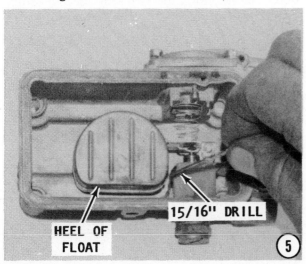

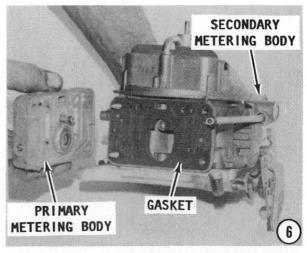

8- Position the fuel transfer tube, if one
is used, so one inch extends beyond the
secondary metering bowl, as shown.

9- Install the accelerating pump discharge needle into the discharge passageway
in the center of the primary venturi.

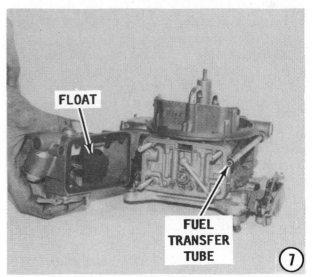

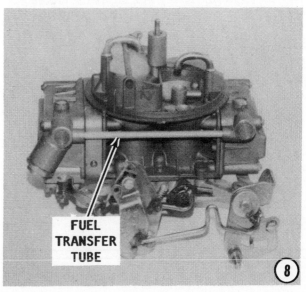

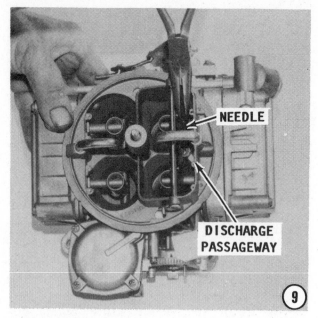

NEEDLE

DISCHARGE PASSAGEWAY

⑨

10- Install the pump discharge nozzle gasket, nozzle, mounting screw, and gasket. Align the notch in the rear of the nozzle with the projection on the boss of the casting. Install the choke valve and shaft.

11- Assemble the secondary throttle-opening diaphragm by first sliding a new diaphragm into the housing. Next, place the diaphragm in position with the vacuum hole in the housing aligned with the vacuum hole in the diaphragm. Now, install the return spring with the coiled end snapped over the button in the cover. Lower the cover into position with the vacuum port in the cover aligned with the port in the housing. Keep the diaphragm flat while the cover is being installed. Install and tighten the attaching

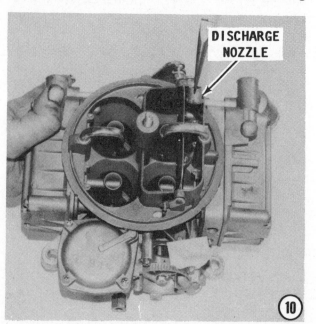

DISCHARGE NOZZLE

⑩

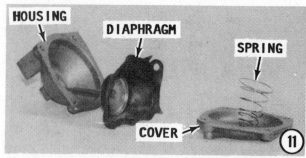

HOUSING DIAPHRAGM SPRING

COVER ⑪

screws. Check the assembly for air leaks by depressing the diaphragm stem and then placing your finger over the port to seal it. If the diaphragm does not remain in the retracted position, there is an air leak.

12- Place a **NEW** gasket in the vacuum passageway recess in the diaphragm housing, then install the secondary diaphragm assembly and at the same time, engage the stem with the secondary stop lever. Install and tighten the attaching screws.

HOLLEY ADJUSTMENTS
TWO BARREL — MODEL 2300C AND
FOUR BARREL — MODEL 4150 AND 4160

Carburetor adjustments **MUST** be performed in the sequence outlined in the following steps.

For best performance and economical results, follow the specifications included in the carburetor repair kit.

Electric Choke Adjustment

1- Loosen the three screws and retainers securing the thermostatic spring housing to the choke housing. Rotate the cover **COUNTERCLOCKWISE** for operation in colder weather. The colder weather will require a higher thermostatic spring temperature to fully open the choke plate.

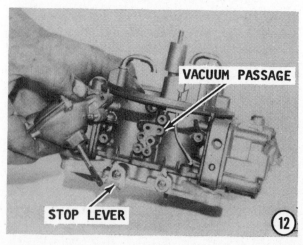

VACUUM PASSAGE

STOP LEVER ⑫

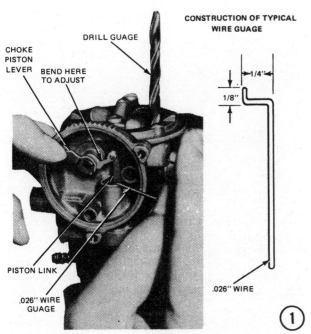

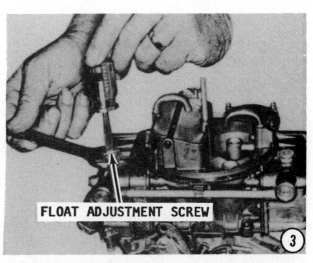

Rotate the cover **CLOCKWISE** for operation in hotter weather. The hotter weather will require a lower thermostatic spring temperature to fully open the choke plate. It is also the leaner setting as indicated by the arrow embossed on the choke housing.

Align the mark on the cover with the specified mark on the housing as an initial setting. If this setting is found unacceptable after a cold start, adjust one notch at a time until a satisfactory setting is obtained.

Vacuum Choke Setting

Obtain a piece of 0.026" wire and make a gauging tool as shown in the accompanying illustration. A standard size paper clip will do.

Insert the bent end into the piston bore until the hooked end engages with the bore slot. Close the choke valve with the choke piston lever until the end of the wire slides to the end of the bore slot.

Hold this position and at the same time measure the clearance between the edge of the choke plate and the inside wall of the air horn with a 9/64" (3.56mm) drill bit.

Adjust the plate clearance to this specified clearance by either bending the pivot arm or bending the stop (if so equipped).

2- A richer or leaner mixture can be obtained for the warm-up period, if desired, by rotating the choke cover. **NEVER** set the index mark on the cover more than two marks from the specified setting.

Fuel Level Adjustment

3- Loosen the locknut and turn the float adjustment screw clockwise to lower the fuel level and counterclockwise to raise the fuel level. A 1/16 turn of the screw will equal approximately 1/16 inch difference in fuel level. After the adjustment has been made, tighten the locknut. On newer model carburetors, this adjustment can only be done with the bowl removed.

4- With the engine approximately level and the engine running, the fuel level in the sight plug must be in line with the threads at the bottom of the plug within 1/32". This sight plug is not used on late model carburetors.

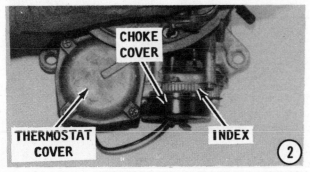

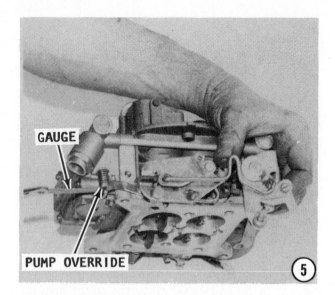

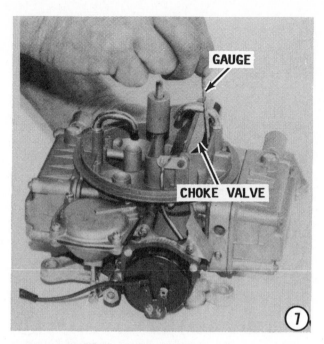

Accelerator Pump Lever Clearance

5- First, move the throttle to the wide-open position. Hold the pump lever down. You should be able to insert a 0.015" gauge between the adjusting nut and the lever. Rotate the pump override screw to obtain the correct clearance. There must be **NO FREE PLAY** with the pump lever at idle.

Choke Control Lever Adjustment

6- First, open the throttle to the mid-position. Next, close the choke valve with a light pressure on the choke control lever. With the carburetor on the engine, measure the distance between the top of the choke rod hole in the control lever and the choke assembly and compare it with 1-11/16". **CAREFULLY** bend the choke shaft rod until the measurement is within 1/64" of the specification given. This adjustment is necessary to ensure a correct relationship between the choke valve, the thermostatic coil spring, and the fast-idle cam.

Choke Unloader Adjustment

7- Hold the throttle in the wide-open position and at the same time insert the specified gauge between the upper edge of the choke valve and the inner wall of the air horn. **CAREFULLY** bend the indicated tang until the required measurement is obtained.

Secondary Throttle Valve Adjustment

8- First, back out the secondary throttle stop screw until the valves are fully closed. Next, turn the stop screw in until it barely contacts the secondary throttle stop lever. Now, turn the adjusting screw in 1/2 turn.

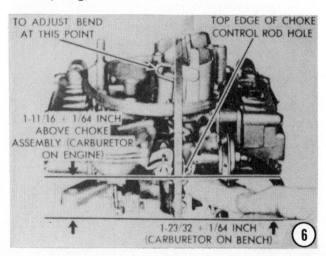

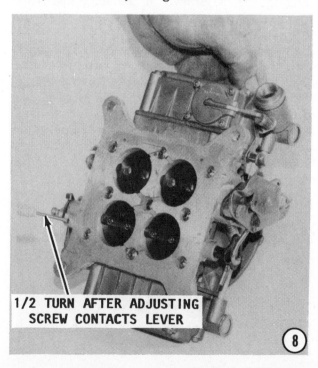

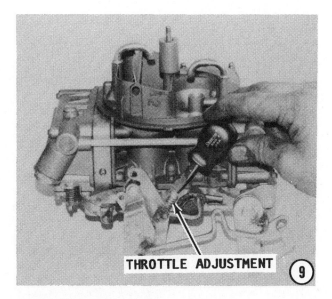

THROTTLE ADJUSTMENT ⑨

Fast Idle Adjustment

9- With the choke plate wide open, the fast-idle adjusting screw should just touch the lowest step of the fast-idle cam. This adjustment **MUST** be done **BEFORE** adjusting the idle needles.

Idle Speed and Mixture Adjustment

10- Turn each idle speed needle in until it barely seats, and then back it out 1-1/2 turns. **TAKE CARE** not to seat the needles tightly because it would groove the tip of the needle. If the needle should become grooved, a smooth idle cannot be obtained. Start the engine and allow it to warm to full operating temperature.

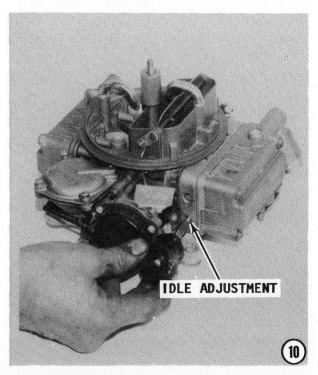

IDLE ADJUSTMENT ⑩

CAUTION

Water must circulate through the stern drive, also to and from the engine, anytime the engine is run to prevent damage to the water pump located in the lower unit (original and Alpha drives); or on the front of the engine (Bravo drive). Just a few seconds without water will damage the water pump.

Adjust the idle-speed screw until the required rpm is reached as given in the Specifications in the Appendix. Now, set one of the two idle adjusting needles for the highest steady manifold vacuum reading. If a vacuum gauge is not available, turn the idle adjusting needle inward until the rpm begins to drop off, then back the needle out over the "highspot" until the rpm again drops off. Set the idle adjusting needle halfway between these two points as an idle speed setting. Repeat the procedure for the other needle. If these adjustments result in an increase in idle rpm, reset the idle speed screw to obtain the required idle rpm, and then adjust the idle adjusting needles again.

11- Install the spacer over the center screw, and then install the flame arrestor. **CHECK TO BE SURE** the flame arrestor is clean because a dirty one will consume too much fuel and cause the engine to run rough at high speeds.

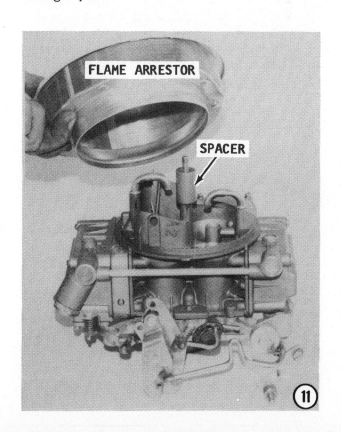

FLAME ARRESTOR

SPACER ⑪

The Holley Model 4011 4-barrel carburetor is used on GMC large block engines.

4-12 HOLLEY CARBURETOR MODEL 4011 4V

INTRODUCTION

See Section 4-10 for detailed description and operation of Holley carburetors. The Model 4011 4V carburetor is used only on the "big block" GMC powerplants, such as the 454 CID unit.

The following procedures pickup the work after the flame arrestor, fuel lines, electric choke harness, and linkage has been disconnected and the carburetor removed from the manifold.

Use a clean suitable work surface, make an attempt to keep parts in a logical order. Be sure to separate parts removed from each side -- primary and secondary -- to ensure all components are returned to their original location.

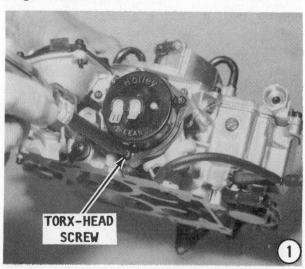

TORX-HEAD SCREW

①

ELECTRIC CHOKE   RETAINER

②

DISASSEMBLING

T-20 and T-25 Torx-head screwdrivers are absolutely necessary when working on this model carburetor.

DO NOT commence disassembling work on the carburetor until a rebuild kit including a full set of gaskets, diaphragms, O-rings and springs has been obtained and is at hand.

1- Remove the three Torx-head screws from the retainer around the electric choke.

2- Remove the retainer from the electric choke.

3- Pull the choke cover free of the housing, and then disengage the loop of the thermostat coil from the tang on the choke arm. Remove and discard the gasket.

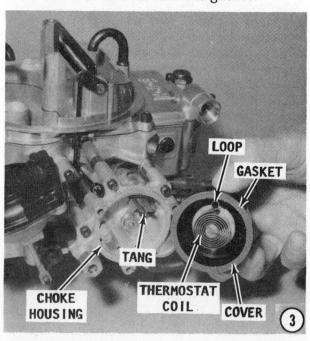

LOOP
GASKET
TANG
CHOKE HOUSING
THERMOSTAT COIL   COVER

③

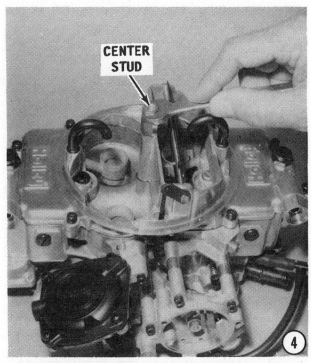

(4)

4- Remove the center stud from the carburetor body.

WARNING

The circlip to be removed in the next step is very small and under considerable tension. Therefore, use utmost caution and cover the circlip with one hand while removing the clip with a pair of needlenose pliers.

5- Using a pair of needlenose pliers, pry the circlip free from the lower end of choke link.

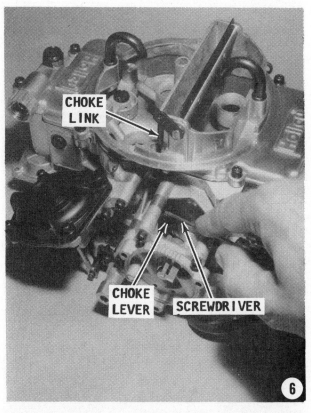

(6)

6- Insert a small screwdriver between the choke link and the choke lever, and then pry them apart.

7- Lift the airhorn assembly up and free of the carburetor body. The choke linkage and floats are attached and will come up with the airhorn. If necessary, rotate the free end of the choke link to allow the end to pass through the hole in the carburetor body.

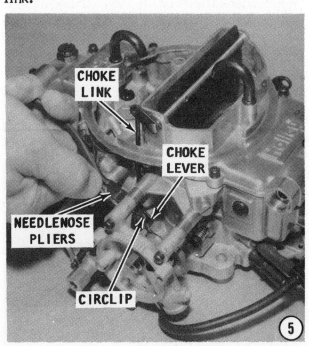

(5)

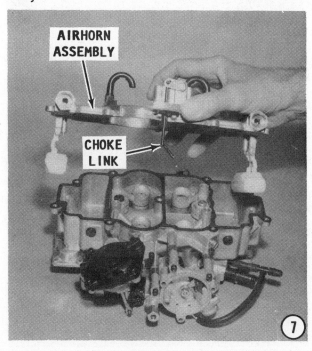

(7)

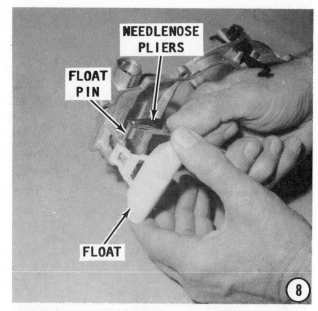

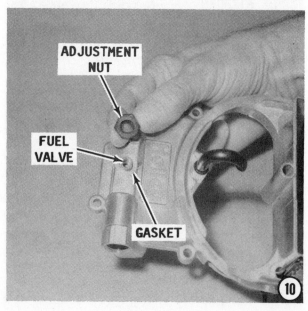

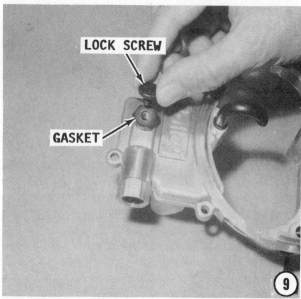

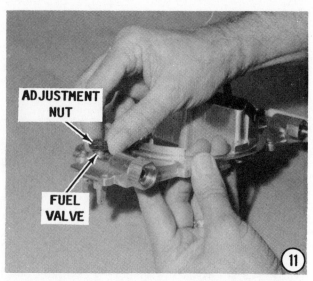

8- Push one of the float pins free of the float boss, and then remove the float. Remove the other float in the same manner.

9- Remove the lock screw and gasket from each side of the airhorn.

10- Remove the adjustment nut and gasket from each side of the airhorn.

11- Using the adjustment nut as a tool, unscrew both fuel valves from the airhorn.

12- Lift off and discard the carburetor main body gasket.

13- Remove the banjo bolt from the primary cluster.

14- Lift out the primary cluster. Remove and discard the gasket.

15- Lift out the weight rod from the center hole on the primary side. Turn the carburetor upside down and **CATCH** the very small check ball as it falls free of the center hole. Use the proper size screwdriver and remove the seat valve for the accelerator pump fuel channel, located under the check ball which was just removed.

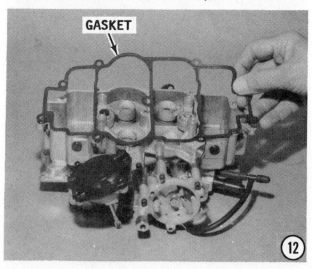

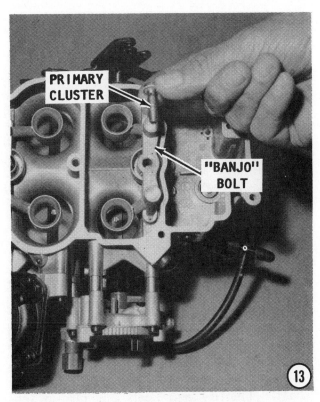

PRIMARY CLUSTER

"BANJO" BOLT

13

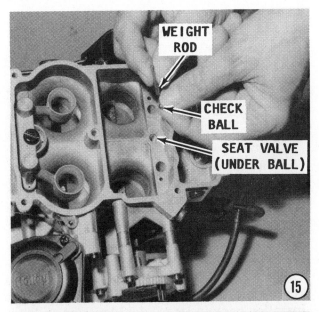

WEIGHT ROD

CHECK BALL

SEAT VALVE (UNDER BALL)

15

16- Remove the bolt from the secondary cluster. Lift out the cluster. Remove and discard the gasket. The secondary cluster **DOES NOT** have a weight rod and check ball.

17- Using the proper size slotted screwdriver, remove a total of four main jets

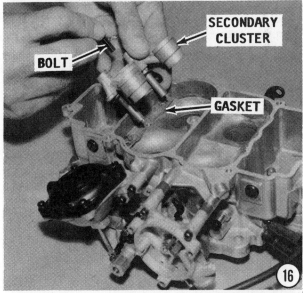

SECONDARY CLUSTER

BOLT

GASKET

16

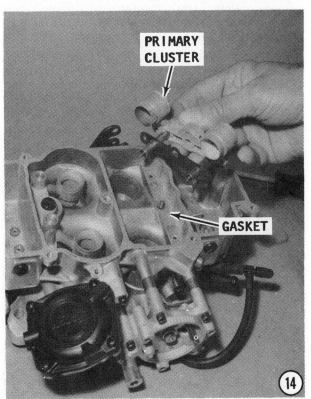

PRIMARY CLUSTER

GASKET

14

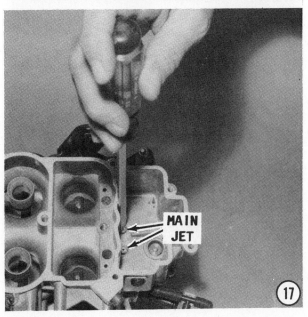

MAIN JET

17

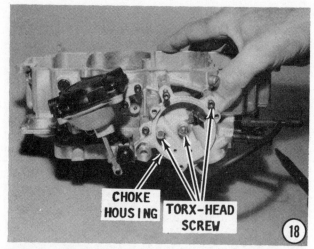

CHOKE HOUSING   TORX-HEAD SCREW

18

from the primary and secondary fuel bowls. Take **EXTRA** precaution to identify and keep the jets separate to ensure each is installed back in its original location.

18- Loosen the three long Torx-head screws securing the choke housing to the carburetor body.

19- Remove each of the Torx-head screws and be sure to **SAVE** the small spacer used on each bolt, as shown.

20- Disconnect the vacuum hose from the fitting under the primary fuel bowl. Set the choke housing with vacuum hose attached, to one side.

WARNING

The circlip to be removed in the next step is very small and under considerable tension. Therefore, use utmost caution and cover the circlip with one hand while removing the clip with a pair of needlenose pliers.

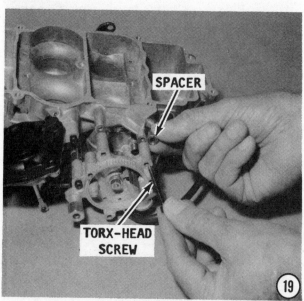

SPACER

TORX-HEAD SCREW

19

VACUUM HOSE

20

21- Using a small screwdriver, pry the circlip free of the secondary diaphragm shaft. Identify and keep this circlip separate from the clip removed earlier in Step 5.

22- Remove the four Torx-head screws securing the cover to the diaphragm housing. Lift off the cover and spring.

23- Lift out the secondary diaphragm from the housing.

24- Turn the carburetor over and **CATCH** the small check valve ball as it falls free of the fifth hole in the housing.

25- Remove one of the idle adjustment needles, and spring as shown, from the side of the primary fuel bowl.

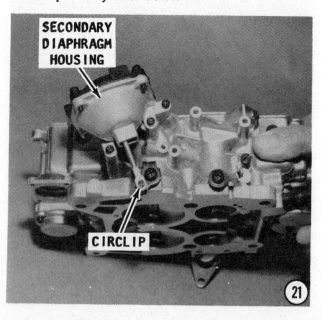

SECONDARY DIAPHRAGM HOUSING

CIRCLIP

21

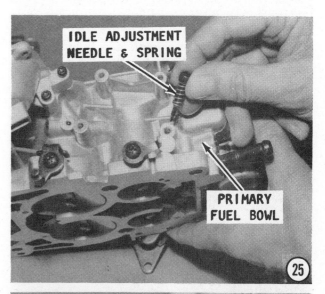

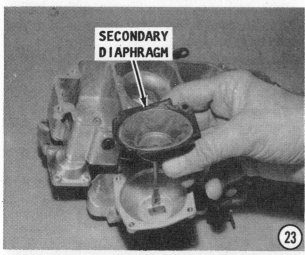

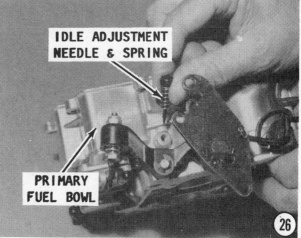

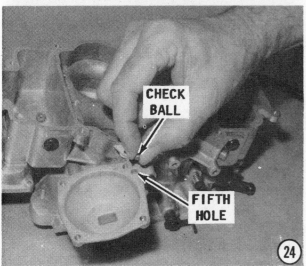

26- Remove the other idle adjustment needle and spring from the other side of the primary fuel bowl.

SPECIAL WORDS

The primary and secondary power valves are identical. Each is removed in the same manner as described in the next three steps.

POWER VALVE

COVER

GASKET

(28)

POWER VALVE

GASKET (HIDDEN)

(29)

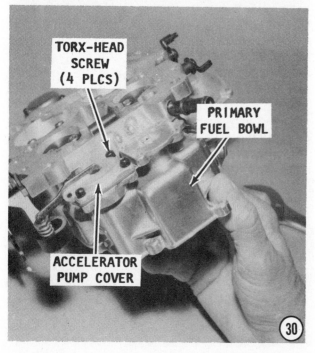

TORX-HEAD SCREW (4 PLCS)

PRIMARY FUEL BOWL

ACCELERATOR PUMP COVER

(30)

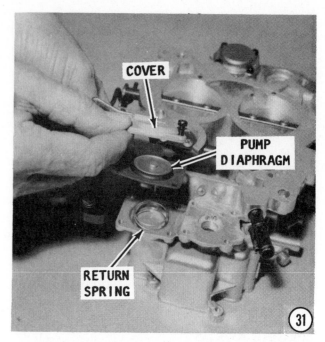

COVER

PUMP DIAPHRAGM

RETURN SPRING

(31)

27– Remove the four Torx-head screws securing the power valve cover to the carburetor body.

28– Lift off the cover, and then remove and discard the gasket.

29– Using the correct size wrench, unscrew the power valve from the body. Remove and discard the gasket. The power valve **CANNOT** be disassembled even though diassembly appears possible. The valve is replaced as a complete unit.

30– Remove the four Torx-head screws securing the accelerator pump cover to the base of the primary fuel bowl.

31– Lift off the cover. Withdraw the pump diaphragm and return spring.

32– Using a small screwdriver, lift up an edge of the check valve diaphragm, and then

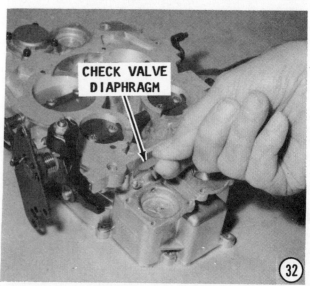

CHECK VALVE DIAPHRAGM

(32)

FUEL PIPE CONNECTION

33

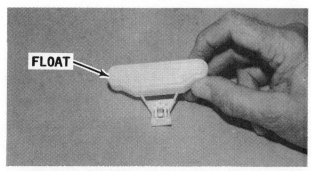

FLOAT

Check the float for any evidence of leakage by shaking it and listening for the sound of fluid movement inside.

pull the diaphragm out by hand. Handle with **CARE,** because the diaphragm can be easily damaged.

33- Using a 3/4" open end wrench, remove the fuel pipe connections on the airhorn.

34- Pull out the gasket and filter screen from the primary inlet and the secondary inlet on the airhorn.

CLEANING AND INSPECTING

NEVER dip rubber parts, plastic parts, or diaphragms in carburetor cleaner. Place all of the metal parts in a screen-type tray and dip them in carburetor solvent until they appear completely clean, then blow them dry with compressed air.

Blow out all of the passages in the castings with compressed air. Check all of the parts and passages to be sure they are not clogged or contain any deposits. **NEVER**

use a piece of wire or any type of pointed instrument to clean drilled passages or calibrated holes in a carburetor.

Inspect the choke and throttle shafts for excessive wear and replace the complete carburetor if the throttle shaft is worn.

Check the floats for leaking by shaking them and listening for fluid movement inside. If a float contains fluid, it must be replaced. Inspect the seat valve contact surface and the float shaft and replace them if either has any grooves worn in it.

ALWAYS replace the valve seat assemblies. These parts receive the most wear in the carburetor and the proper fuel level cannot be maintained if they are worn.

Check the secondary vacuum diaphragm by first depressing the diaphragm stem, and then placing a finger over the vacuum fitting to seal the opening - the diaphragm housing must be removed from the carburetor to perform this test and completely assembled. Now, release the diaphragm stem. If the stem moves out more than 1/16" in ten seconds, the diaphragm has an internal leak and it should be replaced.

Most of the parts that should be replaced during a carburetor overhaul, including the latest updated parts are found in a carburetor kit available at your local marine dealer. In addition to the parts, most kits include the latest specifications which are so important when making adjustments.

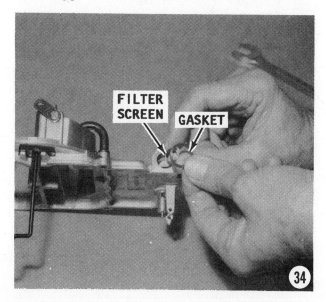

FILTER SCREEN   GASKET

34

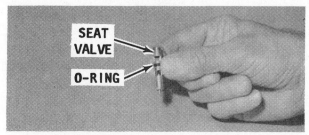

SEAT VALVE

O-RING

Proper fuel level in the fuel bowls cannot be maintained if the valve seat assemblies are worn.

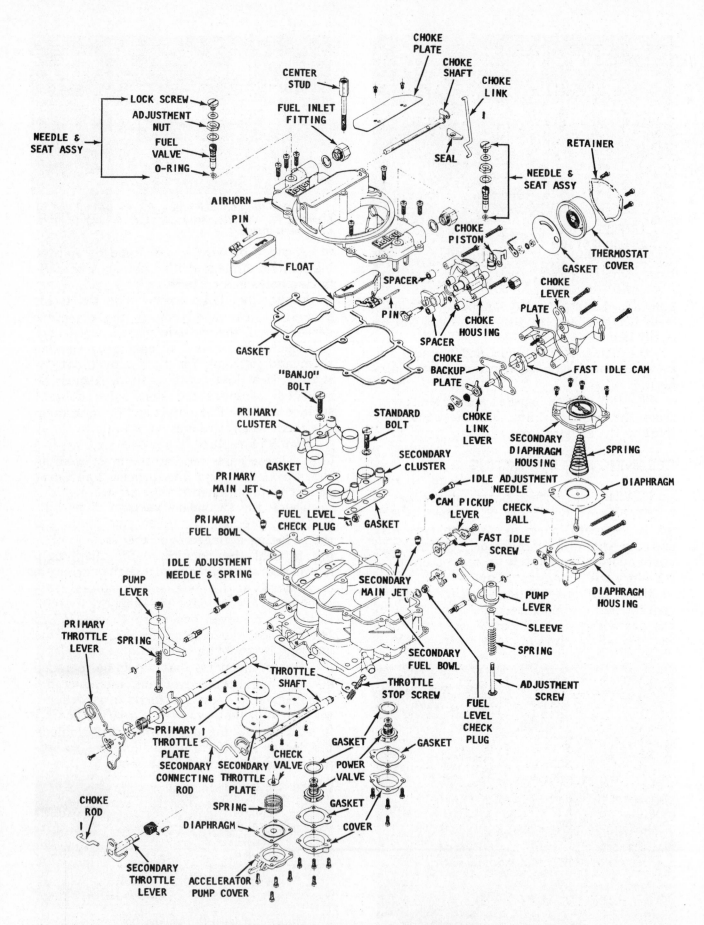

Exploded drawing of the Holley Model 4011 carburetor with major parts identified.

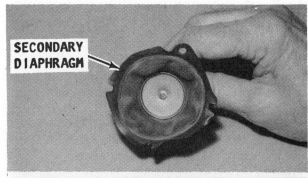

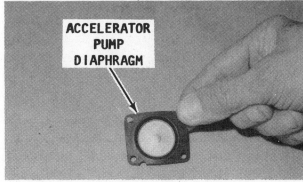

*Check the secondary diaphragm (above) and the accelerator pump diaphragm (below), for pin holes or tears. A damaged diaphragm **MUST** be replaced.*

ASSEMBLING HOLLEY 4011

1- Slide the primary filter screen into the airhorn opening. Position a **NEW** gasket in place over the opening. Install the secondary filter screen and gasket.

2- Install the two fuel pipe connections and tighten them securely.

3- Insert the end of the check valve diaphragm into the primary fuel bowl.

4- Hold the diaphragm in place with a finger and at the same time turn the carburetor body over. Now, using a pair of needlenose pliers, pull the protruding end of

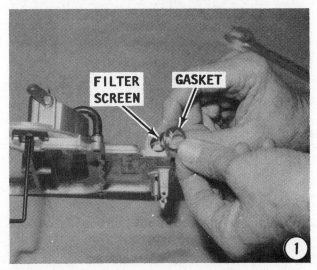

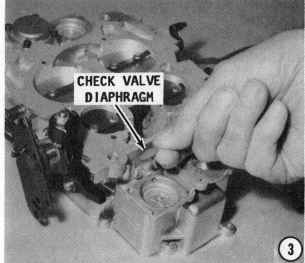

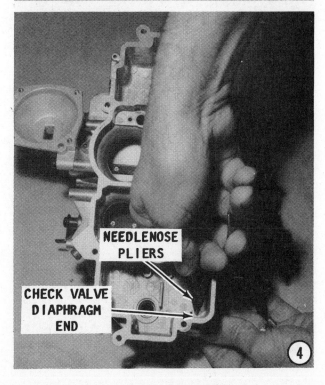

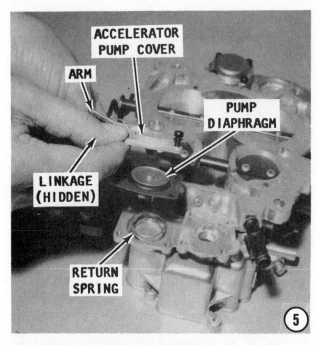

ACCELERATOR PUMP COVER

ARM

PUMP DIAPHRAGM

LINKAGE (HIDDEN)

RETURN SPRING

5

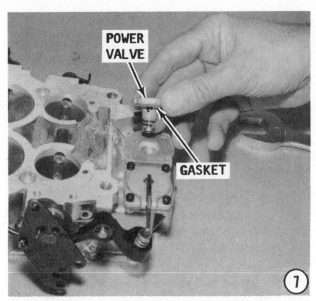

POWER VALVE

GASKET

7

the check valve diaphragm up until the diaphragm snaps into place.

5- Position the return spring in place over the diaphragm. Place the pump diaphragm over the spring with the tab on the diaphragm facing **DOWN**. Install the accelerator pump cover over the pump diaphragm with the pump arm extending towards the linkage, as shown.

6- Secure the cover in place with the four Torx-head screws.

7- Install one of the power valves, using a **NEW** gasket. Tighten the valve to a torque value of 8 ft lbs (11 Nm). Install the other power valve in the same manner.

8- Install the power valve covers using **NEW** gaskets. Tighten the Torx-head screws alternately and securely.

9- Slide a spring onto one of the idle adjustment needles and then thread the needle into the side of the primary fuel bowl

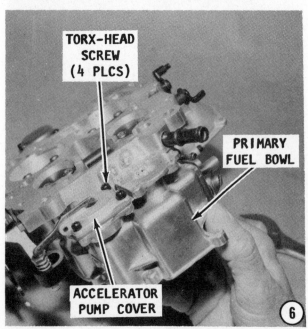

TORX-HEAD SCREW (4 PLCS)

PRIMARY FUEL BOWL

ACCELERATOR PUMP COVER

6

POWER VALVE

COVER

GASKET

8

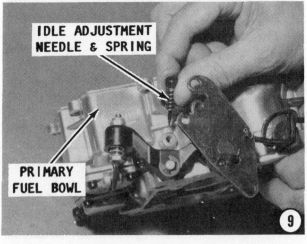

IDLE ADJUSTMENT NEEDLE & SPRING

PRIMARY FUEL BOWL

9

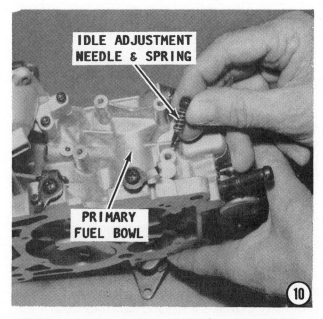

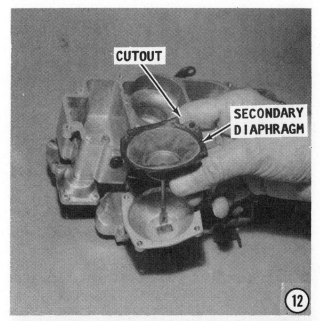

until it barely seats, and then back it out 3/4 turn (counterclockwise) as a preliminary adjustment at this time.

10- Slide a spring onto the other idle adjustment needle; thread it into the other side of the primary fuel bowl until it barely seats; and then back it out 3/4 turn.

11- Drop the check ball into the fifth hole of the secondary diaphragm housing.

12- Insert the secondary diaphragm into the housing. Check to be **SURE** the hole in the diaphragm aligns with the fifth hole in the housing.

13- Position the spring over the diaphragm with the taper facing **UP**. Install the diaphragm cover over the spring with the hole next to one of the cover shoulders aligned with the fifth hole in the housing, as shown in the accompanying illustration. Only one shoulder has a hole next to it.

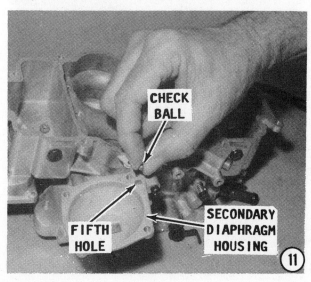

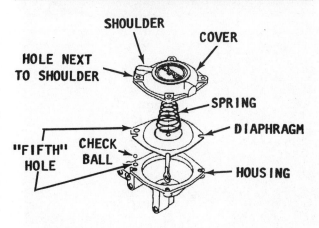

Exploded drawing of the secondary diaphragm housing with major parts including the "fifth" hole identified.

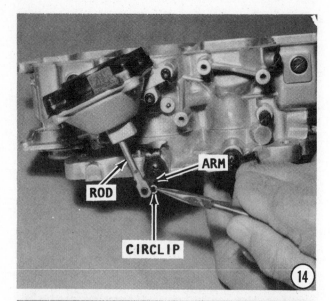

ROD   ARM

CIRCLIP

14

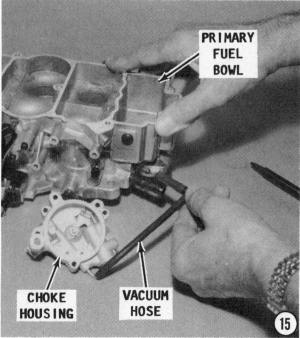

PRIMARY
FUEL
BOWL

CHOKE
HOUSING

VACUUM
HOSE

15

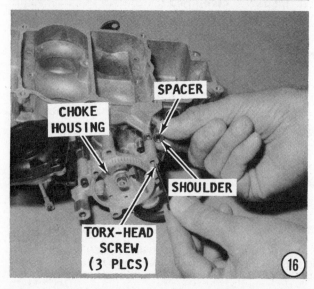

SPACER

CHOKE
HOUSING

SHOULDER

TORX-HEAD
SCREW
(3 PLCS)

16

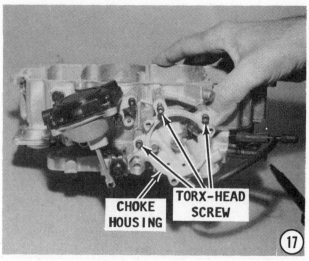

CHOKE
HOUSING

TORX-HEAD
SCREW

17

14- Guide the diaphragm rod over the arm on the secondary throttle shaft and secure it in place with the small circlip.

15- Push the choke housing vacuum hose onto the fitting under the primary fuel bowl.

16- Lightly secure the choke housing to the carburetor body with the three long Torx-head screws. A spacer is used at each mounting boss and installed with the shoulder on the spacer facing the choke housing.

17- Check behind the choke housing to be sure the choke actuating rod is above the fast idle cam. Tighten the three Torx-head screws securely.

18- Install the primary and secondary main jets into their original locations. Standard primary jets are embossed with the number **"64"** and standard secondary jets are embossed with the number **"68"**.

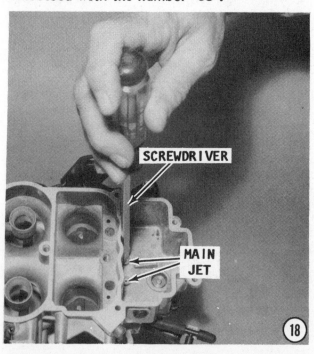

SCREWDRIVER

MAIN
JET

18

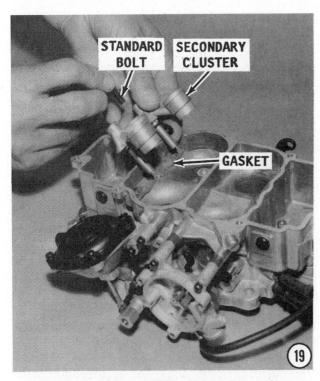

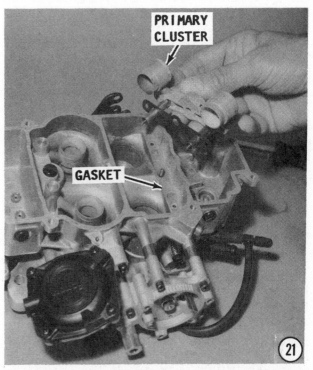

19- Position a **NEW** gasket into the secondary fuel bowl. Lower the secondary cluster into place. The secondary cluster usually has no "shooter" holes in the chamfered edge of the hole for the bolt. The primary cluster will have "shooter" holes in this location. Install and tighten the securing bolt. This bolt is normally black and solid to identify it from the primary bolt which is usually silver and of the "banjo" type - hollow, with holes at the top to allow fuel to pass through the bolt. No check ball and weight rod is used on the secondary side under the cluster.

20- Use the proper size screwdriver and install the seat valve for the accelerator pump fuel channel, located at the base of the threaded hole in the center of the primary fuel bowl. Drop the check ball followed by the weight rod down the banjo bolt hole in the primary fuel bowl.

21- Position a **NEW** gasket in the primary fuel bowl. Lower the primary cluster into place over the gasket.

22- Install and tighten the silver banjo bolt securing the primary cluster to the carburetor.

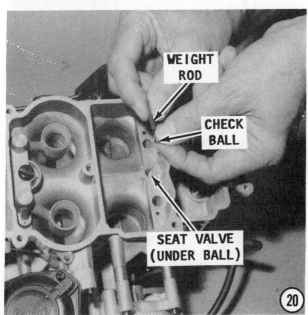

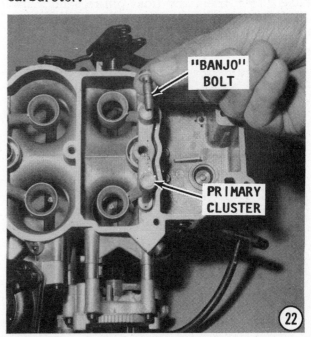

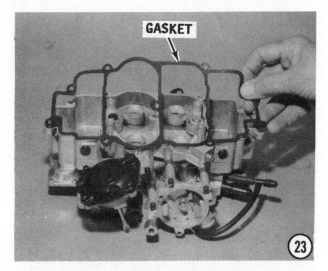

23- Install a **NEW** gasket onto the carburetor body. This gasket can only be installed one way.

24- Lubricate the **NEW O**-ring on each seat valve with engine oil. Install the valves into the primary and secondary sides of the airhorn using the adjustment nut as a tool. The precise depth to which each valve is threaded into the airhorn can only be determined during engine operation. However, a preliminary adjustment is covered in Carburetor Adjustments, Step 2 following the assembling procedures. The seat valve must be temporarily installed to a depth enabling the float - installed in the following step -to hang freely and parallel to the surface of the airhorn.

25- Position one of the floats between the airhorn bosses. Using a pair of needle-nose pliers insert the float pin to secure the float between the two mounting bosses. Center the pin. Install the other float in the same manner.

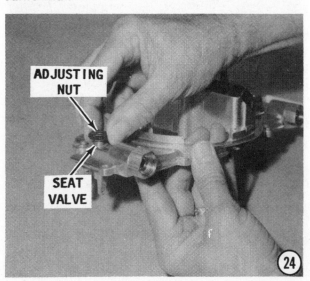

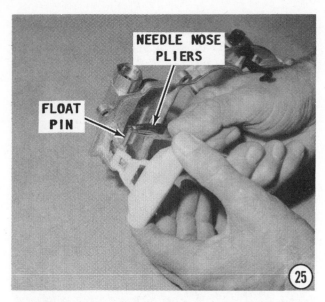

26- Invert the airhorn and allow both floats to hang freely. The upper surface of each float should be parallel to the horizontal surface of the airhorn, as shown. Adjust the depth of the seat valve threaded into the airhorn to obtain the proper alignment for each float.

27- Place **NEW** gaskets above and below the adjustment nut. The larger gasket is

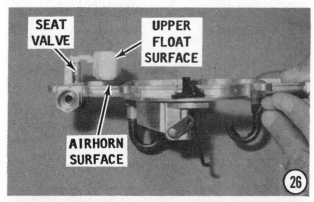

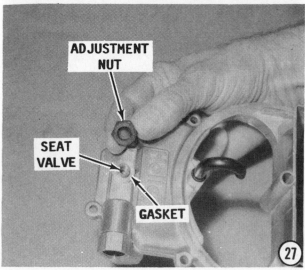

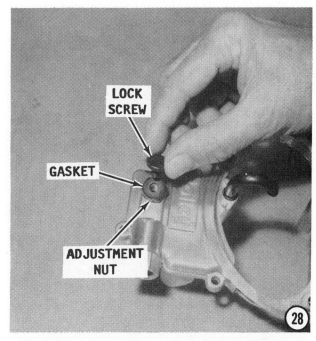

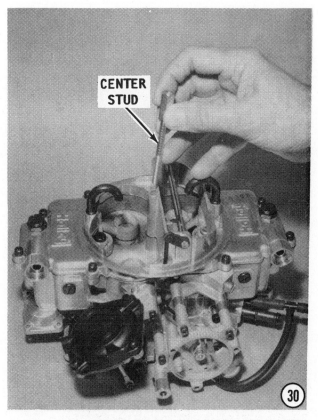

used below the nut. Install the adjustment nuts, one on each seat valve.

28- Use the correct size wrench to prevent the adjustment nut from rotating, and install the lock screw into the seat valve. **TAKE CARE** not to disturb the seat valve setting. Install the other adjustment nut in the same manner. Repeat Step 26 on the previous page, to check the float settings before proceeding with the work.

29- Lower the assembled airhorn down over the fuel bowls. Guide the choke link rod through the hole in the top of the carburetor body and into place close to the choke linkage. Make sure the lower end of the choke rod faces outward, away from the carburetor.

30- Install the center stud through the airhorn and tighten it securely.

31- Using a small screwdriver, guide the lower end of the choke link through the hole in the choke lever.

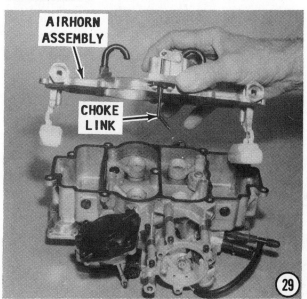

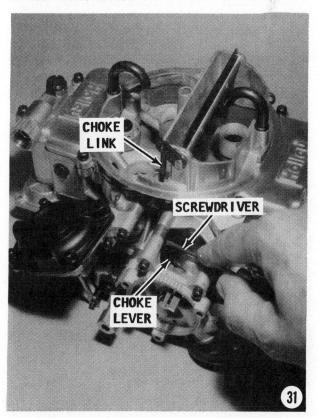

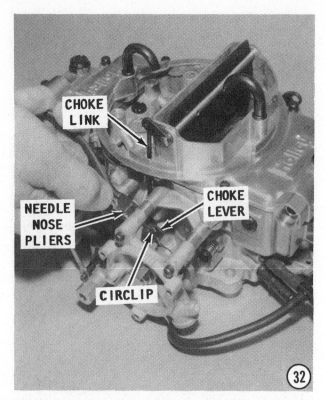

32- Install the small circlip securing the choke link to the choke lever.

33- Install the choke thermostat cover into the choke housing using a **NEW** gasket. The small tang attached to the choke piston must index into the loop on the thermostat.

34- Place the retainer over the choke cover with the convex side against the cover.

35- Install, but **DO NOT** tighten the three Torx-head screws securing the retainer to the housing, at this time. A choke adjustment follows.

CARBURETOR ADJUSTMENTS
MODEL 4011 — FOUR BARREL

CRITICAL WORDS

Carburetor adjustments **MUST** be performed in the sequence outlined in the following steps. Unfortunately, this means the carburetor is installed on the intake manifold; the engine is operated at idle speed while some adjustments are made; the carburetor is removed for other adjustments; and finally, the carburetor is installed a second time.

For best performance and economy results, follow the specifications included in the carburetor repair kit.

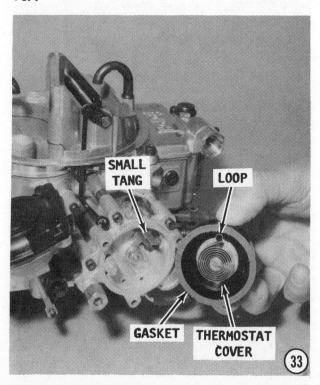

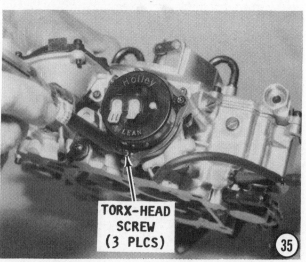

Automatic Choke Adjustment

1- The automatic choke is set at the factory to give maximum performance under all weather conditions. If necessary, the choke can be returned to its original position by aligning the index mark with the proper specification. Loosen the three Torx-head screws around the retainer. Rotate the choke cover to align the mark on the cover with the sixth mark from the left on the choke housing. Tighten the three Torx-head screws.

A richer or leaner mixture can be obtained for the warm-up period, if desired, by rotating the choke cover. **NEVER** set the index mark on the cover more than two marks from the specified setting.

Fuel Level Adjustment

2- Start the engine and allow it to warm to full operating temperature.

CAUTION: Water must circulate through the lower unit to the engine any time the engine is run to prevent damage to the water pump mounted on the engine.

Place a shop towel beneath the level plug to catch any spilled fuel. With the engine approximately level and running at idle speed, remove the level plug on the primary fuel bowl. The fuel level in the bowl must be in line with the threads at the

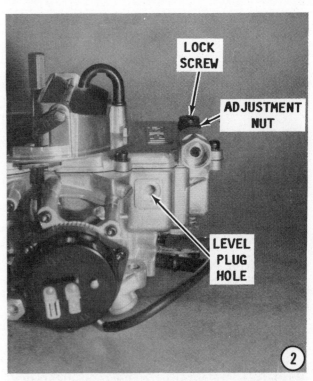

bottom of the plug. To adjust the level: Loosen the lock screw and turn the adjusting nut clockwise to lower the fuel level and counterclockwise to raise the fuel level. A 1/16 turn of the nut will equal approximately 1/16 inch difference in fuel level. After the adjustment has been made, tighten the lock screw and install the level plug. Repeat this procedure for the secondary fuel bowl.

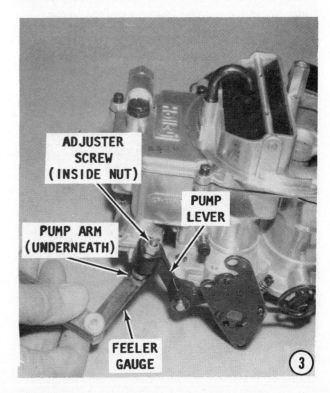

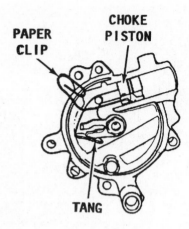

PAPER
CLIP

CHOKE
PISTON

TANG

④

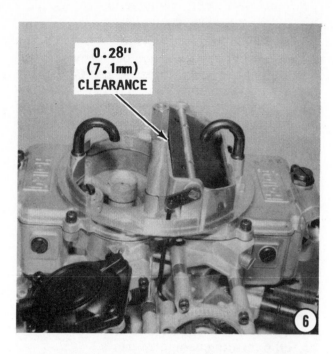

0.28"
(7.1mm)
CLEARANCE

⑥

Accelerator Pump Lever Clearance

3- First, move the throttle to the wide-open position. Hold the pump lever down. Insert a 0.015" gauge between the bolt head under the pump lever and the pump arm. Loosen the top locknut and rotate the pump adjuster screw -- a slot cut through the bolt threads -- to obtain the correct clearance. There must be **NO FREE PLAY** with the pump lever at idle.

Choke Control Lever Adjustment

4- Open out a small size paper clip and bend over the last 1/8" (3mm) to form a right angle. Insert the bent end onto the choke piston and hook it onto the piston edge. Pull out the clip and piston as far as possible. Next, close the choke valve with a light pressure on the choke control lever.

5- Insert a 1/4" (7mm) drill bit between the upper edge of the choke valve and the inner wall of the air horn. **CAREFULLY** bend the indicated tang (shown in illustration No. 4), with needlenose pliers until the required measurement is obtained.

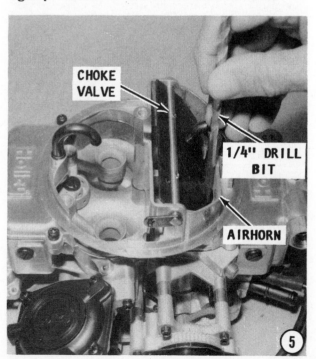

CHOKE
VALVE

1/4" DRILL
BIT

AIRHORN

⑤

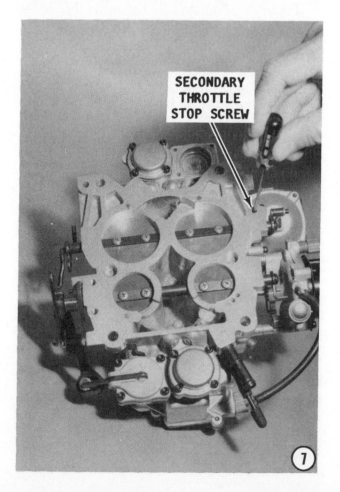

SECONDARY
THROTTLE
STOP SCREW

⑦

Choke Unloader Adjustment

6- Hold the throttle in the wide-open position and at the same time insert a 1/4" drill bit (7mm) between the upper edge of the choke valve and the inner wall of the air horn. **CAREFULLY** bend the tang, located on the choke lever under the choke housing, using needlenose pliers until the required measurement is obtained.

Secondary Throttle Valve Stop Adjustment

7- This adjustment is made with the carburetor removed from the manifold. First, back out the secondary throttle stop screw until the valves are fully closed. Next, turn the stop screw in until it barely contacts the secondary throttle stop lever. Now, turn the adjusting screw in 1/8 turn.

Secondary Throttle Valve Opening Adjustment

8- Set the secondary lever to the idle position. Measure the clearance, to the right, between the link arm and the oval hole. This clearance shuld be no more than 0.020" (0.5mm). Bend the link arm if necessary.

Idle Speed and Mixture Adjustment

9- Turn each idle adjustment needle - one on each side of the primary fuel bowl -

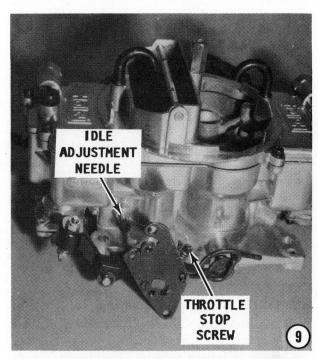

in, until it barely seats, and then back it out 3/4 turn. **TAKE CARE** not to seat the needles tightly, because it would groove the tip of the needle. If the needle should become grooved, a smooth idle cannot be obtained. Start the engine and allow it to warm to full operating temperature.

CAUTION: Water must circulate through the lower unit to the engine any time the engine is run to prevent damage to the water pump mounted on the engine.

Adjust the throttle stop screw on the throttle valve lever until the tachometer indicates 750 rpm.

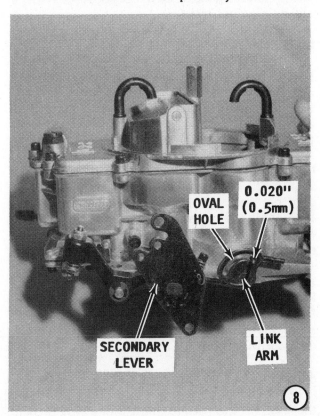

Holley Model 4011 4-barrel carburetor ready for service.

4-13 ELECTRONIC FUEL INJECTION (EFI) DESCRIPTION AND OPERATION

From 1987 to 1989, some GMC V8 engines were equipped with an Electronic Fuel Injection (EFI) system. This fuel distribution system is computer controlled and includes:

The fuel tank.
An electric fuel pump (replacing any type of mechanical fuel pump).
A fuel pressure regulator.
Two injector fuel rails -- one for each bank of cylinders and each supporting four fuel injectors.
Eight fuel injectors -- four mounted on each rail -- one for each cylinder and each a mini solenoid.
An Electronic Control Unit (ECU) -- actually a "black box".
Seven different sensors providing information to the ECU on:
  Engine rpm.
  Throttle setting.
  Manifold air temperature.
  Manifold pressure -- automatically adjusts to barometric pressure.
  Oil pressure.
  Fuel pressure.
  Coolant temperature.

GENERAL INFORMATION

The advantages of a fuel injection system over a carbureted fuel system are:

a- More efficient engine operation.
b- Thorough atomization of the injected mixture.
c- Increased fuel economy.
d- Fuel efficiency -- complete burning of the injected mixture.
e- Better "volumetric efficiency" (the ability of the engine to breathe air), due to the absence of a venturi.
f- Less harmful exhaust gases.

The first electronically controlled fuel injection system was introduced over 50 years ago -- in 1932 on diesel truck engines. Today, state-of-the-art microprocessors (commonly referred to as "computer chips") have lowered the cost of electronically controlled fuel injection systems. The price of EFI is now very close to the cost of modern carbureted systems.

EFI SYSTEM DESCRIPTION

The type of fuel injection used on some late model GMC V8 engines, is called "Indirect Multi-Port Fuel Injection", because the fuel is injected into the intake manifold, before entering the combustion chamber. By design, the method of injection is also termed "Port Tuned Injection" -- each port has minimum and equal restriction to ensure all ports pass the same amount of air into the cylinders.

A computer housed in the Electronic Control Unit (ECU) accepts data from a number of sensors. Based on the information received, the ECU signals the fuel injectors to inject a precise and correct amount of fuel.

A fuel injection system must provide the correct air/fuel ratio for all engine loads, rpm, and temperature conditions.

Manifold Pressure

Under normal operating conditions, an ideal air/fuel ratio is maintained. When the operator places a power demand on the engine by advancing the throttle, intake manifold pressure is reduced. Conversely, a reduction in intake manifold pressure, indicates an additional load has been placed on the engine. This added demand requires a richer air/fuel mixture.

ECU (Electronic Control Unit)

The ECU computes the new ideal air/fuel ratio and the fuel injectors deliver the correct amount of fuel. When the engine load demands stabilize, the ECU again receives data from the sensors and readjusts the ratio to the new stabilized condition.

Fuel Pump & Routing

An EFI engine is equipped with an electric fuel pump. Fuel drawn from the fuel tank, first passes through a fuel filter and then a water separating filter before reaching and passing through the pump.

A fuel pressure regulator, set at the factory, restricts fuel pressure to 39psi. From the pump, fuel is then routed to the fuel rails, one for each cylinder bank, where it is distributed to the injectors, one for each cylinder, and finally into the cylinder.

Air Temperature Sensor

An air temperature sensor is located on the side of the intake manifold. Two Brown leads connect the sensor to the ECU, (one lead junctions with a Black lead from the air valve to the ECU). The sensor measures the ambient air temperature and conducts this information in the form of an electrical signal to the ECU. As the air temperature changes, the amount of oxygen per cubic foot also changes. The quantity of available oxygen has an affect on combustion and therefore must be taken into account when computing the ideal air/fuel ratio.

Auxiliary Air Valve

An auxiliary air valve is mounted at the rear of the top cover. This valve monitors and controls air volumn to the engine, at engine start and during warmup. The auxiliary air valve is connected between the ECU and the water temperature sensor and fuel pressure sensor -- together. At engine temperatures below 120°F (66°C), the coolant temperature sensor signals the ECU to provide the engine with additional fuel. Not only is additional fuel needed to start a cold engine, but at low temperatures additional air is also required to increase power output at idle speeds and to overcome the increased friction of a cold engine. Therefore, the enriched air and fuel supply will provide a steady idle speed and a smooth warmup period.

The correct starting procedure for an engine equipped with EFI, is to rotate the ignition key to the **ON** position and **WAIT** for a full minute before advancing the key to the **START** position. This one minute delay gives the auxiliary air valve time to reduce the amount of air in the system prior to engine start and to allow the fuel pressure in the fuel rails to rise to at least 25 psi.

Manifold Absolute Pressure (MAP) Sensor

A MAP sensor is located inside the ECU and connected to the intake manifold by a vacuum hose. The MAP sensor is a flexible type resistor. This sensor monitors changes in altitude and barometric pressure. As the pressure changes, the resistor flexes and its resistance and the voltage applied across the sensor changes. This change is registered at the ECU.

Two conditions can affect the pressure in the manifold. The first, and most common condition, is a reduction in manifold pressure when load on the engine is increased. As mentioned earlier under "Manifold Pressure", when the operator places a power demand on the engine by advancing the throttle, intake manifold pressure is reduced. Conversely, a reduction in intake manifold pressure, indicates an additional load has been placed on the engine.

The second condition which may affect the manifold pressure is operation at high altitudes. The MAP sensor compensates for both barometric pressures and operation at high altitudes.

Coolant Temperature Sensor

The coolant temperature sensor is located in the cylinder head. This sensor is a thermistor -- an electronic device which functions in the opposite manner of a resistor.

A resistor increases resistance with temperature (decreases voltage), with an increase in temperature. A thermistor varies resistance (increases or decreases voltage), with a change in temperature. When this voltage is received by the ECU, the information is used to determine injector pulse widths and spark advance. The temperature information is also used to determine if an extra charge of fuel is necessary for a cold engine. Engine temperature information is used by the ECU to assist in starting a cold engine by automatic fuel enrichment. In this manner, a choke is not required. Once the engine has reached operating temperature, enrichment is no longer required.

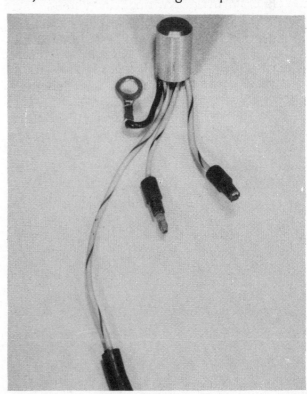

An EFI coolant temperature sensor.

The engine temperature sensor has a Green and a Purple lead. Indirectly, both leads connect back to the ECU.

Throttle Position Sensor

The throttle position sensor is mounted on a throttle shaft and **MUST** have the two securing bolts "SAFETY WIRED" to prevent the bolts loosening due to vibration. The sensor is an encased potentiometer sending a signal to the ECU indicating engine load for a specific rpm. Engine load is influenced by the load in the boat, by the propeller type and size used with the stern drive unit, performance demands imposed by the operator, and other operational factors.

The body of the sensor is stationary with a small shaft emerging from the center of the sensor. The shaft is connected to the throttle valve. As the throttle is advanced, movement is transferred to the sensor and the resistance changes. Therefore, the variable voltage signal sent to the ECU is directly proportional to the throttle position.

The operation of this sensor can be compared to the volume knob on a radio. If the listener has a favorite position for the knob, in time a "flat spot" will wear on the shaft and interference may result.

If the operator of the stern drive has a favorite throttle position, in time a "flat spot" will wear on the sensor shaft and not provide the ECU with accurate information. This condition could lead to ignition timing which varies by as much as 20° with no change in throttle position and incorrect operation of the EFI.

The throttle position sensor has three leads -- a Tan, a Light Blue and an Orange lead. All three leads are connected to the ECU.

An EFI throttle position sensor.

CRITICAL WORDS ON SENSORS AND SENSOR LEADS

The sensors in an EFI system send signals to the ECU via wires of predetermined length, material and therefore, resistance. These signals consist of minute changes in either resistance or voltage. Any tampering with either the wire length or any kind of splicing will alter the resistance and send a misleading signal to the ECU. Even sensors which send voltage signals are affected by resistance changes, according to Ohm's law.

An incorrect signal will result in corrective action by the ECU. This change will be a richer or leaner mixture to compensate for an erroneously indicated physical condition. The culprit actually is a man made change in resistance! Such a created problem is not easy to trace even by the most experienced of factory trained mechanics. Therefore, heed these words: **DO NOT** tamper with sensor wires.

Fuel Injectors

The eight fuel injectors, one for each cylinder, force fuel under pressure into the intake manifold. Each injector, four on each rail, is mounted under the fuel rail, next to an intake valve.

The injectors used in this type fuel injection system are solenoid operated. Each consists of a valve body, a needle valve, and valve seat. A small voltage is sent from the ECU to each injector. When this voltage is applied to the windings of the solenoid, a magnetic field is induced around the needle valve. The valve lifts off its seat and fuel is allowed to pass between the needle valve and the needle seat. Because the fuel is pressurized, a spray emerges from the injector nozzle.

The time interval for the injector to be open and emitting fuel is called the "pulse width". The actual "pulse width" for the injector is controlled by the ECU and must be measured in **MICROSECONDS**.

A small return spring seats the needle valve back onto the seat, the instant the voltage is removed.

The nozzle spray angle of each injector remains constant and is the same for all injectors.

The injectors are timed by the ECU to correspond with engine rpm, not to piston position, as in a diesel engine equipped with EFI.

Two O-rings are used to secure each injector in the fuel rail. One O-ring provides a seal between the injector nozzle and the intake manifold. The other O-ring provides a seal between the injector and the fuel inlet connection. Both O-rings prevent excessive injector vibration. These O-rings are replaceable, but at press time for this manual, a little difficult to obtain. They may only be purchased directly from the engine manufacturer or from Bosch -- the fuel injector manufacturer.

Fuel Rail

A fuel rail, one for each cylinder bank, extends the length of the intake manifold. The fuel injectors, four for each rail, are mounted to the underside of the rail. Fuel enters the rail under pressure at one end and is then evenly distributed to the injectors. A fuel pressure test point (actually a plug), is provided at the opposite end of the rail for troubleshooting purposes.

Fuel Pressure Regulator

An electro-mechanical fuel pressure regulator is connected to the fuel rail. This regulator maintains constant fuel pressure to ensure a uniform spray from the injectors into the cylinders. The electric fuel pump delivers fuel under pressure to the fuel rail. Usually an excessive amount of fuel is delivered to the injectors. When fuel pressure exceeds the regulator setting, excess fuel is routed back to the fuel tank.

The fuel regulator is pre-set at the factory to operate at 39 psi under **ALL** operating conditions.

The fuel pressure sensor has an adjustment screw preset by the factory and **MUST NOT** be altered.

Oil Pressure Sensor

An oil pressure sensor is included in the EFI package as a safety feature. If this sensor detects a dangerously low or no oil pressure, the sensor sends a signal to the ECU. When this signal is received, the ECU immediately cuts power to the electric fuel pump, thus disabling the entire EFI system.

The oil pressure sensor also has an adjustment screw preset at the factory and **MUST NOT** be altered.

Throttle Shutter

The throttle shutter is a butterfly valve mounted at the air intake port of each cylinder. This valve controls the amount of air entering the intake manifold. The valve is controlled by the operator by the throttle setting. There are eight such valves, one per cylinder and each valve opens simultaneously and equally. The throttle position sensor is mounted on the throttle shaft of one of the butterfly valves.

The throttle position sensor reports the movement of the throttle shutter valve as an electronic signal to the ECU.

Electric High Pressure Fuel Pump

An electric fuel pump replaces the mechanically operated fuel pump, and delivers the fuel to the injectors under constant pressure of 39psi.

The electric fuel pump is a roller-cell type and is **NOT SERVICEABLE.**

The accompanying illustration shows the internal parts of an electric fuel pump. **DO NOT ATTEMPT TO OPEN THE PUMP ON THE ENGINE BEING SERVICED.**

Once opened, the airtight seal is lost and the integrity of the seal **CANNOT** be regained. The pump **MUST** be replaced.

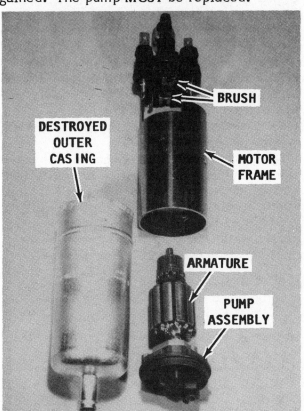

*Illustration to show a destroyed electric fuel pump. An operational pump is an air tight sealed unit and must **NEVER** be opened. Air inside the pump would create a highly volatile (explosive) atmosphere in the pump as explained in the text.*

This electric pump is unique because the pump and electric motor are housed together in one permanently sealed case, constantly surrounded with fuel. **Yes!,** the fuel actually flows past the electric motor brushes. If the case remains **ABSOLUTELY** airtight, there is **NO DANGER OF EXPLOSION.** However, if the case develops a crack, or is deformed in any way, there is the distinct possibility air may enter the case and together with the fuel form an **EXPLOSIVE MIXTURE!** Fuel, by itself will not ignite while exposed to a spark from the electric motor brushes, because there is no oxygen present for combustion. A mixture of 0.7 parts air to 1.3 parts fuel becomes explosive.

Pump action will commence the moment the ignition key switch is rotated to the **ON** position and will continue as long as there is adequate oil pressure in the engine. If the oil pressure sensor detects a low or no oil condition, the ECU will immediately cut power to the pump to avoid damage to the engine.

CRITICAL SAFETY WORDS

The engine must never be cranked without an adequate supply of fuel to the electric fuel pump. If the electric pump draws air into the pump case, an explosive mixture may then form. On the other hand, a fuel injection system cannot be flooded. The fuel pressure regulator will activate the return circuit, and excess fuel will not flood the engine as in a conventional carbureted engine.

The correct starting procedure for an engine equipped with EFI, is to rotate the ignition key to the **ON** position and **WAIT** for a full minute before advancing the ignition key to the **START** position. This time delay gives the auxiliary air valve time to purge air from the system before engine start.

Water Separating Filter

A canister type filter is installed between the fuel tank and the electric fuel pump. As the name suggests, this filter separates water from the fuel. To comply with U.S.C.G. regulations, the electric fuel pump and the water separator **MUST** be mounted within 12" (30cm) of the engine. The presence of water in the fuel will alter the proportion of air/fuel mixture to the "lean" side, resulting in a higher operating temperature. If the condition is not cor-

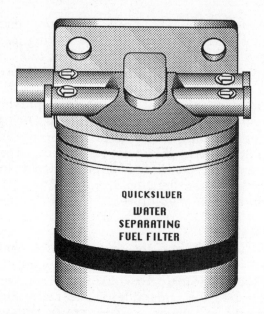

*The water separating fuel filter is a disposable canister type filter. Notice the four fuel connectors. Fuel lines **MUST** be hooked up correctly as explained in the text.*

rected, serious damage could be caused to the pistons. This filter may be installed incorrectly due to the four fittings on the filter top. Refer to Page 4-2 for installation instructions.

RPM Sensor

An engine rpm sensor is housed within the ECU. Electronic pulses from the ignition coil are received by the rpm sensor. This signal indicates engine rpm. A Purple lead connects the ignition coil to the ECU

Electronic Control Unit (ECU)

The fuel injection system is controlled by the ECU -- an onboard computer mounted as far from heat and vibration as possible. The computer is a sealed "black box" type unit and is in no way serviceable. The ECU receives signals from seven sensors on the engine. From the signals received, the ECU determines the amount of fuel to be injected. This computer also determines the timing of the spark at the spark plugs.

The amount of fuel injected is determined by how long each injector nozzle remains open -- the "pulse width". The nozzle opens and closes in response to signals from the ECU.

The ECU receives two types of input signals.

Analog signals change with changing conditions. For example: the coolant temperature sensor will have more electrical resistance when the engine is cold, than when the engine is hot.

Digital signals are a series of on and off pulses. These pulses are counted by the computer to determine a condition. For example: the signal sent from the ignition coil will provide the computer with information on engine rpm.

Sensors located on the engine or inside the computer housing provide information to the computer on engine load, rpm, temperature and other conditions affecting operation.

The computer is **programmed** or provided with instructions, to produce correct air/-fuel mixtures and throttle openings for varying conditions.

"Limp Home" Mode

The computer is programed to accept signals from sensors within a predicted range. If one sensor should fail and send a signal which is totally out of the acceptable range, the computer recognizes this signal as a sensor failure and not as a true reading of engine condition. The computer then goes into what the manufacturer terms the "limp home" mode. The engine will not respond to an increase in throttle. The computer will continue to monitor all sensor input. However, the computer will shut down the engine if a determination is made that a harmful condition exists, if engine operation is continued.

4-12 TROUBLESHOOTING EFI

If a problem is encountered with engine startup or operation, the determination must be made: the problem is in the ignition system or the fuel system is at fault.

The first and easiest test to determine an ignition problem is to perform a spark test as outlined in Chapter 5. If the ignition system checks out and the preliminary troubleshooting narrows the area to the fuel system, proceed with the following fuel system tests.

VERY IMPORTANT WORDS

The following tests were designed for the individual working with a Volt/Ohm/Ammeter only. All dealerships will have a Quicksilver Fuel Injection Tester. Seloc strongly suggests a certified marine mechanic be contacted to check a suspected EFI electrical component. The part may be removed and bench tested in the shop or left installed on the engine and tested in a shop having professional test equipment. Proper testing and accurate diagnosis is necessary and **CRITICAL** before purchasing an expensive **NON RETURNABLE** electronic component, especially a **NEW ECU.**

Installed Injector Test

1- Disconnect the ignition coil high tension lead from the distributor center tower. Obtain a jumper lead and ground the coil lead to a suitable spot on the engine. Grounding the ignition coil is necessary to prevent the engine from accidentally starting while tests are being performed and also to protect the coil from trying to arc to ground. Keep the coil grounded with the jumper cable in place while performing Steps 2 and 3, then remove the cable.

2- Rotate the ignition key to the **ON** position, and then wait for a full minute. Move the switch to the **START** position and crank the engine through several revolutions. Rotate the key switch back to the **OFF** position. Remove the spark plugs one by one and inspect the electrode end. A damp electrode will indicate the injector is functioning for the cylinder being checked.

If all spark plugs are damp, indicating the presence of fuel, all injectors are spraying fuel into their respective cylinders. If some spark plugs are damp with fuel, while others are dry, the electric fuel pump is probably functioning correctly. However, either the injectors of the "dry" cylinders have a restricted fuel flow **OR** the dry injectors have an electrical problem. If one fuel rail exhibits damp spark plugs while the other rail is "dry", the problem can be isolated to a clogged fuel rail, the Red/Yellow lead, the Red lead, the Red lead connection to the starter slave solenoid, the Yellow/Red lead, the Purple/White lead, or the ECU.

Clean the connection between the Red lead and the starter slave solenoid, then using a ohmmeter check each lead for continuity. If everything checks out OK, the ECU, still installed, should be checked at your local Mercury dealer using the proper test equipment, before a replacement is purchased.

If all spark plugs are dry -- none are damp with fuel, a problem with fuel delivery exists. The problem may be:

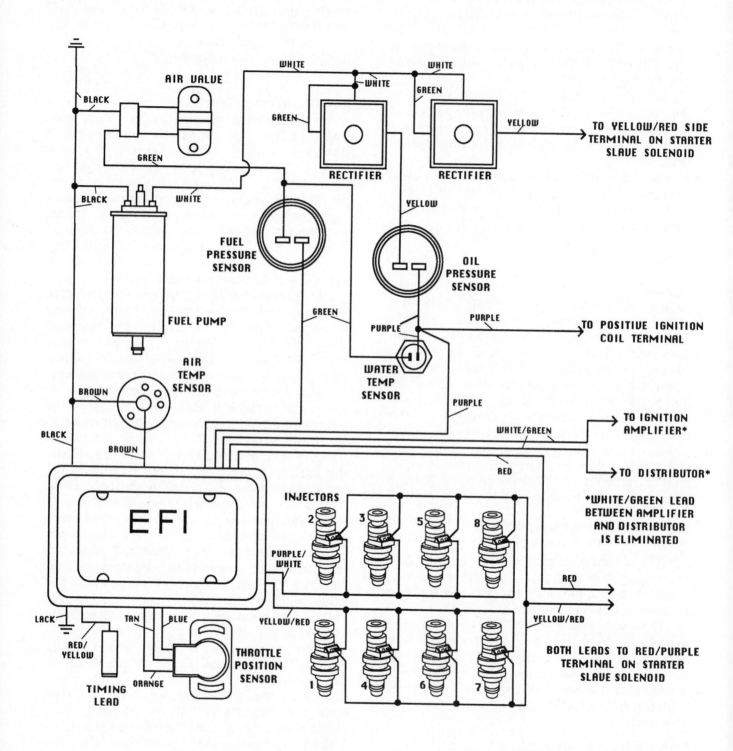

A wiring diagram for the EFI System. This diagram will be most helpful when trouble shooting electrical connections and components as explained in the text.

Main fuel line blockage.

Faulty electrical component affecting operation of the entire system, such as

· Defective electric fuel pump.

Defective harness plug.

Defective ECU component.

Voltage to Electric Fuel Pump Test

FIRST, THESE WORDS

The electric fuel pump is supplied with power from the starter slave solenoid, independant of the ECU. The electric fuel pump is **GROUNDED** through the ECU. Therefore, if the pump fails to operate because of a defective ground, the grounding lead to the ECU, or the ECU is at fault.

3- Leave one end of the jumper cable connected to a good ground on the engine. Disconnect the other end of the cable from the ignition coil. This connection was made in Step 1. Connect the high tension lead from the ignition coil back into the distributor center tower. Now, connect the free end of the jumper cable to the negative terminal on the electric fuel pump -- the terminal with a Black lead still attached.

CAUTION

Water must circulate through the stern drive, also to and from the engine, anytime the engine is run to prevent damage to the water pump located in the lower unit (original and Alpha drives); or on the front of the engine (Bravo drive). Just a few seconds without water will damage the water pump.

Attempt to start the engine. If the electric fuel pump now operates and the engine starts, the problem is a bad ground connection to the pump. Unfortunately this is provided by the ECU. To obtain a satisfactory ground and still retain all the necessary safety features provided by the ECU, check the Black lead grounding the ECU to the engine. If this grounding connection is good, the reader must decide if replacement of the ECU is warranted to retain the safety features the unit provides **OR** if the temporary jumper cable will be replaced with a new permanent ground lead and all safety features are to be sacrificed.

If the engine did not start, remove the temporary jumper cable from the electric fuel pump terminal and connect it back to the high tension lead from the ignition coil as directed in Step 1.

Obtain a voltmeter. Select the 12V DC scale. Rotate the ignition key to the **ON** position, with the jumper cable connected to the ignition coil still in place.

Make contact with the Black meter lead to a suitable ground on the engine throughout the following series of checks.

a- Make contact with the Red meter lead to the **SIDE** terminal on the starter slave solenoid with the Yellow/Red lead attached.

CAUTION

There is also a front terminal with a Yellow/Red lead attached. Be sure to use the **SIDE** terminal.

The meter should register 12 volts. If the meter registers 12 volts proceed to test **b**.

If the meter fails to register 12 volts, the problem lies in one or more of the following areas: the battery, cables, connections, harness plug, or ignition switch.

b- Trace the Yellow lead from the starter slave solenoid to one of the two rectifiers. Make contact with the Red meter lead to the terminal on the rectifier with the Yellow lead attached. The meter should register 12 volts.

If the meter registers 12 volts proceed to test **c**.

If the meter fails to register 12 volts, the problem lies in the Yellow lead or the connection of the lead to the solenoid terminal.

c- Make contact with the Red meter lead to the rectifier terminal with the White lead attached. The meter should register 12 volts.

If the meter registers 12 volts proceed to test **d**.

If the meter fails to register 12 volts, the rectifer is defective.

d- Make contact with the Red meter lead to the positive terminal (terminal with White lead attached), on the pump. The meter should register 12 volts.

If the meter registers 12 volts the electric fuel pump is receiving the correct voltage. If the pump still fails to operate when the ignition coil is reconnected and the ground circuit is bypassed, as explained in the beginning of this step, the pump is defective and must be replaced. The pump is **not** a servicable item.

If the meter fails to register 12 volts, either the White lead or the lead connections between the pump and the rectifier is at fault.

Testing Electric Fuel Pump Operating Pressure

4- Obtain a pressure gauge capable of registering 50psi. Remove the plug from the pressure port located above the fuel pressure regulator. Connect the gauge to the pressure port. Rotate the ignition key to the **ON** position and **WAIT** for one full minute, then crank the engine for about 15 seconds, with the jumper cable still in place grounding the ignition coil. Note the reading on the gauge. Normal engine operation requires pressure of 39psi (268.9kPa).

If the reading is low, the cause may be a restriction in a fuel line. If there is no reading, the pump is defective and must be replaced. The pump is a sealed unit and **CANNOT** be serviced.

Testing Injector Wiring Harness

5- Disconnect both the White and Black leads at the fuel pump, to prevent fuel being pumped during this test. Position the White "hot" lead in such a manner to prevent it from making contact with any part of the engine. Be sure the jumper cable is still in position on the ignition coil to prevent the engine from cranking.

6- To test for voltage to the injectors, the injector harness connector plug must be removed from each injector at the fuel rail.

Once the connector has been removed from the injector, obtain a voltmeter and set the scale to 12V DC.

A fuel injector may be tested using a 9 volt battery as explained in the text.

Rotate the ignition key to the **ON** position with the jumper cable still in place on the ignition coil.

Connect the Red meter lead to the Red-/Yellow injector lead and connect the Black meter lead to the Yellow/Red or Purple/-White lead at each injector in turn.

The meter should register at least 9V at each injector.

If voltage is present at each injector harness plug the injector harness is satisfactory.

If one or more injectors is not receiving the correct voltage, the ECU or the injector harness may be at fault. Remove the ECU and take the "Black Box" to the dealer for testing with a Quicksilver Fuel Injection Tester before purchasing a new unit.

Reconnect the White and Black leads at the electric fuel pump.

The jumper cable, temporarily grounding the ignition coil may now be removed.

Testing Fuel Injectors

7- Follow the instructions given on Page 4-74 to remove the injectors from the fuel rails. Place each injector on the workbench in sequence as it is removed for identification purposes. Once the injectors are unplugged from their wiring harness and removed from the fuel rail, a simple test with a 9V battery can be performed to determine if the solenoid inside the injector is functioning properly.

Obtain a 9V battery, such as those used in radios and calculators. Obtain two jumper leads with small alligator clips at each end.

Momentarily connect the battery across the two terminals inside the injector connector, making sure the two leads do not contact each other. For these tests the jumper cables may be connected in either way, because polarity does not affect the tests.

As soon as contact is made, the injector should emit a sound as the solenoid is energized and the needle is pulled off its seat. When contact is broken, the injector should again emit a sound as the needle returns under spring pressure back to its original position.

Perform this test on each injector in turn. If any injector fails this test, there is an internal defect in the injector. The needle is stuck or there is bad electrical connection. The injector must be replaced.

A quick test to determine which of the two listed reasons prevent the injector from working properly is to perform a resistance test across the two injector terminals using a **DIGITAL** ohmmeter set to the Rx1 scale. Ohmmeter leads across the two injector terminals should register a reading of 1.1 ohm.

A clogged injector may possibly be cleaned by purchasing a fuel additive from the local automotive parts supply store and following the directions on the can.

Air Temperature Sensor Resistance Check

The air temperature sensor need not be removed from the engine for testing purposes.

8- Disconnect the two Brown leads from the air temperature sensor at their quick disconnect fittings. Obtain an ohmmeter and select the Rx1000 ohm scale. Make contact with the two meter leads across the two Brown leads. The meter should register approximately 8, (in reality 8,000 ohms). If the meter registers zero or infinity, the sensor should be replaced. If the meter registers in the kilo-ohm range, the sensor is probaby okay.

Coolant Temperature Sensor
Resistance Test

9- Identify the Green and Purple leads from the coolant temperature sensor located in the head. Obtain an ohmmeter and select the Rx1000 ohm scale. Make contact with the two meter leads across the two sensor leads. The meter should register approximately 1, (in reality 1,000 ohms). If the meter registers zero or infinity, the sensor should be replaced. If the meter registers in the kilo-ohm range, the sensor is probably satisfactory.

Throttle Position Sensor
Resistance Test

10- Mark the original location of the throttle position sensor **BEFORE** performing any tests on the sensor which may lead to disturbing its original alignment. If this sensor is disturbed during testing or service procedures, a **DIGITAL** type voltmeter is needed to correctly reset the sensor on its mounting bracket, because voltages in the 1/10 range must be accurately read. A misaligned sensor could send misleading signals to the ECU and consequently affect the fuel delivery and ignition timing.

Start the engine and allow it to run at idle speed until warmed to operating temperature.

CAUTION
Water must circulate through the stern drive, also to and from the engine, anytime the engine is run to prevent damage to the water pump located in the lower unit (original and Alpha drives); or on the front of the engine (Bravo drive). Just a few seconds without water will damage the water pump.

DO NOT remove or change the throttle sensor position during the following test. Disconnect the wire harness from the sensor consisting of Orange, Blue and Tan leads at the harness connector. Obtain an ohmmeter and select the Rx1000 scale. Make contact with the Red meter lead to the Orange sensor terminal. Make contact with the Black meter lead to the Tan sensor terminal. The meter should register between 800 and 1200 ohms at idle speed. Retain the Red meter lead in place, and move the Black meter lead to the Blue sensor terminal. The meter should register the same reading.

The lower the resistance reading, the leaner the air/fuel mixture. If the engine is overheating at idle speeds, perhaps due to an excessively lean mixture, the problem may lie in the throttle position sensor. Conversely, if the engine is smoking at idle speeds, due to an overrich mixture, again the problem may lie in a misaligned throttle position sensor.

4-13 EFI MAINTENANCE

Preventative maintenance of the EFI system includes checking, repairing or replacing:

Fuel filter and water separating filter.

Loose, corroded, or grounded electrical connections.

Loose, or improperly supported fuel rails, lines, and injectors.

Replacing "old" fuel with "new". Even with the high price of fuel, removing gasoline that has been standing unused over a long period of time, is still the easiest and least expensive preventative maintenance possible. In most cases this old gas can be used without harmful effects in an automobile using regular gasoline.

LEADED GASOLINE AND GASOHOL

The manufacturer of the units covered in this manual normally recommends using either regular unleaded or regular leaded gasoline having a minimum octane rating of 86 or higher for all carbureted engines. However, the manufacturer recommends the use of premium unleaded or regular leaded gasoline with a minimum octane rating of 87 or higher for those engines equipped with EFI.

In the United States, the Environmental Protection Agency (EPA) has slated a proposed national phase-out of leaded fuel, "Regular" gasoline, by 1990. Lead in gasoline boosts the octane rating (energy). Therefore, if the lead is removed, it must be replaced with another agent. Unknown to the general public, many refineries are adding alcohol in an effort to hold the octane rating.

Alcohol in gasoline can have a deteriorating effect on certain fuel system parts. Seals can swell, pump check valves can swell, diaphragms distort, and other rubber or neoprene composition parts in the fuel system can be affected.

Since about 1981, all manufacturers have made every effort to use materials that will resist the alcohol being added to fuels.

Fuels containing alcohol will slowly absorb moisture from the air. Once the moisture content in the fuel exceeds about 1%, it will separate from the fuel taking the alcohol with it. This water/alcohol mixture will settle to the bottom of the fuel tank. The engine will fail to operate. Therefore, storage of this type of gasoline for use in marine engines is not recommended for more than just a few days.

One temporary, but aggravating, solution to increase the octane of "unleaded" fuel is to purchase some aviation fuel from the local airport. Add about 10 to 15 percent of the tank's capacity to the unleaded fuel.

4-14 SERVICING EFI

FIRST, THESE SAFETY WORDS

Before loosening any fuel fitting on the fuel rails or electric fuel pump, position the wrench in place over the fitting and wrap a shop towel around both the wrench and fitting. The towel will absorb any fuel which may spray from the fitting when it is first cracked open. The fuel injection sys-

tem operates under a constant pressure of 39 psi. When the engine is shut down, the fuel pressure, after a few minutes, gradually drops down to atmospheric pressure. However, some residual pressure may remain in the system. Therefore the use of a shop towel will help contain the spraying fuel.

Removal

1- Obtain some masking tape. Wrap a piece of tape around each fuel line on both fuel rails. Identify the fuel lines as an assist in later assembling. Disconnect all fuel lines from both fuel rails, taking adequate precautions with a shop towel against spraying fuel, as described in the paragraph above.

2- Remove the four long bolts securing each fuel rail to the manifold and remove the supports between pairs of bolts.

3- Use the masking tape once more to identify each injector harness lead, so they may be installed in their original positions. Disconnect the harness plugs from the injectors.

4- SLOWLY pull each injector straight up and free of the manifold. The injectors are held in the manifold only by rubber sealing rings. These rings may harden or become gummy, depending on use, making removal of the injectors difficult.

Observe the injector harness connector. Some models may have a wire retaining clip securing the connector to the injector. Other models have no wire retainer, the connector simply pulls off from the injector.

If servicing a unit equipped with wire retaining clips: Use a small slotted screwdriver and pry the retaining clip from the injector harness connector.

All models: Disconnect the harness from the injector. Place the injectors on the work bench in sequence for identification.

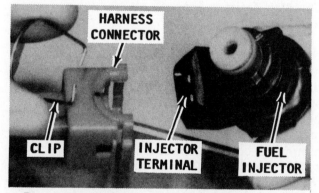

Disconnecting the harness plug on a fuel injector. This type injector uses a wire retaining clip.

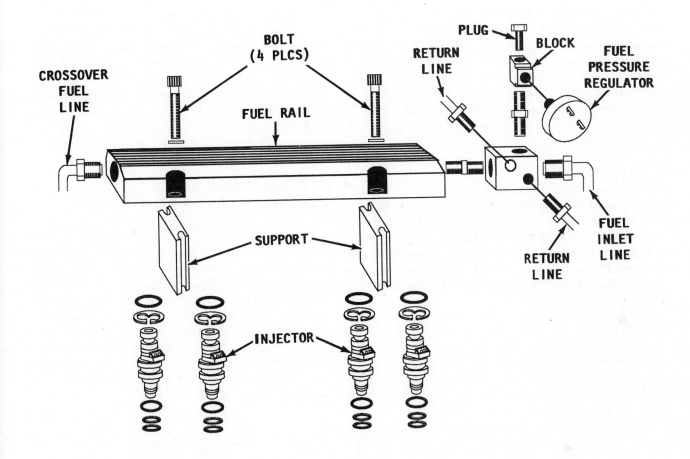

Functional diagram to depict the EFI fuel rail and fuel injectors on a single bank of the engine.

CLEANING AND INSPECTING

A simple test on the internal fuel injector solenoids may be performed as described in Step 7 on Page 4-72.

Two **O**-rings are used to secure each injector in the fuel rail. One **O**-ring provides a seal between the injector nozzle and the intake manifold. The other **O**-ring provides a seal between the injector and the fuel inlet connection. Both **O**-rings prevent excessive injector vibration. These **O**-rings are replaceable and are included in an injector overhaul kit.

Inspect the sensors as far as possible without disturbing them. Check the wire harnesses for signs of chafing, cracks, or corroded connections.

INSTALLATION

1- Obtain Insulating Compound P/N 92-41669-1 and apply this compound to the injector terminals.

If servicing a model equipped with wire retaining clips: Position the retaining clip around the injector wire harness. Push the injector harness connector firmly over the injector terminal. The harness can only be connected one way. Slide the clip into place to secure the harness connector to the injector.

All other models: Push the injector harness connector firmly over the injector terminal. The harness can only be connected one way.

2- Apply a light coating of clean engine oil to the entire surface area of the **NEW O**-rings. The use of engine oil on these **O**-rings will prevent galling. Push the injector into place to seat firmly into the intake manifold. Repeat Step 1 and this step for all injectors.

3- Work on one fuel rail at a time. Position the supports in place between the bolt holes on the manifold. Lower the fuel rail down over the installed injectors and guide each injector top into its respective hole in the fuel rail. Pass the long bolts and washers through the fuel rail and supports and into the intake manifold. Once all four bolts are in place, tighten alternately and evenly to a torque value of 30 in lb (3Nm).

4- Coat the threads of all fuel lines with Loctite Pipe Sealant with Teflon. **DO NOT** wrap the threads with Teflon tape. Install all fuel lines to the fuel rail and tighten them securely.

CAUTION

Water must circulate through the stern drive, also to and from the engine, anytime the engine is run to prevent damage to the water pump located in the lower unit (original and Alpha drives); or on the front of the engine (Bravo drive). Just a few seconds without water will damage the water pump.

Start the engine and check for fuel leaks.

CRITICAL WORDS

The fuel system will be pressurized almost instantly -- as soon as the engine is cranked. Watch for leaks around fuel line fittings and injector O-rings. Because the system is pressurized, a leak will not appear as drips. Instead, fuel will be **SPRAYED** all over the engine. Should this occur, shut down the engine immediately. **DO NOT** forget the system is pressurized. Therefore precautions must be taken before making necessary repairs.

SPECIAL TIMING WORDS
ENGINES EQUIPPED WITH EFI

Identify the Red/Yellow lead at the ECU. Normally this lead is not connected to anything. However, when timing an engine equipped with EFI, this Red/Yellow lead **MUST** be grounded to the engine block. Failure to ground this lead will produce a retarded timing adjustment. After the timing procedure has been completed this lead **MUST** be disconnected and the lead protected to prevent accidental contact (grounding) on the engine block.

5
IGNITION

5-1 DESCRIPTION

STANDARD IGNITION SYSTEM

Engine performance and efficiency are, to a large degree, governed by how fine the engine is tuned to factory specifications as determined by the designers. The service work outlined in this chapter must be performed in the sequence given and to the Specifications listed in the Appendix.

The ignition system consists of a primary and a secondary circuit. The low-voltage current of the ignition system is carried by the primary circuit. Parts of the primary circuit include the ignition switch, ballast resistor, neutral-safety switch, primary winding of the ignition coil, contact points in the distributor, condenser, and the low-tension wiring.

The secondary circuit carries the high-voltage surges from the ignition coil which result in a high-voltage spark between the electrodes of each spark plug. The secondary circuit includes the secondary winding of the coil, coil-to-distributor high-tension lead, distributor rotor and cap, ignition cables, and the spark plugs.

When the contact points are closed and the ignition switch is on, current from the battery or from the alternator flows through the primary winding of the coil, through the contact points to ground. The current flowing through the primary winding of the coil creates a magnetic field around the coil windings and energy is stored in the coil. Now, when the contact points are opened by rotation of the distributor cam, the primary circuit is broken. The current attempts to surge across the gap as the points begin to open, but the condenser absorbs the current. In so doing, the condenser creates a sharp break in the current flow and a rapid collapse of the magnetic field in the coil. This sudden change in the strength of the magnetic field causes a voltage to be induced in each turn of the secondary windings in the coil.

The ratio of secondary windings to the primary windings in the coil increases the voltage to about 20,000 volts. This high voltage travels through a cable to the center of the distributor cap, through the rotor to an adjacent distributor cap contact point, and then on through one of the ignition wires to a spark plug.

When the high-voltage surge reaches the spark plug it jumps the gap between the insulated center electrode and the grounded side electrode. This high voltage jump across the electrodes produces the energy required to ignite the compressed air/fuel mixture in the cylinder.

The entire electrical build-up, breakdown, and transfer of voltage is repeated as each lobe of the distributor cam passes the rubbing block on the contact breaker arm, causing the contact points to open and close. At high engine rpm operation, the number of times this sequence of actions take place is staggering.

BALLAST RESISTOR

Beginning at the key switch, current flows to the ballast resistor and then to the positive side of the coil. When the resistor is cold its resistance is approximately one ohm. The resistance increases in proportion to the resistor's rise in temperature.

While the engine is operating at idle or slow speed, the cam on the distributor shaft revolves at a relatively slow rate. Therefore, the breaker points remain closed for a slightly longer period of time. Because the points remain closed longer, more current is allowed to flow and this currrent flow heats the ballast resistor and increases its resistance to cut down on current flow thereby reducing burning of the contact points.

During high rpm engine operation, the reduced current flow allows the resistor to cool enough to reduce resistance, thus increasing the current flow and effectiveness of the ignition system for high-speed performance.

The voltage drops about 25% during engine cranking due to the heavy current demands of the starter. These demands reduce the voltage available for the ignition system. In order to reduce the problem of less voltage, the ballast resistor is by-passed during cranking. This releases full battery voltage to the ignition system.

SHIFT CUTOUT SWITCH

The shift cutout switch is connected between the primary side of the ignition coil and ground. This switch is normally open. The function of this switch is to ground the ignition system during a shift to neutral. By grounding the ignition system, gear pressure is released and the shift is made much easier. Obviously, if the ignition system is grounded, the cylinders will not fire during this period. In actual practice only a few cylinders fail to fire and it is usually not noticeable. The shift cutout switch is mounted on the transom and is activated by the remote control shift cable.

TAKE NOTE: If this switch is not adjusted properly, or if it shorts out (is grounded) then the primary side of the ignition coil will be grounded and the engine will not start.

IGNITION TIMING

In order to obtain the maximum performance from the engine, the timing of the spark must vary to meet operating conditions. For idle, the spark advance should be as low as possible. During high-speed operation, the spark must occur sooner, to give the air/fuel mixture enough time to ignite, burn, and deliver power to the piston for the power stroke.

Manual setting and centrifugal advance are the two methods of obtaining the constantly changing demands of the engine. The manual setting is made at idle speed. This setting allows the contact points to

open at a specified position of the piston in the same manner as with conventional ignition systems. Idle timing must be set below speeds of 900 rpm.

The ignition amplifier provides for ignition advance. After initial timing, the advance starts at 1000 rpm and increases to a maximum of 24 degrees advance between 3600 and 3800 rpm. If the key is left in the **ON** position when the engine is at rest, the coil primary current will automatically turn off in a short time to protect the ignition coil.

The ignition amplifier will withstand a reverse battery connection for only about a minute.

THUNDERBOLT IV IGNITION SYSTEM

The Thunderbolt IV is a battery powered High Energy Ignition (HEI) system. The system is quite different from the standard ignition system discussed in the previous paragraphs. This ignition system is used on the MCM200, 230, 300, 320 (with EFI), 330, 370, 420, 575, 898R, 228R and 260R, the 5.0 Litre, the 5.7 Litre and the 350 and 454 Magnum. This ignition system is also used on the MIE230, 260, and 340 engines. The system does not utilize breaker points or a mechanical advance mechanism. It was designed to be almost maintenance free without the requirement for periodic adjustment.

In addition to a conventional-type distributor shaft, housing, cap, and rotor, the system consists of four major components, namely an ignition coil, ignition amplifier, ignition sensor, and a sensor wheel. A 12-volt source of power is used for operation.

The sensor wheel is mounted near the top of the distributor shaft under a conventional rotor. The ignition sensor is a device that acts similar to a switch when subjected to a magnetic field. The sensor is mounted on the inside of the distributor body and used to trigger the ignition. This sensor senses the toothed sensor wheel as the wheel rotates. Action of the sensor is not a factor of speed and it will trigger at almost zero rpm. Sensor triggering accuracy is not affected by trigger air gap or end play in the distributor shaft. The distributor rotor and cap distribute current to the proper spark plug and do not determine engine timing advance. Initial ignition timing is adjusted by rotating the distributor housing

NO ADVANCE   **FULL ADVANCE**

Movement of the centrifugal advance mechanism. All parts must be free to move properly or the ignition timing will not advance with engine speed and all phases of performance will suffer, as explained in the text.

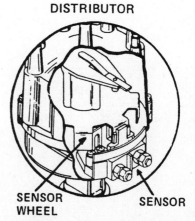

DISTRIBUTOR

SENSOR   SENSOR
WHEEL

The Thunderbolt IV ignition system distributor has a sensor and sensor wheel instead of points, as explained in the text.

in the same manner as with conventional ignition systems. Idle timing must be set below speeds of 900 rpm.

The ignition amplifier provides for ignition advance. After initial timing, the advance starts at 1000 rpm and increases to a maximum of 24 degrees advance between 3600 and 3800 rpm. If the key is left in the **ON** position when the engine is at rest, the coil primary current will automatically turn off in a short time to protect the ignition coil.

The ignition amplifier will withstand a reverse battery connection for only about a minute.

The ignition coil appears to be a standard size and shape. **HOWEVER**, the HEI coil has a special winding and core. If a standard coil is used, the ignition amplifier will not be damaged, but it will only supply a low output and will overheat.

Standard tachometers and most synchronizers monitoring ignition impulses at the negative terminal of the coil will operate satisfactorily with the Thunderbolt IV ignition system.

DIGITAL DISTRIBUTORLESS IGNITION SYSTEM

In 1990, a Digital Distributorless Ignition System (DDIS), was installed as standard equipment on the 181 CID 3.0 LX 4-cylinder in-line engine. This system was also made available as "optional" equipment for the 181 CID 3.0 L. As the name suggests, this ignition system does not have a distributor.

This ignition system is described in Section 5-11, beginning on Page 5-30.

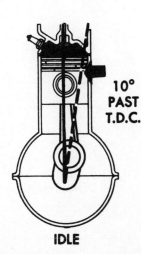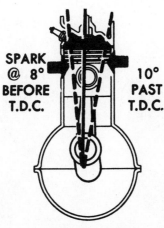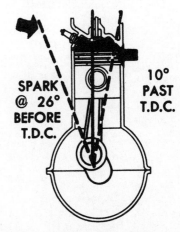

10°
PAST
T.D.C.

SPARK
@ 8°
BEFORE
T.D.C.

10°
PAST
T.D.C.

SPARK
@ 26°
BEFORE
T.D.C.

10°
PAST
T.D.C.

IDLE   **1,000 ENG. RPM**   **2,000 ENG. RPM**

The mechanical spark advance mechanism advances ignition timing when the engine speed increases, as shown in these three drawings. Maximum cylinder pressure must occur by 10 degrees ATDC, at all speeds. Therefore the spark must occur at TDC for idle speeds (left); at approximately 8 degrees BTDC for 1,000 rpm (center); and at 26 degrees BTDC for 2,000 rpm (right).

Cylinder pressure should not vary between cylinders by more then 15 psi for the engine to run smoothly. A variation between the cylinders is much more important than the actual individual readings.

5-2 IGNITION TROUBLESHOOTING

COMPRESSION
ALL MODELS

Before spending time and money seeking a problem in the ignition system, a compression check should be made of each cylinder. Without adequate compression, your efforts in the ignition system will not give the desired results.

Remove each spark plug in turn; insert a compression gauge in the hole; crank the engine several times; and note the reading.

A variation between the cylinders is more important than the actual reading. A variation of more than 20 psi indicates either a ring or valve problem.

To determine which is defective, squirt about a teaspoonful of oil into the cylinder that has the low reading. Crank the engine a few times to distribute the oil. Now, recheck the compression and note the difference from the first reading. If the pressure increased, the compression loss is past the piston rings; if no change is noted, the loss is past a burned valve.

IGNITION SYSTEM TESTS
ALL MODELS EXCEPT THUNDERBOLT IV
AND DDIS

Any problem in the ignition system must first be localized to the primary or secondary circuit before the defective part can be identified.

1- GENERAL TESTS

Disconnect the wire from the center of the distributor cap and hold it about 1/4" from a good ground.   Turn the ignition switch to **START**, and crank the engine with the starter.   If you observe a good spark, go to Test 5 (Secondary Circuit Test).   If you do not have a good spark, go to Test 2 (Primary Circuit Test).

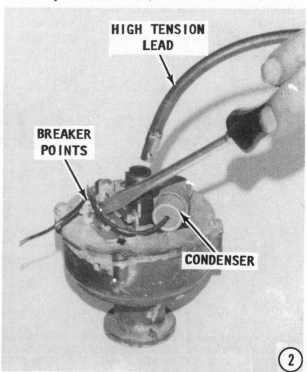

2- PRIMARY CIRCUIT TEST

Remove the distributor cap; lift off the rotor; and then turn the crankshaft until the contact points close.   Turn the ignition switch on, and open and close the contact points using a small screwdriver or a non-metallic object. Hold the high-tension coil wire about 1/4" from a good ground. If you observe a good spark jump from the wire to the ground, the primary circuit checks out. Go to Test 5, (Secondary Circuit Test). If there is no spark, go to Test 3, (Contact Point Test).

3- CONTACT POINT TEST

Remove the distributor cap and rotor. Turn the crankshaft until the points are open, and then insert some type of insulator between the points.   Now, hold the high-tension coil wire about 1/4" from a good ground, and at the same time move a small screwdriver up and down with the screwdriver shaft touching the moveable point and the tip making intermittent contact with the contact point base plate. In this manner, you are using the screwdirver for a set of contact points. If you get a spark from the high-tension wire to ground, then the problem is in the contact points. Replace the set of points. If there is no spark from the high-tension wire to ground the

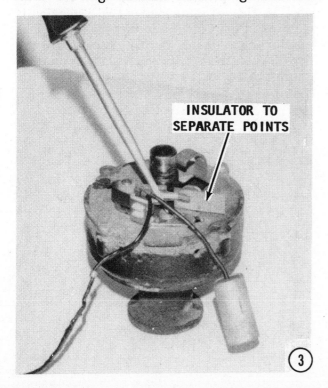

INSULATOR TO
SEPARATE POINTS

③

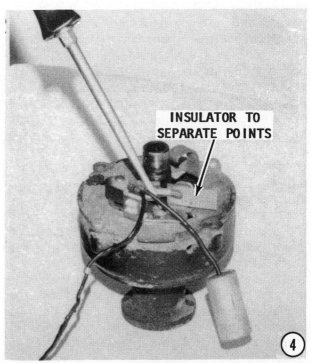

INSULATOR TO
SEPARATE POINTS

④

problem is either a defective coil or condenser.   To test the condenser, go to Test 4 (Condenser Test),

4- CONDENSER TEST

Condensers seldom cause a problem. However, there is always the possibility one may short out and ground the primary circuit. Before testing the condenser, check to be sure one of the primary wires or connections inside the distributor has not shorted out to ground.

The most accurate method of testing a condenser is with an instrument manufactured for that purpose. However, seldom is one available, especially during an emergency.   Therefore, the following procedure is outlined for emergency troubleshooting the condenser and the primary circuit insulation for a short.

First, remove the condenser from the system. TAKE CARE that the metallic case of the condenser does not touch any part of the distributor. Next, insert a piece of insulating material between the contact points. Now, move the blade of a small screwdriver up and down with the shaft of the screwdriver making contact with the movable contact point and the tip making and breaking contact with the contact point base plate. Observe for a low-tension spark between the tip of the screwdriver and the contact point base plate as you make and break the contact with the screwdriver tip. You should observe a spark during this test and it will prove the primary circuit complete through the neutral-safety switch, the primary side of the ignition coil, the ballast resistor, the shift cutout switch, and the primary wiring inside the distributor. If you have a spark, reconnect the condenser and again make the same test with the screwdriver. If you do not get a spark, either the condenser is defective and should be replaced, or the shift control switch should be adjusted or replaced.

If you were unable to get a spark with the condenser disconnected, it means no current is flowing to this point, or there is a short circuit to ground. Use a continuity tester and check each part in turn to ground in the same manner as you did at the movable contact point. If you get a spark, indicating current flow, at one terminal of the part, but not at the other, then you have isolated the defective unit.

5- SECONDARY CIRCUIT TEST

The secondary circuit cannot be tested using emergency troubleshooting procedures **UNLESS** the primary circuit has been tested and proven satisfactory, or any problems discovered in the primary ciruucit have been corrected.

If the primary circuit tests are satisfactory, use the same procedures as outlined in Test 2, Primary Circuit Test, to check the secondary circuit. Hold the high-tension coil lead about 1/4" from a good ground and at the same time, use a small screwdriver to open and close the contact points. A spark at the high-tension lead proves the ignition

coil is good. However, if the engine still fails to start and the problem has been traced to the ignition system, then the defective part or the problem must be in the secondary circuit.

The distributor cap, the rotor, high-tension leads, or the plugs may require attention or replacement. To test the rotor, go to Test 6 (Rotor Test). If you were unable to observe a spark during the secondary circuit test just described, the ignition coil is defective and must be replaced.

IGNITION VOLTAGE TESTS

Many times hard starting and misfiring problems are caused by defective or corroded connections. Such a condition can lower the available voltage to the ignition coil. Therefore, make voltage tests at critical points to isolate such a problem. Move the voltmeter test probes from point-to-point in the following order.

6- ROTOR TEST

With the distributor cap removed and the rotor in place on the distributor shaft, hold the high-tension coil lead about 1/4" from the rotor contact spring, and at the same time crank the engine with the ignition switch turned **ON**. If a spark jumps to the rotor, it means the rotor is shorted to ground and must be replaced. If there is no spark to the rotor, it means the insulation is good and the problem is either in the distributor cap (check it for cracks), in the high-tension leads (check for poor insulation or replace it), or in the spark plugs (replace them).

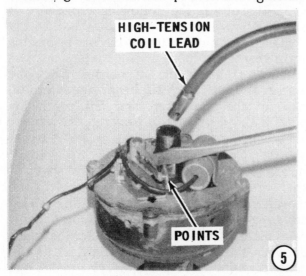

HIGH-TENSION COIL LEAD

POINTS

5

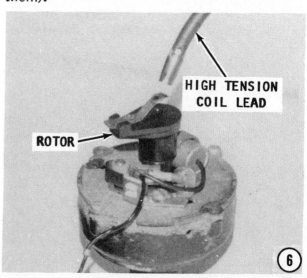

HIGH TENSION COIL LEAD

ROTOR

6

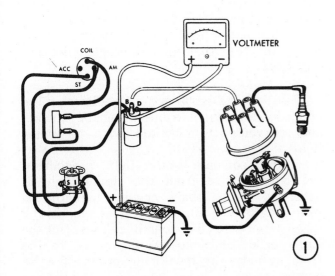

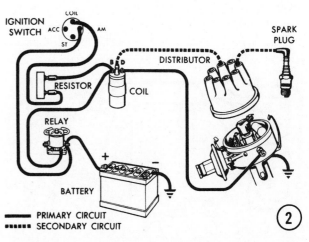

PRIMARY CIRCUIT
SECONDARY CIRCUIT

TEST 1 Voltage Loss Across Entire Ignition Circuit: Connect a voltmeter between the battery side of the ignition coil and the positive post of the battery, as shown in the Test 1 illustration. Crank the engine until the contact points are closed. Turn the ignition switch to ON. The voltage loss should not exceed 3.2 volts. This figure allows for a 0.2 loss across each of the connections in the circuit, plus a calibrated 2.4 volt drop through the ballast resistor. If the total voltage loss exceeds 3.2 volts, then it will necessary to isolate the corroded connection in the circuit of the key resistor, the wiring, or at the battery.

TEST 2 Cranking System: First disconnect the high-tension wire and ground it securely to minimize the danger of sparks and a possible fire. Next, connect a voltmeter between the battery side of the ignition coil and ground, as shown in Test 2 illustration. Now crank the engine and check the voltage. A normal system should have a reading of 8.0 volts. If the voltage is lower, the battery is not fully charged or the starter is drawing too much current.

TEST 3 Contact Points and Condenser: Measure the voltage between the distributor primary terminal and ground, as shown in Test 3 illustration. Crank the engine until the contact points are closed. Turn the ignition switch to the ON position. The voltage reading must be less than 0.2 volts. A higher reading indicates the contact points are oxidized and must be replaced. To check the condenser, crank the engine until the contact points are open, and then take a voltage reading. If the reading is not equal to the battery voltage, the condenser is shorted to ground. Check the condenser installation or replace the condenser.

"R" TERMINAL ON STARTER SOLENOID

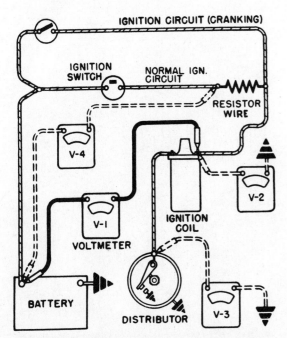

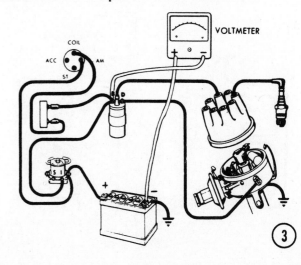

Functional diagram showing voltmeter hookups to test the ignition system.

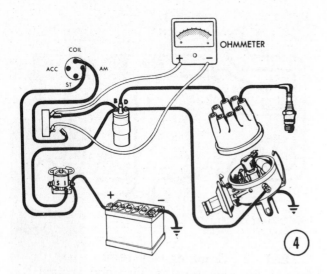

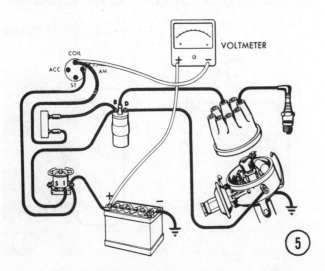

TEST 4 Primary Resistor: Disconnect the battery wire at the primary resistor to prevent damage to the ohmmeter. Connect an ohmmeter across the terminals of the resistor, as shown in Test 4 illustration. The specified resistance is between 1.3 and 1.4 ohms. If the reading does not fall within this range, replace the resistor.

TEST 5 Voltage Loss in the ignition switch, ammeter, and battery cable: Crank the engine until the points are closed. Connect the voltmeter to the battery post (not the cable terminal) and to the load side of the ignition switch, as shown in Test 5 illustration. Now, turn the ignition switch to the ON position and note the voltage reading. The meter reading should not be more than 0.8 volts. A 0.2 volt drop across each of the connections is permitted.

If the voltage drop is more than 0.8 volts, move the test probe to the "hot" side of the ammeter. If the reading is 0.4 volts, the ignition switch is satisfactory. Once the

corroded connection has been located, remove the nut, clean the wire terminal and connector, and then tighten the connection securely.

TEST 6- Distributor Condition: The condition of the distributor can be quickly and conveniently checked with a timing light.

Under normal timing light procedures the trigger wire from the timing light is attached to the spark or plug wire of the No. 1 cylinder. In this test, connect the trigger wire to the fourth cylinder in the firing order of an in-line engine or to the fifth cylinder in the firing order of a **V8** engine.

The timing mark and the pointer should align in the same position as it did with the number one cylinder. If there is a variation of a few degrees, the distributor shaft bushings or cam lobes may be worn and the condition will have to be corrected.

Before setting the timing, make sure the point dwell is correct. **TAKE CARE** to aim the timing light straight at the mark. Sighting from an angle may cause an error of two or three degrees.

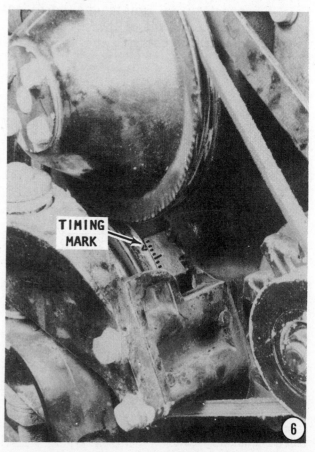

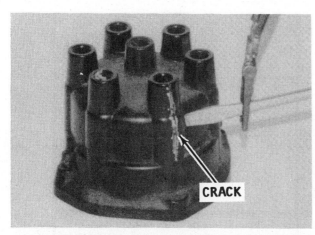

A cracked distributor cap should be replaced. For emergency use, such a cap may be tested as follows: When installed, with high tension leads attached, a distributor cap may be tested by using a grounded screwdriver blade and a spark plug tester as shown. If there is evidence of arcing between the cap and screwdriver blade, when the engine is cranked, the cap is no longer fit for service.

THUNDERBOLT IV TROUBLESHOOTING

The only equipment needed to trouble-shoot the Thunderbolt IV ignition system is a voltmeter and a spark gap tester.

WARNING

Check to be sure the engine compartment is well ventilated and free of any gasoline vapors before starting any of the following tests. A spark will be generated creating a potential fire hazard if fuel vapors are present.

1- Check to be sure the battery is up to a full charge. If not, correct the condition by charging the battery or making a substitution.

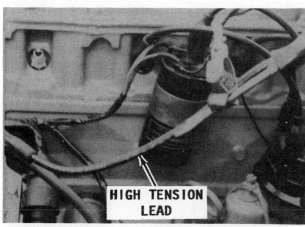

Hookup for making a voltage check of high tension spark plug wires. One end of a jumper wire is clipped to the screwdriver; the other end is grounded; and the spark plug wire is disconnected from the plug. As a preventative measure against hard starting, the high tension wires should be replaced as a set every two years.

2- Check the terminal connections at the distributor, the ignition amplifier and at the ignition coil.

3- Turn the ignition key to the **RUN** position. Attach one lead of the voltmeter to the positive terminal of the ignition coil. The voltmeter should indicate 12-volts. If no voltage is present, check the engine and instrument wiring harness, battery cables, and the key switch. If 12-volts is present proceed to step 4.

4- Leave the one lead of the voltmeter attached to ground on the engine. Make contact with the other lead to the **WHT/RED** terminal on the distributor. The voltmeter should indicate 12-volts. If no voltage is present, continue with this step. If 12-volts is present proceed to step 5. Disconnect the **WHT/RED** lead from the distributor terminal. Check for 12-volts on this lead. If no voltage is present, replace the ignition amplifier. If 12-volts is present replace the ignition sensor in the distributor.

5- Remove the high tension lead from the distributor to the coil. Insert a spark gap tester from coil tower to ground. Remove the **WHT/GRN** lead from the distributor terminal. Turn the ignition key to the RUN position. Strike the terminal on the **WHT/GRN** lead against a good ground on the engine. If there is a spark at the coil, proceed to Step 6. If there is no spark at the coil, proceed to Step 7.

6- Replace the ignition sensor in the distributor.

7- Substitute a **NEW** ignition coil and repeat the test in Step 5. If there is now spark at the coil, install a **NEW** ignition coil. If there is no spark at the coil, replace the ignition amplifier.

5-3 SPARK PLUG TROUBLESHOOTING

The following conditions are keyed to the illustrations with the same number. Refer to the illustration for a visual indication of which plugs are involved.

Condition 1: One spark is overheated. Check the firing order. If the burned plug is the second of two adjacent, and consecutive-firing plugs, the overheating may be the result of crossfire. If you found the spark plug of No. 7 cylinder was overheated, and the firing order is 1-8-4-3-6-5-7-2, the crossfire might result because cylinders No. 5 and No. 7 are adjacent to each other physically and No. 7 follows No. 5 in firing order.   Separate the high-tension leads to these two plugs and the problem may be corrected.

Condition 2: Four spark plugs are fouled in the unusual pattern shown in the No. 2 illustration.   This pattern follows the usual fuel flow in a V8 engine.   This condition may indicate one barrel of the carburetor is running too rich.

Condition 3: The four rear spark plugs are overheated.   This condition indicates a problem in the cooling system.   A reverse-flush of the engine may restore circulation to the rear of the cylinder heads.

Condition 4: Two adjacent plugs are fouled.   Check the high-tension leads to the

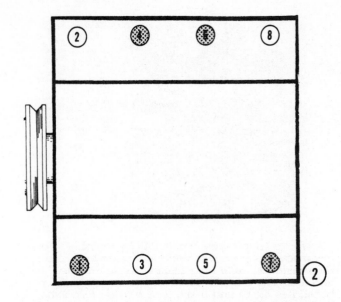

spark plugs to be sure they are connected in the proper sequence and leading to the correct plug for firing order.   If the high-tension wires are all in good order and

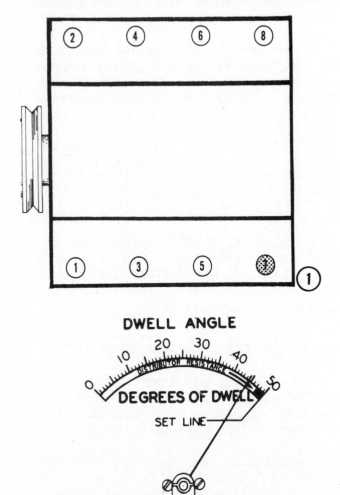

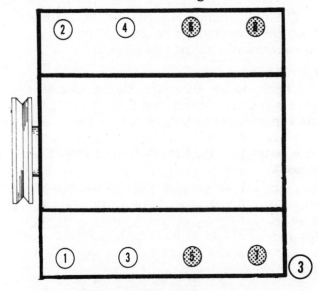

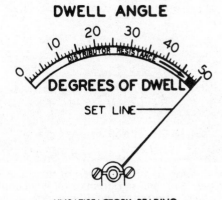

Dwell meter reading with a satisfactory set of contact points (left) adjusted at 46 degrees.   The same set of points with an unsatisfactory reading (right) at 50 degrees.

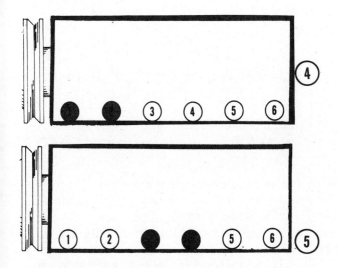

connected properly, then check for a blown cylinder head gasket, refer to Chapter 3.

Condition 5: The two center plugs of a 6-cylinder engine are fouled. The cause may be raw fuel boiling out of the carburetor into the intake manifold after the engine is shut down. This condition can be the result of fuel percolating in the carburetor, a leaking intake valve and seat, a heavy float, or a too-high fuel level in the carburetor bowl. Refer to Chapter 4.

SPARK PLUG EVALUATION

Removal: Remove the spark plug wires by pulling and twisting on only the molded cap. **NEVER** pull on the wire or the connection inside the cap may become separated or the boot damaged. Remove the spark plugs and keep them in order. **TAKE CARE** not to tilt the socket as you remove the plug or the insulator may be cracked.

Examine: Line the plugs in order of removal and carefully examine the firing end to determine the firing conditions in each cylinder.

Correct Color: A proper firing plug should be dry and powdery. Hard deposits inside the shell indicate the engine is starting to use some oil, but not enough to cause concern. The most important evidence is the light gray color of the porcelain, which is an indication this plug has been running at the correct temperature. This means the plug is one with the correct heat range and also that the air-fuel mixture is correct.

Overheating: A dead white or gray insulator, which is generally blistered, is an indication of overheating and pre-ignition.

The electrode gap wear rate will be more than normal and in the case of pre-ignition, will actually cause the electrodes to melt as shown in this illustration. Overheating and pre-ignition are usually caused by over-advanced timing, detonation from using too-low an octane rating fuel, an excessively lean air-fuel mixture, or problems in the cooling system.

Fouled: A fouled spark plug may be caused by the wet oily deposits on the insulator shorting the high-tension current to ground inside the shell. The condition may also be caused by ignition problems which prevent a high-tension pulse to be delivered to the spark plug.

Carbon Deposits: Heavy carbon-like deposits are an indication of excessive oil consumption. This condition may be the result of worn piston rings, worn valve guides, or from a valve seal that is either worn or was incorrectly installed.

Deposits formed only on the shell is an indication the low-speed air-fuel mixture is too rich. At high speeds with the correct mixture, the temperature in the combustion chamber is high enough to burn off the deposits on the insulator.

Too Cool: A dark insulator, with very few deposits, indicates the plug is running too cool. This condition can be caused by low compression or by using a spark plug of an incorrect heat range. If this condition shows on only one plug it is most usually caused by low compression in that cylinder. If all of the plugs have this appearance, then it is probably due to the plugs having a too-low heat range.

Plug with too-cold a rating.

Rich Mixture: A black, sooty condition on both the spark plug shell and the porcelain is caused by an excessively rich air-fuel mixture, both at low and high speeds. The rich mixture lowers the combustion temperature so the spark plug does not run hot enough to burn off the deposits.

Electrode Wear: Electrode wear results in a wide gap and if the electrode becomes carbonized it will form a high-resistance path for the spark to jump across. Such a condition will cause the engine to misfire during acceleration. If all of the plugs are in this condition, it can cause an increase in fuel consumption and very poor performance at high-speed operation. The solution is to replace the spark plugs with a rating in the proper heat range and gapped to specification.

Red rust-colored deposits on the entire firing end of a spark plug can be caused by water in the cylinder combustion chamber. This can be the first evidence of water entering the cylinders through the exhaust manifold because of an accumulation of scale or defective exhaust shutter. This condition **MUST** be corrected at the first opportunity. Refer to Chapter 9, Cooling System Service.

POLARITY CHECK

Coil polarity is extremely important for proper battery ignition system operation. If a coil is connected with reverse polarity, the spark plugs may demand from 30 to 40 percent more voltage to fire. Under such demand conditions, in a very short time the coil would be unable to supply enough voltage to fire the plugs. Any one of the following three methods may be used to quickly determine coil polarity.

1- The polarity of the coil can be checked using an ordinary D.C. voltmeter. Connect the postive lead to a good ground.

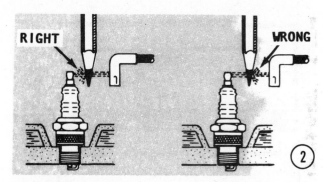

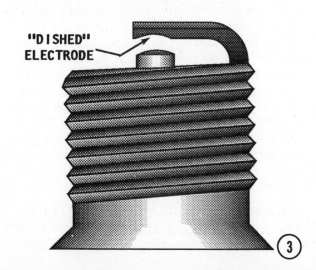

With the engine running, momentarily touch the negative lead to a spark plug terminal. The needle should swing upscale. If the needle swings downscale, the polarity is reversed.

2- If a voltmeter is not available, a pencil may be used in the following manner: Disconnect a spark plug wire and hold the metal connector at the end of the cable about 1/4" from the spark plug terminal. Now, insert an ordinary pencil tip between the terminal and the connector. Crank the engine with the ignition switch ON. If the spark feathers on the plug side and has a slight orange tinge, the polarity is correct. If the spark feathers on the cable connector side, the polarity is reversed.

3- The firing end of a used spark plug can give a clue to coil polarity. If the ground electrode is "dished", it may mean polarity is reversed.

5-4 DISTRIBUTOR SERVICE

The four- and six-cylinder in-line engines all have a Delco-Remy distributor. The V8 engines may have one of several makes: A Delco-Remy, Mallory, Mercury Marine, or an Autoline/Prestolite.

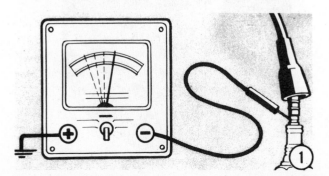

"DISHED" ELECTRODE

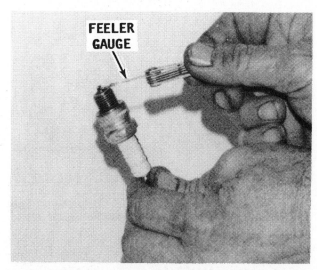

A rounded wire-type feeler gauge should always be used to check the spark plug gap. Bend the side electrode slightly to make an adjustment.

Service procedures for these distributors will be given in the above listed order.

The distributor should be removed for any service including the installation and adjustment of the breaker points. New breaker points can be installed with the distributor in place but the position of the distributor usually makes it awkward to do a perfect job of aligning the points and making the point gap setting.

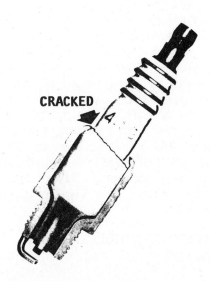

A cracked insulator is always caused by tilting the socket during removal or installation.

DISTRIBUTOR REMOVAL

Remove the distributor cap by turning the cap retainer screws 90 degrees and then lifting off the cap. Some distributor caps have one retainer screw and others have two screws.

Turn the crankshaft until the rotor points to the front of the engine. Loosen the retaining screw and disconnect the con-

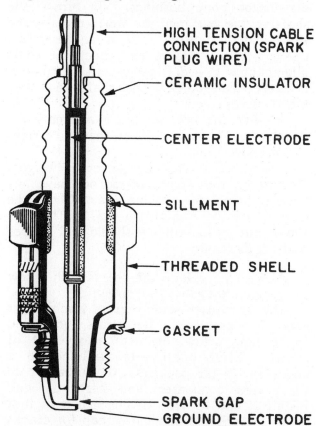

HIGH TENSION CABLE CONNECTION (SPARK PLUG WIRE)

CERAMIC INSULATOR

CENTER ELECTRODE

SILLMENT

THREADED SHELL

GASKET

SPARK GAP
GROUND ELECTRODE

Cross-section view of a spark plug.

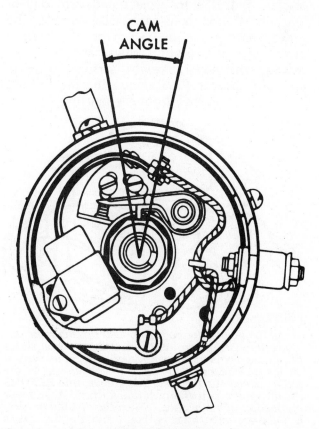

CAM ANGLE

Cam angle of closed points.

denser and primary leads from the terminal. Remove the hold-down clamp bolt, and then **SLOWLY** begin to lift the distributor straight up. As the distributor is pulled up, the rotor will rotate through an arc as the distributor gear slides free of the camshaft gear. Scribe a mark on the distributor body which is aligned with the rotor and a matching mark on the block as soon as the gears disengage and **BEFORE** the distributor is withdrawn any further. After the marks have been scribed, remove the distributor from the block.

5-5 SERVICING A DELCO-REMY DISTRIBUTOR

DISASSEMBLING
FROM A 4- OR 6-CYLINDER ENGINE

Remove the rotor. Remove the breaker point set and the condenser by first disconnecting the primary and condenser leads from the contact point quick-disconnect terminal. Remove the attaching screws, and then remove the breaker plate. Further disassembly of the breaker plate is not necessary.

Remove the pin securing the main shaft drive gear, and then slide the gear off the shaft. Pull the mainshaft and cam out of the distributor housing. Disassemble the main shaft parts by removing the weight cover and stop-plate screws, and then removing the cover, weight springs, weights and cam assembly. Remove the main shaft bushing felt washer from the housing.

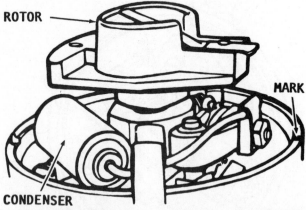

*As the distributor body is pulled up, the rotor will rotate through an arc as the distributor gear slides free of the camshaft gear. A mark should be scribed on the distributor body and a matching one on the block as soon as the gears disengage and **BEFORE** the distributor is withdrawn any further. Align these marks during installation. When the distributor body is lowered down into the block and the two gears mesh, the rotor will swing back to its original position.*

NO ADVANCE   **FULL ADVANCE**

Increased engine speed causes the centrifugal weights to be thrown outward to advance the timing.

The main shaft bushings in the housing are not serviced individually. The housing and bushings **MUST** be serviced together as an assembly.

CLEANING AND INSPECTING

NEVER wash the distributor cap, rotor, condenser, or breaker plate assembly of a distributor in any type of cleaning solvent. Such compounds may damage the insulation of these parts or, in the case of the breaker plate assembly, saturate the lubricating felt.

Check the shaft for wear and fit in the distributor body bushings. If either the shaft or the bushings are worn, replace the shaft and distributor body as an assembly. Use a set of V-blocks and check the shaft alignment with a dial gauge. If the run-out is more than 0.002", the shaft and body **MUST** be replaced.

Inspect the breaker plate assembly for damage and replace it if is there are signs of excessive wear.

Check to be sure the governor weights fit free on their pins and do not have any burrs or signs of excessive wear. Check the cam fit on the end of the shaft. The cam should not fit loose but it should still be free without binding.

ALWAYS replace the points with a new set during a distributor overhaul.

The condenser seldom gives trouble, but good shop practice a few years ago called for a new condenser with a new set of points. Some point sets still include a condenser in the package. If you have paid for a new condenser, you might as well install it and be free of concern over that part. Inspect the distributor cap for cracks or damage. Check the spark plug wires.

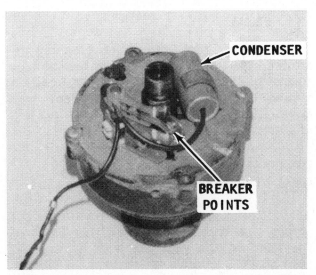

Breaker point parts of a GMC in-line engine distributor.

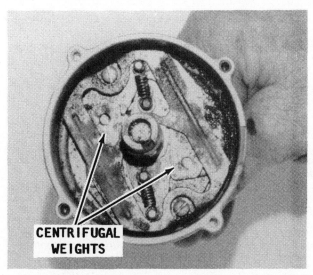

Centrifugal weights and the weight cover installed in the distributor housing.

1. Cap and Button Assy.
2. Lead - Coil
3. Rotor
4. Contact Points Set
5. Condenser
6. Breaker Plate
7. Housing
8. Gear
9. Pin
10. Spring, Weight
11. Lockwasher
12. Weight Hold Down
13. Cam Assembly
14. Weight
15. Main Shaft
16. Screw (#8-32)
    Screw (#10-32)
17. Screw
18. Felt Lubricator
19. Grommet
20. Lead, Primary
21. Retainer
22. Lead - Spark Plug
23. Grommets

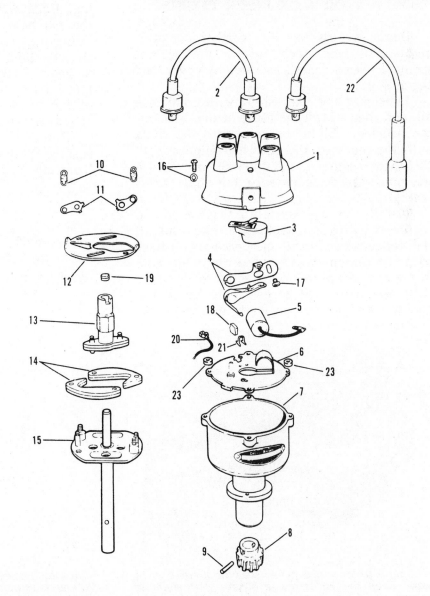

Breaker point parts of an in-line engine distributor.

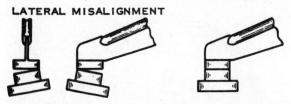

LATERAL MISALIGNMENT   PROPER LATERAL ALIGNMENT

New contact points must be aligned by bending the fixed contact support. *CAUTION: Never bend the breaker lever.*

Before setting the breaker point gap, the points must be properly aligned (right). **ALWAYS** *bend the stationary point,* **NEVER** *the breaker lever. Attempting to adjust an old worn set of points is not practical, because oxidation and pitting of the points will always give a false reading.*

ASSEMBLING THE DELCO-REMY FROM A 4- OR 6-CYLINDER ENGINE

Begin, by placing the weights on their pivot pins and then install the weight springs. Next, install the weight cover and the stop plate.

Lubricate the main shaft with crankcase oil, and then install the shaft in the distributor housing. Slide the gear onto the shaft with the mark on the gear hub aligned with the rotor segment, and then secure it with the pin. Check to be sure the shaft turns freely.

Install the breaker plate assembly and secure it with the attaching screws. Install the condenser. Install the contact point set with the pilot indexed in the matching hole in the breaker plate. Secure the set in place with the attaching screws. Connect the primary and condenser leads to the quick-disconnect terminal.

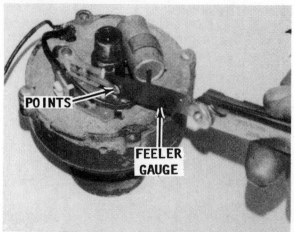

POINTS

FEELER GAUGE

Using a feeler gauge to measure the breaker point gap. Keep the feeler gauge clean. The slightest amount of oil film transferred from the blade to the points will cause oxidation and hard starting.

GAPPING CONTACT POINTS

After the points have been properly installed, the gap must be accurately set, or the dwell adjusted, and then the ignition timing set. These two adjustments are covered in detail in the last part of this chapter.

DISASSEMBLING THE DELCO-REMY GMC V8 ENGINE

Pull off the rotor. Remove both weight springs and both advance weights. Remove the head of the gear pin with a file, and then drive the pin out of the distributor shaft. Remove the gear from the shaft. **BEFORE** removing the shaft from the housing, **REMOVE ANY BURRS** from around the pin hole in the shaft with a flat file. Failure to remove these burrs will damage the shoulder bushing when the shaft is removed. After the burrs are removed, pull the shaft-and-cam weight base assembly out of the housing. Remove the breaker cam-and-weight base assembly from the housing.

Remove the felt washer from around the bushing in the housing. Remove the gasket from the shaft housing.

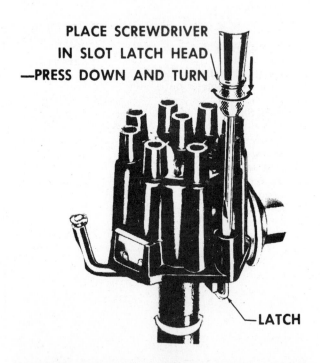

PLACE SCREWDRIVER IN SLOT LATCH HEAD —PRESS DOWN AND TURN

LATCH

Two latch screws hold the distributor cap on a GMC V8 engine in place. To release the cap, simply press down on each screw with a screwdriver and rotate the screw one half turn.

A crack in the distributor will cause misfiring and hard starting.

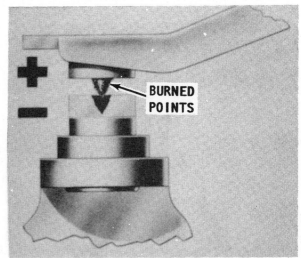

Defective contact points will cause misfire at high speed and hard starting.

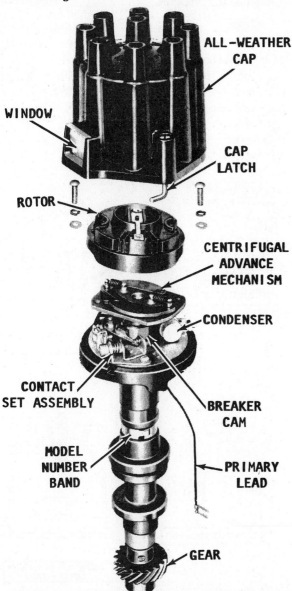

The rotor is secured in place with two screws. Two attaching screws are then removed to release the contact point assembly.

CLEANING AND INSPECTING

NEVER wash the distributor cap, rotor, condenser, or breaker plate assembly of a distributor in any type of cleaning solvent. Such compounds may damage the insulation of these parts or, in the case of the breaker plate assembly, saturate the lubricating felt.

Check the shaft for wear and fit in the distributor body bushings. If either the shaft or the bushings are worn, replace the shaft and distributor body as an assembly. Use a set of V-blocks and check the shaft alignment with a dial gauge. If the run-out is more than 0.002", the shaft and body **MUST** be replaced.

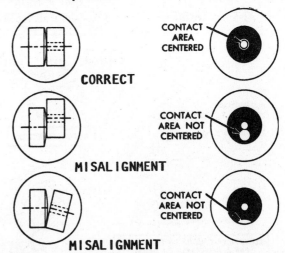

Lateral misalignment of the two point surfaces will result in the contact areas being off center and wearing unevenly. The contact surface of a new breaker lever is always convex to allow for some flexing of the breaker lever and to ensure contact of both stationary surface and the breaker lever surface through all rpm ranges.

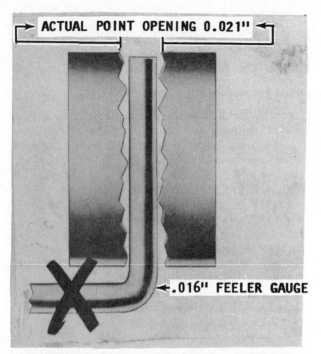

ACTUAL POINT OPENING 0.021"

←.016" FEELER GAUGE

*An exaggerated drawing to illustrate a set of burned points. Notice how the feeler gauge will only register the gap between the high section of the points. Naturally this gives a completely false reading. Therefore, a set of burned points should **ALWAYS** be replaced instead of making an attempt to clean and adjust.*

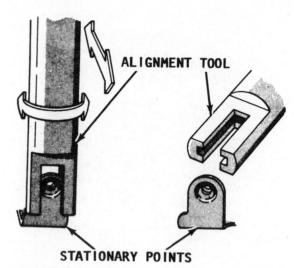

ALIGNMENT TOOL

STATIONARY POINTS

Method of bending the stationary points to align the contact points properly.

Inspect the breaker plate assembly for damage and replace it if is there are signs of excessive wear.

Check to be sure the governor weights fit free on their pins and do not have any burrs or signs of excessive wear. Check the cam fit on the end of the shaft. The cam should not fit loose but it should still be free without binding.

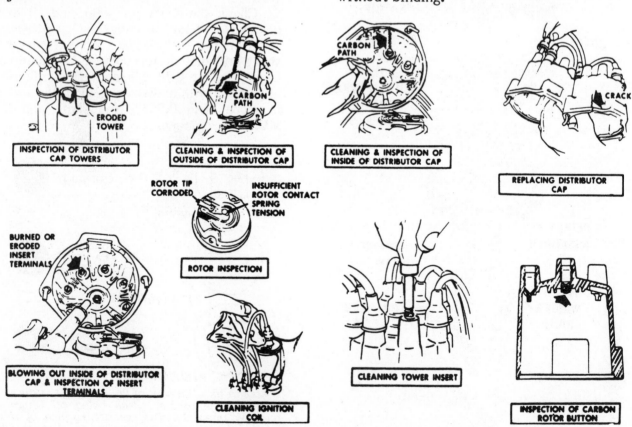

ERODED TOWER

INSPECTION OF DISTRIBUTOR CAP TOWERS

CARBON PATH

CLEANING & INSPECTION OF OUTSIDE OF DISTRIBUTOR CAP

CARBON PATH

CLEANING & INSPECTION OF INSIDE OF DISTRIBUTOR CAP

CRACK

REPLACING DISTRIBUTOR CAP

BURNED OR ERODED INSERT TERMINALS

ROTOR TIP CORRODED

INSUFFICIENT ROTOR CONTACT SPRING TENSION

ROTOR INSPECTION

BLOWING OUT INSIDE OF DISTRIBUTOR CAP & INSPECTION OF INSERT TERMINALS

CLEANING IGNITION COIL

CLEANING TOWER INSERT

INSPECTION OF CARBON ROTOR BUTTON

Cleaning and inspecting a distributor cap.

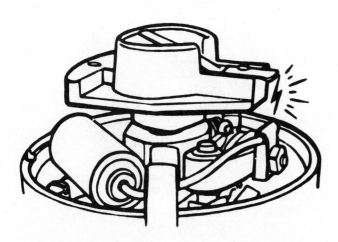

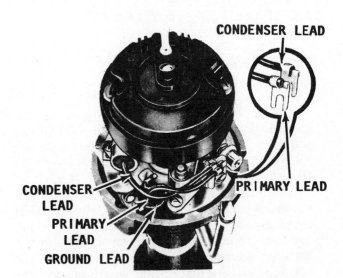

If the primary leads are not properly insulated, a high tension spark can jump from the rotor to the primary wires, as shown.

*Arrangement of the condenser and primary leads. The leads **MUST** be properly insulated or the ignition system will be shorted out and the engine will fail to start.*

ALWAYS replace the points with a new set during a distributor overhaul.

The condenser seldom gives trouble, but good shop practice a few years ago called for a new condenser with a new set of points. Some point sets still include a condenser in the package. If you have paid for a new condenser, you might as well install it and be free of concern over that part.

Inspect the distributor cap for cracks or damage. Check the spark plug wires.

ASSEMBLING THE DELCO-REMY FROM A GMC V8 ENGINE

Install the gasket into the shaft housing. Place the felt washer around the bushing in the housing.

Install the breaker plate in the housing, and then the spring retainer onto the upper bushing. Place the condenser in position and secure it with the attaching screw. Install the breaker point set and secure it with the attaching screws.

Place the cam-and-weight base assembly onto the shaft. If the lubrication in the grooves at the top of the shaft was cleaned out, use Plastilube No. 2 or the equivalent. Install the shaft-and-cam weight assembly in the distributor housing. Install the gear onto the shaft and secure it in place with the pin. Install the advance weights and springs.

Install a new cam lubricator, and then the rotor.

GAPPING CONTACT POINTS

After the points have been properly installed, the gap must be accurately set, or the dwell adjusted, and then the ignition timing set. These two adjustments are covered in detail in the last part of this chapter.

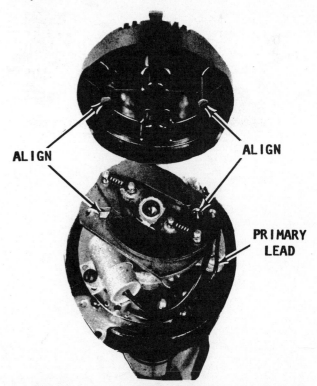

CHECK *to be sure the primary lead is properly connected and safely tucked back to prevent being pinched when the cap is installed. If the lead is pinched and cut, the ignition system will be shorted out and the engine will fail to start.*

5-6 SERVICING A MALLORY DISTRIBUTOR

DISASSEMBLING FROM A V8 ENGINE

Remove the distributor cap and the cap gasket. Pull the rotor off of the distributor shaft.

Remove the condenser and bracket by first disconnecting the condenser lead from the primary terminal, and then removing the retaining screw. Disconnect the breaker assembly lead from the primary terminal, and then remove the retaining screw and the breaker assembly. Scribe a mark on the housing to indicate the position of the breaker plate as an aid to installation. Remove the plate retaining screws, and then the plate. Remove the oiler wick. The oiler wick is a press fit in the housing and need not be removed.

DO NOT disassemble the distributor further because the oil seal may be damaged.

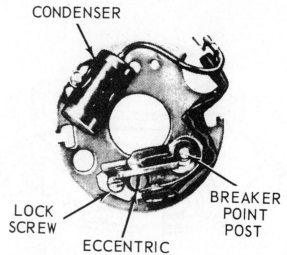

Arrangement of principle parts of a FORD V8 engine distributor.

Drive the retaining pin from the drive gear, and then remove the gear from the shaft. Drive the roll pin out of the distributor shaft collar. Remove the collar, washer, distributor housing, and washer. Remove the vent screen from the housing.

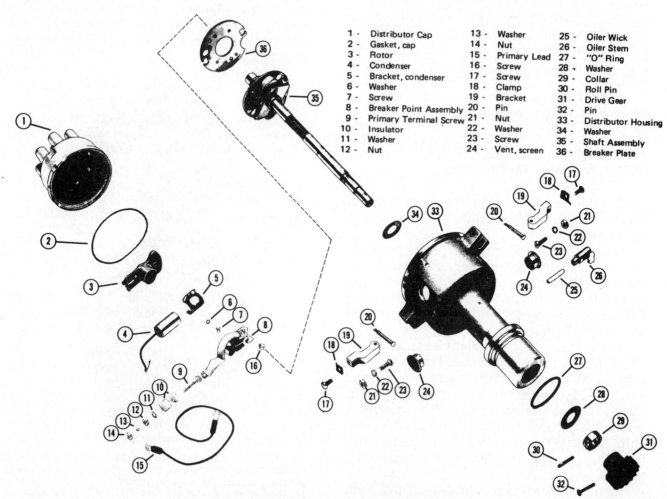

1 -	Distributor Cap	13 -	Washer	25 -	Oiler Wick
2 -	Gasket, cap	14 -	Nut	26 -	Oiler Stem
3 -	Rotor	15 -	Primary Lead	27 -	"O" Ring
4 -	Condenser	16 -	Screw	28 -	Washer
5 -	Bracket, condenser	17 -	Screw	29 -	Collar
6 -	Washer	18 -	Clamp	30 -	Roll Pin
7 -	Screw	19 -	Bracket	31 -	Drive Gear
8 -	Breaker Point Assembly	20 -	Pin	32 -	Pin
9 -	Primary Terminal Screw	21 -	Nut	33 -	Distributor Housing
10 -	Insulator	22 -	Washer	34 -	Washer
11 -	Washer	23 -	Screw	35 -	Shaft Assembly
12 -	Nut	24 -	Vent, screen	36 -	Breaker Plate

Exploded view showing all parts of a Mallory distributor.

CLEANING AND INSPECTING

NEVER wash the distributor cap, rotor, condenser, or breaker plate assembly of a distributor in any type of cleaning solvent. Such compounds may damage the insulation of these parts or, in the case of the breaker plate assembly, saturate the lubricating felt.

Check the shaft for wear and fit in the distributor body bushings. If either the shaft or the bushings are worn, replace the shaft and distributor body as an assembly. Use a set of V-blocks and check the shaft alignment with a dial gauge. If the run-out is more than 0.002", the shaft and body **MUST** be replaced.

Inspect the breaker plate assembly for damage and replace it if is there are signs of excessive wear.

Check to be sure the governor weights fit free on their pins and do not have any burrs or signs of excessive wear. Check the cam fit on the end of the shaft. The cam should not fit loose but it should still be free without binding.

ALWAYS replace the points with a new set during a distributor overhaul.

The condenser seldom gives trouble, but good shop practice a few years ago called for a new condenser with a new set of points. Some point sets still include a condenser in the package. If you have paid for a new condenser, you might as well

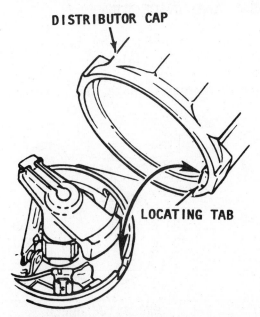

DISTRIBUTOR CAP

LOCATING TAB

*Locating tab on the distributor cap and matching cutout in the rim of the distributor housing. The tab **MUST** index in the cutout or the tip of the rotor will strike one segment of the cap and crack it.*

install it and be free of concern over that part.

Inspect the distributor cap for cracks or damage. Check the spark plug wires.

ASSEMBLING THE MALLORY FROM A V8 ENGINE

Install the vent screen in the distributor housing vent hole. Secure the screen in place by crimping the inside flange.

Slide a washer onto the distributor shaft, and then install the shaft through the housing. Install the washer and collar on the shaft and secure the collar in place with a **NEW** pin. Check the shaft for end play which should be between 0.008" and 0.010". Install the drive gear on the distributor shaft with the shoulder going on first and secure the gear in place with a **NEW** pin through the shoulder and shaft. Peen both ends of the pin to prevent it from coming out.

Insert the oiler wick into the oiler stem. Install the breaker plate screws, the plate, washers, and nuts. **DO NOT** tighten the nuts at this time.

Install the breaker plate with the marks made during dissasembly aligned, and then tighten the screws. Install the breaker assembly and connect the breaker assembly lead to the distributor primary terminal.

Install and secure the condenser with the bracket and attaching screw. Connect the condenser lead to the distributor primary terminal.

Slide the rotor onto the distributor shaft. Install a **NEW** gasket on the distributor cap with the notch in the cap aligned with the locating pin on the housing. Install the distributor cap with the tab on the inside rim of the cap aligned with the notch in the housing. If the cap is not properly positioned, as described, the rotor will strike one of the segments in the cap and crack it. Secure the cap in place by tightening the two screws.

GAPPING CONTACT POINTS

After the points have been properly installed, the gap must be accurately set, or the dwell adjusted, and then the ignition timing set. These two adjustments are covered in detail in the last part of this chapter.

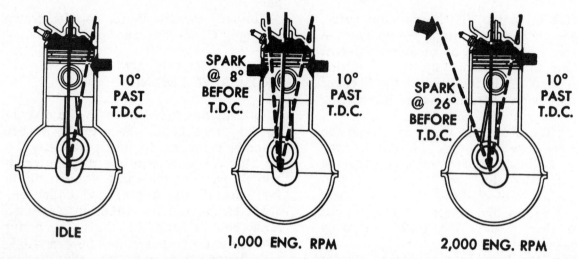

The mechanical spark advance mechanism advances the ignition timing when the engine speed increases, as shown in these three drawings. Maximum pressure must occur by 10 degrees ATDC, at all speeds. Therefore, the spark must occur at TDC for idle speeds (left); at approximately 8 degrees BTDC for 1,000 rpm (center); and at 26 degrees BTDC for 2,000 rpm (right).

5-7 SERVICING A MERCURY MARINE DISTRIBUTOR

DISASSEMBLING FROM A V8 ENGINE

Remove the two distributor cap screws, and then remove the cap. Lift the rotor from the distributor shaft.

Disconnect the condenser lead from the primary terminal. Removing the condenser bracket attaching screw, then the bracket and condenser. Disconnect the coil wire from the primary terminal. Remove the breaker assembly retaining screw, and then the breaker assembly. **FURTHER DISASSEMBLY** of this part of the distributor is not necessary.

Use a drift punch to remove the pin from the drive gear and shaft, then slide the gear off of the shaft. Remove the vent screen from the distributor housing.

Individual parts are not available for the pilot-and-shaft assembly. Therefore, further disassembly of these parts is not recommended.

CLEANING AND INSPECTING

NEVER wash the distributor cap, rotor, condenser, or breaker plate assembly of a distributor in any type of cleaning solvent. Such compounds may damage the insulation

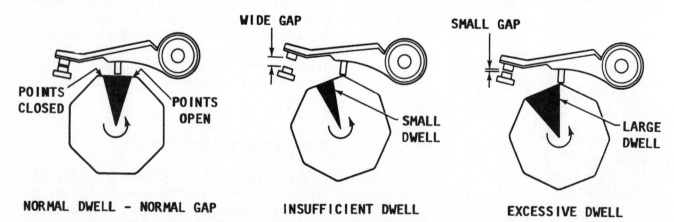

Three different point gap/dwell angle conditions. A normal gap (and dwell angle) is shown in the left view. The center view illustrates an excessive gap with too small a dwell angle. The right illustration depicts too small a gap with the resulting dwell angle too great. If the gap is too small, the ignition is retarded, causing loss of power.

of these parts or, in the case of the breaker plate assembly, saturate the lubricating felt.

Check the shaft for wear and fit in the distributor body bushings. If either the shaft or the bushings are worn, replace the shaft and distributor body as an assembly. Use a set of V-blocks and check the shaft alignment with a dial gauge. If the run-out is more than 0.002", the shaft and body **MUST** be replaced.

Inspect the breaker plate assembly for damage and replace it if there are signs of excessive wear.

Check to be sure the advance weights fit free on their pins and do not have any burrs or signs of excessive wear. Check the cam fit on the end of the shaft. The cam should not fit loose but it should still be free without binding.

ALWAYS replace the points with a new set during a distributor overhaul.

The condenser seldom gives trouble, but good shop practice a few years ago called for a new condenser with a new set of points. Some point sets still include a condenser in the package. If you have paid for a new condenser, you might as well install it and be free of concern over that part.

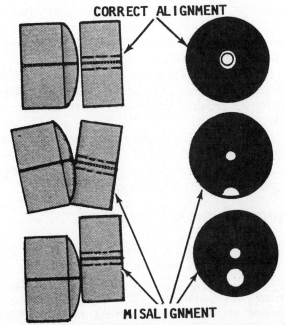

Lateral misalignment of the two point surfaces will result in the contact areas being off center and wearing unevenly. The contact surface of a new breaker lever is always convex to allow for some flexing of the breaker lever and to ensure contact of both the stationary surface and the breaker lever surface through all rpm ranges.

Inspect the distributor cap for cracks or damage. Check the spark plug wires.

ASSEMBLING THE MERCURY MARINE DISTRIBUTOR FROM A V8 ENGINE

Position the vent screen in the distributor housing vent hole, and then secure it in place with the keeper.

Install the drive gear onto the distributor shaft with the shoulder end of the gear facing the housing. Secure the gear on the shaft with a **NEW** pin through the gear shoulder and the shaft.

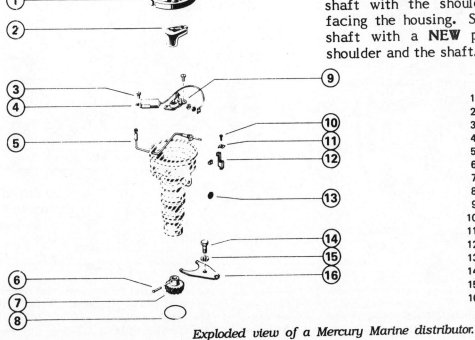

1 - Distributor Cap
2 - Rotor
3 - Screw (3)
4 - Condenser
5 - Primary Wire
6 - Drive Gear Pin
7 - Drive Gear
8 - "O" Ring
9 - Breaker Assembly
10 - Screw (2)
11 - Lockwasher (2)
12 - Distributor Cap Clamp (2)
13 - Screen (2)
14 - Screw
15 - Lockwasher
16 - Clamp

Exploded view of a Mercury Marine distributor.

Install the breaker assembly in the housing and secure it in place with the attaching screws. Connect the coil wire to the breaker primary terminal.

Install and secure the condenser with the bracket and attaching screw. Connect the condenser lead to the breaker primary terminal.

Slide the rotor onto the end of the distributor shaft. Install the distributor cap with the notch in the cap aligned with the locating pin on the housing.

GAPPING CONTACT POINTS

After the points have been properly installed, the gap must be accurately set, or the dwell adjusted, and then the ignition timing set. These two adjustments are covered in detail in the last part of this chapter.

5-8 SERVICING AN AUTOLITE/PRESTOLITE DISTRIBUTOR

DISASSEMBLING FROM A FORD V8 ENGINE

Remove the distributor cap, and then remove the rotor.

Disconnect the primary wire and the condenser lead from the breaker point assembly terminal. Remove the breaker point

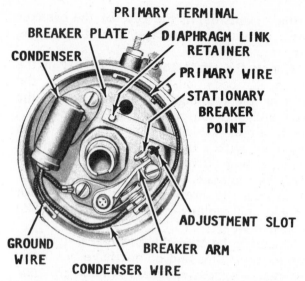

Internal view of an Autolite/Prestolite distributor. Notice the routing of the primary lead and the condenser wire.

assembly by removing the two attaching screws. Remove the condenser attaching screw and the condenser.

Pull the primary lead through the opening in the housing. Remove the two breaker plate attaching screws, and then remove the breaker plate.

Identify one of the distributor weight springs and its bracket with a mark. Mark one of the weights and its pivot pin. CAREFULLY unhook and remove the weight springs. Pull the lubricating wick out of the cam assembly. Remove the cam assembly by first removing the retainer, and then lifting the assembly off the distributor shaft.

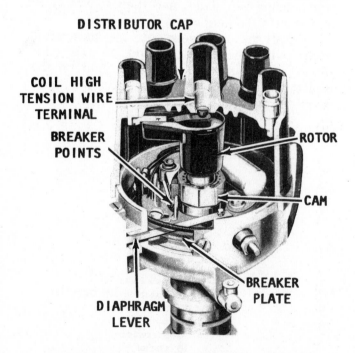

Cutaway view of an Autolite/Prestolite distributor with major parts identified.

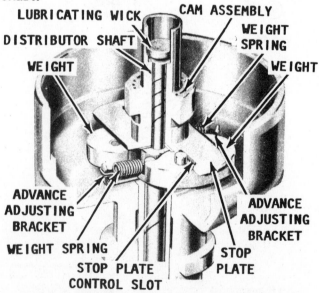

Cutaway view of a distributor with the breaker plate and associated parts removed. Notice the location of the two weight springs and the relationship of other major parts.

Remove the thrust washer installed only on counterclockwise rotating engines.

Remove the weight retainers, and then remove the weights.

Remove the distributor cap clamps. Scribe a mark on the gear and a matching mark on the distributor shaft as an aid in locating the pin holes during assembly. Place the distributor shaft in a V-block, and then use a drift punch to remove the roll pin. Remove the gear from the shaft. Remove the shaft collar roll pin.

CLEANING AND INSPECTING

NEVER wash the distributor cap, rotor, condenser, or breaker plate assembly of a distributor in any type of cleaning solvent. Such compounds may damage the insulation of these parts or, in the case of the breaker plate assembly, saturate the lubricating felt.

Check the shaft for wear and fit in the distributor body bushings. If either the shaft or the bushings are worn, replace the shaft and distributor body as an assembly.

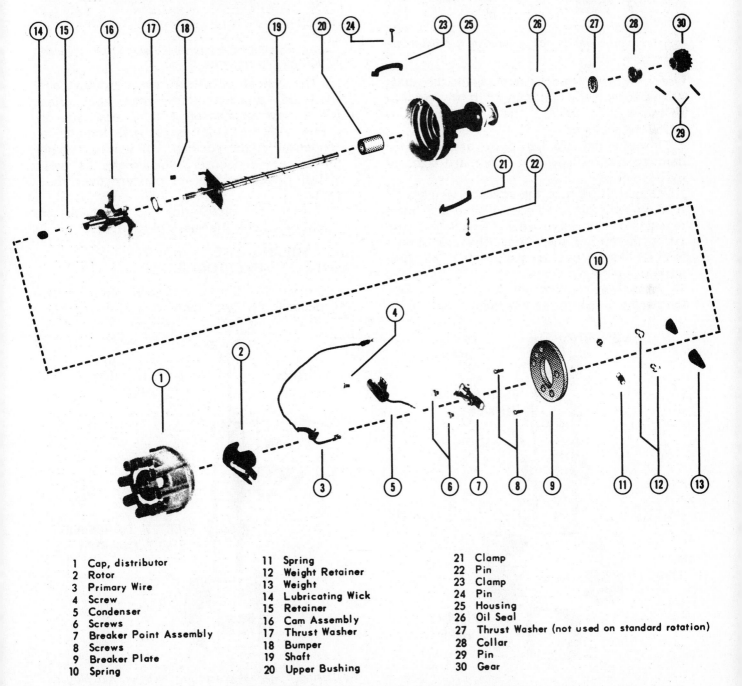

1	Cap, distributor
2	Rotor
3	Primary Wire
4	Screw
5	Condenser
6	Screws
7	Breaker Point Assembly
8	Screws
9	Breaker Plate
10	Spring

11	Spring
12	Weight Retainer
13	Weight
14	Lubricating Wick
15	Retainer
16	Cam Assembly
17	Thrust Washer
18	Bumper
19	Shaft
20	Upper Bushing

21	Clamp
22	Pin
23	Clamp
24	Pin
25	Housing
26	Oil Seal
27	Thrust Washer (not used on standard rotation)
28	Collar
29	Pin
30	Gear

Exploded drawing of an Autolite/Prestolite distributor.

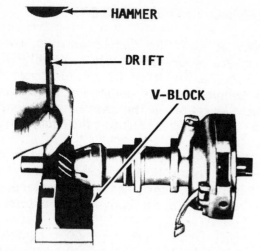

Setup to drive out the roll pin holding the distributor gear in place.

Use a set of V-blocks and check the shaft alignment with a dial gauge. If the run-out is more than 0.002", the shaft and body **MUST** be replaced.

Inspect the breaker plate assembly for damage and replace it if there are signs of excessive wear.

Check to be sure the advance weights fit free on their pins and do not have any burrs or signs of excessive wear. Check the cam fit on the end of the shaft. The cam should not fit loose but it should still be free without binding.

ALWAYS replace the points with a new set during a distributor overhaul.

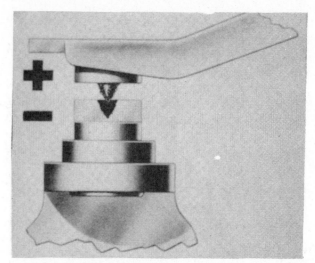

Oxidized contact points which should be replaced for satisfactory performance.

The condenser seldom gives trouble, but good shop practice a few years ago called for a new condenser with a new set of points. Some point sets still include a condenser in the package. If you have paid for a new condenser, you might as well install it and be free of concern over that part.

Inspect the distributor cap for cracks or damage. Check the spark plug wires.

ASSEMBLING THE AUTOLITE/PRESTOLITE

Lubricate the distributor shaft with crankcase oil, and then slide it into the

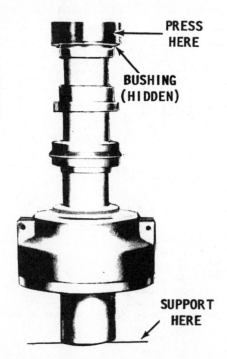

Setup to install a new upper bushing.

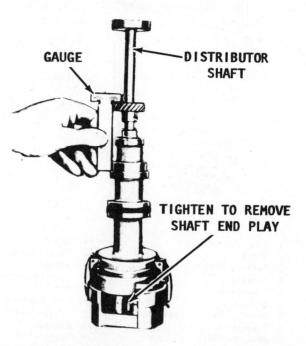

Using a special gauge to check for end play as described in the text.

distributor body. Slide the collar onto the shaft; align the holes in the collar with the hole in the shaft; and then install a **NEW** pin.

Install the distributor cap clamps. Left-hand rotating engine distributor assemblies have an additional thrust washer between the collar and the base. Use a feeler gauge between the collar and the distributor base to check the shaft end play. The end play should be between 0.024" and 0.035"

Install the gear onto the shaft with the marks on the gear and the shaft you made during disassembly aligned. The holes through the gear and the shaft should be aligned after the gear is installed. Install the gear roll pin.

Fill the grooves in the weight pivot pins with distributor cam lubricant. Position the weights in the distributor with the weight you identified with a mark during disassembly matched with the marked pivot pin. Secure the weights in place with the retainers. Slide the thrust washer onto the shaft. Fill the upper distributor shaft grooves with distributor cam lubricant.

Install the cam assembly with the marked spring bracket near the marked spring bracket on the stop plate. If a new cam assembly is being installed, **TAKE CARE** to be sure the cam is installed with the hypalon-covered stop in the correct cam plate control slot. The proper slot can be determined by measuring the length of the slot used on the old cam, and then using the corresponding slot on the new cam. Some new cams will have the size of the slot

stamped in degrees near the slot. If the **WRONG** slot is used, the maximum advance will not be correct.

Coat the distributor cam lobes with a light film of distributor cam lubricant. Install the retainer and wick. Use a few drops of SAE 10W engine oil on the wick. Install the weight springs with the spring and bracket you marked during disassembly matched.

Place the breaker plate in position, and then secure with the attached screws.

Push the primary wire through the opening in the distributor. Place the breaker point assembly and the condenser in position and secure them in place with the attaching screws. Connect the primary wire and the condenser lead to the breaker point primary terminal.

Adjust the point gap and dwell as outlined in the following procedures, then install the rotor and the distributor cap.

5-9 ADJUSTING THE POINT GAP ALL DISTRIBUTORS

A feeler gauge or a dwell meter may be used to adjust the contact points. However, due to the rough surface of used points, a feeler gauge will never provide an accurate setting. The feeler gauge can give satisfactory results if a new set of contact points is being adjusted or if a dwell meter is not available. The feeler gauge is used when the points are adjusted with the distributor out of the engine and the dwell meter when the distributor is installed and the engine is running.

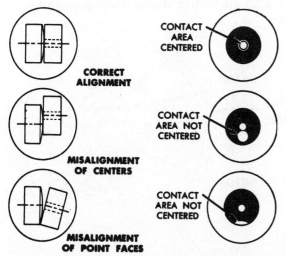

*Before setting the breaker point gap, the points must be properly aligned (top). **ALWAYS** bend the stationary point, **NEVER** the breaker lever. Attempting to adjust an old worn set of points is not practical, because of oxidation and pitting of the points will always give a false reading.*

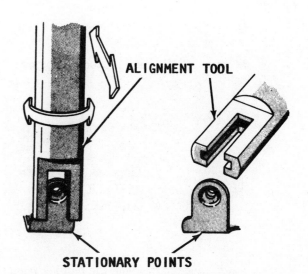

Method of bending the stationary point to align the contact points properly.

USING A FEELER GAUGE

If the distributor is not installed, rotate the shaft until the fiber rubbing block is on the high point of the cam. If the distributor is installed, crank the engine until the rubbing block is on the high point of the cam. Adjust the gap to the specification given in the Appendix by turning the eccentric adjuster on the stationary contact point or moving the point with a screwdriver in the base slot. Rotate the distributor shaft until the points are closed. Check to see if the points are properly aligned, as shown in the accompanying illustration.   If necessary, use a pair of needle-nose pliers, or a contact point alignment tool, to bend the **STATIONARY** point bracket until the points are aligned, as shown.

ALWAYS use a clean feeler gauge to make the final adjustment or you may leave a thin coating of oil on the points. Any oil on the points will oxidize in a short time and cause problems. **TAKE CARE** when making the gap measurement with the feeler gauge not to twist or cock the gauge. If the guage is not inserted square with the points, you will not get an accurate measurement.

Adjust the point gap about 0.003" wider than the specification to allow for initial rubbing block wear. Keep the contact point retaining screw snug while making the adjustment to keep the gap from changing when the screw is finally tightened.

After the proper gap has been obtained, tighten the retaining screw, and then recheck the gap to be sure the setting has not changed.

Coat the distributor cam with a light film of heavy grease, and then turn the distributor shaft in the normal direction of rotation to wipe the lubricant off against the back of the rubbing block. The lubricant will remain there as a reservoir while the rubbing block wears. Wipe any excess lubricant from the cam.

5-10 DISTRIBUTOR INSTALLATION SPECIAL WORDS FOR MCM185 229 CID V6

The MCM185 229 CID V6 engine is considered an uneven firing engine because the crankshaft angle alternately changes from 132° to 108° through the firing cycle. To obtain this angle change, the cam in the distributor has alternating sharp and rounded lobes.   Care must be exercised when installing the distributor.

Crank the engine until the timing mark on the crankshaft balancer is aligned with the TDC-0 mark on the the timing tab **AND** the No. 1 cylinder is in the firing position.

Install the distributor into the engine with the rotor aligned with the No. 1 spark plug tower and the rubbing block of the breaker points is positioned on a rounded lobe.

In order for the spark plug wires to fit properly, select the rounded lobe which positions the distributor body with the lead from the distributor to the coil coming out the aft side of the distributor, as shown in the accompanying illustration.

Secure the distributor with the clamp.

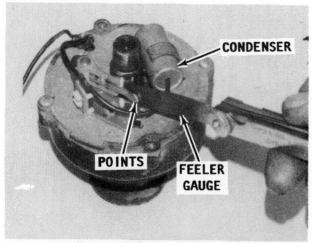

Using a feeler gauge to measure the breaker point gap. Keep the feeler gauge blade clean. The slightest amount of oil film transferred from the blade to the points will cause oxidation and hard starting.

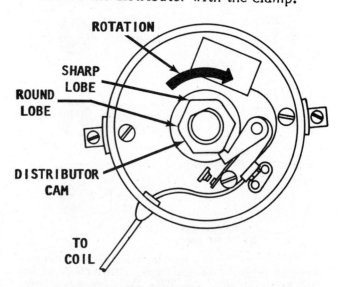

The cam for the MCM 185 229 CID, V6 has alternating rounded and sharp lobes as explained in the text.

As the distributor body is pulled up, the rotor will rotate through an arc as the distributor gear slides free of the camshaft gear. A mark should be scribed on the distributor body and a matching one on the block AFTER the distributor is removed, but BEFORE it is lifted clear. Align these marks during installation. When the distributor body is lowered down into the block and the two gears mesh, the rotor will swing back to its original position.

DISTRIBUTOR INSTALLATION
ALL OTHER ENGINES

During removal a mark was scribed on the distributor housing and a matching one on the block as an aid to installation. Now, slide the distributor shaft into the block with the rotor pointing toward the front of the engine and with the marks you scribed during removal aligned as close as possible. As the distributor is lowered down into the block, the rotor will swing through a small arc as the gear on the end of the shaft indexes with the camshaft gear, which will place the rotor in the same position as before removal.

If the crankshaft was turned for any reason while the distributor was removed, the timing was lost and it will be necessary to retime the engine.

To time the engine, first remove the rocker arm cover. Next, rotate the crankshaft in the normal direction with a wrench on the harmonic balancer bolt until both valves for No. 1 cylinder are closed and the timing mark on the balancer is aligned with the "0" on the timing indicator. NEVER rotate the crankshaft in the opposite direction from the normal or the water pump in the stern drive will be damaged. Now, with both valves closed, and the timing mark aligned with the timing indicator, the No. 1 cylinder is in firing position.

Align the rotor with the No. 1 cylinder wire terminal in the distributor cap, and then install the distributor in the block. If the distributor will not seat fully in the block, press down lightly on the housing while a partner turns the crankshaft slowly until the distributor tang snaps into the oil pump shaft slot and the distributor moves into its full seated position. Tighten the distributor hold-down bolt.

Ignition fine-tuning will be accomplished after the engine is running.

Wipe the distributor cap and the coil of any moisture to be sure it does not cause a leakage path.

ADJUSTING THE DWELL

A dwell meter accurately measures the length of time the points are closed, as shown in the accompanying illustration. Connect one lead of the dwell meter to the negative side of the ignition coil and the other lead to a good ground. Start the engine and adjust the dwell to the Specifications in the Appendix.

CAUTION
Water must circulate through the stern

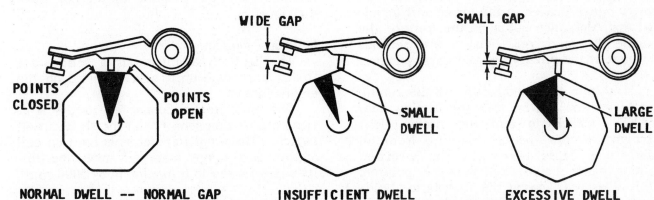

Three different point gap/dwell angle conditions. A normal gap (and dwell angle) is shown in the left view. The center view illustrates an excessive gap with too small a dwell angle. The right illustration depicts too small a gap with the resulting dwell angle too great. If the gap is too small, the ignition is retarded, causing loss of power.

drive, also to and from the engine, anytime the engine is run to prevent damage to the water pump located in the lower unit (original and Alpha drives); or on the front of the engine (Bravo drive). Just a few seconds without water will damage the water pump.

ADJUSTING IGNITION TIMING

The breaker point gap or the dwell must be accurately set before attempting to fine-tune the engine because the point gap directly affects the timing.

Check the timing mark on the balancer or pulley and the lines on the timing tab. If they are hard to see, mark them with paint or chalk. Connect a timing light to an adaptor for No. 1 spark plug. **NEVER** puncture the high-tension wire the core will be damaged.

SPECIAL TIMING WORDS
ENGINES EQUIPPED WITH EFI

Identify the Red/Yellow lead at the ECU. Normally this lead is not connected to anything. However, when timing an engine equipped with EFI, this Red/Yellow lead **MUST** be grounded to the engine block. Failure to ground this lead will produce a retarded timing adjustment. After the timing procedure has been completed this lead **MUST** be disconnected and the lead protected to prevent accidental contact (grounding) on the engine block.

Start the engine and adjust the idle speed to specification.

CAUTION

Water must circulate through the stern drive, also to and from the engine, anytime the engine is run to prevent damage to the water pump located in the lower unit (original and Alpha drives); or on the front of the engine (Bravo drive). Just a few seconds without water will damage the water pump.

Now, aim the timing light at the timing mark on the pulley and the timing tab. **TAKE CARE** to aim the timing light straight at the mark. Sighting from an angle may cause an error. The specified timing mark should align with the pointer. If it does not, loosen the distributor hold-down bolt and rotate the distributor until it is aligned. Tighten the hold-down bolt, and then check the marks again with the light.

Check operation of the centrifugal advance mechanism by accelerating the engine and checking the position of the timing mark with the light. The mark should advance on the pulley if the advance mechanism is operating properly.

5-11 DIGITAL DISTRIBUTORLESS IGNITION SYSTEM

In 1990, a Digital Distributorless Ignition System (DDIS), was installed as standard equipment on the 181 CID 3.0 LX 4-cylinder in-line engine. This system was also made available as "optional" equipment for the 181 CID 3.0 L. As the name suggests, this ignition system does not have a distributor.

The block is identical for both model engines. The 3.0 L with standard Thunderbolt IV ignition has a conventional electronic distributor. The 3.0 LX with Digital Distributorless Ignition (DDIS) has a motion sensor located where the conventional distributor would be. The engine firing order for the 3.0 LX with DDIS is 1-3-4-2.

The system consists of the following components:

Two ignition coils
An ignition amplifier
A motion sensor
A shift cutout switch

A brief description of each unit is presented in the following paragraphs.

Ignition Coils
Two ignition coils are mounted on the block. Each coil simultaneuously provides the spark for two cylinders. As explained later, the spark to one cylinder is wasted. As a result, this system is sometimes referred to as the "waste spark" method of spark distribution.

Ignition Amplifier
The ignition amplifier is a solid state "black box" type unit, also mounted on the engine block.

The amplifier is connected to the ignition coil, motion sensor, and the shift cutout switch. The amplifier controls ignition coil output, determines spark advance, and limits engine speed to a maximum of 5000 rpm.

Motion Sensor
The motion sensor is located next to the oil filter. The lower portion of the motion

sensor resembles a conventional distributor. The spiral gear on the sensor shaft is driven by a matching gear on the camshaft in the block. The lower end of the shaft is slotted to drive the oil pump.

The upper portion of the motion sensor is entirely different from a conventional distributor. A close tolerance machined reluctor disc is mounted on top of the sensor shaft and rotates at engine speed. Six "high points" are machined into the reluctor disc. Four of the spaces between the "high points" are quite large, and two spaces are small, as shown in the illustration on the following page. A crankshaft position sensor is mounted to one side of the reluctor disc. This sensor is connected to the ignition amplifier.

Shift Cutout Switch
This switch is described in detail on Page 5-2.

Cylinder "Pair"
On a four cylinder engine, each cylinder is paired with another to form two cylinder pairs. Usually, No. 1 is paired with No. 4, and No. 2 is paired with No. 3. One ignition coil provides spark to a pair of cylinders, and the other ignition coil provides spark for the other cylinder pair.

Each spark plug in a cylinder pair fires simultaneously. One cylinder plug fires on the compression stroke and the other cylinder fires on the exhaust stroke.

At this time, spark does not occur in the other cylinder pair, as the piston comes down on the intake or power stroke. When the spark plug fires in the cylinder on the exhaust stroke, very little energy is drawn from the coil. Therefore, most of the energy is directed to the cylinder on the compression stroke.

DDIS Operation
The reluctor disc at the upper end of the motion sensor shaft rotates at engine speed. Six "high points" are machined into the reluctor disc. Four of the spaces in between the "high points" are quite large and are spaced approximately 90° apart. The small spaces are 180° -- opposite each other. As each space passes the crankshaft position sensor, an induced voltage pulse is created. The sensor monitors crankshaft position and engine speed. The sensor sends signals to the ignition amplifier.

The ignition amplifier compares time intervals between pulses. As a small space sweeps past the crank position sensor, the ignition amplifier sends a signal to energize the first ignition coil and fires one of the two cylinder pair. After the crankshaft has rotated through 180°, the second small space sweeps past the crankshaft position sensor. The ignition amplifier again sends a signal to energize the second ignition coil and fires the other cylinder pair.

TROUBLESHOOTING
The tests on the following page **MUST** be performed at room temperature -- 68°F (20°C). If the engine has just been operated, allow time for the ignition components to cool. To accurately determine the problem area, the following tests **MUST** be performed in the sequence listed. If a component fails any test, it **MUST** be replaced. DDIS components do not have servicable parts.

Motion Sensor Service
No service is possible. The following instructions include removal of a defective sensor and installation of a new sensor.

First, disconnect the harness lead to the motion sensor. Next, crank the engine until the timing mark is at TDC. Now, remove the bolt and clamp securing the sensor to the engine block. Lift out the sensor. Remove and discard the gasket.

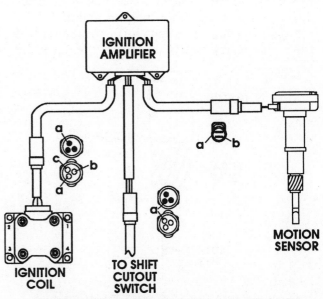

Connector terminal identification used for testing, as explained in the text.

COMPONENT BEING TESTED	IGNITION AMPLIFIER	IGNITION AMPLIFIER	IGNITION COIL (PRIMARY RESISTANCE)	IGNITION COIL (PRIMARY RESISTANCE)	IGNITION COIL (SECONDARY RESISTANCE)	IGNITION COIL (SECONDARY RESISTANCE)	IGNITION COIL (INSULATION RESISTANCE)	MOTION SENSOR	SHIFT CUTOUT SWITCH (IN NEUTRAL GEAR)
IGNITION KEY POSITION	RUN	RUN	OFF	OFF	OFF	OFF	OFF	OFF	OFF
METER SCALE	15 V DC	15 V DC	5 OHMS	5 OHMS	20K OHMS	20K OHMS	INFINITY	250 OHMS	INFINITY
DISCONNECT LEAD FROM	IGNITION COIL	SHIFT CUTOUT SWITCH	IGNITION AMPLIFIER	IGNITION AMPLIFIER	SPARK PLUGS	SPARK PLUGS	IGNITION AMPLIFIER	IGNITION AMPLIFIER	IGNITION AMPLIFIER
RED METER LEAD TO	TERMINAL "a"	TERMINAL "a"	TERMINAL "a"	TERMINAL "a"	TOWER 1	TOWER 4	TERMINAL "a", "b", THEN "c"	TERMINAL "a"	TERMINAL "a"
BLACK METER LEAD TO	ENGINE GROUND	ENGINE GROUND	TERMINAL "b"	TERMINAL "c"	TOWER 2	TOWER 3	ENGINE GROUND	TERMINAL "b"	ENGINE GROUND
SPECIFIED METER READING	12 V DC	12 V DC	1.9 - 2.5 OHMS	1.9 - 2.5 OHMS	11.3 - 15.5 OHMS	11.3 - 15.5 OHMS	MINIMUM 10 MILLION OHMS	140 - 180 OHMS	CONTINUITY

Component testing table. Component being tested -- read across. How component is tested -- read down.

Remove the cover of the new motion sensor. Check to be sure the timing mark has not moved from the TDC position. If necessary, crank the engine to align the marks. The No. 1 piston may be at the top of compression or exhaust stroke -- TDC -- it makes no difference with the DDIS system -- because of the "twin" firing.

Align the hole -- not the roll pin hole -- at the base of the gear with the notch on the lower housing. Install a new gasket. Lower the sensor into the block. If the slotted end of the sensor fails in index with the oil pump, use a long slotted screwdriver to realign the oil pump.

The sensor is correctly installed when the small drain hole in the housing faces directly away from the engine and the dimple on the reluctor aligns with the center of the crankshaft position sensor. Secure the

sensor in place with the clamp and bolt. Tighten the bolt to a torque value of 20 ft lb (27Nm). Install the cover over the motion sensor and tighten the bolts securely.

DDIS Timing

Initial ignition timing is adjusted by loosening the clamp bolt and rotating the motion sensor in the same manner as with conventional ignition systems. Hook a timing light pickup around the No. 1 spark plug lead. Aim the light at the timing mark and rotate the motion sensor until the timing mark aligns with 8° BTDC at 700 rpm. Tighten the clamp bolt. The electronic characteristics of this ignition system will advance the timing to 32° at 2200 rpm.

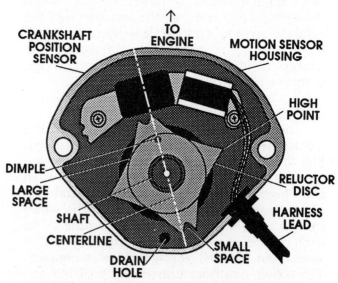

Notch and hole alignment, prior to motion sensor installation.

Reluctor disc and crankshaft position sensor alignment, prior to motion sensor installation.

6
ELECTRICAL

6-1 INTRODUCTION

The battery, gauges and horns, charging system, and the cranking system are all considered subsystems of the electrical system. Each of these units or subsystems will be covered in detail in this chapter beginning with the battery.

6-2 BATTERIES

The battery is one of the most important parts of the electrical system. In addition to providing electrical power to start the engine, it also provides power for operation of the the running lights, radio, electrical accessories, and possibly the pump for a bait tank.

Because of its job and the consequences if it should fail in an emergency, the best advice is to purchase a well-known brand with an extended warranty period from a reputable dealer.

The usual warranty covers a prorated replacement policy which means you would be entitled to a consideration for the time left on the warranty period if the battery should prove defective before its time.

Do not consider a battery of less than 70-ampere hour capacity. If in doubt as to how large your boat requires, make a liberal estimate and then purchase the one with the next higher ampere rating.

MARINE BATTERIES

Because marine batteries are required to perform under much more rigorous conditions than automotive batteries, they are constructed much differently than those used in automobiles or trucks. Therefore, a marine battery should always be the No. 1

unit for the boat and other types of batteries used only in an emergency.

Marine batteries have a much heavier exterior case to withstand the violent pounding and shocks imposed on it as the boat moves through rough water and in extremely tight turns.

The plates in marine batteries are thicker than in automotive batteries and each plate is securely anchored within the battery case to ensure extended life.

The caps of marine batteries are "spill proof" to prevent acid from spilling into the bilges when the boat heels to one side in a tight turn or is moving through rough water.

Because of these features, the marine battery will recover from a low charge condition and give satisfactory service over a much longer period of time than any type of automotive-type unit.

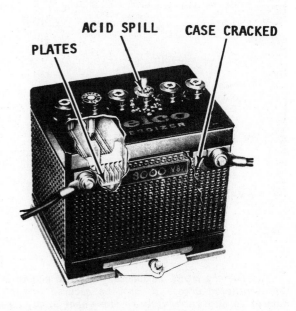

Principle parts of a modern battery.

BATTERY CONSTRUCTION

A battery consists of a number of positive and negative plates immersed in a solution of dilute sulfuric acid. The plates contain dissimilar active materials and are kept apart by separators. The plates are grouped into what are termed elements. Plate straps on top of each element connect all of the positive plates and all of the negative plates into the groups. The battery is divided into cells which hold a number of the elements apart from the others. The entire arrangement is contained within a hard-rubber case. The top is a one-piece cover and contains the filler caps for each cell. The terminal posts protrude through the top where the battery connections for the boat are made. Each of the cells are connected to each other in a positive-to-negative manner with a heavy strap called the cell connector.

BATTERY RATINGS

Two ratings are used to classify batteries: One is a 20-hour rating at 80°F and the other is a cold rating at 0°F. This second

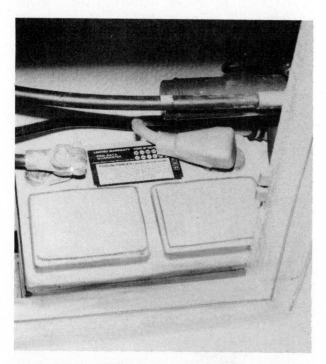

The battery **MUST** be located near the engine in a well-ventilated area. It must be secured in such a manner to allow absolutely no movement in any direction under the most violent action of the boat.

figure indicates the cranking load capacity and is refered to as the Peak Watt Rating of a battery. This Peak Watt Raing (PWR) has been developed as a measure of the batteries cold-cranking ability. The numerical rating is embossed on each battery case at the base and is determined by multiplying the maximum current by the maximum voltage.

The ampere-hour rating of a battery is its capacity to furnish a given amount of amperes over a period of time at a cell voltage of 1.5. Therefore, a battery with a capacity of maintaining 3 amperes for 20 hours at 1.5 volts would be classified as a 60-ampere hour battery.

Do not confuse the ampere-hour rating with the PWR because they are two unrelated figures used for different purposes.

A replacement battery should have a power rating equal or as close to the old unit as possible.

BATTERY LOCATION

Every battery installed in a boat must be secured in a well-protected ventilated area. If the battery area is not well ventilated, hydrogen gas which is given off during charging could become very explosive if the gas is concentrated and confined. Because of its size, weight, and acid content, the battery must be well-secured. If the battery should break loose during rough boat maneuvers, considerable damage could be done, including damage to the hull.

BATTERY SERVICE

The battery requires periodic servicing and a definite maintenance program will ensure extended life. If the battery should test satisfactorily, but still fail to perform properly, one of four problems could be the cause.

1- An accessory might have accidently been left on overnight or for a long period during the day. Such an oversight would result in a discharged battery.

2- Slow speed engine operation for long periods of time resulting in an undercharged condition.

3- Using more electrical power than the alternator can replace would result in an undercharged condition.

4- A defect in the charging system. A slipping fan belt, a defective voltage regulator, a faulty alternator, or high resistance somewhere in the system could cause the battery to become undercharged.

5- Failure to maintain the battery in good order. This might include a low level of electrolyte in the cells; loose or dirty cable connections at the battery terminals; or possibly an excessive dirty battery top.

Electrolyte Level

The most common practice of checking the electrolyte level in a battery is to remove the cell cap and visually observe the level in the vent well. The bottom of each vent well has a split vent which will cause the surface of the electrolyte to appear distorted when it makes contact. When the distortion first appears at the bottom of the split vent, the electrolyte level is correct.

Some late-model batteries have an electrolyte-level indicator installed which operates in the following manner: A transparent rod extends through the center of one of the cell caps. The lower tip of the rod is immersed in the electrolyte when the level

is correct. If the level should drop below normal, the lower tip of the rod is exposed and the upper end glows as a warning to add water. Such a device is only necessary on one cell cap because if the electrolyte is low in one cell it is also low in the other cells. **BE SURE** to replace the cap with the indicator onto the second cell from the positive terminal.

During hot weather and periods of heavy use, the electrolyte level should be checked more often than during normal operation. Add colorless, odorless, drinking water to bring the level of electrolyte in each cell to the proper level. **TAKE CARE** not to overfill because it will cause loss of electrolyte and any loss will result in poor performance, short battery life, and will contribute quickly to corrosion. **NEVER** add electrolyte from another battery. Use only clean pure water.

Cleaning

Dirt and corrosion should be cleaned from the battery just as soon as it is discovered. Any accumulation of acid film or dirt will permit current to flow between the terminals. Such a current flow will drain the battery over a period of time.

Clean the exterior of the battery with a solution of diluted ammonia or a soda solution to neutralize any acid which may be present. Flush the cleaning solution off with clean water. **TAKE CARE** to prevent any of the neutralizing solution from entering the cells by keeping the caps tight.

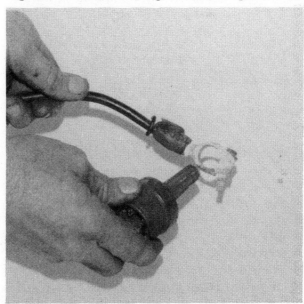

An inexpensive brush can be purchased and used to clean the battery terminals to ensure a proper connection.

One of the most effective means of cleaning the battery terminals is by using a wire brush with holder designed for this specific purpose.

A poor contact at the terminals will add resistance to the charging circuit. This resistance will cause the voltage regulator to register a fully charged battery, and thus cut down on the alternator output adding to the low battery charge problem.

Scrape the battery posts clean with a suitable tool or with a stiff wire brush. Clean the inside of the cable clamps to be sure they do not cause any resistance in the circuit.

Battery Testing

A hydrometer is a device to measure the percentage of sulfuric acid in the battery electrolyte in terms of specific gravity. When the condition of the battery drops from fully charged to discharged, the acid leaves the solution and enters the plates, causing the specific gravity of the electrolyte to drop.

The following six points should be observed when using a hydrometer.

1- NEVER attempt to take a reading immediately after adding water to the battery. Allow at least 1/4 hour of charging at a high rate to thoroughly mix the electrolyte with the new water and to cause vigorous gassing.

2- ALWAYS be sure the hydrometer is clean inside and out as a precaution against contaminating the electrolyte.

3- If a thermometer is an integral part of the hydrometer, draw liquid into it several times to ensure the correct temperature before taking a reading.

Corroded battery terminals such as these result in high resistance at the connections. Such corrosion places a strain on any and all electrically operated devices on the boat and causes hard engine starting.

4- BE SURE to hold the hydrometer vertically and suck up liquid only until the float is free and floating.

5- ALWAYS hold the hydrometer at eye level and take the reading at the surface of the liquid with the float free and floating.

Disregard the light curvature appearing where the liquid rises against the float stem due to surface tension.

6- DO NOT drop any of the battery fluid on the boat or on your clothing, because it is extremely caustic. Use water and baking soda to neutralize any battery liquid that does drop.

After withdrawing electrolyte from the battery cell until the float is barely free, note the level of the liquid inside the hydrometer. If the level is within the green band range, the condition of the battery is satisfactory. If the level is within the white

A check of the electrolyte in the battery should be a regular part of the maintenance schedule on any boat. A hydrometer reading of 1.300 or in the green band, indicates the battery is in satisfactory condition. If the reading is 1.150 or in the red band, the battery needs to be charged. Observe the six points listed in the text when using a hydrometer.

band, the battery is in fair condition, and if the level is in the red band, it needs charging badly or is dead and should be replaced. If level fails to rise above the red band after charging, the only answer is to replace the battery.

JUMPER CABLES

If booster batteries are used for starting an engine the jumper cables must be connected correctly and in the proper sequence to prevent damage to either battery, or to the alternator diodes.

ALWAYS connect a cable from the positive terminal of the dead battery to the positive terminal of the good battery **FIRST**. **NEXT**, connect one end of the other cable to the negative terminal of the good battery and the other end to the **ENGINE** for a good ground. By making the ground connection on the engine, if there is an arc when you make the connection it will not be near the battery. An arc near the battery could cause an explosion, destroying the battery and causing serious personal injury.

DISCONNECT the battery ground cable before replacing an alternator or before connecting any type of meter to the alternator.

If it is necessary to use a fast-charger on a dead battery, **ALWAYS** disconnect one of the boat cables from the battery first, to prevent burning out the diodes in the alternator.

NEVER use a fast charger as a booster to start the engine because the diodes in the alternator will be **DAMAGED**.

STORAGE

If the boat is to be laid up for the winter or for more than a few weeks, special attention must be given to the battery to prevent complete discharge or possible damage to the terminals and wiring. Before putting the boat in storage, disconnect and remove the batteries. Clean them thoroughly of any dirt or corrosion, and then charge them to full specific gravity reading. After they are fully charged, store them in a clean cool dry place where they will not be damaged or knocked over.

NEVER store the battery with anything on top of it or cover the battery in such a manner as to prevent air from circulating around the fillercaps. All batteries, both new and old, will discharge during periods of storage, more so if they are hot than if they remain cool. Therefore, the electrolyte level and the specific gravity should be checked at regular intervals. A drop in the specific gravity reading is cause to charge them back to a full reading.

In cold climates, care should be excused in selecting the battery storage area. A fully-charged battery will freeze at about 60 degrees below zero. A discharged battery, almost dead, will have ice forming at about 19 degrees above zero.

DUAL BATTERY INSTALLATION

Three methods are available for utilizing a dual-battery hook-up.

1- A high-capacity switch can be used to connect the two batteries. The accompanying illustration details the connections

An explosive hydrogen gas is released from the cells when the caps are removed. This battery exploded when the gas ignited from smoking in the area with the caps removed, or possibly from a spark at the terminal post.

The charging system output can be determined while the engine is running at a fast idle speed by holding an induction-type ammeter over the main wire.

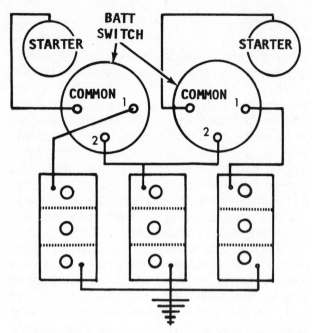

Schematic diagram for a three battery, two engine hookup.

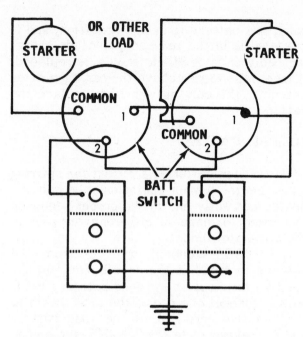

Schematic diagram for a two battery, two engine hookup.

for installation of such a switch. This type of switch installation has the advantage of being simple, inexpensive, and easy to mount and hookup. However, if the switch is forgotten in the closed position, it will let the convenience loads run down both batteries and the advantage of the dual installation is lost. However, the switch may be closed intentionally to take advantage of the extra capacity of the two batteries, or it may be temporarily closed to help start the engine under adverse conditions.

2- A relay, can be connected into the ignition circuit to enable both batteries to be automatically put in parallel for charging or to isolate them for ignition use during engine cranking and start. By connecting the relay coil to the ignition terminal of the ignition-starting switch, the relay will close during the start to aid the starting battery. If the second battery is allowed to run down, this arrangement can be a disadvantage since it will draw a load from the starting battery while cranking the engine. One way to avoid such a condition is to connect the relay coil to the ignition switch accessory

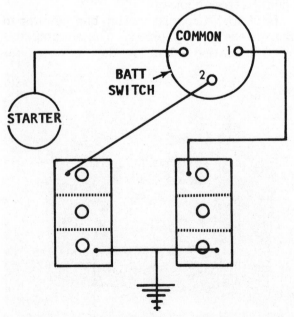

Schematic diagram for a two battery, one engine hookup.

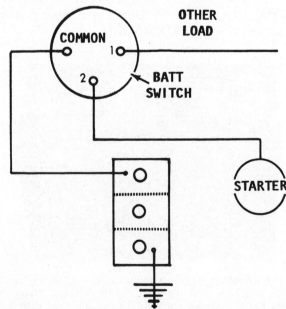

Schematic diagram for a single battery, one engine hookup.

terminal. When connected in this manner, while the engine is being cranked, the relay is open, but when the engine is running with the ignition switch in the normal position, the relay is closed, and the second battery is being charged at the same time as the starting battery.

3- A heavy duty switch installed as close to the batteries as possible can be connected between them. If such an arrangement is used it must meet be the standards of the Amnerican Boat and Yacht Council, INC. or the Fire Protection Standard for Motor Craft, N.F.P.A. No. 302.

6-3 GAUGES AND HORNS

Gauges or lights are installed to warn the opertor of a condition in the cooling and lubrication systems that may need attention. The fuel gauge gives an indication of the amount of fuel in the tank. If the engine overheats or the oil pressure drops to low for safety, a gauge or warning light reminds the operator to shut down the engine and check the cause of the warning before serious damage is done.

CONSTANT-VOLTAGE SYSTEM

In order for gauges to register properly, they must be supplied with a steady voltage. The voltage variations produced by the engine charging system would cause erratic gauge operation, too high when the alternator voltage is high and too low when the alternator is not charging. To remedy this problem, a constant-voltage system is used to reduce the 12-14 volts of the electrical system to an average of 5 volts. This steady 5 volts ensures the gauges will read accurately under varying conditions from the electrical system.

SERVICE PROCEDURES

Systems utilizing warning lights do not require a constant-voltage system, therefore, this service is not needed.

Service procedures for checking the gauges and their sending units is detailed in the following sections.

6-4 OIL AND TEMPERATURE GAUGES

The body of oil and temperature gauges must be grounded and they must be supplied with 12 volts. Many gauges have a terminal on the mounting bracket for attaching a ground wire. A tang from the mounting bracket makes contact with the gauge. **CHECK** to be sure the tang does make good contact with the gauge.

Ground the wire to the sending unit and the needle of the gauge should move to the full right position indicating the gauge is in serviceable condition.

Check Gauge Sender for a defective motor temperature warning system: Remove the sender from the engine. Connect the sender terminals to an ohmmeter. Submerge the sender in a container of oil with a thermometer. Heat the oil over a fireless heating element. Now, observe the meter and thermometer readings. The meter should read 450 ohms +5% at 220°F.

Check The Sender for a defective oil pressure warning system: Substitute a new sender unit of the correct value. Start the engine with the propeller in the water. Observe the gauge. If the reding is still unsatisfactory, replace the original gauge and test again. If the reading is still unsatisfactory, the problem may be in the engine lubrication system and due to worn bearings. **NEVER** attempt to interchange the sending unit from a system using a gauge with a unit from one using a warning light.

6-5 WARNING LIGHTS

If a problem arises on a boat equipped with water, temperature, and oil pressure

The indicator and control panel should be kept clean and protected from water spray, especially when operating in a salt water atmosphere.

lights, the first area to check is the light assembly for loose wires or burned-out bulbs.

When the ignition key is turned on, the light assembly is supplied with 12 volts and grounded through the sending unit mounted on the engine. When the sending unit makes contact because the water temperature is too hot or the oil pressure is too low, the circuit to ground is completed and the lamp should light.

Check The Bulb: Turn the ignition switch on. Disconnect the wire at the engine sending unit, and then ground the wire. The lamp on the dash should light. If it does not light, check for a burned-out bulb or a break in the wiring to the light.

Check The Sender for a defective motor temperature warning system: Remove the sending unit from the engine and connect the terminals to an ohmmeter. Submerge the sender in a container of oil with a thermometer. Heat the oil over a fireless heating element. Observe the thermometer readings. The meter should indicate an open circuit until the temperature reaches $200+5^{\circ}$F. If the circuit does not close at the specified temperature, replace the unit.

Check The Sender for a defective oil pressure warning system: Disconnect the electrical lead at the sending unit. Connect an ohmmeter between the terminal and ground. Turn the ignition switch on and the meter should indicate a complete circuit. Start the engine.

CAUTION
Water must circulate through the stern drive, also to and from the engine, anytime the engine is run to prevent damage to the water pump located in the lower unit (original and Alpha drives); or on the front of the engine (Bravo drive). Just a few seconds without water will damage the water pump.

Increase engine rpm and the meter should then indicate an open circuit. If it does not, replace the sender. If a new sender still fails to open the circuit, the problem may be in the engine lubricating system or because of worn bearings.

THERMOMELT STICKS

Thermomelt sticks are an easy method

of determining if the engine is running at the proper temperature. Thermomelt sticks are not exsspensive and are available at your local marine dealer.

Start the engine with the propeller in the water and run it for about 5 minutes at about 3000 rpm.

CAUTION
Water must circulate through the stern drive, also to and from the engine, anytime the engine is run to prevent damage to the water pump located in the lower unit (original and Alpha drives); or on the front of the engine (Bravo drive). Just a few seconds without water will damage the water pump.

The 140 degree stick should melt when you touch it to the lower thermostat housing. If it does not melt, the thermostat is stuck in the open position and the engine temperature is too low.

Touch the 170 degree stick to the same spot on the lower thermostat housing and it should not melt. If it does, the thermostat is stuck in the closed position or the water pump is not operating properly because the engine is running too hot. For service procedures on the cooling system, see Chapter 9.

6-6 FUEL GAUGES

The fuel gauge is intended to indicate the quantity of fuel in the tank. As the experienced boatman has learned, the gauge reading is seldom an accurate report of the fuel available in the tank. The main reason for this false reading is because the boat is rarely on an even keel. A considerable difference in fuel quantity will be indicated by the gauge if the bow or stern is heavy and if the boat has a list to port or starboard.

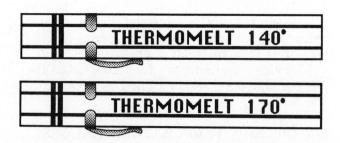

A "Thermomelt" stick is a quick, simple, and fairly accurate method to determine engine operating temperature.

Therefore, the reading is usually low. The amount of fuel drawn from the tank is dependent on the location of the fuel pickup tube in the tank.

The engine may cutout while cruising because the pickup tube is clear of the fuel level. Instead of assuming the tank is empty, shift weight in the boat to change the trim and the problem may be solved until you are able to take on more fuel.

FUEL GAUGE HOOKUP

The Boating Industry Association recommends the following color coding be used on all fuel gauge installations:

Black — for all grounded current-carrying conductors.

Pink -- insulated wire for the fuel gauge sending unit to the gauge.

Red -- insulated wire for a connection from the positive side of the battery to any electrical equipment.

Connect one end of a pink insulated wire to the terminal on the gauge marked **TANK** and the other end to the terminal on top of the tank unit.

Connect one end of a black wire to the terminal on the fuel gauge marked **IGN** and the other end to the ignition switch.

Connect one end of a second black wire to the fuel gauge terminal marked **GRD** and the other end to a good ground. It is important for the fuel gauge case to have a good common ground with the tank unit. Aboard an all-metal boat, this ground wire is not necessary. However, if the dashboard is insulated, or made of wood or plastic, a wire **MUST** be run from the gauge ground terminal to one of the bolts securing the sending unit in the fuel tank, and then from there to the **NEGATIVE** side of the battery.

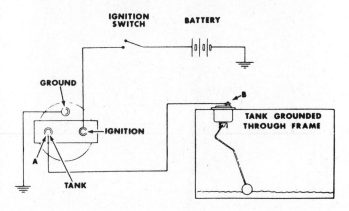

Schematic diagram for safe fuel tank hookup.

FUEL GAUGE TROUBLESHOOTING

In order for the fuel gauge to operate properly the sending unit and the receiving unit must be of the same type and preferably of the same make.

The following symptoms and possible corrective actions will be helpful in restoring a faulty fuel gauge circuit to proper operation.

If you suspect the gauge is not operating properly, the first area to check is all electrical connections from one end to the other. Be sure they are clean and tight.

Next, check the common gound wire between the negative side of the battery, the fuel tank, and the gauge on the dash.

If all wires and connections in the circuit are in good condition, remove the sending unit from the tank. Run a wire from the gauge mounting flange on the tank to the flange of the sending unit. Now, move the float up-and-down to determine if the receiving unit operates. If the sending unit does not appear to operate, move the float to the midway point of its travel and see if the receiving unit indicates half full.

If the pointer does not move from the **EMPTY** position one of four faults could be to blame:

1- The dash receiving unit is not properly grounded.

2- No voltage at the dash receiving unit.

3- Negative meter connections are on a positive grounded system.

4- Positive meter connections are on a negative grounded system.

If the pointer fails to move from the FULL position, the problem could be one of three faults.

1- The tank sending unit is not properly grounded.

2- Improper connection between the tank sending unit and the receiving unit on the dash.

3- The wire from the gauge to the ignition switch is connected at the wrong terminal.

If the pointer remains at the 3/4 full mark, it indicates a six-volt gauge is installed in a 12-volt system.

If the pointer remains at about 3/8 full, it indicates a 12-volt gauge is installed in a six-volt system.

Erratic Fuel Gauge Readings

Inspect all of the wiring in the circuit for possible damage to the insulation or conductor. Carefully check:

1- Ground connections at the receiving unit on the dash.

2- Harness connector to the dash unit.

3- Body harness connector to the chassis harness.

4- Ground connection from the fuel tank to the trunk floor pan.

5- Feed wire connection at the tank sending unit.

GAUGE ALWAYS READS FULL , when the ignition switch is ON:

1- Check the electrical connections at the receiving unit on the dash; the body harness connector to chassis harness connector; and the tank unit connector in the tank.

2- Make a continuity check of the ground wire from the tank to the trank floor pan.

3- Connect a known good tank unit to the tank feed wire and the ground lead. Raise and lower the float and observe the receiving unit on the dash. If the dash unit follows the arm movement, replace the tank sending unit.

GAUGE ALWAYS READS EMPTY, when the ignition switch is ON:

Disconnect the tank unit feed wire and do not allow the wire terminal to ground. The gauge on the dash should read FULL.

If Gauge Reads Empty:

1- Connect a spare dash unit into the dash unit harness connector and ground the unit. If spare unit reads FULL, the original unit is shorted and must be replaced.

2- A reading of EMPTY indicate a short in the harness between the tank sending unit and the gauge on the dash.

If Gauge Reads Full:

1- Connect a known good tank sending unit to the tank feed wire and the ground lead.

2- Raise and lower the float while observing the dash gauge. If dash gauge follows movement of the float, replace the tank sending unit.

GAUGE NEVER INDICATES FULL

This test requires shop test equipment.

1- Disconnect the feed wire to the tank unit and connect the wire to ground thru a variable resistor or thru a spare tank unit.

2- Observe the dash gauge reading. The reading should be FULL when resistance is increased to about 90 ohms. This resistance would simulate a full tank.

3- If the check indicates the dash gauge is operating properly, the the trouble is either in the tank sending unit rheostat being shorter, or the float is binding. The arm could be bent, or the tank may be deformed. Inspect and correct the problem.

6-7 TACHOMETER

An accurate tachometer can be installed on any engine. Such an instrument provides an indication of engine speed in revolutions per minute (rpm). This is accomplished by measuring the number of electrical pulses per minute generated in the primary circuit of the ignition system.

The meter readings range from 0 to 6,000 rpm, in increments of 100. Tachometers have solid-state electronic ciruits which eliminates the need for relays or batteries and contributes to their accuracy. The electronic parts of the tachometer susceptible to moisture are coated to prolong their life.

Maximum engine performance can only be obtained through proper tuning using a tachometer.

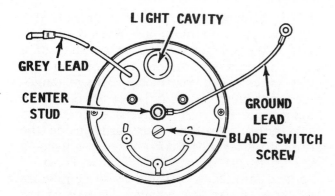

A blade-type tachometer can be installed for use with a four, six, or eight cylinder engine. Simply shift the blade to the connector matching the number of engine cylinders.

6-8 HORNS

The only reason for servicing a horn is because it fails to operate properly or because it is out of tune. In most cases, the problem can be traced to an open circuit in the wiring or to a defective relay.

Cleaning: Crocus cloth and carbon tetrachloride should be used to clean the contact points. **NEVER** force the contacts apart or you will bend the contact spring and change the operating tension.

Check The Relay and Wiring: Connect a wire from the battery to the horn terminal. If the horn operates, the problem is in the relay or in the horn wiring. If both of these appear satisfactory, the horn is defective and needs to be replaced.

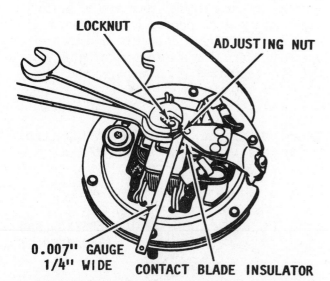

The tone of a horn can be adjusted with a 0.007" feeler gauge, as described in the text. TAKE CARE to prevent the feeler gauge from making contact with the case, or the circuit will be shorted out.

Before replacing the horn however, connect a second jumper wire from the horn frame to ground to check the ground connection.

Test the winding for an open circuit, faulty insulation, or poor ground. Check the resistor with an ohmmeter, or test the condenser for capacity, ground, and leakage. Inspect the diaphragm for cracks.

Adjust Horn Tone: Loosen the locknut, and then rotate the adjusting screw until the desired tone is reached. On a dual horn installation, disconnect one horn and adjust each one-at-a-time. The contact point adjustment is made by inserting a 0.007" feeler gauge blade between the adjusting nut and the contact blade insulator. **TAKE CARE** not to allow the feeler gauge to touch the metallicparts of the contact points because it would short them out. Now, loosen the locknut and turn the adjusting nut down until the horn fails to sound. Loosen the adjusting nut slowly until the horn barely sounds. The locknut **MUST** be tightened after each test. When the feeler gauge is withdrawn the horn will operate properly and the current draw will be satisfactory.

6-9 CHARGING SYSTEM

The alternator, regulator, battery, ammeter, and the necessary wiring to connect it all together comprise the charging system.

Before the alternator is blamed for battery problems, consider other areas which may be the cause:

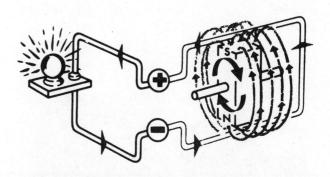

Simplified drawing to illustrate voltage generation in its most basic form. The stator has many coils of wire, shown here as a single loop. The rotor spins; cuts the magnetic field; an electrical voltage is produced in the loop; the voltage moves out of the loop and into the system. Higher revolutions per minute of the loop generates more voltage.

1- Excessive use of lights and accessories while the engine is shut down or operating at low speed for short periods.

2- Voltage losses and the cause.

3- Corroded battery cables, connectors, and terminals.

4- Low electrolyte level in the battery cells.

5- Prolonged disuse of the battery causing a self-discharged condition.

6-10 ALTERNATOR
DESCRIPTION AND OPERATION

The alternator used on Mercruiser engines is an AC alternator replacing the conventional DC generator. The unit installed may be a Delco-Remy, Motorola, Autolite/Prestolite, or a Mando. In very general terms, all alternators operate the same. The alternator has four distinct advantages over the generator:

1- A higher charging rate.

2- A lower cut-in charging speed.

3- A lighter weight.

4- Longer trouble-free service.

The alternator operates on a different principle from the conventional generator. The armature of the alternator is the stationary part and is called the **STATOR**; and the field is the rotating part and is called the **ROTOR**. With this arrangement, the higher current carried by the stator is conducted to the external circuit through fixed leads. This method results in trouble-free operation of a small current supplied to the fields to be conducted through small brushes and rotating slip rings. This is in contrast to the DC generator where the current is carried through a rotating commutator and brushes.

The alternator has a three-phase stator winding. The windings are phased electrically 120° apart. The rotor consists of a field coil encased between two four- or six-pole, interleaved sections. This arrangement produces an eight- or twelve-pole magnetic field, with alternate north and south poles. As the rotor turns inside the stator an alternating current (AC) is induced in the stator windings. This current is changed into DC (rectified) by silicon diodes and then sent to the output terminals of the alternator.

Consider the silicon diode rectifiers as electrical one-way valves. Three of the diodes are polarized one way and are pressed into an aluminum heat sink which is grounded to the slip ring end head. The other three diodes are polarized the opposite way and are pressed into a similar heat sink, which is insulated from the end head and connected to the alternator output terminal. Since a diode has high resistance to the flow of electricity in one direction and passes current with very little resistance in the opposite direction, it is connected in a manner which allows current to flow from the alternator to the battery in a low-resistance direction.

The high resistance in the opposite direction prevents battery current from flowing to the alternator, therefore, no circuit breaker (cutout) is required between the alternator and the battery.

The magnetism left in the rotor field poles is negligible. Thus, the field must be excited by an external source, the battery. The battery is connected to the field winding through the ignition switch and the voltage regulator. The alternator charging voltage is regulated by varying the field strength. This is controlled by the voltage

The Mando alternator (right), looks very much like the Motorola (left), but is tested like a Delco-Remy.

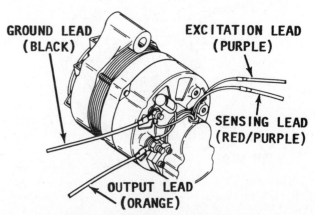

Wire identification for a Mando alternator.

regulator. No current regulator is required since the alternator has self-limiting current characteristics.

ALTERNATOR PROTECTION DURING SERVICE

The alternator is an important and expensive piece of equipment on the boat. Unless certain precautions are taken before and during servicing, parts of the alternator or the charging circuit may be damaged. **TAKE TIME** to review the following points **BEFORE** servicing or troubleshooting the charging system.

1- If the battery connections are reversed, the diodes in the alternator will be damaged. Check the battery polarity with a voltmeter before making any connections to be sure the connections correspond to the battery ground polarity.

2- The field circuit between the alternator and the regulator **MUST NEVER BE GROUNDED.** If this circuit is grounded the regulator will be damaged.

3- NEVER ground the alternator ouput terminal while the engine is running or shut down because no circuit breaker is used, and battery current is applied to the alternator output terminal at all times.

4- NEVER operate the alternator on an open circuit with the field winding energized, or the unit will be **DAMAGED.**

5- NEVER attempt to polarize the alternator because it is never necessary. Any attempt to polarize the alternator will result in **DAMAGE** to the alternator, the regulator, and the wiring.

6- NEVER short the bending tool to the regulator base while adjusting the voltage, or the unit will be **DAMAGED.** The bending tool should be **INSULATED** with tape or an insulating plastic sleeve while making the adjustment.

7- ALWAYS disconnect the battery ground strap before replacing an alternator or connecting any meter to it.

8- If booster batteries are used for starting an engine the jumper cables must be connected correctly and in the proper sequence to prevent damage to either battery, or to the alternator diodes.

ALWAYS connect a cable from the positive terminal of the dead battery to the positive terminal of the good battery **FIRST. NEXT,** connect one end of the other cable to the negative terminal of the good battery and the other end to the **ENGINE** for a good ground. By making the ground connection on the engine, if there is an arc when you make the connection it will not be near the battery. An arc near the battery could cause an explosion, destroying the battery and causing serious personal injury.

9- If it is necessary to use a fast charger on a dead battery, **ALWAYS** disconnect one of the boat cables from the battery first, to prevent burning out the diodes in the alternator.

10- NEVER use a fast charger as a booster to start the engine because the diodes in the alternator will be **DAMAGED.**

CHARGING SYSTEM TROUBLESHOOTING

The following symptoms and possible corrective actions will be helpful in restoring a faulty charging system to proper operation.

Alternator Fails To Charge

1- Drive belt loose or broken. Replace and/or adjust.

2- Corroded or loose wires or connection in the charging circuit. Inspect, clean, and tighten.

3- Worn brushes or slip rings. Replace as required.

4- Sticking brushes.

5- Open field circuit. Trace and repair.

6- Open charging circuit. Trace and repair.

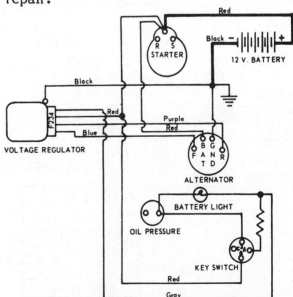

Schematic diagram for a charging system with an indicator light on the dashboard and a relay-type voltage regulator installed.

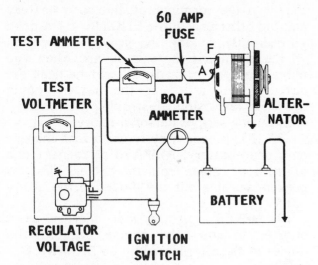

Functional diagram for hookup to test the electrical generating system.

7- Open circuit in the stator windings.

8- Open rectifier. Replace the unit.

9- Defective regulator.   Replace the unit.

Alternator Charges Low or Unsteady

1- Drive belt loose or broken. Replace and/or adjust.

2- Battery charge too low. Charge or replace the battery.

3- Defective regulator.   Replace the unit.

4- High resistance at the battery terminals. Remove cables, clean connectors and battery posts, replace and tighten.

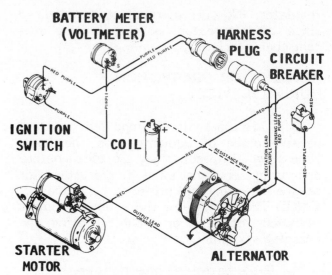

Functional diagram -- Mando alternator.

5- High resistance in the charging circuit. Trace and repair.

6- Open stator winding. Replace alternator.

7- High resistance in body-to-ground lead. Trace and repair.

Alternator Output Low and a Low Battery

1- Drive belt slipping. Adjust.

2- High resistance in the charging circuit. Trace and repair.

3- Shorted or open-circuited rectifier. Replace rectifier.

4- Grounded stator windings. Replace alternator.

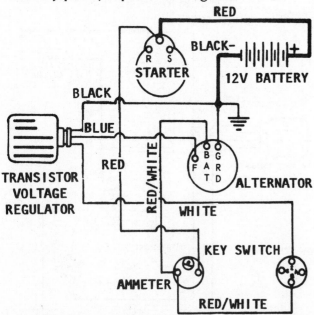

Schematic diagram for a charging system with an ammeter on the dashboard and a relay-type voltage regulator installed.

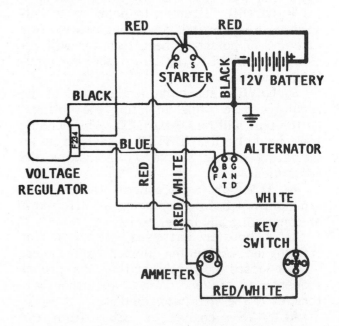

Schematic diagram for a charging system with an ammeter on the dashboard and a transistor-type regulator installed.

Alternator Output Too High — Battery Overcharged

1- Faulty voltage regulator. Replace.

2- Regulator base not grounded properly. Correct condition to make good ground.

3- Faulty ignition switch. Replace.

Alternator Too Noisy

1- Worn, loose, or frayed drive belt. Replace belt and adjust properly.

2- Alternator mounting loose. Tighten all mounting hardware securely.

3- Worn alternator bearings. Replace.

4- Interference between rotor fan and stator leads or rectifier. Check and correct.

5- Rotor or fan damaged. Replace.

6- Open or shorted rectifier. Replace.

7- Open or shorted winding in stator. Replace.

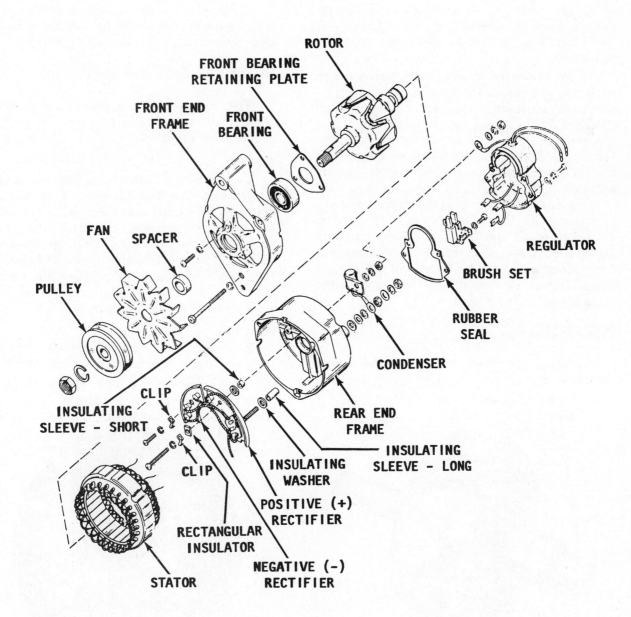

Exploded view of the Prestolite alternator with principle parts identified.

Ammeter Fluctuates Constantly

1- High resistance connection in the alternator or voltage regulator circuit. Trace and repair.

6-11 SERVICING CHARGING SYSTEM WITH SEPARATE REGULATOR

Trouble between the alternator and the regulator may be isolated as follows:

First, connect one end of a jumper wire to the **BAT** terminal and the other end to the F (field) terminal.

If the charging system works with the jumper wire in place, but not without it, then the regulator is at fault and must be replaced.

If the charging system does not operate with the jumper in place, then the alternator is at fault and must be replaced. The **REGULATOR MUST** be replaced any time the alternator is replaced, because the regulator will also be defective as evidenced by its failure to control the alternator output properly.

Regulator trouble is often caused by high resistance in the charging circuit at the ammeter or at the battery. Check both of these areas, and if corrosion is discovered, the connection must be disassembled, cleaned, and then tightened securely to prevent recurrence of the problem.

DISASSEMBLING

Remove the nuts, washers, and cover from the top of the brush holder. Remove the studs and the brush-and-spring assemblies. Remove the screws securing the brush holder to the end frame, and then remove the brush holder.

Scribe a mark on both end frames and matching marks on the stator as an aid to properly assembling the parts.

Remove the four through-bolts. Separate the drive end frame and rotor assembly from the stator by **CAREFULLY** prying with a screwdriver at the slot in the stator frame. **NEVER** pry anywhere except at the slot or the castings will be damaged.

Cover both sides of the bearing on the end frame and the bearing surface with tape as a prevention from damaging them.

CLEANING AND TESTING

The following procedures are keyed by number to matching illustrations as an aid in performing the work.

ROTOR

1- Check the rotor windings for a short or ground: Connect an ohmmeter between one brush slip ring and the shaft. The meter must indicate an open circuit on the R-100 scale. Check the rotor windings for continuity: Connect the ohmmeter between the two slip rings, and the reading should be 6 ohms for a 42-ampere alternator rotor, or 4.5 ohms for a 32-ampere rotor.

DIODES

2- Each diode must be tested to be sure it is not open or shorted. Check the three diodes in the heat sink: Disconnect the

(CHECK FOR GROUNDS)

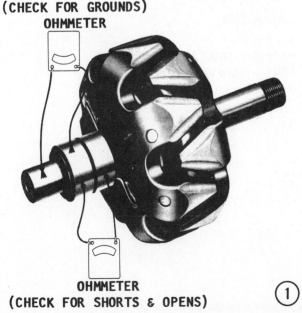

OHMMETER

OHMMETER
(CHECK FOR SHORTS & OPENS)   ①

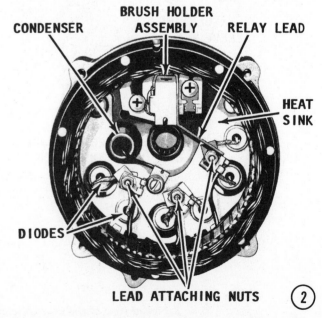

CONDENSER   BRUSH HOLDER
ASSEMBLY   RELAY LEAD

HEAT SINK

DIODES

LEAD ATTACHING NUTS   ②

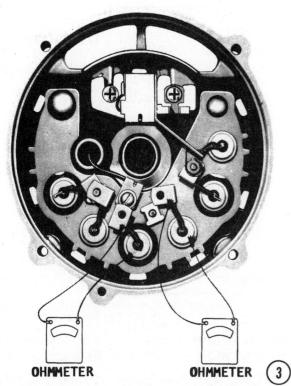

OHMMETER   OHMMETER   ③

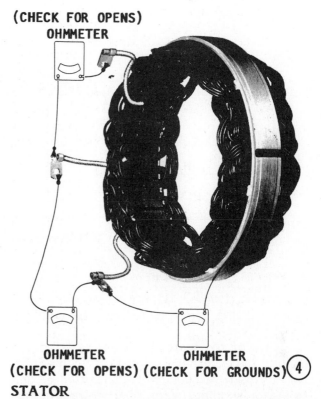

(CHECK FOR OPENS)
OHMMETER

OHMMETER   OHMMETER   ④
(CHECK FOR OPENS) (CHECK FOR GROUNDS)

STATER

stator leads and test each diode with one ohmmeter probe on the heat sink and the other on the disconnected diode terminal. Note the reading, and then reverse the test leads and again note the reading. If one reading is high and the other low, the diode is satisfactory. If both readings are zero or infinity, the diode is defective and must be replaced.

3- Check the three diodes in the end frame: Connect the ohmmeter in the same manner as for the previous test **EXCEPT** connect one test lead to the end frame instead of the heat sink. Note the reading, then reverse the leads and again note the reading. The results should be the same as the previous test.

STATOR

4- The stator is not checked for shorts due to the very low resistance of the windings. Neither are checks of the delta stator for opens because the windings are connected in parallel.

Checks that are made on the stator must be made with all of the diodes disconnected from the stator.

Check the Y-connected stator for open circuits: Connect an ohmmeter or test light across any two pair of terminals. If the ohmmeter reading is high, or if the test light fails to come on, there is an open in the winding.

Both types of stator winding can be checked for grounds by connecting an ohmmeter or test light from either terminal to

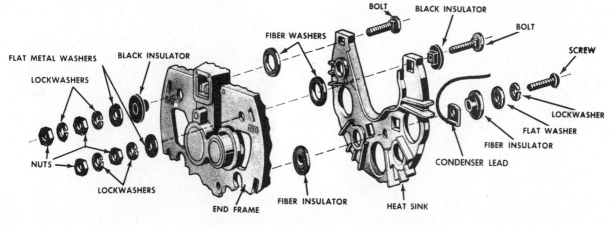

Exploded view showing all parts of the heat sink.

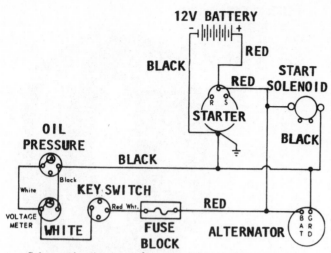

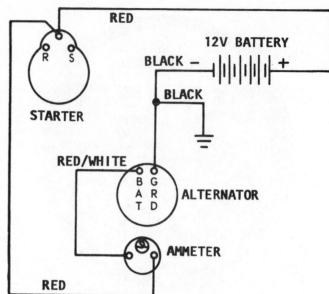

Schematic diagram of an alternator with an integral regulator and a harness length over 20 feet.

the stator frame. If the ohmmeter reads low or if the lamp comes on, the windings are grounded.

If all of the tests check out satisfactorily, but the alternator still fails to meet its rated output, a shorted Y-connected or delta stator winding, or an open delta winding, may be the cause.

6-12 SERVICING DELCOTRON CHARGING SYSTEM

Alternators with an integral solid-state regulator installed, such as the Delcotron, are easily identified by an explosion-resistant screen at each end frame.

Trouble between the alternator and the regulator may be isolated as follows:

Start the engine.

CAUTION

Water must circulate through the stern drive, also to and from the engine, anytime the engine is run to prevent damage to the water pump located in the lower unit (original and Alpha drives); or on the front of the

Schematic diagram of an alternator with an integral regulator and a harness length less than 20 feet.

engine (Bravo drive). Just a few seconds without water will damage the water pump.

Now, with the engine running, **CAREFULLY AND SLOWLY** insert a small bladed screwdriver **NO MORE THAN ONE-INCH** into the test hole in the slip ring end frame. **NEVER** insert the screwdriver more than 1", because you will contact the rotor and **DESTROY THE ALTERNATOR.**

With the screwdriver in place, the field winding is grounded and the regulator is bypassed. If the alternator charges with the screwdriver in place, the regulator is defective and must be replaced. If the alternator does not charge with the screwdriver in place, then the alternator is defective and must be repaired or replaced.

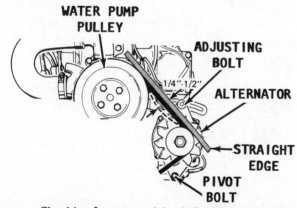

Checking for proper drive belt tension.

Exterior view of an alternator with an integral regulator.

Removing the alternator pulley.

WHENEVER the alternator is repaired or replaced, the regulator must be replaced because it is also defective, as evidenced by its failure to control the alternator output properly.

Many times, the cause of a defective regulator is a high resistance connection in the charging circuit, at the ammeter, or at the battery. Each of these areas should be checked and serviced thoroughly to prevent repetition of the trouble. Disassemble the connection, clean, and then tighten it.

DISASSEMBLING

Remove the four through-bolts, and then take off the cover. Secure a 5/16" Allen wrench in a vise, as shown in the accompanying illustration.

Now, slide the alternator shaft over the exposed end of the Allen wrench. Use a 15/16" box-end wrench to loosen, and then remove the end nut, washer, pulley, fan, collar, drive-end frame, and collar from the rotor shaft.

CAREFULLY pull the rotor assembly

from the slip-ring end frame, and **TAKE CARE** not to lose the brush springs. Cover the slip-ring end frame with tape to keep dirt out of the bearing. Cover the slip rings with tape also as a precaution against scoring them. **NEVER** use friction tape, which leaves a gummy deposit which is very difficult to remove.

CLEANING AND TESTING

The following procedures are keyed by number to matching illustrations as an aid in performing the work.

ROTOR

1- Check the rotor windings for a short or ground: Connect an ohmmeter between one brush slip ring and the shaft. If the meter does not show an open circuit on the R-100 scale there is a short in the rotor windings. Check the rotor windings for continuity: Connect the ohmmeter between the two slip rings, and a normal reading of 2.6 ohms will indicate the rotor coil is serviceable.

STATOR

2- The stator is not checked for shorts due to the very low resistance of the windings. Neither are checks of the delta stator for opens because the windings are connected in parallel.

Checks that are made on the stator must be made with all of the diodes disconnected from the stator.

(CHECK FOR OPENS)
OHMMETER

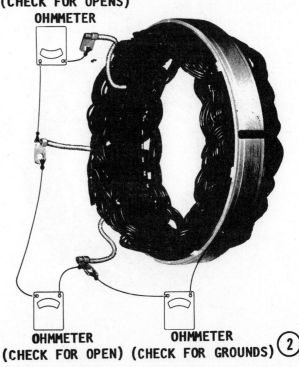

OHMMETER   OHMMETER ②
(CHECK FOR OPEN) (CHECK FOR GROUNDS)

(CHECK FOR GROUNDS)
OHMMETER

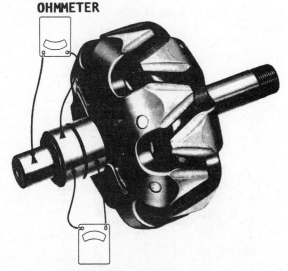

OHMMETER ①
(CHECK FOR SHORTS & OPENS)

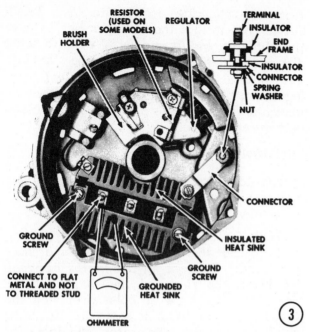

Check the Y-connected stator for open circuits: Connect an ohmmeter or test light across any two pair of terminals. If the ohmmeter reading is high, or if the test light fails to come on, there is an open in the winding.

Both types of stator winding can be checked for grounds by connecting an ohmmeter or test light from either terminal to the stator frame. If the ohmmeter reads low or if the lamp comes on, the windings are grounded.

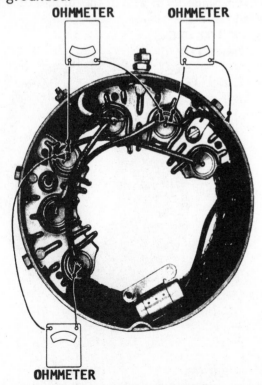

Ohmmeter hookups for testing.

If all of the tests check out satisfactorily, but the alternator still fails to meet its rated output, a shorted Y-connected or delta stator winding, or an open delta winding, may be the cause.

RECTIFIER BRIDGE

3- First, set the ohmmeter scale to R-100. Next, connect one test lead to the grounded heat sink and the other test lead to one of the three terminals. Now, note the reading on the scale. Reverse the test leads, and again note the reading. If both readings are zero or very high, the rectifier bridge is defective. If one of the readings is high and the other low, the rectifier checks out for service.

Repeat the tests between the grounded heat sink and the other two terminals, and then between the insulated heat sink and each of the three terminals. When you are finished, you should have made a total of six tests, with two readings during each test.

DIODE TRIO

4- Remove the diode trio by removing the attaching bolts and nuts. Set the ohmmeter scale to R-100. Check each diode for a short or open by first connecting one test lead to a single connector, of the three connectors, and then noting the reading on the scale. Now, reverse the test leads, and again note the reading. If both readings are zero or very high, the diode is defective. If one reading is low and the other is high, the diode checks out for service.

Repeat each test for each of the other two connectors, taking two readings during each test for a total of six readings. Replace the diode trio if any one diode fails to check out.

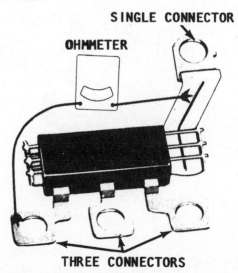

6-13 SERVICING VALEO ALTERNATOR

TROUBLESHOOTING

A great many charging system problems may be traced to a defective battery. Before attempting any troubleshooting of the charging system make a thorough check of the battery condition as described in Section 6-1.

Drive Belt Check

Before attempting troubleshooting procedures, first check the alternator belt for correct tension. A loose belt will slip, reducing alternator output. Such a condition will, in time, lead to an undercharged battery. A belt which is overtightened will place a great strain on the bearings and bushings inside the alternator, leading to premature failure of these components.

Replace the drive belt if it is cracked, worn or saturated with oil.

Check the drive belt tension by depressing the belt at a point midway between pulleys. Heavy thumb pressure on the belt should depress it approximately 1/4" (6mm). To adjust tension on the belt, loosen the alternator pivot and mounting bolts, and then tighten them again finger tight. Move the alternator inboard or outboard, until the correct belt tension is obtained. Tighten the mounting bolt first and then the pivot bolt.

Alternator Output Test

Before beginning charging system tests, the alternator output should be tested to verify a charging problem exists.

The alternator must be tested at operating temperature. Some electrical problems only occur when the temperature rises and increased resistance occurs. Operate the engine for approximately 3 minutes at 2,000 rpm before making any tests.

CAUTION

DO NOT disturb any connections for the alternator, regulator, or battery while the engine is operating. Disconnecting any portion of the charging circuit while it is operating, may cause the electrical current to reverse itself and burn out diodes.

Take care to keep measuring meter cables, hair, clothing, etc., well clear of the drive belt while the engine is operating.

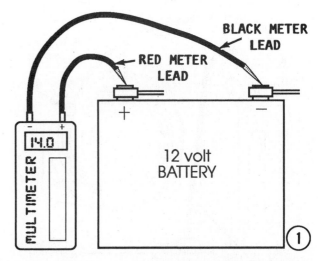

1- Obtain a multimeter, preferably a digital multimeter. This type meter gives a very high degree of accuracy and is easier to read.

CAUTION

Water must circulate through the stern drive, also to and from the engine, anytime the engine is run to prevent damage to the water pump located in the lower unit (original and Alpha drives); or on the front of the engine (Bravo drive). Just a few seconds without water will damage the water pump.

Start the engine and allow it to reach normal operating temperature at idle speed.

Select a meter scale capable of registering more than 14 volts. Make contact with the positive Red meter lead to the positive battery terminal. Make contact with the negative Black meter lead to the negative battery terminal. The meter should register approximately 14 volts. Make a note of the reading for future reference. If the reading is less than 14 volts, proceed to Step 2. If the reading is greater than 14.4 volts, proceed to Removal and Disassembling and

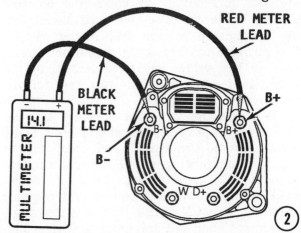

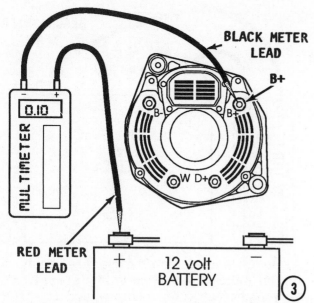

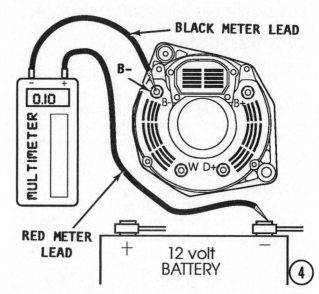

further testing of the internal components of the alternator.

2- Use the same multimeter scale and make contact with the Red meter lead to the **B+** alternator terminal. Make contact with the Black meter lead to the **B-** alternator terminal. Keep the engine idle at 2,000 rpm. The meter should register between 14.0-14.4 volts **AND** only be 0.4 volts less than the reading obtained in the previous step. If the voltage drop is less than 0.2 volts, both the battery cables are satisfactory. If the voltage drop exceeds 0.3 volts, proceed to Step 3 to check the positive battery cable.

3- Select a meter scale capable of registering 0.00 to 0.50 volts. Keep the engine idle at 2,000 rpm.

Make contact with the Red meter lead to the positive battery terminal. Make contact with the Black meter lead to **B+** alternator terminal. The meter should register no more than 0.2 volts. If the meter registers more than 0.2 volts, the positive battery cable has excessive resistance causing too great a voltage drop in the circuit. The positive battery cable should be replaced. If the meter registers less than 0.2 volts, proceed to the following test for the negative battery cable.

4- Keep the same meter scale and engine rpm as in the preceding step. Make contact with the Red meter lead to the negative battery terminal. Make contact with the Black meter lead to the **B-** alternator terminal. The meter should register no more than 0.2 volts. If the meter registers more than 0.2 volts, the positive battery cable has excessive resistance causing too great a voltage drop in the circuit. The positive battery cable should be replaced.

If the meter registers less than 0.2 volts, and has passed the preceding tests and still fails to produce the necessary output; it will have to be disassembled and bench tested component by component.

SPECIAL TESTING WORDS

Accurate bench testing of the regulator can only be accomplished with a factory regulator tester. The following tests are limited to those which can be performed using a multimeter with a diode testing function. If the rotor windings, stator windings, and diode bridge are found to be satisfactory, then replace the regulator.

REMOVAL

Disconnect both battery cables from the battery. **REMEMBER**, even with the ignition key in the **OFF** position, the **B+** terminal may still be "hot". Therefore, the battery cables **MUST** be disconnected before starting work on the charging system. Remove the protective boots from the electrical connections on the back of the alternator and then remove the electrical connections from the **B+**, **B-**, and **D+** alternator terminals. Tag each wire **BEFORE** removal, as an assist in later work.

Loosen the pivot and mounting bolts. Move the alternator inboard and slide the drive belt from the pulley. Remove the bolts, spacers (if used), washers and bolts securing the alternator to the engine. Make a note of the location of any spacers used.

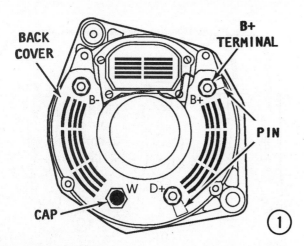

Fig. 1

DISASSEMBLING

1- With the alternator on a suitable workbench, remove the nut on the **B+** alternator terminal and slide the lead from the regulator free of the terminal. Remove the connecting pin from the **D+** terminal. Pry the protective plastic cap from the **W** terminal. Remove the back cover.

2- Remove the four securing screws from the regulator. Insert a flat blade screwdriver under the securing lug - one on each side of the regulator, and pry outward to release the regulator.

Use a pair of needlenose pliers to disconnect the two regulator leads from the terminals on the diode bridge. **DO NOT** pull the leads free of the terminals, because the leads may be pulled away from their connectors. The regulator will still be connected to the brush holder. Therefore set the regulator to one side for later work.

3- Remove the two screws securing the brush holder to the alternator. Lift out the regulator and brush holder. To separate the regulator from the brush holder, unsolder the two leads joining the two components.

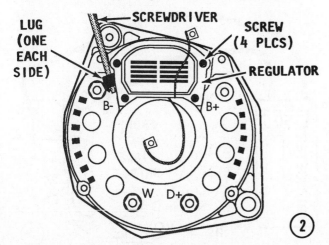

Fig. 2

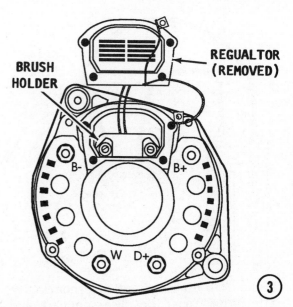

Fig. 3

CLEANING AND TESTING BRUSHES

Clean the brush holder assembly with a brush or compressed air. Use a soft cloth dipped in a mild solvent to wipe the brushes clean, then blow them dry with compressed air. Inspect the brushes for cracks, grooves, and oil saturation.

The original brush set may be reused, if the brushes are 5/16" (8mm) or longer. The brushes and holders are replaced as a unit. If the brush holder is to be replaced, unsolder the two regulator leads from the holder. Solder the two leads back onto the new brush holder using acid free solder.

Alternator Testing

1- Select a meter scale capable of registering 10.0 ohms. Connect the meter leads across the two slip rings to test the rotor windings, one meter lead probe on each ring. Make sure each probe makes good contact with the metal ring. The meter should register 4 - 6 ohms. Keep one meter probe on a slip ring. Move the other meter probe to the alternator frame. The meter should register **NO CONTINUITY**. Repeat this test for the other ring. If one ring shows continuity, the other ring will

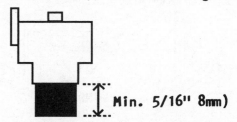

Min. 5/16" 8mm)

Replace the brush holder if the brushes are worn beyond the specified limit.

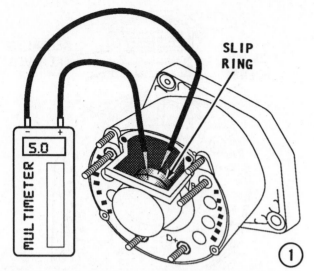

SLIP RING

①

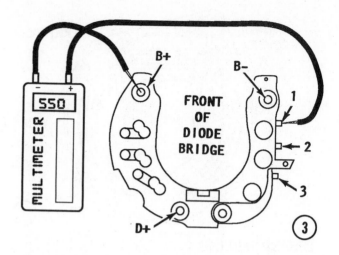

B+ B-

FRONT OF DIODE BRIDGE

1
2
3

D+

③

also. If both rings indicate continuity, the alternator must be disassembled and the grounded part located and insulated.

2– The diode bridge must be removed to check the diodes and the stator windings. Carefully unsolder the three stator winding leads. Do not overheat the solder connections, because excessive heat will damage the diodes. Remove the remaining nuts, spacers and washers securing the diode bridge to the alternator.

Testing Diodes

If the reason for testing the charging system is because the battery apparently discharges itself overnight, the cause is probably a bad diode. At the begining of these procedures, it was stated that the **B+** alternator terminal remained "hot" even though the ignition switch was in the **OFF** position. A bad diode will allow the battery to discharge overnight.

The three diodes on the diode bridge can only be tested if the multimeter being used has a diode test function. If the meter is not equipped with this feature, take the diode bridge together with these instruct-

ions, to your nearest electronic equipment repair shop.

A diode will only pass current in one direction. The multimeter applies a small voltage to the diode in a specified direction. If the diode passes the voltage in this specified direction, the diode is considered good. The meter leads are then changed around and the voltage is applied in the opposite direction. A good diode will not allow the voltage to pass. A bad diode will allow the voltage to pass in both directions. If any one of the three diodes on the diode bridge fails the test, the bridge, as a unit, must be replaced.

3– Select the diode test function on the multimeter. Make contact with the Red meter lead to the No. 1 stator winding connection, identified in the accompanying illustration. Make contact with the Black meter lead to the **B+** terminal on the diode bridge. The meter should register 450 - 650 mV (milli-volt - 1/1000 volt). Reverse the meter leads. The meter needle should swing

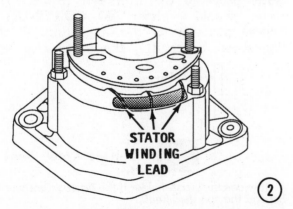

STATOR WINDING LEAD

②

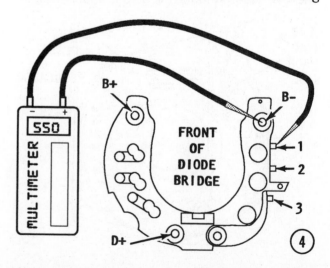

B+ B-

FRONT OF DIODE BRIDGE

1
2
3

D+

④

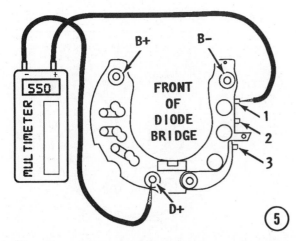

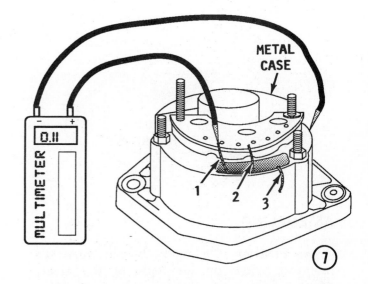

to the far left. If the diode passes both tests, it can be considered good. If the diode fails one or both tests, the entire diode bridge must be replaced. Repeat the test for the No. 2 and No. 3 stator winding connections.

4- Keep the meter on the diode test function. Make contact with the Red meter lead to the No. 1 stator winding connection, identified in the accompanying illustration. Make contact with the Black meter lead to the **B-** terminal on the diode bridge. The meter should register 450 - 650 mV. Reverse the meter leads. The meter needle should swing to the far left. If the diode passes both tests, it can be considered good. If the diode fails one or both tests, the entire diode bridge must be replaced. Repeat the test for the No. 2 and No. 3 stator winding connections.

5- Keep the meter on the diode test function. Make contact with the Red meter lead to the No. 1 stator winding connection, identified in the accompanying illustration. Make contact with the Black meter lead to the **D+** terminal on the diode bridge. The meter should register 450 - 650 mV. Re-

verse the meter leads. The meter needle should swing to the far left. If the diode passes both tests, it can be considered good. If the diode fails one or both tests, the entire diode bridge must be replaced. Repeat the test for the No. 2 and No. 3 stator winding connections.

6- Select a meter scale capable of registering 0.00 - 1.00 ohms. Touch both meter leads together and make a note of the resistance of the measuring cables (e.g. 0.10 ohm). This value must be deducted from all test measurements taken in this step.

The three stator leads **MUST** be unsoldered, as directed previously in Step 2, for the test to be valid. Identify the leads as No. 1, No. 2, and No. 3. Determine the resistance as follows:

No. 1 and No. 2
No. 1 and No. 3
No. 2 and No. 3

The meter should register 0.11 - 0.15 ohms for each test **AFTER** the resistance of the measuring cables has been subtracted from the reading. If the resistance is not within the limits shown, the stator must be replaced.

7- Keep the meter on the same scale. Make contact with one meter lead to the alternator metal case. Make contact with the other meter lead to each of the three stator winding leads in turn. All three meter readings should show **NO CONTINU-ITY**, indicating there is no internal shorting of the stator to the case. If the meter indicates a short in the stator windings, the stator must be replaced.

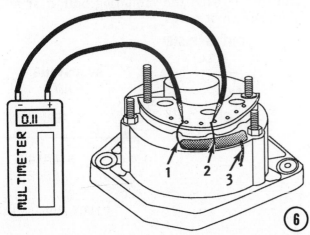

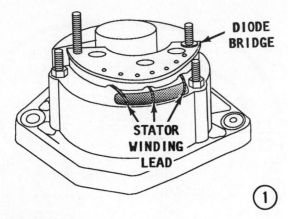

1

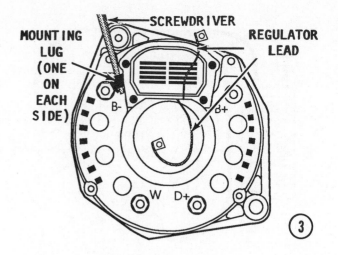

3

ASSEMBLING

1- Position the diode bridge over the mounting studs on the back of the alternator. Solder the three stator windings to the three terminals on the diode bridge using acid free solder. Secure the bridge with the attaching hardware.

2- Solder the two regulator leads to the brush holder using acid free solder. Insert the brush holder into the alternator and secure it in place with two screws. Rotate the alternator pulley and listen for any noises which may indicate the brushes are not seated correctly. The pulley should rotate smoothly.

3- Place the regulator over the brush holder and snap the mounting lugs into place. Secure the regulator with the four attaching screws. Install the two regulator leads onto their respective terminals on the diode bridge.

4- Install the back cover and secure it with the attaching hardware. Install the plastic protective cap over the **W** terminal.

INSTALLATION

5- Mount the alternator and secure it with the pivot bolt and mounting bolt. Tighten the bolts fingertight. Slip the drive

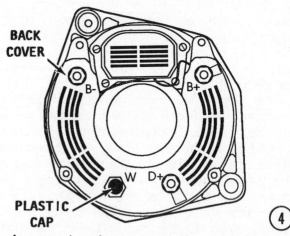

4

belt over the alternator pulley. Tighten the belt by moving the alternator outboard and first tightening the mounting bolt, then the pivot bolt. Check the drive belt tension by depressing the belt at a point midway between pulleys. The belt should be able to be depressed with thumb pressure approximately 1/4" (6mm). To adjust the tension on the belt, loosen the alternator pivot and mounting bolts, and then tighten them fingertight. Move the alternator inboard or outboard, until the correct belt tension is obtained. Tighten the mounting bolt first and then the pivot bolt.

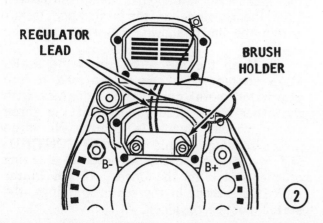

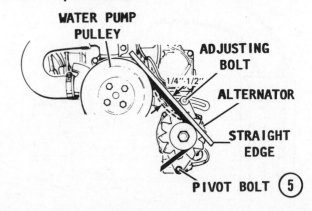

2

5

6-14 CRANKING SYSTEM

The cranking system includes the starter and drive, battery, solenoid, ignition switch, and the required cables and wires to connect the parts for efficient operation. A neutral-start switch to prevent opertion of the starter unless the shift selector lever is in the **NEUTRAL** position, is installed in the shift box on all boats.

The Delco-Remy, Autolite, and Prestolite cranking motors (starters) are used on most marine engine installations. The Delco-Remy starter has a solenoid mounted on the field coil housing. The Autolite and Prestolite units have separate solenoids.

Detailed, illustrated service procedures are given in the following sections for each starter.

6-15 CRANKING SYSTEM TROUBLESHOOTING

Regardless of how or where the solenoid is mounted, the basic circuits of the starting system on all makes of cranking motors are the same and similar tests apply. In the following testing and troubleshooting procedures, the differences are noted.

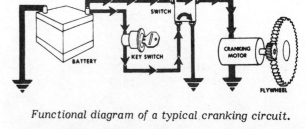

Functional diagram of a typical cranking circuit.

ALWAYS TAKE TIME TO VENT THE BILGE WHEN MAKING ANY OF THE TESTS AS A PREVENTION AGAINST IGNITING ANY FUMES ACCUMULATED IN THAT AREA. AS A FURTHER PRECAUTION, REMOVE THE HIGH-TENSION WIRE FROM THE CENTER OF THE DISTRIBUTOR CAP AND GROUND IT SECURELY TO PREVENT SPARKS.

All starter problems fall into one of three problem areas:

1- The starter fails to turn.
2- The starter spins rapidly, but does not crank the engine.
3- The starter cranks the engine, but too slowly.

The following paragraphs provide a logical sequence of tests designed to isolate a problem in the cranking system.

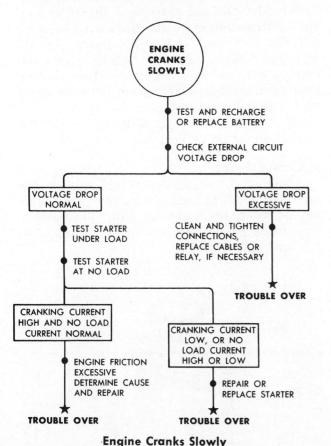

Engine Cranks Slowly

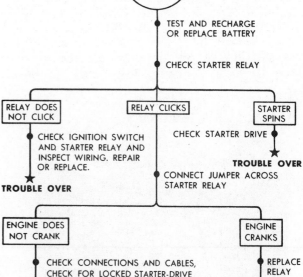

Engine Will Not Crank

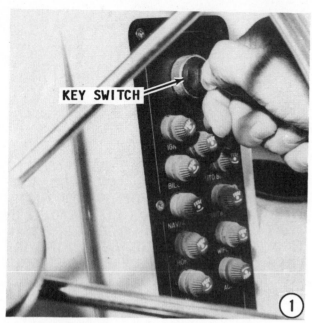

KEY SWITCH

①

The procedures are keyed by number to matching numbered illustrations as an aid in performing the work.

BATTERY TEST

1- Turn on several of the cabin lights. Now, turn the ignition switch to the **START** position and note the reaction to the brightness of the lights.

With a normal electrical system, the light will dim slightly and the starter will crank the engine at a reasonable rate. If the lights dim considerably and the engine does not turn over, one of several causes may be at fault.

A- If the lights go out completely, or dim considerably, the battery charge is low or almost dead. The obvious remedy is to charge the battery; switch over to a secondary battery if one is available; or to replace it with a known fully charged one.

BATTERY
TERMINALS

②

B- If the starting relay clicks, sounding similar to a machine gun firing, the battery charge is too low to keep the starting relay engaged when the starter load is brought into the circuit.

C- If the starter spins without cranking the engine, the drive is broken. The starter will have to be removed for repairs.

D- If the lights do not dim, and the starter does not operate, then there is an open circuit. Proceed to Test 2, Cable Connection Test.

CABLE CONNECTION TEST

2- If the starter fails to operate and the lights do not dim when the ignition switch is turned to **START**, the first area to check is the connections at the battery, starting relay, starter, and neutral-safety switch.

First, remove the cables at the battery; clean the connectors and posts; replace the cables; and tighten them securely.

Now, try the starter. If it still fails to crank the engine, try moving the shift box selector lever from **NEUTRAL** to **FORWARD** to determine if the neutral-safety switch is out of adjustment or the electrical connections need attention.

Sometimes, after working the shift lever back-and-forth and perhaps a bit sideways, the neutral-switch connections may be temporarily restored and the engine can be started. Disconnect the leads; clean the connectors and terminals on the switch; replace the leads; and tighten them securely at the first opportunity.

If the starter still fails to crank the engine, move on to Test 3, Solenoid Test.

SOLENOID TEST

3- The solenoid, commonly called the starting relay, is checked by directly bridging between the terminal from the battery (the large heavy one) to the terminal from the ignition switch.

TEST CABLE

③

TAKE EVERY PRECAUTION TO ENSURE THERE ARE NO GASOLINE FUMES IN THE BILGE BEFORE MAKING THESE TESTS.

If a bilge blower is installed, operate it for at least five minutes to clear any fumes accumulated in the bilge.

Turn the ignition switch to the **START** position. Now, bridge between the battery lead terminal and the ignition lead terminal with a very heavy piece of wire. If the relay operates, the trouble is in the circuit to the ignition switch. If the starter still fails to operate, continue with Test 4, Current Draw Test.

CURRENT DRAW TEST

4- Lay an amperage gauge on the cable between the battery and the starter. Attempt to crank the engine and note the current draw reading of the amperage gauge under load. The current draw should not exceed 190 amperes.

CHECKING CRANKING CIRCUIT RESISTANCE

If the starter turns very slowly or not at all, or if the solenoid fails to engage the starter with the flywheel, the cause may be

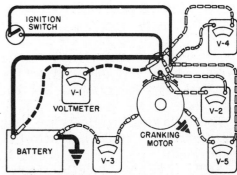

Functional diagram of hookup for various voltage tests outlined in the text.

excessive resistance in the cranking circuit.

The following checks can be performed with the starter installed on the engine in the boat.

1- Test the battery and bring it up to a full charge, if necessary.
GROUND the distributor primary lead to prevent the engine from firing during the following checks.

2- Measure the voltage drop during cranking between the positive battery post and the battery lead terminal of the solenoid.

3- Measure the voltage drop during cranking between the battery lead terminal of the solenoid and the motor lead terminal of the solenoid.

4- Measure the voltage drop during cranking between the negative battery post and the starter motor frame.

If the voltage drop during any of the previous three tests is more than 0.2 volt, excessive resistance is indicated in the circuit being checked. Trace and correct the cause of the resistance.

If the solenoid fails to pull in, the problem may be due to excessive high voltage drop in the solenoid circuit. To check voltage drop in this circuit, measure the voltage drop during cranking, between the battery terminal of the solenoid and the switch terminal of the solenoid. If the voltage drop is more than 2.5 volts, the resistance is excessive in the solenoid circuit.

If the voltage drop is not more than 2.5 volts and the solenoid does not pull in, measure the voltage available at the switch terminal of the solenoid. The solenoid should pull in with 8.0 volts at temperatures up to 200°F. If it does not pull in, remove the starter motor and test the solenoid.

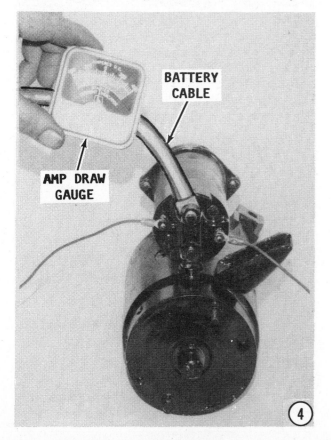

BATTERY CABLE

AMP DRAW GAUGE

④

AUTOLITE AND PRESTOLITE CRANKING SYSTEM VOLTAGE TESTS

Even though the starter cannot be tested accurately while it is mounted on the engine, several tests can be made for excessive resistance in the cranking system circuits. The following four tests are designed to isolate excessive resistance with the system under load.

TAKE EVERY PRECAUTION TO ENSURE THERE ARE NO GASOLINE FUMES IN THE BILGE BEFORE MAKING THESE TESTS.

If a bilge blower is installed, operate it for at least five minutes to clear any fumes accumulated in the bilge.

REMOVE THE HIGH-TENSION WIRE FROM THE CENTER OF THE DISTRIBUTOR AND GROUND IT BEFORE MAKING ANY OF THE FOLOWING TESTS.

The circled numbers on the accompanying illustration are keyed to the tests and indicate the connection for the voltmeter leads.

TEST 1

Voltage Drop Across the Battery Terminal Cable and Post:

Connect the positive lead of the voltmeter to the positive post of the battery and the negative lead of the voltmeter to the battery terminal of the starter. Have a partner turn the ignition switch to **START** and note the reading of the voltmeter. If a partner is not available, connect a remote starter cable switch between the S and battery terminals of the starter relay.

If the reading is more the 0.5 volt, there is a high-resistance connection between the battery post and the cable. The battery post is the first place for corrosion to form due to the high corrosive nature of the battery electrolyte (acid).

Remove both battery cable connectors; clean the connectors and the battery posts thoroughly; replace the connectors; and tighten them securely.

TEST 2

Voltage Drop In the Battery Cable to Starter Relay:

Leave the positive lead of the voltmeter connected to the positive post of the battery. Connect the negative lead of the voltmeter to the battery terminal of the starter relay. Do not connect it to the cable, but directly to the terminal.

Attempt to crank the engine as described in Test 1, and note the voltage drop. If the voltage drops more than 0.1 volt, the starter is drawing too much current because the cable is too thin, or the connection at the relay is corroded. Clean and tighten the connection or replace the cable.

TEST 3

Voltage Drop Across the Contacts Inside the Starter Relay:

Leave the positive lead of the voltmeter connected to the positive battery post. Connect the negative lead of the voltmeter to the starter terminal of the starter relay, as shown. Attempt to crank the engine as described in Test 1 and note the voltage drop. If the voltage drop is more than 0.3 volt, the starter relay must be replaced because the contacts are burned.

TEST 4

Voltage Drop Between the Negative Battery Post and the **Engine Ground:**

Connect the positive lead of the voltmeter to a **GOOD** engine ground. Do not attempt to obtain a good ground on a painted surface. Connect the negative lead of the voltmeter to the negative battery post.

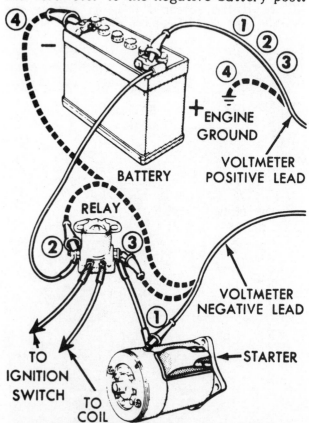

Functional diagram of hookup for various tests of an Autolite or Prestolite starting system. The numbers are identified in the text.

Attempt to crank the engine as described in Test 1 and note the voltage drop. If the voltage drop is more than 0.1 volt, the negative battery post or the cable connector is corroded and must be cleaned and the connector tightened securely. One other possibility is paint between the engine and the cable. Remove the cable, scrape the paint away to bare metal, connect the cable again and repeat the test.

6-16 DELCO-REMY STARTER DESCRIPTION AND OPERATION

Delco-Remy starters consist of a set of field coils positioned over pole pieces, which are attached to the inside of a heavy iron frame. An armature, an overrunning clutch drive mechanism, and a solenoid are included inside the iron frame.

The armature consists of a series of iron laminations placed over a steel shaft, a commutator, and the armature winding. The windings are heavy copper ribbons assembled into slots in the iron laminations. The ends of the windings are soldered or welded to the commutator bars. These bars are electrically insulated from each other and from the iron shaft.

An overrunning clutch drive arrangement is installed near one end of the starter shaft. This clutch drive assembly contains a pinion which is made to move along the

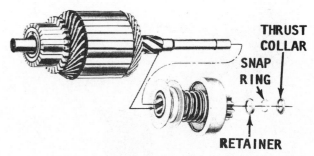

Starter armature and clutch assembly.

shaft by means of a shift lever to engage the engine ring gear for cranking. The relationship between the pinion gear and the ring gear on the engine flywheel provides sufficient gear reduction to meet cranking requirement speed for starting.

The overrunning clutch drive has a shell and sleeve assembly, which is splined internally to match the spiral splines on the armature shaft. The pinion is located inside the shell. Spring-loaded rollers are also inside the shell and they are wedged against the pinion and a taper inside the shell. Some starters use helical springs and others use accordion type springs. Four rollers are used. A collar and spring, located over a sleeve completes the major parts of the clutch mechanism.

When the solenoid is energized and the shift lever operates, it moves the collar endwise along the shaft. The spring assists

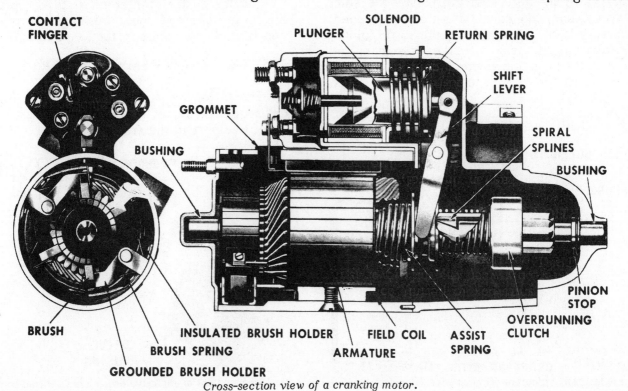

Cross-section view of a cranking motor.

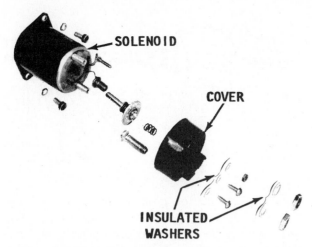

Arrangement of external parts of a starter solenoid.

movement of the pinion into mesh with the ring gear on the flywheel. If the teeth on the pinion fail to mesh for just an instant with the teeth on the ring gear, the spring compresses until the solenoid switch is closed; current flows to the armature; the armature rotates; the spring is still pushing on the pinion; the pinion teeth mesh with the ring gear; and cranking begins.

Torque is transfered from the shell to the pinion by the rollers, which are wedged tightly between the pinion and the taper cut into the the inside of the shell. When the engine starts, the ring gear drives the pinion faster than the armature; the rollers move away from the taper; the pinion overruns the shell; the return spring moves the shift lever back; the solenoid switch is opened; current is cutoff to the armature; the pinion moves out of mesh with the ring gear; and the cranking cycle is completed. The start switch should be opened immediately when

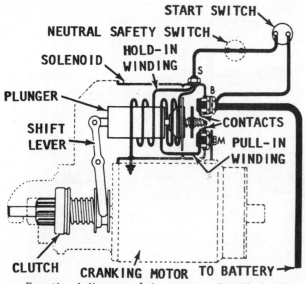

Functional diagram of the starter solenoid circuit.

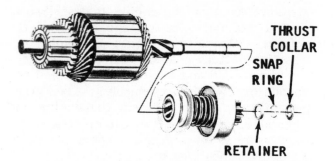

*Principle parts of the starter armature and clutch assembly. **NEVER** wash the clutch assembly in solvent or the lubricant will dissolve causing the unit to seize.*

the engine starts to prevent prolonged over-run.

SERVICING DELCO-REMY STARTERS

The starter is usually installed in an inaccessible location. For this reason the starter drive-end bushing receives little or no lubrication. For lack of lubrication, the drive-end bushing wears; the armature drops down and rubs against the field pole pieces; internal drag is created; and the output torque of the starter is reduced.

The commutator-end bushing is more accessible and therefore, may receive too much lubrication. The commutator becomes covered with oil; this oil insulates the commutator from the brushes; resistance is increased; and the efficiency of the cranking system is reduced.

Service on the starter consists of replacing defective switches, bushings, brushes, and turning the commutator to make it true.

TESTING DELCO-REMY STARTERS

The following paragraphs provide a logical sequence of tests designed to isolate a defective part in the starter.

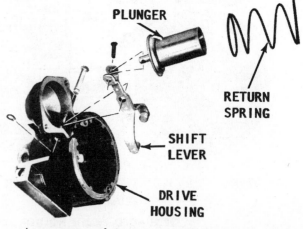

Arrangement of the drive housing and related parts.

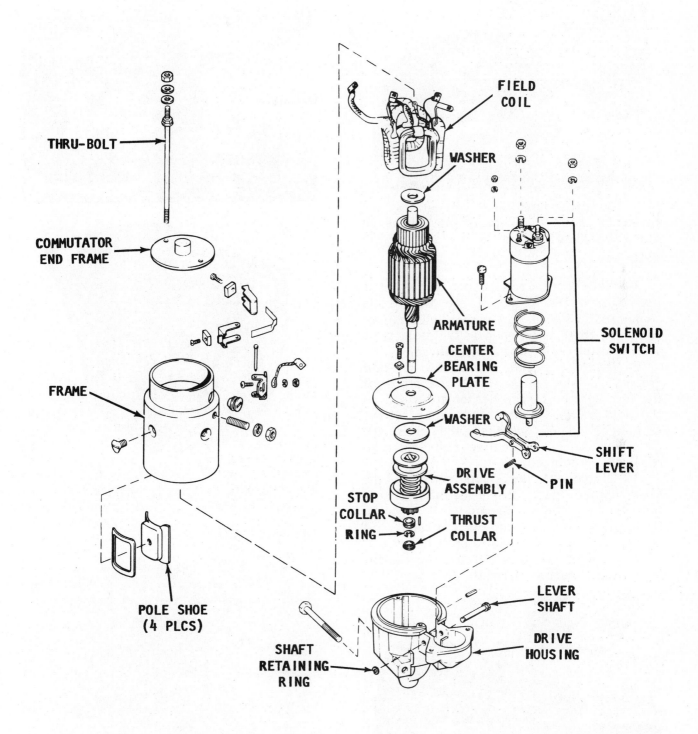

THRU-BOLT

COMMUTATOR
END FRAME

FRAME

POLE SHOE
(4 PLCS)

FIELD
COIL

WASHER

ARMATURE

CENTER
BEARING
PLATE

WASHER

DRIVE
ASSEMBLY

STOP
COLLAR

RING

THRUST
COLLAR

SOLENOID
SWITCH

SHIFT
LEVER

PIN

SHAFT
RETAINING
RING

LEVER
SHAFT

DRIVE
HOUSING

Exploded view of the Delco-Remy starter with principle parts identified.

FIELD COILS

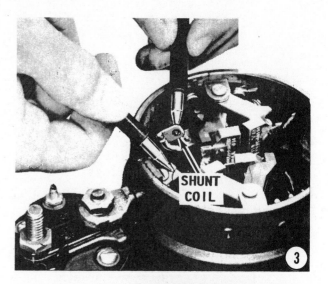

SHUNT COIL

The procedures and suggestions are keyed by number to matching numbered illustrations as an aid in performing the work.

1- Make contact with one probe of a test light on each end of the field coils connected in series. If the test light fails to come on, there is an open in the field coils and repair or replacement is required.

2- Disconnect the shunt coil or coil ground. Make contact with one probe of the test light on the connector strap and on the field frame with the other probe. If the test light comes on, the field coils are grounded and the defective coils must be repaired or replaced.

3- Disconnect the shunt coil grounds. Make contact with one probe of the test light on each end of the shunt coil, or coils. If the light fails to come on, the shunt coil is open and must be repaired or replaced.

4- ALWAYS replace the drive-end bushing during a starter overhaul.

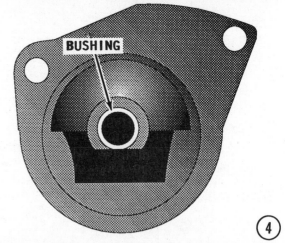

BUSHING

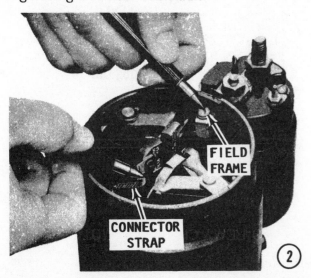

ARMATURE

FIELD FRAME

CONNECTOR STRAP

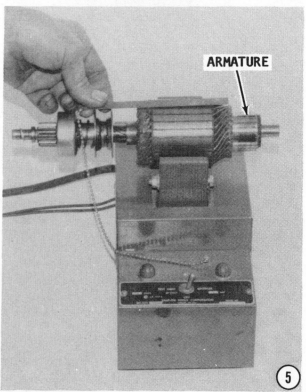

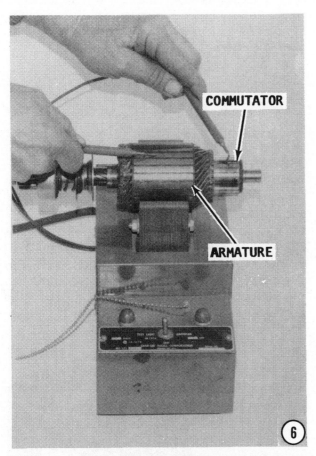

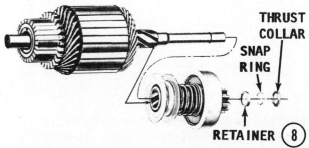

5- True the commutator, if necessary, in a lathe. **NEVER** undercut the mica because the brushes are harder than the insulation. Check the armature for a short circuit by placing it on a growler and holding a hack saw blade over the armature core while the armature is rotated. If the saw blade vibrates, the armature is shorted. Clean between the commutator bars, and then check again on the growler. If the saw blade still vibrates, the armature must be replaced.

6- Make contact with one probe of the test light on the armature core or shaft and the other probe on the commutator. If the light comes on, the armature is grounded and must be replaced.

7- Wash the brush holder in solvent, and then blow it dry with compressed air. Use the test light to verify two of the brush holders are grounded and two are insulated.

8- The overrunning clutch is secured to the armature by a snap ring. This ring may be removed if the clutch requires replacement. **NEVER** wash an overrunning clutch in solvent or the lubricant will be disolved; the clutch will fail; the engine will drive the cranking motor armature at high speed; the windings will be thrown out by centrifugal force; and the armature will be destroyed.

9- Check the clearance between the pinion and the retainer. **NEVER** use a 12-volt battery while making this check or the armature will turn. Using a **6-VOLT** battery, energize the solenoid coil; push the pinion

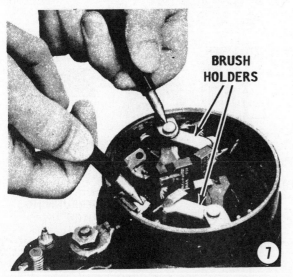

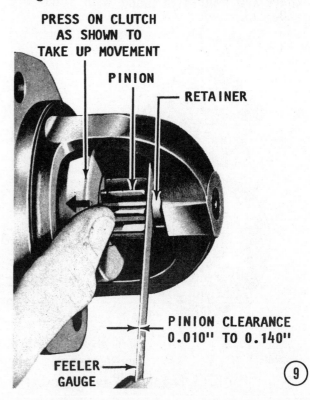

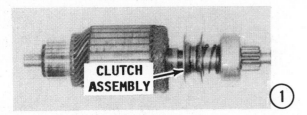

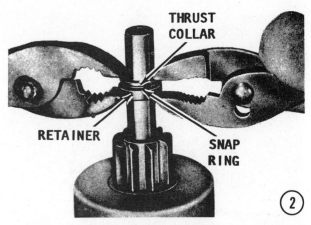

back to take up the slack; and measure the clearance with a feeler gauge. If the clearance is not between 0.010" and 0.140", the solenoid is not properly installed, or the linkage is worn.

ASSEMBLING A DELCO REMY

After all parts have been cleaned, tested, and replacements obtained, the starter is ready to be assembled.

The following instructions are numbered and matched to numbered illustrations as an aid in performing the work.

1- Lubricate the drive end of the armature shaft with a thin coating of silicone lubricant. Slide the clutch assembly onto the armature shaft with the pinion facing outward. Slide the retainer onto the shaft with the cupped surface facing the end of the shaft (away from the pinion).

2- Stand the armature on end on a wooden surface with the commutator down. Position the snap ring on the upper end of the shaft and hold it in place with a block of wood.

Now, tap on the wood block with a hammer to force the snap ring over the end of the shaft. Slide the snap ring down into the groove. Assemble the thrust collar onto the shaft with the shoulder next to the snap ring.

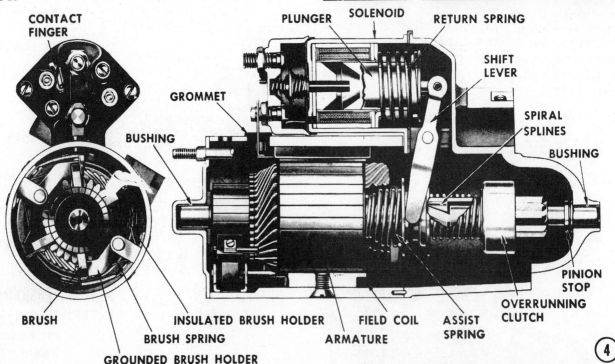

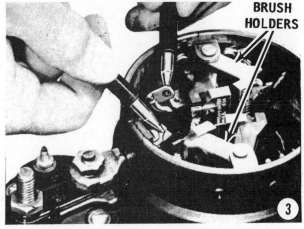

Place the armature flat on the work bench, and then position the retainer and thrust collar next to the snap ring. Next, using two pair of pliers at the same time (one pair on each side of the shaft), grip the retainer and thrust collar and squeeze until the snap ring is forced into the retainer.

Lubricate the drive housing bushing with a thin coating of silicone lubricant. **MAKE SURE** the thrust collar is in place against the snap ring and the retainer.

3- Install the brushes into the brush holders. Assemble the insulated and grounded brush holders together with the **"V"** spring and position them as a unit on the support pin. Push the holders and springs to the bottom of the support, then rotate the springs to engage the **"V"** in the support.

Attach the ground wire to the grounded brush and the field lead wire to the insulated brush.

4- Slide the armature and clutch assembly into place in the drive housing engaging the shift lever with the clutch.

Position the field frame over the armature and apply a thin coating of liquid neoprene (Gaco) between the frame and the solenoid case. Place the frame in position against the drive housing. **TAKE CARE** not to damage the brushes.

Apply a coating of silicone lubricant to the bushing in the commutator end frame. Place the leather brake washer onto the armature shaft and slide the commutator

end frame onto the shaft. Connect the field coil connectors to the **MOTOR** solenoid terminal.

Pinion Clearance Check:

After the starter has been assembled, check the pinion clearance as follows:

1- Connect a battery, of the same voltage as the solenoid, from the solenoid switch terminal to the solenoid frame or ground terminal. **DISCONNECT** the motor field coil connector for the test.

Momentarily make contact with a jumper lead from the solenoid motor terminal to the solenoid frame or ground terminal. The pinion will now shift into the cranking position and it will remain there until the battery is disconnected.

2- Push the pinion back towards the commutator end to eliminate any slack movement. Now, measure the distance between the pinion and the pinion stop with a feeler gauge. The clearance should be between 0.10" and 0.040". If the clearance is not within these limits, it may indicate excessive wear of the solenoid linkage shift lever yoke buttons or improper assembly of the shift lever mechanism. Any worn or defective parts should be replaced.

6-17 AUTOLITE CRANKING SYSTEM

The Autolite cranking system consists of the starter, relay, ignition switch, battery,

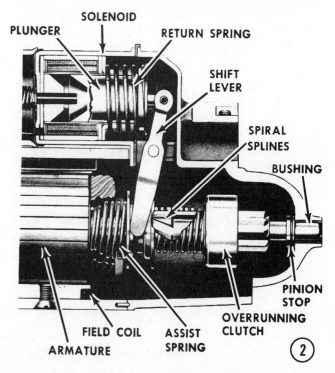

cables, and necessary wiring for efficient operation. The starter has a special moveable pole shoe within the field housing to engage the drive assembly. As on all marine installations, a neutral-start switch is installed in the shift box to permit operation of the starter only when the shift lever is in the **NEUTRAL** position.

AUTOLITE STARTERS
DESCRIPTION AND OPERATION

The starter has an integral, positive-engagement drive mechanism. The sequence of events in the cranking operation is as follows: The ignition key is turned to the **START** position; current flows to the relay; the relay makes contact through heavy-current type contacts; current is directed through the grounded field coil; the moveable pole shoe is activated; the special pole shoe is attached to the starter drive plunger lever which forces the drive (with the overrunning clutch) to engage the ring gear of the engine flywheel; after the shoe is fully seated, it opens the field coil grounding contacts; and the starter is in normal cranking operation. While the starter is turning the engine flywheel, a holding coil holds the moveable pole shoe in the fully seated position.

DISASSEMBLING AN
AUTOLITE STARTER

Loosen the retaining screw, and then slide the brush cover band off the starter.

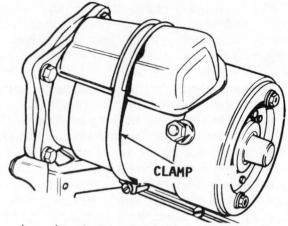

A marine starter must have a clamp installed, as shown. The starter cap coming off during engine operation could cause a fire.

Remove the starter drive plunger cover and gasket. Take note of the lead position as an aid to assembling. Remove the commutator brushes from the brush holders.

Remove the long through bolts. Separate the drive end housing from the starter. Remove the plunger return spring. Drive out the pivot pin retaining the starter gear plunger lever, and then remove the lever and the armature.

Remove the stop ring retainer from the shaft, and then remove and discard the stop ring. Slide the starter drive gear assembly off the shaft.

Remove the retaining screws, and then the brush end plate. Remove the screws retaining the ground brushes to the frame.

CAREFULLY bend the tab on the field coil retaining sleeve up, and then remove the sleeve.

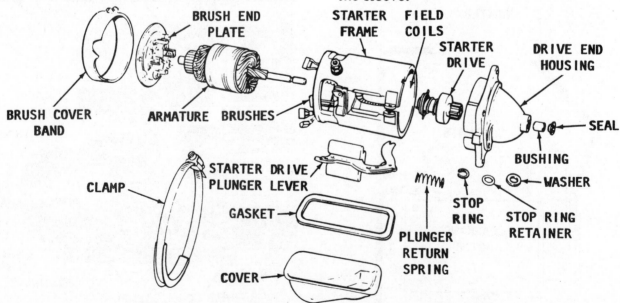

Exploded view of an Autolite starter with principle parts identified.

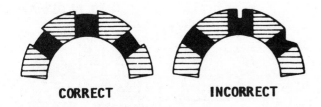

CORRECT   INCORRECT

Armature segments properly cleaned (left) and improperly cleaned (right).

CLEANING AND INSPECTING

Clean the field coils, armature, commutator, armature shaft, brush-end plate and drive-end housing with a brush or compressed air. Wash all other parts in solvent and blow them dry with compressed air.

Inspect the insulation and the unsoldered connections of the armature windings for breaks or burns.

Perform electrical tests on any part suspected of defect, according to the procedures outlined in the Testing Sections of this Chapter.

Check the commutator for run-out. Inspect the armature shaft and both bearings for scoring.

Turn the commutator in a lathe if it is out-of-round by more than 0.005".

Check the springs in the brush holder to be sure none are broken. Check the spring tension and replace if the tension is not 32-40 ounces. Check the insulated brush holders for shorts to ground. If the brushes are worn down to 1/4" or less, they must be replaced.

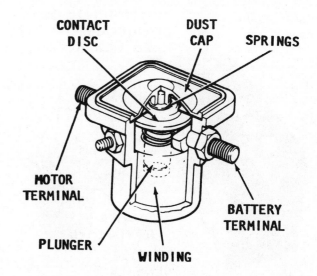

Cutaway view of a starter solenoid showing major parts.

Check the field brush connections and lead insulation. A brush kit and a contact kit are available at your local marine dealer, but all other assemblies must be replaced rather than repaired.

BENCH TESTING AUTOLITE STARTERS

The following paragraphs provide a logical sequence of bench tests designed to isolate a defective part in the starter.

The procedures and suggestions are keyed by number to matching numbered illustrations as an aid in performing the work.

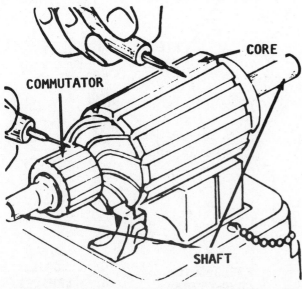

Armature check for a short: one test light lead on each commutator segment, alternately, and the other lead on the armature core. No continuity.

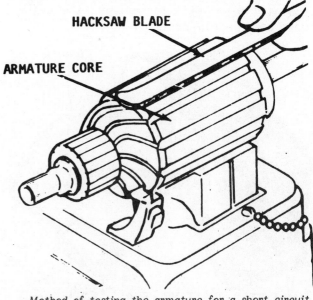

Method of testing the armature for a short circuit using a growler and hacksaw blade. If the blade vibrates, the mica must be cleaned out or the armature replaced.

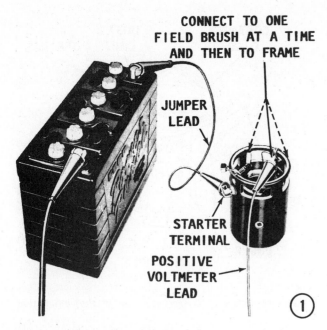

CONNECT TO ONE FIELD BRUSH AT A TIME AND THEN TO FRAME

JUMPER LEAD

STARTER TERMINAL

POSITIVE VOLTMETER LEAD

①

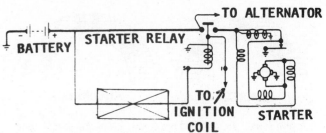

Schematic diagram of the Ford cranking system.

Armature and Field Open Circuit Test

1- Examine the commutator for any evidence of burning. A spot burned on the commutator is caused by an arc formed every time the commutator segment connected to the open-circuit winding passes under a brush.

Connect a jumper wire from the positive terminal of a battery to the starter terminal. Connect the negative lead from a voltmeter to the negative battery terminal. Connect the positive voltmeter lead to one field brush, as shown. Since the starter has three field windings, it will be necesasary to check each of the windings separately. If the voltmeter fails to register, the coil is open and must be replaced.

Armature Grounded Circuit Test

2- This test will determine if the winding insulation has failed, permitting a conductor to touch the frame or armature core.

Connect a jumper wire from the positive terminal of a battery to the end of the starter shaft opposite the commutator, as shown. Connect the negative lead from a voltmeter to the negative battery terminal. Make contact with the positive voltmeter lead to the commutator and check for voltage. If the voltmeter fails to register, the windings are grounded and must be replaced.

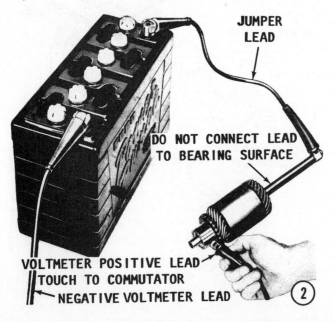

JUMPER LEAD

DO NOT CONNECT LEAD TO BEARING SURFACE

VOLTMETER POSITIVE LEAD TOUCH TO COMMUTATOR

NEGATIVE VOLTMETER LEAD

②

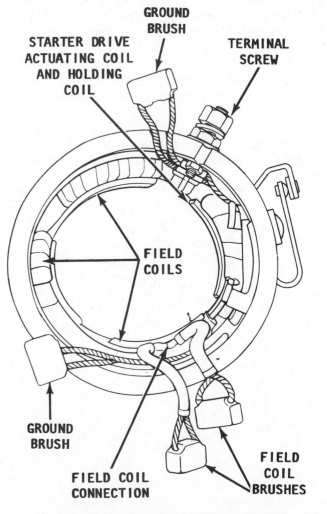

GROUND BRUSH

STARTER DRIVE ACTUATING COIL AND HOLDING COIL

TERMINAL SCREW

FIELD COILS

GROUND BRUSH

FIELD COIL CONNECTION

FIELD COIL BRUSHES

Principle parts of the field coil and brushes.

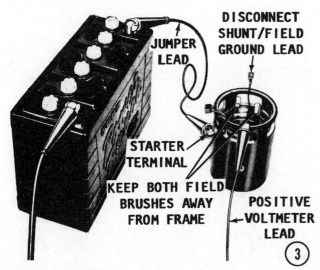

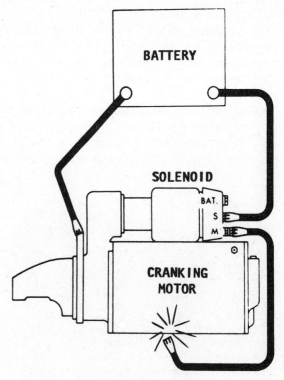

Field Ground
Circuit Test

3- Grounded field windings can be detected using a voltmeter and battery. First, **DISCONNECT** the shunt-field ground lead. Next, connect a jumper lead from the positive lead of the battery to the starter terminal. Connect the negative lead from the voltmeter to the negative battery terminal. Now, keep both field brushes away from the starter frame, and make contact with the positive lead from the voltmeter to the starter frame, as shown. If the voltmeter indicates any voltage, the field windings are grounded and must be replaced.

ASSEMBLING AN
AUTOLITE STARTER

Position the new insulated field brushes lead onto the field coil terminal and secure it with the clip. To ensure extended service, solder the lead, clip and terminal together, using resin core solder.

Place the solenoid coil ground terminal over the nearest ground screw hole. Place each ground brush in position to the starter

frame and secure them with the retaining screws.

Place the brush end plate into position on the starter frame with the boss on the plate indexed in the slot on the frame.

Coat the armature shaft splines with a thin layer of Lubriplate. Slide the starter drive assembly onto the armature shaft, and secure it in place with a **NEW** stop ring and stop-ring retainer.

Slide the fiber thrust washer onto the commutator end of the shaft, then insert the armature shaft into the starter frame. (The thrust washer is not used with molded commutator armatures.)

Place the drive gear plunger lever in position in the frame and the starter drive assembly, and then install the pivot pin.

Fill the drive-end housing bearing bore **ONLY 1/4 FULL** with lubrication. Insert the starter drive plunger lever return spring in position, and then mate the drive-end housing with the starter frame. Install and tighten the through bolts to a torque value of 55-75 in-lbs. **TAKE CARE** not to pinch the brush leads between the brush plate and the frame. **CHECK TO BE SURE** the stop ring retainer is properly seated in the drive housing.

Install the brushes in the brush holder with the **SPRINGS CENTERED** on the brushes.

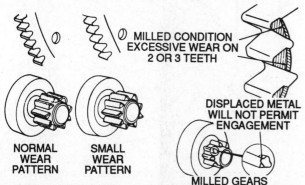

Wear pattern on a starter drive gear and on the flywheel ring gear.

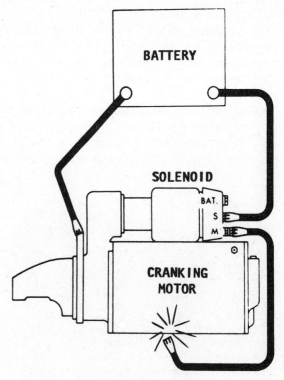

Bench testing a starter.

Check the brush spring tension by pulling on a line with a scale. The line should be hooked under the brush spring near the brush and the pull should be parallel to the face of the brush. Take the reading just as the spring leaves the brush. The reading should be from 32 to 40 ozs.

Finally, position the gasket and drive gear plunger lever cover in place, then slide the brush cover band in place and secure it with the retaining screw.

If the engine should start, but fail to reach the predetermined rpm to disengage the detent pin, a ratchet type clutch on the screw shaft allows the pinion to overrun the armature shaft and prevent damage to the starter. When this overrun occurs, a light buzzing sound is audible from the clutch ratcheting. When the engine is accelerated, the drive will release and the buzzing sound will stop.

6-18 PRESTOLITE CRANKING SYSTEM

DESCRIPTION AND OPERATION

The Prestolite cranking system includes the starter, relay, ignition switch, battery, cables, wiring, and a special starter relay. As with all marine installations, a safety-start switch is installed in the shift box. This switch prevents operation of the cranking system unless the shift lever is in the **NEUTRAL** position.

The starter motor has a Bendix Folo-Thru type drive designed to overcome disengagement of the flywheel ring gear when engine speed has reached a predetermined rpm.

After the pinion engages the flywheel ring gear, a spring-loaded detent pin in the pinion gear assembly indexes in a notch in the screw shaft. The pinion then remains

Prestolite cranking motor which may be installed on some stern drive powerplants.

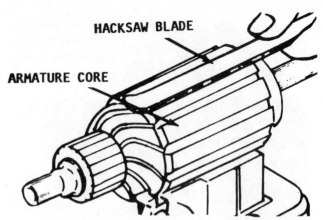

Method of testing the armature for a short circuit using a growler and hacksaw blade. If the blade vibrates, the mica must be cleaned out or the armature replaced.

locked in the engaged position until the engine rpm reaches a predetermined speed. At that point, centrifugal force moves the detent pin out of the notch in the shaft and allows the pinion to disengage the flywheel.

If, during the starting operation, the engine fails to continue running, movement of the pinion in a disengaging direction is prevented by the pin indexed in the screw shaft. For this reason, if the starting motor is re-engaged while the engine kicks back, the starter will not be damaged.

DISASSEMBLING A PRESTOLITE STARTER

First, remove the two through-bolts, and then separate the cover assembly from the commutator end. Next, remove the starter housing from the armature and the end frame.

Now, remove the two screws securing the bearing assembly to the end frame, and then remove the end frame assembly.

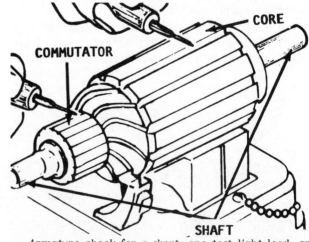

Armature check for a short: one test light lead on each commutator segment, alternately, and the other lead on the armature core. No continuity.

Remove the pin and the Bendix drive. Remove the bearing assembly from the armature shaft. Remove the three attaching screws, and then the brush plate. Remove the springs and brushes.

If the field is to be removed, disconnect the field wire from the terminal stud; the four screws; field assmbly; and pole shoes.

CLEANING AND INSPECTING

Clean the field coils, armature, commutator, brushes, and bushings with a brush or compressed air. Wash all other parts in solvent and blow them dry with compressed air. **NEVER** use a grease dissolving solvent to clean electrical parts and bushings, because the solvent would damage the insulation and remove the lubricating qualities from the bushings.

Perform electrical tests on any part suspected of defect, according to the procedures outlined in the Testing Sections of this Chapter.

Check the armature shaft to be sure it is not worn or bent. Check the other armature parts: Commutator worn; laminations core scored; or connections requiring attention.

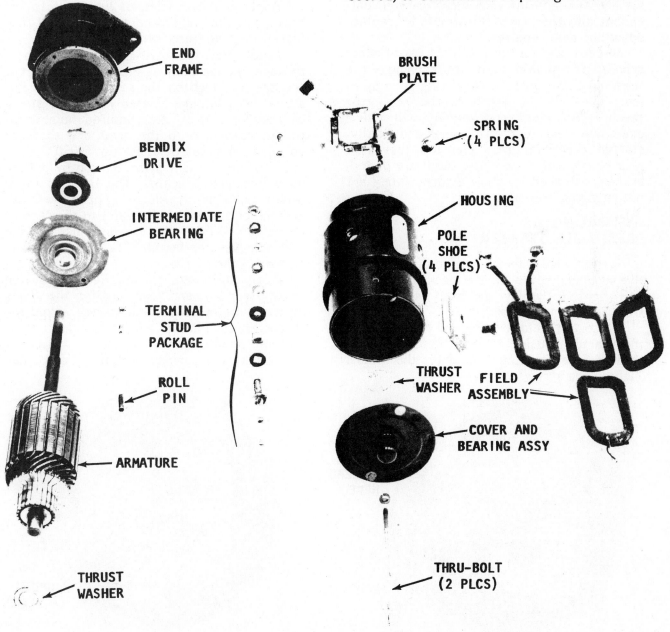

Prestolite starter motor with major parts identified.

Inspect the armature shaft and bearings for scoring or excessive wear. If the commutator is rough, burned, out-of-round, or has high mica on it, the commutator must be turned on a lathe. The out-of-round limit is 0.005".

Check the springs in the brush holder to be sure none are broken. Check the spring tension and replace if the tension is not 32-40 ounces. Check the insulated brush holders for shorts to ground. If the brushes are worn down to 1/4" or less, they must be replaced.

Check the field brush connections and lead insulation. A brush kit and a contact kit are available at your local marine dealer, but all other assemblies must be replaced rather than repaired.

Inspect the drive teeth of the Bendix drive. The pinion teeth must engage the teeth on the engine flywheel ring gear by at least one-half the depth of the ring gear teeth. Any less engagement will cause excessive wear to the ring gear and finally, starter drive failure. Replace the drive gear or the ring gear if the teeth are pitted, broken, damaged, or show evidence they are not engaging properly.

ASSEMBLING A PRESTOLITE STARTER

Install a **NEW** terminal stud according to the sequence given on the package. Use a test light to verify the stud is insulated from the starter housing.

Place the field winding and pole shoes in the starter housing. Slide the four srews through the housing and into the poles. Tighten the screws firmly.

Position the brush plate assembly in place and install the three screws and lockwashers through the starter housing and into the plate assembly, then tighten them firmly.

Connect the shunt ground wire to the brush plate assembly and install the brush springs.

Apply a thin coating of lubricant to the bearing surface of the bearing assembly, and then slide the asembly into place on the armature shaft. Apply a thin coating of lubricant to the drive-end of the armature shaft. Slide the Bendix drive assembly onto the shaft, and then install a **NEW** roll pin.

Apply a thin coating of lubricant to the bearing in the end frame asesmbly, and then install the armature into the end frame.

Install the two screws and lockwashers through the bearing assembly and into the end frame. Tighten the screws firmly. Position the armature assembly into the starter housing. Slide the required number of thrust washers onto the armature shaft to obtain an end play of 0.005" to 0.030".

Check the brush spring tension by pulling on a line with a scale. The line should be hooked under the brush spring near the brush and the pull should be parallel to the face of the brush. Take the reading just as the spring leaves the brush. The reading should be from 32 to 40 ozs.

Apply lubricant to the bearing in the cover assembly, and then position the cover on the armature shaft and starter. Slide the through-bolts through the housing and tighten them securely.

Check the armature shaft end play again.

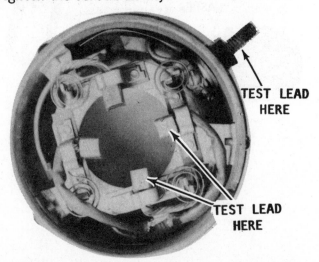

Points to attach test leads when testing a Prestolite starter.

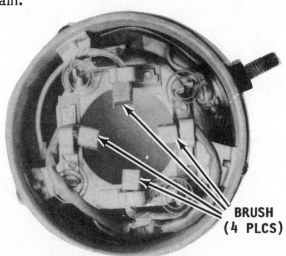

Prestolite starter with the brushes installed.

7
REMOTE CONTROLS

7-1 INTRODUCTION

Boat accessories are seldom obtained from the original equipment manufacturer. Shift boxes, steering, bilge pumps, blowers, and other similar equipment may be added by the boat manufacturer. Because of the wide assortment, styles, and price ranges of such accessories, the boat manufacturer, or customer has a wide selection from which to draw, when outfitting the boat.

Therefore, the procedures and suggestions in this chapter are general in nature in order to cover as many units as possible, but still specific and in enough detail to allow you to troubleshoot, repair, and adjust each of these accessories for maximum comfort, performance and safety.

7-2 SHIFT/STEERING CABLE SERVICE

The shift and steering cables MUST be properly adjusted whenever they have been reconnected for any reason, particularly after the stern drive has been replaced.

After each shift cable adjustment, the following checks and adjustments MUST be made to ensure proper performance:

Shift cut-out switch lever position
Reverse lock
Trim-limit switch (Since 1975)

Early type shift/steering cable installations differ considerably from those used by MerCruiser in later years. The change was affected during the 1973 model year. For this reason, the service procedures and adjustments for each type will be covered in separate sections.

The trim-limit switch MUST be checked and adjusted after the shift cables have been installed and the switches adjusted. Proper adjustment of the trim-limit switch is necessary to ensure safe operation of the boat at high speed and to prevent damage to the drive unit. Therefore, the simple procedure to make the adjustment begins on Page 8-15.

STARBOARD SIDE

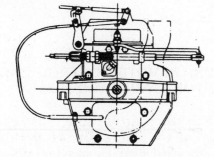

PORT SIDE

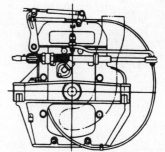

The shift cables may be routed for the starboard side (left drawing) or for the port side (right drawing).

7-3 EARLY-TYPE — PRIOR TO 1973 SHIFT CABLE SERVICE

The procedures in these sections cover service on the early-type shift cable installations from the first Mercruiser stern drive units to the change made sometime during the 1973 model year.

ATTACHING STERN DRIVE SHIFT CABLE MOUNTED ON STARBOARD SIDE

1- On early production models, the drive unit shift cable was mounted on the starboard side of the boat. The shift cable on this type of installation is connected by first pushing the Nylon tube (from the shift cable) into the recess in the transom plate, and then anchoring the tube and cable to the inner transom plate with a tab washer and flat-head screw.

2- To properly set the brass barrel on early model installations, move the remote control unit into the full **FORWARD** position and at the same time have an assistant rotate the propeller shaft counterclockwise until it stops, to ensure the clutch is fully engaged in **FORWARD** gear. In this position, the inner core wire **MUST** extend exactly 1-3/8" from the end of the cable guide insert.

3- Later production models of this early type have cables that require installation of the support tube and the shift cable inner core wire. On this installation, slide the support tube over the core wire until 1/2" of the core wire is exposed. Secure the tube to the core wire by crimping. Next, slide the cable and guide over the end of the cable through the cable anchor; tighten one anchor screw just snug, and then tighten the other screw to secure the core wire.

DRIVE UNIT SHIFT CABLE

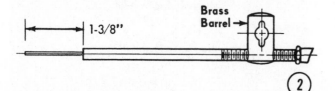

1-3/8"   Brass Barrel

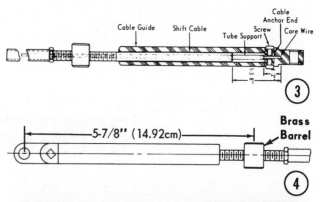

Cable Guide   Shift Cable   Cable Anchor End   Screw   Core Wire   Tube Support

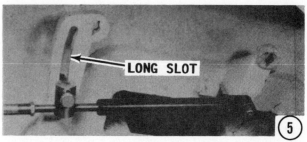

Brass Barrel

5-7/8" (14.92cm)

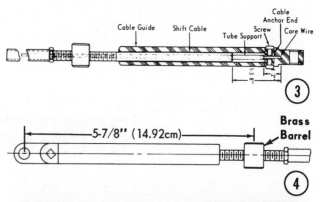

LONG SLOT

4- After the cable end guide has been installed, shift the drive unit into the full **FORWARD** position and at the **SAME TIME** have an assistant rotate the propeller shaft counterclockwise until it stops, to ensure the clutch is fully engaged in FORWARD gear. Measure the distance from the centerline of the brass barrel to the centerline of the cable end guide mounting hole. This measurement **MUST** be exactly 5-7/8". Adjust the brass barrel until this measurement is obtained.

5- Secure the cable guide to the mounting bracket stud on the transom plate, but do not tighten the nut securely because the end guide must be able to pivot freely on the mounting stud. The brass barrel mounting screw **MUST** be located at the bottom of the curved slot with a thick washer installed on both sides of the slot in the shift lever. Attach the shift cables to the anchor and the adjusting stud.

ADJUSTING REMOTE-CONTROL CABLE MOUNTED ON STARBOARD SIDE

1- After the control cable is attached in the bottom of the slot, shift the stern drive unit into forward gear by moving the shift lever toward the port side as far as possible and at the **SAME TIME** have an assistant

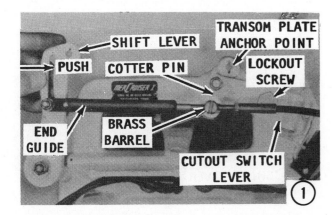

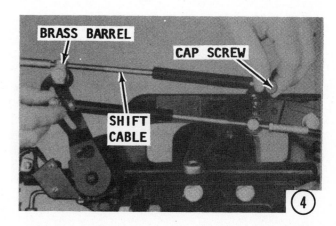

turn the propeller shaft counterclockwise until it stops to ensure the clutch is fully engaged. **ONE WORD:** On E-Z Shift models through 160, place the lockout screw through the cut-out switch lever.

2- Anchor the remote control shift cable end guide to the transom plate anchor point. Remove all slack in the cable by pulling the outer guide away from the cable guide. Now, adjust the brass barrel on the remote control shift cable until the barrel aligns with the mounting hole in the shift lever.

3- Fasten the brass barrel and spacer to the shift lever, and tighten the cap screw securely.

4- Disconnect the shift cable end guide from the transom plate anchor point. Shift the stern drive into reverse gear by moving the shift lever toward the starboard side as far as possible and at the **SAME TIME** have

an assistant rotate the propeller shaft clockwise until it stops to ensure the clutch is fully engaged. Loosen, and then move the end guide of the shift cable up the slot in the shift lever until the cable end guide can be reinstalled to the transom anchor point. Secure the stern drive cable end guide to the shift lever.

ATTACHING STERN DRIVE SHIFT CABLE MOUNTED ON PORT SIDE

1- On early production models of the early-type shift cables, the brass barrel on the shift cable is secured to the transom plate with a spacer, a washer, and an elastic-stop nut, as shown.

2- On later models of the early-type shift cable, the brass barrel of the shift cable fits into the recess of the transom plate and is secured with a cotter pin.

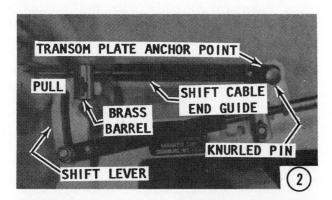

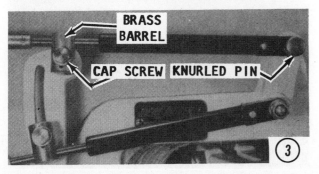

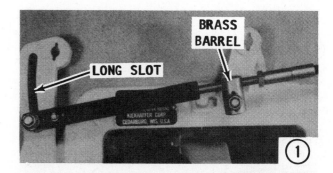

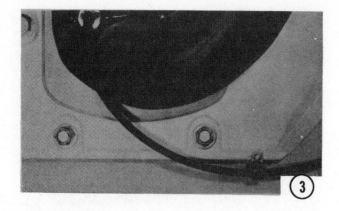

③

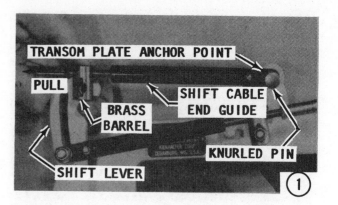

①

3- Anchor the shift cable to the inner transom plate with a clamp and screw.

4- After the cable end guide has been installed, shift the drive unit into the full forward position and at the **SAME TIME** have an assistant turn the propeller shaft counterclockwise until it stops to ensure the clutch is fully engaged. Measure the distance from the centerline of the brass barrel to the centerline of the cable end guide mounting hole. This measurement **MUST** be exactly 5-7/8".

5- Adjust the brass barrel until this measurement is obtained. Secure the cable guide to the mounting bracket stud, but do not tighten the nut securely because the end guide must be able to pivot freely on the mounting stud. Secure the brass barrel to the transom plate. Attach the shift cables to the anchor and the adjusting stud.

ADJUSTING REMOTE CONTROL CABLE MOUNTED ON PORT SIDE

1- On the E-Z Shift models through the 160, insert the lockout screw through the cut-out switch lever. With the shift cable attached in the bottom of the slot, shift the stern drive unit into forward by moving the shift lever toward the port side as far as possible. At the **SAME TIME** have an assistant rotate the propeller shaft counterclockwise until it stops to ensure the clutch is fully engaged in **FORWARD** gear. Move the remote control handle to the full forward position. Next, anchor the shift cable end guide to the shift lever. Remove any slack in the cable by pulling the outer guide away from the cable guide. Now, adjust the brass barrel on the shift cable to align with the transom plate anchor point.

2- Secure the brass barrel, with a spacer, to the shift lever. Tighten the cap screw securely.

3- Disconnect the shift cable end guide from the shift lever. Move the remote control handle to the full reverse position. Manually, move the shift lever to the starboard side as far as possible and at the **SAME TIME** have an assistant rotate the propeller shaft clockwise to ensure the clutch is fully engaged.

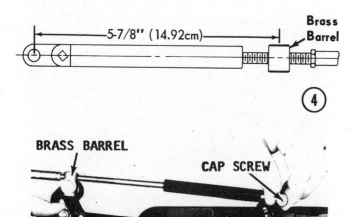

④

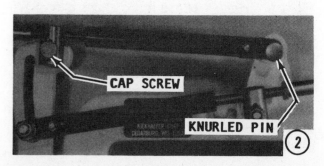

②

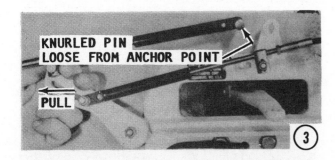

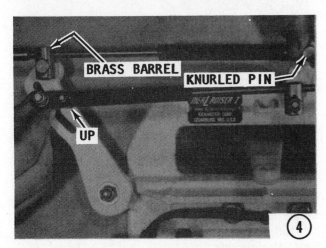

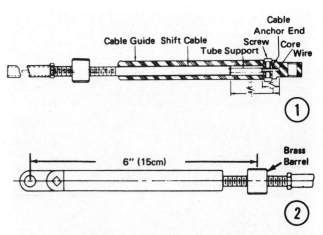

4- Loosen and move the end guide of the stern drive shift cable up the slot in the shift lever until the cable end guide can be reinstalled on the shift lever. Securely tighten the nut on the stud to secure the stern drive cable end guide to the shift lever.

7-4 SERVICING LATER STYLE — 1973 to 1982 SHIFT CABLES

The procedures in this section cover service on later-style shift cable installations since the change made sometime during the 1973 model year and used through most of the 1982 model year.

INSTALLATION DIMENSIONS

1- Before making any measurements of the remote control shift cable, shift the unit into forward and at the **SAME TIME** have an assistant turn the propeller shaft counterclockwise until the shaft stops to ensure the clutch is fully engaged. Measure the amount the inner core of the drive unit shift cable extends from the end of the cable guide insert. This measurement **MUST** be exactly 1-3/8". Slide the support tube over the core wire until 1/2" of the core wire is exposed. Crimp the wire to secure it. Slide

the cable end guide over the end of the cable and through the cable anchor. Secure the core wire by bringing one anchor screw up snug, and then tighten the other screw.

2- After the cable end guide is properly installed, and the shift unit is in full forward gear, measure the distance from the centerline of the brass barrel to the centerline of the cable end guide mounting hole. This measurement **MUST** be exactly 6".

STERN DRIVE SHIFT CABLE INSTALLATION

3- Coat the recess of the cut-out switch lever and shift cable anchor points with Multi-purpose Lubricant, or equivalent. Route the shift cable in the cable clamp (located on the engine flywheel housing cover on six-cylinder engines and under the right exhaust elbow or through the J-clip on a V8 engine), and then bend the clamp shut. Remove the cotter pin from the shift cut-out lever recess and remove the elastic stop nut and flat washer from the shift lever adjustable stud. Place the brass barrel of the drive unit shift cable into the recess of the shift cut-out lever, and then install the cable end guide on the shift lever stud. Insert a **NEW** cotter pin through the cut-out switch lever, and then spread the ends. Place a flat washer on the shift lever stud, and then thread on the elastic-stop nut. **TAKE CARE** not to tighten the elastic-stop nut securely because the end guide must pivot freely on the stud.

REMOTE CONTROL SHIFT CABLE ADJUSTMENT

4- Place the stern drive unit in the full forward gear position, and at the **SAME**

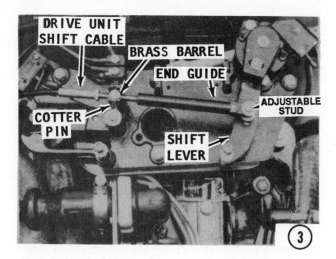

TIME have an assistant rotate the propeller shaft counterclockwise until it stops, to ensure the clutch is fully engaged in FORWARD gear. GENTLY pull the brass barrel away from the remote control shift cable end guide to remove any slack. TAKE NOTE: The adjustable stud anchor point for the drive unit shift cable MUST be to the bottom of the slot in the shift lever, which is toward the reverse lock valve. Now, adjust the brass barrel on the remote control shift cable until the brass barrel guide aligns with the anchor stud on the shift lever and the cable end guide with the anchor stud on the shift plate. Back the brass barrel off exactly FOUR TURNS away from the cable end guide. TAKE CARE not to tighten the elastic-stop nuts against the cable because the end guide and brass barrel MUST pivot freely on the anchor studs.

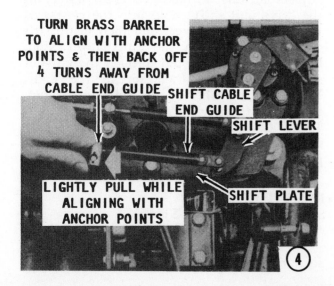

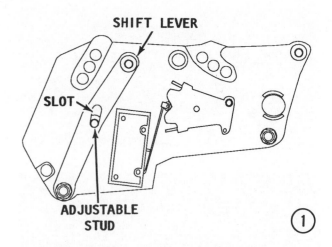

7-5 SERVICING SHIFT CABLES DRIVE MODELS — R, MR, AND ALPHA SINCE 1983

SHORT INTRODUCTION

Two shift cables are connected to the shift plate on the engine. The lower cable is routed from the stern drive unit to the plate and the upper cable is routed from the remote control shift box to the plate.

STERN DRIVE (LOWER) SHIFT CABLE INSTALLATION

1- Move the adjustable stud to the lowest position in the slot on the shift lever.

2- Shift the stern drive into FORWARD gear by pushing the cable end guide, at the end of the stern drive shift cable, towards the barrel, as shown in the accompanying

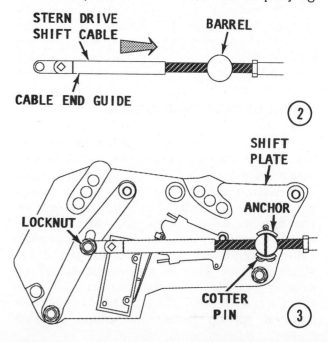

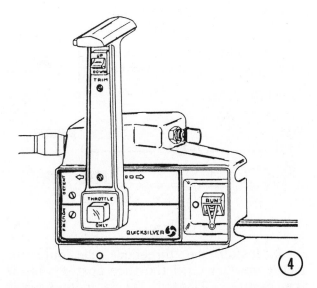

④

illustration. At the same time, have an assistant rotate the propeller shaft **COUNTERCLOCKWISE** until the shaft stops to ensure the clutch is fully engaged.

3- Install the stern drive shift cable onto the shift plate by first hooking the eye on the guide onto the stud on the shift lever, and then rotating the brass barrel along the threaded portion of the cable until the barrel may be positioned in the anchor with the stud remaining in position at the bottom of the shift lever slot.

Place two washers over the stud. Thread the locknut over the stud and tighten the nut securely. Now, back off the locknut one half turn. Push the brass barrel into the anchor until the top and bottom holes on the anchor are clearly visible. Slide the cotter pin through the two holes in the anchor and spread both ends to secure the barrel in the anchor.

4- Shift the remote control lever into **FORWARD** gear, and at the same time have

an assistant rotate the propeller shaft **COUNTERCLOCKWISE** until the shaft stops to ensure the clutch is fully engaged. If difficulty in shifting is encountered, then pull the installed stern drive shift cable in the same direction as described in Step 1, until the clutch engages. From this point on, **TAKE CARE** not to disturb the position of the shift lever on the the shift plate.

REMOTE CONTROL (UPPER) SHIFT CABLE INSTALLATION

5- Hold the remote control shift cable up to the shift plate and visually align the eye at the end of the cable guide with the stud at the top of the shift lever.

Adjust the position of the brass barrel along the threaded portion of the cable until the hole in the barrel indexes with the upper stud on the shift plate. Pull the barrel slightly away from the cable guide while making this adjustment. Correct installation is with the cable positioned **ABOVE** the stud, not below the stud.

After the cable eye and the barrel hole are aligned, rotate the brass barrel through **FOUR** complete turns **CLOCKWISE**, towards the eye end of the cable. This action places the correct amount of preload on the cable.

6- Now, slide the cable eye over the shift lever stud and the barrel over the shift plate stud. Install locknuts over the two studs and tighten the locknuts securely.

SERVICING SHIFT CABLES BRAVO STERN DRIVE

SPECIAL WORDS

If the stern drive is removed from the boat and the only cable disturbed was the stern drive shift cable

or

If the stern drive cable has been replaced, then perform only Step 1 thru Step 6.

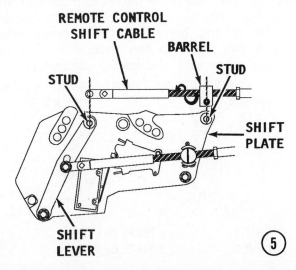

REMOTE CONTROL SHIFT CABLE
BARREL
STUD
STUD
SHIFT PLATE
SHIFT LEVER

⑤

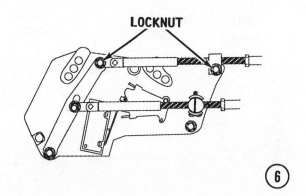

LOCKNUT

⑥

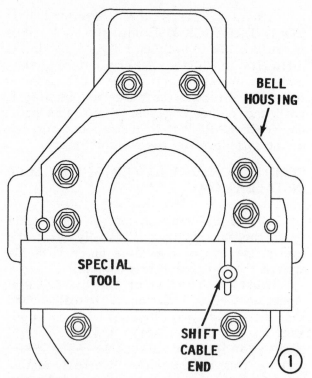

BELL
HOUSING

SPECIAL
TOOL

SHIFT
CABLE
END

1

If both the stern drive cables and the remote control cables need adjustment, then proceed directly to Step 7, on the next page.

Stern Drive Shift Cable Adjustment
Stern Drive Removed from Boat

1- Obtain Core Wire Locating Tool P/N 91-17263 and Shift Cable Anchor Adjustment Tool P/N 91-17262. If these tools are not available, an alternate method of achieving the correct cable measurements is given in the following procedures.

Place the core wire locating tool against the bell housing and slide the end of the shift cable through the slot in the tool, as shown. The thickness of this tool is 1/4". Therefore, if this tool is not available obtain a piece of metal 1/4" thick, which will extend across the bell housing. Cut a slot in the appropriate place for the cable end to slide through.

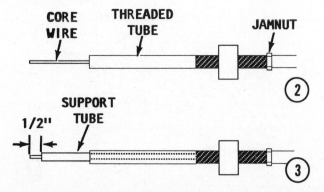

CORE   THREADED
WIRE   TUBE   JAMNUT

2

1/2"   SUPPORT
        TUBE

3

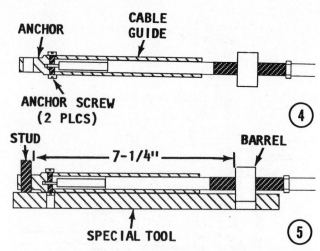

ANCHOR   CABLE
              GUIDE

ANCHOR SCREW
(2 PLCS)

4

STUD   BARREL

|← 7-1/4" →|

SPECIAL TOOL

5

2- Install the threaded tube onto the free core wire until the tube bottoms inside the casing. Tighten the jamnut securely against the casing.

3- Slide the support down over the cable core and into the threaded tube. Position the visible end of the support tube 1/2" from the end of the exposed core wire. Crimp the support tube over the core wire to hold it in place.

4- Slide the cable end guide over the core wire. Make sure the core wire end passes through the two anchor screws and into the hole inside the cable anchor at the end of the cable guide. Tighten the two anchor screws securely. These two screws "pinch" the cable and attach the cable end guide to the wire core.

5- Lightly pull the cable guide away from the brass barrel to take up any slack on the cable end which rests against the special tool at the bell housing.

Slide the anchor eye over the stud on the shift cable anchor adjustment tool. Rotate the brass barrel along the threaded portion of the tube until the barrel can be slid into the hole of the special tool. This action positions the barrel in the correct place along the tube. If the special tool is not available, rotate the barrel along the tube

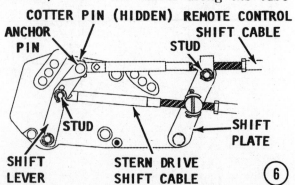

COTTER PIN (HIDDEN) REMOTE CONTROL
ANCHOR   SHIFT CABLE
PIN   STUD

                                    SHIFT
                                    PLATE

SHIFT   STERN DRIVE
LEVER   STUD SHIFT CABLE

6

threads until there is a distance of 7-1/4"
between the edge of the barrel and the edge
of the anchor eye. This distance is edge-to-
edge and **NOT** center-to-center. **TAKE
CARE** not to disturb the position of the
barrel during installation of the cable to the
shift plate. Remove the special tool from
the cable end at the bell housing.

6- Shift the remote control lever to the
NEUTRAL position.

GOOD WORDS

Unlike the R, MR, and Alpha One stern
drive units, the upper cable on a Bravo
connects the remote control shift box to the
shift plate and the lower cable connects the
shift plate to the stern drive.

Align the anchor eye with the hole at the
top of the shift lever. Slide the anchor pin
through the shift lever and anchor eye.
Install the cotter pin through the anchor pin
and spread both ends to secure the pin. This
is not an easy task because very little
clearance exists between the shift plate and
the shift lever. The anchor pin head may be
rotated to bring the ends to the top. In this
position, the ends are visible.

Slide the barrel over the upper stud on
the shift plate. The barrel **MUST** be installed with the cable **ABOVE** the stud and not
below the stud. If the barrel will not align
with the upper shift plate stud, loosen the
stud in the slot of the shift lever and move
the lever slightly until the barrel can be
installed over the shift plate stud. Tighten
the shift lever stud to hold the adjustment.

Shift Cables Installation and Adjustment
Stern Drive Installed
(Bravo Stern Drive)

FIRST, THESE WORDS

Shift Cable Adjustment Tool 91-12427
must be obtained to correctly install both
shift cables. Without the use of this tool,
the proper adjustment can only be obtained
through a time consuming trial and error
method.

Propeller Rotation

Installation of the control cable at this
point will determine rotation of the propeller -- either right or left hand.

After installation, if a **RIGHT** hand pro-

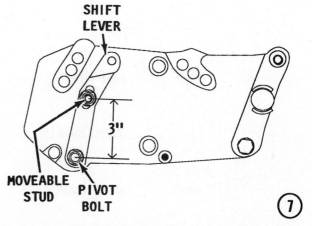

peller motion is required, the remote control cable end **MUST** move **TOWARD** the
barrel, when the cable is activated.

After installation, if a **LEFT** hand propeller motion is required, the remote control cable end **MUST** move **AWAY** from the
barrel, when the cable is activated.

The inner notch cut in the lower edge of
the tool is used to align the moveable stud
on the shift lever to permit right hand
rotation of the propeller. This position
allows the cable end to move toward the
brass barrel when the unit is shifted into
FORWARD gear.

The outer notch cut in the lower edge of
the tool is used to align the moveable stud
on the shift lever to permit left hand rotation of the propeller. This allows the cable
end to move away from the brass barrel
when the unit is shifted into **FORWARD**
gear.

7- Loosen the moveable stud in the slot
on the shift lever. Move the stud until a
distance of three inches can be measured
from the center of the stud to the center of
the pivot bolt on the shift lever, as shown.
Tighten the securing nut just **SNUG** to hold
this adjustment.

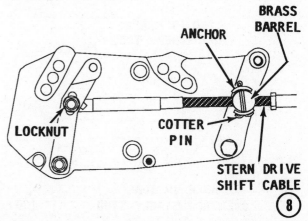

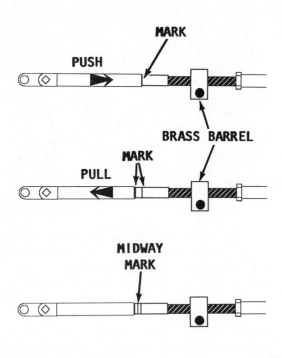

PUSH ← MARK

PULL ← MARK

BRASS BARREL

MIDWAY MARK

(9)

8- Hook the stern drive shift cable eye over the moveable stud. Rotate the brass barrel along the threaded portion of the cable until the barrel may be positioned in the anchor. Place the two washers over the stud. Thread the locknut over the stud. Tighten the nut securely, and then back off the locknut one half turn. Push the brass barrel into the anchor until the top and bottom holes are clearly visible. Slide the cotter pin through the two holes in the anchor and spread both ends to secure the barrel in the anchor.

Remote Control Shift Cable Installation (Bravo Stern Drive)

9- Place the remote control lever on the shift box in the **NEUTRAL** position. Obtain

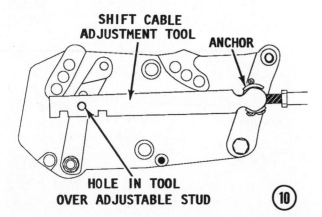

SHIFT CABLE ADJUSTMENT TOOL   ANCHOR

HOLE IN TOOL OVER ADJUSTABLE STUD (10)

a suitable marker to mark the cable. **CAREFULLY** push the end of the remote control cable toward the brass barrel. Make a mark on the tube at the point where the tube emerges from the cable guide. Next, **CAREFULLY** pull the end of the remote control cable away from the brass barrel. Again, make a mark on the tube as before.

Now, measure the distance between the two marks and make a third mark midway between the two marks. This midway mark is the true neutral position with the cable play/slack having been removed.

10- Place Shift Cable Adjustment Tool P/N 91-12427 in place over the installed stern drive shift cable, with the rounded end of the tool indexed with the anchor and the hole in the tool indexed over the adjustable stud. If necessary, place a piece of tape over the rounded end of the special tool to keep it in place within the anchor.

11- Hold the remote control shift cable up to the shift plate and visually align the anchor eye at the end of the cable guide with the hole at the top of the shift lever.

Hold the cable guide at the midway mark on the tube, and at the same time adjust the

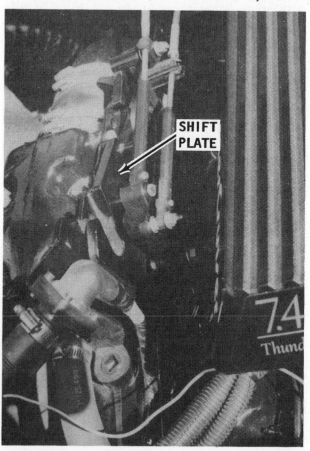

The shift plate and shift cables installed on a 7.4 Litre Bravo powerplant.

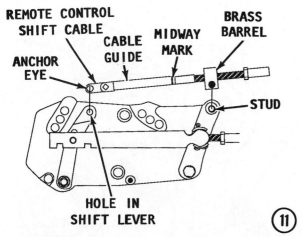

REMOTE CONTROL SHIFT CABLE — CABLE GUIDE — MIDWAY MARK — BRASS BARREL — ANCHOR EYE — STUD — HOLE IN SHIFT LEVER — ⑪

position of the brass barrel along the threaded portion of the cable until the hole in the barrel indexes with the upper stud on the shift plate. Correct installation is with the cable **ABOVE** the stud.

12- Align the anchor eye with the hole at the top of the shift lever. Slide the anchor pin through the shift lever and anchor eye. Install the cotter pin through the anchor pin and spread both ends to secure the pin. This is not an easy task due to the small amount of clearance between the shift plate and the shift lever. The anchor pin head may be rotated to bring the ends to the top. In this position, the ends are visible.

Slide the barrel over the upper stud on the shift plate. Correct installation is with the barrel **ABOVE** the stud. Install the locknut over the stud and tighten the nut securely.

13- Remove the special tool. Shift the remote control lever at the shift box into **FORWARD** gear. Install the rounded end of the special tool, just removed, back over the brass barrel of the stern drive shift cable.

If a **RIGHT** hand propeller motion is desired, the inner slot on the lower edge of the tool **MUST** fit over the shift lever stud.

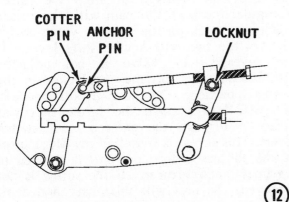

COTTER PIN — ANCHOR PIN — LOCKNUT — ⑫

RIGHT HAND PROPELLER INSTALLATION

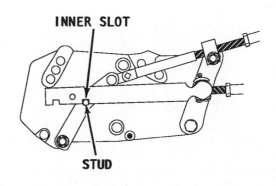

INNER SLOT — STUD

LEFT HAND PROPELLER INSTALLATION

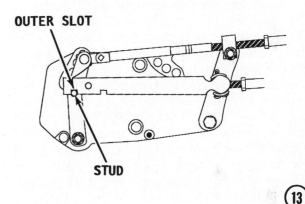

OUTER SLOT — STUD — ⑬

If a **LEFT** hand propeller motion is desired, the outer slot on the lower edge of the tool **MUST** fit over the shift lever stud.

If the appropriate slot will not align with the stud, loosen the nut at the base of the stud and move the shift lever slightly until the appropriate slot indexes with the stud. Tighten the nut to hold the adjustment.

CUT-OUT SWITCH LEVER POSITION ADJUSTMENT

FIRST, THESE WORDS

The cutout switch performs a very important function in the shifting process. When the shift movement begins, the cutout switch grounds the ignition for a split second. This momentary delay relieves the torque on the gear and clutch, allowing the operator to shift with ease. This action is so close to being instantaneous, it is impossible for the operator to notice. If the stern drive is continually operated without the cutout switch properly adjusted, severe

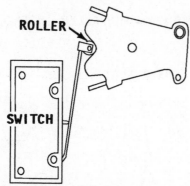

An ignition cutout switch is mounted on the shift plate on most late model stern drive units.

damage could be caused to the shift cables, clutch dog, stern drive gears, and shifting mechanism in the lower unit. **THEREFORE,** at the first indication of hard shifting when coming out of gear, the cutout switch adjustment should be checked.

Move the remote control shift handle to the full reverse position and at the **SAME TIME** have an assistant turn the propeller shaft clockwise until the shaft stops to ensure the clutch is fully engaged. When the remote control shift handle is in the full reverse position and the clutch is fully engaged, the switch lever **MUST** be in the center of the slot for the neutral position. Adjust the remote control cable brass barrel one turn at a time (four turns maximum) toward the end guide until the shift cut-out switch lever is positioned in the center of the actuating arm slot.

GOOD WORDS

If a standard MerControl side mount or extra long remote control cable is used, or if there are a large number of bends in the remote control shift cable, an additional adjustment may be necessary.

This adjustment is made as follows: With the boat out of the water and the engine shut down, move the remote control shift handle into the reverse position until the throttle **JUST BEGINS TO OPEN.** Next, turn the propeller shaft clockwise; the clutch should engage and cause the propeller to lock. If the clutch does not engage, loosen the adjustable stud which anchors the drive unit shift cable to the shift lever, and then move it forward in the shift lever slot until the clutch is firmly engaged. Again, check the shift cut-out lever position which **MUST** be in the center of the slot for the neutral position. Tighten the adjustable anchor stud at the necessary location of the shift lever slot.

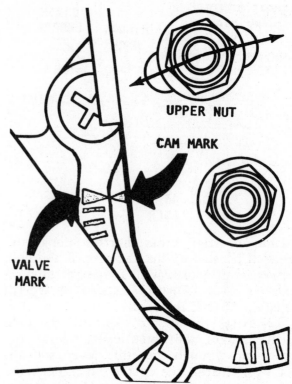

Making the reverse lock valve adjustment, as described in the text.

REVERSE LOCK VALVE ADJUSTMENT

GOOD WORDS

The lock out valve cuts the circuit to the tilt pump when the stern drive is in reverse gear. If the tilt switch could be operated with the stern drive in reverse gear, severe damage would be caused to the pump or the tilt cylinders.

From the neutral position, move the shift lever to the full reverse position, and at the **SAME TIME** have an assistant rotate the propeller shaft clockwise until it stops, to ensure the clutch is fully engaged in **REVERSE** gear. Loosen the two nuts on the shift lever and move the upper nut in the direction required to align the raised triangular mark on the cam with the raised triangular mark on the reverse valve cover. Tighten the two nuts on the shift lever. If the reverse lock valve is not adjusted **PROPERLY**, the unit may be locked in the neutral position. On the E-Z Shift models, **BE SURE** to remove the screw that was attached through the shift cut-out switch lever. This screw is installed my Mercruiser at the factory before the unit is shipped to the boat manufacturer. If the screw is not removed, the unit will shift hard and damage will be caused to the stern drive unit.

SPARK PLUG

7-6 COMMANDER CONTROL SHIFT BOX SERVICE

REMOVAL AND DISASSEMBLING

The following detailed instructions cover removal and disassembly of the "Commander" control shift box from the mounting panel in the boat.

1- Turn the ignition key to the **OFF** position. Disconnect the high tension leads from the spark plugs, with a twisting motion.

TACHOMETER CONNECTION

2- Disconnect the remote control wiring harness plug from the outboard trim/tilt motor and pump assembly.

3- Disconnect the tachometer wiring plug from the forward end of the control housing.

4- Remove the three locknuts, flat washers, and bolts securing the control housing to the mounting panel. One is located next to the **RUN** button (the ignition safety stop switch), and the second is beneath the control handle on the lower portion of the plastic case. The third is located behind the control handle when the handle is in the **NEUTRAL** position. Shift the handle into **FORWARD** or **REVERSE** position to remove the bolt, then shift it back into the **NEUTRAL** position for the following steps.

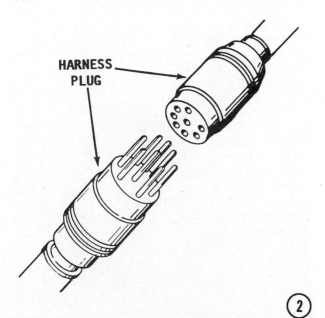

HARNESS PLUG

SECURING BOLT (HIDDEN IN RECESS)

SECURING BOLT

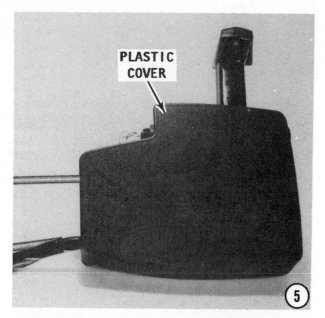

PLASTIC
COVER

⑤

5- Pull the remote control housing away and free of the mounting panel. Remove the plastic cover from the back of the housing. Lift off the access cover from the housing. (Some "Commander" remote control units do not have an access cover.)

6- Remove the two screws securing the cable retainer over the throttle cable, wiring harness, and shift cable. Unscrew the two Phillips-head screws securing the back cover to the control module, and then lift off the cover.

Throttle Cable Removal

7- Loosen the cable retaining nut and raise the cable fastener enough to free the throttle cable from the pin. Lift the cable from the anchor barrel recess.

8- Remove the grommet.

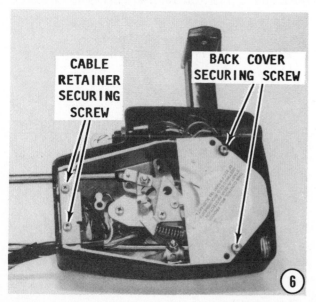

CABLE
RETAINER
SECURING
SCREW

BACK COVER
SECURING SCREW

⑥

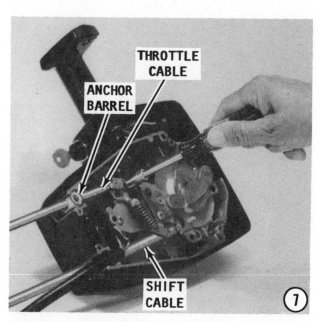

THROTTLE
CABLE

ANCHOR
BARREL

SHIFT
CABLE

⑦

Shift Cable Removal

9- Shift the outboard unit into **REVERSE** gear by depressing the neutral lock bar on the control handle and moving the control handle into the **REVERSE** position. **LOOSEN,** but do not remove, the shift cable retainer nut with a 3/8" deep socket as far as it will go without removing it. Raise the shift cable fastener enough to free the shift cable from the pin.

DO NOT attempt to shift into **REVERSE** while the cable fastener is loose. An attempt to shift may cause the cable fastener to strike the neutral safety microswitch and cause it damage.

Lift the wiring harness out of the cable anchor barrel recess and remove the shift cable from the control housing.

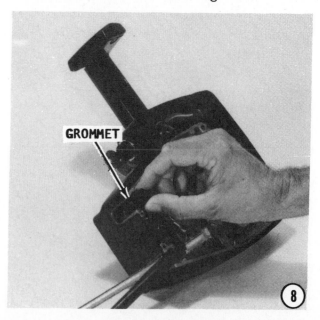

GROMMET

⑧

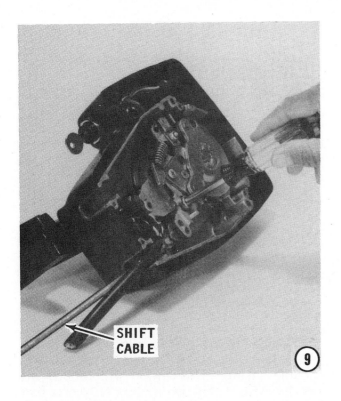

SHIFT
CABLE

⑨

Control Handle Removal
With Toggle Trim Switch
Or Push-Button Trim Switch

10- Depress the **NEUTRAL** lock bar on the control handle and shift the control handle back to the **NEUTRAL** position. Remove the two Phillips head screws which secure the cover to the handle, and then lift off the cover. The push button trim switch will come free with the cover, the toggle trim switch will stay in the handle body.

Unsnap and then remove the wire retainer. Carefully unplug the trim wires and straighten them out from the control panel hub for ease of removal later.

11- Back-off the set screw at the base of the control handle to allow the handle to be removed from the splined control shaft.

12- Grasp the "throttle only" button and pull it off the shaft.

SPECIAL WORDS

Take care not to damage the trim wires when removing the control handle.

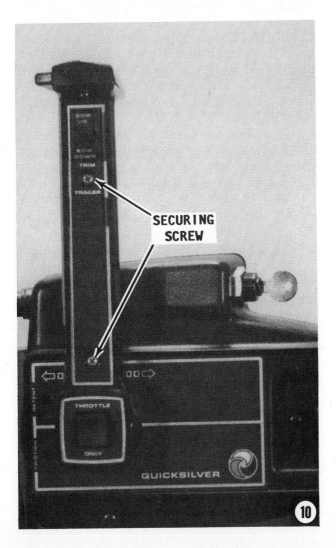

SECURING
SCREW

QUICKSILVER

⑩

BASE
SET SCREW

⑪

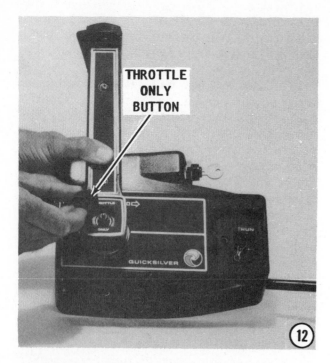

THROTTLE ONLY BUTTON

⑫

13– Remove the control handle.

14– Lift the neutral lockring from the control housing.

TAKE CARE to support the weight of the control housing to avoid placing any unnecessary stress on the control shaft during the following disassembling steps.

15– Remove the three Phillips-head screws securing the control module to the plastic case. Two are located on either side of the bearing plate and one is in the recess where the throttle cable enters the control housing.

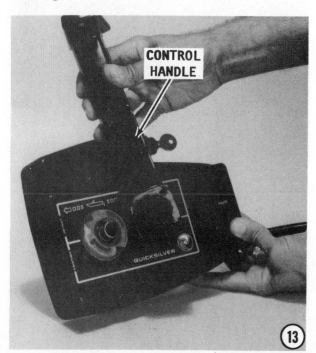

CONTROL HANDLE

⑬

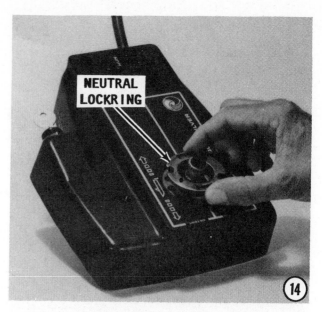

NEUTRAL LOCKRING

⑭

16– Back-out the detent adjustment screw and the control handle friction screw until their heads are flush with the control module casing. This action will reduce the pre-load from the two springs on the detent ball for later removal.

GOOD WORDS

As this next step is performed, count the number of turns for each screw as they are backed-out and record the figure somewhere. This will be a tremendous aid during assembling.

17– Remove the two locknuts securing the neutral safety switch to the plate assembly and lift out the micro-switch from the recess in the assembly.

18– Remove the Phillips-head screw securing the retaining clip to the control module.

⑮

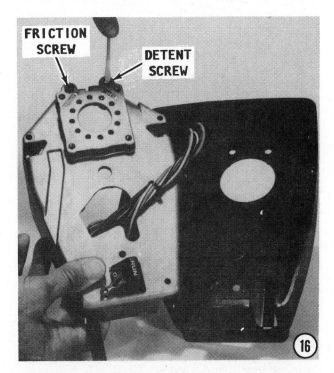

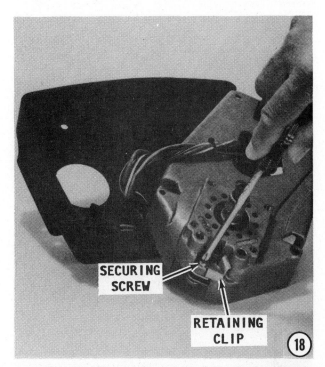

19- Support the module in your hand and tilt it until the shift gear spring, shift nylon pin (earlier models have a ball), shift gear pin, another ball the shift gear ball (inner), fall out from their recess. If the parts do not fall out into your hand, attach the control handle and ensure the unit is in the **NEUTRAL** position. The parts should come free when the handle is in the **NEUTRAL** position.

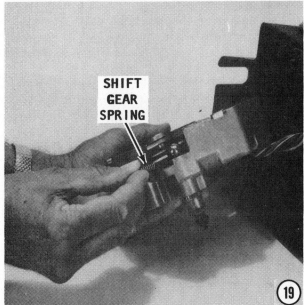

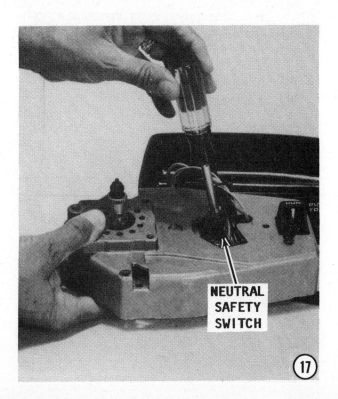

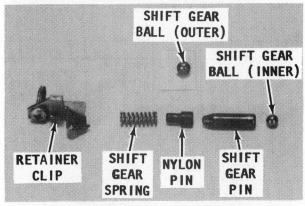

Arrangement of parts from the control module recess. As the parts are removed and cleaned, keep them in order, ready for installation.

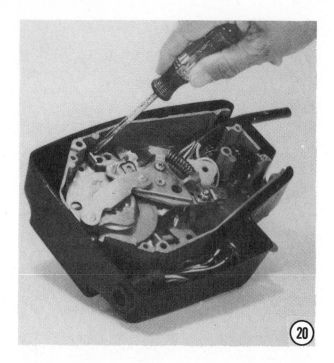

(20)

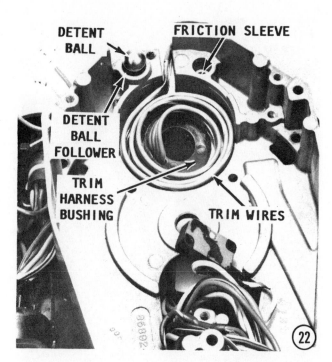

(22)

20– Remove the three Phillips-head screws securing the bearing plate assembly to the control module housing.

21– Lift out the bearing plate assembly from the control module housing.

22– Uncoil the trim wires from the recess in the remote control module housing and lift them away with the trim harness bushing attached.

23– Remove the detent ball, the detent ball follower, and the two compression springs (located under the follower), from

their recess in the control module housing.

24– If it is not part of the friction pad, remove the control handle friction sleeve from the recess in the control module housing.

25– Pull the throttle link assembly from the module. Remove the compression spring from the throttle lever. It is not necessary to remove this spring unless there is cause

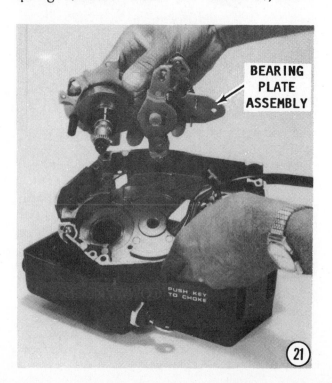

BEARING PLATE ASSEMBLY

(21)

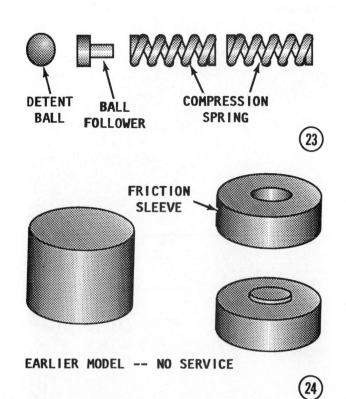

(23)

(24)

EARLIER MODEL -- NO SERVICE

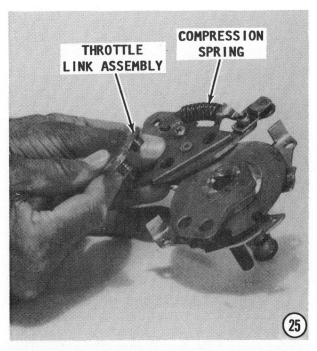

THROTTLE LINK ASSEMBLY

COMPRESSION SPRING

(25)

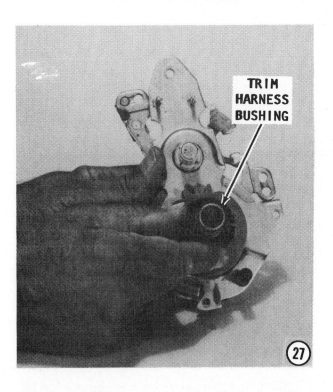

TRIM HARNESS BUSHING

(27)

least amount of tension on the compression spring, therefore, now would be the time to replace it, if required.

26- Lift the shift pinion gear (with attached shift lever), off the pin on the bearing plate. The nylon bushing may come away with the shift lever or stay on the pin. Remove the shift lever and shift pinion gear as an assembly. **DO NOT** attempt to separate them. Both are replaced if one is worn.

27- Remove the trim harness bushing and wiring harness retainer from the control shaft, if they are used.

28- Remove the shift gear retaining ring from its groove with a pair of circlip pliers.

SPECIAL WORDS

If the circlip slipped out of its groove, this would allow the shift gear to ride up on the shaft and cause damage to the small parts contained in its recess. The shift gear ball (inner), the shift gear pin, the shift gear ball (outer), or the nylon pin and **PARTICULARLY** the shift gear spring **MUST** be inspected closely.

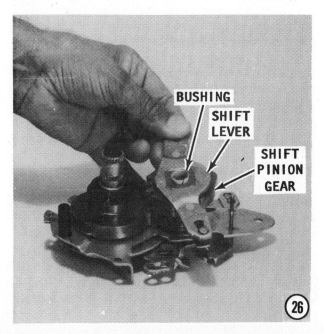

BUSHING

SHIFT LEVER

SHIFT PINION GEAR

(26)

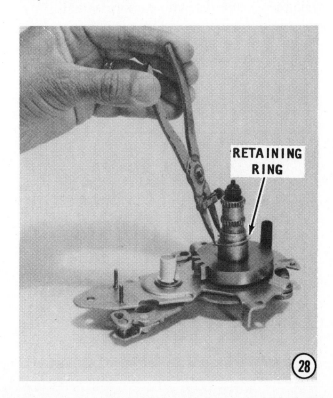

RETAINING RING

(28)

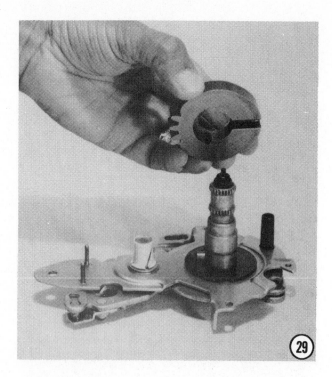

29- Lift the gear from the control shaft.

30- Remove the "throttle only" shaft pin and "throttle only" shaft from the control shaft.

31- Remove the step washer from the base of the bearing plate.

CLEANING AND INSPECTING

Clean all metal parts with solvent, and then blow them dry with compressed air.

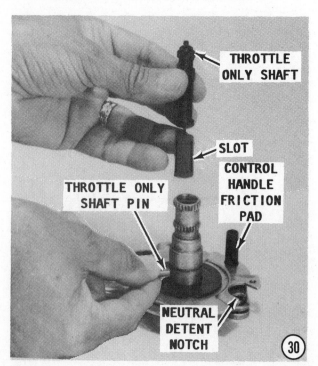

NEVER allow nylon bushings, plastic washers, nylon pins, wiring harness retainers, and the like, to remain submerged in solvent more than just a few moments. The solvent will cause these type parts to expand slightly. They are already considered a "tight fit" and even the slightest amount of expansion would make them very difficult to install. If force is used, the part is most likely to be distorted.

Inspect the control housing plastic case for cracks or other damage allowing moisture to enter and cause problems with the mechanism.

Carefully check the teeth on the shift gear and shift lever for signs of wear. Inspect all ball bearings for nicks or grooves which would cause them to bind and fail to move freely.

Closely inspect the condition of all wires and their protective insulation. Look for exposed wires caused by the insulation rubbing on a moving part, cuts and nicks in the insulation and severe kinking which could cause internal breakage of the wires.

Inspect the surface area above the groove in which the circlip is positioned for signs of the circlip rising out of the groove. This would occur if the clip had lost its "spring" or worn away the top surface of the groove as mentioned previously in Step 29. If the circlip slipped out of its groove, this would allow the shift gear to ride up on the shaft and cause damage to the small parts contained in its recess. The shift gear ball (inner), the shift gear pin, the shift gear ball (outer), or the nylon pin and **PARTICULARLY** the shift gear spring **MUST** be inspected closely.

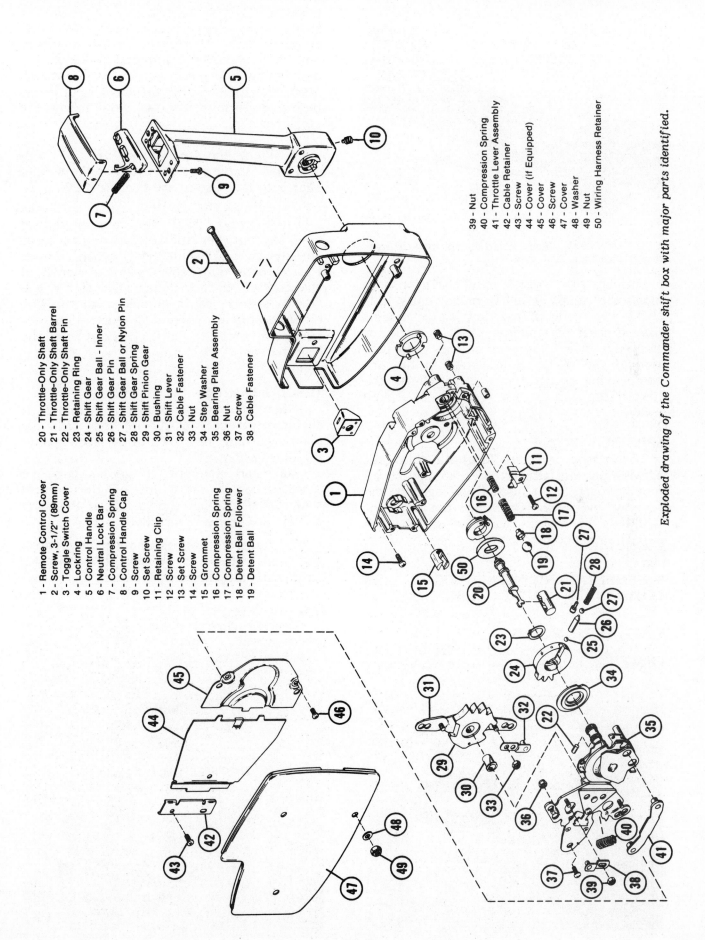

1 - Remote Control Cover
2 - Screw, 3-1/2" (89mm)
3 - Toggle Switch Cover
4 - Lockring
5 - Control Handle
6 - Neutral Lock Bar
7 - Compression Spring
8 - Control Handle Cap
9 - Screw
10 - Set Screw
11 - Retaining Clip
12 - Screw
13 - Set Screw
14 - Screw
15 - Grommet
16 - Compression Spring
17 - Compression Spring
18 - Detent Ball Follower
19 - Detent Ball

20 - Throttle-Only Shaft
21 - Throttle-Only Shaft Barrel
22 - Throttle-Only Shaft Pin
23 - Retaining Ring
24 - Shift Gear
25 - Shift Gear Ball - Inner
26 - Shift Gear Pin
27 - Shift Gear Ball or Nylon Pin
28 - Shift Gear Spring
29 - Shift Pinion Gear
30 - Bushing
31 - Shift Lever
32 - Cable Fastener
33 - Nut
34 - Step Washer
35 - Bearing Plate Assembly
36 - Nut
37 - Screw
38 - Cable Fastener

39 - Nut
40 - Compression Spring
41 - Throttle Lever Assembly
42 - Cable Retainer
43 - Screw
44 - Cover (if Equipped)
45 - Cover
46 - Screw
47 - Cover
48 - Washer
49 - Nut
50 - Wiring Harness Retainer

Exploded drawing of the Commander shift box with major parts identified.

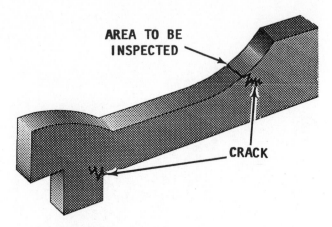

The throttle only shaft should be inspected for wear along the ramp, as indicated.

Inspect the "throttle only" shaft for wear along the ramp. In early model units, this shaft was made of plastic. Later models have a shaft of stainless steel. Check for excessive wear or cracks on the ramp portion of the shaft, as indicated in the accompanying illustration. Also check the lower "stop" tab to be sure it has not broken away.

SPECIAL WORDS

Good shop practice dictates a thin coat of Multipurpose Lubricant be applied to all moving parts as a precaution against the "enemy" moisture. Of course the lubricant will help to ensure continued satisfactory operation of the mechanism.

ASSEMBLING AND INSTALLATION COMMANDER CONTROL SHIFT BOX

FIRST, THESE WORDS

The Commander control shift box, like others, has a number of small parts that MUST be assembled in only one order -- the proper order. Therefore, the work should not be "rushed" or attempted if the person assembling the unit is "under pressure". Work slowly, exercise patience, read ahead before performing the task, and follow the steps closely.

1- Place the step washer over the control shaft and ensure the steps of the washer seat onto the base of the bearing plate.

2- Rotate the control shaft until the "throttle only" shaft pin hole is aligned centrally between the neutral detent notch and the control handle friction pad. Lower the "throttle only" shaft into the barrel of the control shaft with the wide slot in the "throttle only" shaft aligned with the line drawn on the accompanying illustration. Secure the shaft in this position with the "throttle only" shaft pin.

SPECIAL WORDS

When the pin is properly installed, it should protrude slightly in line with the plastic bushing, as shown in the accompanying illustration.

Make an attempt to gently pull the "throttle only" shaft out of the control shaft. The attempt should fail, if the shaft and pin are properly installed.

3- Place the shift gear over the control shaft, and check to be sure the "throttle only" shaft pin clears the gear.

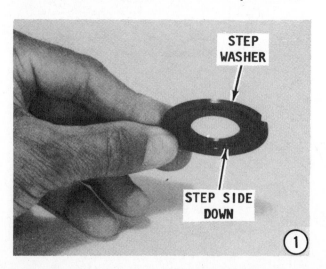

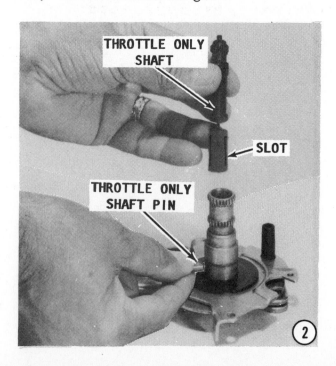

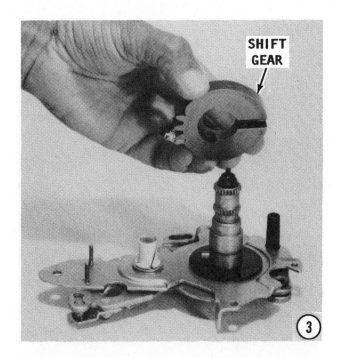

SHIFT GEAR

③

4- Install the retaining ring over the control shaft with a pair of circlip pliers. Check to be sure the ring snaps into place within the groove.

5- Slide the wiring harness retainer (if a retainer is used), and then slide the trim harness bushing over the control shaft. The trim harness bushing is placed "stepped side" **UP** and the notched side toward the forward side of the control housing.

6- Carefully coil the wires around the trim harness bushing, as shown in the accompanying illustration. Ensure the black

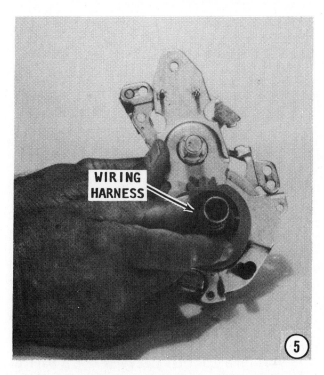

WIRING HARNESS

⑤

line on the trim harness is positioned at the exact point shown for correct installation. The purpose of the coil is to allow slack in the wiring harness when the control handle is shifted through a full cycle. The bushing and wires move with the handle.

7- Position the bushing, shift pinion gear and shift lever onto the pin on the bearing plate, with the shift gear indexing with the shift pinion gear.

8- Install the two compression springs,

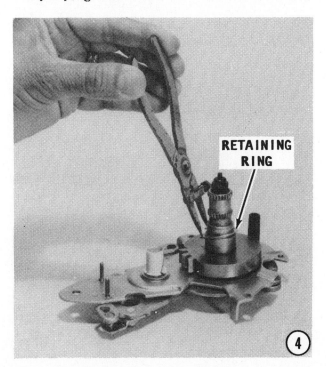

RETAINING RING

④

POSITION OF BLACK LINE ON TRIM HARNESS

TRIM HARNESS SLOT

TRIM HARNESS BUSHING

⑥

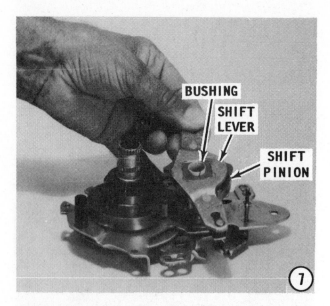

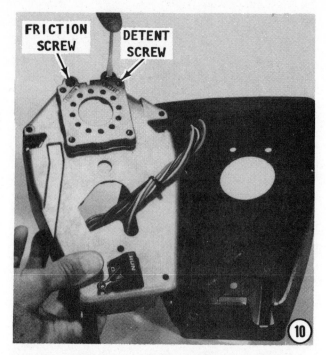

the detent ball follower and the detent ball into their recess in the control module housing.

9- If the friction sleeve is not a part of the friction pad, then place the control handle friction sleeve into its recess in the control module housing.

SPECIAL WORDS

In Step 16 of the disassembling procedures, instructions were given to count the number of turns required to remove the detent adjustment screw and the control handle friction screw. The number of turns is now necessary for ease in performing the next step.

10- Thread the detent adjustment screw and the control handle friction screw the exact number of turns as recorded during Step 16 of the disassembling procedures. A fine adjustment may be necessary after the unit is completely assembled.

11- Place the compression spring (if removed) in position on the shift lever and shift pinion gear assembly against the bearing plate, as shown. Use a rubber band to secure the shift pinion gear to the bearing plate. Lower the complete bearing plate assembly into the control module housing.

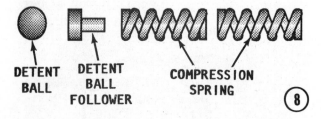

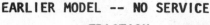

EARLIER MODEL -- NO SERVICE

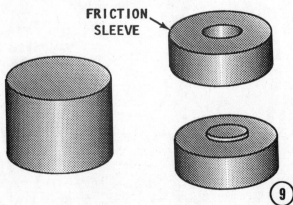

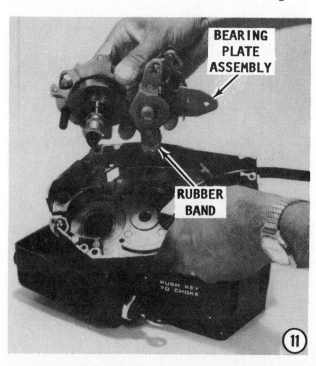

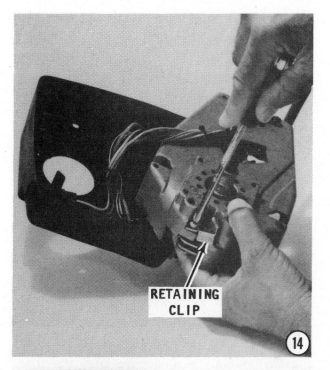

12- Secure the bearing plate assembly to the control module housing with the three Phillips head screws, remove the rubber band.

13- Insert the gear shift ball (inner) into the recess of the shift gear and hole in the "throttle only" shaft barrel. Now, insert the shift gear pin into the recess with the rounded end of the pin away from the control shaft. Insert the nylon pin or shift gear ball (outer) into the same recess. Next, insert the gear shift spring.

14- Hold these small parts in place and at the same time secure them with the retaining clip and the Phillip head screw. This retaining clip secures the trim wire to the control module also.

15- Insert the neutral safety micro-switch into the recess of the plate assembly and secure it with the two locknuts.

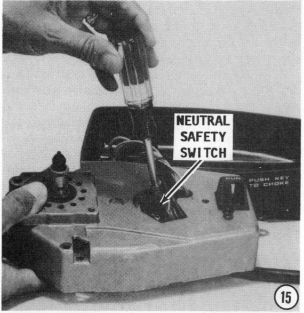

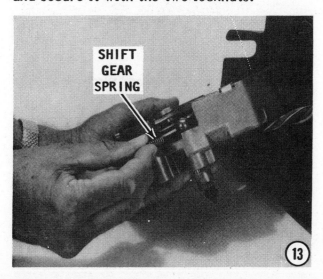

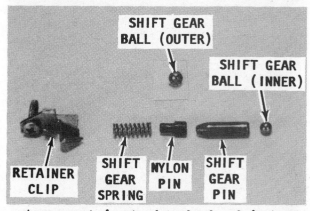

Arrangement of parts, cleaned and ready for installation into the control module recess.

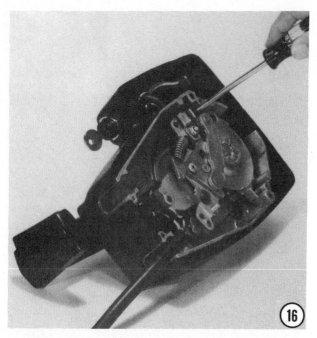

16- Secure the control module to the plastic control housing case with the three Phillips head screws. Two are located on either side of the bearing plate and the third in the recess where the throttle cable enters the control housing.

17- Temporarily install the control handle onto the control shaft. Shift the unit into forward detent **ONLY**, not full forward, to align the holes for installation of the throttle link. After the holes are aligned, remove the handle. Install the throttle link.

18- Again, temporarily install the control handle onto the control shaft. This time shift the unit into the **NEUTRAL** position, and then remove the handle. Place the neutral lockring over the control shaft, with

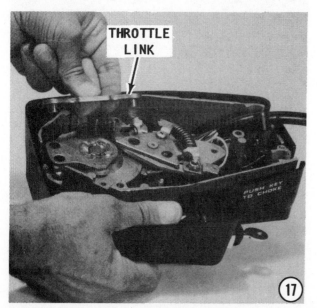

the index mark directly beneath the small boot on the front face of the cover.

19- Install the control handle onto the splines of the control shaft. **TAKE CARE** not to cut, pinch, or damage the trim wires on the power trim/tilt unit.

CRITICAL WORDS

When positioning the control handle, ensure the trim wire bushing is aligned with its locating pin against the corresponding slot in the control handle. If this bushing is **NOT** installed correctly, it will not move with the control handle as it is designed to move -- when shifted. This may pinch or cut the trim wires, causing serious problems.

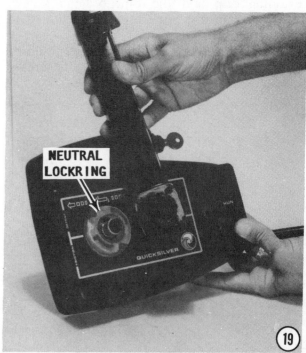

Misplacement of this bushing (it is possible to install this bushing upside down) will not allow the control handle to seat properly against the lockring and housing. This situation will lead to the Allen screw at the base of the control handle to be incorrectly tightened to seat against the splines on the control shaft, instead of gripping the smooth portion of the shaft. Subsequently the control handle will feel "sloppy" and could cause the neutral lock to be ineffective.

WARNING
IF THIS HANDLE IS NOT SEATED PROPERLY, A SLIGHT PRESSURE ON THE HANDLE COULD THROW THE LOWER UNIT INTO GEAR, CAUSING SERIOUS INJURY TO CREW, PASSENGERS, AND THE BOAT.

20- Push the "throttle only" button in place on the control shaft.
21- Ensure the control handle has seated properly, and then tighten the set screw at the base of the handle to a torque value of 70 in. lbs (7.9Nm).

SAFETY WORDS
FAILURE to tighten the set screw to the required torque value, could allow the handle to disengage with a loss of throttle and shift control. An extremely **DANGEROUS** condition.

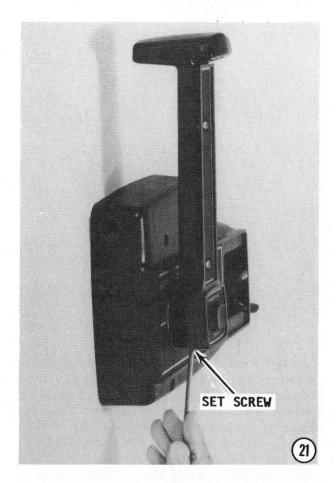

SET SCREW

㉑

22- Slide the hooked end of the neutral lock rod into the slot in the neutral lock release. Route the trim wires in the control handle in their original locations. Connect them with the wires remaining in the handle and secure the connections with the wire retainer. Install the handle cover and tighten the two Phillips head screws.

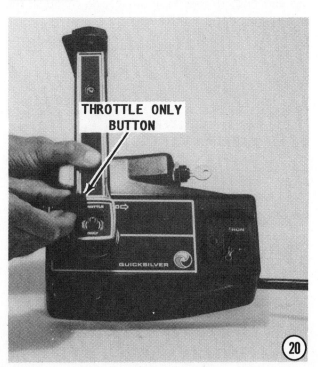

THROTTLE ONLY BUTTON

⑳

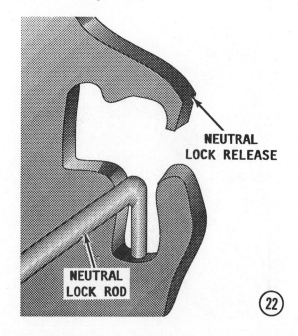

NEUTRAL LOCK RELEASE

NEUTRAL LOCK ROD

㉒

23

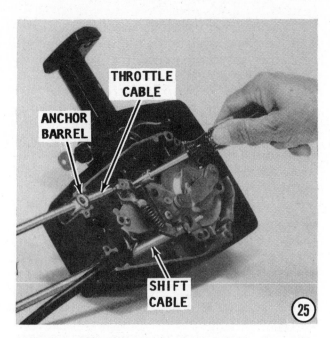

25

23– Move the wiring harness clear of the barrel recess. Thread the shift cable anchor barrel to the end of the threads, away from the cable converter, and place it into the recess. Hook the pin on the end of the cable fastener through the outer hole in the shift lever. Depress the **NEUTRAL** lock bar on the control handle and shift the handle into the **REVERSE** position. **TAKE CARE** to ensure the cable fastener will clear the neutral safety micro-switch. The access hole is now aligned with the locknut.

STOP

Check to be sure the pin on the cable fastener is all the way through the cable end

and the shift lever. A pin partially engaging the cable and the shift lever may cause the cable fastener to **BEND** when the nut is tightened.

Tighten the locknut with a 3/8" deep socket to a torque value of 20 to 25 in. lbs (2.26 to 2.82 Nm). Position the wiring harness over the installed shift cable.

24– Install the grommet into the throttle cable recess.

25– Thread the throttle cable anchor barrel to the end of the threads, away from the cable connector, and then place it into the recess over the grommet. Hook the pin on the end of the cable fastener through the outer hole in the shift lever.

24

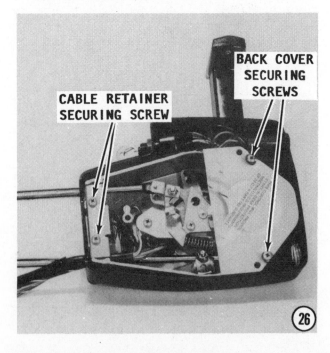

26

STOP AGAIN

Check to be sure the pin on the cable fastener is all the way through the cable end and the throttle lever. A pin partially engaging the cable and throttle lever may cause the cable fastener to **BEND** when the nut is tightened.

Tighten the locknut to a torque value of 20 to 25 in lbs (2.26 to 2.82 Nm).

26- Position the control module back cover in place and secure it with the two Phillips-head screws. Tighten the screws to a torque value of 60 in lbs (6.78 Nm). Install the cable retainer plate over the two cables and secure it in place with the two Phillip-head screws.

27- Place the plastic access cover over the control housing.

28- Position the control housing in place on the mounting panel and secure it with the three long (3-1/2") bolts, flat washers, and locknuts. One is located next to the **RUN** button (the ignition safety stop switch). The second is beneath the control handle on the power portion of the plastic case. The third bolt goes in behind the control handle when the handle is in the **NEUTRAL** position.

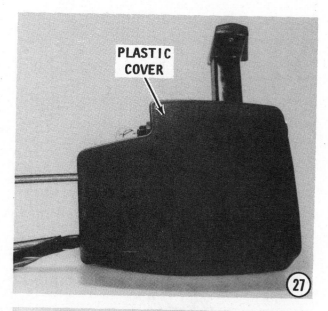

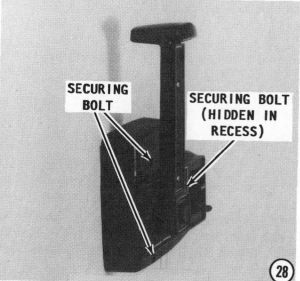

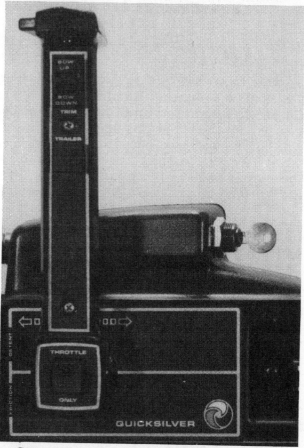

Commander shift box ready for installation into the boat.

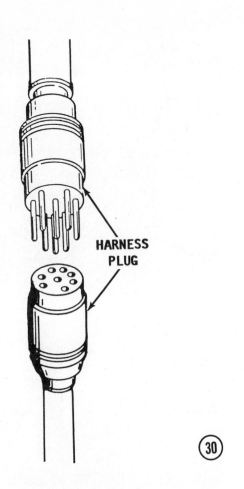

Therefore, in order to install this bolt, shift the handle into the **FORWARD** or the **RE-VERSE** position, and then install the bolt. After the bolt is secure, shift the handle back to the **NEUTRAL** position for the next few steps.

29- Connect the tachometer wiring plug to the forward end of the control housing.

30- Connect the remote control wiring harness plug from the outboard trim/tilt motor and pump assembly.

31- Install the high-tension leads to their respective spark plugs.

32- Route the wiring harness alongside the boat and fasten with the "Sta-Straps". Check to be sure the wiring will not be pinched or chafe on any moving part and will not come in contact with water in the bilge. Route the shift and throttle cables the best possible way to make large bends and as few as possible. Secure the cables approximately every three feet (one meter).

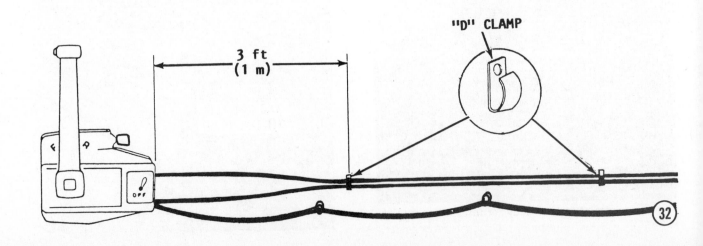

STANDARD NEUTRAL POSITION

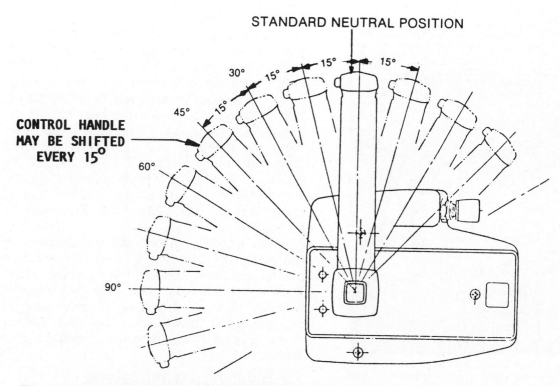

CONTROL HANDLE MAY BE SHIFTED EVERY 15°

The neutral position of the remote control handle may be changed to any one of a number of convenient angles to meet the owner's preference. The change is accomplished by shifting the handle one spline on the shaft at a time. Each spline about 15° of arc, as shown. The procedures on Page 7-26 explain the positioning in detail.

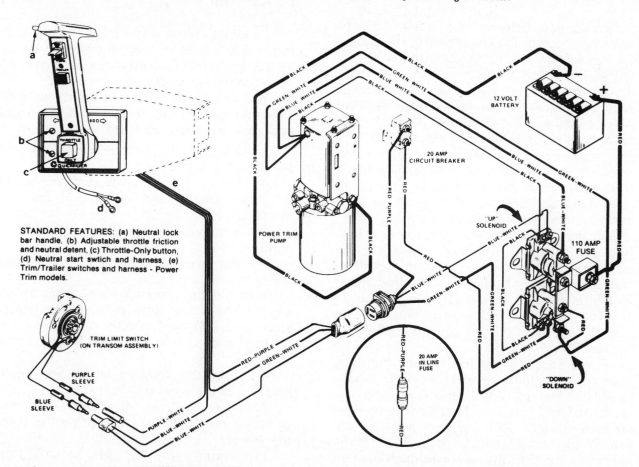

STANDARD FEATURES: (a) Neutral lock bar handle, (b) Adjustable throttle friction and neutral detent, (c) Throttle-Only button, (d) Neutral start switch and harness, (e) Trim/Trailer switches and harness - Power Trim models.

Functional diagram of the Commander remote control shift box and other related electrical components.

The steering system should be checked each season by moving the steering wheel from hardover to hardover -- port and starboard, several times without evidence of stiffness.

7-7 STEERING SYSTEMS

Would you believe: Probably 90% of steering cable problems are directly caused by the system not being operated, just sitting idle during the off-season. Without movement, all steering cables have a tendency to "freeze". **Would you also believe:** Service shops report almost 50% of boat cables are replaced every year, due to lack of movement. Therefore, during off-season when the boat is laid up in a yard, or on a trailer alongside the house, take time to go aboard and operate the steering wheel from hard-over to hard-over several times.

Service procedures for the steering system are divided into two major section according to the type of system installed. Instructions for working on the Directional Indicator installation are given in Section 7-5. The Rotary Steering with 90°, 20° and Tilt Wheel Mount procedures are outlined in Section 7-6.

These sections provide step-by-step detailed instructions for the complete disassembly, cleaning and inspection, and assembly of each system. Disassembly may be stopped at any point desired and the assembly process begun at that point. However, for best results and maximum performance, the entire system should be serviced if any one part is disassembled for repair.

Steering Check:

The steering system may be checked by moving the steering wheel from hard-over to hard-over several times. The stern drive unit should move without any sign of stiffness. If binding or stiffness is encountered, the cause may be defective bearings in the gimbal housing ring. To replace these bearings, see Chapter 11. Stiffness of stern drive movement may also be caused by frozen U-joints. To replace the U-joints, see Chapter 10. To correct either one of these defects, the stern drive unit must be removed.

The steering cable may be checked by removing the steering bolt at the transom plate, and then turning the steering wheel back-and-forth from hard-over to hard-over several times. If there is any sign of stiffness, the cables may be corroded or there may be a defect in the steering mechanism.

7-8 DIRECTIONAL INDICATOR SYSTEM SERVICE

DISASSEMBLY

To assist you in performing the work, the items mentioned in the following steps are keyed by number with those shown in the accompanying exploded drawing. This method of presentation will help you to see the relationship of the various items as parts are removed and installed.

1- Remove the center button (2) from the steering wheel (1).

2- Use a 7/16" socket to remove the bolt (8), lockwasher (9) and washer (10), from the inside hub (4).

3- Remove the hub (4) from inside the center of the steering wheel.

4- Remove the bushing (5) from inside the hub.

5- Hold the steering wheel from turning and at the same time remove the elastic stop nut (28) and washer (29) from the bottom end of the steering wheel center bolt (12).

6- Lift the steering wheel and center bolt off of the steering shaft (16).

7- Remove the directional indicator nylon pinion shaft (11) from inside the steering wheel.

8- Remove the nut (35), bolt (32), and spacer (33) from the gear rack assembly to steering mount (31).

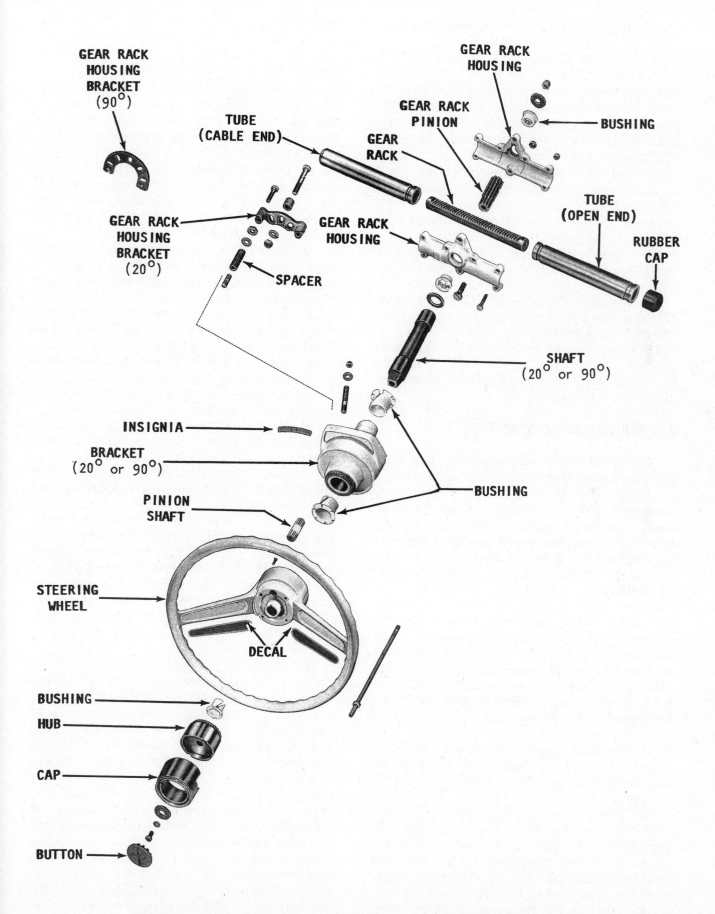

GEAR RACK
HOUSING
BRACKET
(90°)

GEAR RACK
HOUSING

GEAR RACK
PINION

BUSHING

TUBE
(CABLE END)

GEAR
RACK

TUBE
(OPEN END)

GEAR RACK
HOUSING
BRACKET
(20°)

GEAR RACK
HOUSING

RUBBER
CAP

SPACER

SHAFT
(20° or 90°)

INSIGNIA

BRACKET
(20° or 90°)

BUSHING

PINION
SHAFT

STEERING
WHEEL

DECAL

BUSHING

HUB

CAP

BUTTON

Exploded drawing of the directional indicator system, with major parts identified.

9- Remove the gear rack assembly (31) from the steering mount (14).

10- Slide the rubber cap (23) from the end of the rack (22), and then move the transom end of the cable to bottom-out inner core cable out of the way. Remove the cotter pin from the locking cap, and then remove the locking cap, if one is installed. Remove the jam-nut and the retainer nut.

11- Loosen the steering cable to gear rack jam-nut, and then turn the gear rack off the steering cable.

12- Remove the pinion gear bushings (18) from the gear rack (31).

13- Remove the upper and lower steering shaft bushings (13) from the steering mount (14).

14- Remove the six nuts (27) and bolts (25) securing the two halves of the gear rack housing (19) together.

15- Separate the gear rack housing halves and remove the tubes (21 and (22) from the housings. Slide the tubes off the rack gear (20).

CLEANING AND INSPECTING

Clean all parts with solvent, and then dry them thoroughly with compressed air.

Inspect the steering shaft, pinion gear, rack gear and the rack gear tubes for wear and/or galled surfaces. Inspect all of the bushings for excessive wear and/or cracks.

Replace any defective or worn parts.

ASSEMBLING

FIRST THIS WORD: Elastic-type locknuts may not be used more than twice, three times at the most before they loose their locking ability. If the locking ability of the elastic locknut is questionable, replace it with a new one. **NEVER** use worn-out locknuts or non-locking nuts.

1- Apply a liberal coating of Universal Joint Lubricant or equivalent to the inside of the rack tubes (21) and (22), and to the rack gear (20).

2- Place the end of the rack gear (20), with the smaller diameter hole, into the rack tube (22) which has the open end.

3- Place the rack tube (21), which has the threaded hole for the steering cable, over the other end of the rack gear (20).

4- Position the rack gear with the tubes into the center housing (19), with the groove in the bottom of the rack gear over the stop in the center housing. **TAKE CARE** to be sure the rack tubes are properly seated into the grooves in the center housing.

5- Mate the other half of the center housing over the rack tubes, and then secure the two halves with the six bolts (25) and nuts (27). Tighten the nuts to a torque value of 70 in.-lbs.

6- Lubricate the outer surfaces of the pinion gear bushings (18) with Universal Joint Lubricant, or equivalent, and then install the bushings into the pinion openings in the gear rack assembly (19).

7- Install the pinion gear (24) into the gear rack assembly with the steering shaft spline (smaller diameter spline) toward the steering mount (14).

8- Thread the gear rack assembly onto the steering cable. **A GOOD WORD: NEVER** thread the steering cable into the gear rack. **ALWAYS** thread the gear rack all the way onto the steering cable to prevent the cable housing from winding up and causing hard steering.

9- Push the transom end of the steering cable inward to bottom-out the inner core cable.

10- Clean any grease from the inner core cable threaded area which is exposed beyond the rack gear.

11- Thread the cable retainer nut onto the inner core cable, and then tighten it to a torque value of 90 in.-lbs. Install the second nut and tighten it securely, but not **OVER** 90 in.-lbs. Install the locking cap over the cable core nuts, and then install the cotter pin, if the installation is equipped for one. Install the rubber cap (23) onto the end of the tube (22).

12- Move the transom end of the steering cable in-and-out several times to ensure there is no binding in the rack assembly.

13- If the steering cable is attached to the drive unit, move the drive unit to the dead-ahead position. Install a large ID washer (17) over the pinion gear steering mount side.

14- Lubricate the steering shaft bushings (13) with Universal Joint Lubricant, or equivalent, and then install the bushings into the steering mount (14).

15- Place the gear rack assembly (19) in position behind the steering mount (14) and secure it with the 3/8" bolt (32), spacer (33), and nut (35). **DO NOT** tighten the nut at this time.

16- Insert the steering shaft (13) into the steering mount (14). Rotate the shaft until the one flat, tapered portion flat side, is horizontal. This position will reference the steering shaft and the steering wheel spoke correctly when the boat is moving dead-ahead.

17- Tighten the 3/8" diameter bolt (32) to a torque value of 25 ft-lbs.

18- Lubricate the directional indicator pinion gear (11) with Universal Joint Lubricant, or equivalent, and then install the gear into the steering sheel (1) with the letter "F" toward the helmsman's position.

19- Slide the steering wheel over the steering shaft (16) with the spokes of the wheel in the horizontal position. Install the center bolt (12) though the steering wheel, shaft and pinion gear. Secure the center bolt with washer (29) and elastic stop nut (30).

20- Tighten the center bolt retainer nut to a torque value of 80 in.-lbs. Tighten the steering cable to gear rack Jam-nut to a torque value of 30 ft-lbs.

21- Lubricate the indicator hub bushing (5) with Universal Joint Lubricant, or equivalent, and then install the bushing into the center of the steering wheel. Place the hub (4) into the center of the steering wheel and turn it to engage the gear teeth on the hub with the teeth on the pinion gear.

7-9 ROTARY STEERING SERVICE

These sections provide step-by-step detailed instructions for the complete disassembly, cleaning and inspection, and assembly of the rotary steering system. Disassembly may be stopped at any point desired and the assembly process begun at that point. However, for best results and maximum performance, the entire system should be serviced if any one part is disassembled for repair.

Steering Check:
The steering system may be checked by moving the steering lever back-and-forth from hard-over to hard-over several times. The stern drive unit should move without

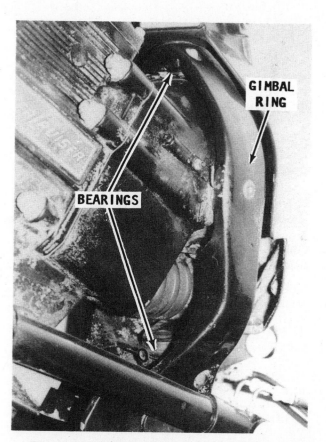

This "frozen" U-joint will prevent movement of the stern drive to port and starboard. Do not let this damage lead to an incorrect conclusion the steering system is at fault.

The upper and lower gimbal ring bearings may be "frozen" and give a false indication the steering mechanism is at fault.

any sign of stiffness. If binding or stiffness is encountered the cause may be defective bearings in the gimbal housing ring. To replace these bearings, see Chapter 11. Stiffness of stern drive movement may also be caused by frozen U-joints. To replace the U-joints, see Chapter 10. To correct either one of these defects, the stern drive unit must be removed.

The steering cable may be checked by first disconnecting the steering cable at the transom plate, and then turning the steering wheel back-and-forth from hard-over to hard-over several times. If there is any sign of stiffness the cables, may be corroded or there may be a defect in the steering mechanism.

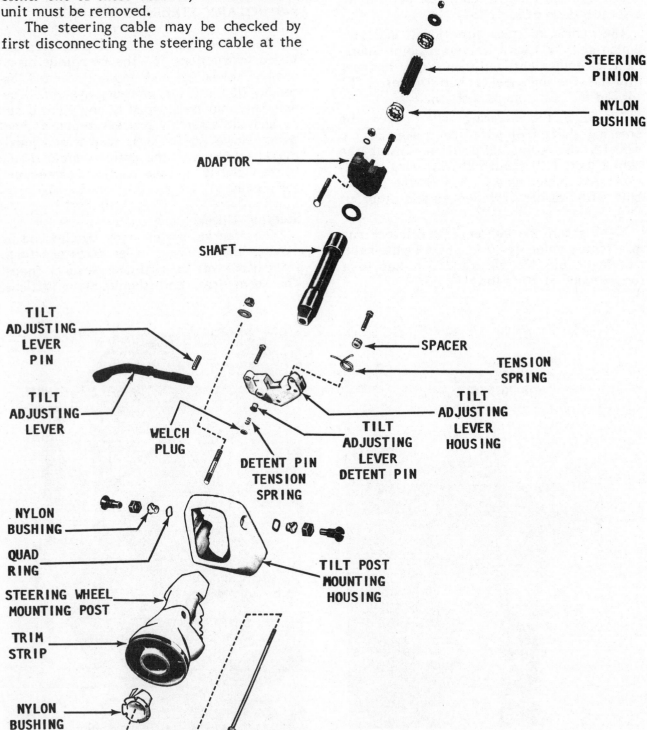

STEERING
PINION

NYLON
BUSHING

ADAPTOR

SHAFT

SPACER

TENSION
SPRING

TILT
ADJUSTING
LEVER
PIN

TILT
ADJUSTING
LEVER

TILT
ADJUSTING
LEVER
HOUSING

WELCH
PLUG

TILT
ADJUSTING
LEVER
DETENT PIN

DETENT PIN
TENSION
SPRING

NYLON
BUSHING

QUAD
RING

TILT POST
MOUNTING
HOUSING

STEERING WHEEL
MOUNTING POST

TRIM
STRIP

NYLON
BUSHING

Exploded drawing of the rotary ride-guide type steering mechanism, with major parts of the steering helm identified.

DISASSEMBLING

To assist you in performing the work, the items mentioned in the following steps are keyed by number with those shown in the accompanying exploded drawing. This method of presentation will help you to see the relationship of the various items as parts are removed and installed.

1- Pop the button out from the center of the steering wheel.

2- Remove the center bolt (4) retainer nut (33), and large OD washer (32), and then lift off the steering wheel and center bolt.

3- Remove the two 3/8" nuts (12) and washers (11) securing the rotary head to the steering mount (5). Remove the rotary head from the mounting bolts.

4- Remove the steering shaft (23) and pinion gear (31) from the rotary head.

5- Remove the two bolts (43) and (44) and the washers (45) holding the housing halves together.

6- Separate the housing halves (38) and (42) by prying them apart with a thin blade screwdriver.

7- Remove the steering cable anchor from the recess in the rotary housing. If necessary, tap **LIGHTLY** with a mallet to loosen the cable from the housing. Work the cable loose, **NEVER** use force.

8- Lift the rotary rack (40) out of the steering control housing (42).

9- Using a thin blade screwdriver, disengage the inner core cable hook from the rotary rack gear (40).

10- Remove the nylon center shaft bushing (39) from the housing hub (38) or from inside the rotary rack gear (40).

CLEANING AND INSPECTING

Clean all parts with solvent, and then dry them thoroughly with compressed air. **NEVER** allow "nyliner" or bushings to remain submerged in solvent for any length of time, because the liquid will cause them to expand.

Inspect the housing halves for dents, cracked casting, or other damage. Check the nyliner and bushings for cracks, broken pieces and wear. Inspect the pinion gear and the rack gear teeth for wear.

Replace any parts that show signs of excessive wear or damage.

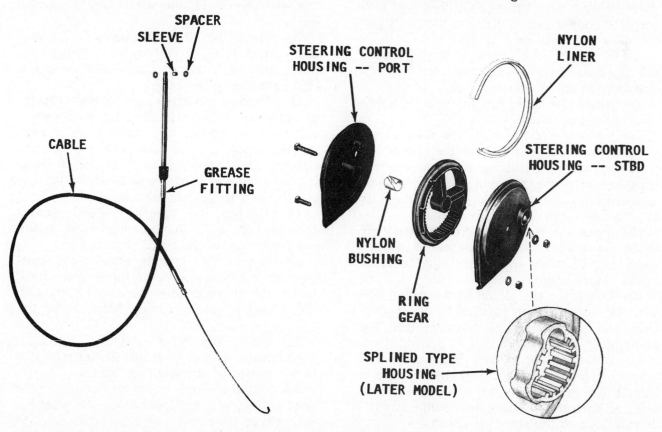

Exploded drawing of the rotary ride-guide cable and drum parts, with a detail of the later model spline-type housing.

ASSEMBLING

FIRST, THESE WORDS

Elastic-type locknuts may not be used more than twice, three times at the most before they loose their locking ability. If the locking ability of the elastic locknut is questionable, replace it with a new one. **NEVER** use worn-out locknuts or non-locking nuts.

1- Grasp the ring gear (40) with the gear teeth toward the left side, as shown. Install the inner core cable hook (34) into the retainer hole in the ring gear (40) with the steering cable extended toward the left (counterclockwise). If necessary, tap **LIGHTLY** on the inner core cable hook to seat the inner core cable into the ring gear recess.

2- Wrap the inner core cable approximately 1/2 turn counterclockwise around the ring gear.

3- Secure the inner core cable using Cable Retainer Tool C-91-54175.

4- Lubricate the teeth of the ring gear (40), the nyliner (41), bushing (39), and the housing halves (38) and (42) with Universal Joint Lubricant, or equivalent.

5- Install the center hub bushing (39) onto the center hub and place the "nyliner" (41) over the inner core cable.

6- Place the ring gear (40) squarely onto the starboard steering control housing (42). Remove the cable retainer tool.

7- Place the cable housing into the starboard steering control housing (42).

8- Place the other half, the port steering control housing (38), over the assembled starboard steering housing and secure them together with the two bolts (43) and (44), washers (45), and nuts (46). Tighten the nuts to a torque value of 10 ft-lbs.

9- Move the transom end of the steering cable in-and-out several times to ensure there is no binding in the steering cable or in the rotary head.

10- Lubricate the pinion bear bushing (30) with Universal Joint Lubricant, or equivalent, and then insert the bushings into the pinion gear openings.

11- Position the rotary head with the correct side toward the steering mount and install the pinion gear (31) into the pinion opening with the steering post spline (smaller end) toward the steering mount.

12- Install the rotary head over the adaptor bolts (26) and secure the head with the washers (27) and elastic stop nuts (28). **DO NOT** tighten the nuts at this time.

13- Lubricate the steering shaft bushing (3) with Universal Joint Lubricant, or equivalent, and then install the bushing into the steering mount (1).

14- Attach the steering cable to the drive unit.

15- Move the drive unit to the dead-ahead position.

16- Slide the steering shaft (23) into the steering mount (1) with one flat of the squared end in the horizontal position.

17- After the steering shaft has been indexed as described in the previous step, hold the pinion gear in position with one hand, and with the other hand engage the steering wheel shaft (23) with the pinion gear spline (31).

18- Observe the steering wheel from the back side. Notice the recess (square opening) in the steering post has one flat side above the spoke in the steering wheel. Now, with the parts aligned as described in steps 15 through 17, install the steering wheel to the steering shaft with the flat areas aligned.

19- Install the center though-bolt (4) thru the steering wheel, the shaft, and the pinion gear.

20- Place the large OD washer (32) over the center through-bolt against the pinion gear, and then thread the elastic stop nut (33) onto the center bolt.

21- Tighten the steering assembly retainer bolts (26) started in Step 12, to a torque value of 25 ft-lbs., if working on a non-splined cable head as shown in the accompanying illustration, or to 12 ft-lbs. if working on a splined cable head as shown in the small insert in the illustration.

22- Tighten the steering wheel center through-bolt to 8 in.-lbs., and then install the cotter pin in the bolt.

23- Rotate the steering wheel from hard-over to hard-over several times and check to be sure there is no evidence of binding in the steering cable, rotary head or in the steering mount.

24- Place the drive unit in the dead-ahead position, and then install the trim cap in the center of the steering wheel.

25- Make a final check to be sure the boat will turn to the port when the wheel is rotated to port and the boat will turn to starboard when the wheel is rotated to starboard.

7-10 POWER STEERING

Description and Operation

The power steering system used with the MerCruiser Alpha and Bravo stern drive units is very similar to other power steering systems used on a wide variety of boats, with different stern drives and powerplants. Sometimes the unit is installed in reverse of what is shown in this chapter. This means the installation would appear as an opposite (mirror image), but the procedures for service will not differ.

A hydraulic pump is mounted to the front of the engine on the port side. The pump is driven by a belt off the crankshaft pulley. Hydraulic lines connect the pump to an actuator valve and actuating cylinder mounted over the engine bell housing. A hydraulic reservoir tank is mounted to the engine above the pump. An oil cooler is connected to the system to prevent the hydraulic fluid from overheating.

Hydraulic Pump

The pump is a rotor vane type with a flow restriction orifice, a pressure relief valve, and a reservoir. The high pressure outlet fitting is of the molded type and capable of withstanding pressures in excess of 1100 psi (7500 kPa). This outlet fitting is located at the bottom rear of the pump and the low pressure return fitting is also at the back above the high pressure fitting. The low pressure hose is secured to the fitting with a hose clamp. The clamp is adequate because the hose only carries fluid at low pressure.

The pump utilizes **METRIC** fittings and a decal affixed to the pump publicizes this fact.

When the engine is operating at idle speed, the pump output pressure is enough to provide the desired power assist for turning the stern drive unit.

During normal operations, hydraulic fluid from the reservoir is provided to the intake port of the pump. High pressure forces the pump vanes against the pump cavity walls. Output flow of the pump is restricted to approximately 2.3 gallons per minute by the flow restriction orifice.

When extreme high steering loads are demanded, the pump is protected by a pressure relief valve. This valve restricts the output pressure of the pump to less than 1100 psi (7500 kPa).

CRITICAL WORDS

The pump can be damaged internally if the system is operated with full load at the relief pressure for more than just a few consecutive seconds.

Steering Cylinder

When there is no steering input from the helm, the stern drive is held in a fixed constant position. The fluid flow is routed through both of the actuator valve chambers and returned through the oil cooler to the hydraulic reservoir.

When the helm is turned to port, hydraulic fluid from the pump is directed to the rod end of the cylinder. Fluid from the base of the cylinder is routed back to the oil cooler and the reservoir. The piston moves to starboard, pulling the steering arm to starboard, swinging the stern drive to port and the bow of the boat swings to port.

When the helm is turned to starboard, hydraulic fluid from the pump is directed to the base end of the cylinder. Fluid from the rod end of the cylinder is routed back to the oil cooler and reservoir. The piston moves to port, pushing the steering arm to port, swinging the stern drive to starboard and the bow of the boat swings to starboard.

Operation Without Pressure

Should hydraulic pressure be lost for any number of reasons, manual steering without the power assist is still possible. Naturally,

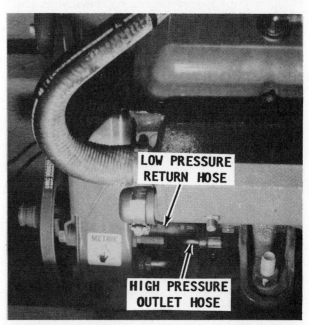

Identification of the two power steering hoses attached to the rear of the power steering pump.

much more steering effort is required due to the lack of the power assist and the movement of fluid through the system when the pump is not operating.

Any manual movement of the steering cable is transmitted to the actuator valve spool which then transmits the full manual effort to the steering arm.

7-11 SYSTEM MAINTENANCE

The power steering system requires very little attention and seldom does it require major maintenance. An exposed "Zerk" fitting on the actuator valve provides the necessary lubrication for the valve. This fitting should be serviced a couple times each season.

The pump cannot be repaired. Therefore, if a malfunction should develop and the determination is made finding the pump at fault, it must be replaced with a new unit.

The hydraulic fluid cooler is located in an exposed position and can be removed and "rodded" out with a stiff piece of welding rod or similar tool.

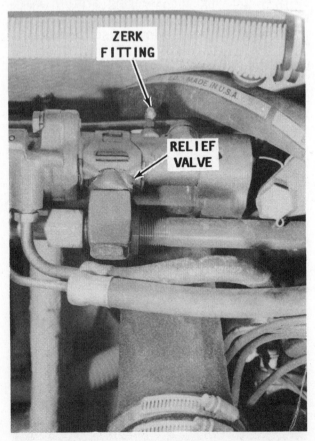

*Force grease into the lubrication zerk fitting using only a **HAND** operated grease gun, until grease appears at the relief valve.*

The actuator valve can be adjusted to correct any adverse steering "drag" to either port or starboard.

ZERK FITTING

The actuator valve is provided with a Zerk fitting for lubrication purposes. This fitting may be difficult to reach because of its location, but it should be serviced at least once every six months. Use marine wheel bearing grease, or equivalent. **TAKE CARE** not to force too much material into the fitting and **DO NOT** use a power grease gun.

ADDING HYDRAULIC FLUID

CRITICAL WORDS
NEVER allow the hydraulic pump to operate for even a couple seconds without fluid in the system. Should such action take place, serious damage would be caused to the pump.

Fluid is added to the system as follows: Start the engine and allow it to warm to operating temperature.

CAUTION
Water must circulate through the stern drive, also to and from the engine, anytime the engine is run to prevent damage to the water pump located in the lower unit (original and Alpha drives); or on the front of the engine (Bravo drive). Just a few seconds without water will damage the water pump.

After the engine has reached normal operating temperature, shut the engine down. Remove the reservoir filler cap and check oil level to the **HOT** mark on the dipstick.

If the fluid level is low, add one of the approved hydraulic fluids listed below to the proper level on the dipstick.

GMC Power Steering Fluid
Automatic Transmission Fluid Dexron
Automatic Transmission Fluid Dexron II
Equivalent Quicksilver product

SAFETY WORDS

The fluid used in the power steering system is the type of motor oil with a high concentration of detergent, Therefore, the fluid is flammable.

TAKE CARE not to overfill the reservoir. Overfilling the reservoir will cause the pump to overflow through the breather cap hole. Such action would spray hot fluid, under high pressure, into the engine compartment. Fluid landing on a hot manifold could cause a very possible **FIRE HAZARD**.

MORE SAFETY WORDS

If the breather hole in the cap should become blocked, excess fluid, under high pressure, will attempt to escape. The fluid will find a weak or loose fitting, worn seal, or other route, even a seam in the pump body, and be released -- a most **DANGEROUS** condition.

BLEEDING SYSTEM

Anytime the hydraulic fluid supply is changed, fluid added to the system, or the system has been opened for any reason, the system **MUST** be "bled" of air. The procedure is simple, no tools are required, and the task can be accomplished in a few minutes.

First, rotate the helm to the hard over port position, with the engine **NOT** running.

Next, add power steering fluid to the reservoir until the level is within the **COLD** range on the dipstick.

After the fluid has been added, start and operate the engine at idle speed for just a short time -- couple of minutes.

CAUTION

Water must circulate through the stern drive, also to and from the engine, anytime the engine is run to prevent damage to the water pump located in the lower unit (original and Alpha drives); or on the front of the engine (Bravo drive). Just a few seconds without water will damage the water pump.

With the engine not running, check the fluid level again and add fluid, if necessary.

Now, start and operate the engine at idle speed. "Bleed" air from the system by turning the helm port and starboard, **BUT** not so far as to hit the stops. Fluid with air will develop "foam" and have a light tan or red appearance.

If difficulty in removing foam is encountered, allow the unit to stand idle, with the cap not secured, for at least an hour and the foam will disappear -- much the same as the head of foam on a glass of cold beer.

Return the helm to the center position and continue to run the engine for two or three minutes. Shut the engine down and water test the boat at the first opportunity. Check for satisfactory performance including quiet operation.

After the engine has been run for some time and the fluid level warmed to approximately 150 to 190°F (65 to 88°C), check the fluid level. The fluid level should be within the **HOT** range on the dipstick.

BELT ADJUSTMENT

Tension of the drive belt should be checked at regular intervals. Correct tension is achieved when finger pressure midway between the pump pulley and the adjacent pulley will deflect the belt from 1/4 to 1/2" (6.4 to 12.7mm).

If the belt tension is not satisfactory, loosen the pump attached bolts and adjust the belt by moving the pump outward, away from the engine.

Checking fluid level at the power steering pump mounted on the powerplant.

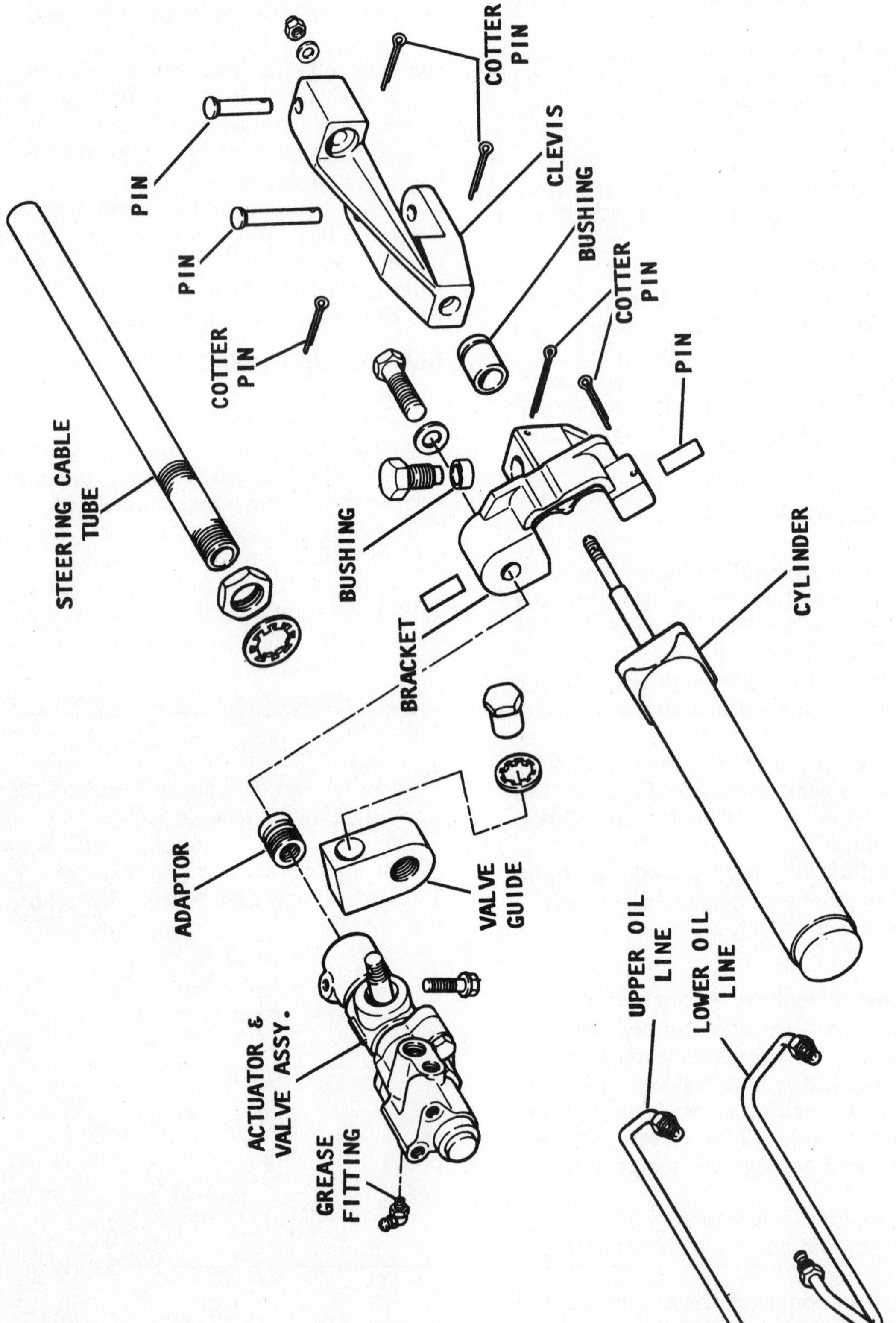

COTTER PIN

COTTER PIN

PIN

PIN

CLEVIS

BUSHING

COTTER PIN

BUSHING

COTTER PIN

PIN

STEERING CABLE TUBE

BUSHING

BRACKET

CYLINDER

ADAPTOR

VALVE GUIDE

ACTUATOR & VALVE ASSY.

GREASE FITTING

UPPER OIL LINE

LOWER OIL LINE

Exploded drawing of the Mercruiser power steering system, with major parts identified.

NEVER pry against the pump reservoir or pull against the filler neck when making a belt adjustment. Instead, insert a 1/2" breaker bar into the hole provided in the pump mounting bracket and work the pump outward.

Tighten the pump mounting bolts and remove the breaker bar.

Belt Too Loose

If the belt tension is too loose, the helmsperson may experience momentary difficulty in steering, or hear chirps or squeals coming from the pump, pulley, or belt. Although not necessarily harmful, these symptoms are annoying.

Belt Too Tight

If the belt tension is too great, no appreciable difference in peformance will be detected until it is too late. Excessive belt tension will shorten the life of the power steering pump bearing. The only warning will be a loud whine from the pump. The whine will indicate the bearing has or is about to fail. If the bearing fails, the pump **MUST** be replaced.

Correct belt tension ensures long life of the power steering pump bearing and efficient system operation.

Removing the end cap from the actuator valve in preparation to making an adjustment.

ACTUATOR VALVE ADJUSTMENT

WARNING

Exercise care when making an adjustment to the actuator valve because the cable end adaptor will move while the adjustment is being made.

First, check the hydraulic fluid level in the reservoir and fill as described in the previous section. If the fluid level was low, the problem may be corrected. If not, disconnect the steering control cable and the link rod from the end adaptor of the cylinder.

Next, remove the protection cap from the end of the actuator valve.

Now, start the engine and allow it to warm to operating temperature. With the engine running at a normal idle rpm, **SLOWLY** rotate the adjusting nut **CLOCKWISE** until the steering cylinder begins to move by itself. As soon as it begins to move, **SLOWLY** rotate the nut **COUNTERCLOCKWISE** counting the number of revolutions, until

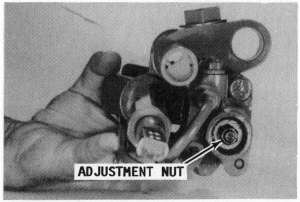

ADJUSTMENT NUT

Close-up end view of the cylinder and actuator valve to show the adjustment nut.

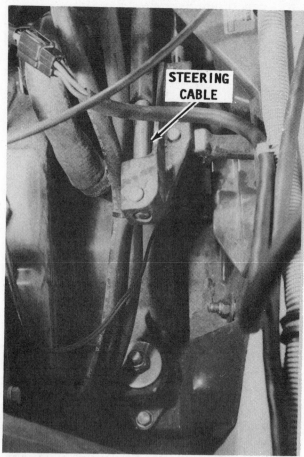

STEERING CABLE

Lubricate the exposed steering cable with Quick-silver Lubricant.

the cylinder just begins to move in the opposite direction.

At this point, rotate the nut again **CLOCKWISE** exactly 1/2 the number of turns counted as the nut was being turned counterclockwise.

This position neutralizes the actuator valve in the center.

CRITICAL WORDS

Do not turn the adjusting nut any more than necessary. Excessive rotation of the nut can cause the nut to lose its locking effect.

After the adjusting nut has been properly positioned, it should be possible to push the plunger rod in and out with hand pressure.

Shut the engine down and connect the steering cable and link rod. Again, start the engine. If the adjustment is correct, the cable end adaptor will not move in either direction.

Coat the adjustment nut with water resistant grease. Replace the cap over the end of the actuator valve covering the adjustment nut.

SPECIAL WORDS

If the actuator valve should be damaged, replacement parts are available at the local dealer. The steering cylinder is also available as a replacement item.

POWER STEERING COOLER

The power steering fluid cooler is a tube type heat exchanger cooler mounted on the port side of the engine, forward. Water from the upper gear housing is circulated through the cooler and then into the thermostat housing. From the thermostat housing, the water is routed through the engine block and finally overboard through the stern drive and propeller.

Hydraulic fluid heat is drawn off by the water passing through the heat exchanger and into the engine cooling system. Therefore, if the cooler should become plugged, the engine could overheat.

The cooler may be easily removed and the tubes rodded out with a heavy piece of welding rod or similar tool. Take care not to use a sharp pointed piece of material which could cause damage to the tubes.

To remove the cooler, first disconnect the inlet and outlet water hoses. Next, disconnect the inlet and outlet hydraulic lines. Remove the attaching hardware and the cooler will come free.

8

TRIM/TILT

8-1 INTRODUCTION

All stern drive engine installations are equipped with some means of raising or lowering the stern drive unit for efficient operation under various load, boat design, and water conditions. The most simple form is a mechanical tilt adjustment consisting of a series of holes in the gimbal ring through which an adjustment stud passes through to secure the stern drive at the desired angle. A second and more modern method is a hydraulically operated system controlled from the helmsman's position.

8-2 MECHANICAL TILT MECHANISM

The mechanical tilt arrangement is found only on the older model units. A change in the tilt angle of the stern drive is accomplished by inserting the tilt adjust-

ment stud through one of a series of holes in the gimbal ring. These holes allow the operator to obtain the desired boat trim under various speeds and loading conditions.

The tilt angle of a stern drive unit is properly set when the anti-cavitation plate is approximately parallel with the bottom of the boat. The boat trim is corrected by stopping the boat, removing the adjustment stud, tilting the drive unit upward or downward, as desired, and then installing the stud through the new hole exposed in the gimbal ring.

To raise the bow of the boat, the drive unit is raised one hole at-a-time until the operator is satisfied with the boat's performance. If the bow is to be lowered, the drive unit is lowered one hole at-a-time.

Performance will generally be improved if the bow is lowered during operation in rough water. The boat should never be operated with the stern drive set at an excessive raised position because such a tilt angle will cause the boat to porpoise through the water, which is very dangerous

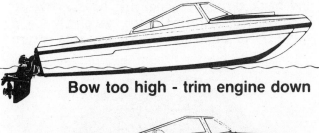

Bow too high - trim engine down

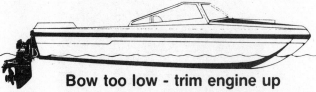

Bow too low - trim engine up

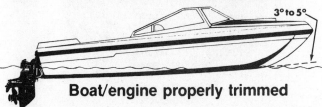

Boat/engine properly trimmed

The tilt position of the stern drive directly affects the bow position and boat performance.

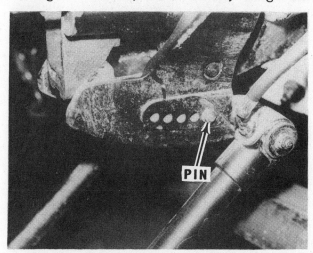

The tilt position, on early model stern drive units, is adjusted by inserting a tilt pin through one of a series of holes in the gimbal housing.

in rough water. Under such conditions the helmsman does not have complete control at all times.

Instead of making extreme changes in the stern drive angle, it is far better to shift passengers and/or the load to obtain proper performance.

8-3 HYDRAULICALLY POWERED TRIM/TILT SYSTEM

The hydraulically powered trim/tilt system permits changing the tilt angle of the stern drive from the helmsman's position. Controls and indicators for the system are located on the control panel.
The angle of the stern drive is properly set when the boat is operating to give maximum performance, including comfort and safety.

The powered tilt system consists of a hydraulic pump, two tilt cylinders, a reverse lock valve, trim indicator sender, controls, an indicator gauge, and associated hoses and fittings.

The hydraulic pump includes a valve body-and-gear assembly, control valve, pump motor, and a reservoir. The controls consist of an **UP** button, an **UP/OUT** button, and an **IN** button. The indicator gauge is installed on the dash next to the control buttons.

The older models with only one hose for each cylinder is identified as a low-pressure system. This is only capable of lifting the stern drive to the full tilt position when the

boat is not moving through the water. The cylinders on these early model installations cannot be rebuilt nor can they be updated to the power trim operation.

Cylinders on the late model units can be rebuilt and returned to service.

SYSTEM OPERATION

The relationship of the various units in the hydraulically powered trim system and the theory of operation are discussed in the following paragraphs. Their operation will be more easily understood if the two functional diagrams in this section are studied while reading how the system operates and before servicing the system. Many of the items mentioned in the text are identified by number. Matching numbers will be found on the functional diagrams.

Hydraulic UP Circuit

When the **UP/OUT** button on the control panel is pushed, hydraulic pressure is created in the **UP** side of the system. This pressure is indicated by the dark shaded area on the **UP** circuit functional diagram. The pressure moves the slide valve (2) and allows hydraulic fluid to enter the hoses through the valve (4) and into the trim cylinder chambers (3).

At this point the pressure pushes the dual-piston hydraulic ram outward. Fluid in the **DOWN** side of the system is forced out of the cylinders by movement of the piston and back through connecting hoses, reverse lock valve, trim-indicator assembly, and valve (7) to the pump reservoir. This return

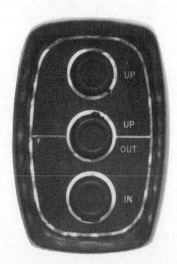

On installations since 1967, three buttons are used to control the tilt adjustment. One button is used to raise the stern drive for trim, another to lower the unit, and the third is used simultaneously with the UP button to position the stern drive for trailering or to launch the boat.

A dash-mounted gauge advises the helmsman at all times of the relative position of the stern drive.

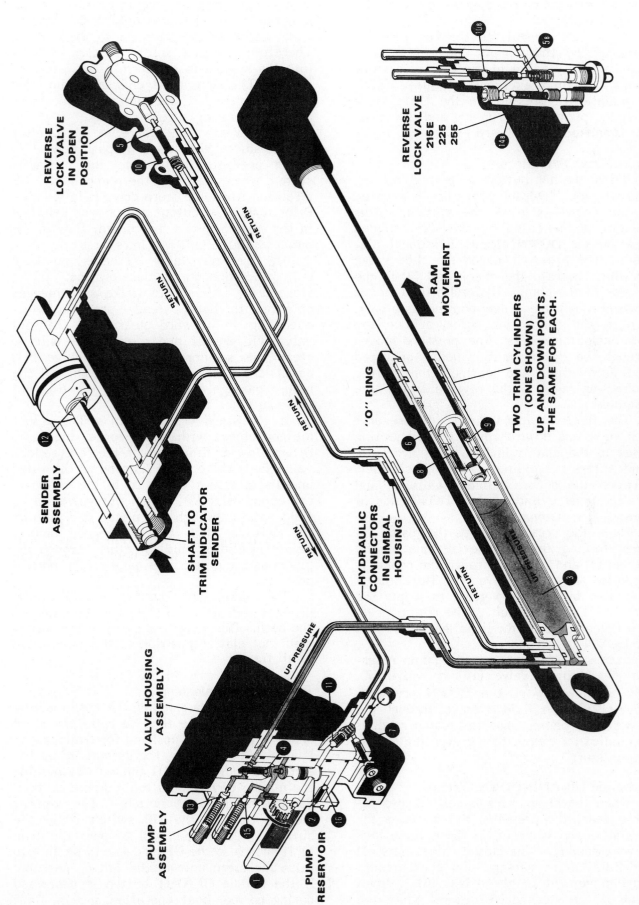

REVERSE LOCK VALVE IN OPEN POSITION

5

10

SENDER ASSEMBLY

SHAFT TO TRIM INDICATOR SENDER

12

RETURN

RETURN

RETURN

RETURN

RETURN

REVERSE LOCK VALVE 215E 225 255

UB

5B

14B

"O" RING

RAM MOVEMENT UP

TWO TRIM CYLINDERS (ONE SHOWN) UP AND DOWN PORTS, THE SAME FOR EACH.

6

9

8

3

UP PRESSURE

HYDRAULIC CONNECTORS IN GIMBAL HOUSING

UP PRESSURE

VALVE HOUSING ASSEMBLY

11

4

1

7

13

15

2

16

1

PUMP ASSEMBLY

PUMP RESERVOIR

Functional diagram of the power tilt & trim UP circuit. The text explains in detail, the operation of this system.

fluid is indicated on the diagram by the light shaded area.

Valve (15) regulates the **UP** pressure developed by the pump. Valve (13) is a back-up safety valve to protect the system from a build up of excessive pressure. Valve (4) maintains pressure in the **UP** system when the pump is not running and the drive unit is shifted into forward gear.

Hydraulic DOWN Circuit

When the **IN** button is pushed on the control panel, hydraulic pressure is created in the **DOWN** side of the system. This pressure is indicated by the dark shaded area on the **DOWN** circuit functional diagram. The pressure moves the slide valve (2) which unseats the check ball (4) in the **UP** side of the pump. Hydraulic fluid moves down through the valve housing assembly, through the connecting hoses, and to the trim indicator sender. The pressure moves through the sending unit, which is operated by fluid pressure, and continues through the connecting hose to the reverse lock valve assembly.

The fluid continues through the reverse lock valve and down through the connecting hoses to the trim cylinders. The hydraulic fluid enters the cylinders through channels between the double cylinder walls (6) and into the inner cylinder on the **DOWN** side of the cylinder pistons.

Hydraulic pressure moves the piston inward, forcing fluid to the **UP** side of the system (3) and returns through the connecting hoses and check valve (4). This fluid is then used by the pump gears to supply the hydraulic **DOWN** circuit. The return pressure is indicated on the light shaded area on the **DOWN** circuit diagram.

Excess fluid returns to the pump reservoir (1) through valve (16) or valve (7). Valve (16) regulates the **DOWN** pressure. Valve (12) allows fluid to travel through the trim sender piston when the piston reaches the end of its travel, thus preventing excessive pressure.

Hydraulic UP-AND-DOWN Circuit

When the stern drive is shifted into reverse gear, the propeller thrust closes the normally-open valve (5) in the reverse lock valve assembly. The closed valve shuts off the fluid return passage and prevents upward movement of the drive unit. Movement of the stern drive upward while the

unit is operating in reverse gear will result in a build up of high pressure. Such pressure could cause serious damage to the system. Therefore, another safety feature is built into the system in the form of an interlock switch. This switch prevents the pump from running in the **UP** direction while the stern drive is operating in reverse gear.

The reverse lock valve on MerCruiser models since 1977, has a by-pass valve (14) which prevents damage to the hydraulic system when operating the pump in the UP direction while the stern drive is in reverse. Prior to 1977, a cutout switch was installed in the system to cut the current flow to the pump while the unit was in reverse.

The trail-out valve (7) holds the drive unit in position during deceleration. If the stern drive unit should strike a submerged object with light, steady pressure while operating in forward gear, the trail-out valve will unseat and permit the unit to clear the submerged object. Valve (11) assists the trail-out valve by closing off the return passage and permitting the trail-out valve to function.

If the stern drive should strike a submerged object with great force, the rapid return flow of fluid through the reverse lock valve seats valve (10). As soon as the valve unseats, pressure builds up in the cylinders and opens valve (8). When this valve opens, fluid passes through the piston and permits the unit to clear the submerged object. Valve (9) allows fluid to return through the piston as the unit comes back to the normal position.

The power trim sender unit transmits pressure indication to the gauge mounted on the dash. This gauge indicates to the helmsman the relative position of the stern drive at all times.

Trailering or Launching

Two control buttons on the dash must be pushed at the same time to raise the stern drive to the full up position for trailering or launching the boat -- the normal **UP** button and the middle **UP/OUT** button. By pushing the middle **UP/OUT** button, current is passed to the top **UP** switch. The current passing through the **UP** switch while the button is depressed, will by-pass the trim limit switch so the **UP** circuit will be able to raise the stern unit to the full up position. If the middle **UP/OUT** button is depressed during normal boat operation, a trim limit

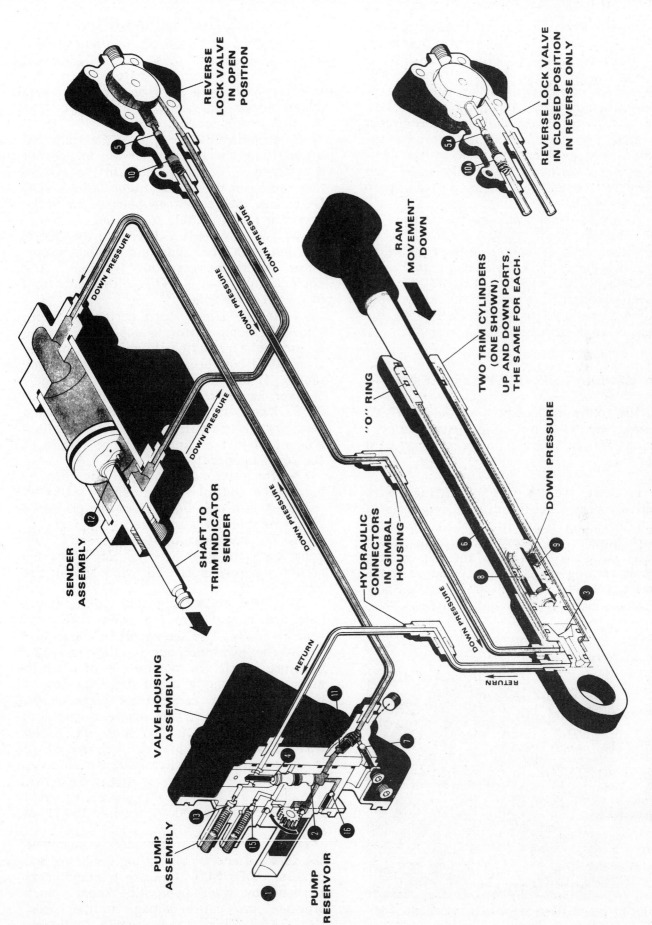

REVERSE LOCK VALVE IN OPEN POSITION

REVERSE LOCK VALVE IN CLOSED POSITION IN REVERSE ONLY

DOWN PRESSURE

DOWN PRESSURE

DOWN PRESSURE

DOWN PRESSURE

DOWN PRESSURE

RAM MOVEMENT DOWN

"O" RING

TWO TRIM CYLINDERS (ONE SHOWN) UP AND DOWN PORTS, THE SAME FOR EACH.

DOWN PRESSURE

HYDRAULIC CONNECTORS IN GIMBAL HOUSING

DOWN PRESSURE

RETURN

RETURN

SENDER ASSEMBLY

SHAFT TO TRIM INDICATOR SENDER

VALVE HOUSING ASSEMBLY

PUMP ASSEMBLY

PUMP RESERVOIR

Functional diagram of the power tilt & trim DOWN circuit. The text explains in detail, the operation of this system.

switch will keep the drive unit from moving out beyond the gimbal ring support guides.

When the hydraulic cylinders reach their full extent of travel, the pump motor will labor if the control buttons are not released. Therefore, to prevent damage to the system, a bimetal switch will open the circuit to stop the pump motor and prevent the motor from overheating. The switch contacts will close automatically after the motor has cooled and the motor can again be operated.

A reverse inter-lock switch is used to open the **UP** circuit and stop the motor when the drive unit is operating in reverse to prevent damage to the hydraulic system.

ONE MORE WORD: Whenever the boat is being trailered, a trailer bracket should **ALWAYS** be used to mechanically lock the stern drive in the up position.

8-4 GENERAL SERVICE INSTRUCTIONS

The following 13 points should **ALWAYS** be observed when installing, testing, or servicing any part of the power tilt or trim system.

1- Coat the threads of fittings with Multi-Purpose Lubricant, or equivalent. **ALWAYS** start hydraulic line fittings with your fingers to prevent cross-threading them. To prevent stripping threads, **NEVER** tighten fittings to a torque value of more than 125 to 150 in.-lbs. Use a special Flare nut open-end wrench to tighten fittings.

High-pressure pump installed on some late model stern drive units.

2- Use clean **MS, SD, SE 20W, OR 5W-30,** or equivalent oil to fill the hydraulic reservoir. Use one of these oils to flush the system. Use automatic automobile transmission fluid, **Type A,** when servicing early model system stamped with a six digit number followed by an "A".

3- **REMEMBER** to remove the vent seal when filling the pump reservoir to prevent overfilling. Too much fluid in the system can cause pump body or motor failure. **DO NOT** close the vent screw because air in the reservoir **MUST** be able to escape.

4- The drive unit **MUST** be in the **DOWN** position with the hydraulic cylinders collapsed when the reservoir is filled. The correct fluid level is **EVEN** with the bottom of the oil filler hole.

5- If the pump stops during long use, allow the pump motor to cool at least one minute before starting it again. An internal thermal circuit breaker protects the pump motor. If the pump will operate, the circuit breaker may be tripped, in the off-position.

6- Keep the work area **CLEAN** when servicing disassembled parts. The smallest amount of dirt or lint can cause failure of the pump to operate.

7- The valve body-and-gear cage assembly, or the control valve assembly **MUST** be replaced as a unit because of the precision fitting of the valves and gears.

8- The can over the inlet screen of the valve body-and-gear assembly **MUST** be installed securely or the oil will foam out the vent.

9- **TAKE CARE** not to allow liquid neoprene to get into the valve body-and-gear assembly. Use the neoprene sparingly.

10- The hydraulic pump motor must be moisture-proof. To moisture-proof a motor after service, apply liquid neoprene around the pump motor commutator plate edges, around the motor frame where it contacts the reservoir, and in the area where the wire leads enter the motor. Any sign of oil in the pump motor indicates either the reservoir seal is damaged or the vent screw was closed. If the vent was closed, air could not escape and oil was forced into the pump motor.

11- **KEEP** the trim cylinders attached to the forward anchor pin during servicing and repairs. **DO NOT** allow the trim cylinders to hang by their hydraulic hoses. Such practice may cause damage to the hoses. Use **CARE** when handling trim cylinders during removal/installation of the stern drive unit. Rough treatment of the hoses could result in a weakened hose, partial separation at the fitting, or bending of the metal tubing, any one of which could restrict the flow of oil to the trim cylinders. For many years, Mercury fought a hard battle with the high-pressure hoses being submerged in water. Now, however, the battle seems to have been won. Mercury engineers have developed a special high-pressure stainless steel hose that will probably result in long life and satisfactory service.

12- **ALWAYS** hold the metal ferrule on the hose with pliers when tightening the fitting. The indicator line on the hose **MUST** follow the hose bend without a twist. The fittings on the cylinder end of the hose **MUST** point approximately 45° from the stern drive centerline toward the transom. A twisted hose will cause severe loads and result in over-stress on the hose and will bend the tubing. On early model units, the hoses are routed from the cylinder to a block between the U-joint and the exhaust bellows. On later model units, the hoses are routed directly from the cylinder to the block under the gimbal housing. **TAKE EXTRA CARE** when attaching the hoses to the block. If the fittings are cross-threaded, the block would have to be replaced. This job would entail removal of the engine.

13- Following service on the system, move the stern drive unit from hard-over to hard-over, port and starboard, and from the full **DOWN** to the full **UP** position, and at the **SAME TIME** check for possible kinks, twists, or severe bends in the hoses or at the fitting ends.

8-5 TROUBLESHOOTING TRIM/TILT SYSTEM

The shift interlock switch prevents the power trim system from operating when the stern drive unit is in **REVERSE** gear. Therefore, if the hydraulic system fails to function properly, check to be sure the stern drive is in **FORWARD**.

Determining if problem is electrical or mechanical:

Disconnect the blue wire in the **UP** circuit or the green wire in the **DOWN** circuit from the pump motor wiring harness. One way to remember the color code is to think: The sky is **UP** and blue; the grass is **DOWN** and green. Connect the wire directly to

the positive terminal of the battery. If the pump motor runs, the problem is in the electrical control circuit. Make a thorough check of all switches on the control panel. Check the wiring for a possible short or open circuit.

If the pump motor fails to run, the problem is hydraulic. Check the lines for proper installation including tight fittings. Verify that the level of the pump reservoir is even with the threads of the fill hole. Check the reverse lock valve and the hydraulic pump.

ALWAYS check to be sure the **UP** and **DOWN** hydraulic hoses are not reversed. Check the exploded view of the cylinder for the correct hose position. After servicing the system and to operate the system for the first time, push the **DOWN** button **FIRST**. If the drive unit rises, the hoses are obviously reversed and the condition **MUST** be corrected.

If reverse lock cover breakage occurs, the control valve on the hydraulic pump **MUST** be replaced. If the O-ring seal comes out of the reverse lock valve, or if the valve leaks, the valve cover must be removed and washer No. C-12-59570 and a **NEW** O-ring seal installed on the valve cover. Use Loctite 35, or equivalent on the threads of the cam-to-cam lever screw. Adjustment of the reverse lock valve is outlined in Section 8-9.

Power Trim—Double Hose

Before troubleshooting the system, verify that the fluid level in the hydraulic reservoir is even with the threads of the fill hole. **Would you believe**, probably **90%** of problems in the hydraulic system can be attributed directly to a low fluid level condition.

MECHANICAL PROBLEMS

Unit will not trim up or out
1- Reverse lock valve improperly adjusted.
2- Stern drive not in neutral or forward gear.
3- Interference between the drive housing and the gimbal ring or tilt pin.
4- Frozen U-joints in the stern drive.

Unit will not tilt to full UP position
1- Battery not up to full charge.
2- Failure of operator to push proper button.

Unit will trim up or out when stern drive is in reverse gear
1– Reverse interlock switch is shorted or switch will not open.

Pump motor fails to run continuously
1– Open in the armature.
2– Loose or open connections.

Pump motor runs only in DOWN position
1– Failure of operator to push proper button.
2– Interlock switch open or disconnected.
3– Wiring not connected properly.
4– Defective solenoid.
5– Defective wiring, connections, or solenoid.
6– Defective key, push button, or rocker switch.
7– Defective pump motor.

Pump motor runs only in UP position
1– Failure of operator to push proper button.
2– Wiring not connected properly.
3– Defective wiring, connections, ground, or solenoid.
4– Defective pump motor.
5– Defective push button or rocker switch.

Pump motor fails to run
1– Battery not up to full charge.
2– Motor overload switch open due to overheating.
3– Defective wiring, connection, ground, or solenoid.
4– Defective key, push button, or rocker switch.
5– Wiring not connected properly.
6– Defective pump motor.

Unit trims out or tilts up while unattended
1– Moisture in key switch.
2– Defective push button or rocker switch.

HYDRAULIC PROBLEMS

Unit will not trim up or out
1– Fluid level low.
2– External leak.
3– Internal cylinder leak.
4– Low pump pressure.

5– Reverse lock inoperative.
6– Internal resistance in one or both cylinders.
7– U-joints "frozen".

Unit fails to reach full UP position
1– Internal resistance in one or both cylinders.
2– Fluid level low.
3– U-joints "frozen".

Unit fails to release from full UP position
1– Internal resistance in one or both cylinders.
2– Fluid level low.
3– Pump pressure low or no pressure.

Unit fails to return to full DOWN position or fails to move smoothly (moves with jerks)
1– Hoses reversed on one cylinder.
2– Air trapped in the system.
3– Pump pressure low or no pressure.
4– Internal resistance in one or both cylinders.

Unit will not hold in reverse
1– Contamination (dirt) in the system.
2– External leaks.
3– Internal cylinder leaks.
4– Defective reverse lock valve.

Stern drive vibrates when shifted
1– Air trapped in hydraulic system.
2– Internal leak in one or both cylinders.
3– Defective control valve (leaks).

Unit trails out when backing off throttle at high speed
1– Air trapped in hyraulic system.
2– Defective control valve assembly (inoperative).
3– Internal leak in one or both cylinders.

Unit will not hold a trimmed position, or fails to remain tilted
1– Contamination (dirt) in hydraulic system.
2– Defective pump check valve (leaks).
3– External leaks in hydraulic system.
4– Internal leak in one or both cylinders

Unit moves up or out slowly or with a jerky motion when tilting or trimming
1– Hoses reversed on one cylinder.

Reverse lock valve broken and leaking
1- Defective control valve assembly (inoperative).
2- Hoses reversed on control valve.
3- Hoses reversed on gimbal housing.

Oil foams from pump vent
1- Fluid level low.
2- Can improperly installed over filter screen.
3- Check valve contains entrapped foreign particles. May be cleared by operating system up-and-down several times while flushing system. If flushing fails to correct problem, check valve is defective.

Oil in the motor
1- Vent screw closed.
2- Seal in pump reservoir damaged or missing.

ELECTRICAL PROBLEMS

Unit fails to trim up or out
1- Failure of operator to push proper button.
2- Limit switch open or disconnected.
3- Defective push button or rocker switch.
4- Defective trim motor solenoid.

Unit fails to tilt to full UP position
1- Battery not up to full charge.
2- Failure of operator to push proper button.

Unit will trim up or out when stern drive is in reverse gear
1- Defective interlock switch (shorted or will not open).

Pump motor fails to run continuously
1- Open or loose connections.
2- Open in armature.

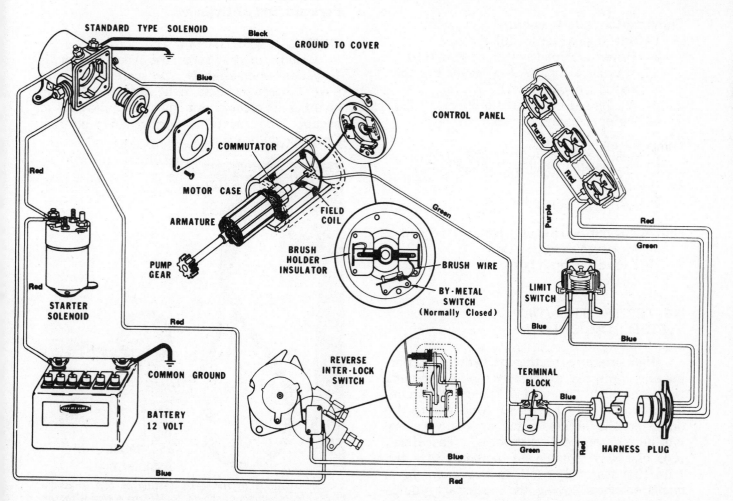

Functional diagram of the power trim electrical system with the three-button control. This illustration may be helpful during troubleshooting of the system.

Pump motor runs only in DOWN direction

1- Failure of operator to push proper button.
2- Wiring not connected properly.
3- Interlock switch open or disconnected.
4- Defective solenoid.
5- High resistance, broken or loose connection in wiring, ground, or solenoid.
6- Defective push button, or rocker switch.
7- Defective pump motor.

Pump motor runs only in UP direction

1- Failure of operator to push proper button
2- Wiring not connected properly.
3- Defective push button or rocker switch.
4- High resistance, broken or loose connection in wiring, ground, or solenoid.
5- Defective pump motor.

Pump motor fails to operate

1- Battery not up to full charge.
2- Motor overload switch open due to overwrheating. (Allow motor to cool, switch should close.)
3- Wiring not connected properly.
4- Defective key, push button, or rocker switch.
5- High resistance, broken or loose connection in wiring, ground, or solenoid.
6- Defective pump motor.

Unit trims out or tilts up when unattended

1- Moisture in key switch.
2- Defective push button or rocker switch.

8-6 TESTING TRIM/TILT SYSTEM COMPONENTS

High pressure testing of hydraulic components requires expensive special gauges, tools and highly trained personnel to accurately interpret test results. A danger to the operator and others in the area always exists during the testing process. Therefore, it is highly recommended the boat, with the trim/tilt system installed, be taken to a qualified shop having the necessary equipment and trained technicians to conduct the testing properly and safely.

Do not attempt to work on hydraulic hose connections without the proper tools designed for that specific purpose. The use of a common box end wrench will quite likely result in "rounding off" the hydraulic fitting. Technicians accustomed to working on hydraulic systems will have a set of "flare" wrenches which will almost guarantee the fitting will not be damaged.

8-7 PURGING AIR ("BLEEDING") TRIM/TILT SYSTEM

Power trim/tilt systems operate with hydraulic fluid. Each time the trim/tilt system is serviced for any reason, including the opening of a hose or fitting, unwanted air enters the system. This air must be purged ("bled") from the system because any air mixed with the hydraulic fluid will cause the system to operate erratically.

**Early Model Systems
Both Hoses Connected to
Forward End of Cylinder**

CRITICAL WORDS

Check to be sure the vent screw is "cracked" open slightly. On systems without a vent screw, check to be sure the fill cap opening is clear. Air must be allowed to escape the system during the purging ("bleeding") process.

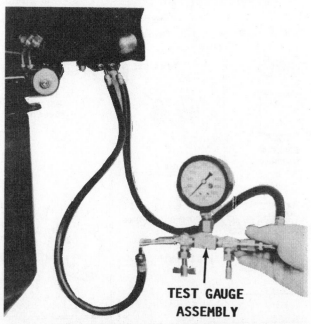

**TEST GAUGE
ASSEMBLY**

As explained in the text, testing of high pressure hydraulic components should be left to shops with the proper equipment, trained personnel, and safety procedures well ingrained in their procedures.

1- Operate the system until the stern drive is in the full **DOWN** position. Remove the fill screw and check the fluid level. If necessary, fill the reservoir to the bottom of the fill screw threads, with **SAE, MS, SD, SE 20W-30**, or **5W-30**. Earlier systems are stamped with a six digit number followed by an "A". On these earlier model systems, use automatic automobile transmission **Type A** fluid. **TAKE CARE** not to overfill the reservoir.

2- **Bleed the DOWN side of the cylinder.** Operate the system until the stern drive is in the full **DOWN** position. Disconnect the aft end of the cylinders from the driveshaft housing. Remove the rear mounting pins. Pass a piece of rope through the lifting eye in the top cover of the driveshaft housing, and then secure each end to each cylinder. Adjust the rope until each cylinder is as near horizontal as possible. With both piston rods fully retracted, check the fluid level.

3- Disconnect the aft hose from one cylinder. Extend the piston rod and purge oil and air from that cylinder by running the pump in the **UP** direction. Operate the pump in the **DOWN** direction for a moment to purge any air in the hose line to the cylinder, and then connect the hose. Bring the fluid level in the reservoir up to the bottom of the fill screw threads. Repeat the bleed procedure for the other cylinder.

4- **Bleed the UP side of the cylinder.** Remove the rope and allow the cylinders to hang. Operate the pump in the **UP** direction until the piston rods are fully extended.

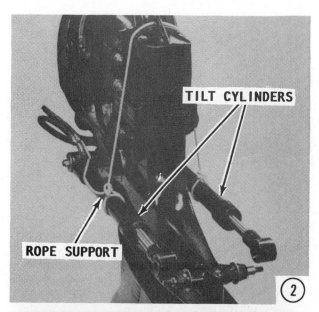

TILT CYLINDERS

ROPE SUPPORT

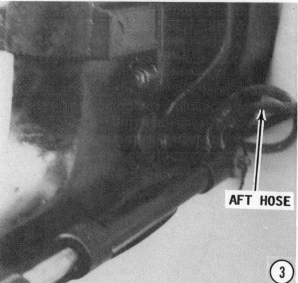

AFT HOSE

VENT SCREW

OIL FILL

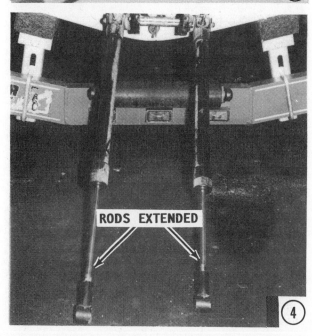

RODS EXTENDED

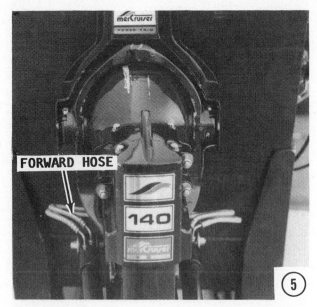

FORWARD HOSE

⑤

5- Disconnect the front hose on one cylinder. Operate the pump in the **DOWN** direction to retract the piston and purge air and oil from the cylinder. Operate the pump in the **UP** direction for a moment to purge air in the hose to the cylinder. Connect the hose and bring the fluid level in the reservoir back up to the bottom of the fill screw threads. Repeat the bleed procedure for the other cylinder. Connect both cylinders to the aft anchor pins.

6- Again fill the reservoir, if necessary, and then install the vent screw. **BE SURE** to leave the vent screw open.

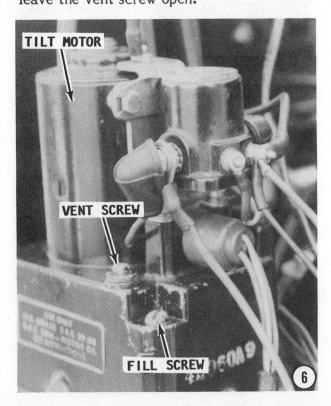

TILT MOTOR

VENT SCREW

FILL SCREW

⑥

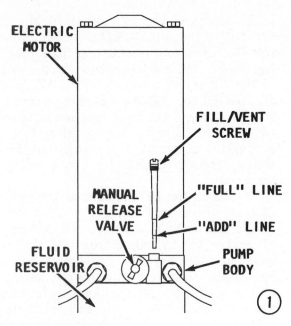

ELECTRIC MOTOR

FILL/VENT SCREW

MANUAL RELEASE VALVE

"FULL" LINE

"ADD" LINE

FLUID RESERVOIR

PUMP BODY

①

Purging ("Bleeding")
Late Model Systems
One Hose Connected to
Each End of Cylinder

The power trim/tilt system is designed to be self-purging with nothing to be done other than raising and lowering the stern drive several times. If, during service work, components were installed already filled with hydraulic fluid, or a line was disconnected and then reconnected, then the self purging feature of this system will adequately eliminate the small amount of unwanted air. However, if components were removed and installed "dry", the following procedures **MUST** be performed to adequately purge air from the system.

1- Check to be sure the fill/vent screw is opened slightly to allow air to escape during the purging process. Operate the system to lower the stern drive to the full **DOWN** position. Remove the fill screw/dipstick and check the fluid level. If necessary, fill the reservoir to the proper level indicated on the dipstick. If both cylinders were replaced and installed without being filled with fluid, fill the reservoir with fluid to the bottom of the dipstick threads. Adding fluid at this time will prevent the pump from running out of fluid during the purging process.

If the hydraulic pump is equipped with a manual release valve, ensure the valve is closed and tightened securely.

2- Obtain a suitable container and a supply of lint free shop towels. Disconnect

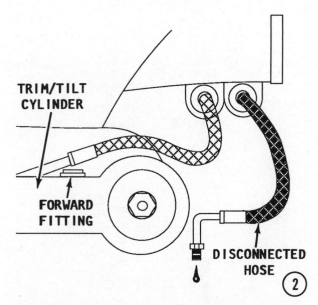

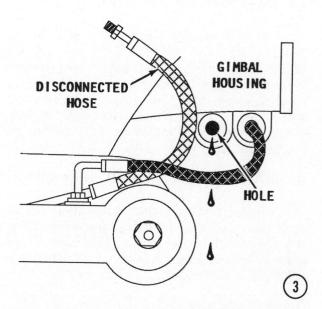

the hydraulic hose from the forward end of the cylinder which was serviced. If both cylinders were serviced, or if the pump assembly was replaced, this procedure may be performed on both cylinders simultaneously, or one after the other. Be sure the container is of sufficient size to accommodate fluid from both hoses.

Direct the end/s of the hose/s into the container. Operate the unit to the full **UP** position until an uninterrupted stream of hydraulic fluid is being discharged from the end/s of the hose/s. Connect and tighten the hoses.

All air should have been discharged after the first raising of the stern drive. If there is any doubt, reconnect the hoses/s; operate the system to lower the stern drive to the full **DOWN** position; disconnect the hose/s; and repeat the procedure just described.

After it has been determined the sytem is free of air, operate the system to lower the stern drive to the full **DOWN** position; check the fluid level at the reservoir; and add fluid to bring the level to the "full" line.

3- Obtain special plug P/N 22-38609. Two plugs are required if both cylinders are to be purged simultaneously. Beginning with the unit in the full **DOWN** position, disconnect the hydraulic hose connected to the gimbal housing from the aft end of the cylinder. **DO NOT** disconnect the hose at the fitting on the cylinder.

Now, use the special plug to plug the hole at the gimbal housing. If the special plug is not available, obtain a suitable re-

placement, bearing in mind the threads of the hole are pipe threads and not standard machine screw threads. If both cylinders are to be purged simultaneously, disconnect the hose and install the plug for the second cylinder.

Position the container under the gimbal housing hole/s. Operate the system until the stern drive is to the full **UP** position. Remove the plug/s from the gimbal housing, and then operate the system to lower the stern drive until an uninterrupted stream of hydraulic fluid is dischared from the holes/s in the gimbal housing. Connect and tighten the hose/s.

Continue to operate the system until the stern drive is in the full **DOWN** position. Check the fluid level at the reservoir. Add fluid as required to bring the level to the "full" line.

All air should have been discharged from the system when the stern drive unit was lowered the first time. However, if there is any doubt, reconnect the hoses/s; operate the system to raise the stern drive to the full **UP** position; disconnect the hose/s; and repeat the procedure just described.

After it has been determined the sytem is free of air, operate the system to lower the stern drive to the full **DOWN** position; check the fluid level at the reservoir; and add fluid to bring the level to the "full" line.

Finally, for good measure, raise and lower the stern drive several time to permit the self purging feature to function, and then make a final check of the fluid level with the stern drive in the full **DOWN** position.

8-8 TRIM SENDER SYSTEMS

The dash-mounted gauge on the control panel indicates to the helmsman the position of the stern drive at all times. Two indicator systems are in common use. The early system, prior to 1975 is a hydraulic/electric unit. Since 1975 the sender unit is only electrical with the indicator trim switch on the gimbal ring.

HYDRAULIC TRIM SENDER SERVICE

This section presents step-by-step instructions for disassembling and assembling the hydraulic sender unit. Procedures are also given to bleed the system after the sender is installed.

DISASSEMBLING

1- Remove the end caps and piston by removing the four through-bolts. Hold the piston shaft end and at the same time remove the nut and washer. Slide the piston assembly from the shaft. Separate the piston halves. **TAKE CARE** not to lose the by-pass valve pins, the check balls, and the spring.

ASSEMBLING

2- Take the extra time required to keep the work area and all parts **CLEAN** as a precaution against contamination entering the system. Lubricate the parts with light oil. Position a **NEW** seal washer and quadring on the piston half. Assemble the piston on the shaft. **TAKE CARE** to replace all of the by-pass valve parts properly. Tighten the nut securely.

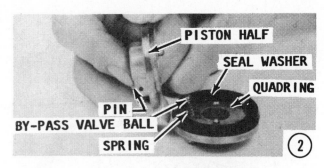

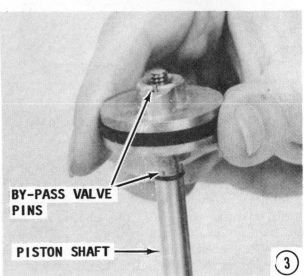

3- Install a **NEW** O-ring seal on the shaft side of the end cap. **DO NOT** disturb the staked washer. Install **NEW** quadrings in both end caps. Assemble the sender and tighten the through-bolts securely.

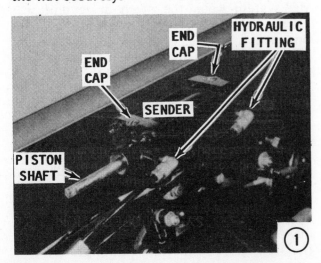

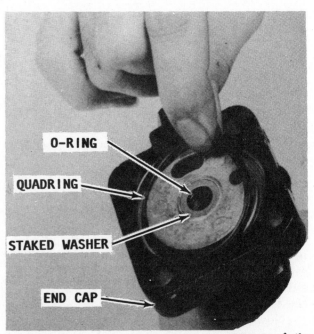

Placement of the O-ring into the end cap of the hydraulic trim sender unit.

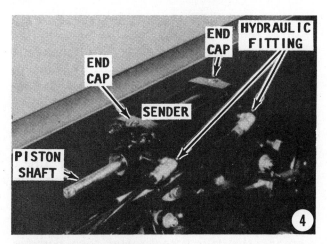

BLEEDING AFTER SENDER INSTALLATION

4- Loosen bleed screw indicated by "B" on the accompaning illustration. Momentarily press the **DOWN** or **IN** button on the control panel until the oil flowing out is free of air, and then tighten the bleed screw. Loosen bleed screw "A" and press the **UP** button for just a second, until the oil flow is free of air. Install and bleed screw. Add fluid to the reservoir, as required. **TAKE CARE** not to lose the O-ring seals from the bleed screws.

GOOD WORDS

Air entering the hydraulic system during installation of the indicator will become trapped in the sender. Therefore, it is not necessary to bleed the system at the cylinders.

ELECTRIC TRIM INDICATOR SENDER AND TRIM LIMIT SWITCH ADJUSTMENT

The trim indicator sender unit is located on the starboard side of stern drive and the trim limit switch is on the port side. Would you believe, one is a sending unit and the other is a switch, **BUT** their appearance on the outside is identical.

Trim Indicator Sender Adjustment

Turn the stern drive hard-over to port in the full **DOWN** trim position. Loosen the two adjusting screws securing the indicator sender housing. Turn the ignition key to the **RUN** position. Rotate the indicator sender housing in the required direction until the panel indicator pointer is at the bottom of the green arc. The sender is now properly adjusted. Tighten the adjusting screws and turn the ignition key to **OFF**.

The trim indicator sender is installed on the starboard side of the stern drive.

Trim Limit Switch Adjustment:

Loosen the two adjusting screws securing the switch housing. Rotate the housing **COUNTERCLOCKWISE** until a distance of 6-1/2" (165mm) can be measured between the outer edge of the end cap and the center

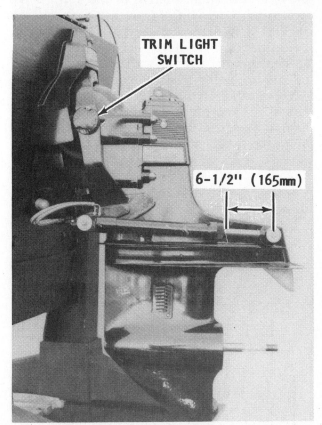

The trim limit switch is installed on the port side of the stern drive.

SHIFT LEVER ATTACHING NUTS

REVERSE LOCK VALVE COVER

SHIFT LEVER

SHORT ADJUSTING SLOT

①

of the aft pivot bolt, as shown in the accompanying illustration. Tighten the screws to hold the adjustment.

8-9 REVERSE LOCK VALVE

DESCRIPTION

The reverse lock valve is installed to prevent upward movement of the stern drive while it is operating in reverse gear. If the reverse lock valve is defective or out of adjustment, the stern drive unit can be forced upward due to the pressure created by the propeller thrust. The position of the reverse lock valve is determined by the shift cables. Therefore, the cables **MUST** be adjusted before the reverse lock valve can be properly adjusted, see Chapter 7.

Separate shift cable systems are used for in-line engine and for V-8 engine installations. Separate instructions for adjustment of the shift cables for each installation follow the service procedures for the lock valve.

DISASSEMBLING

1- Disconnect the two hydraulic lines. Loosen the two hex-head screws on the switch, and then disconnect the electrical leads from the switch. Back off the two nuts, and then remove the cam assembly. Remove the two screws, and then remove the reverse lock valve. Remove the four Phillips-head screws, and then remove the cover-and-cam assembly. Remove the cam-to-cam lever screw. **TAKE CARE** not to lose any of the 30 ball bearings. Remove the **O**-ring seal and the wavy washer.

CLEANING AND INSPECTING

Carefully check each part for any sign of damage.
Obtain an ohmmeter and connect one lead of the meter to each switch terminal. Operate the switch and check the meter reading. The reading should go from zero to infinity. Replace the switch if it fails the test. **ALWAYS** replace all **O**-ring seals.

ASSEMBLING

The accompanying exploded illustration showing all parts of the reverse lock valve

1 - Reverse Lock Valve Body
2 - Elbow
3 - Connector, elbow
4 - Follower, cam
5 - Washer, wave
6 - "O" Ring
7 - Cam, reverse lock valve
8 - Screw, cam to cam lever
9 - Lockwasher, cam screw
10 - Washer, retaining
11 - Ball, valve cover
12 - "O" Ring, valve cover
13 - Cover, reverse lock valve
14 - Screw
15 - Nut, cover screw
16 - Plug, pipe
17 - Lever Assembly, cam
18 - Stud
19 - Screw
20 - Lockwasher
21 - Switch Assembly, cut-out
22 - Shield, cut-out switch
23 - Screw
24 - Clamp
25 - Screw
26 - "D" Washer

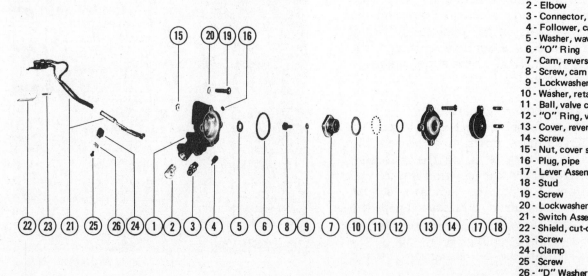

②

(15) (20)(19)(16)

(22)(23)(21)(25)(26)(24)(1)(2)(3)(4)(5)(6)(8)(9)(7)(10)(11)(12)(13)(14)(17)(18)

will be helpful in following the sequence of assembly and the relationship of each part to the others.

2- Count to be sure you have 30 ball bearings for installation. Attach the cam, retaining washer, the ball bearings, and a **NEW** O-ring seal to the cover. Secure these parts in place with a lockwasher and screw. Install the Belleville (wavy) washer and a **NEW** O-ring seal in the valve body. Install the cover assembly to the body and secure it with the four Phillips-head screws. Install the valve assembly onto the engine. **ONE MORE WORD:** If the reverse lock valve leaks, or if the O-ring seal comes out of the valve, the complete reverse lock valve assembly **MUST** be replaced.

REVERSE LOCK ADJUSTMENT

3- Move the remote control shift handle to the full **REVERSE** position and at the same time have an assistant turn the propeller shaft clockwise until the shaft stops, to ensure the clutch is fully engaged. Loosen the two shift lever-to-reverse lock valve attaching nuts. Rotate the forward nut in a direction to align the raised triangular mark on the cam with the matching raised mark on the reverse lock valve cover. Tighten the two shift lever attaching nuts. After the boat is in the water and with the engine running, operate the stern drive in reverse gear. If the unit trails out (fails to hold in the reverse position), adjust the reverse lock valve one mark at-a-time until it holds in the reverse gear position. **ONE MORE WORD:** The drive unit will be locked in neutral if the reverse lock is overadjusted.

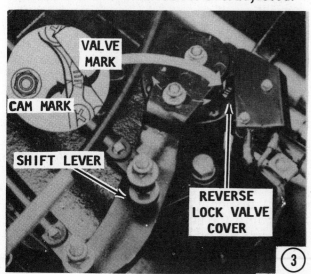

The tilt system will not operate properly if the battery is not up to a full charge.

8-10 PUMP MOTOR

Several different different pump motors are used with MerCruiser stern drive units covered in this manual. They are very similar. Therefore, only one will be disassembled and tested in this section.

TROUBLESHOOTING

1- Before going directly to the pump as a source of trouble, check the battery to be sure it is up to a full charge; inspect the wiring for loose connections corrosion and

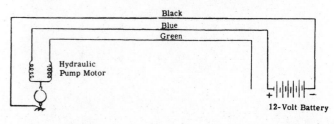

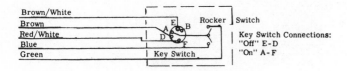

TO TILT SWITCH
TERMINAL BLOCK   ②

the like; and take a good look at the control switches and connections for evidence of trouble. The control switches may be eliminated as a source of trouble by connecting the pump directly to the battery for testing purposes. This can be accomplished by disconnecting the black and the blue wires from the terminal block, and then connecting the black wire to the negative (–) battery terminal and the blue wire to the positive (+) terminal. If the pump motor does not run with this direct connection, either the pump motor or the pump-and-valve assembly is defective. If the pump motor does operate, the problem is in the wiring or the panel switches.

2– Check rocker-type panel switches and wiring on models through 1966: Disconnect the red/white and the blue wires in the control panel harness from the control panel terminal block. Connect one test lead from a continuity meter to the terminal for the red/white and the other test lead to the terminal for the blue wire. Turn the control panel switch to the **ON** position. Depress the rocker switch and note the meter. The

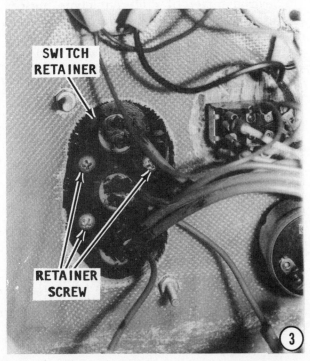

meter **MUST** indicate continuity for a satisfactory switch. If there is no continuity, connect one meter test lead to the **"F"** terminal on the back of the control panel key switch and connect the other test lead to the **"A"** terminal. With the control panel switch still **ON**, the meter **MUST** indicate continuity if the switch is satisfactory. If there is no continuity, remove the test lead from the **"F"** terminal and connect it to the terminal on the back of the rocker switch with the blue wire soldered to it. Depress the top of the rocker switch and note the meter reading. The meter **MUST** indicate continuity or the rocker switch is defective. If these checks indicate the switch is satisfactory, the trouble **MUST** be an open circuit in the red/white or the blue wires.

3– Test panel switches since 1967: Tests for these switches must be done at the back of the switches because the wires are soldered to the switch terminals and cannot be disconnected. Actually, the test procedures are the same as for the other type of switches. Unplug the trim harness from the trim pump. Connect a continuity meter between the terminals on the back of the switch being tested. The meter **MUST** indicate an open circuit with the button in the free position and **MUST** indicate continuity with the button depressed.

8-10 PUMP MOTOR SERVICE

DISASSEMBLING

1– Remove the two screws from the top of the motor and reservoir assembly. Scribe a mark on the reservoir and a matching mark on the motor housing as an aid to assembling. Separate the reservoir from the motor. **TAKE CARE** not to lose the spacers from the armature. Pull the armature out of the frame. Disconnect the ground wire from the upper end cap. Remove the end cap from the frame-and-field assembly.

CLEANING AND TESTING

Any sign of oil in the pump motor indicates either the reservoir seal is damaged or the vent screw was closed. If the vent was closed, air could not escape and oil was forced into the pump motor.

Check the amount of wear to the brushes. If they are worn to half their original length, they should be replaced. If the

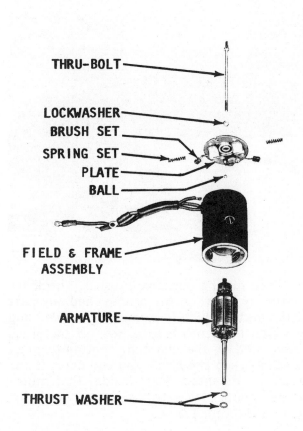

THRU-BOLT

LOCKWASHER
BRUSH SET
SPRING SET
PLATE
BALL

FIELD & FRAME
ASSEMBLY

ARMATURE

THRUST WASHER

①

②

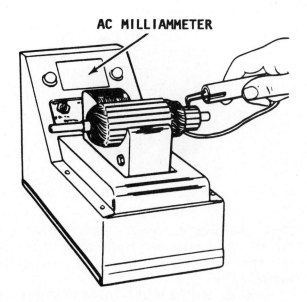

AC MILLIAMMETER

Testing the armature segments for possible short between the segments or to the armature core.

commutator is worn, true it on a lathe, and undercut the mica.

2- Check the armature on a growler for shorts, open windings, or shorted windings.

3- Check the resistance of the thermal switch. If the switch has high resistance, it **MUST** be replaced.

4- Check each positive and negative brush as described in the following paragraphs under "Good Words".

GOOD WORDS ABOUT BRUSHES

Depending on the model being serviced, the brushes may differ in number and loca-

Testing for a short between a brush pigtail and the end cap.

End cap of a tilt motor showing worn brushes and corrosion.

tion. All brushes sweep across the surface of the commutator, but, they may be mounted on the end cap or just inside the motor frame.

Brushes are **ALWAYS** found in pairs, one negative and the other positive, or two negative and two positive.

Positive brushes are **ALWAYS** insulated, usually in some form of plastic or fiber casing. Positive brushes are **ALWAYS** attached to the field windings and must **NEVER** be allowed to contact the motor frame, the end cap, or any part of the electric motor which is normally grounded. Such contact would cause a short.

When testing positive brushes, check for continuity between the brushes and the two field leads. If no continuity exists, there is an open in the circuit, and therefore the field and frame assembly is defective.

Negative brushes usually have bright braided leads and are **ALWAYS** grounded to the end cap or the motor frame.

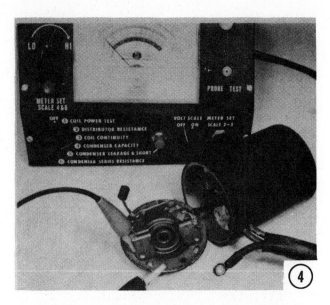

When testing negative brushes, check for continuity between the brushes and any part of the motor which is normally grounded and to which the brush is attached. If the brush is mounted on the end cap, continuity must exist between the brush and end cap. If the brush is mounted just inside the motor frame, continuity must exist between the brush and motor frame.

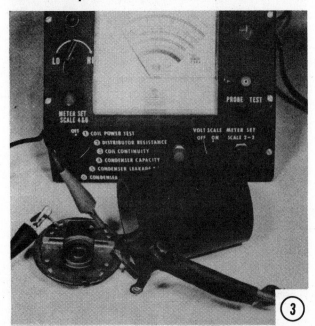

Also test for **NO** continuity between the negative brushes and the two field leads, if the brushes are mounted inside the motor frame. If continuity is found here, the field and frame assembly is grounded and defective.

GOOD WORDS

If there is any measureable high resistance in any of the tests in Step 5, the frame-and-field assembly **MUST** be replaced.

5- Check the field coils for a short, ground, or excessive resistance in the windings. With the tester on scale No. 2, check for resistance between the green wire and the black jumper wire. Check for resistance between the blue wire and the black jumper wire. Check the black ground wire for resistance.

Move the tester to scale No. 3, and check for a short between the black jumper wire and ground. A short is indicated if the needle moves to the right. The assembly **MUST** be replaced.

6- Install the armature into the frame-and-field assembly. Place the ball bearing on top of the armature shaft, if such a bearing is used. Depress the brushes into place, and slide them over the commutator bars. **TAKE CARE** not to mar the brushes. Position the assembly onto the reservoir housing and work the shaft into the pump. Install the screws into the cap. Be sure to install the ground wire onto one of the screws securing the cap. Tighten the two bolts to the reservoir housing.

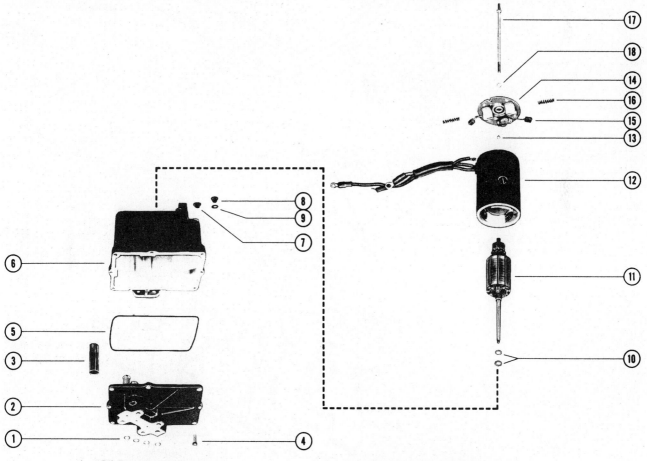

1 - "O" Ring, hydraulic pump
2 - Valve Body and Gear Assembly
3 - Can, valve body
4 - Screw, valve body to reservoir
5 - Seal, reservoir to valve body and gear
6 - Reservoir Assembly
7 - Screw, vent - hydraulic pump
8 - Screw, filler - hydraulic pump
9 - Washer, sealing - filler screw

10 - Washer, thrust - armature
11 - Armature, field and frame
12 - Field and Frame Assembly
13 - Ball, commutator end plate
14 - Plate Assembly, commutator end
15 - Brush Set, commutator end plate
16 - Spring Set, brush tension
17 - Bolt Set, thru - plate and frame to reservoir
18 - Lockwasher, thru bolt

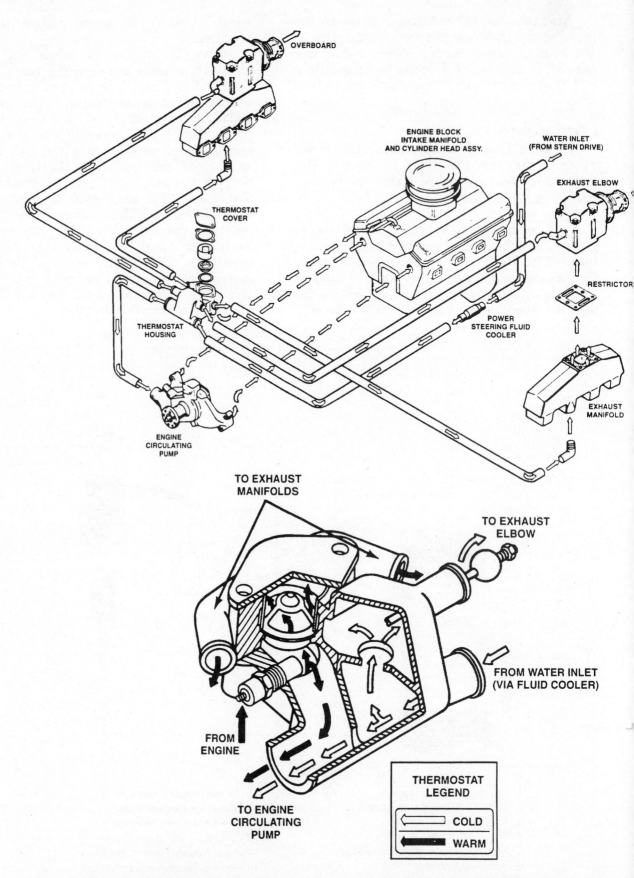

Functional diagram to illustrate water flow for the 5.0 Litre, 5.0 Litre LX, 5.7 Litre, and 350 Magnum (305/350 CID), V8 engines. The cutaway drawing in the lower portion depicts the interior of the thermostat housing and the water routing through the housing.

9
COOLING

9-1 DESCRIPTION

MerCruiser engines are water cooled by one of two methods: One is an external water system which circulates external water through the block for cooling purposes. The other is a system utilizing fresh water circulating through the block to cool the engine block and head. In this second system, the fresh water is cooled by external water moving through a heat exchanger.

The standard external water cooling system has two pumps. On all models except the Bravo, one pump is located in the stern drive and pulls the external water up to the engine from outside the boat. However, on the Bravo stern drive, this pump is mounted

on the powerplant and is driven by the crankshaft drive belt. The second pump, on all models, circulates the water through the cylinder head and engine block. A thermostat is installed in the routing to close off water circulation in a cold engine for faster warm up.

An optional fresh-water cooling system, standard on Model 470 installations because of the aluminum block, is a closed-type system. Fresh water circulates through the cylinder head and engine block through a closed circuit. The water never leaves the system. A portion of the routing is through a heat exchanger where the fresh circulating water is cooled by external water pumped from the stern drive through the heat

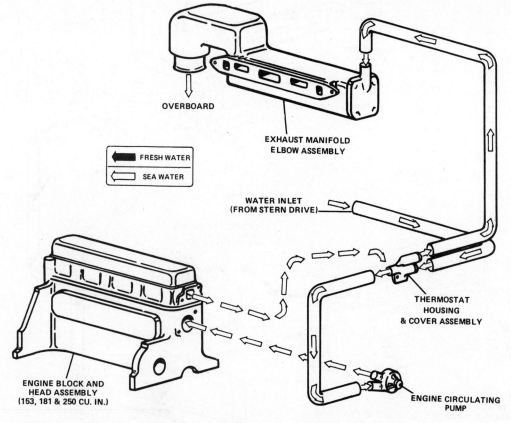

OVERBOARD

EXHAUST MANIFOLD ELBOW ASSEMBLY

◄— FRESH WATER

⇐ SEA WATER

WATER INLET (FROM STERN DRIVE)

THERMOSTAT HOUSING & COVER ASSEMBLY

ENGINE BLOCK AND HEAD ASSEMBLY (153, 181 & 250 CU. IN.)

ENGINE CIRCULATING PUMP

Water flow routing through a standard engine installation.

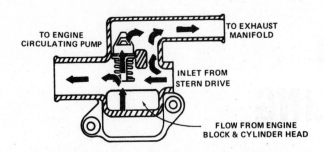

Water flow through the thermostat housing on a standard in-line engine installation, when the thermostat is OPEN.

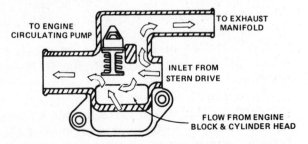

Water flow through the thermostat housing on a standard in-line engine installation, when the thermostat is CLOSED.

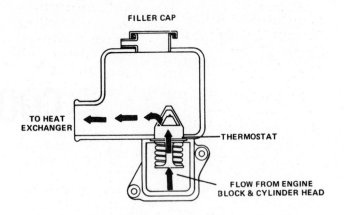

Water flow through the thermostat on an engine installation with a closed water system.

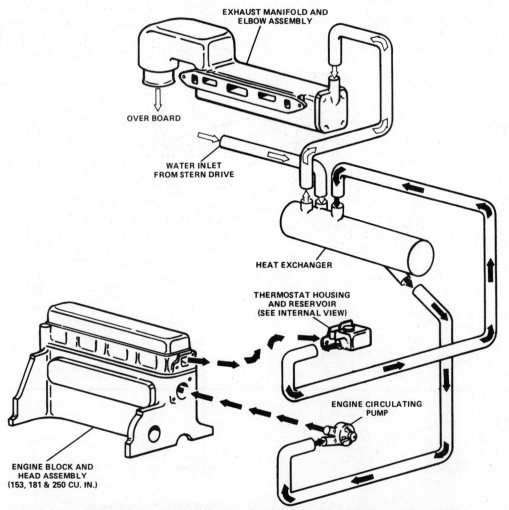

Water flow routing for an in-line engine installation with a closed-water system.

exchanger and then overboard. The fresh water section of the cooling system is pressurized, and the coolant temperature is thermostatically controlled.

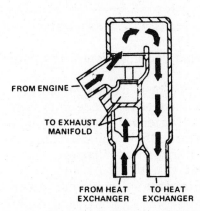

Water flow through the thermostat housing on a 470 engine installation with a closed water system, when the thermostat is OPEN.

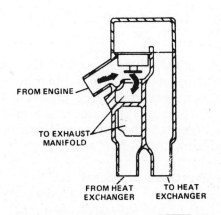

Water flow through the thermostat housing on a 470 engine installation with a closed water system, when the thermostat is CLOSED.

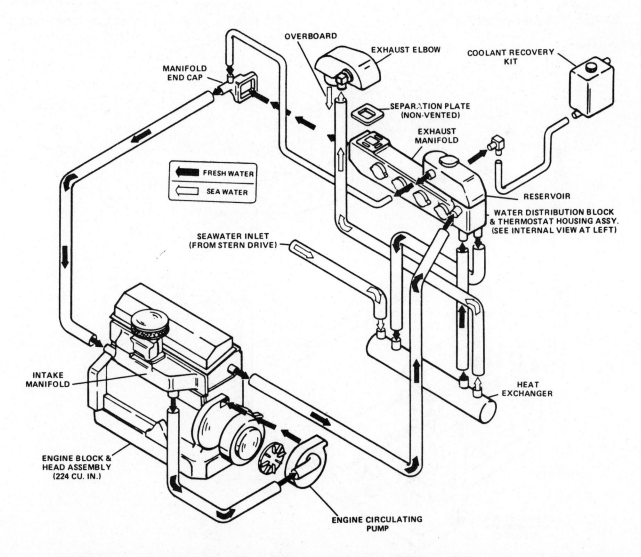

Water flow routing through a 470 engine installation, Serial No. 4886619 and below, with closed water system.

9-2 TROUBLESHOOTING THE COOLING SYSTEM

The following paragraphs list troubles encountered in the various portions of the system with accompanying probable causes for that problem. The causes are given in a logical order of checking until the problem is corrected.

Engine Overheats—General
1- Loose or broken circulating pump belt or pickup water pump belt
2- Inaccurate temperature gauge or sender
3- An accessory or barnacles in front of water pickup causing turbulance
4- Defective stern drive water pump
5- Loose hose connections between pick-up and pump (sucking air)
6- Pump fails to hold prime due to air leaks
7- Ice in water passages
8- Defective engine circulating pump
9- Defective thermostat
10- Plugged water passages in the exhaust manifold or elbows

Engine overheats with fresh-water-cooling in addition to above
1- Closed cooling system reservoir level low
2- Plugged heat exchanger cores
3- Too much anti-freeze in system
4- Fresh water kit improperly installed
5- Exhaust elbow dump fittings bottomed on inner water jacket

Engine overheats — problems other than cooling system
1- Incorrect ignition timing
2- Spark plug wires crossed
3- Lean air-fuel mixture
4- Preignition -- spark plugs are wrong heat range
5- Engine laboring -- engine rpm below specs at WOT

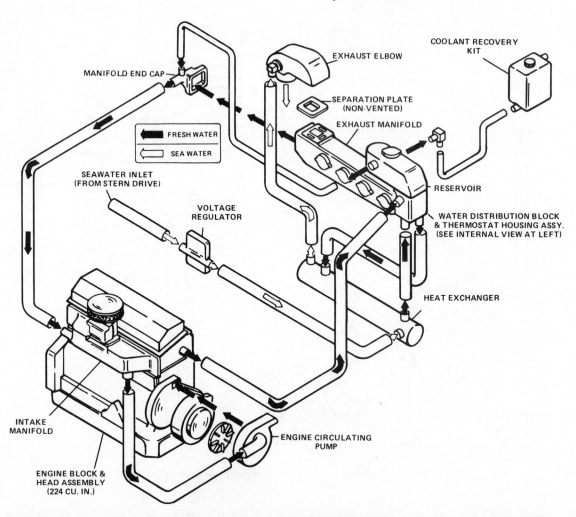

Water flow routing for a 470 engine installation (Serial No. 4886620 and above) with a closed water system.

6- Poor lubrication
7- Water in cylinders due to warped cylinder head
8- Sticking distributor mechanical advance mechanism
9- Clogged exhaust elbows
10- Exhaust flappers stuck closed
11- Cabin hot-water heater incorrectly connected to engine

Water in cylinders

1- Backwash through exhaust system
2- Loose cylinder head bolts
3- Blown cylinder head gasket
4- Warped cylinder head
5- Cracked or corroded-out exhaust manifold
6- Cracked block in valve lifter area on V8 engine
7- Improper engine or exhaust hose installation
8- Cracked cylinder wall

Water in crankcase oil

1- Backwash through exhaust system
2- Water seeping past piston rings from flooded combustion chamber
3- Thermostat stuck open or missing -- condensation forms because of engine running too cool
4- Cracked cylinder block
5- Intake manifold water passage leak on V8 engines only

9-3 COOLING SYSTEM SERVICE

FIRST, READ the most important words in this book: The engine cannot be run for even five seconds without water moving through the stern drive water pump or the pump impeller will be damaged. Therefore, **NEVER** start the engine, even for testing purposes without the boat being in the water, or provision having been made for water to pass through the stern drive.

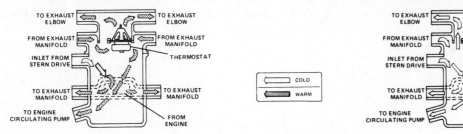

Water flow through the thermostat housing on Models 888/225-S/233 with standard cooling, when the thermostat is OPEN.

Water flow through the thermostat housing on Models 888/225-S/233 with standard cooling, when the thermostat is CLOSED.

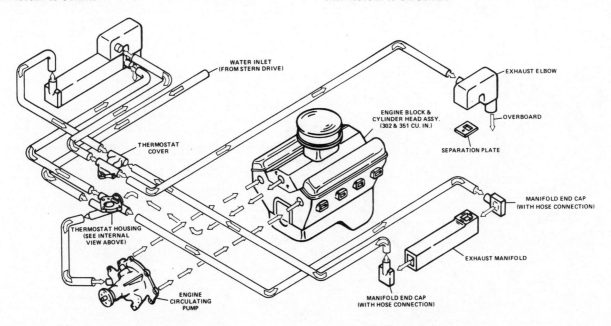

Water flow routing for Models 888/225-S/233 with standard cooling.

The service procedures given in this chapter provide step-by-step instructions for replacing a defective heat exchanger, water pump or thermostat. Replacement of these parts requires complete draining and filling of the cooling system.

Instructions are also given for replacing the water pump in the stern drive unit.

SPECIAL WORDS

Marine thermostats are rated at 143° to 160°F. An automotive-type thermostat must **NEVER** be used because of the higher temperaure ratings. Such a high rating would cause the engine to run much hotter than normal.

Procedures for removal and installation of the lower unit will be given in this chapter, in order to service the external water pump housed in the lower gear case of most stern drive units. Complete service instructions for the stern drive will be found in Chapter 10.

If the water pump impeller burns out due to a clogged water intake or improper installation, the exhaust shutters in the exhaust elbow will not be cooled properly and the shutters will also burn out.

Detailed procedures are also given in this chapter for servicing the exhaust manifold and the exhaust shutters. These procedures must be performed periodically to clean out the scale that accumulates in these areas. An increased temperature reading on the dash gauge indicates the need for this service.

9-4 DRAINING THE COOLING SYSTEM

General

The cooling system should be drained, cleaned, and refilled each season. The bow of the boat **MUST** be higher than the stern to properly drain the cooling system. If the bow is not higher than the stern, water will remain in the cylinder block and in the exhaust manifold. Insert a piece of wire into the drain holes, but **NOT** in the petcock,

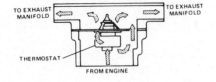

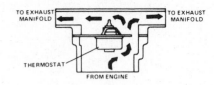

Water flow through the thermostat housing on Models 898/228/250/260 with a closed water system, when the thermostat is **OPEN**.

Water flow through the thermostat housing on Models 898/228/250/260 with a closed water system, when the thermostat is **CLOSED**.

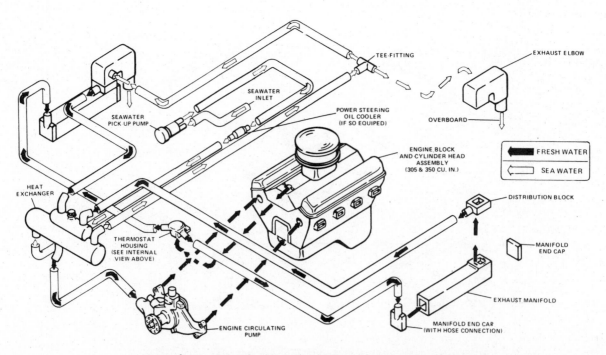

Water flow routing for Models 898/228/250/260 with closed water system.

to be sure sand, silt, or other foreign material is not blocking the drain hole.

If the engine is not completely drained for winter storage, trapped water can freeze and cause severe damage. The water in the oil cooler, on boats equipped with power steering, **MUST** also be drained.

For complete storage or pre-season preparation procedures, see Chapter 14.

Safety precautions **MUST** be observed when removing the filler cap on a closed-type cooling system because these systems are pressurized to 14 psi and sudden escaping scalding water could result in nasty burns.   To properly remove the reservoir cap, first allow the engine to cool down. After the engine is cool, turn the pressure cap 1/4 turn counterclockwise to allow any remaining pressure to escape slowly.   After the pressure has been released, the cap can be safely removed.

ONE MORE WORD: When running the

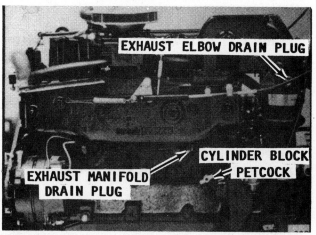

Location of water drains on a standard in-line engine.

engine in freezing weather with fresh water cooling and anti-freeze, the engine external water system **MUST** be drained following each engine operation.

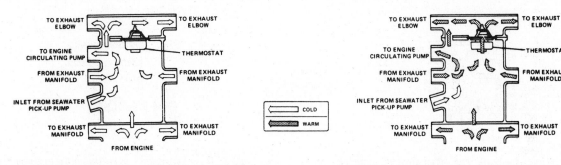

Water flow through the thermostat housing on Models 898/228/250/260 with standard cooling, when the thermostat is OPEN.

Water flow through the thermostat housing on Models 898/228/250/260 with standard cooling, when the thermostat is CLOSED.

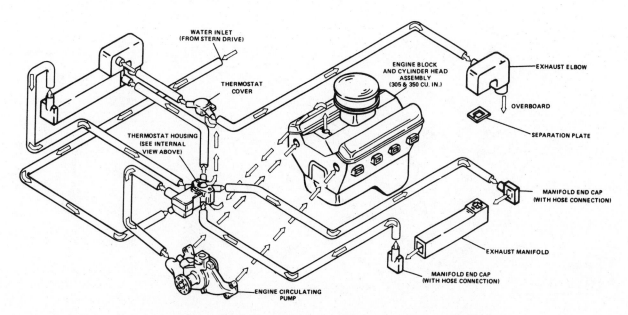

Water flow routing for Models 898/228/250/260 with standard cooling.

Draining In-line Engines with Standard Cooling System:

Place suitable containers under the drain plugs and petcock to prevent water from draining into the bilge. Remove the drain plugs and open the petcock. After the water has completely drained, apply Perfect Seal, or equivalent, to the threads of the drain plugs, and then install them. Close the petcock securely.

Draining In-line Engines with a Closed-Type Cooling System:

Place suitable containers under the drain plugs and petcock to prevent water from draining into the bilge. Remove the external water drain plugs. Remove the zinc electrode from the heat exchanger and inspect it for erosion. If erosion appears to be less than 25%, continued use of the zinc is permitted. If erosion appears to be more than 25%, the zinc **should** be replaced. Remove the external water inlet hose clamp at the heat exchanger, and then remove the

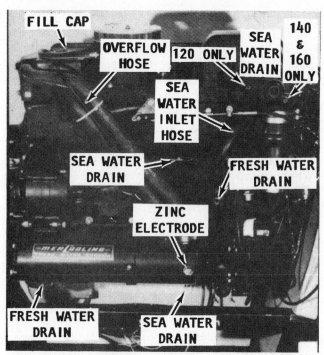

Location of water drains on an in-line engine installation with closed water system.

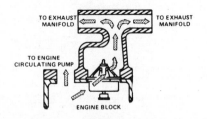

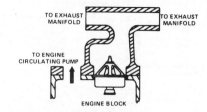

Water flow through the thermostat housing on Models 888/225-S/233 with closed water system, when the thermostat is OPEN.

Water flow through the thermostat housing on Models 888/225-S/233 with closed water system, when the thermostat is CLOSED.

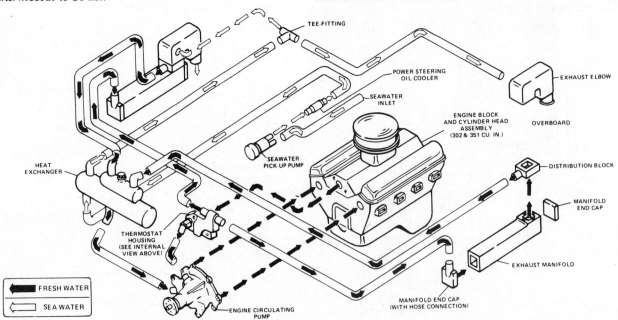

Water flow routing for Models 888/225-S/233 with closed water system.

a- PRESSURE FILL CAP
b- EXHAUST ELBOW DRAIN PLUG
c- EXHAUST MANIFOLD DRAIN PLUG
d- WATER-COOLED VOLTAGE REGULATOR
e- CYLINDER BLOCK DRAIN PLUG (HIDDEN)
f- SEA WATER INLET HOSE
g- FRESH WATER DRAIN PLUG
h- SEA WATER DRAIN PLUG
i- HEAT EXCHANGER

Location of water drains on a 470 engine installation with closed water system.

a- EXHAUST ELBOW DRAIN PLUG (HIDDEN)
b- EXHAUST MANIFOLD DRAIN PLUG
c- THERMOSTAT HOUSING WATER INLET HOSE
d- ENGINE CYLINDER BLOCK PETCOCK
   (HIDDEN)
e- THERMOSTAT HOUSING TO CIRCULATING
   PUMP HOSE

Location of water drains, starboard side, V8 engine with standard water cooling.

hose. After the water has drained completely, apply Perfect Seal, or equivalent, to the threads of the drain plugs. Install and tighten the drain plugs securely. Install the external water inlet hose and clamp.

a- PRESSURE FILL CAP
b- EXHAUST ELBOW DRAIN PLUG
c- EXHAUST MANIFOLD DRAIN PLUG
d- CYLINDER BLOCK DRAIN PLUG
e- FRESH WATER DRAIN PLUG
f- EXTERNAL WATER DRAIN PLUG
g- HEAT EXCHANGER

Location of water drains on a V8 engine installation with closed water system.

a- EXHAUST ELBOW DRAIN PLUG
b- EXHAUST MANIFOLD DRAIN PLUG
c- ENGINE CYLINDER BLOCK PETCOCK

Location of water drains on Model 470 (Serial No. 4886619 and below) with closed water system.

a- EXHAUST ELBOW DRAIN PLUG
b- EXHAUST MANIFOLD DRAIN PLUG
c- ENGINE CYLINDER BLOCK PETCOCK

Location of water drains, port side, V8 engine, with standard water cooling.

a- EXHAUST ELBOW DRAIN PLUG (HIDDEN)
b- EXHAUST MANIFOLD DRAIN PLUG
c- THERMOSTAT HOUSING WATER INLET HOSE
d- THERMOSTAT HOUSING TO CIRCULATING PUMP HOSE
e- CYLINDER BLOCK PETCOCK

Location of water drains, port side, V8 engine, with closed water system.

Draining V8 Engines with Standard Cooling System:

Place suitable containers under the drain plugs and petcocks to prevent water from draining into the bilge. Remove the cooling

a- EXHAUST ELBOW DRAIN PLUG
b- EXHAUST MANIFOLD DRAIN PLUG
c- CYLINDER BLOCK PETCOCK

Drain plug locations, port side, on a V8 engine.

a- REMOVE HOSE IF NO OIL COOLER DRAIN
b- DRAIN PLUG
c- POWER STEERING OIL COOLER

Location of oil cooler water drains on a V8 engine.

a- POWER STEERING OIL COOLER
b- DRAIN PLUG

Power steering oil cooler drains on a V8 engine installation.

system drain plugs and open the petcocks. Disconnect the water inlet hose from the thermostat housing. Lower the hose to drain any water in the hose. Connect the hose to the thermostat housing and tighten the clamp. Apply Perfect Seal, or equivalent, to the threads of the drain plugs. Install and tighten the plugs securely. Close the petcocks.

Draining V8 Engines with a Closed-Type Cooling System:

Place suitable containers under the drains and petcocks to prevent water from draining into the bilge. Remove the external water drain plugs. Disconnect the external water pump hoses. Remove the drain plug from the pump.

Remove the closed cooling system drain plugs and open the petcocks. Remove the coil high-tension lead wire from the distributor cap and ground the lead wire. Now, crank the engine with the starter motor to drain the pump completely. **TAKE CARE** to ensure the engine compartment is well ventilated and no gasoline vapors are present, as a precaution against fire or possible explosion.

After the cooling system has drained completely, apply Perfect Seal, or equivalent, to the threads of the drain plugs. Install and tighten the drain plugs securely. Close the petcocks. Connect the external water pump hoses and tighten the clamps. Install the coil high-tension lead wire to the distributor cap.

9-5 ENGINE WATER PUMP SERVICE

REMOVAL

1- After the the cooling system has been drained, loosen the pulley retaining bolts. Loosen the alternator upper brace and swivel bolt. Shift the alternator inward, and then remove the drive belt. Disconnect the large hose to the water pump behind the

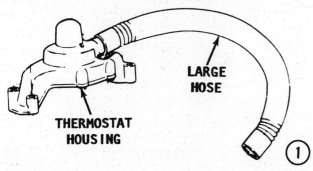

pulley. Remove the attaching bolts, and then remove the pump.

INSTALLATION

2- Coat a **NEW** water pump gasket with Perfect Seal, or equivalent. Install the new pump assembly onto the cylinder block with the new gasket. Secure the pump in place with the retainer bolts and tighten them to a torque value of 15 ft-lbs. Install the pump pulley and bolts. Tightening the pulley bolts will be much easier **AFTER** the alternator belt is installed.

3- Install the alternator drive belt and adjust the belt tension to 45 lbs. with a tension gauge. Tighten the adjustment brace bolt. Tighten the alternator pivot bolt. If a belt tension gauge is not available, the belt tension can be adjusted by pivoting the alternator away from the engine until the belt deflection, halfway between the circulating pump pulley and the alternator pulley, is 1/4" when a slight

downward pressure is applied to the belt, as shown. Tighten the circulating pump pulley bolts. Install the hoses and secure them in place with the clamps. If the engine is equipped with a closed cooling system, it must be filled according to the procedures given in Section 9-9. Start the engine and check for leaks.

CAUTION

Water must circulate through the stern drive, also to and from the engine, anytime the engine is run to prevent damage to the water pump located in the lower unit (original and Alpha drives); or on the front of the engine (Bravo drive). Just a few seconds without water will damage the water pump.

9-6 THERMOSTAT SERVICE

REMOVAL

Drain the water from the cooling system, see Section 9-5 and the particular steps for your engine. Refer to the accompanying illustrations for thermostat location on the most popular engine installations. Remove the thermostat cover and lift out the thermostat.

TESTING

1- Inspect the thermostat at room temperature. If the thermostat is fully open, it is defective and must be replaced. Hold the thermostat up to the light and check it for leaks. A light leak around the perimeter indicates the thermostat is not closing, and therefore, it **MUST** be replaced.

2- Check the thermostat opening temperature by first opening the thermostat valve and inserting a piece of thread to keep the valve from seating. This will ensure some flow of water during the test. Now, suspend the thermostat and a thermometer

inside a tester. Take care to be sure neither the thermostat or the thermometer touches the container. If either one does touch the container, the metal of the container will influence the test. Fill the tester with enough water to cover the thermostat. Plug in the tester and observe the temperature at which the thermostat starts to open. If a tester is not available, suspend the thermostat and thermometer in a pan of water on the stove and allow the water to heat rather slowly and observe the thermometer and thermostat. The thermometer reading **MUST** agree with the rating stamped on the thermostat. If the unit fails the test, it **MUST** be replaced.

INSTALLATION

3- Apply a coating of Perfect Seal, or equivalent, to both sides of a **NEW** gasket. Install the thermostat and gasket, as shown. Position the thermostat cover over the housing, and secure it in place with the attaching bolts. Tighten the cover bolts on in-line engines to a torque value of 30 ft lbs. Tighten the cover bolts on V8 engines to a torque value of 15 ft lbs. If a closed water system is used, fill the system with a solution of fresh water and rust inhibitor or antifreeze. See Section 9-9.

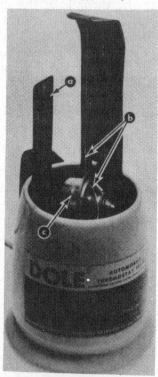

a- THERMOMETER
b- NYLON STRING
c- THERMOSTAT ②

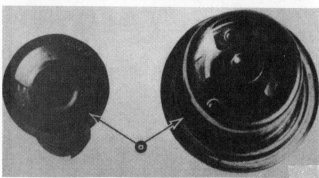

a- CHECK FOR LIGHT LEAKAGE
AROUND PERIMETER OF VALVE ①

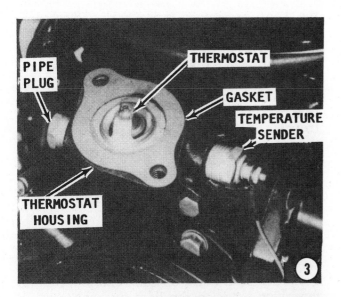

Thermostat location on a MerCruiser 470 engine with a closed water system.

Thermostat location on a 120, 140, and 165 engine with standard cooling system.

Thermostat location on a 120, 140, and 165 engine with a closed water system.

Thermostat location on a Ford V8 engine with closed water cooling.

Thermostat location on a Ford V8 engine with standard water cooling system.

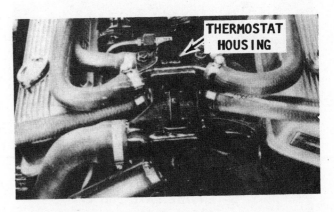

Thermostat location on a GMC V8 engine with standard water cooling system.

9-7 HEAT EXCHANGER SERVICE IN-LINE ENGINES

1- Drain the external and fresh water portions of the cooling system, see Section 9-4. Disconnect all hoses from the heat exchanger. Remove the two bolts from the front and the one bolt from the rear of the heat exchanger mount brackets, and then remove the heat exchanger.

Scribe a mark on the heat exchanger end plate and a matching mark on the the heat exchanger. Remove the bolt securing the end plate to the heat exchanger, and then remove the end plate. Remove the seal washer from the retainer bolt. Clean any deposits that may have formed from the water passages, in the heat exchanger, with a suitable wire brush. Remove the zinc electrode from the heat exchanger and inspect it for erosion. If erosion appears to be less than 25%, continued use of the zinc is permitted. If erosion appears to be more than 25%, the zinc **should** be replaced.

Hold the heat exchanger in the vertical position (upright) with the water passages toward the bottom. Use compressed air to blow all loose particles out of the water passages.

Apply a coating of Perfect Seal, or equivalent, to both sides of a **NEW** end plate gasket and position the gasket in place on the heat exchanger. Install a **NEW** seal washer on the end cap retainer bolt. Now, align the scribe marks on the end plate and

the heat exchanger made during disassembling. Install the heat exchanger onto the engine and secure it in place with the two bolts through the front bracket and the one bolt through the rear brace.

Fill the closed water system with a solution of water and rust inhibitor or antifreeze. See Section 9-9.

9-8 HEAT EXCHANGER SERVICE V8 ENGINES

1- Drain the external and fresh water sections of the cooling system, see Section 9-4. Scribe a mark on each end plate and a matching mark on the heat exchanger as an aid during installation. Remove each end plate retaining bolt, the seal washer from the bolt, and then the end plates. Clean the gasket material from the end plates and from the heat exchanger. Remove the zinc electrode from the heat exchanger and inspect it for erosion. If erosion appears to be less than 25%, continued use of the zinc is permitted. If erosion appears to be more than 25%, the zinc **should** be replaced. Clean all deposit that may have formed in the water passages with a suitable brush. Use compressed air to blow loose particles out of the water passages. Apply a coating of Perfect Seal, or equivalent, to both sides of **NEW** end plate gaskets. Position the gaskets in place on the heat exchanger. Install the end plates onto the heat exchanger with the marks made during disassembly aligned. Install **NEW** seal washers and end plate retaining bolts. Tighten the bolts securely. Fill the closed water system with a solution of water and rust inhibitor or antifreeze. See Section 9-9.

a- PRESSURE CAP
b- ENGINE WATER HOSE
c- DRAIN PLUG
d- OUTLET HOSE   ①

a- THERMOSTAT COVER
b- HEAT EXCHANGER   ①

9-9 FILLING THE COOLING SYSTEM

Check to be sure the petcocks are closed tightly and the the drain plugs are installed securely.

Remove the zinc electrode from the heat exchanger and inspect it for erosion. If erosion appears to be less than 25%, continued use of the zinc is permitted. If erosion appears to be more than 25%, the zinc **should** be replaced.

Engines NOT Exposed to Freezing Temperatures: Fill the closed cooling system with a mixture of clean, soft water and an automotive cooling system rust inhibitor. Take time to read the recommendations of the manufacturer and fill with the proper mixture of water and inhibitor.

Engines Exposed to Freezing Temperatures: Fill the closed cooling system with a mixture of clean, soft water and a permanent-type anti-freeze.Take time to read the manufacturer's recommendations, and then mix in the proper proportions to protect the engine to the lowest expected weather temperature. If soft water is not available in your area, hard water may be used. However, in many areas hard water contains a high mineral content which will cause scale and deposits to form rapidly in the cooling system.

Start the engine and allow it to run at idle speed for at least five mintues.

CAUTION

Water must circulate through the stern drive, also to and from the engine, anytime the engine is run to prevent damage to the water pump located in the lower unit (original and Alpha drives); or on the front of the engine (Bravo drive). Just a few seconds without water will damage the water pump.

Add a mixture of water and anti-freeze to the heat exchanger slowly while the engine is running to maintain the coolant level. The coolant should fill the heat exchanger 1" below the filler neck. Install the pressure cap. Allow the engine to continue running and check all hose connections, fittings, plugs, the drain petcock, and gaskets for leaks. Check the engine temperature gauge on the dash. If the temperature indication enters the red portion of the gauge, stop the engine and determine the cause, see Section 9-2.

Check the coolant level following the first run of the boat at full throttle.

If an anti-freeze solution is added to the coolant system, the solution should be removed if the boat is no longer to be operated in freezing temperatures. Fill the system following the instructions for operation in non-freezing temperatures. **REMEMBER,** an anti-freeze solution is not as effective a cooling agent as water and a rust inhibitor solution. The anti-freeze solution will cause the engine to operate at a higher temperature.

9-10 EXHAUST MANIFOLD MAINTENANCE

All MerCruiser cooling systems circulate external water from outside the boat to cool the engine directly, or indirectly through a heat exchanger, to cool the water which circulates through the cylinder head and engine block. In either case, after the water leaves the engine, or heat exchanger, it is routed through a hollow jacket surrounding the exhaust manifold. The water cools the exhaust manifold and then passes through the exhaust elbow and stern drive unit to the propeller, where it is discharged with the exhaust gases. This arrangement of exhaust gases and water is peculiar to marine engine installations and is known as "propeller exhaust".

Exhaust manifolds are subjected to considerable rusting and scale build-up due to the mineral content in the water in which the boat is operating, expecially in salt water. These deposits **MUST** be rodded out at regular intervals or the flow of coolant will be restricted and the engine will overheat. An engine operating continuously at higher than normal temperatures will develop serious and expensive mechanical damage. If the movement of exhaust gases through the exhaust manifold is restricted, the engine cannot "breathe" properly and the gases back up into the engine affecting performance. Deposit build up in the manifold is a gradual process over a period of time. Therefore, the problem is not noticeable until it becomes serious. For this reason exhaust manifold service **MUST** be performed on a regular basis. As a general "ball park" figure, service should be performed after 300 hours of operation, possibly sooner if the boat is operating in high mineral content waters such as salt water. If the water contains a heavy concentration of salt, such as the Salton Sea in California, or in areas with high marine plankton growth, the service should be performed every six months.

If the exhaust manifold rusts through, external water can enter the exhaust passageways and pass into the combustion chamber while the engine is shut down and the boat is in the water. An extremely small amount of water in the combustion chamber can cause severe metallic corrosion and contaminate the crankcase oil. The rusting process is a slow one over a long period of time. Therefore, the initial seepage from the exhaust manifold may not be noticed until serious corrosive damage has occurred. Checking the spark plugs, for any sign of rust, on a regular schedule is one way to detect such a condition in its early stages.

The operating habits of the skipper and the installation of the power plant are unique marine problems that affect the life of the exhaust manifold. In some boats the engine may be mounted too low and under certain operating conditions, external water can back-up into the exhaust manifolds. If the boat capacity is exceeded, the hull will ride too low in the water, allowing external water to enter the exhaust manifold. This may be possible when the boat is not underway. Two courses of action are available. One is to not overload the boat, and the other is to install an elevated elbow riser

kit, if the loading is marginal and a common practice. Such a riser kit will prevent water from entering the exhaust manifold when the boat is low in the water.

Another cause of water entering the exhaust manifold is a malfunction in the exhaust shutters. These shutters are installed in the exhaust pipe just below the exhaust rubber hose. If the exhaust shutters become burned, they cannot seal the flow of external water when the boat is operating in reverse. In some cases the shutters may fail to seal when the throttle is backed-off to idle speed while the boat is under way. The backwash can flood the exhaust manifold with external water, and the lowered exhaust pressure of an idling engine does not purge the water from the manifold. A "worse condition" is when the engine is shut down and the boat is riding in a heavy rolling sea. Under such a condition, it is possible for water to enter the exhaust manifolds.

A third cause of water entering the exhaust manifold is a misfiring spark plug. If one engine cylinder is misfiring or firing with a weak spark, a vacuum may exist during the valve overlap period. Such a vacuum can suck external water into the defective cylinder.

1 - Bolt (4)
2 - Washer (4)
3 - Screw
4 - Nut (2)
5 - Washer (2)
6 - Bolt (4)
7 - End Cap (Front)
8 - Gasket
9 - Exhaust Heat Tube
10 - Stud (2)
11 - Gasket
12 - Plug (4)
13 - Water Drain Plug
14 - Lockwasher (4)
15 - Nut (4)
16 - Exhaust Elbow
17 - Water Separation Tube
18 - Gasket
19 - Stud (4)
20 - Exhaust Manifold
21 - Gasket
22 - End Cap (rear)
23 - Bolt (4)
24 - Water Drain Plug
25 - Plug (2)
26 - Wiring Harness Clamp
27 - Lockwasher (1)
28 - Screw (1)
29 - Plug

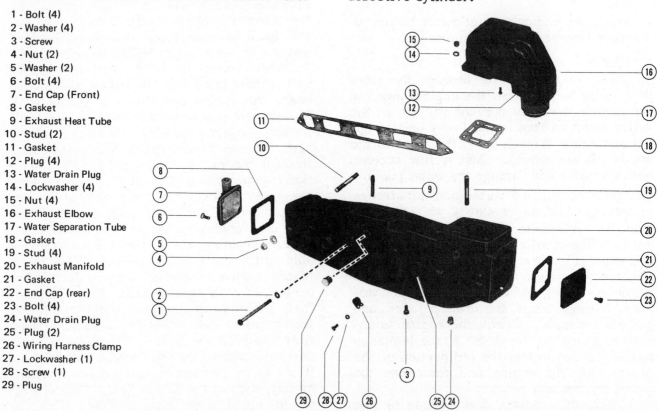

Arrangement of exhaust manifold and exhaust elbow parts on an in-line engine installation.

Another serious condition is the following set of circumstances in the order presented. The engine has been shut down for a short time while the boat is in the water. Water is in the exhaust manifold due to either defective exhaust shutters or overloading. The engine is started again. If it backfires and runs backwards for just one or two turns, each of the exhaust valves becomes an intake valve to suck in water from the flooded exhaust manifold. The created hydrostatic lock can destroy the pistons and rods.

9-11 EXHAUST MANIFOLD SERVICE GM IN-LINE ENGINES

The intake and exhaust manifolds on in-line engines are one unit.

Disconnect the battery cable at the battery. Drain the water from the engine block, exhaust manifold, exhaust elbow, and heat exchanger, if one is installed, see Section 9-4. Disconnect the throttle cable at the carburetor. Disconnect the exhaust hose and the cooling hose. Disconnect the fuel line at the carburetor and plug the line to prevent fuel from siphoning out of the fuel tank into the bilge. Disconnect the crankcase ventilation hose at the rocker arm cover. Remove the wiring harness clamps from the manifold. Disconnect the alternator bracket from the manifold. Remove the exhaust manifold fasteners. Remove the manifold unit and **DISCARD** the gaskets.

CLEANING AND INSPECTING

Carefully check the engine exhaust ports for any sign of rust or indication water has entered the ports from the exhaust manifold. If there is any evidence of porosity, or that water has passed between the water passages and the exhaust chambers, the manifold being inspected **MUST** be replaced. Water leaking from the water passages into the exhaust system could cause water to pass into the engine through the exhaust valves.

Check the exhaust elbow water passage. Blow air or force water through the passage to verify the passage is not clogged. Check the inside of the exhaust hose closely for any sign of burning. Any evidence of burning indicates a lack of water passing through the manifold or elbow.

The manifolds and elbows should be cleaned and rodded out every 300 to 400 hours of operation, or every 3-1/2 to 4 years, whichever occurs first.

Soak the manifold and elbow, **CAST IRON ONLY**, for about 1-1/2 hours in Muriatic Acid, obtainable from any swimming pool service outlet. **NEVER** use any kind of acid on **ALUMINUM** parts. After they have soaked, wash them thoroughly with water and then rod out the passages with a stiff rod to break loose any debris in the passage. Wash them again with water and use the rod a second time.

Clean the mating surfaces of the exhaust manifold, elbow, and the block.

INSTALLATION

Place a **NEW** gasket on the engine head, and then carefully move the manifold into position. Take care to ensure the gasket remains in the proper position. If in doubt as to the sealing ability of the gasket, apply a coating of Permatex "Form-A-Gasket" to the mating surfaces of the engine head and manifold in addition to the gasket.

Insert the elbow end into the exhaust hose, and then hold the manifold in place and install the fasteners. Tighten the fasteners evenly by starting in the center of the manifold and working toward both ends. The fasteners loosening from compression of the manifold gasket, will cause a noticeable vibration. Therefore, make sure they are tightened securely. Slide the exhaust hose clamp into place, and then tighten it securely.

Connect the crankcase ventilation hose at the rocker arm cover. Install the alternator bracket to the manifold. Install the wiring harness and clamps to the manifold.

If a new exhaust manifold is being installed, transfer the intake manifold and carburetor to the new exhaust manifold. Connect the throttle cable at the carburetor. Connect the fuel line and choke tube. Install the exhaust hose and the cooling hose.

Connect the battery cables at the battery. Close the water drain valves. Fill the closed water system with a solution of clean water and rust inhibitor or antifreeze.

Start the engine and check for leaks.

CAUTION

Water must circulate through the stern drive, also to and from the engine, anytime the engine is run to prevent damage to the

water pump located in the lower unit (original and Alpha drives); or on the front of the engine (Bravo drive). Just a few seconds without water will damage the water pump.

Adjust the carburetor idle speed and mixture, as required.

9-12 EXHAUST MANIFOLD SERVICE 470 ENGINE

REMOVAL

Drain the external water system and the closed-circuit cooling system, see Section 9-4. Disconnect the wire from the temperature sending unit. The unit is installed on the thermostat/distribution housing. Disconnect all hoses and the exhaust bellows from the manifold assembly. Drain the oil from the engine crankcase, and then remove the oil dipstick tube.

Support the manifold and remove the exhaust manifold fasteners. After all fasteners have been removed, lift the manifold assembly clear of the engine. DISCARD the gasket.

DISASSEMBLING

Remove the closed-circuit cooling reservoir from the thermostat/distribution housing. Remove the thermostat.

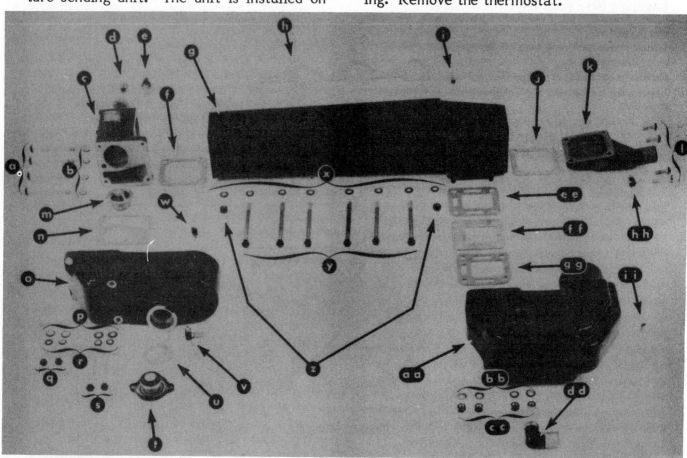

a- BOLT (4)	m- THERMOSTAT	y- HEAD BOLT (6)
b- WASHER (4)	n- GASKET	z- HEAD NUT (2)
c- THERMOSTAT HOUSING	o- RESERVOIR	aa- EXHAUST ELBOW
d- BRASS PLUG	p- WASHER (4)	bb- LOCKWASHER (4)
e- TEMPERATURE SENDER	q- SHORT BOLT (2)	cc- BODY NUT (4)
f- GASKET	r- LOCKWASHER (4)	dd- ELBOW
g- MANIFOLD BODY	s- LONG BOLT (2)	ee- GASKET
h- GASKET	t- RESERVOIR CAP	ff- PLATE
i- DRAIN PLUG	u- GASKET	gg- GASKET
j- GASKET	v- OVERFLOW FITTING	hh- FITTING
k- END CAP	w- FITTING	ii- DRAIN PLUG
l- BOLT (4)	x- LOCKWASHER (8)	

Arrangement of exhaust manifold and exhaust elbow parts on a 470 engine installation.

Remove the thermostat/distribution housing from the manifold body. Remove the temperature sender and the brass plug from the thermostat/distribution housing.

Remove the exhaust elbow from the manifold body. Remove the end cap from the aft end of the manifold body.

CLEANING AND INSPECTING

Carefully check the engine exhaust ports for any sign of rust or indication water has entered the ports from the exhaust manifold. If there is any evidence of porosity, or that water has passed between the water passages and the exhaust chambers, the manifold being inspected **MUST** be replaced. Water leaking from the water passages into the exhaust system could cause water to pass into the engine through the exhaust valves.

Check the exhaust elbow water passage. Blow air or force water through the passage to verify the passage is not clogged. Check the inside of the exhaust hose closely for any sign of burning. Any evidence of burning indicates a lack of water passing through the manifold or elbow.

Some models have a small 0.060" hole in the corner of the stainless steel plate between the gaskets at the manifold and elbow joint. Clean the hole -- vent hole -- to prevent steam pockets from forming.

The manifolds and elbows should be cleaned and rodded out every 300 to 400 hours of operation, or every 3-1/2 to 4 years, whichever occurs first.

Soak the manifold and elbow, **CAST IRON ONLY, NO ALUMINUM** for about 1-1/2 hours in Muriatic Acid, obtainable from any swimming pool service outlet. After they have soaked, wash them thoroughly with water and then rod out the passages with a stiff rod to break loose any debris in the passage. Wash them again with water and use the rod a second time.

Clean the mating surfaces of the exhaust manifold, elbow, and the block.

ASSEMBLING

Position a **NEW** gasket in place. Install the end cap on the aft end of the manifold body, with the hose connector facing toward the engine side of the manifold. Install a **NEW** gasket, plate, second **NEW** gasket, and exhaust elbow onto the manifold body, as shown. Install the lockwashers and nuts.

Coat the threads of the temperature sender and the brass plug with Perfect Seal, or equivalent, and then install these two items. Install a **NEW** gasket, the thermostat/distribution housing, and attaching bolts. Use copper washers on these bolts.

Insert the thermostat into its housing cavity. Position a **NEW** gasket in place, and then install the closed-circuit cooling system reservoir. Secure it in place with the attaching bolts, flat washers, and lockwashers.

INSTALLATION

Position a **NEW** manifold gasket in place, and then move the manifold assembly into place on the cylinder head. Secure the manifold with the bolts, washers, and nuts.

Install the dipstick tube to the oil pan and secure it in place with the support clamp on the exhaust manifold fastener bolt.

Connect all hoses and the exhaust bellows to the manifold assembly. Connect the engine harness wire to the temperature sender on the thermostat/distribution housing.

Fill the crankcase with the recommended amount and weight oil. Fill the closed-circuit cooling system to the proper fluid level. Obtain an automotive-type radiator cap tester and pressure-check the cooling system.

9-13 EXHAUST MANIFOLD SERVICE FORD V8 ENGINE

REMOVAL

Drain the water from the engine and manifolds, see Section 9-4. Disconnect the cooling hose from the manifold. Remove the dipstick bracket and fuel filter mount from the port manifold. Remove the solenoid bracket.

Disconnect the exhaust hoses. Drain the water from the manifold housing and elbow. Remove the attaching bolts and washers, and then the manifold.

CLEANING AND INSPECTING

Carefully check the engine exhaust ports for any sign of rust or indication water has entered the ports from the exhaust manifold. If there is any evidence of porosity, or that water has passed between the water passages and the exhaust chambers, the manifold being inspected **MUST** be replaced. Water leaking from the water passages into the exhaust system could cause water to pass into the engine through the exhaust valves.

Check the exhaust elbow water passage. Blow air or force water through the passage to verify the passage is not clogged. Check the inside of the exhaust hose closely for any sign of burning. Any evidence of burning indicates a lack of water passing through the manifold or elbow.

The manifolds and elbows should be cleaned and rodded out every 300 to 400 hours of operation, or every 3-1/2 to 4 years, whichever occurs first.

Soak the manifold and elbow, **CAST IRON ONLY, NO ALUMINUM** for about 1-1/2 hours hours in Muriatic Acid, obtainable from any swimming pool service outlet. After they have soaked, wash them thoroughly with water and then rod out the passages with a stiff rod to break loose any debris in the passage. Wash them again with water and use the rod a second time.

Clean the mating surfaces of the exhaust manifold, elbow, and the block.

INSTALLATION

Install the exhaust manifold to the cylinder head with a **NEW** gasket. Tighten the bolts to a torque value of 25 ft-lbs. On the starboard manifold, install the dipstick tube support bracket with the proper manifold bolt. Install the exhaust tubes and water hoses. On the starboard manifold, install the water-separating fuel filter and connect the fuel lines.

Start the engine and check for leaks.

CAUTION

Water must circulate through the stern drive, also to and from the engine, anytime the engine is run to prevent damage to the water pump located in the lower unit (original and Alpha drives); or on the front of the engine (Bravo drive). Just a few seconds without water will damage the water pump.

9-14 EXHAUST MANIFOLD SERVICE GM V8 ENGINE
REMOVAL

Port manifold: Drain the engine cooling system, see Section 9-4. Remove the exhaust tube and water hoses from the port manifold. Remove the bolts securing the manifold to the cylinder head.

1 - Screw
2 - Clip
3 - End Cap
4 - Gasket
5 - Water Drain Plug
6 - Screw (8)
7 - Washer (8)
8 - Plug (2)
9 - Exhaust Manifold
10 - Gasket
11 - End Cap
12 - Screw
13 - Screw (4)
14 - Lifting Eye
15 - Screw (2)
16 - Gasket
17 - Stud (4)
18 - Exhaust Elbow Plate
19 - Gasket (2)
20 - Water Drain Plug
21 - Exhaust Elbow
22 - Lockwasher (4)
23 - Nut (4)
24 - Fitting
25 - Plug
26 - Screw (4)
27 - Clamp (4)
28 - Exhaust Bellows
29 - Exhaust Elbow

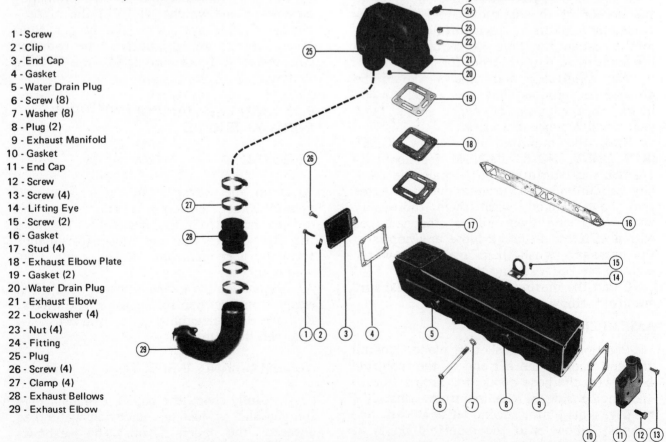

Arrangement of exhaust manifold and exhaust elbow parts, starboard side, on a Ford V8 engine installation.

Starboard manifold: Drain the engine cooling system, see Section 9-4. Remove the exhaust tube and water hoses from the manifold. Disconnect the fuel lines from the water-separating fuel filter, and then remove the fuel filter assembly. Remove the bolts securing the manifold to the cylinder head.

CLEANING AND INSPECTING

Carefully check the engine exhaust ports for any sign of rust or indication water has entered the ports from the exhaust manifold. If there is any evidence of porosity, or that water has passed between the water passages and the exhaust chambers, the manifold being inspected **MUST** be replaced. Water leaking from the water passages into the exhaust system could cause water to pass into the engine through the exhaust valves.

Check the exhaust elbow water passage. Blow air or force water through the passage to verify the passage is not clogged. Check the inside of the exhaust hose closely for any sign of burning. Any evidence of burning indicates a lack of water passing through the manifold or elbow.

The manifolds and elbows should be cleaned and rodded out every 300 to 400 hours of operation, or every 3-1/2 to 4 years, whichever occurs first.

Soak the manifold and elbow, **CAST IRON ONLY,** for about 1-1/2 hours in Muriatic Acid, obtainable from any swimming pool service outlet. **NEVER** use any kind of acid on **ALUMINUM** parts. After they have soaked, wash them thoroughly with water and then rod out the passages with a stiff rod to break loose any debris in the passage. Wash them again with water and use the rod a second time.

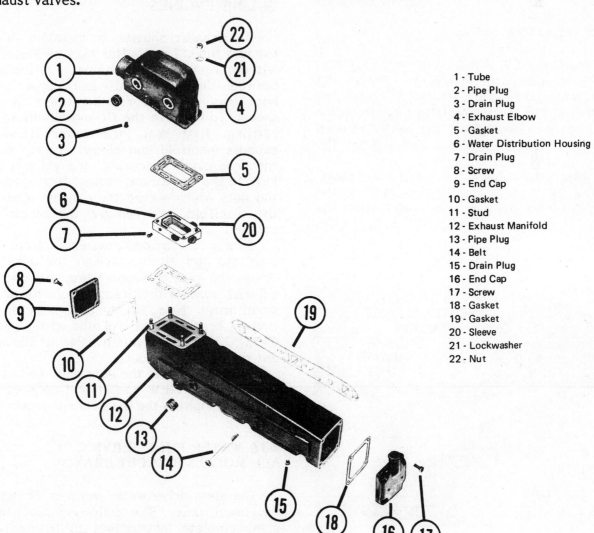

1 - Tube
2 - Pipe Plug
3 - Drain Plug
4 - Exhaust Elbow
5 - Gasket
6 - Water Distribution Housing
7 - Drain Plug
8 - Screw
9 - End Cap
10 - Gasket
11 - Stud
12 - Exhaust Manifold
13 - Pipe Plug
14 - Belt
15 - Drain Plug
16 - End Cap
17 - Screw
18 - Gasket
19 - Gasket
20 - Sleeve
21 - Lockwasher
22 - Nut

Arrangement of exhaust manifold and exhaust elbow parts, starboard side on a GMC V8 engine installation.

Check the exhaust elbow water passage. Blow air or force water through the passage to verify the passage is not clogged. Check the inside of the exhaust hose closely for any sign of burning. Any evidence of burning indicates a lack of water passing through the manifold or elbow.

The manifolds and elbows should be cleaned and rodded out every 300 to 400 hours of operation, or every 3-1/2 to 4 years, whichever occurs first.

Soak the manifold and elbow, **CAST IRON ONLY, NO ALUMINUM,** for about 1-1/2 hours in Muriatic Acid, obtainable from any swimming pool service outlet. After they have soaked, wash them thoroughly with water and then rod out the passages with a stiff rod to break loose any debris in the passage. Wash them again with water and use the rod a second time.

Clean the mating surfaces of the exhaust manifold, elbow, and the block.

INSTALLATION

Position **NEW** manifold gaskets in place, and then install the manifolds. Secure the manifolds with the attaching bolts and washers. Tighten the bolts to a torque value of 25 ft lbs. Work from the center of each manifold toward the ends and tighten the bolts evenly.

Install the exhaust hoses and drain plugs. Install the dipstick housing bracket and the fuel filter on the port manifold. Install the solenoid bracket. Connect the cooling inlet hose.

Start the engine and check for leaks.

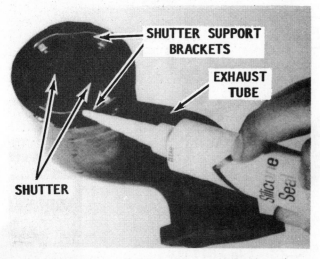

Applying Form-A-Gasket, or an equivalent silicone sealant, to the shutter support brackets.

CAUTION

Water must circulate through the stern drive, also to and from the engine, anytime the engine is run to prevent damage to the water pump located in the lower unit (original and Alpha drives); or on the front of the engine (Bravo drive). Just a few seconds without water will damage the water pump.

9-15 EXHAUST SHUTTER SERVICE

V8 ENGINES

The exhaust shutters on stern drive units with **V8** engines are installed in the exhaust separator. The engine must be removed to gain access to the shutters. See Chapter 13, Engine Removal/Installation. See Chapter 12 Gimbal Housing/Transom Plate to service the shutters.

IN-LINE ENGINES

An exhaust shutter is installed in the exhaust tube. The shutter may be inspected without removing the engine. A loose fit between the shutter shaft and support brackets may cause a rattling noise at idle speed. To improve the fit and eliminate the rattling, first drain the water from the exhaust manifold and elbow. Next, loosen the hose clamps securing the exhaust bellows to the the exhaust tube. Next, remove the nuts which secure the exhaust elbow to the manifold, and remove the elbow and bellows.

Scrape all gasket material from the manifold and elbow. Clean the area between the shutter support brackets and the exhaust tube with cleaning solvent and a small brush. Now, fill the area between the support brackets and the tube with Sealant, C-92-52238, or equivalent. Permit the sealant to cure for 24 hours.

Alright, now its the next day. Install the elbow with a **NEW** gasket and connect the bellows. Tighten the hose clamps securely.

9-16 WATER PUMP SERVICE ALL MODELS (EXCEPT BRAVO)

The stern drive water pump is located in the lower unit. The following procedures give complete instructions to remove and install the water pump. Additional procedures are given in Chapter 10.

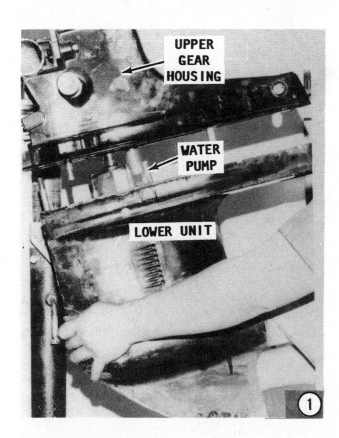

Worn blades on a water pump impeller. A worn impeller MUST be replaced to obtain satisfactory service and adequate cooling water to the upper gear housing and the engine.

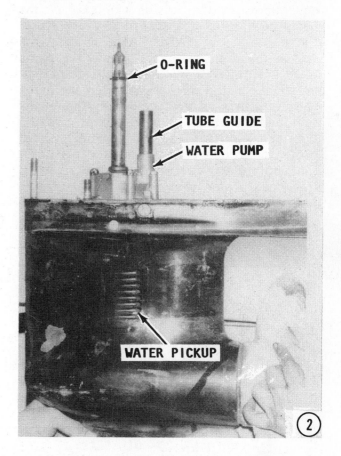

Lower Unit Removal

1- The lower unit **MUST** be removed in order to service the water pump. This unit may be removed without disturbing the upper gear housing. Assuming only the lower unit is to be removed, first, shift the unit into **FORWARD** gear by moving the shift lever at the helm to the full **FORWARD** position and at the same time, have an assistant turn the propeller **COUNTER-CLOCKWISE** until a definite "click" sound is heard, indicating the unit has slipped all the way into forward gear. (See Section 10-6, Step 1, for detailed instructions to remove the propeller, if desired.) Mark the position of the trim tab as an aid to installing it back in the same position. Remove the trim tab. Remove the 5/8" hex nuts located in the center bottom side of the anti-cavitation plate. Remove the 5/8" hex nut from the leading edge of the driveshaft housing. Remove the remaining 5/8" hex nuts from both sides of the housing. Remove the Allen screw from the cavity for the trim tab. Separate the lower unit from the upper gear housing.

2- Remove the water pump guide and the O-ring seal on the upper end of the driveshaft. This O-ring is used to keep exhaust gases and water from washing the lubricant off the driveshaft splines. Re-
move the one screw and three nuts securing the water pump body in place. Now, lift off the pump housing, impeller, drive pin, gasket, base plate, pump base, and gasket. On early pumps, the flush screw **MUST** be removed before the pump base can be removed. This screw is located on the port side close to the oil vent screw.

CLEANING AND INSPECTING

Wash all parts in solvent, and then blow them dry with compressed air. Remove all traces of the old seal and gasket from all mating surfaces. Blow all water passageways and screw holes clean with compressed air.

Water Pump Installation

3- Place a **NEW** water pump base oil seal in position, with the lip of the seal facing toward the impeller side of the base. Install a **NEW** larger oil seal with the lip facing the gear housing. Install a **NEW** O-ring seal on the water pump base. Press a **NEW** oil seal into the pump body, with the lip facing upward. **NOTE:** Late model drive units do not have this oil seal in the pump body. Apply a coating of Perfect Seal, or equivalent, to the outside diameter of the pump housing insert. Install the insert into the pump body. **TAKE CARE** to align the tab on the insert with the hole in the pump body. **REMOVE** any excess Perfect Seal material. Install a **NEW** gasket on the pump base, and then lubricate the base, the **O**-ring seal, and the oil seal lips with Multi-purpose Lubricant, or equivalent. Install a seal protector to cover the splines on the drive-shaft to prevent damaging the new oil seals. If a seal protector is not available, wrap tape around the driveshaft splines. After

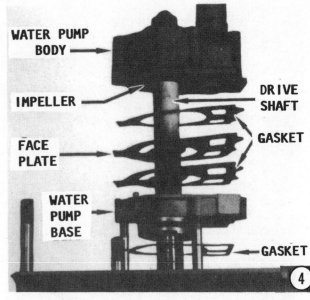

the driveshaft splines have been covered, install the base assembly.

4- This step calls for the installation of two gaskets, an upper one and a lower one. Take time to identify each gasket and then be sure to install them in the proper position. Place a **NEW** lower gasket in position. Install the face plate with the lip facing down. Install a **NEW** upper gasket. Install the drive key in the slot in the driveshaft.

13- DRIVE SHAFT PIN
14- COMPRESSION SPRING
15- UPPER O-RING
20- GASKET
21- O-RING
22- WATER PUMP BASE ASSY
23- OIL SEAL
24- OIL SEAL
25- DOWEL PIN
26- LOWER GASKET
27- FACE PLATE
28- UPPER GASKET
29- WATER PUMP BODY ASSY
30- INSERT
31- OIL SEAL
32- RUBBER SEAL
33- WATER PUMP IMPELLER
34- IMPELLER DRIVE PIN
35- SCREW
36- LOCKWASHER
37- NUT
38- NUT
39- WASHER
40- RUBBER RING
41- GUIDE SLEEVE

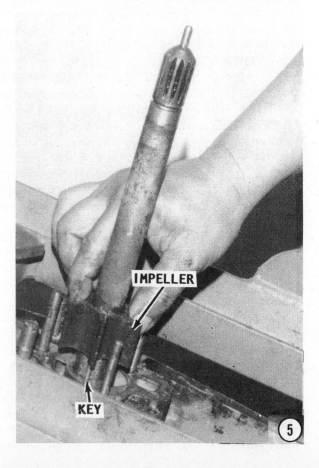

Use Multi-purpose Lubricant, or equivalent, to hold the key in place until the impeller is installed.

5- Slide the impeller down the drive-shaft with the key slot aligned with the key on the driveshaft. Push down on the impeller to seat it securely on the face plate.

6- Before installing the water pump body: Lubricate the oil seal with Multi-purpose Lubricant, or equivalent. Cover the pump insert with a mixture of soap and water and at the same time, place the pump body over the impeller. Push the body down and at the same time, turn the driveshaft clockwise to assist the impeller to enter the pump body without cutting it. Install the washers, screw, and nuts. Tighten the nuts to the torque value given in the accompanying illustration. **DO NOT** overtighten the nuts because the added strain could cause the pump to fail. Now, install the centrifugal slinger and a **NEW** driveshaft O-ring seal. Install the water inlet tube guide. Install the shift shaft seal and flat washer.

ALRIGHT, the lower unit is now assembled and ready to be installed. See the next section to properly install the lower unit.

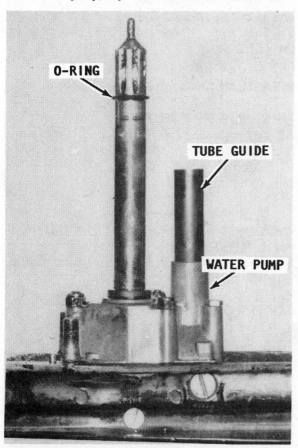

Assembled water pump. The lower unit is now ready for installation to the upper gear housing.

Lower Unit Installation

This section provides instructions only to install the lower unit. For detailed service procedures, see Chapter 10, Stern Drive.

1- Check to be sure the remote control shift handle at the helmsman's position is still in **FORWARD** gear. If the shift is not

in the forward gear position, move the handle to the forward position. Place the lower unit in forward gear. If difficulty is encountered, see Section 10-25, Step 4. Check to be sure the trim tab bolt is in place in the aft section of the upper gear housing. Lift the lower unit upward into place against the upper gear housing, and at the same time **CHECK** three important areas: The water pump tube **MUST** start into the water pump guide tube properly; the male splines of the driveshaft **MUST** index with the female splines of the upper gear housing; and the male and female splines on the shift **MUST** index with each other.

2- Start the two 5/8" hex nuts on both sides of the upper gear housing, but **DO NOT** tighten them at this time. Install the 5/8" hex nut from the leading edge of the housing.

Install the hex head nuts into the bottom of the anti-cavitation plate. Install the allen-head screw into the cavity for the trim tab. Now, tighten all the 5/8" nuts on both sides and inside the trim cavity of the housing a little-at-a-time until the lower unit is snug against the upper gear housing. Clean the inside of the tab cavity to ensure a good ground to the stern drive. **NEVER** paint the tab or the cavity. Install the trim tab, with the marks made during removal aligned. Use an Allen wrench to tighten the Allen screws. Install the trim tab plastic cover plug.

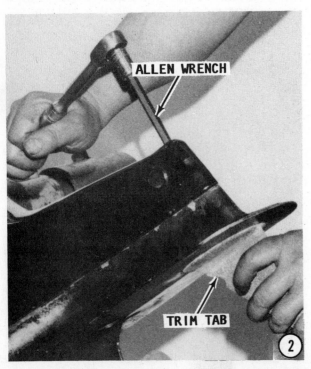

Start the engine and check the shift mechanism.

CAUTION

Water must circulate through the stern drive, also to and from the engine, anytime the engine is run to prevent damage to the water pump located in the lower unit (original and Alpha drives); or on the front of the engine (Bravo drive). Just a few seconds without water will damage the water pump.

9-17 WATER PUMP SERVICE BRAVO STERN DRIVE

INTRODUCTION

Powerplants mated with a Bravo stern drive are equipped with a belt driven water pump mounted **ON** the forward end of the engine. This location results in much easier service work than the involved procedures for servicing a water pump installed inside the lower unit.

This water pump is equipped with a pulley and is driven by the engine drive belt. The lower hose on the pump is the **INLET** and is connected to the water pickup.

The upper hose on the pump is the **OUTLET** hose.

DISASSEMBLING

FIRST, THESE WORDS
The drawing on Page 9-28 showing all parts of the water pump will be most helpful in performing the following steps to service the water pump.

1- If the engine being serviced is installed in the boat in the water, the water inlet valve **MUST** be closed before the inlet hose is removed to prevent water from draining into the boat.

GOOD WORDS
On some powerplants equipped with this type water pump, it is possible to remove the water pump without removing the mounting bracket. If this is the case, simply disregard the mounting bracket removal and installation instructions.

MORE GOOD WORDS
If the engine being serviced is equipped with power steering, the power steering

drive belt **MUST** be removed before the water pump drive belt can be removed.

2- Loosen the screws on both hose clamps on the inlet and outlet hoses. Slide both hose clamps a few inches down the hoses, and then lightly tighten each clamp screw to prevent the clamps from coming free of the hoses. Pry the inlet and outlet hoses free of the pump.

3- If the pump is to be serviced, "break" loose the four front bolts attaching the pulley to the pump. **DO NOT** remove the bolts at this time. If the water pump is being removed because of scheduled work on the engine, leave these four bolts undisturbed. Loosen the pivot bolt on the mounting bracket and swing the pump toward the engine to relieve belt tension. Slide the belt free of the pulley.

4- Remove the four pulley attaching screws and the clamping ring (some models only). Lift the pulley from the pump studs.

5- Remove the bolt securing the pump mounting bracket to the engine, and then lift the bracket and pump free as an assembly. On some models, it may not be necessary to remove both the mounting bracket and the pump. To remove the pump from the mounting bracket, loosen, but **DO NOT** remove, the clamping bolt. Spread the

clamping boss apart to allow the pump shaft to slide free.

6- Remove the five long screws and washers securing the cover to the pump body. Remove the cover, gasket, outer wear plate, and second gasket from the pump. Discard both gaskets. Slide the pump body from the pump shaft. The small rubber plug and impeller will remain in the pump body. Remove the rubber plug from the center of the impeller, and then push the impeller from the pump body. Remove the Woodruff key from the slot in the pump shaft.

7- Remove the gasket, inner wear plate, and O-ring from the housing. Discard the gasket and O-ring.

8- If the front oil seal or either ball bearing needs to be replaced, the pulley hub must first be pressed from the pump shaft. Obtain a universal bearing separator tool. Position the bearing separator under the hub. Rest the separator tool on an arbor press and apply pressure to the pump shaft to press it down and free of the hub.

9- "Puncture" the upper oil seal with an awl, and then pry the seal up from the housing.

10- Obtain a pair of snap ring pliers and remove the snap ring retaining the ball bearing in place. The shaft, two ball bearings and the two aft oil seals are removed as an assembly from the housing. This assembly is a "slide fit" inside the housing. However, on an older unit, it may be necessary to press this assembly free of the housing using an arbor press. Apply force on the aft end of the shaft to push the assembly out the front (pulley) end of the housing.

11- Use the same setup in the arbor press and press the two ball bearings and the two oil seals free of the shaft.

CLEANING AND INSPECTING

Wash all metal water pump parts in solvent, and then dry them with compressed air.

Clean all traces of sealer and gasket material from all sealing surfaces.

Inspect the inside surface of the housing where the outer races of both ball bearings make contact with the housing. Look for evidence of the races spinning within the housing. Inspect the surface of the pump shaft where the inner ball bearing races make contact with the pump shaft. Check

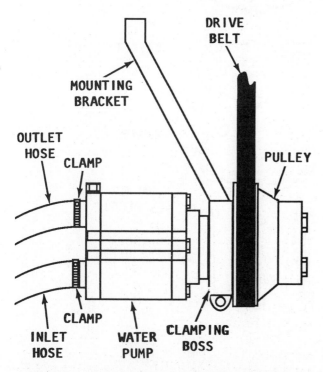

Belt driven water pump installed on an engine mated with a Bravo stern drive.

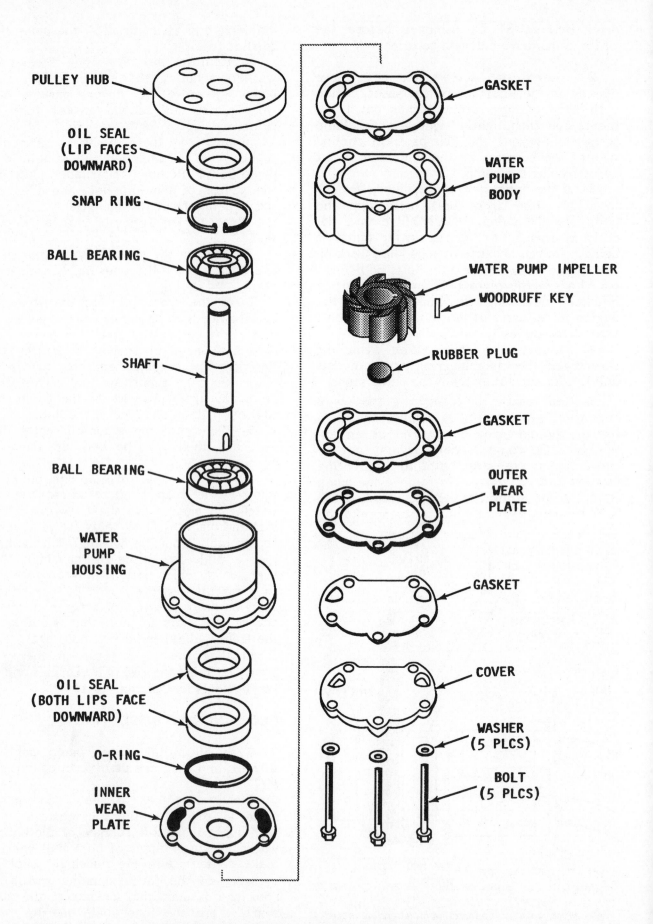

PULLEY HUB

OIL SEAL
(LIP FACES
DOWNWARD)

SNAP RING

BALL BEARING

SHAFT

BALL BEARING

WATER
PUMP
HOUSING

OIL SEAL
(BOTH LIPS FACE
DOWNWARD)

O-RING

INNER
WEAR
PLATE

GASKET

WATER
PUMP
BODY

WATER PUMP IMPELLER

WOODRUFF KEY

RUBBER PLUG

GASKET

OUTER
WEAR
PLATE

GASKET

COVER

WASHER
(5 PLCS)

BOLT
(5 PLCS)

Exploded drawing of the Bravo engine mounted water pump.

for evidence of the races spinning on the shaft.

Inspect the cover and both wear plates for distortion, rough surfaces, and cracks, possibly caused by overheating.

If possible, **ALWAYS** install a new water pump impeller. A new impeller will ensure extended satisfactory service and give "peace of mind" to the owner. Inspect the blades for cracks in the area where they flex. If the impeller blades have taken a set, remain in a curved position, the impeller **MUST** be replaced. If the old impeller must be returned to service, **NEVER** install it in reverse to the original direction of rotation. Installation in reverse will most surely cause premature impeller failure.

ASSEMBLING

1- Obtain Shell Alvanina No. 2 grease or equivalent lubricant suitable for continuous high speed heavy duty operation. The manufacturer states 2-4-C Marine Lubricant may be substituted in an emergency, but its use is not recommended. Pack the lips of all three oil seals with the No. 2 grease.

2- The manufacturer orders a light coating of Loctite Type "A" be applied to the inside diameter of the two rear oil seals. Obtain a suitable driver and press both oil seals into the aft end of the housing. **BOTH** oil seal lips face aft -- toward the impeller. These oil seals are **NOT** installed back-to-back, as in other cases with double seals. Press the first oil seal in until the seal seats within the housing. Press the second seal in until it is flush with the housing.

3- Obtain a suitable driver which will contact **ONLY** the ball bearing inner race. Using an arbor press and the driver, press the two ball bearings, one at each end of the shaft, until they seat against their respective shoulders on the shaft. Pack both bearings with the No. 2 grease. Slide the assembled shaft into the housing with the slotted end of the shaft (slotted for the Woodruff key), at the flanged end of the housing. Using the ring pliers, install the snap ring into the groove at the pulley end of the housing to secure the shaft assembly within the housing.

4- Pack the oil seal lip with the No. 2 grease and apply a light coating of Loctite Type "A" to the outer circumference. Press in the front oil seal over the installed snap ring with the oil seal lip facing downward -- toward the housing.

5- Obtain Quicksilver Special Lubricant 101 (P/N 92-79214A1), or equivalent heavy duty water resistant lubricant. Apply a light coating of the lubricant to the forward end of the shaft. Place the shaft in an arbor press with the shaft standing on end (the aft end), and install the hub onto the forward end. **DO NOT** rest the shaft on the installed ball bearings during this installation procedure. Such action will place an excessive load on the bearings. Press the hub squarely into place until 1/4" (6mm) of the shaft extends beyond the top surface of the hub. This dimension **MUST** be observed to establish correct drive belt alignment.

6- Clamp the pump housing in a vise equipped with soft jaws with the flanged (aft) end facing upward. Coat a new **O**-ring with 2-4-C Marine Lubricant, or equivalent, and install the **O**-ring into the housing groove. Position the inner wear plate onto the housing with the holes in the plate aligned with the holes in the housing. Apply a light coating of Perfect Seal to both sides of a **NEW** gasket and place the gasket over the wear plate with all holes aligned.

7- Install the impeller into the rear of the pump body. Rotate the impeller **COUNTERCLOCKWISE** and at the same time, push it into place. All impeller blades must face the same direction when installed.

8- Apply a dab of grease to the slot in the shaft to help hold the Woodruff key in place during assembling. "Stick" the key to the shaft. Slide the impeller and pump body

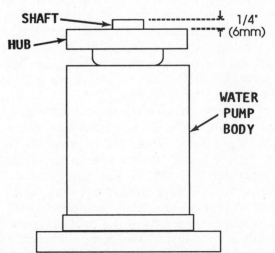

The shaft must be pressed through the hub and extend, as shown, in order to ensure correct drive belt alignment.

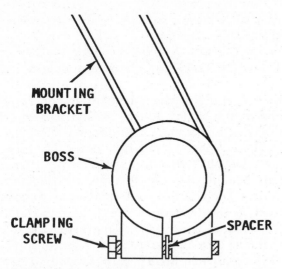

A spacer, between the two mounting bracket clamping bosses, prevents damage to the pump if the clamping screw is overtightened.

over the shaft with the Woodruff key index-ing into the impeller groove. Align the holes in the pump body with the holes in the installed gasket, inner wear plate and hous-ing. Install the small rubber plug into the end of the impeller.

9- Apply a light coating of Perfect Seal to both sides of the two remaining gaskets. Refer to the drawing on Page 9-28 for correct positioning of the two gaskets. In-stall the outer wear plate, with the new gaskets, one on each side, onto the water pump body. Align all holes. Finally, posi-tion the back cover over the water pump and secure the cover with the five washers and long bolts. Tighten the bolts alternately and evenly to a torque value of 10 ft lb (14Nm).

10- Slide the hub through the mounting bracket boss. Rotate the water pump within the boss until the brass plug on the outlet fitting is uppermost. This will position the outlet fitting at the top of the pump and the inlet fitting at the bottom of the pump. Incorrect positioning of the pump will lead to overheating problems.

Check to be sure the spacer between the two mounting bracket clamping bosses is still in position. This spacer prevents dam-age to the pump if the clamping screw is overtightened. Install the pulley over the hub with the four bolts and lockwashers. Some models are also equipped with a clamping ring. Tighten the bolts fingertight at this time. Install the pump and mounting bracket to the engine with the attaching hardware. Tighten these bolts also just fingertight, at this time.

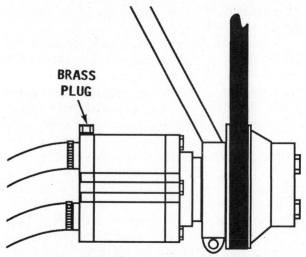

The brass plug MUST be positioned uppermost when installing the water pump for a Bravo stern drive.

Check to be sure the installed pulley is optically aligned with the crankshaft pulley.

Check to be sure the brass plug is still positioned at the top of the installed water pump. Make any necessary adjustments before tightening the bolt at the base of the boss to a torque value of 30 ft lb (41Nm).

Drive Belt Installation

11- Install the drive belt around the crankshaft pulley and the water pump pul-ley. Pivot the water pump away from the engine until the correct belt tension is ob-tained. A deflection of about 1/4" (6mm) should be obtained at a point midway be-tween the two pulleys.

12- Tighten the mounting bracket secur-ing the bolt to a torque value of 30 ft lb (41Nm). Install the inlet hose to the lower fitting on the back of the water pump. Install the outlet hose to the upper fitting on the back of the water pump. Tighten both hose clamps securely. On models equipped with power steering, install and tighten the power steering belt.

CAUTION

Water must circulate through the stern drive, also to and from the engine, anytime the engine is run to prevent damage to the water pump located in the lower unit (origi-nal and Alpha drives); or on the front of the engine (Bravo drive). Just a few seconds without water will damage the water pump.

13- Start the engine and check the com-pleted work. Check for leaks. Run the engine for about five minutes and then shut it down. Recheck the tension on the drive belt/s and make any necessary adjustments.

10
STERN DRIVE

10-1 INTRODUCTION

Early Models

Early model stern drive units include the 1, 1A, 1B, and the 1C.

Later Models

Later model Mercruiser stern drive units are classified as Type I units and include the 120, 140, 160, 165, 470, 225S, 228, 233, 250, 260, 888, and the 898. The model number of each unit is displayed prominently by decal on the upper gear housing.

"MR" AND Alpha Units

In early 1985, a new stern drive unit was introduced and identified to the trade as an "MR" unit. In all sales literature, the unit was named "Alpha". **THEREFORE**, any and all references in this manual to the "MR" unit are also valid for the "Alpha" drive. These units are identified by a decal affixed to the stern drive unit. Serial numbers and gear ratios are listed on Page 14-3.

When procedures differ between early model, late model or the "MR" ("Alpha") model, these differences will be noted in the text and supported by separate illustrations.

Bravo Units

An advanced engineered stern drive identified as Bravo "One" was introduced at the 1988 IMTEC Boat Show in Chicago. A huskier version of the Bravo One was made available the following year, identified as Bravo "Two". A complete separate section, 10-29, beginning on Page 10-59 presents procedures to service the Bravo units.

Coverage — All Except Bravo

The detailed procedures in this chapter contain complete instructions for trouble-shooting, disassembling, cleaning and inspecting, and assembling these units.

The service procedures give complete illustrated instructions for the entire stern drive unit. The information is divided into these broad areas:

Operation

Troubleshooting

Precautions

Stern drive removal

Upper gear housing service

U-joint, drive gear & bearing service

Lower unit, including water pump service

Assembling the various units

Stern drive installation

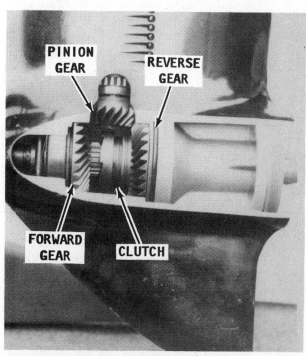

Cutaway view of a Type I, MR, or Alpha lower unit showing the gear arrangement. The clutch dog is engaged with the reverse gear.

10-2 OPERATION

The upper gear housing transfers engine power to the lower unit through a driveshaft. Splines on the forward end of the drive shaft mate with matching splines of a coupler attached to the engine flywheel. The aft end of the driveshaft is connected to a universal joint and the upper gear assembly. The pinion gear and drive gear changes the direction of the power train from horizontal to vertical. The power is transferred from the upper gear housing down to the lower unit through a vertical driveshaft. A water pump installed in the stern drive is constantly driven by the vertical driveshaft. This pump supplies cooling water directly to the engine, or indirectly through a heat exchanger.

CAUTION

Water must circulate through the stern drive, also to and from the engine, anytime the engine is run to prevent damage to the water pump located in the lower unit (original and Alpha drives); or on the front of the engine (Bravo drive). Just a few seconds without water will damage the water pump.

The forward and reverse gears are contained within the lower unit. A sliding clutch, actuated by the shift linkage, engages a forward gear and creates a direct coupling for transmitting power through the pinion gear and forward gear to the propeller shaft. The reverse gear utilizes the same mechanics as the forward gear, except that the clutch is coupled with the reverse gear.

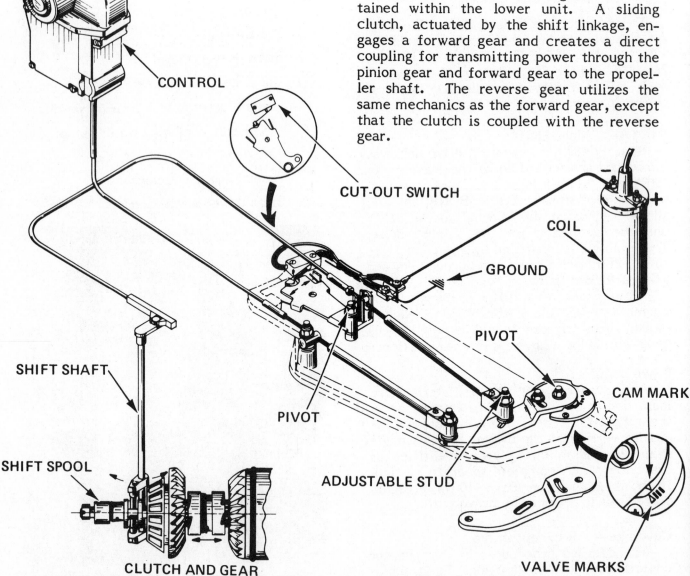

CONTROL

CUT-OUT SWITCH

COIL

GROUND

PIVOT

SHIFT SHAFT

CAM MARK

PIVOT

SHIFT SPOOL

ADJUSTABLE STUD

CLUTCH AND GEAR

VALVE MARKS

Functional diagram of an EZ-shift system showing all major parts. Such a drawing is useful in understanding the working relationship of the system parts and their operation.

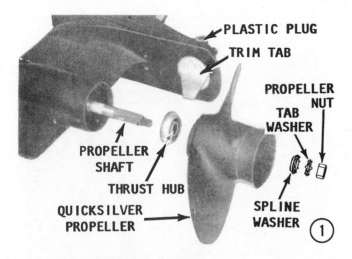

PLASTIC PLUG

TRIM TAB

PROPELLER NUT

PROPELLER TAB WASHER

PROPELLER SHAFT

THRUST HUB

QUICKSILVER PROPELLER

SPLINE WASHER

①

10-3 TROUBLESHOOTING

Troubleshooting **MUST** be done before the unit is removed from the boat to permit isolating the problem to one area.

1- Check the propeller and the rubber hub. See if the hub is shredded. If the propeller has been subjected to many strikes with underwater objects, it could slip on its hub. If the hub appears to be damaged, replace it with a **NEW** hub. Replacement of the hub must be done by a propeller rebuilding shop equipped with the proper tools and experience for such work.

2- **Shift mechanism check:** Verify that the ignition switch is **OFF**, to prevent possible personal injury, should the engine start. Shift the unit into **REVERSE** gear and at the same time have an assistant turn the propeller shaft to ensure the clutch is fully engaged. If the shift handle is hard to move, the trouble may be in the stern drive unit, transom shift cable, remote control cable, or the shift box.

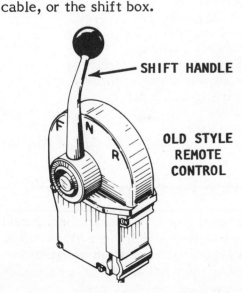

SHIFT HANDLE

OLD STYLE REMOTE CONTROL

②

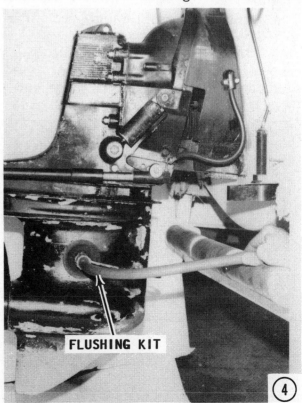

NUT AND WASHER

END GUIDE

COTTER PIN

JAM NUT

SET SCREW

CABLE END

③

3- **Isolate the problem:** Disconnect the remote-control cable at the transom plate, by first removing the two nuts, and then lifting off the remote-control shift cable. Operate the shift lever. If shifting is still hard, the problem is in the shift cable or control box. If the shifting feels normal with the remote-control cable disconnected, the problem must be in the stern drive. To verify the problem is in the stern drive unit, have an assistant turn the propeller and at the same time move the shift cable between the transom plate and stern drive back-and-forth. Determine if the clutch engages properly. Most of hard shifting problems are caused because this cable is not moved during the off-season. Water entering the cable, especially salt water, will cause rapid corrosion and hard shifting. If the cable

FLUSHING KIT

④

requires replacement, see Chapter 11. If the cable moves freely, then reconnect the control cables at the transom plate.

4- Stern drive noise check: FIRST, a word about stern drive noise: When the stern drive is positioned for dead-ahead operation, the U-joints will make very little noise. However, as the stern drive is moved to port or starboard, the noise will increase, due to the working of the U-joints. You will be accustomed to the normal noise of the U-joints. This test and procedure is for abnormal noises.

Attach a Flush-Test device to the stern drive and turn on the water. Start the engine and shift into gear.

CAUTION

Water must circulate through the stern drive, also to and from the engine, anytime the engine is run to prevent damage to the water pump located in the lower unit (original and Alpha drives); or on the front of the engine (Bravo drive). Just a few seconds without water will damage the water pump.

5- Run the engine at idle speed. Turn the stern drive slightly to port and then slightly to starboard, and at the same time listen at the upper driveshaft housing. An unusual noise is an indication the U-joints are worn or the gimbal housing bearing is defective. If the bearing is defective, see Chapter 11.

Stop the engine. Turn off the water. Disconnect the Flush-Test device.

6- Trim/Tilt system check: Operate the control buttons on the dashboard, or on the shift control lever. Check to determine if the stern drive unit moves to the full up and

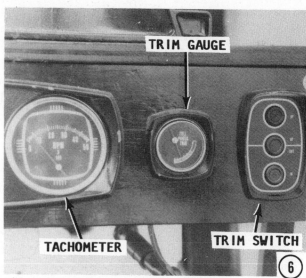

and full down positions. If the drive unit fails to move, the U-joints may be frozen due to corrosion from water entering through the U-joint bellows; lack of lubrication; or from non-operation over a long period of time. If the stern drive does not move properly, the trim/tilt mechanism may need service, see Chapter 8.

TROUBLESHOOTING
STERN DRIVE NOISES

Noise in lower unit
1- Propeller not installed properly
2- Propeller shaft bent
3- Worn, damaged, or loose parts
4- Metal particles in lubricant
5- Gear alignment incorrect
6- Incorrect shimming

Noise in upper gear housing
1- Oil level low
2- Worn U-joints
3- Worn bearings
4- Incorrect shimming
5- Gear alignment incorrect
6- Engine not aligned properly
7- Worn engine coupler
8- Transom too light for stern drive installation
9- Worn, damaged, or loose internal parts

10-4 SERVICE PRECAUTIONS

The following safety precautions should be observed while servicing the stern drive. The list is presented here as an assist to you in preventing personal injury or exspensive damage to parts.

1- Whenever possible, use special tools when they are specified. Always use the proper size common tool when removing or installing attaching parts.

2- Threaded parts are right-hand, unless otherwise indicated.

3- Clean parts in regular cleaning solvent and blow them dry with compressed air. Disassembly and assembly should be done on a clean work bench to prevent contamination (foreign matter, dirt) from entering the hydraulic system, or from becoming attached to parts. Such material can result in false shimming indications and will lead to premature bearing failure.

4- If bearings are removed from assemblies, keep them **TOGETHER**, with the original shims as an aid during asembling. If new parts are used with the assembly, the unit will have to be reshimmed in order to obtain the proper bearing preload or gear lash.

5- Work slowly when removing shims and take time to note the number and location of the shims. Follow shimming instructions to the letter. Gears must be mounted to the correct feeler gauge reading and with the correct gear backlash to avoid noisy performance and possibly early failure. Check the specifications for the proper preload. This is the **ONLY** way satisfactory service may be obtained from the unit after the work has been completed. There are **NO** short cuts.

6- Always keep bearing assemblies and spacers together and note the order of removal as an aid during installation.

7- The proper mandrels and supports **MUST** be used to remove pressed-on parts, such as bearings and gears. Therefore, do not use force on the bearing cages, mating surfaces, or edges which could be damaged or cause failure.

8- Lubricate parts with oil or grease to ease installation, retard rust, and prevent corrosion. One exception is the installation of oil seals. These seals must be coated with Loctite "A" to prevent lubrication seepage and to hold the seal in position.

9- When it is necessary to hold a part or housing in a vise, **ALWAYS** use soft brass jaws or wooden blocks to prevent marring or damaging the part.

10-5 PROPELLER REMOVAL

1- Bend the locking tabs forward out of the locking washer. **NEVER** pry on the edge of the propeller. Any small distortion will affect propeller performance. Place a block of wood between one blade of the propeller and the anti-cavitation plate to keep the shaft from turning. Use a socket and breaker bar and loosen the retaining nut. Remove the nut, tab washer, splined washer, and then the propeller.

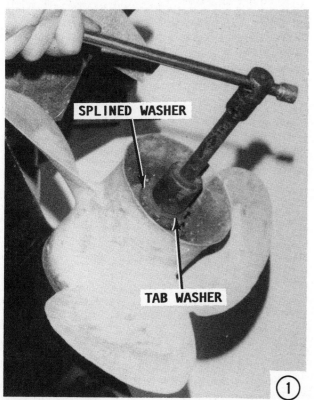

SPLINED WASHER

TAB WASHER

①

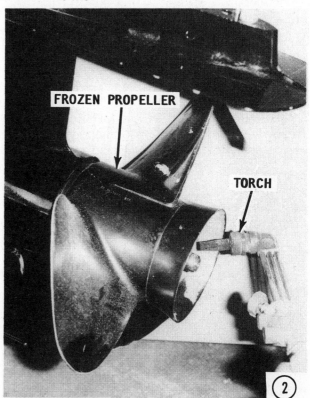

FROZEN PROPELLER

TORCH

②

2- If the propeller is frozen to the shaft, heat must be applied to the shaft to melt out the rubber inside the hub. Using heat will destroy the hub, but there is no other way. As heat is applied, the rubber will expand and the propeller will actually be blown from the shaft. Therefore, **STAND CLEAR** to avoid personal injury.

3- Use a knife and cut the hub off the inner sleeve.

4- The sleeve can be removed by cutting it off with a hacksaw, or it can be removed with a puller. Again, if the sleeve is frozen, it may be necessary to apply heat.

5- Remove the thrust hub from the propeller shaft.

Procedures for propeller installation are given at the end of this chapter, after the stern drive has been installed.

10-6 TRIM CYLINDER REMOVAL

1- Remove the two 9/16" nuts on the power-trim cylinder on the aft end of the stern drive unit. If the aft anchor pin turns in the housing, one side must be removed at

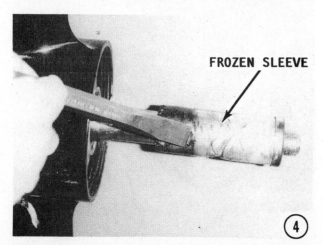

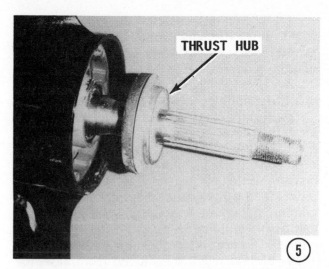

a time. Hold the pin with clamping-type pliers, and then remove the nut from the other side.

2- After the nut has been removed, take off the washer and rubber bushings on both sides. Remove the trim cylinder from the anchor pin by pulling it straight out. Repeat the procedure for the other cylinder. Secure the cylinders up out of the way as a prevention against damaging them during work on the stern drive unit.

3- Remove the spiral spring, rubber bushing, washer, and anchor pin. The spiral spring electrically grounds the trim cylinders. One is installed on each side, with these exceptions: On models between 1975 and 1978, the cylinders were coated with Nylon, as a substitute for the spiral springs. However, the coating was not satisfactory. The coating should be removed and the springs installed. Since 1979, the hydraulic hoses are made of stainless steel to provide

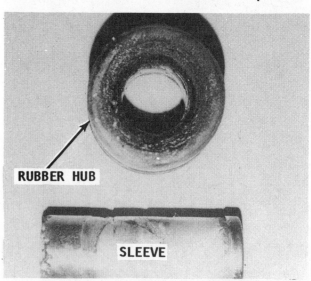

Worn propeller hub (top) and sleeve (bottom). Such damaged parts must be replaced for satisfactory service.

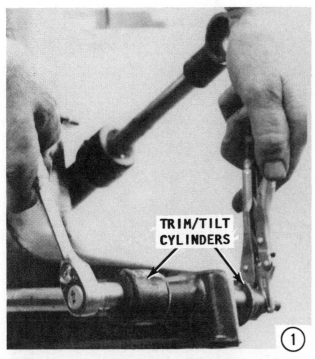

TRIM/TILT
CYLINDERS

①

a ground between the cylinders and the stern drive. Even with these new hoses, the springs should **STILL** be used to ensure a satisfactory ground.

10-7 STERN DRIVE REMOVAL

1- Remove the **FILL** plug and washer at the bottom of the lower unit. Remove the **VENT** screw and washer from the upper gear housing. This vent screw **MUST** be removed in order to release a possible air lock. Such an air lock could prevent the lubricant from draining. On models prior to approximately 1973, the two reservoirs are independent and must be drained separately. **NEVER** remove the vent or filler plugs when the drive unit is hot. Expanded lubricant would

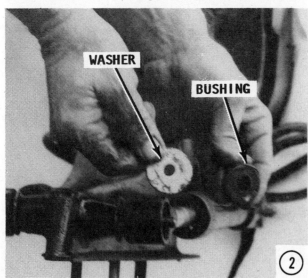

WASHER

BUSHING

②

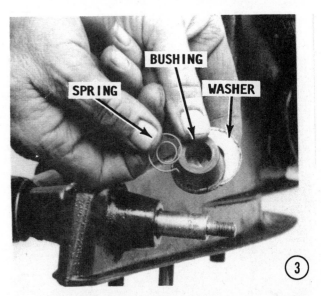

SPRING

BUSHING

WASHER

③

be released through the plug hole. Check the drained lubricant carefully to determine if it contains any water or metal particles. Rub some of the lubricant between your fingers. Any metal particles in the lubricant will thus be evident. Do not be mislead if the color of the lubricant is metal colored. This is not harmful and is caused by the lubricant used in the unit the first time after manufacture.

2- Shift the unit into **FORWARD** gear. Remove the 5/8" nuts and the washers securing the stern drive to the bell housing.

3- Remove the stern drive unit by pulling it straight back and free of the upper gear housing. Remove and **DISCARD** the gasket.

4- If the stern drive unit is difficult to remove, the driveshaft splines may be frozen in the engine coupler or the shaft may be frozen onto the gimbal bearing. One of two methods may be used to break the stern

GEAR OIL DRAINING

①

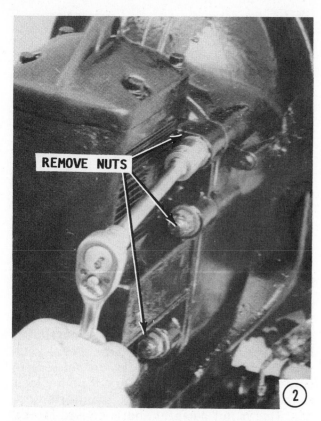

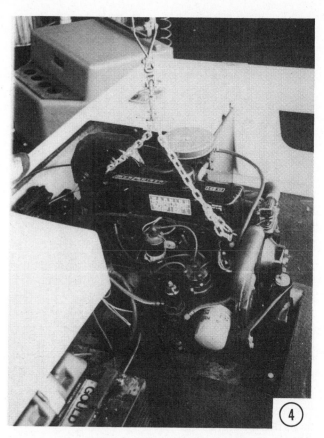

drive loose from the coupler. One involves disconnecting the engine mounts and then moving the engine forward slightly to enable the driveshaft to be pryed from the splines.

5- The second method is to remove the

top cover plate and then drive off the universal joint nut with a punch and hammer. The stern drive can then be removed, leaving the frozen shaft in the bell housing.

6- Remove the bell housing, see Chapter 11. Cut or remove the **U**-joint and exhaust bellows. **A GOOD WORD:** The shift cable does not have to be removed. Set the bell housing aside.

7- Remove the snap ring securing the gimbal bearing. This can be accomplished with a pair of truarc pliers, as shown. The

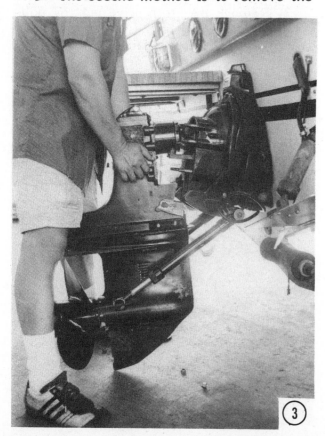

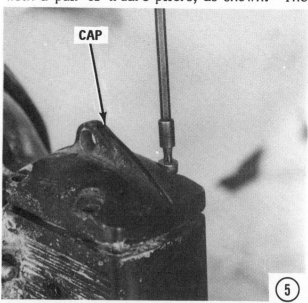

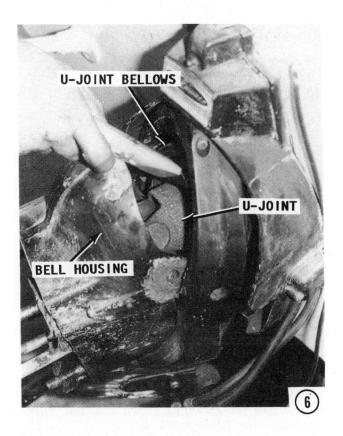

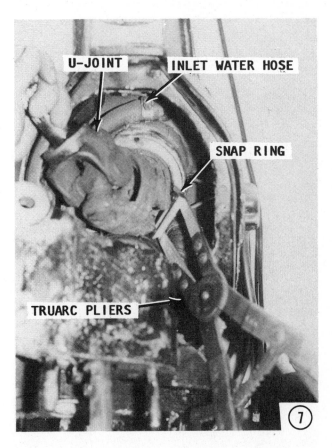

U-joints are in the way, but the snap ring can be removed, with skill and patience.

8- Use a slide hammer and remove the U-joint and gimbal bearing. If the bearing refuses to release, use a torch to heat the bearing and the gimbal housing, while using the slide hammer. To replace the gimbal bearing, see Chapter 12.

9- Check the shift cable to be sure it is not frozen. If problems are indicated in this area, see Chapter 7.

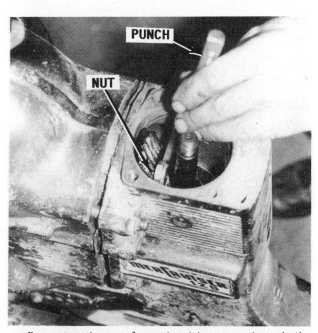

Removing the nut from the drive gear through the upper gear housing, as described in the text.

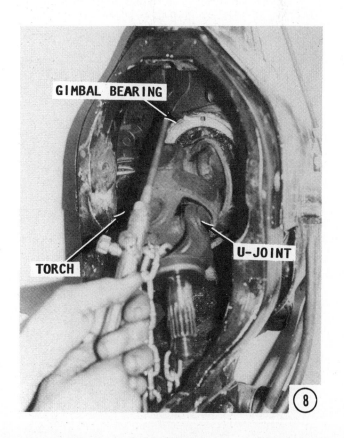

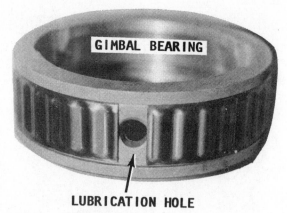

A new gimbal bearing retainer. The hole shown **MUST** be positioned at 3 o'clock to align with the tube in the gimbal housing. The hole and tube permit lubrication of the bearing through the housing.

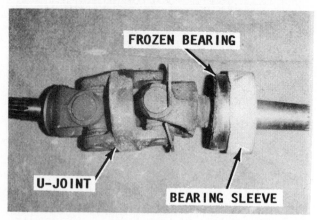

Gimbal bearing frozen to the universal driveshaft. In this case the gimbal bearing and U-joints must be replaced. Leaking U-joint bellows caused this damage.

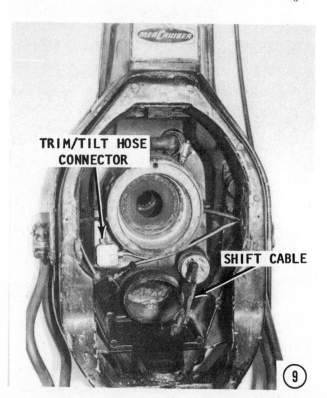

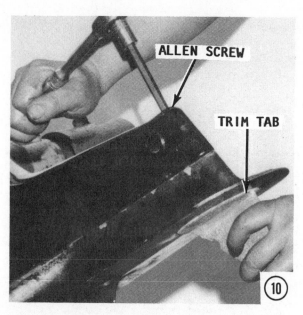

Separating the housing:

10- Assuming the propeller has been removed, place the assembly in a holding fixture. Mark the location of the trim tab trailing edge on the anti-cavitation plate as an aid to assembling. Remove the plug. Remove the Allen-head set screw directly above the trim tab. Remove the trim tab.

11- Remove the 5/8" hex nuts located in the center bottom side of the anti-cavitation plate. Remove the 5/8" hex nut from

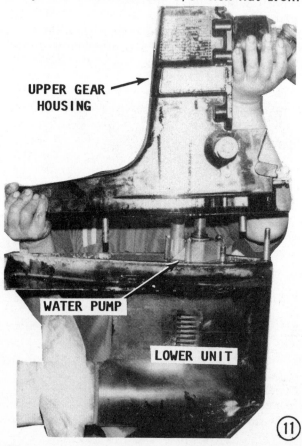

the leading edge of the driveshaft housing. Remove the remaining two 5/8" hex nuts on both sides of the housing. Remove the Allen screw from the cavity for the trim tab. Separate the housings. In some cases the housings may be difficult to separate because the water tubes and the driveshaft are frozen in the upper gear housing.

10-8 UPPER GEAR HOUSING DISASSEMBLING

As an assist, the photographs in this section were taken of a sectioned unit. Thus, you are able to see more clearly exactly what is happening inside the unit as the work progresses.

1- Remove the four screws securing the cover, and then lift off the cover. Remove the tapered roller bearing cup, shims, and O-ring seal from inside the cover. **SAVE** and **TAG** the shims as an aid during assembling. Some models have a spacer and free-load spring behind the bearing cup instead of the shims.

2- **BEFORE** removing the retainer, scribe a mark on the retainer and a matching mark on the housing as an aid to assembling. Loosen the U-joint roller bearing retainer with special wrench C-91-36235. Work the tool counterclockwise to remove the retainer. Pull the U-joint assembly from the housing. Reach inside the housing bore and remove the shims. **SAVE** and **TAG** the shims as an aid to assembling.

3- Lift the upper driveshaft assembly from the housing. Set the assembly up in an arbor press, and then remove the tapered roller bearing and gear. Remove the upper tapered roller bearing with the arbor press.

4- If the bearing cup is defective, remove it, using a slide hammer and universal puller. Remove, **SAVE**, and **TAG** the bearing

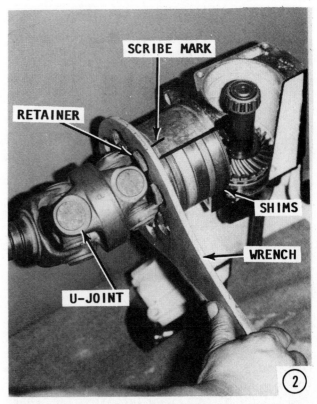

shims. Pull out the upper oil seal using the slide hammer and universal puller.

5- Drive out the lower driveshaft oil seal. Remove the intermediate shift shaft retaining cotter pin, washer, and shift shaft. Remove the water tube and **DISCARD** the seal under the tube.

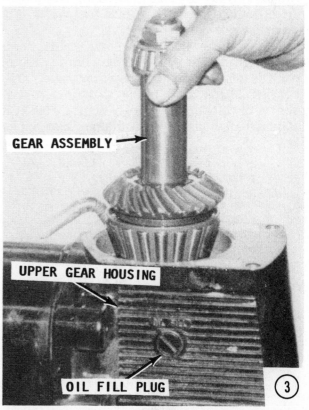

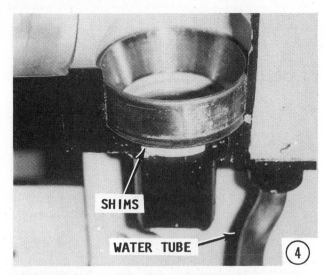

CLEANING AND INSPECTING

Inspect the gearcase, housing, and covers inside and out for cracks. Check carefully around screw and shaft holes. Inspect around machined faces and holes for burrs. Check to be sure all old gasket material has been removed. Inspect for stripped threads.

Check O-ring seal grooves for sharp edges, which could cut a new seal. Check all oil holes.

Inspect the bearing surfaces of the shafts, splines, and keyways for wear and burrs. Check for evidence of an inner bearing race turning on the shaft. Measure the runout on all shafts to detect any bent condition. If possible, check the shafts in a lathe for out-of-roundness.

Inspect the gear teeth and shaft holes for wear and burrs. Hold the center race of each bearing and turn the outer race. The bearing **MUST** turn freely without any evidence of binding or rough spots. **NEVER** spin a ball bearing with compressed air or

Cut-a-way view of a late model upper gear housing, showing the relationship of the driven gear and associated parts.

the bearing will be ruined. Inspect the balls and rollers for pitting and flat spots. Inspect the outside diameter of the outer races and the inside diameter of the inner races for evidence of turning in the housing or on a shaft. Deep discoloration and scores are evidence the bearing has been overheating.

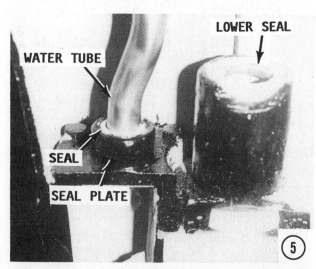

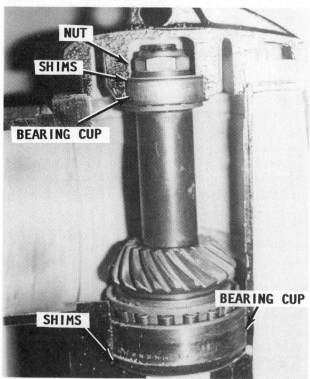

Cut-a-way view of an early model upper gear housing, showing the nut securing the upper bearing in place. The driven gear and associated parts are also visable.

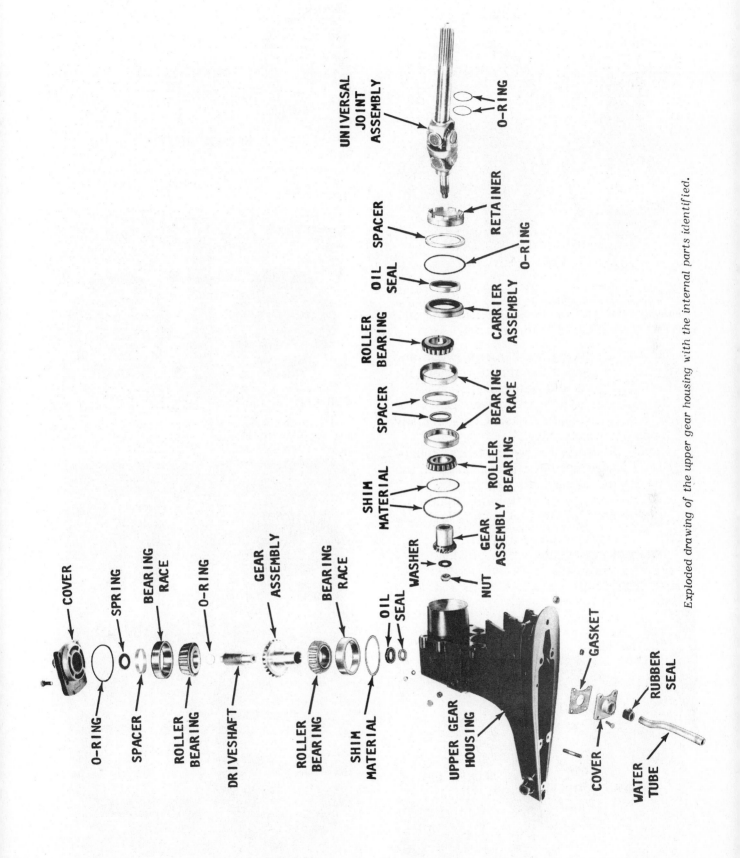

Exploded drawing of the upper gear housing with the internal parts identified.

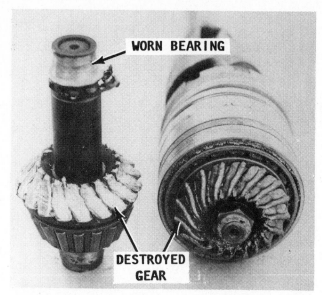

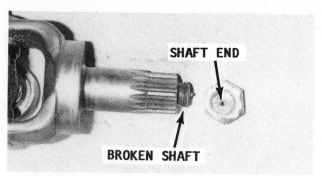

The end of this shaft broke off because the driven gear nut was tightened to a higher torque value than recommended.

Damaged driven gear (left) and drive gear (right). This expensive replacement job resulted from improper shimming. The two gears worked against each other and were destroyed in a short time.

Inspect all thrust washers for wear and distortion. Measure the thrust washers for uniform thickness and flatness.

Inspect the pinion and drive gears for distortion, burrs, and cracks. Check for any evidence of overheating.

Check oil seals for wear, tears, roughness, and proper spring tension. Inspect the O-rings for nicks, hardness, and cracks.

For confidence in obtaining maximum service, replace all seals, O-rings, and gaskets.

A damaged upper gear housing cap (left), alongside a new cap for comparison. The cap was destroyed because it was not shimmed properly.

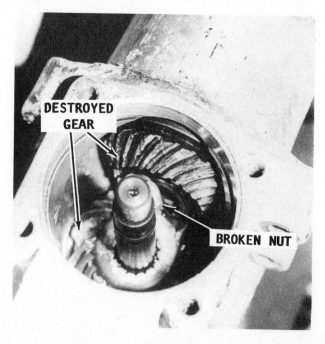

Drive and driven gears destroyed because the driven gear nut was not tightened to the proper torque value.

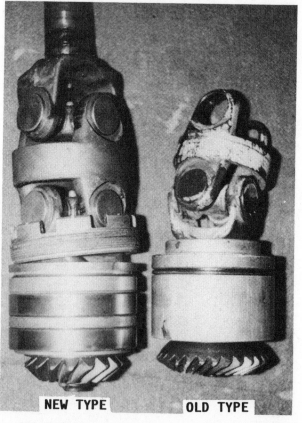

A new style (left) and an old style (right) driven gear bearing assembly.

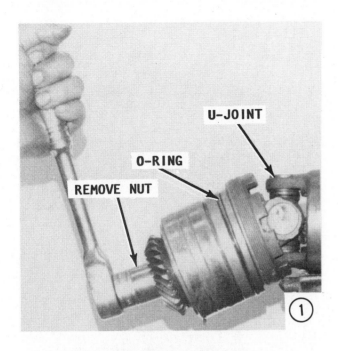

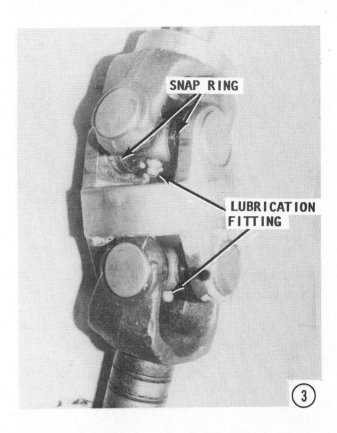

10-9 U-JOINT, DRIVE GEAR AND BEARING DISASSEMBLING

1- Clamp one end of a punch in a vise and the other end through the U-joint to prevent it from turning. Remove the nut securing the parts together.

2- Slide the parts off and note the order as follows: brass retainer, retainer ring, O-ring, oil seal with the oil seal carrier, tapered roller bearing, bearing cup, inner pre-load spacer, outer spacer, second bearing cup with tapered roller bearing, drive gear, washer, and nut. These parts **MUST** be installed in the proper order. Therefore, **KEEP** the parts in this order as an assist during assembling.

3- Drive the eight snap rings off the U-joint bearings with a punch and hammer.

4- Obtain a suitable adaptor to support the U-joint yoke. Press one bearing until the opposite bearing is pressed into the adaptor. Remove the two loose bearings. Rotate the U-joint assembly 180°. Use the adaptor and press on the bearing cross-member and remove the bearings. Remove the cross-member from the yoke. Press the other four bearings out in the same manner as described in the previous step and in the first part of this step. Remove the O-rings on the yoke shaft.

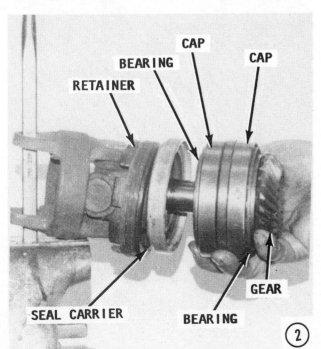

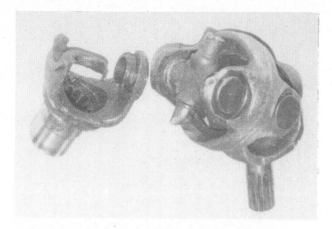

This U-joint was damaged from lack of lubrication.

10-10 U-JOINT, DRIVE GEAR AND BEARING ASSEMBLING

1- Lubricate the **U**-joint bearing cups with a liberal amount of Universal Joint Lubricant, C-92-58229. Place the lubricated cups in the yoke and start them on the bearing cross-members. Press the bearings through the yoke onto the cross-members. as shown.

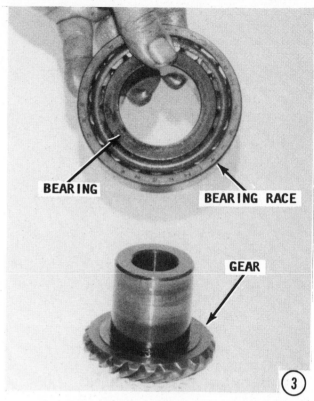

2- Drive the bearing cup retaining snap rings into place with a hammer. Lubricate and install the other sets of bearings in the same manner.

3- Position the tapered roller bearing and race onto the drive gear, with the **LETTERED** side of the bearing race facing **AWAY** from the gear.

4- Place the small spacer over the gear. The spacer is no longer used with press-fit bearings.

⑤

5- Install the large spacer.

6- Install the second tapered roller bearing, with the **LETTERED** side of the inner cone facing **UP.** The lettered side of the outer race **MUST** face **DOWN.**

⑥

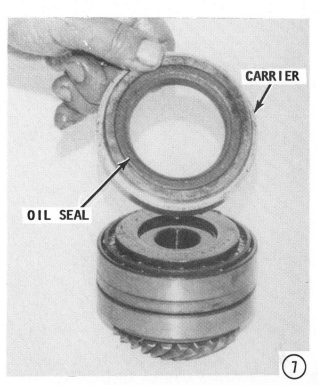

⑦

7-Press a **NEW** oil seal into the oil seal carrier, with the lip of the seal facing the **CONCAVE** side of the carrier, until the seal is **FLUSH** with the carrier. Coat the lip of the seal with Multi-purpose Lubricant. Install the oil seal carrier, with the **LIP** of the oil seal facing **DOWN,** toward the drive gear. The seal prevents lubricant from flowing out of the upper housing and onto the U-joints.

8- Install the outer ring, and then place the threaded bearing retainer in position on the stack.

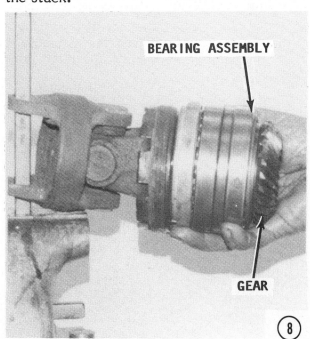

⑧

OUTER RING

O-RING GROOVE

⑨

9- Clamp a suitable pry bar in a vise, with the sharp end through the **U-joint**. This arrangement will hold the parts as the work progresses. Notice how the threaded bearing retainer is positioned to the left of the pry bar as you slide the bearing pack over the **U-joint** output shaft.

10- Install the flat washer, with the **CONCAVE** side **FACING** the drive gear. Thread the locking nut onto the end of the U-joint shaft. Tighten an automotive-type piston ring compressor over the bearing pack to align the parts for ease in installing the unit into the housing. Tighten the self-locking nut to a torque value of 85 ft-lbs. The torque value is **IMPORTANT** because the shaft nut applies the correct bearing preload to the drive gear bearing assembly. Set the completed work aside for later installation.

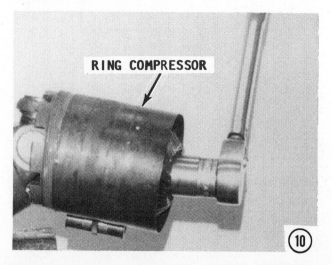

RING COMPRESSOR

⑩

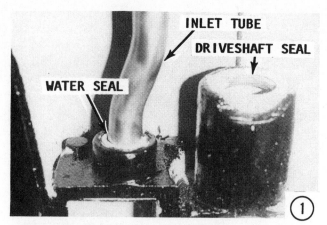

INLET TUBE

DRIVESHAFT SEAL

WATER SEAL

①

10-11 UPPER GEAR HOUSING ASSEMBLING

1- Coat the rubber groument with a small amount of Perfect Seal. Install the groument into the water pickup. Coat the outside metal case of the lower driveshaft oil seal with a small amount of Loctite. Drive the seal into the cavity with the **LIP** of the seal facing the **DRIVEN** gear. Continue to drive the seal into place until it is flush with the housing.

To ease the installation of the upper driveshaft, lubricate the lip of the seal with Multi-purpose Lubricant. This oil seal prevents exhaust gases and water from entering the splined area of the driveshaft. **DO NOT** install the upper driveshaft oil seal until the upper driveshaft bearing preload has been measured.

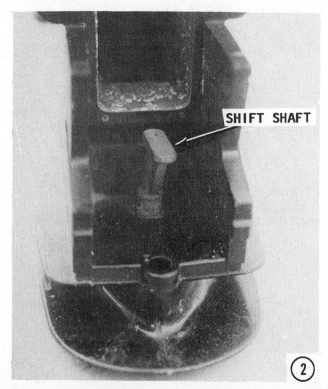

SHIFT SHAFT

②

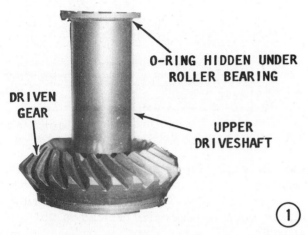

O-RING HIDDEN UNDER
ROLLER BEARING

DRIVEN
GEAR

UPPER
DRIVESHAFT

①

2- Install the intermediate shift shaft. Install the retaining washer. Secure the washer in place with a **NEW** cotter pin.

10-12 UPPER DRIVESHAFT AND DRIVEN GEAR ASSEMBLING

1- Install a **NEW** O-ring seal over the driven gear end of the upper driveshaft. Push the O-ring into place up against the roller bearing. Press the driveshaft into the driven gear using an arbor press and suitable mandrel until the shoulder on the shaft seats against the gear collar.

2- Turn the driveshaft over. Position

SHIMS

SHIMS

②

the tapered roller bearing cup over the bearing to protect the bearing. Press on the other tapered roller bearing. On early model stern drive units, Models 1, 1A, 1B, and 1C, **DO NOT** install the tapered roller bearing or cup because the shimming tool for determining the drive gear position replaces this bearing and cup.

UPPER DRIVESHAFT ADJUSTMENTS

The following adjustments are for late-model stern drive units, Models 120-889. The sequence of instructions are divided into three sections and they all **MUST** be performed in this order:

Upper driveshaft bearing preload adjustment, Section 10-13.
Driven gear shimming, Section 10-14.
Drive gear shimming, Section 10-16.

For early style stern drive units, Models 1, 1A, 1B, and 1C, the adjustment sequence **MUST** be:

Drive gear shimming, Section 10-16.
Upper driveshaft bearing preload adjustment, Section 10-13.
Driven gear shimming, Section 10-14.

10-13 UPPER DRIVESHAFT BEARING PRELOAD ADJUSTMENT

1- Install the shims, removed during disassembly, under the bottom bearing race. Install the upper driveshaft. If the upper oil seal is in place, it **MUST** be **REMOVED**

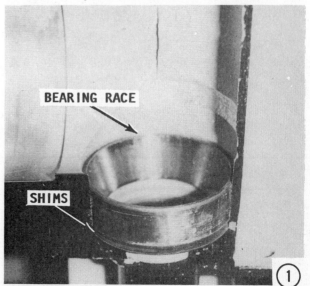

BEARING RACE

SHIMS

①

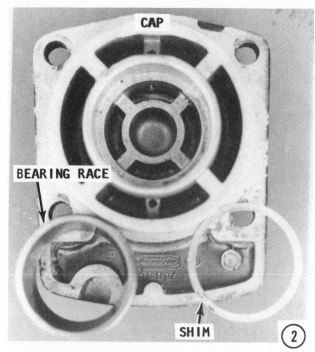

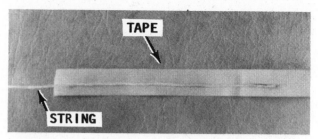

Masking tape attached to a section of fish line in preparation to making a preload test of the driven gear assembly.

before replacing the driveshaft. This seal must be removed in order to obtain an accurate bearing preload measurement. The friction a seal exerts on the shaft would give a false measurement.

2- Install the same number of shims under the bearing race of the top cover as were removed during disassembling. Press in the top tapered bearing race. Install the top cover temporarily, **WITHOUT** the O-ring seal in place. A gasket is **NOT** used under

the top cap. Now, tighten the four nuts, securing the top cap in place, to a torque value of 20 ft-lbs. On models with a spring-loaded bearing race, the preload is not adjustable.

3- Insert an old gear housing driveshaft, with a pinion gear retaining nut, in place into the upper driveshaft splines. Determine the effort required to turn the upper driveshaft. This may be done with a torque wrench calibrated in in-lbs. The torque value should be 2-6 in-lbs. If the reading is too high, remove shims from under the bearing cup in the top cover. If the torque value is less than 2 in-lbs, add shims. If the shim pack under the top cover is changed, the nuts securing the cover in place **MUST** be tightened again to the 20 ft-lbs torque value.

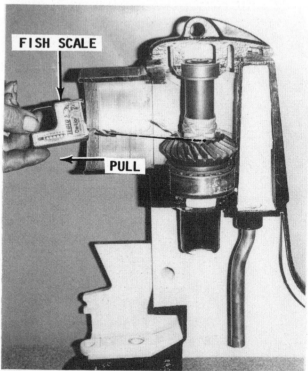

Cut-a-way view of the upper gear housing with a 3-foot section of fish line wrapped around the driven gear shaft and a ounce fish scale attached to the line. The preload test of the driven gear assembly can now be made, as described in the text.

10-14 DRIVEN GEAR SHIMMING

These instructions pick up the work **AFTER** the driveshaft bearing preload has been properly adjusted, as described in the previous section. The measurements in this section **MUST** been done slowly and precisely, to avoid problems that will develop if inaccurate measurements are taken.

4- Install shimming gauge tool, C-91-60526 (or C-91-36384 for early models), into the drive gear cavity, with the proper face, **X, Y,** or **Z,** according to the chart in the Appendix, aligned with three teeth of the driven gear. Measure the clearance between the gear face and the shimming gauge. This clearance should be 0.025". Most feeler gauges do not have a long enough blade to make this clearance measurement. A hacksaw blade is usually about 0.023 in. thick. Therefore, if the teeth of the blade are ground off, it may then be used to check the clearance. It would be best to check the thickness of the hacksaw blade with a micrometer before grinding the teeth down.

Rotate the shimming tool slightly each way to obtain a slight drag on the feeler gauge when it is aligned with one outside tooth of the three aligned teeth. Now, without moving the shimming tool, insert the feeler gauge between the shimming tool and the other outside tooth of the three aligned gear teeth. **A WORD:** This procedure is necessary to align the face of the gauge parallel to the driven gear teeth. Accurately determine the actual clearance. If the measured clearance is **LESS** than 0.025", subtract the measured clearance from the specified 0.025", and then remove shims of that thickness from under the driven gear tapered roller bearing race, but add an equal amount of shims under the upper

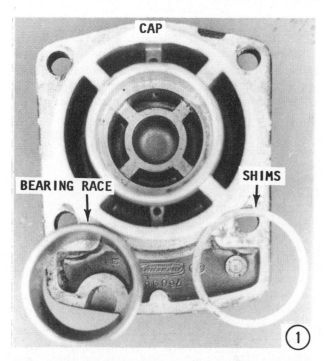

driveshaft tapered roller bearing race to maintain the previously corrected bearing preload.

Check the clearance a second time. If the clearance is correct, coat the outside metal case of the upper driveshaft oil seal with a thin layer of Loctite. Apply a thin coating of Multi-purpose Lubricant to the lip of the seal to ease installation of the upper driveshaft. This seal **MUST** be removed when measuring the driveshaft bearing preload because of the extra drag a seal places on a rotating shaft. This seal prevents lubricants from working from the driveshaft housing into the exhaust chamber.

10-15 U-JOINT BEARING AND BEARING ASSEMBLY INSTALLATION

1- Remove the top cover. Install the same number of bearing-to-driveshaft housing shims that were removed and tagged during disassembling.

2- First, a good word: Most MerCruiser drive and driven gears have matching marks. These marks **MUST** be aligned when the gears are meshed in the gear housing. However, on some models, the gears do not have

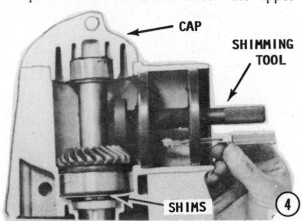

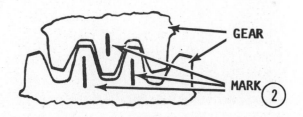

these marks. In this case, the gears may be installed with any gear mesh. Position the driven and drive gears to mesh properly before installing the U-joint shaft assembly.

3– Install the O-ring seal. Insert the assembled shaft, with the gears properly meshed into the upper gear housing. Tighten the bearing retainer nut to a torque value of 200 ft-lbs. It is very difficult to tighten the retainer nut to the proper torque value. An alternate method is to tighten the nut securely and then to bring it around to the mark you made during disassembling. From this point tighten it another 1/4" past the mark.

10-16 DRIVE GEAR SHIMMING

This procedure **MUST** be followed precisely as described in order to obtain a correct reading of the clearance between the driving gear and the shimming tool. The measurement is needed to ensure the shimming tool face is parallel with the face of the gear.

4– Insert shimming tool, C-91-60523, into the driveshaft housing top cover cavity. Align the proper tool face, **X**, **Y**, or **Z**, according to the chart in the Appendix, with three teeth of the drive gear. Insert a 0.025" feeler gauge blade between one of the outside teeth of the three aligned gear teeth, and the shimming tool. Rotate the tool to obtain a slight drag on the feeler

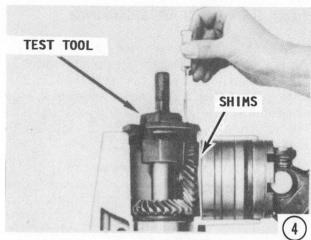

gauge blade, and then, without moving the shimming tool, insert the gauge between the tool and the other outside tooth, of the three aligned teeth. If the clearance is greater than the gauge thickness, repeat the measuring procedure with a thicker gauge until the same clearance is obtained between both outside gear teeth and the tool. If the clearance is less, use a thinner thickness gauge blade and repeat the measuring procedures described. Now, if the measurement between the tool and the gear is less than 0.025", subtract the reading obtained from 0.025" and **ADD** shims of that thickness between the U-joint tapered roller bearing race and the driveshaft housing. If the final measured clearance is more than 0.025" subtract 0.025" from the measurement and **REMOVE** shims of that thickness. Check the clearance a second time after making changes to the shims. Remove the shimming tool.

5– After the shimming procedure is complete, position a **NEW** O-ring seal on the upper cover. Install the cover and tighten

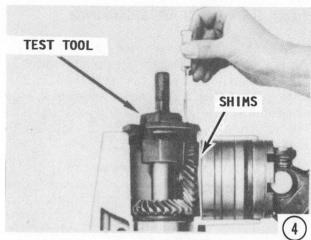

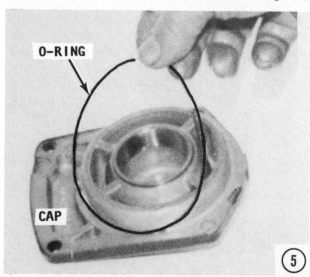

the retaining nuts to a torque value of 20 ft-lbs. A gasket under the cover is **NOT** needed.

The upper driveshaft housing assembly is now completely rebuilt and ready to be assembled to the lower unit, if the lower unit does not require service. For assembling and installation procedures see Section 10-25. See Section 10-26 to install the stern drive.

10-17 LOWER UNIT HOUSING REMOVAL

If only the lower unit requires servicing, this unit may be removed without disturbing the upper gear housing.

1- Assuming the propeller has been removed, secure the assembly in a holding fixture. Mark the location of the trim tab trailing edge on the anti-cavitation plate as an aid during assembling. Remove the plug. Remove the Allen-head set screw directly above the trim tab. Remove the trim tab. Remove the 5/8" hex nut from the leading edge of the driveshaft housing. Remove the remaining two 5/8" hex nuts on both sides of the housing. Remove the Allen-head screw from the cavity for the trim tab.

Alpha Generation II Words

The Generation II lower unit has an odometer installed. The lead must be disconnected before the lower unit is separated from the intermediate housing. This connection is a simple "push-twist-pull" type.

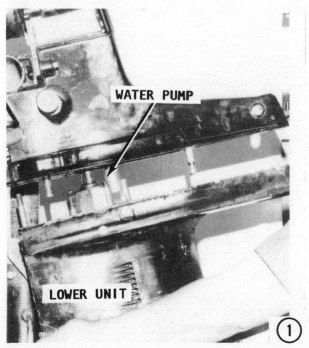

Separate the housings. In some cases the housing may be difficult to separate because the water tubes and the driveshaft are "frozen" in the upper gear housing.

LOWER UNIT DISASSEMBLING UNITS UP THROUGH ORIGINAL ALPHA DRIVE

2- Remove the water pump guide and **O**-ring seal on the upper end of the driveshaft. The **O**-ring prevents exhaust gases and water from washing the lubricant off the driveshaft splines. Remove the one bolt and three nuts securing the water pump body in place.

3- Lift off the pump housing, impeller, drive pin, gasket, base plate, pump base, and gasket.

4- On early model pumps, the flush screw located on the port side, close to the oil vent screw, must be removed first in order to permit removal of the pump base.

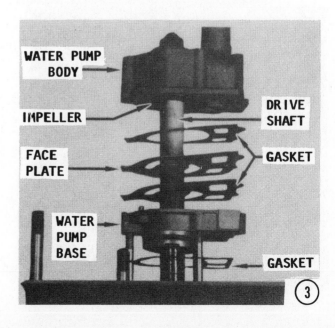

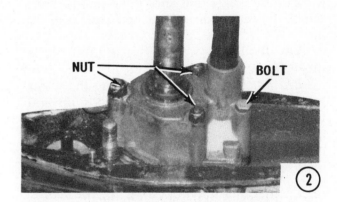

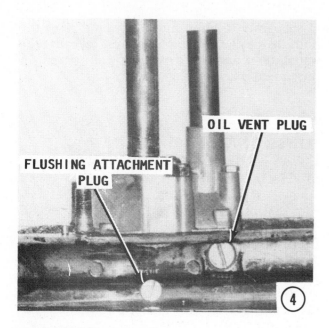

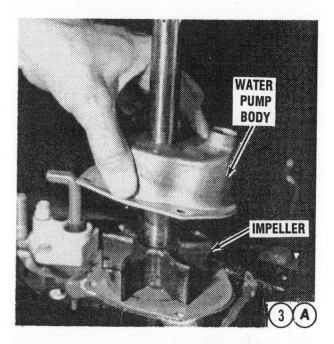

ALPHA STERN DRIVE
GENERATION II

The following "A" steps and supporting "A" illustrations cover removal and disassembly of the Generation II water pump. To continue work on the lower unit after the pump is disassembled, proceed directly to Step 5 on Page 10-25.

2A- Pull the water seal up and free of the driveshaft.

3A- Remove the bolts securing the water pump body, and then slide the body up and free of the driveshaft. It may be necessary to use a couple screwdrivers -- one on each mounting flange -- to gently persuade the pump body to "break" loose from the plate. Slide the impeller up and free of the driveshaft. Remove the Woodruff key from the cutout in the driveshaft.

4A- Remove the gasket from the underside of the body. This gasket may remain on the plate when the body is removed.

5A- Slide the plate and gasket up and free of the driveshaft.

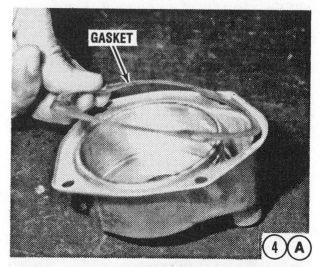

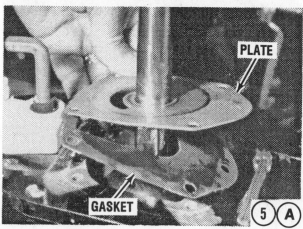

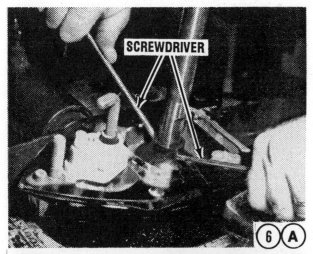

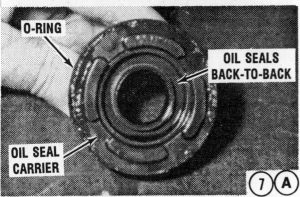

Use of an electric shop drill to drill out a frozen bearing carrier nut.

6A- Use two screwdrivers -- one on each side as shown, to pry the oil seal carrier free. Remove the carrier.

7A- If the oil seals are damaged, they may be pried out of the carrier with a screwdriver after the carrier has been clamped in a vise. Be sure replacements have been obtained and are at hand, because removing the seals will destroy their sealing qualities.

FURTHER DISASSEMBLING OF LOWER UNIT CONTINUES

5- To remove the reverse bearing carrier, first straighten the locking tab on the tab washer. (This washer is not used on early models.) Loosen the cover nut by turning it **COUNTER-CLOCKWISE** with special wrench C-91-61069. Remove the cover nut and tab washer. Remove the aligning key.

6- Pull out the reverse gear bearing carrier assembly using a bearing carrier puller. The thrust hub for the propeller, which was removed previously, can be used to keep the puller jaws in place.

7- If the reverse gear bearing carrier is corroded in place, apply heat to the housing to loosen the corrosion. While the housing is still

A drilled bearing carrier nut after removal.

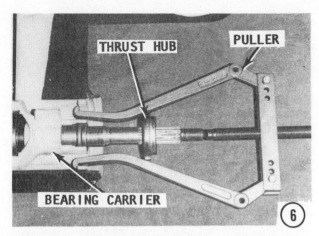

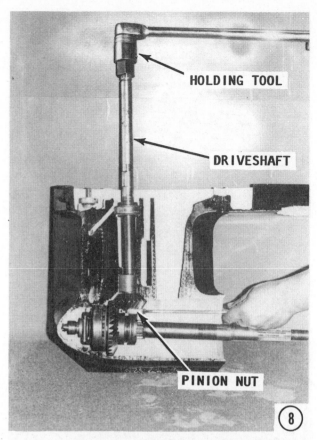

hot, use a bearing carrier puller to remove the assembly. **DO NOT** overheat the housing or it will become distorted and useless. Reach in and remove the thrust washer-to-gear housing shim material. Tag and wire them together, as an aid during assembling.

8-   Remove the driveshaft pinion nut by sliding the pinion nut holding tool, C-91-61067A1, over the propeller shaft. If working on an **MR** lower unit, use holding tool C-91-61067A2.   Turn the driveshaft **COUNTER-CLOCKWISE** with the special splined wrench adapter, C-91-56775, (or C-91-34377A1 for coarse splines), and a break-over handle. If the pinion nut holding tool is not available, a substitute can be made from a box-end wrench of the proper size for the nut.   Grind the wrench down on both sides until it will clear the clutch dog and the pinion gear. Place the wrench on

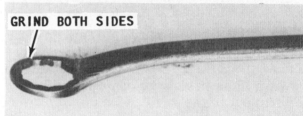

A modified box-end wrench used to remove/install the pinion gear nut, if the special tool shown below is not available.

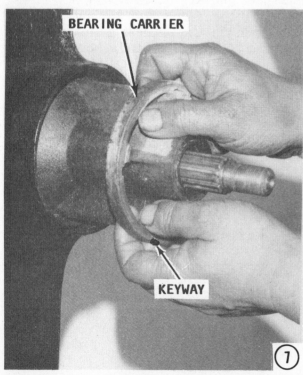

*New pinion nut holding tool, left, alongside an old one, right. The new tool has three cutouts and may be used on all lower unit models. The old one can only be used on models prior to the **MR** unit.*

the nut and proceed with disassembling. Remove the nut.

SPECIAL WORDS FOR MR UNITS

Remove the outer race retaining nut in order for the driveshaft to clear the housing. To remove the nut, obtain driveshaft bearing retainer nut tool 91-43506. Slide the tool down the driveshaft until the tool is indexed over the nut, as shown in illustration **8A**. Now, use a breaker bar on the special tool and loosen the nut, as shown in illustration **8B**. Slide the special tool free of the driveshaft, and then back the nut off and clear of the driveshaft.

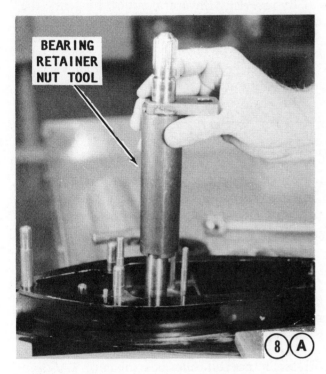

BEARING RETAINER NUT TOOL

8 A

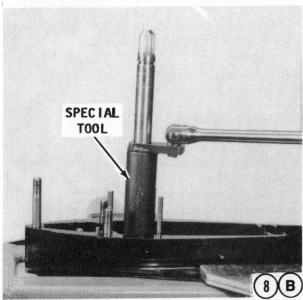

SPECIAL TOOL

8 B

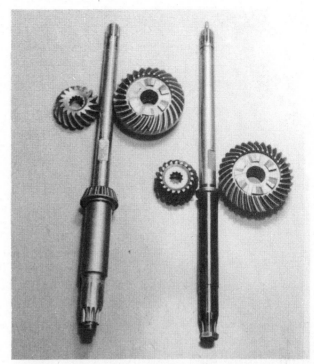

A new driveshaft, pinion gear, and driven gear, for the **MR** *model stern drive, alongside the old. Differences are explained in the paragraph below.*

ALL UNITS

9- Clamp the driveshaft in a vise equipped with brass soft jaws. If soft jaws are not available, clamp the driveshaft on a cast surface, away from the machine splines. Use a block of wood to protect the housing from damage when hammering on it. Drive the housing downward by striking the block of wood with a hammer. Once the driveshaft is free of the pinion gear, slide the housing clear

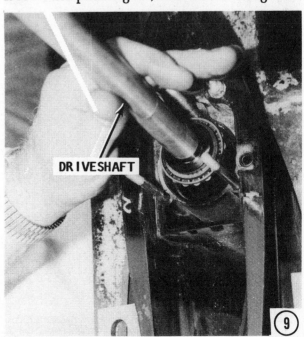

DRIVESHAFT

9

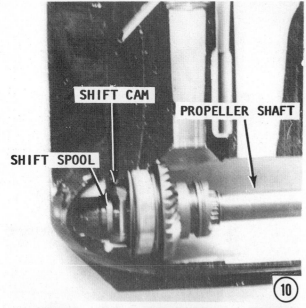

SHIFT CAM

PROPELLER SHAFT

SHIFT SPOOL

⑩

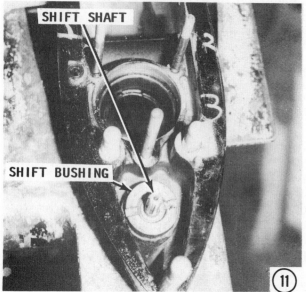

SHIFT SHAFT

SHIFT BUSHING

⑪

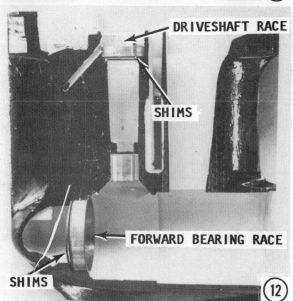

DRIVESHAFT RACE

SHIMS

FORWARD BEARING RACE

SHIMS

⑫

of the driveshaft. The pinion gear can be removed from the housing after the driveshaft has been removed.

MORE WORDS FOR MR UNITS

One of the major differences and improvements to the MR unit, is the increased strength built into the driveshaft and the reverse spiral cut of both the pinion gear and driven gear as indicated in the accompanying illustration on the previous page. The new shaft and gear set is shown on the left, and the old on the right. Notice the absence of the preload pin at the top of the new driveshaft.

By reversing the spiral cut of the gears, the driveshaft is constantly being drawn upward, thereby placing a preload on the driveshaft bearing at all times. This type of preload provides a more constant backlash under varying load conditions.

10-   Move the propeller shaft to the starboard side of the lower unit to pass by the shift cam. Pull out the propeller shaft assembly and forward gear.

11-   Remove the metal washer and the rubber washer from the shift shaft. Remove the shift shaft bushing using special tool C-91-31107. Lift the shift shaft from the gear housing. Remove the shift crank.

GOOD WORDS

Step 12 is not required, unless the bearings are to be replaced. Any time the bearings are replaced, they **MUST** be shimmed properly or their service life will be drastically reduced.

12-   Using a slide hammer, remove the driveshaft tapered roller bearing cup. **TAG** and wire the shim material from under the bearing cup together, as an aid during assembling. Use a slide hammer and remove the propeller shaft tapered roller bearing cup. **TAG** and wire the shim material pieces from behind the bearing cup together, as an aid during assembling.

CLEANING AND INSPECTING

Wash all parts in solvent and blow them dry with compressed air. Remove all traces of seal and gasket material from all mating surfaces. Blow all water, oil passageways, and screw holes clean with compressed air. After cleaning, apply a light coating of engine oil to the bright surfaces of all gears, bearings, and shafts as a prevention against rusting and corrosion. Check the shift crank to be sure it is not bent.

Inspect the water pump impeller plate for wear and corrosion. Replace any worn, corroded, or damaged parts. Use a fine file to remove burrs. **REPLACE** all O-rings, gaskets, and seals to ensure satisfactory service from the unit. Clean the corrosion from inside the housing where the bearing carrier was removed.

Check to be sure the water intake is clean and free of any foreign material.

Inspect the gearcase, housings, and covers inside and out for cracks. Check carefully around screw and shaft holes. Check for burrs around machined faces and holes. Check for stripped threads in screw holes and traces of gasket material remaining on mating surfaces.

Check O-ring seal grooves for sharp edges, which could cut a new seal. Check all oil holes.

Inspect the bearing surfaces of the shafts, splines, and keyways for wear and burrs. Look for evidence of an inner bearing race turning on the shaft. Check for damaged threads. Measure the runout on all shafts to detect any bent condition. If possible, check the shafts in a lathe for out-of-roundness.

Inspect the gear teeth and shaft holes for wear and burrs. Hold the center race of each bearing and turn the outer race. The bearing must turn freely without binding or evidence of rough spots. **NEVER** spin a ball bearing with compressed air or it will be ruined. Inspect the outside diameter of the outer races and the inside diameter of the inner races for evidence of turning in the housing or on a shaft. Deep discoloration and scores are evidence of overheating.

Inspect the thrust washers for wear and distortion. Measure the washers for uniform thickness and flatness.

Inspect the forward and reverse gears for distortion, burrs, and cracks. Check for any sign of discoloration, which means the gears are running too hot.

Replace all seals, O-rings, and gaskets to ensure maximum service after the work is completed.

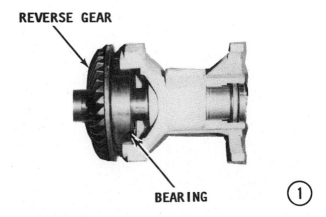

REVERSE GEAR

BEARING

①

10-18 REVERSE BEARING CARRIER DISASSEMBLING

1- Clamp the reverse gear bearing carrier in a vise equipped with soft jaws. Use a slide hammer and pull the reverse gear. Use a universal puller and slide hammer to pull out the ball bearing and thrust ring from the bearing carrier.

2- Remove the two propeller shaft oil seals from the bearing carrier. Press the needle bearing from the carrier, using an arbor press.

CLEANING AND INSPECTING

Wash all parts in solvent and blow them dry with compressed air. Remove any corrosion from the outside surface of the bearing carrier. Remove any seal and gasket material from all mating surfaces. Blow all water, oil passageways, and screw holes clean with compressed air. After all parts are clean and dry, apply a light coating of engine oil to the bright surfaces of all gears,

SEALS

②

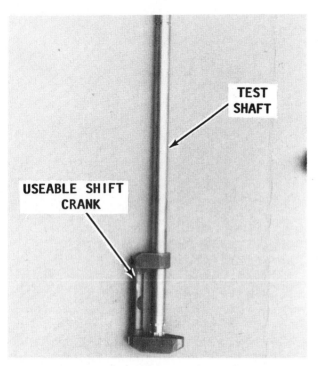

A single test rod inserted through the shift crank proving the crank is not bent.

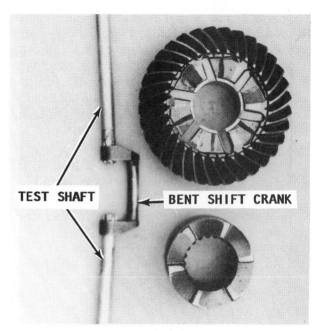

A shift crank, clutch dog, and forward gear. A test rod has been inserted into each side of the crank to indicate how much the crank has been bent.

A bearing carrier damaged during removal to prevent injury to the much more expensive lower unit housing.

A lower unit destroyed in order to remove a frozen bearing carrier.

bearings, and shafts as a prevention against rusting.

Inspect the gears for any sign of excessive wear, nicks, or broken teeth. Check all splines to be sure they have not been damaged. Inspect the bearing sets for pits, grooves, or uneven wear. If the bearing carrier is worn or damaged, it may be replaced with a new assembly at the same time the bearings and seals are installed.

10-19 REVERSE GEAR BEARING CARRIER ASSEMBLING

1- Install the new bearing into the bearing carrier with the numbers on the bearing **FACING** the installing tool, as shown.

2- Press two **NEW** oil seals into the front of the reverse bearing carrier. Seat the first one with the lip facing **IN** to hold the lubrication within the housing. Seat the second seal with the lip facing **OUT** to prevent water from entering the housing.

TAKE TIME to obtain the proper mandrel for the ball bearing to ensure the force will be applied to the inner race of the bearing. Press the reverse gear ball bearing and thrust washer into place with the washer **FACING** the reverse gear. Turn the bearing carrier over so the seal side is **DOWN**, and then press the reverse gear and bearing assembly into the carrier. **ALWAYS** use an adapter to protect the reverse gear. Set the assembly aside for later installation.

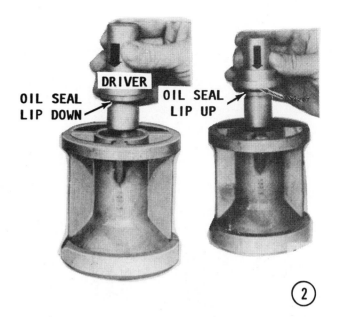

OIL SEAL LIP DOWN   DRIVER   OIL SEAL LIP UP

②

10-20 PROPELLER SHAFT SERVICE

DISASSEMBLING

1- With a small screwdriver, unwind the spring from around the sliding clutch. **TAKE CARE** not to damage the spring by bending it out of shape. Remove the cross-pin. Slide the shift spool and the shift-actuating shaft out. Remove the forward gear-and-bearing assembly.

2- Remove the cotter pin and nut, and then pull the spool off the shift-actuating shaft. Remove the bearings from the forward gear by first splitting the needle bearing inside the gear with a chisel and hammer. Remove the bearings. Use a universal puller plate to remove the tapered roller bearing. Press the gear from the bearing.

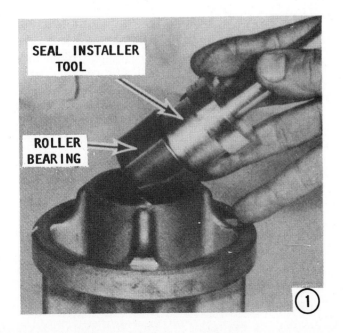

SEAL INSTALLER TOOL

ROLLER BEARING

①

FORWARD GEAR

SPRING

SHIFT SPOOL

CLUTCH DOG

CROSS PIN

①

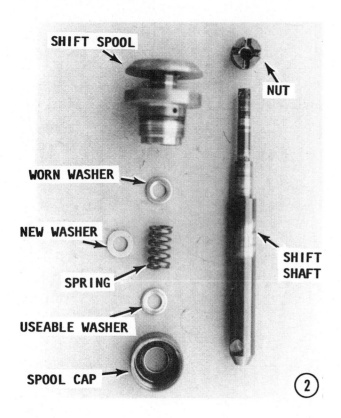

SHIFT SPOOL

NUT

WORN WASHER

NEW WASHER

SPRING

USEABLE WASHER

SPOOL CAP

SHIFT SHAFT

②

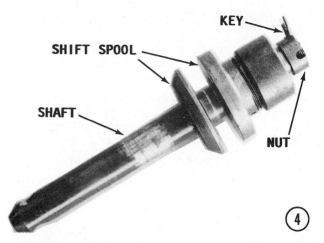

KEY

SHIFT SPOOL

SHAFT

NUT

④

CLEANING AND INSPECTING

Wash all parts in solvent and blow them dry with compressed air. Remove any seal and gasket material from all mating surfaces. Blow all water, oil passageways, and screw holes clean with compressed air. After all parts are clean and dry, apply a light coating of engine oil to the bright surfaces of all gears, bearings, and shafts as a prevention against rusting.

Inspect the gears for any sign of excessive wear, nicks, or broken teeth. Check all splines to be sure they have not been damaged. Inspect the bearing sets for pits, grooves, or uneven wear.

Use a file, and clean the grooves inside the propeller hub. Inspect the propeller and remove any nicks and burrs with a file. **TAKE CARE** not to remove any more material than is absolutely necessary.

Inspect the propeller for cracks, damage, or bent condition. Roll the propeller shaft on a flat surface and check for a bent condition. If the shaft is bent over 0.015", it **MUST** be replaced.

Check the clutch dog and gears for worn lugs. If the lugs are worn, the clutch dog or gear **MUST** be replaced. If the lugs show wear, check the shift cable between the inside of the boat and the stern drive. If the cable requires service, see Chapter 11.

ASSEMBLING

3- Assemble the parts of the shift clutch actuating shaft. Place the first washer, spring, and second washer into the brass shift actuating spool, and then thread on the retainer. Tighten the retainer securely.

4- Slide the spool assembly over the clutch actuating shaft and thread the adjusting sleeve on fingertight. Now, back off the adjusting sleeve until the cotter key slides through the first hole.

5- Use a suitable adapter and press the tapered roller bearing onto the forward gear. The force must be applied only a-

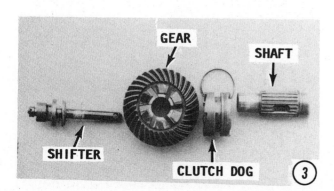

GEAR

SHAFT

SHIFTER

CLUTCH DOG

③

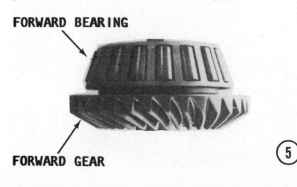

FORWARD BEARING

FORWARD GEAR

⑤

gainst the center race. Seat the bearing against the shoulder of the forward gear. Use the proper adapter and press the forward gear needle bearing into the gear from the **NUMBERED** side. Seat the bearing against the inner gear shoulder.

6- Place the sliding clutch on the propeller shaft, with the grooves in the clutch toward the reverse gear. The clutch on early models do not have grooves. Instead, the clutch has a copper coating on only one side. The face **WITHOUT** the copper coating **MUST** face toward the reverse gear. Install the forward gear-and-bearing assembly, with the gear **FACING** the sliding clutch. Insert the clutch actuating shaft assembly into the hole in the forward end of the propeller shaft. Rotate the shaft assembly

until the hole in the shaft is aligned with the holes in the sliding clutch and the propeller shaft. Insert the cross-pin through the sliding clutch, propeller shaft, and actuating shaft. Install the cross-pin retainer spring by winding it through the sliding clutch in the clutch retainer groove. **DO NOT** over-stretch the spring. Set the assembly aside for later installation.

10-21 LOWER UNIT ASSEMBLING

1- READ AND BELIEVE: The driveshaft bearing has a slight taper. Therefore, if the driveshaft roller bearing is reversed during installation, the driveshaft will fail. The only way the bearing can be installed properly, is with the **NUMBERED** end of the bearing case facing **UP** and **AWAY** from the pinion gear and **TOWARD** the anti-cavitation plate. Coat the driveshaft roller bearing bore in the housing with a thin layer of Multi-purpose Lubricant. Use Bearing-and-Removal Kit, C-91-31229A5, and install the driveshaft roller bearing, in the manner described. Pull the bearing up until it bottoms on the gear housing shoulder.

2- If the driveshaft bearing was removed, install it using the same number of shims removed during disassembly. If the number of shims was not recorded; if some were lost; or if a new assembly is being installed; begin with 0.015" thickness of shim material. To replace the upper driveshaft bearing races: place the tapered roller bearing shims

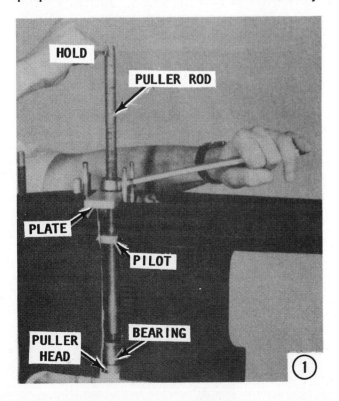

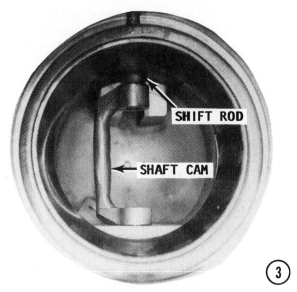

that were tagged and wired together during disassembly, into the bearing recess. Now, use a bearing race driver, driver rod, and mallet to drive the tapered roller bearing race into the bearing race cavity.

3- Pick up the shift-actuating crank. Now, reach all the way into the lower unit to the forward end and install the crank on the locating pin, with the throw side facing the oil fill hole side of the gear housing. On the non **E-Z-shift** type, install the shift cam with the **NOTCHES** to the **LEFT**.

4- Install a **NEW** seal in the shift shaft bushing, with the lip facing **UP**.

5- Install the shift shaft, with the retaining clip in position, as you slide the male splines of the shaft down to index with the female splines of the shift-actuating crank. Install the bushing to hold the shaft in place. **ALWAYS** install a **NEW** O-ring and seal to ensure satisfactory service. The seal and O-ring prevent exhaust gases and water from entering the lower unit and at the same time the seal holds the lubricant inside the housing.

Removing the seal in the shift shaft bushing.

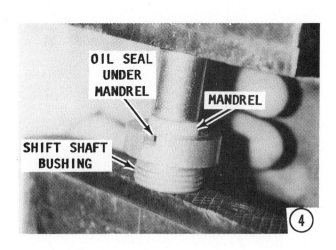

6- If the forward gear bearing race was removed, install it using the same number of shims removed during disassembly. If the number of shims was not recorded; if some were lost; or if a new assembly is being installed, begin with 0.015" thickness of shim material. Place the forward gear bearing race shims that were tagged and wired together during disassembly, into the forward gear bearing race cavity. Position the forward gear bearing race on the shoulder in the gear housing, with the **SMALL** inside diameter of the bearing race **TOWARD** the shift crank. Drive the bearing race into position.

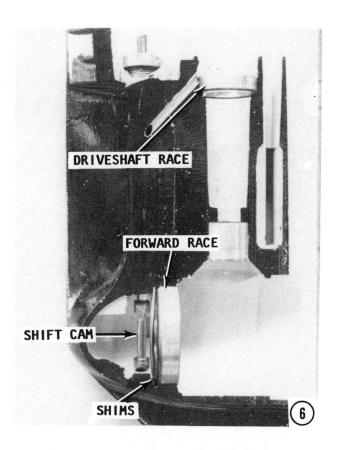

DRIVESHAFT RACE

FORWARD RACE

SHIFT CAM

SHIMS

6

STOP AND READ: Two methods are available to properly shim the lower unit. The first method involves shimming and checking the backlash before the unit is assembled. The second method is the factory approved procedure which involves assembling the unit; taking the required clear-

ance measurements; and then disassembling the unit if necessary, in order to change the number of shims installed. This second method requires the use of special tools obtained only from a MerCruiser dealer, while the alternate procedure can be accomplished using standard tools or modified standard tools.

Three steps are necessary to properly shim the lower unit. First the pinion gear must be shimmed to the correct depth. Secondly, the forward gear must be shimmed to the pinion gear for the proper backlash. And finally, the reverse gear is shimmed to the pinion gear with the correct backlash.

The following procedures shim the lower unit **WITHOUT** the use of factory special shimming tools.

7- Paint the forward and pinion gear teeth with Dyken machine dye, or equivalent substance.

8- Assemble the forward bearing onto the forward gear, and then lower the assembly into the lower unit.

ALL UNITS EXCEPT MR

9- With one hand, lower the driveshaft into the top of the lower unit; with the other hand, hold the pinion gear up in the lower unit just forward of the forward gear. Lower the driveshaft with the splines indexing with the splines of the pinion gear. Use an old pinion nut for this step, because if a

MACHINE DYE

7

FORWARD GEAR

8

PINION GEAR

NUT

9

new nut is used, it will lose its locking ability. Slide a driveshaft adaptor onto the end of the driveshaft. Now, turn the drive-shaft with a wrench on the adaptor, and at the same time thread the old pinion nut onto

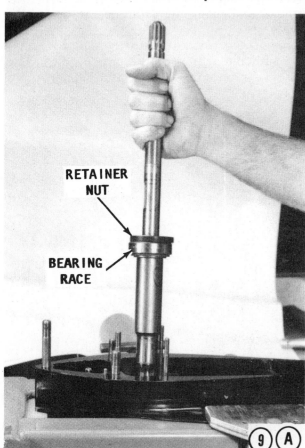

RETAINER NUT

BEARING RACE

9 **A**

the lower end of the driveshaft below the pinion gear. Hold the pinion nut with a box-end wrench, and at the same time turn the adaptor with a torque wrench until the nut has been tightened to the specification given in the Appendix.

SPECIAL PROCEDURE FOR MR UNIT

9A- Slide the outer driveshaft bearing race and retainer nut down the driveshaft over the bearing. Insert the driveshaft into the lower unit with the outer bearing race seating in the cavity of the lower unit. Secure the bearing in place with the retain-er nut. Tighten the nut to a torque value of 100 ft-lbs. Install the propeller shaft. Push in on the propeller shaft and at the same time, pull **UPWARD** on the driveshaft, then rotate the driveshaft clockwise from 25 to 30 turns. This action will establish a for-ward and pinion gear wear pattern for the MR unit.

ALL OTHER UNITS

Install the propeller shaft. Push inward on the propeller shaft and at the same time push **DOWNWARD** on the driveshaft, then rotate the driveshaft clockwise from 25 to

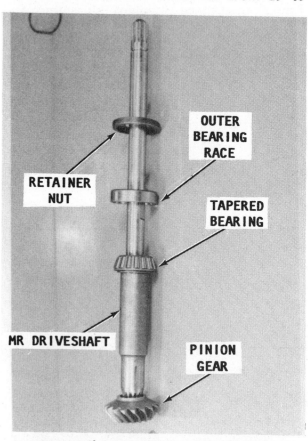

OUTER BEARING RACE

RETAINER NUT

TAPERED BEARING

MR DRIVESHAFT

PINION GEAR

Arrangement of parts on the new MR driveshaft.

30 turns. This action will establish a forward and pinion gear wear pattern for all except the MR unit.

10- Checking forward gear backlash:

Push hard on the propeller shaft to exert pressure onto the forward gear and to keep it from turning. Now, while exerting this force on the propeller shaft, turn the driveshaft back-and-forth with a light pressure. The movement of the driveshaft is the approximate backlash of the forward gear. The amount of play should be according to the specifications in the Appendix. A dial indicator is necessary to accurately determine the amount of backlash. If the backlash is too tight, remove shims from behind the forward gear bearing race. If the backlash is too great, add shim material behind the forward gear bearing race. As a

general guide: for each 0.003" of backlash, add or remove 0.002" of shim material.

11- Hold onto the driveshaft and remove the pinion nut, pinion gear, and the forward gear assembly. Examine the wear pattern on the forward gear and on the pinion gear. The pattern **MUST** be within 1/16" of the bottom of the gear cut onto the tooth. Also, the wear pattern should start about 1/8" in on the tooth and end about 1/8" from the end of the tooth. To change the wear pattern, the pinion gear must be raised or lowered by adding or removing shim material. If the backlash was satisfactory, but it is necessary to change the shim material from under the drivehshaft bearing, the backlash should be checked a second time. Once the pinion depth is established for the forward gear, or backlash, it **CANNOT** be changed for reverse gear. The reverse gear will be checked only for proper backlash.

12- Checking reverse gear backlash:

Disregard the propeller shaft and forward gear assembly in the accompanying illustration. Use an old pinion nut for this step, because if a new nut is used, it will loose its locking ability. With one hand, lower the driveshaft into the top of the lower unit, and with the other hand hold the pinion gear up in the lower unit. Lower the driveshaft with the splines indexing with the splines of the pinion gear. Slide a drive-

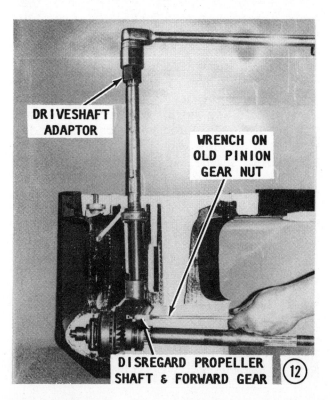

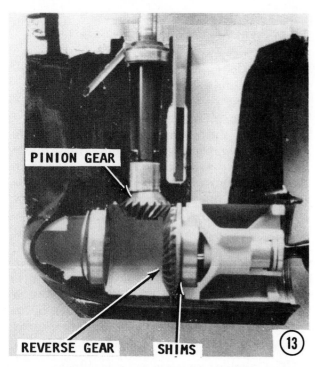

PINION GEAR

REVERSE GEAR   SHIMS   (13)

shaft adaptor onto the end of the driveshaft. Now, turn the driveshaft with a wrench on the adaptor, and at the same time thread the old pinion nut onto the lower end of the driveshaft below the pinion gear. Hold the pinion nut with a modified box-end wrench (used during disassembly), and at the same time turn the adaptor with a torque wrench until the nut has been tightened to the specification given in the Appendix.

13- Install the thrust washer onto the reverse gear, and then install the assembly into the bearing carrier. Install the same number of reverse gear shims into the lower unit housing as was removed during disassembly. If the number was not recorded; if shims were lost; or if a new assembly is being installed, begin with 0.015" total thickness of shim material. Coat the inside surface of the lower unit with oil as an aid to installing the bearing carrier.

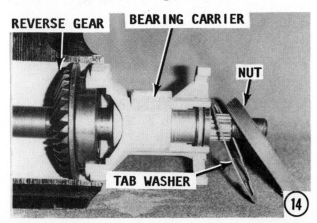

REVERSE GEAR   BEARING CARRIER

NUT

TAB WASHER   (14)

Leave the O-ring off the bearing carrier, as an aid to installing and removing the carrier at this time. Slide the bearing carrier into place. Install the tab washer and retainer nut. Tighten the nut to the specification given in the Appendix. **TAKE CARE** when tightening the nut not to jam the teeth of the gears together. This can be accomplished by continually moving the driveshaft to ensure the proper amount of backlash is being maintained while the nut is being tightened. Insert a long-shanked screwdriver through the propeller shaft hole in the bearing carrier. Exert a force on the screwdriver to prevent the reverse gear from turning. While exerting the force on the reverse gear, rotate the driveshaft lightly back-and-forth and determine the amount of backlash.

SPECIAL WORDS FOR MR UNIT

While exerting the force on the reverse gear, pull **UPWARD** on the driveshaft and at the same time rotate the driveshaft back-and-forth to determine the amount of backlash.

ALL UNITS

The movement of the driveshaft is the approximate backlash of the reverse gear. This amount of play should be according to the specifications in the Appendix. A dial indicator is necessary to accurately determine the amount of backlash. If the backlash is too tight, add shims to the reverse gear. If the backlash is too great, remove shims from the reverse gear. As a general guide: For each 0.003" of backlash, add or remove 0.002" of shim material.

14- Remove the retainer nut, tab washer, and bearing carrier. Remove the pinion nut, pinion gear and driveshaft. **DISCARD** the old pinion nut because it has lost its locking ability. **ALWAYS** use a **NEW** pinion nut during final assembling.

END SHIMMING W/O SPECIAL TOOLS

This completes the procedure of shimming without the use of special factory shimming tools.

SPECIAL WORDS

The next numbered step sequence continues with the work of assembling the lower unit. When shimming procedures are required, factory approved shimming tools

PROPELLER SHAFT ASSEMBLY

⑮

are used. However, if you have performed the preceding shimming instructions, then disregard the references to shimming in the following steps. **UNDERSTAND**, the assembling instructions are to be followed closely, as with any other procedure.

ALL UNITS

15- Install the assembled propeller shaft into the gear housing. This is accomplished by tilting the propeller end of the shaft toward the oil fill hole side of the gear housing to allow the actuating shaft-and-spool to engage the shift crank. Straighten the propeller shaft and operate the shift shaft. The sliding clutch should not move unless the shift crank and spool are moved.

16- If the tapered roller bearing was pressed off the driveshaft, install the bearing onto the driveshaft. Insert the drive-

shaft into a suitable adapter, against the bearing shoulder on the shaft, and then press against the inner bearing race to seat the bearing. Use a wrench adapter on the upper end of the driveshaft to protect the splines.

17- Lower the driveshaft pinion gear into the cavity. Move the propeller shaft slightly to allow the pinion gear to drop into position and mesh with the forward gear. Secure the lower unit housing in the horizontal position. Install the driveshaft, with the threaded end through the two bearings and the pinion gear. Coat the driveshaft locknut threads with a small amount of Loctite. Place the nut in the pinion nut holding tool, C-91-61067A2. If a pinion nut holding tool is not available, a substitute can be made from a box-end wrench, as described in the disassembly procedures. Place the nut in the wrench and proceed with the installation. Slide the tool into position under the driveshaft, and then thread the driveshaft into position, using driveshaft adaptor C-91-56775, over the splines.

MR MODEL

17A On the **MR** model, a thin washer is used to more evenly distribute the load of the nut against the pinion gear. When the washer is new, both sides are flat. However, after use, one side becomes slightly

TAPERED ROLLER BEARING   ADAPTOR

⑯

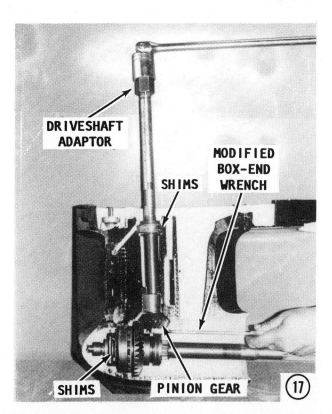

DRIVESHAFT ADAPTOR

SHIMS

MODIFIED BOX-END WRENCH

SHIMS   PINION GEAR

⑰

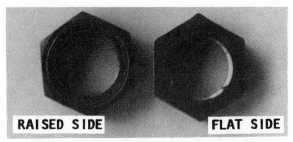

Both sides of a pinion gear nut. The raised portion, shown in the left view, MUST face the pinion gear.

concave, the other side raised. When installing a used washer, the raised side **MUST** face toward the pinion gear.

17B– After the driveshaft is in place, slide the outer bearing race and retainer nut down the driveshaft and over the bearing.

17C– Secure the retaining nut by tightening it with special tool 91-43506 to a torque value of 100 ft-lbs.

PINION GEAR SHIMMING PRIOR TO MR MODEL

1– Insert pinion gear shimming tool, C-91-56048, over the propeller shaft and into the housing. The design of this tool will allow the flat portion at the front of the gauge to clear the pinion gear. After the tool is inserted all the way past the pinion gear, turn the tool until the flat on the tool is **AWAY** from the pinion gear. Now, with the tool held in this position, insert a 0.025" feeler gauge between the bottom face of the pinion gear and the rounded part of the

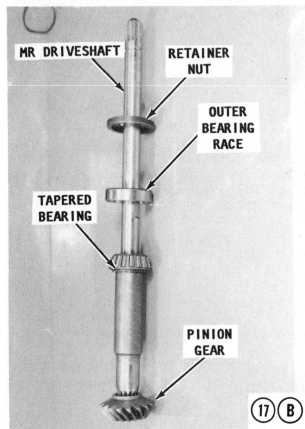

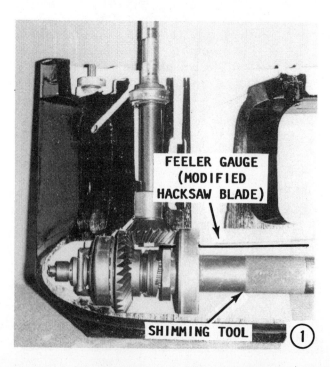

gauge. Most feeler gauges are not long enough to make this measurement properly. However, a suitable gauge can be made by grinding the teeth off a hacksaw blade and then using it to take the clearance measurement. The feeler gauge in the accompanying illustration is a modified hacksaw blade and is twisted 90° from its gauging position for photographic clarity. Twist the feeler gauge a quarter turn, and then insert it between the gauging point and the bottom face of the pinion gear. While taking the measurement, push down on the driveshaft to seat its roller bearing. If the clearance is over 0.025", **REMOVE** shims from under the upper bearing cup. If the clearance is under 0.025", **ADD** shims under the bearing cup.

PINION GEAR SHIMMING
MR UNITS

1A- Obtain special pre-load tool No. 91-44307A1. This tool consists of a spring that will seat on top of the driveshaft retaining nut; a plate that fits over the driveshaft and the three water pump studs; a thrust bearing to ride on top of the plate; a collar with Allen screw that slides down the driveshaft

and the screw bears on a flat on the driveshaft; and three nuts for the water pump studs.

Install the spring, plate, spacers, and nuts of the tool as described and shown. One of the nuts takes a 1/2" wrench and the other two a 7/16" wrench. Now, tighten the nuts alternately and evenly --moving the plate down the driveshaft as the spring is compressed. Continue tightening the nuts until the 1/2" nut "bottoms out" (all the threads on the stud are used. Check to be sure the plate is fairly level with the surface of the lower unit.

Slide the thrust bearing down the driveshaft and onto the surface of the plate. The thrust bearing may be installed with either side facing down. Check to be sure the Allen set screw in the collar is backed out to allow the collar to slide down the driveshaft. Move the collar down the driveshaft with the Allen screw end going down first, as shown and positioned to allow the Allen set screw to bear onto the water pump impeller key flat.

1B- Tighten the Allen set screw securely, and then back-off the three nuts

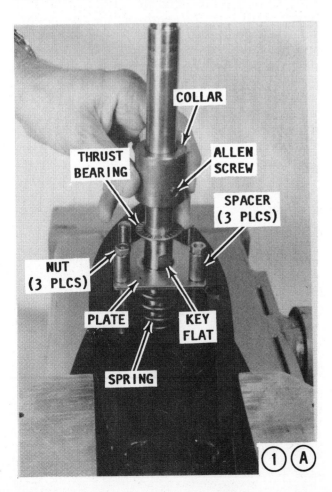

COLLAR

THRUST BEARING

ALLEN SCREW

SPACER (3 PLCS)

NUT (3 PLCS)

PLATE

KEY FLAT

SPRING

1 A

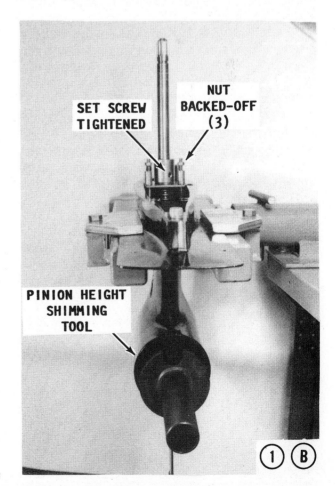

SET SCREW TIGHTENED

NUT BACKED-OFF (3)

PINION HEIGHT SHIMMING TOOL

1 B

on the water pump studs. The driveshaft has now been forced **UPWARD** placing an **UPWARD** load on the driveshaft bearing.

Insert pinion gear shimming tool, C-91-56048, over the propeller shaft and into the housing. The design of this tool will allow the flat portion of the front of the gauge to clear the pinion gear. After the tool is inserted all the way past the pinion gear, turn the tool until the flat on the tool is **AWAY** from the pinion gear. Now, with the tool held in this position, insert a 0.025" feeler gauge between the bottom face of the pinion gear and the rounded part of the gauge. Most feeler gauges are not long enough to make this measurement properly. However, a suitable gauge can be made by grinding the teeth off a hacksaw blade and then using it to take the clearance measurement. If the clearance is over 0.025", **REMOVE** shims from under the upper bearing cup. If the clearance is under 0.025", **ADD** shims under the bearing cup. Leave the preload shimming tool installed for forward and reverse gear shimming procedures.

ALL MODELS

2- Install the carrier shims, which were tagged and wired together during disassembly. Install a **NEW O**-ring seal on the carrier between the thrust washer and the carrier housing. Coat the surfaces of the **O**-ring seal and carrier with anti-corrosion grease. Insert the carrier assembly.

3- Insert the bearing carrier-to-gear housing alignment key. If the housing has two notches, use the top notch. Install the bearing carrier retainer tabbed washer, with the **V**-tab aligned with the **V**-notch on the bearing carrier.

4- Thread the gear housing cover into place **BY HAND** to prevent cross-threading. Tighten it a couple of turns by hand. Use bearing carrier wrench C-91-61069 and

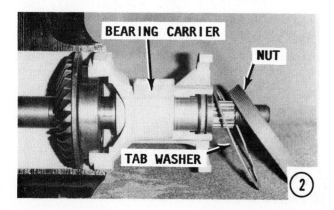

tighten the cover to a torque value of 150 ft-lbs, if the locking washer has a locking tab, or to 250 ft-lbs, if the washer does not have a locking tab. **DO NOT** secure the tabbed washer at this time. The tab is not secured until **AFTER** the forward and reverse gear backlash has been adjusted.

Shimming Forward Gear
To Correct Backlash
Prior to MR Model

5- Install bearing carrier puller with the arms of the puller on the carrier and the center bolt on the end of the propeller shaft. Tighten the puller center bolt to a torque value of 45 in.-lbs. This action places a preload on the forward gear, pushing the forward gear into the forward gear bearing. Rotate the driveshaft about three full revolutions, then recheck the torque value on the center puller bolt. Attach the backlash dial indicator rod 91-53459 to the driveshaft. Position the dial indicator shaft to the line marked "I" on the dial indicator rod. Exert a **DOWNWARD** force and at the same time rotate the driveshaft back-and-

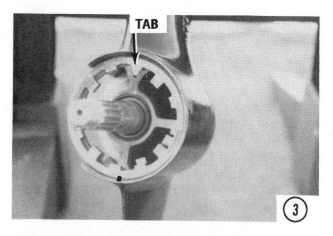

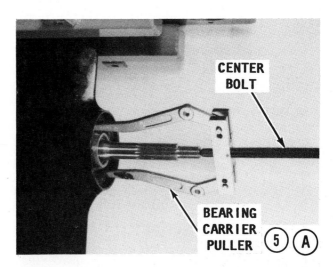

Cut-a-way view of the lower unit showing the clutch dog engaged with the forward gear.

forth and observe movement of the dial indicator. Total movement is the forward gear backlash. Check the listings in the Appendix for proper amount of backlash permissible for the unit being serviced. If the backlash is too much, add shim material behind the forward gear bearing race. If the backlash is too small, remove shim material from behind the forward gear bearing race. For each 0.003" of backlash, add or remove 0.002" shim material.

MR MODEL

SPECIAL NOTE: The driveshaft preload tool will remain installed for both the forward and reverse gear shimming procedures.

5A- Install bearing carrier puller with the arms of the puller on the carrier and the center bolt on the end of the propeller shaft. Tighten the puller center bolt to a

torque value of 45 in.-lbs. This action places a preload on the forward gear, pushing the forward gear into the forward gear bearing. Rotate the driveshaft about three full revolutions, then recheck the torque value on the center puller bolt.

5B- Attach the backlash dial indicator rod 91-53459 to the driveshaft. Position the dial indicator shaft to the line marked "I" on the dial indicator rod. Exert an **UPWARD** force on the driveshaft, and at the same time, rotate the driveshaft back-and-forth and observe movement of the dial indicator. Total movement is the forward gear backlash.

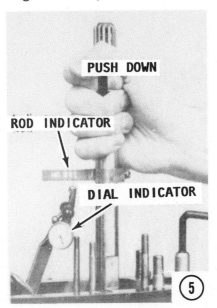

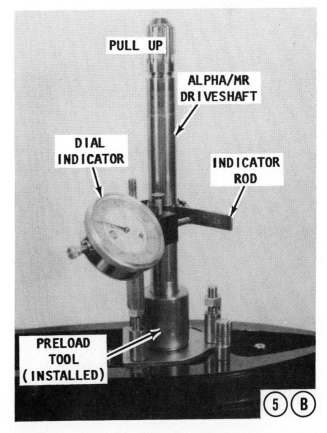

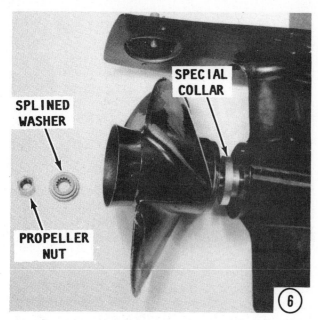

Check the listings in the Appendix for proper amount of backlash permissible for the unit being serviced. If the backlash is too much, add shim material behind the forward gear bearing race. If the backlash is too small, remove shim material from behind the forward gear bearing race. For each 0.003" of backlash, add or remove 0.002" shim material.

**Shimming Reverse Gear
To Correct Backlash**

ALL MODELS

6- The reverse gear backlash adjustment outlined in this step should be checked before correcting the forward gear backlash. In this sequence, the changing of shims can be accomplished in one disassembly job.

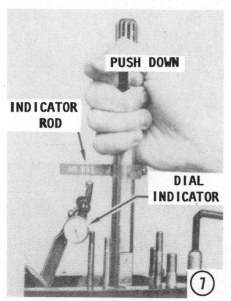

Slide the special collar 90-30366 onto the propeller shaft with the smaller end of the collar **TOWARDS** the bearing carrier. Install the propeller, then the splined washer with the splined side going on first -- to seat against the propeller. Thread the propeller nut onto the shaft, and then tighten the nut to a torque value of 45 in.-lbs. This action will place a preload onto the reverse gear -- pushing it into the reverse gear bearing.

7- Attach the backlash dial indicator rod 91-53459 to the driveshaft. Position the dial indicator shaft to the line marked "I" on the dial indicator rod.

PRIOR TO MR MODELS

Hold the propeller firmly to prevent the propeller shaft from turning. At the same time, exert a **DOWNWARD** force and rotate the driveshaft back-and-forth. Observe movement of the dial indicator. Total movement is the reverse gear backlash. Check the listings in the Appendix for proper amount of reverse gear backlash permissible for the unit being serviced. If the backlash is too much, remove shim material from in front of the bearing carrier thrust washer. If the backlash is too small, add shim material in front of the bearing carrier thrust washer. For each 0.003" of backlash, add or remove 0.002" shim material.

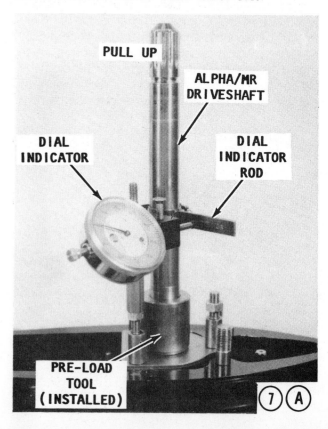

Cut-a-way view of the lower unit showing the clutch dog engaged with the reverse gear.

MR MODEL

7A- Hold the propeller firmly to prevent the propeller shaft from turning. Exert and **UPWARD** force on the driveshaft, and at the same time, rotate the driveshaft back-and-forth and observe movement of the dial indicator. Total movement is the reverse gear backlash. Check the listings in the Appendix for proper amount of reverse gear backlash permissible for the unit being serviced. If the backlash is too much, remove shim material from in front of the bearing carrier thrust washer. If the backlash is too small, add shim material in front of the bearing carrier thrust washer. For each 0.003" of backlash, add or remove 0.002" shim material.

ALL MODELS

8- After correcting the forward gear backlash by changing the thickness of shims

Pinion gear **PROPERLY** engaged with the teeth of the reverse gear. The text explains in detail how this proper adjustment is obtained.

behind the forward gear cup, **OR** the reverse gear backlash by changing the thickness of shim material between the reverse gear and reverse gear bearing carrier assembly; **REMOVE** the bearing carrier assembly and clean it of all lubricant used during assembling. After the assembly is clean, apply a liberal coating of Perfect Seal to the outer diameters of the carrier that contact the gear housing. **DO NOT** allow any sealer to enter the ball bearing or the reverse gear. Coat the threads of the retainer with the sealer, and then thread the retainer by **HAND**, to prevent any possibility of cross-threading.

9- Tighten the gear housing cover to the required torque value. Bend one of the tabs on the locking washer into one of the slots in the cover.

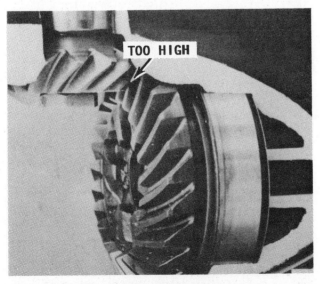

A pinion gear **NOT** engaged properly with the reverse gear. The pinion gear is riding too high on the reverse gear teeth.

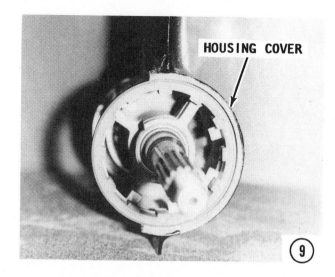

Set-up to show the clutch dog engaged with the *FORWARD* gear.

Instructional-type set-up showing the clutch dog in the NEUTRAL position.

10-22 WATER PUMP INSTALLATION

FIRST, A GOOD WORD: The water pump **MUST** be in very good condition for satisfactory service. The pump performs an extremely important function by supplying enough water to properly cool the stern drive and the power plant. Therefore, in most cases, it is advisable to replace the complete water pump assembly while the lower unit is disassembled.

WATER PUMP THROUGH STANDARD APLPHA DRIVE

1-   Install a **NEW** water pump base oil seal, with the lip facing **TOWARD** the impeller side of the base. Install a **NEW** larger oil seal with the lip facing the **GEAR** housing. Install a **NEW** O-ring seal on the water pump base. Press a **NEW** oil seal into the pump body, with the lip facing **UPWARD**. Late model drive units do **NOT** have this larger oil seal in the pump body.

Set-up to show the clutch dog engaged with *REVERSE* gear.

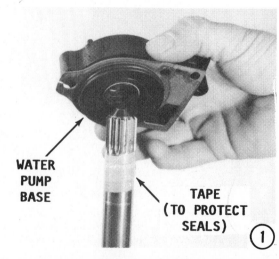

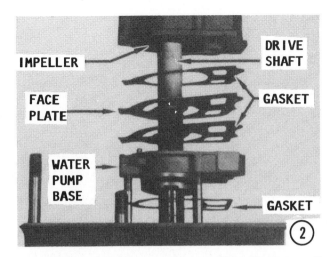

IMPELLER
FACE PLATE
WATER PUMP BASE
DRIVE SHAFT
GASKET
GASKET

②

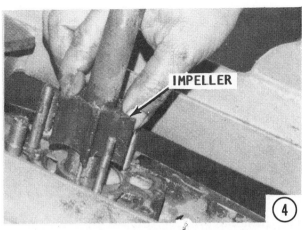

IMPELLER

④

Apply Perfect Seal to the outside diameter of the pump housing insert. Install the insert into the pump body, with the tab on the insert aligned with the hole in the pump body. **REMOVE** any excess Perfect Seal. Position a **NEW** gasket on the pump base. Apply a coating of Multi-purpose lubricant to the base, **O**-ring seal, and the oil seal lips. Install a seal protector to cover the splines on the driveshaft as a prevention against damaging the new oil seals. If a seal protector is not available, the driveshaft splines may be wrapped with tape to protect the seals from the splines. Install the base assembly.

2- Slide a **NEW** lower gasket into place on the water pump base. Install the face plate, with the lip facing **DOWN,** and then install a **NEW** upper gasket.

3- Coat the drive key with Multi-purpose Lubricant to hold it in place, and then insert the key in the slot on the driveshaft.

4- Slide the impeller over the driveshaft, with the slot in the impeller aligned to engage over the key on the driveshaft. Push down on the impeller to seat it securely on the face plate.

5- Before installing the water pump body, first coat the oil seal lip with Multi-purpose Lubricant and cover the pump insert with a

mixture of soap and water. Install the pump body down over the impeller. Push down on the body and at the same time turn the driveshaft **CLOCKWISE** to assist the impeller to enter the pump body without cutting it. Install the washers, screw, and nuts.

6- Tighten the screw and nuts to the torque value given the accompanying illustration. **OVERTIGHTENING** the screw or nuts could result in pump **FAILURE.** Install the centrifugal slinger and a **NEW** driveshaft **O**-ring

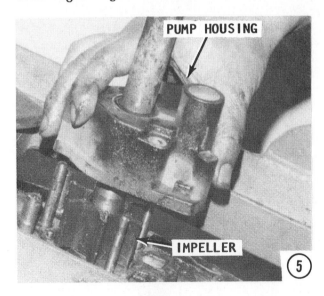

PUMP HOUSING

IMPELLER

⑤

PUMP KEY

③

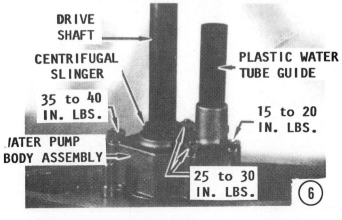

DRIVE SHAFT

CENTRIFUGAL SLINGER

35 to 40 IN. LBS.

WATER PUMP BODY ASSEMBLY

PLASTIC WATER TUBE GUIDE

15 to 20 IN. LBS.

25 to 30 IN. LBS.

⑥

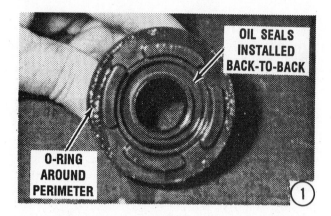

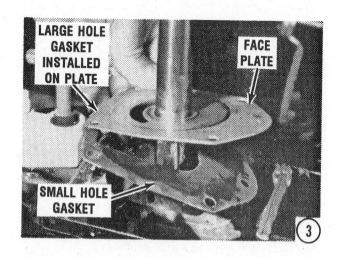

seal. Install the water inlet tube guide. Replace the shift shaft seal and flat washer.

WATER PUMP -- ALPHA DRIVE GENERATION II

1- If the seals in the seal carrier were removed, apply a thin coating of Quicksilver Perfect Seal, P/N 92-34227-1 to the oil seal bore, and then press the seals in back-to-back with the small seal going in first, with the lip facing down -- away from driver P/N 91-817569. The seal is properly installed when the driver bottoms against the carrier. **DO NOT** press further -- the carrier may be damaged. Press the large seal in with the lip facing up, until the driver bottoms on the carrier. This places the seals in the carrier back-to-back. Install a new **O**-ring around the perimeter of the carrier.

2- Apply a light coating of Quicksilver 1-4-C Marine Lubricant P/N 92-90018A12, or equivalent, to the lips of the oil seals and to the **O**-ring. Slide the carrier down the driveshaft and into the lower unit opening. **DO NOT** use a hammer to seat the carrier. Use only **HAND** pressure.

3- Slide the small hole gasket down the driveshaft, followed by the face plate and the large hole gasket. Holes in the gaskets and plate will only align with each other and the holes in the lower unit **ONE** way. If the holes do not align one or more of the items is upside down. Correct the situation by turning one or more of the items over.

4- Apply just a "dab" of grease to the Woodruff key, and then place in the driveshaft keyway. Slide the impeller down the driveshaft and onto the face plate with the cutout in the impeller indexed over the Woodruff key. If an old impeller with a "set" to the blades is being installed, face the curl of the blades in a **COUNTERCLOCKWISE** direction. If the direction is reversed, premature impeller failure will surely occur.

Slide the pump body down the driveshaft and just to the top of the impeller. Keeping the gasket, impeller, and body mounting holes aligned is not an easy task. However, if a couple of pins are inserted down through just two opposite holes, as shown, the holes will stay

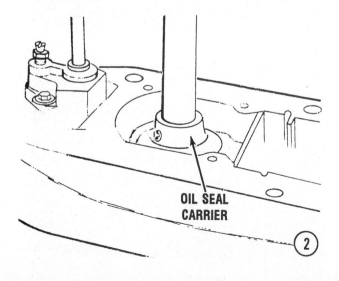

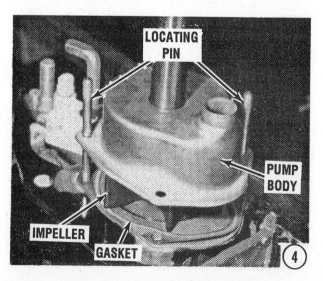

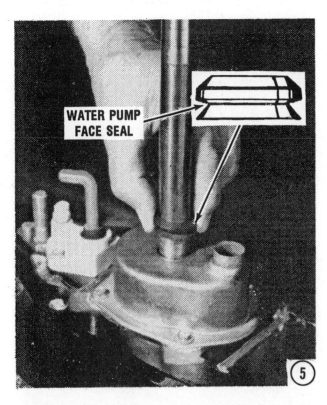

WATER PUMP FACE SEAL

(5)

FORWARD GEAR

SHIFT CAM

CLUTCH DOG

(1)

aligned while the body is worked down over the impeller. The pins can be old drill bits, small diameter bolts, rod, whatever is handy. Exert some downward pressure on the pump body and at the same time rotate the driveshaft **CLOCKWISE** and the impeller blades will "set" in the proper direction.

Start a couple of the body mounting bolts, and then remove the two pins. Install the remaining mounting bolts, and tighten them to a torque value of 60 in lb (7.9Nm).

5- Apply a light coating of lubricant to the driveshaft. Move the water pump face seal down the driveshaft until it is a few inches above the pump body, as shown. Obtain special seal setting tool P/N 92-90018A12. If the tool is not available, a collar or large washer with an inside diameter slightly larger than the driveshaft could be used. Push the seal down onto the water pump body, and then remove the tool.

10-23 LOWER UNIT INSTALLATION

This procedure provides instructions for proper installation of the lower unit, if the upper gear housing was not removed. For instructions to install the upper housing to the lower unit, see Section 10-24.

1- Verify that the helm shift handle is in **FORWARD** and the lower unit is in **FORWARD** gear.

2- Verify that the trim tab bolt is in place in the aft section of the lower unit housing.

3- Check to be sure the O-ring is in place on the passageway for the gear oil.

4- Observe closely that the stainless steel special washer is installed properly over the shift shaft and is seated onto the shift shaft bushing. This washer **MUST** be made of stainless steel. **NEVER** attempt to replace it with an iron or brass washer, because the washer prevents the shift shaft in the upper gear housing from working up-and-down as it turns. An iron or brass washer would give very limited service. Check to be sure the plastic guide tube is installed into the water pump housing. Verify that the thick rubber gasket is fully positioned on top of the water pump. This gasket prevents exhaust gases from entering the water system. Verify that the O-ring is installed onto the driveshaft. This O-ring is

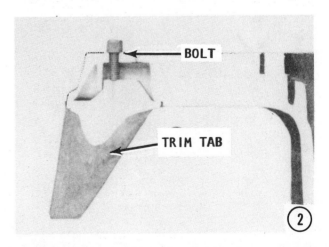

BOLT

TRIM TAB

(2)

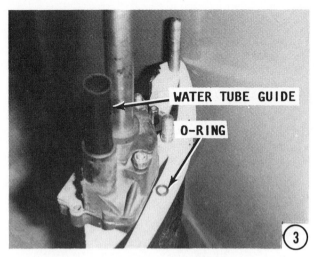

extremely important to prevent water from reaching the driveshaft splines.

5- Now, raise the lower unit to mate with the upper gear housing, and at the same time check to be sure the water tube starts into the water pump guide tube and the male splines of the shift shaft indexes with the female splines of the upper gear housing. It may be necessary to slowly turn the propeller shaft counterclockwise until the splines index.

6- Start the two 5/8" hex nuts on both sides of the housing, to hold the lower unit in place. Install the 5/8" hex nut from the leading edge of the housing. Next, install the Allen-head screw into the trim tab cavity. Tighten the four 5/8" nuts alternately, and a little at-a-time, until the lower unit is tight against the upper gear housing.

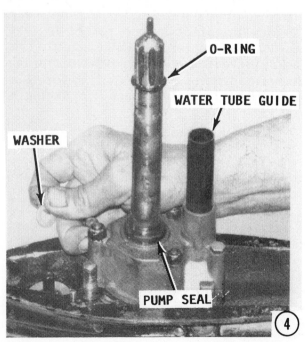

The short arm of the shift shaft which must face forward when mating the upper gear housing to the lower unit.

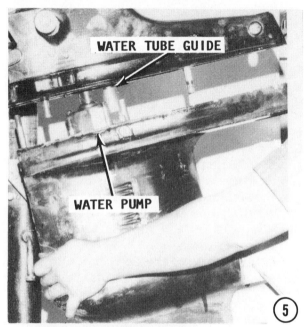

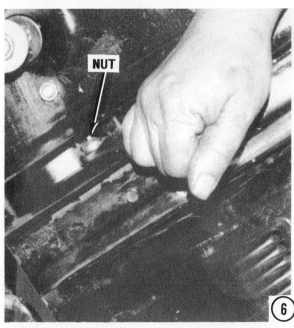

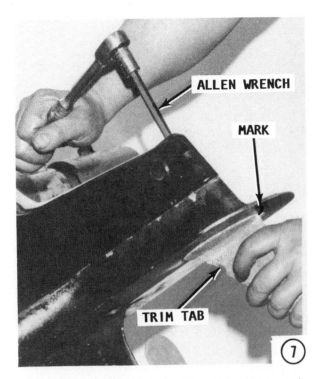

7- Install the trim tab with the marks made during disassembly aligned. Tighten the Allen-head screws securely.

GOOD WORDS:

The trim tab performs two very important jobs, one of which you may not realize. First, the tab compensates for steering torque. If the boat continually seems to move to port or starboard while the helmsman is on a straight course, the trim tab can be adjusted to the side of the pull. The tab also prevents electrolysis from damaging expensive parts. The tab is not expensive; it should show signs of electrolytic action; and should **ALWAYS** be replaced after some of the material has been eaten away or is pitted. **NOW**, if the tab shows no signs of electrolytic action after the boat has been in use over a period of time, the grounding should be checked to ensure more expensive parts are not being damaged. Install the trim tab plastic cover plug.

10-24 UPPER GEAR HOUSING TO LOWER UNIT ASSEMBLING

This section provides detailed instructions to mate the upper gear housing with the lower unit. The assembled stern drive unit is then installed onto the boat, see Section 10-25.

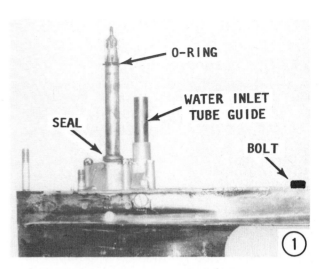

1- Place the trim tab mounting Allen-head screw into the hole provided.

2- On late model units with interconnected oil chambers, install a **NEW O-ring** seal coated with Perfect Seal. Check to be **SURE** the driveshaft O-ring seal is in place.

3- Observe closely that the stainless steel special washer is installed properly over the shift shaft and is seated onto the shift shaft bushing. This washer **MUST** be made of stainless steel. **NEVER** attempt to replace it with an iron or brass washer, because the washer prevents the shift shaft in the upper gear housing from working up-and-down as it turns. An iron or brass washer would give very limited service. Check to be sure the plastic guide tube is installed into the water pump housing. Verify that the thick rubber gasket is fully

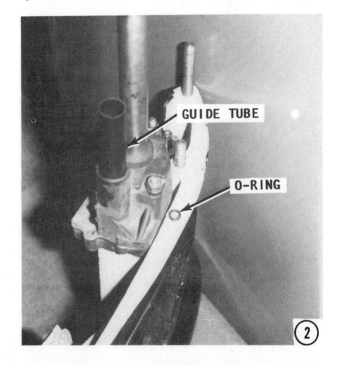

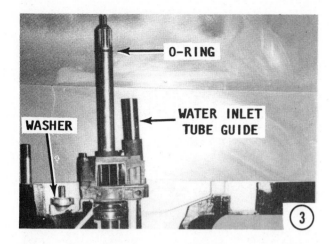

positioned on top of the water pump. This gasket prevent exhaust gases from entering the water system. Verify that the O-ring is installed onto the driveshaft. This O-ring is extremely important to prevent water from reaching the driveshaft splines.

4- Shift the lower unit into full **FORWARD** gear. Move the shift shaft **CLOCKWISE** as far as it will go, and at the same time, turn the propeller shaft counterclockwise until it stops to ensure the clutch is fully engaged. Coat the driveshaft splines with Multi-purpose lubricant.

5- Check the shift shaft extending out of the upper gear housing. The small arm of the shaft **MUST** face directly **FORWARD** in

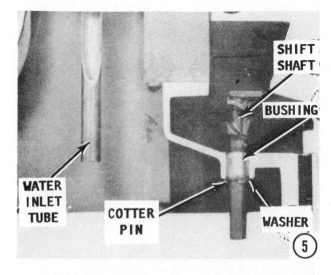

relation to the upper gear housing. In this position, the upper gear housing is in **FORWARD** gear. The lower unit is in **FORWARD** gear, as verified in the previous step. Therefore, both sections of the shaft will mate properly. **BAD NEWS:** If the shaft is not positioned properly (is off by just one spline) the shift cannot be adjusted properly.

6- Insert the driveshaft into the housing with the water tube aligned with the water

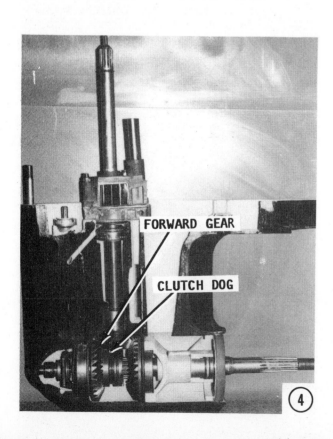

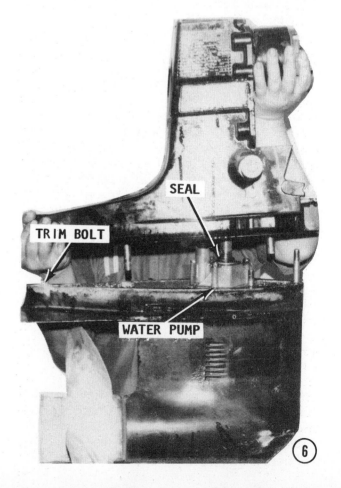

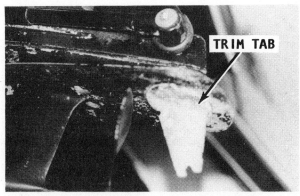

Corroded trim tab which must be replaced. It is perfectly natural for the tab to suffer the effects of electrolysis. This is one of its major functions - to protect more expensive parts.

tube guide and the driveshaft splines indexing with the upper driveshaft splines. If necessary, slowly rotate the propeller shaft counterclockwise to assist the lower driveshaft splines to index with the splines of the upper driveshaft. The splines of the upper and lower driveshafts **MUST** be fully indexed, or the units cannot be properly assembled.

7- Start one hex-head nut on each side of the driveshaft housing. Tighten the nuts one turn at-a-time until the two machined surfaces come together. Install the Allen-head screw into the trim tab cavity and the hex-head nuts on the bottom of the anti-cavitation plate. Install the nut on top of the leading edge. Tighten all nuts to a torque value of 35 ft-lbs.

8- Install the trim tab with the marks made during disassembly aligned. Tighten the Allen-head screws securely.

GOOD WORDS: The trim tab performs two very important jobs, one of which you may

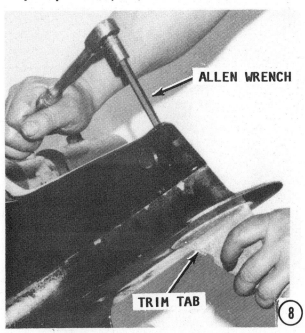

not realize. First, the tab compensates for steering torque. If the boat continually seems to move to port or starboard while the helmsman is on a straight course, the trim tab can be adjusted to the side of the pull. The tab also prevents electrolysis from damaging expensive parts. The tab is not expensive; it should show signs of electrolytic action; and should **ALWAYS** be replaced after some of the material has been eaten away or if it is pitted. **NOW**, if the tab shows no signs of electrolytic action after the boat has been in use over a period of time, the grounding should be checked to ensure more expensive parts are not being damaged. Install the trim tab plastic cover plug.

10-25 STERN DRIVE INSTALLATION

1- **READ AND BELIEVE:** The engine **MUST** be properly aligned or the female splines in the engine coupler and the male splines of the driveshaft will be destroyed in a short time. If troubleshooting indicates the coupler is damaged and requires replacement, remove the engine, see the appropriate section in Chapter 3, and then replace the coupler. The coupler is simply secured to the flywheel with attaching bolts. If the necessary tools are not available for proper engine alignment, the boat **SHOULD** be taken to a marine dealer. Now, from outside the boat, insert the small diameter end of Alignment Shaft, C-91-48247, through the U-joint bellows and into the gimbal bearing. If this tool is not available, use the end of Alignmment Tool, C-91-57797A3, which does not have a threaded hole.

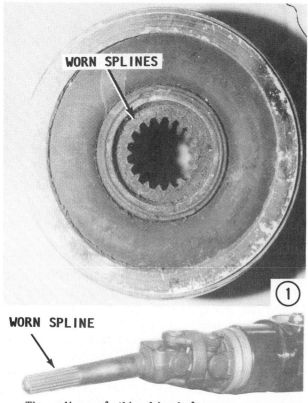

The splines of this driveshaft were prematurely worn due to incorrect engine alignment.

2- Move the gimbal bearing with the shaft to index the shaft with the engine coupling splines. The alignment is correct if the shaft enters the coupling freely, with no pressure. **AS AN AID**, apply a small amount of grease onto the end of the alignment tool. Insert the tool through the U-joint bellows, the gimbal bearing, and into the coupler until it is fully seated. Now, carefully remove the tool **WITHOUT** turning it and inpsect the end with the grease. If the grooves are deep on the top side of the tool and light on the bottom, the indication would be that the engine is too high.

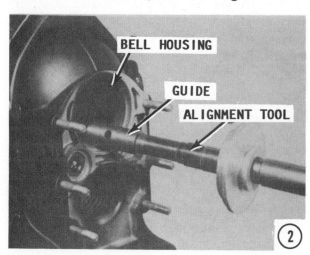

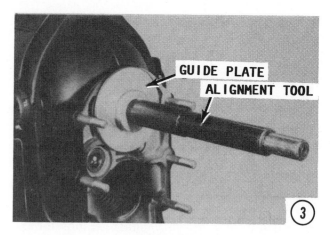

3- If the groove marks are reversed, the front of the engine is too low. If the alignment is not correct, and difficulty is encountered in indexing the splines, loosen the front mounting bracket adjusting nuts and rotate them to raise or lower the front of the engine until the alignment shaft slides into the engine coupling splines with no resistance. Once the correct alignment shaft fit is obtained, tighten the lower adjusting nut securely. Again, check the alignment shaft fit. Secure the mounting bracket to the boat.

4- Check the bell housing bearing to be sure it turns freely without binding or rough spots. If the bearing needs to be replaced, see Chapter 11.

5- Install a **NEW** O-ring in the water passageway between the two housings and the rubber sealing ring inside the bell housing. Install a **NEW** gasket on the face of the bell housing.

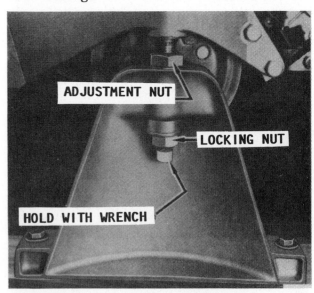

Front engine mount adjustment on an in-line engine. This adjustment is used when aligning the driveshafts of the engine and the upper gear housing.

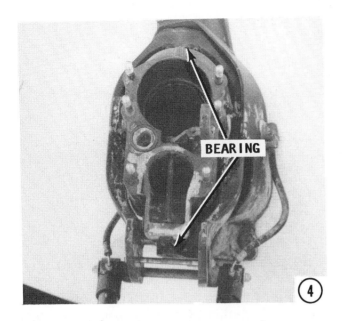

(4)

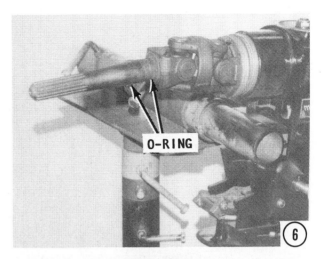

(6)

6- Oil the shaft to prevent damage to the seals, and then slide two **NEW** O-ring seals over the splines of the universal joint shaft, and into the grooves provided.

7- Coat the splines of the driveshaft and the driveshaft housing pilot, with Multi-purpose Lubricant. As an aid to installation, apply a light coating of Multi-purpose Lubricant to the inside surfaces of the bell housing bore.

8- Apply a coating of Multi-purpose Lubricant onto the shift lever coupler, and then align the slot straight by moving the lever to the right. Lubricate the slot and the reverse lock roller.

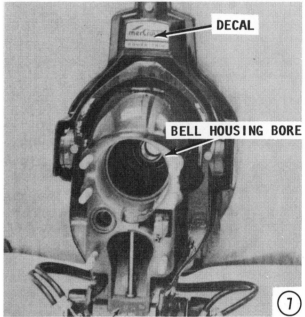

(7)

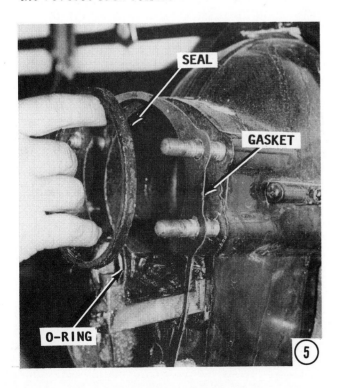

(5)

(8)

9- Check to be sure the stern drive unit is in **FORWARD** gear. Rotate the propeller shaft to align the intermediate shift shaft coupling with the upper shift shaft slot. **BAD NEWS:** If the coupler of the intermediate shift shaft in the driveshaft housing and the locating slot of the shift lever in the bell housing are not aligned properly, the bell housing and shift shaft couplers will be damaged when the drive unit is installed.

10- Coat the **U-joint O-rings** and gimbal housing ball bearing with Multi-purpose Lubricant, to assist installation. Install the

drive unit into the bell housing bore. If the driveshaft splines do not index with the splines of the engine coupler, slide the propeller onto the propeller shaft, and then **SLOWLY** rotate the shaft counterclockwise until the drive unit can be pushed completely into position. While performing this installation maneuver, make sure the shifting slide assembly does not turn out of position. If the stern drive will not move completely into place, check underneath and observe if the shift cam is properly aligned in the groove on the stern drive.

11- Secure the driveshaft housing to the bell housing with the six elastic-stop nuts with flat washers. Tighten the nuts alternately and evenly to a torque value of 50 ft-lbs.

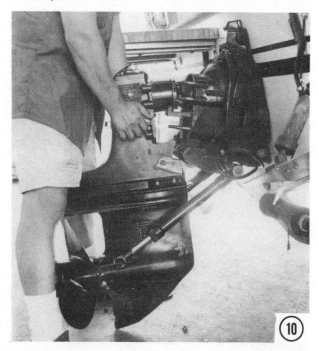

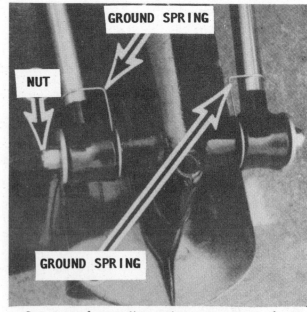

One type of grounding spring arrangement for the trim/tilt cylinders.

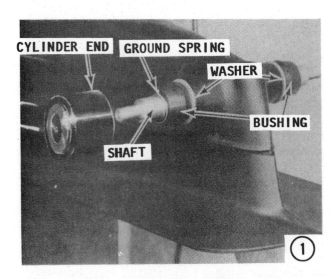

CYLINDER END GROUND SPRING
WASHER
BUSHING
SHAFT
①

10-26 TRIM CYLINDER INSTALLATION

1- On each side of the lower unit: Loosen the nut securing the hydraulic cylinder to the forward anchor pin. Insert the aft anchor pin through the hole in the driveshaft housing. Slide one large washer, one rubber bushing, and one spiral spring onto each aft anchor pin. The spiral springs between the rubber bushings form an electrical ground between the cylinder and the stern drive to reduce electrolysis. On models between 1975 and 1978, the cylinders were coated with Nylon, as a substitute for the spiral springs. However, the coating was not satisfactory. The coating should be removed and the springs installed. On late model units, since 1979, the hydraulic hoses are made of stainless steel, to form a ground between the cylinders and the stern drive. Even with these new hoses, the springs should **STILL** be used to ensure a satisfactory ground.

2- Install the cylinders onto the aft anchor pins. Pick up the rubber bushings and observe that one end has a smaller diameter than the other. The small diameter end **MUST FACE** the driveshaft housing. Push a rubber bushing onto the aft anchor pin and into the mounting hole of the cylinder with the small diameter end facing the housing. Slide one flat washer onto each aft anchor pin. Thread a retaining nut onto each anchor pin and tighten the nuts securely.

10-27 PROPELLER INSTALLATION

1- Check to be sure the splines and the threaded end of the propeller shaft are clean and in good condition. Install the propeller thrust hub. Coat the propeller shaft splines with Perfect Seal No. 4, and the rest of the shaft with a good grade of anti-corrosion lubricant.

2- Install the propeller, and then the splined washer, tab washer, and propeller nut.

3- Position a block of wood between the propeller and the anti-cavitation tab to keep the propeller from turning. Tighten the propeller nut to a torque value of 35-45 ft-lbs. Adjust the nut to fit the tab lock space.

4- Bend three of the tab washer tabs into the spline washer using a punch and hammer. The tabs will prevent the nut from backing out.

10-28 STERN DRIVE LUBRICATION

Filling Stern Drive Unit

1- Remove the vent screw in the upper gear housing. Place the stern drive in an

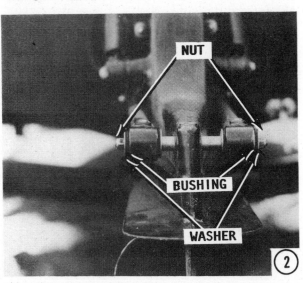

NUT
BUSHING
WASHER
②

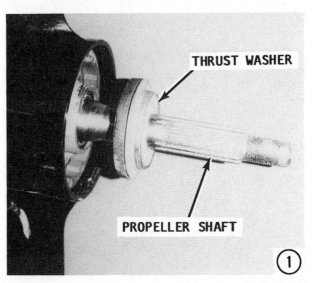

THRUST WASHER
PROPELLER SHAFT
①

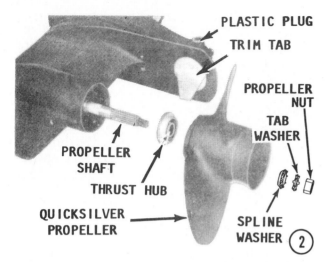

Figure 2

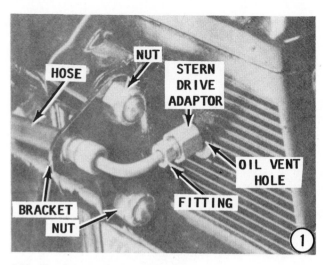

Figure 1

approximate vertical position without a list to port or starboard. Insert a lubricant tube into the **FILL** plug opening in the lower unit. Inject lubricant until the excess starts to flow out the **VENT** hole in the upper gear housing. After the unit is full of lubricant, install the **VENT** and **FILL** plugs, with **NEW** washers. As the plugs are tightened, check to be sure the washers remain in place to prevent water from leaking into the housing.

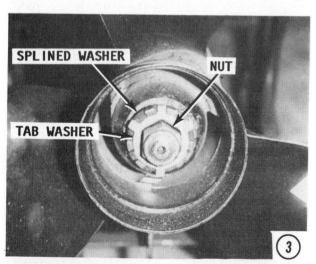

Figure 3

Figure 2

MerCruiser Model 120—260

2- A visual oil reservoir kit is available from the local MerCruiser marine dealer. This kit includes a see-through transparent lubricant reservoir and the necessary hose and fittings for installation inside the boat in any convenient location. After the kit is installed, the reservoir is filled to the **FULL** mark with regular Quicksilver Super Duty Lower Unit Lubricant. The reservoir allows the lubricant to "breathe" as the running components create channels in the lubricant. Any loss of lubricant can be detected by a change in the level of lubricant in the transparent reservoir after the unit has cooled. Lubricant can be added at any time.

A consistent loss of lubricant indicates the stern drive unit is in need of close inspection to determine the cause. Normal maintenance calls for changing the lower unit lubricant every 100 hours of operation, but with the reservoir kit installed, this period recommendation is extended to 200 hours, provided there has not been a consistant loss or discoloration of lubricant.

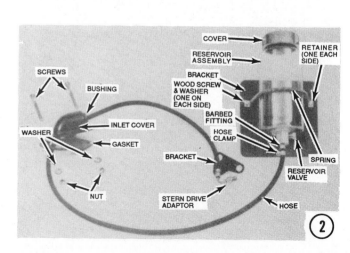

Figure 2

10-29 BRAVO STERN DRIVE

INTRODUCTION

Since its introduction in 1988, the Bravo stern drive has been mated with the GMC V8 350 CID and 454 CID engines. At press time, the manufacturer announced plans to expand the availability to the full range of MerCruiser powerplants.

The Bravo can be easily identified from its older brothers, the MC and Alpha units, by the different shape gearcase and an engine driven water pump.

The stern drive unit serial number and gear ratio will be found on a decal affixed to the port side of the upper gear housing.

No gear ratio change is available (at press time) for the Bravo when operating at high altitudes. However, the manufacturer recommends a propeller change for the higher elevations. Consult the local dealer for his propeller suggestions. The standard gear ratio for all Bravo units is 1.5:1.

The service procedures in this chapter provide complete illustrated instructions to service the entire stern drive unit. Some areas require special tools to remove, install, or adjust certain parts. Therefore, remove only those parts unfit for further service or parts which must be removed to gain access to a faulty unit. Be sure the special tools are available before work is started.

The chapter is divided into the following broad areas:
Operation.
Troubleshooting.
Precautions.
Removal of the stern drive from the bell housing.
Separating the lower unit from the upper gear housing.
Upper gear housing service.
Lower unit service.
Installing the lower unit to the upper gear housing.
Stern drive to bell housing installation.

OPERATION

The upper gear housing transfers engine power to the lower unit through a driveshaft. Splines on the forward end of the U-joint shaft mate with matching splines of a coupler attached to the engine flywheel. The aft end of this shaft is connected to a drive gear which meshes with two clutch gear assemblies. The drive gear and two clutch gears change the direction of the power train from horizontal to vertical. The power is transferred from the upper gear housing down to the lower unit through a vertical driveshaft. A water pump installed on the engine is belt driven by the crankshaft pulley. This pump pulls water through the stern drive. See Chapter 9 for water pump service.

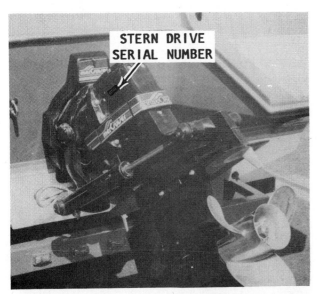

Location of the Bravo stern drive serial number. This number should always be given when ordering replacement parts for the stern drive. This serial number should not be confused with the transom serial number shown and described in the adjacent column.

Location of the Bravo transom serial number. This number should always be used when ordering transom replacement parts. Do not confuse this number with the stern drive serial number shown in the adjacent column.

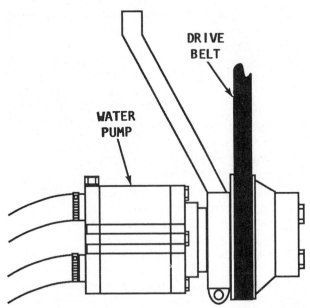

Line drawing to depict a Bravo water pump installed on the front of the engine and driven by a belt off the crankshaft pulley.

All shifting is accomplished inside the upper gear housing. The shift cable actuates a yoke and cam assembly via linkage. The yoke rides in the center of a clutch spool between two brass collars of the clutch gears. The contact surfaces of the collars are not perfectly horizontal. By pivoting left or right the yoke is able to block one of the clutch gears and permit the clutch shaft to rotate in only one direction.

The lower splines of the short clutch shaft are coupled with the upper splines of the driveshaft.

The pinion gear and single driven gears are contained within the lower unit. Unlike many other stern drive units, the lower unit is free of any shifting mechanisms and water pump components.

TROUBLESHOOTING

Troubleshooting **MUST** be done before the unit is removed from the boat to permit isolating the problem to one area.

1- Check the propeller and the rubber hub for shredding. If the propeller has been subjected to many strikes with underwater objects, it could slip on its hub. If the hub appears to be damaged, replace it with a **NEW** hub. Replacement of the hub must be done by a propeller rebuilding shop equipped with the proper tools and experience for such work.

2- Shift mechanism check: Verify that the ignition switch is **OFF**, to prevent pos-

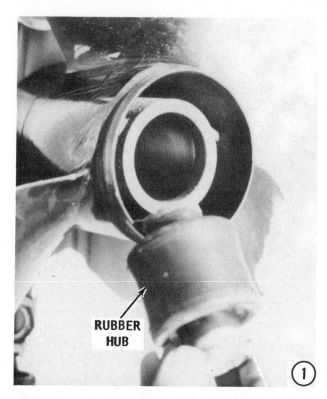

sible personal injury, should the engine start. Shift the unit into **REVERSE** gear and at the same time have an assistant turn the propeller shaft to ensure the clutch is fully engaged. If the shift handle is hard to move, the trouble may be in the stern drive unit, transom shift cable, remote control cable, or the shift box.

3- Isolate the problem: Disconnect the remote-control cable at the transom plate, by first removing the two nuts, and then lifting off the remote-control shift cable. Operate the shift lever. If shifting is still hard, the problem is in the shift cable or control box. If the shifting feels normal with the remote-control cable disconnected,

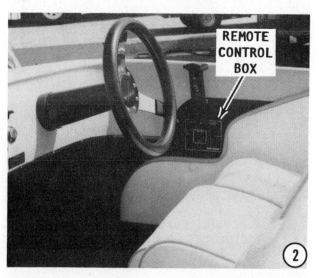

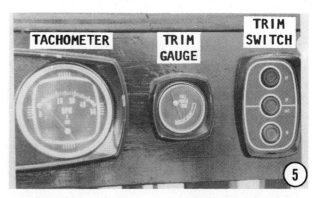

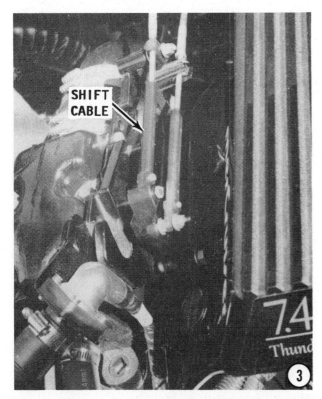

the problem must be in the stern drive. To verify the problem is in the stern drive unit, have an assistant turn the propeller and at the same time move the shift cable between the transom plate and stern drive back-and-forth. Determine if the clutch engages properly. Most of hard shifting problems are caused because this cable is not moved during the off-season. Water entering the cable, especially salt water, will cause rapid corrosion and hard shifting. If the cable requires replacement, see Chapter 11. If the cable moves freely, then reconnect the control cables at the transom plate.

4- Stern drive noise check: FIRST, a word about stern drive noise: When the stern drive is positioned for dead-ahead operation, the U-joints will make very little noise. However, as the stern drive is moved

to port or starboard, the noise will increase, due to the working of the U-joints. This test and procedure is for abnormal noises.

Attach a Flush-Test device to the stern drive and turn on the water. Start the engine and shift into gear.

CAUTION
Water must circulate through the stern drive, also to and from the engine, anytime the engine is run to prevent damage to the water pump located in the lower unit (original and Alpha drives); or on the front of the engine (Bravo drive). Just a few seconds without water will damage the water pump.

Operate the engine at idle speed. Turn the stern drive slightly to port and then slightly to starboard, and at the same time

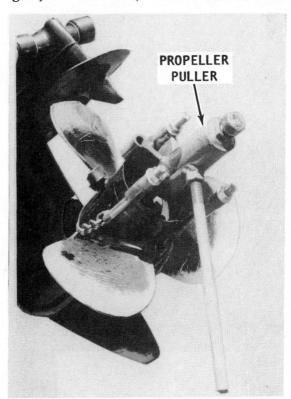

Use of a propeller puller as an aid in removing a propeller "frozen" to the propeller shaft.

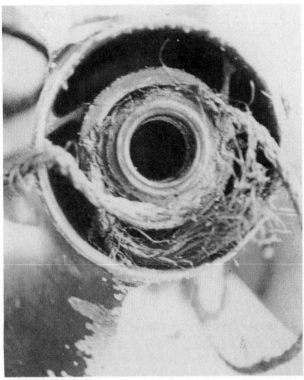

Excellent view of rope and fish line entangled behind the propeller. Entangled fish line can actually cut through the seals allowing water to enter and oil to escape from the lower unit. Check this area constantly.

RUBBER HUB

Cutaway view of a propeller to expose the rubber hub.

listen at the upper driveshaft housing. An unusual noise is an indication the U-joints are worn or the gimbal housing bearing is defective. If the bearing is defective, see Chapter 11.

Shut down the engine. Turn off the water. Disconnect the Flush-Test device.

5- Tilt/Trim system check: Operate the control buttons on the control panel or on the shift control lever. Check to determine if the stern drive unit moves to the full up and full down positions. If the drive unit fails to move, the U-joints may be frozen due to corrosion from water entering through the U-joint bellows; lack of lubrication; or from non-operation over a long period of time. If the stern drive does not move properly, the trim/tilt mechanism may need service, see Chapter 8.

STERN DRIVE NOISES

Noise in upper gear housing
1- Oil level low
2- Worn U-joints
3- Worn bearings
4- Incorrect thrust washer placement
5- Gear timing incorrect
6- Engine not aligned properly
7- Worn engine coupler
8- Transom too light for stern drive installation
9- Worn, damaged, or loose internal parts

Noise in lower unit
1- Propeller not installed properly
2- Propeller shaft bent
3- Worn, damaged, or loose internal parts
4- Metal particles in lubricant
5- Blocked oil passage between upper gear housing and lower unit
6- Worn bearings
7- Incorrect gear alignment
8- Incorrect shimming

SERVICE PRECAUTIONS

The following safety precautions should be observed while servicing the stern drive. The list is presented here as an assist in preventing personal injury or expensive damage to parts.

1- Whenever possible, use special tools when they are specified. Always use the proper size common tool when removing or installing attaching parts.

2- Threaded parts are right-hand, unless otherwise indicated.

3- Clean parts in regular cleaning solvent and blow them dry with compressed air. Disassembly and assembly should be done on a clean work bench to prevent contamination (foreign matter, dirt) from entering the oil passageways or from becoming attached to parts. Such material can result in false shimming indications and will lead to premature bearing failure.

4- If bearings are removed from assemblies, keep them **TOGETHER**, with the original shims as an aid during assembling. If new parts are used with the assembly, the unit will have to be reshimmed in order to obtain the proper bearing preload or gear lash. Save and install the original amount of shim material as a starting point in determining the correct shim material thickness.

5- Work slowly when removing shims and take time to note the number and location of the shims. Follow shimming instructions to the letter. Correct gear backlash will avoid noisy performance and possibly early failure of gears. Check the specifications for the proper **U**-joint shaft and driveshaft preload. This is the **ONLY** way satisfactory service may be obtained from the unit after the work has been completed. There are **NO** short cuts.

6- Always keep thrust bearing and thrust washer assemblies together and note the order of removal as an aid during installation.

7- The proper mandrels and supports **MUST** be used to remove pressed-on parts, such as bearings and gears for safety and preservation of parts which must be reused. Therefore, do not use force on the bearing cages, mating surfaces, or edges which could be damaged or cause failure.

8- Lubricate parts with oil or grease to ease installation, retard rust, and prevent corrosion. Oil seals must be coated with Quicksilver 2-4-C to prevent lubrication seepage and to hold the seal in position during installation.

9- When it is necessary to hold a part or housing in a vise, **ALWAYS** use soft brass jaws or wooden blocks to prevent marring or damaging the part.

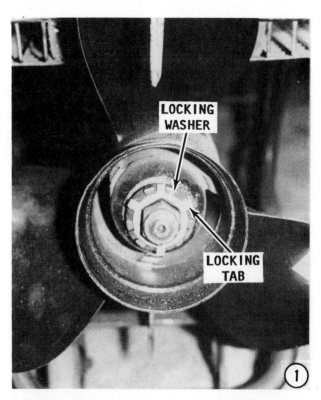

10-30 SERVICE BRAVO STERN DRIVE

PROPELLER REMOVAL

1- Bend the locking tabs forward out of the locking washer. **NEVER** pry on the edge of the propeller. Any small distortion will affect propeller performance. Place a block of wood between one blade of the propeller and the anti-cavitation plate to keep the shaft from turning. Use a socket and breaker bar and loosen the retaining nut. Remove the nut, tab washer, splined washer, continuity washer, propeller and then the thrust hub.

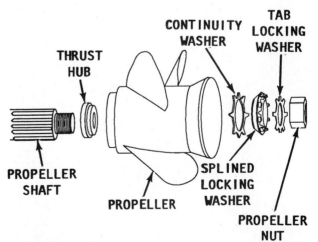

Line drawing to show arrangement of parts used to secure the propeller to the shaft.

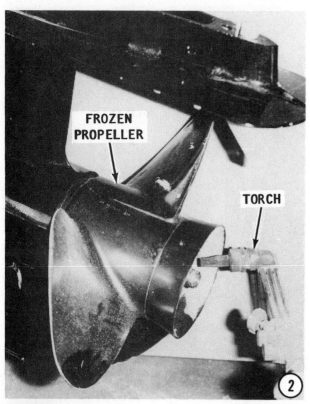

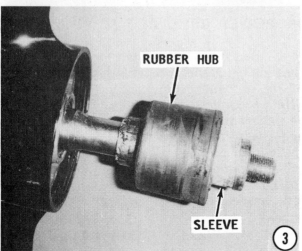

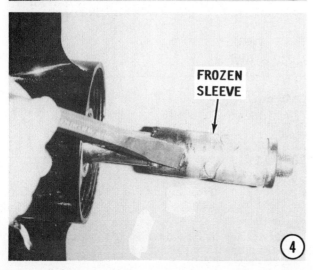

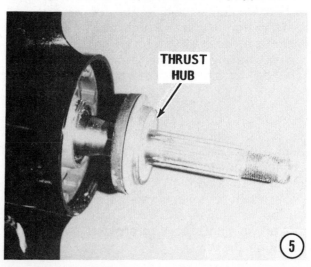

2- If the propeller is frozen to the shaft, heat must be applied to the shaft to melt out the rubber inside the hub. Using heat will destroy the hub, but there is no other way. As heat is applied, the rubber will expand and the propeller will actually be blown from the shaft. Therefore, **STAND CLEAR** to avoid personal injury.

3- Use a knife and cut the hub off the inner sleeve.

4- The sleeve can be removed by cutting it off with a hacksaw, or it can be removed with a puller. Again, if the sleeve is frozen, it may be necessary to apply heat.

5- Remove the thrust hub from the propeller shaft.

Procedures for propeller installation are given at the end of this chapter, after the stern drive has been installed.

DRAINING LUBRICANT FROM UPPER GEARCASE HOUSING AND LOWER UNIT

CRITICAL LUBRICATION WORDS

On the Bravo stern drive, the upper and lower lubricant reservoirs are connected by an oil passageway. Therefore, the same gearcase lubricant is shared between the upper gearcase housing and the lower unit. Because of this arrangement, both upper and lower housings can be drained and filled from the lower unit **DRAIN/FILL** hole located at the base of the lower unit housing.

Trapped Air

NEVER fill the unit at the vent hole in the upper gearcase housing, because trapped air inside both housings will give a false reading on the dipstick (if equipped), or cause the oil to overflow at the vent hole. These trapped air pockets will eventually work their way upward into the upper gearcase. When this happens all the oil will find its way down into the lower unit, leaving the upper gearcase short of lubricant. Such a condition can and will cause extensive and expensive damage to moving parts in the upper gearcase.

Inadequate Lubricant

An inadequate amount of lubricant in the stern drive will cause the oil to whip and foam, greatly reducing its lubricating qualities. Such a condition could lead to disaster -- serious and expensive damage to moving parts.

Adequate Lubricant

When the stern drive is properly filled from the **DRAIN/FILL** hole, any air pockets will eventually work their way upward to the top of the upper gearcase housing, and then escape through the vent hole.

Draining Procedure

1- Move the stern drive to the full up position.

2- Place a suitable container under the lower unit.

3- Remove the **VENT** plug and washer from the starboard side of the upper gearcase housing.

4- Remove the **DRAIN/FILL** plug from the lower unit and allow the fluid to drain from the upper housing and lower unit.

On a Bravo stern drive, the vent plug is located on the starboard side of the upper gearcase housing. The plug MUST be removed before the lower unit can be drained.

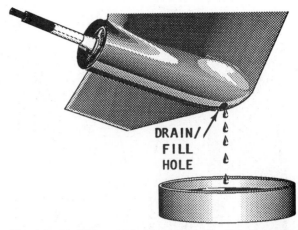

The stern drive must be raised to the full UP position before the lower unit lubricant can be drained into a suitable container.

NEVER remove the vent or drain/fill plugs when the drive unit is hot. Expanded lubricant would be released through the plug hole. Check the lubricant level after the unit has been allowed to cool. Add only Quicksilver Super-Duty Gear Lubricant. **NEVER** use regular automotive-type grease in the stern drive because it expands and foams too much. Stern drive units do not have provisions to accommodate such expansion.

If the lubricant appears milky brown, or if large amounts of lubricant must be added to bring the lubricant up to the full mark, a thorough check should be made to determine the cause of the loss.

SEPARATING STERN DRIVE FROM BELL HOUSING

1- Shift the unit into **NEUTRAL** gear. Pry the decorative/protective caps from the trim cylinder aft pivot bolts. Remove the

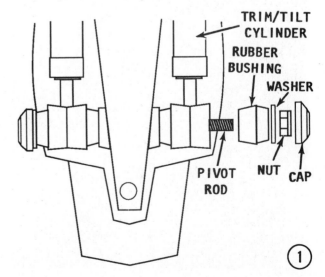

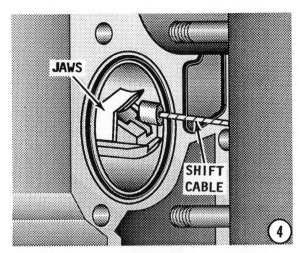

nut, washer and rubber bushing from each end of the pivot rod. Remove the rod, and then secure the cylinders up out of the way as a prevention against damaging them during work on the stern drive unit.

2- Disconnect the speedometer hose fitting at the forward end of the upper gearcase housing.

3- Remove the six locknuts, five washers and one ground plate, securing the upper gear housing to the tramsom assembly.

4- Support, while sliding, the stern drive unit from the six studs until a gap of an inch or so appears between the surfaces.

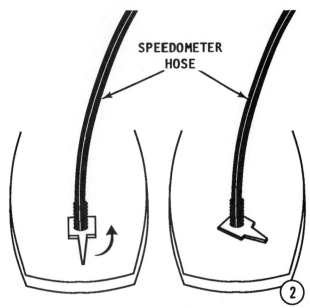

Release the shift cable from the shift linkage jaws by gently pulling out the linkage using a pair of needle nose pliers and opening the jaws to free the cable end.

Now the stern drive is free to be moved to the workbench. Do not allow the splined end of the U-joint shaft to fall as it is released from the engine coupler onto the bellows. The U-joint bellows between the bell housing and the driveshaft housing forms an air tight seal, easily damaged by a heavy blow from the shaft splines.

SEPARATING UPPER GEAR CASE HOUSING FROM LOWER UNIT

1- Place the stern drive in a holding fixture. Mark the position of the trim tab.

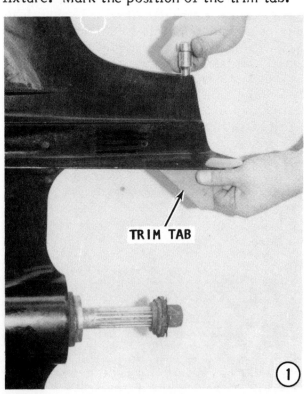

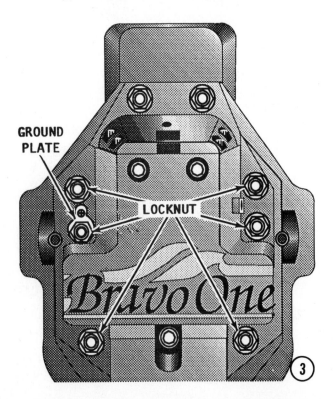

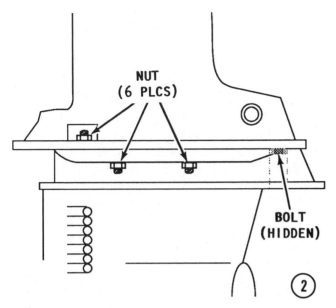

NUT
(6 PLCS)

BOLT
(HIDDEN)

②

GOOD WORDS

If the unit being serviced is equipped with power steering, the trim tab should be centered upon installation. Any steering tendencies should be corrected by adjustment of the steering rod, not the trim tab position.

Remove the small plastic trim tab plug from the upper surface of the lower unit. Use a 1/2" socket and a long extension to loosen and remove the bolt passing through the upper gearcase, lower unit and into the trim tab. Remove the tab.

2- Reach into the trim tab cavity and remove the single bolt, them remove the six securing nuts and washers, two located above the anti-cavitation plate and four located below the plate.

3- Separate the upper gearcase from the lower unit. Slide the retainer and driveshaft coupler from the top of the driveshaft.

MORE GOOD WORDS

There are no locating/indexing pins between the two mating surfaces. No type of gasket is used between the two units. Sealing between the two surfaces is accomplished by two **O**-rings and a seal. One **O**-ring is installed around the driveshaft spacer and a second **O**-ring around the water passageway. A tubular seal is installed in the speedometer water tube bore.

As the two housings separate, watch for the spacer and **O**-ring around the driveshaft. These two items may be in place on top of the lower unit or may remain up inside the upper gear housing cavity. If this is the

case, do not use a screwdriver to dislodge these two items. The lower surface of this spacer is used in a critical measurement during the driveshaft preload shimming procedures and if the surface is scratched, an accurate measurement cannot be made. If this spacer remains in the upper gearcase after separation, use a slide hammer with a jaw attachment. Hook the jaws inside the spacer and pull it down to remove it.

If the only work to be peformed is servicing the upper gearcase, proceed to the following section. If the lower unit is to be serviced, proceed to Section 10-32 on Page 10-90.

10-31 UPPER GEAR CASE SERVICE

WORDS OF WISDOM

Procedural steps are given to remove all items in the upper gearcase. However, do **NOT** remove bearings, bushings or seals if a determination can be made the item is fit for further service. As the work progresses,

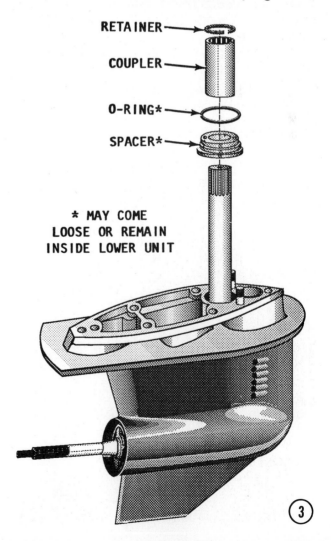

RETAINER

COUPLER

O-RING*

SPACER*

* MAY COME
LOOSE OR REMAIN
INSIDE LOWER UNIT

③

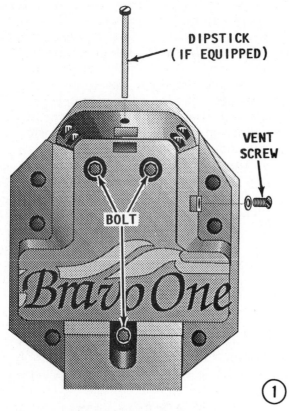

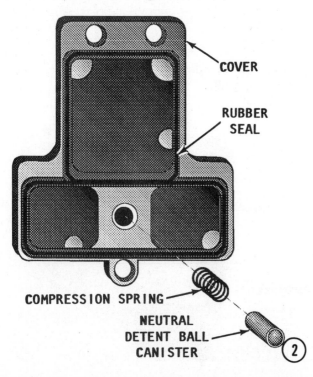

simply skip the steps involving these parts and continue with the required work.

Rear and Top Cover Removal

1- Remove the vent screw, if not already removed. Remove the dipstick. This **MUST** be removed now, it could be bent or broken if left in the gearcase.

Remove the three bolts and washers securing the rear cover to the upper gearcase housing.

2- Remove the neutral detent ball canister and compression spring from the inside of the rear cover. Remove and discard the rubber seal from the groove in the cover.

3- Before further disassembly, rotate the U-joint shaft and observe the brass collars around the two gears visible through the rear opening. The collars will rotate in opposite directions. As they rotate identify a "+" and "-" sign embossed on each of the collars. The "+" on the top collar must be directly above the "-" on the bottom collar as the gears rotate.

Further rotation of the U-joint shaft in the same direction will show a "-" on the top collar directly above the "+" on the bottom collar. The "+" and "-" signs **WILL** align in this manner when the unit is correctly timed. Two "+" signs or two "-" signs must **NEVER** align.

If the "+" and "-" signs do not align, the gears were not timed correctly during assembly. Depending on the degree of misstiming, damage may be caused to the three meshing gears, or at the very least, difficulty in shifting. Most certainly, excessive wear will be caused to shift linkage components.

CRITICAL THRUST WASHER WORDS

Two large thrust washers are used in the upper gearcase. During initial assembly at

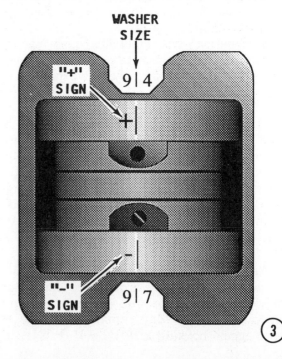

the factory, the thickness of these two washers was determined from dimensions taken in the upper gearcase housing. The thickness of these thrust washers was embossed on the upper gearcase housing-to-rear cover mating surface. The top number represents the thickness of the upper thrust washer in thousands of an inch and the bottom number represents the thickness of the lower thrust washer in thousands of an inch.

Only three washer sizes were used and and the set was patriotically color coded:

91 = 0.091"   Red
94 = 0.094"   White
97 = 0.097"   Blue

There are also two small thrust washers in the upper gearcase. The above mentioned measurements and color coding does not apply to these small washers.

As the two large thrust washers are removed during disassembling, identify the top and bottom washer. If the washers are accidentally mixed - no problem - their thicknesses may be measured with a micrometer.

If the measured thicknesses do not agree with the embossed numbers on the gearcase, obtain the correct thrust washers from the local dealer.

4- Remove the four bolts and washers securing the top cover to the upper gear case. Remove and discard the **O**-ring. Do not remove the small Allen screw on the

front edge of the cover. This Allen screw serves as a plug for an oil passage drilled during manufacture to lubricate the center bearing.

5- Lift out the upper thrust washer and thrust bearing **TOGETHER**. Because a wear pattern has been established, the thrust washer side contacting the thrust bearing **MUST** remain the same during installation. If the washer and bearing are not installed with original contact surfaces against each other, premature and excessive wear will occur. Therefore, keep the upper thrust washer and thrust bearing together at all times and identify them as the top set. The same will hold true when removing the bottom set in a later step.

Remove and discard the two small **O**-rings on the upper surface of the gear housing, one from the shift linkage cavity and the other from the water passage.

Top Cover Bearing Sleeve and Center Needle Bearing Removal

6- Obtain the following special tools:

Puller Jaws - two pieces - P/N 91-90777A1 and P/N 91-90778
Puller Guide - P/N 91-90774
Puller Bolt - P/N 10-90775
Driver Guide - P/N 92-90244

Position the two halves of the puller jaws around the bearing sleeve. Slide the driver guide over the puller guide with the flanged end of the driver guide facing the same direction as the open end of the puller guide. Slide both these pieces over the

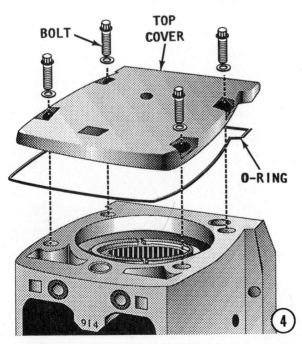

BOLT

TOP COVER

O-RING

④

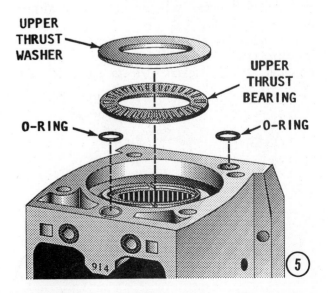

UPPER THRUST WASHER

UPPER THRUST BEARING

O-RING

O-RING

⑤

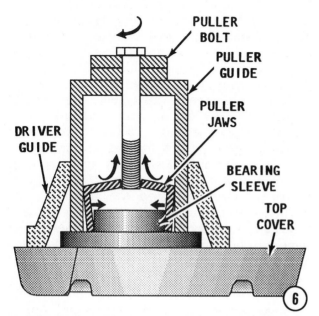

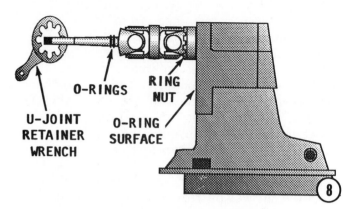

puller jaws. Install the bolt through the hole in the puller guide and thread the bolt into the puller jaws. Continue to rotate the bolt **CLOCKWISE** as the bearing sleeve is drawn up and free of the top cover.

7- Clamp the top cover in a vise equipped with soft jaws. Obtain a slide hammer with jaw attachment P/N 91-34569A1. Hook the jaws under the needle bearing cage and pull the bearing from the top cover.

U-Joint Shaft Removal

8- Remove and discard the **O**-rings around the shift linkage passage and the sealing ring around the water passage.

Remove and discard the two **O**-rings around the **U**-joint shaft. Obtain **U**-joint Retainer Wrench P/N 91-36235. Slide the wrench over the **U**-joint assembly and index the tool with the notches on the retaining ring nut.

Loosen the ring nut and gently pull the **U**-joint assembly **STRAIGHT** out of the housing. Take care not to "tweak" the

assembly as it comes out of the housing or the assembly may jam in the housing.

9- Clamp the **U**-joint assembly in a vice equipped with soft jaws. Remove and discard the **O**-ring between the oil seal carrier and the sealing ring. Loosen and remove the nut and washer from the shaft. Slide the gear assembly from the shaft and then remove the following components. Pay particular **ATTENTION** to their positions to ensure correct installation:

A sealing ring - with the taper facing the **U**-joints.

A large beveled washer - with the bevel facing the **U**-joints.

An oil seal carrier, containing an oil seal - with the seal lip facing away from the **U**-joints and with the carrier bevel facing away from the **U**-joints.

A ring nut - with the threaded end facing away from the **U**-joints.

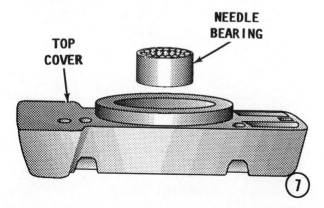

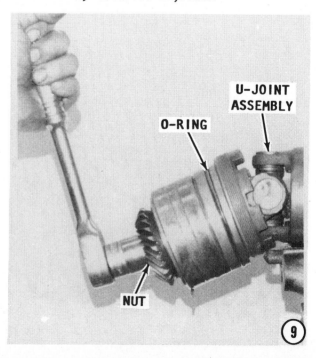

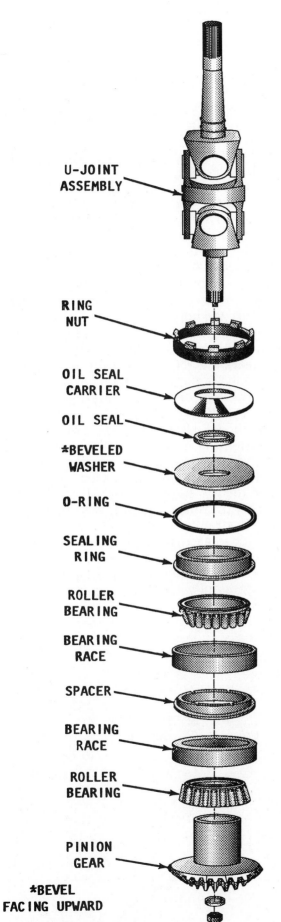

U-JOINT ASSEMBLY

RING NUT

OIL SEAL CARRIER

OIL SEAL

*BEVELED WASHER

O-RING

SEALING RING

ROLLER BEARING

BEARING RACE

SPACER

BEARING RACE

ROLLER BEARING

PINION GEAR

*BEVEL FACING UPWARD

Exploded drawing to show arrangement and identification of parts on the U-joint shaft.

To disassemble the gear assembly: position a universal bearing separator between the gear and the first bearing. Using a suitable mandrel in an arbor press, press on the gear shaft to separate the gear from the bearings. Both bearings will be distorted during the removal procedure. Therefore, the bearings should not be used again.

The spacer between the bearings must be **SAVED** for installation during assembling. Take notice how the washer is installed. When installed correctly, the flat face of the spacer will face the gear.

If necessary, the oil seal in the oil seal carrier may be removed, using a punch and hammer. Work the seal out to the flat side of the carrier. The seal may also be removed by using a suitable mandrel and pressing it out to the flat side of the carrier.

Inspect the condition of the two U-joints to determine if replacement is required. If so, use a pointed punch and hammer and drive off the two snap rings from the cup groves in each U-joint.

Obtain adaptor P/N 91-38756 or a suitably sized socket and a U-joint press. Use the press to remove each joint from its yoke.

Shift Linkage Removal

10- Remove the shift cam cap screw from the lower hole in the shift cam. There

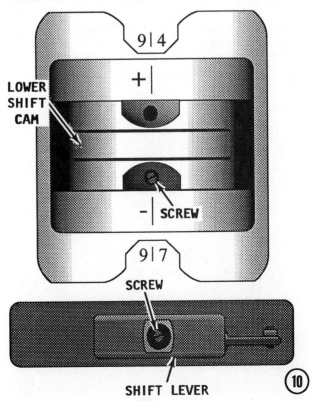

LOWER SHIFT CAM

SCREW

SCREW

SHIFT LEVER

10

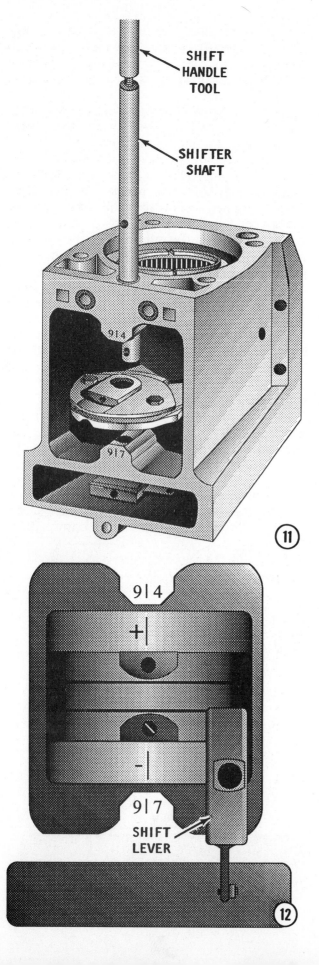

is no cap screw in the hole of the upper shift cam even though there is a hole in the exact location as the lower shift cam. The two halves of the shift cam are manufactured from the same casting, but only one screw is needed to secure the shift cam to the shifter shaft.

Remove the shift linkage cap screw from the shift lever.

11- Obtain Shift Handle tool P/N 91-17302. Thread the tool into the top of the shifter shaft, and then use the tool to pull the shaft up out of the housing.

If the special tool is not available, obtain a bolt at least 4" long with the same threads as the two cap screws removed in Step 10. The two cap screws are too short to be used in place of the bolt.

12- Grasp the shift lever and pull it out from the back of the housing until the pivot joint clears the housing. Rotate the lever 1/4 turn about the pivot point in a **CLOCKWISE** direction. Pull the shift lever to the left as the linkage is pulled out. If the linkage binds, study the path the linkage takes inside the housing and try again. This is a tricky maneuver, but it will come out. Actually, there is no real reason to remove it other than to inspect the components. Do not force the linkage as it may distort or break. If a real problem is encountered and there is no reason to suspect the linkage as being faulty, "let a sleeping dog lie" and proceed directly to Step 17. If the linkage was successfully removed, proceed with Steps 13 thru 16.

13- Remove and discard the cotter pin **ONLY** if a stainless steel pin is available as a replacement. Lift out the securing pin

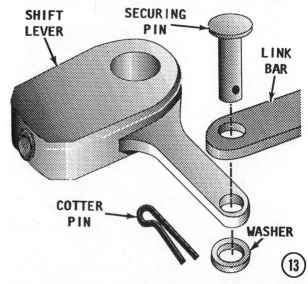

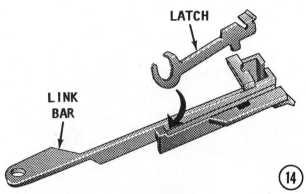

and separate the shift lever from the link bar.

14- Detach the latch from the link bar.

15- Slide the yoke and cam assembly from the clutch assembly.

16- Remove and discard the two lock-nuts. Lift off the top shift cam from the yoke. Remove the spacers from the two bolts and slide the bolts free of the bottom shift cam.

GOOD WORDS

The clutch shaft assembly consists of two identical gears with their related components - one gear at each end of the shaft.

These components appear to be interchangable. Yet, the thrust bearings, thrust washers, gears and the clutch spool, all develop wear patterns on their frictional contact surfaces. These patterns, once established, **MUST** be matched during installation. The only way to ensure they are matched is to lay the components out in order of disassembly. Therefore, unless a component is replaced, they will return to their original locations with their original contact surfaces bearing against each other.

If original friction surfaces are installed bearing against each other, a new wear pattern will be established. This new friction wear will be greatly accelerated and the servicable life of the component is therefore considerably shortened.

The moral to this tale is: Work slowly and with patience to **ENSURE** all used components are installed back in their exact original positions, even though some may appear to be interchangeable.

WORDS ON ECONOMICS

Some mechanics might advise installing interchangable parts in a manner to even out the wear.

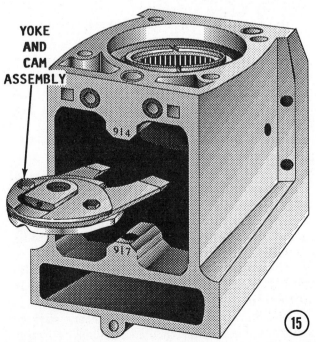

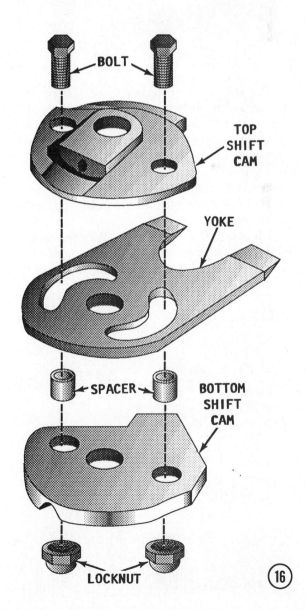

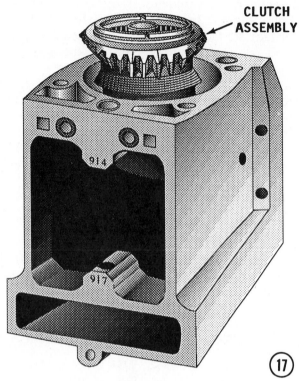

CLUTCH
ASSEMBLY

⑰

The authors strongly advise the reader to take into account the economics of reinstalling a worn part in a location where the labor cost exceeds the replacement cost of the part.

Clutch Disassembling

17- Lift the clutch assembly up and out of the gearcase.

18- Reach inside the gearcase and lift out the lower thrust bearing and thrust washer -- as a pair with contact surfaces together, if the parts are to be used again.

19- Obtain holding fixture P/N 91-17301A1. Slide the clutch shaft splined end

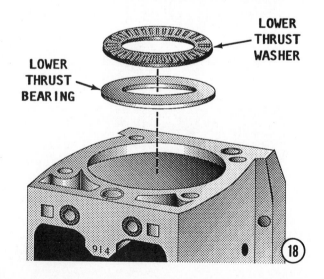

LOWER
THRUST
WASHER

LOWER
THRUST
BEARING

⑱

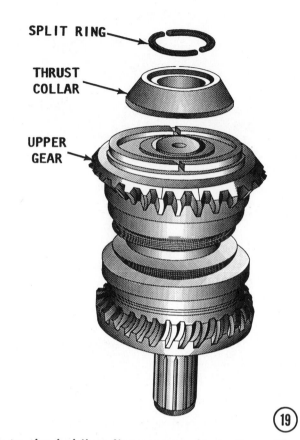

SPLIT RING

THRUST
COLLAR

UPPER
GEAR

⑲

into the holding fixture, or clamp the splined end in a vise equipped with soft jaws.

Push down on the upper gear and upper thrust collar until the collar clears the split ring. Remove the two halves of the split ring from the groove in the clutch shaft.

Slide the thrust collar and gear free of the shaft.

SPECIAL GEAR WORDS

If either the gear or the internal needle bearing is unfit for further service, both are replaced as a set. Both gears, as removed, are not servicable, even though it appears a circlip holds the needle bearing within the gear. The gear and the needle bearing are purchased together under the same part number. Therefore, seperating them would be a pointless task.

20- Lift out the upper thrust bearing, thrust washer and garter spring. Stack these items closely together if they are to be reused afer inspection. Identify the stack as the **UPPER** set.

21- Rotate the clutch spool **COUNTER-CLOCKWISE**, up over the spiral worm gear on the clutch shaft.

22- Remove the lower garter spring, thrust washer and thrust bearing. Stack

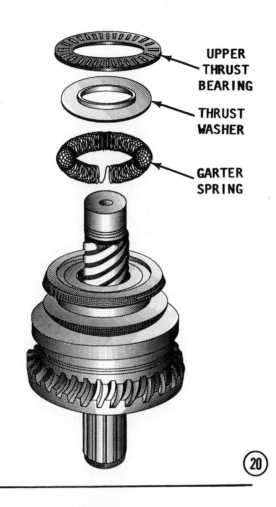

UPPER
THRUST
BEARING

THRUST
WASHER

GARTER
SPRING

(20)

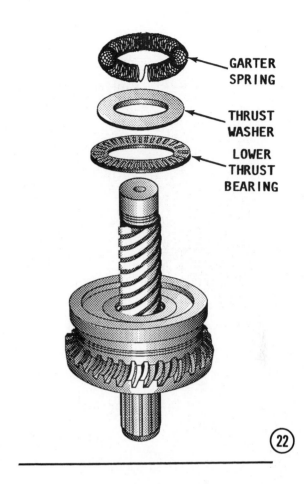

GARTER
SPRING

THRUST
WASHER

LOWER
THRUST
BEARING

(22)

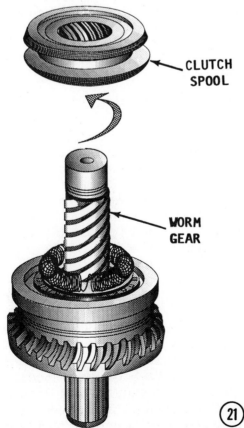

CLUTCH
SPOOL

WORM
GEAR

(21)

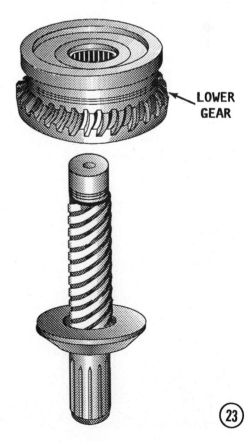

LOWER
GEAR

(23)

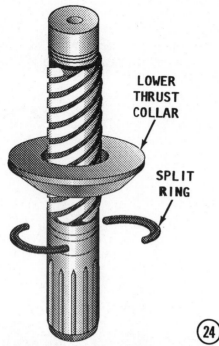

LOWER THRUST COLLAR

SPLIT RING

24

these items closely together, if they are to reused after inspection. Identify the stack as the **LOWER** set.

23- Lift the lower gear from the clutch shaft.

24- Pull up on the lower thrust collar until the collar clears the split ring. Remove the two halves of the split ring and then slide the collar up and free of the shaft.

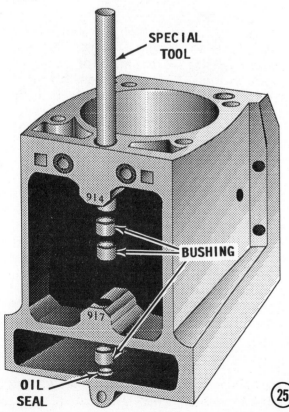

SPECIAL TOOL

914

917

BUSHING

OIL SEAL

25

Shifter Shaft Bushings and Oil Seal Removal

SPECIAL BUSHING & SEAL WORDS

If any of the three shifter shaft bushings or the oil seal need replacement, then all four items **MUST** be replaced. The bushings cannot be removed and replaced without disturbing the oil seal and vise versa.

If the oil seal needs replacement, but the three bushings show little sign of wear, all four items must still be replaced, but the original shifter shaft may be reused.

If any of the bushings show excessive signs of wear, this is most certainly reflected in the shifter shaft. Therefore, the oil seal, three bushings and shifter shaft must be replaced.

Two bushings are located in the casting above the yoke and cam assembly (when installed), and a single bushing and oil seal are located below the yoke and cam assembly. The three bushings are **NOT** identical. Therefore, identify each one as it is removed and match it in length with a new one to be installed.

25- Obtain special tool P/N 91-17273 or a suitable driver. Use the tool or driver and a ball peen hammer to tap the top bushing down onto the center bushing, and then on to remove both bushings from the housing.

Place the tool or driver down through the top of the casting onto the third bushing and drive the bushing and oil seal out in a similar manner.

Driveshaft Housing Bearing Sleeve and Needle Bearing Removal

26- Obtain the following special tools:

Puller Jaws - two pieces - P/N 91-90777A1 and P/N 91-90778
Puller Guide - P/N 91-90774
Puller Bolt - P/N 10-90775
Driver Guide - P/N 92-90244

Position the two halves of the puller jaws around the bearing sleeve. Slide the driver guide over the puller guide and down onto the sealing surface of the housing, with the flanged end of the driver guide facing upward. Slide both these pieces over the puller jaws. Install the bolt through the hole in the puller guide and thread the bolt into the puller jaws. Continue to rotate the bolt

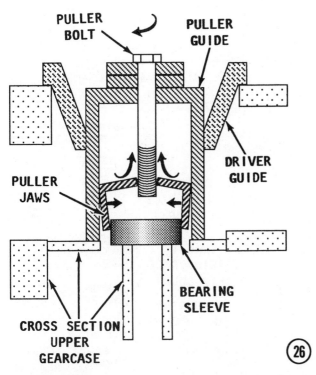

PULLER BOLT

PULLER GUIDE

DRIVER GUIDE

PULLER JAWS

BEARING SLEEVE

CROSS SECTION UPPER GEARCASE

(26)

race. For installation and correct placement of this roller bearing a number of special tools are required. If the necessity arises to remove and replace this roller bearing without the aid of special tools, first look closely at the positioning of the outer bearing race in relation to the housing **BEFORE** disturbing the bearing. Make a mark around the visible portion of the race which may be matched up and transferred to the new bearing race for accurate placement.

27- Obtain a suitable driver and tap the roller bearing down into the oil cavity. If the cage disintegrates, make sure all the old rollers are accounted for.

CLEANING AND INSPECTING

No gaskets are used in the upper gear case housing. All sealing is accomplished with oil seals and O-rings.

The purpose of the "+" and "-" marks embossed on the brass collars of the clutch gears are to identify the high and low areas of the undersurface of the collars. If placed on a flat surface, the top machined surface of each gear will not be parallel to the flat surface. This is quite normal, and as the brass surface wears, so does the shifting efficiency of the unit.

CLOCKWISE as the bearing sleeve is drawn up and free of the housing.

SPECIAL ROLLER BEARING WORDS

Removal of roller bearing destroys the cage. No special tools are required for removal, other than a socket just a fraction smaller in diameter than the outer bearing

If either the gear or the internal needle bearing is unfit for further service, both are replaced. The reason being, both gears, as removed, are not serviceable even though it appears a circlip holds the needle bearing within the gear. The gear and the needle bearing are purchased together under the same part number, therefore separating them would be a pointless task.

The top gear assembly is identical to the bottom gear assembly and in an emergency, they are interchangeable.

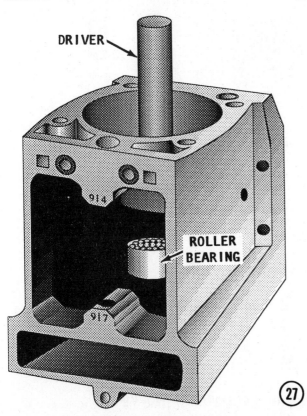

DRIVER

ROLLER BEARING

(27)

The embossed lines around the clutch spool should be cleaned with a soft brass brush or glass beading. Yes, glass beading is a procedure recommended by the manufacturer for cleaning the clutch spool. Evidently, the process is not so harsh as to remove material from the spool surface. Do not use a regular wire brush or lapp the lines with rubbing compound, as this **WILL** wear away the lines which are an essential part of the clutch mechanism. Replace the clutch spool if the lines are no longer visible.

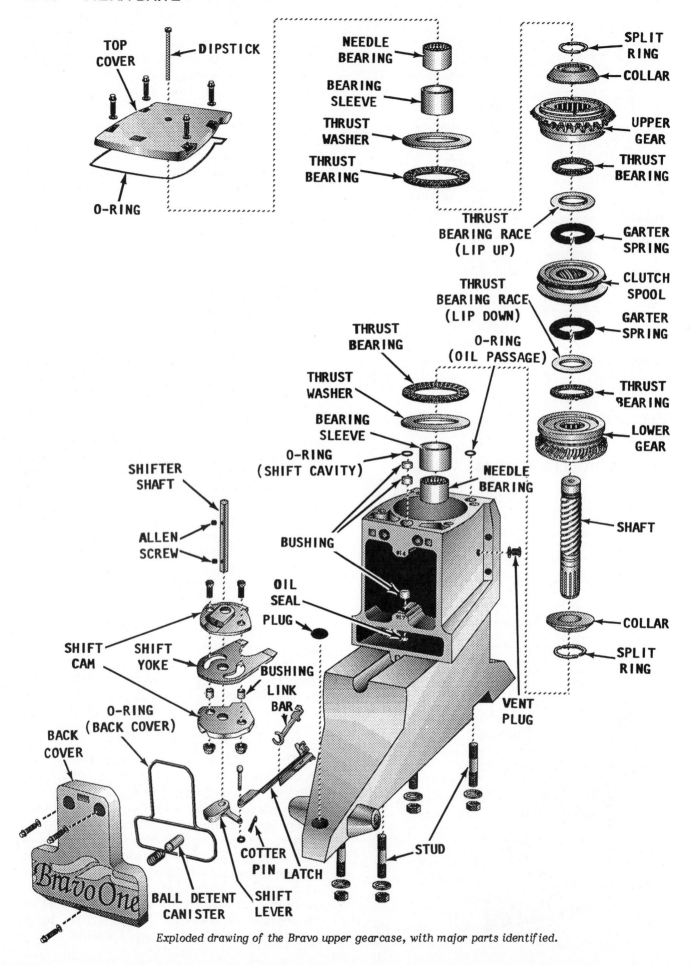

Exploded drawing of the Bravo upper gearcase, with major parts identified.

The function of the two garter springs is to help shift the unit out of gear. These springs should have enough tension in them, to require stretching to fit around the inner collar of the clutch spool. If they fit without stretching, they must be replaced.

If the thrust washers show signs of a "washboard" pattern, they should be replaced. This is the first sign of inadequate lubrication.

Examine the discarded sealing rings and O-rings for signs of cracks and distortion to identify the cause of the problem.

If any of the three shifter shaft bushings or oil seal need replacement, then all four parts **MUST** be replaced. The bushings cannot be removed and replaced without disturbing the oil seal and vise versa.

If the oil seal needs replacement, but the three bushings show little sign of wear, all four parts must still be replaced, but the original shifter shaft may be reused.

If any of the bushings show excessive signs of wear, this is most certainly reflected in the shifter shaft. Therefore, the oil seal, three bushings and shifter shaft must be replaced.

The converse also holds true. If the shifter shaft shows excessive wear in the areas of the bushings, then the shaft, bushings and oil seal must be replaced.

Where nylon locknuts are used and removed, they should be replaced with new locknuts. The nylon insert is intentionally deformed on initial installation to prevent the nut from backing off. Therefore, these nuts loose their locking characteristics if used more than once.

Inspect the cam spacers. Both must be equal in height and long enough to permit the yoke to move independently within the two halves of the cam.

Inspect the neutral detent ball cannister. The cannister must contain a spring confined within its length applying pressure against the ball bearing.

The shift lever and link bar, if functioning correctly, must be able to secure the shift cable end within its jaws, while installed in the housing passage. If the cable end can be pulled free of the jaws while the shift lever and link bar are still confined in the housing passage, the lever and link must be replaced.

No gear ratio change is available (at press time) for the Bravo when operating at high altitudes. However, the manufacturer recommends a propeller change for the higher elevations. Consult the local dealer for his propeller suggestions. The standard gear ratio for all Bravo units is 1.5:1.

ASSEMBLING

Driveshaft Housing Needle Bearing and Sleeve Installation

1- Perform this step only if the driveshaft needle bearing and bearing sleeve were removed in Step 26.

Obtain the following special tools:

Driver Head P/N 91-90773
Puller Guide P/N 91-90774
Bolt P/N 10-90775
Driver Guide P/N 91-90244

These tools require some assembling before the roller bearing can be installed. Slide the driver head onto the puller guide and secure the two together with the bolt. Position the new roller bearing on the base of the driver head. Place the driver guide onto the top of the driveshaft housing.

Lower the assembled guide tools down into the driver guide. Use a ball peen hammer and tap on the head of the bolt until the puller guide seats, positioning the bearing at the correct location in the housing.

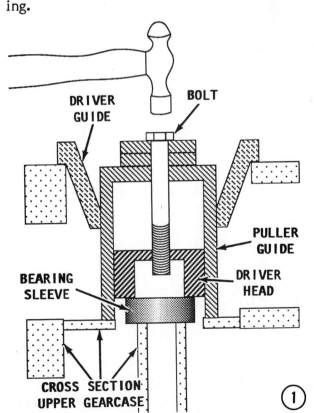

DRIVER GUIDE   BOLT

PULLER GUIDE

DRIVER HEAD

BEARING SLEEVE

CROSS SECTION UPPER GEARCASE

1

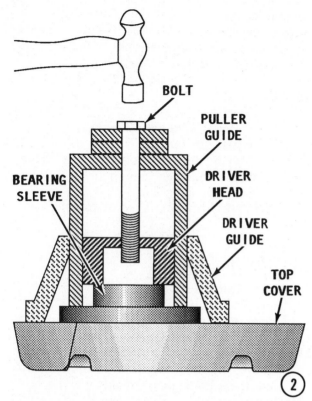

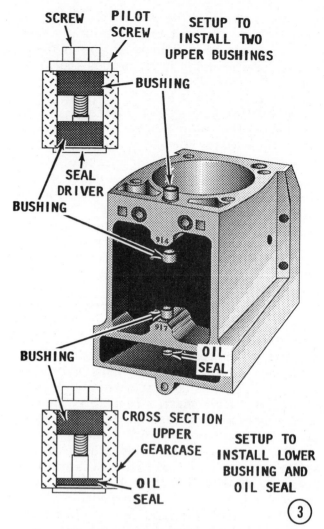

Use the same tool set up as in the beginning of this step, but place the bearing sleeve against the edge of the driver. Install the sleeve into the housing.

Top Cover Needle Bearing and Bearing Sleeve Installation

2- Perform this step only if the top cover needle bearing and bearing sleeve were removed in Steps 6 and 7 of Disassembling.

Again, use the same tool set up as in the previous step, but this time install the driver guide onto the puller guide backwards, with the flanged end facing away from the head of the bolt. Position the flange over the top cover. Tap the head of the bolt until the bearing seats in the cover.

Once more, use the same tool set up as in Step 1, but now place the bearing sleeve against the edge of the driver. Install the sleeve by tapping on the head of the bolt until the sleeve is seated.

Shifter Shaft Bushing and Oil Seal Installation

3- Perform this step and the next step only if the three shifter shaft bushings and oil seal were removed in Step 25 of Disassembling.

Identify each of the new bushings by comparison to the used ones. Obtain seal

driver P/N 91-17275, screw pilot P/N 91-17274 and screw P/N 10-20784.

Pack the oil seal lip with Quicksilver 2-4-C Marine Lubrucant and coat the outside perimeter of the new oil seal with Loctite Type A. Push the lower bushing into the bore by hand as far as it will go. Hold the oil seal up to the bottom of the bore with the lip of the seal facing upward. Place the seal driver portion of the tool under the oil seal. Place the screw pilot into the top of the bushing. Pass the screw through the hole in the screw pilot and thread it into the seal driver.

Tighten the screw. As the screw is tightened, the bushing and oil seal will be drawn into the housing until each is flush with the housing surface. Remove the tool.

By hand, push the new upper and lower bushings into the housing from the top and the bottom as far as they will go. Install the bushings in the same manner as described for the bushing and oil seal in the previous step.

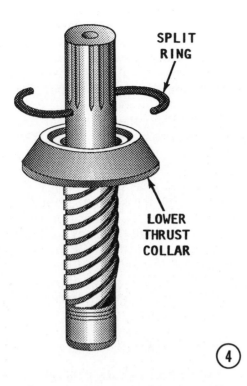

SPLIT
RING

LOWER
THRUST
COLLAR

④

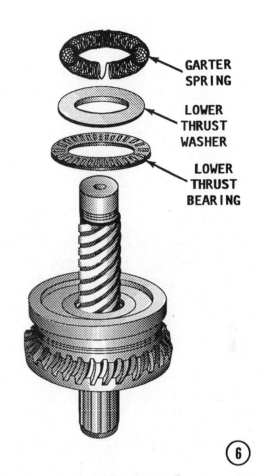

GARTER
SPRING

LOWER
THRUST
WASHER

LOWER
THRUST
BEARING

⑥

SPECIAL CONTACT SURFACE WORDS

The contact surfaces of the thrust bearings and thrust washers removed in Step 5 and Step 18 of Disassembling, must face the same direction as before they were removed to minimize wear.

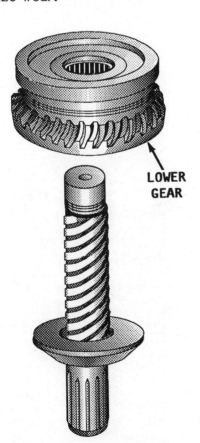

LOWER
GEAR

⑤

Clutch Assembly

4- Slide the lower thrust collar over the splined end of the clutch shaft, with the taper (smaller diameter), facing **UPWARD.** Hold the collar down and slide the two halves of the split ring into the grooves in the shaft. Snap the collar upward to retain the ring in the groove.

5- Invert the shaft and apply a good grade of gear oil to the splines. Slide the lower gear assembly down over the splines with the brass collar facing **UPWARD.** The gear assembly will rotate clockwise as it follows the curved splines going down the shaft and will seat against the lower collar.

6- Slide the lower thrust bearing down over the shaft, followed by the lower thrust washer, with the washer lip facing **DOWNWARD.** Stretch the lower garter spring over the thrust washer. The garter spring will not lie flat. However if it does, the spring has lost too much tension and needs to be replaced.

7- Lower the clutch spool down onto the shaft. The spool will rotate clockwise as it follows the curved splines downward to rest against the lower garter spring.

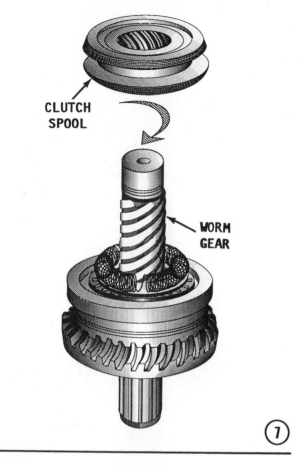

CLUTCH SPOOL

WORM GEAR

⑦

8- Slide the upper garter spring, upper thrust washer, with the washer lip facing upwards and finally the upper thrust bearing over the shaft. At this stage, these items will not appear to be seated. Fear not, for this is a normal condition.

9- Position the upper gear over the shaft with the brass collar facing **DOWNWARD.** Install the upper thrust collar, with the taper (smaller diameter), facing upward.

Snap the gear down sharply to spread and seat both garter springs around each end of the clutch spool. This action will expose the upper groove in the shaft. Hold the gear down and insert the two halves of the split ring around the groove in the shaft. Release pressure on the gear allowing the upper thrust collar and gear to snap up and retain the split ring in the groove.

U-Joint Shaft Assembly

FIRST, THESE GOOD WORDS

If only Step 9 was performed during disassembling, then proceed directly to Step 13 -- checking the rolling torque of the driveshaft.

If the **U**-joint was completely disassembled, as in Step 9 of disassembling, perform Steps 10 thru 12, then Step 13 and on.

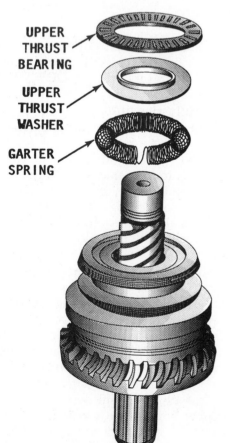

UPPER THRUST BEARING

UPPER THRUST WASHER

GARTER SPRING

⑧

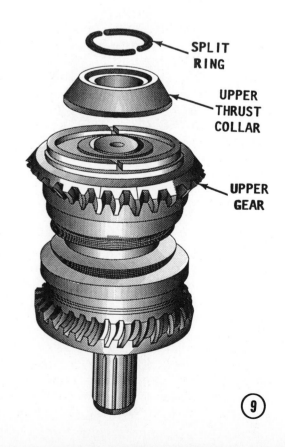

SPLIT RING

UPPER THRUST COLLAR

UPPER GEAR

⑨

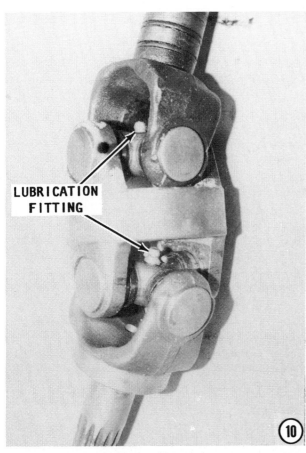

LUBRICATION
FITTING

⑩

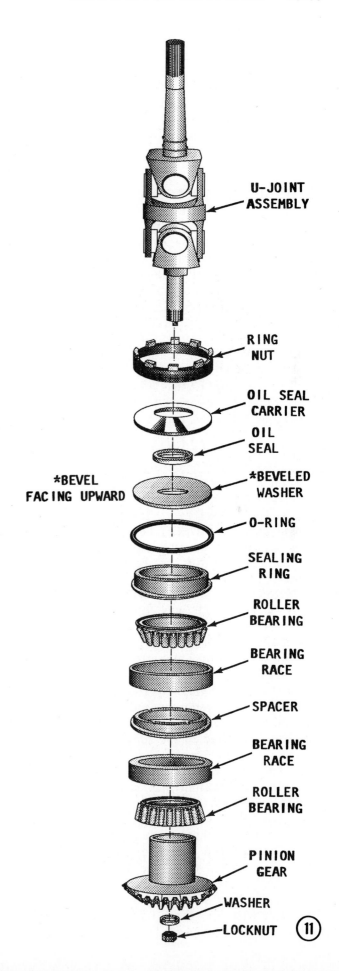

U-JOINT
ASSEMBLY

RING
NUT

OIL SEAL
CARRIER

OIL
SEAL

*BEVEL
FACING UPWARD

*BEVELED
WASHER

O-RING

SEALING
RING

ROLLER
BEARING

BEARING
RACE

SPACER

BEARING
RACE

ROLLER
BEARING

PINION
GEAR

WASHER

LOCKNUT ⑪

10- Lay the two U-joints out on the workbench. Identify and remove all caps. This is **MOST** important, especially if the old U-joints are to be re-used. Once a wear pattern has been established, it is best to maintain the existing pattern. Inject Quicksilver 2-4-C Marine Lubricant into the lubrication fitting at the cross, until the lubricant oozes from each arm. Replace the caps. Repeat for the other U-joint.

Use Adaptor P/N 91-38756, or a suitable sized socket and a U-joint installer. Install both U-joints, with their lubrication fittings facing toward the longer coupler yoke after installation.

Install the bearing cup retaining snap rings.

Apply a coat of Loctite Type A onto the outer diameter of the oil seal. Position the oil seal into the concave (cupped) side of the oil seal carrier with the lip facing downward. Obtain Bearing Driver P/N 91-89868 and tap the seal down until it seats within the carrier. Pack the seal lip with Quicksilver 2-4-C Marine Lubricant.

11- Place the drive gear, teeth down, onto a piece of aluminum or very hard wood. Place the smaller diameter bearing over the gear shaft with the bearing taper (the smaller diameter), facing upward. Obtain bearing installation tool P/N 91-90774, or a suitable mandrel which will contact only the inner race. Press the bearing onto the gear. Slide the outer bearing race onto the installed bearing, and then slide the spacer over the gear shaft, with the wide flat surface of the spacer **TOWARD** the gear.

CRITICAL SPACER WORDS

The spacer **MUST** be installed correctly in the next part of this step. Incorrect installation of the spacer will result in misplacement of the pinion gear. If the pinion gear is out of position, severe damage to the upper gearcase housing could, and most likely would, occur when the retaining ring nut is tightened to specifications.

Place the larger diameter bearing race over the gear shaft, with the flat surface of the race facing **DOWNWARD**. Place the larger diameter bearing into the race. Use a suitable mandrel and press on the inner race of the bearing, until the bearing rollers **BARELY** make contact with the outer race.

Check to be absolutely sure the spacer between the bearings is free to move. If the spacer is held tightly between the two bearings, it must be freed. Install a universal bearing puller between the spacer and the larger diameter bearing and **lightly** tap on the gear teeth with a soft head mallet.

MORE SPECIAL SPACER WORDS

When the unit was disassembled, the spacer between the two bearings was **NOT** free to move. This is normal and correct. The pressure with which the spacer is held, is determined by checking the rolling torque of the U-joint shaft. By tightening the nut retaining the driven gear and double bearing assembly onto the U-joint shaft, the correct pressure is applied to the spacer.

Step 11 Continues

Slide the retainer nut onto the shorter U-joint yoke shaft, with the castellated end facing the U-joints. Next, slide the oil seal carrier onto the shaft with the bevelled side facing away from the U-joints. Now, slide the large washer over the shaft with the taper (smaller diameter), facing the U-joints. Finally, slide the sealing ring over the shaft with the taper (smaller diameter), facing the U-joints.

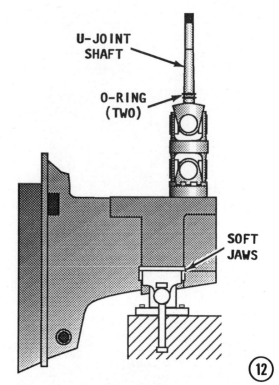

Position a new **O**-ring between the oil seal carrier and the sealing ring.

Install the gear and bearing assembly onto the shaft. Slide the washer onto the shaft and install a **NEW** locknut. Tighten the locknut **ONLY** by hand at this time.

12- Position and block the upper gear case housing on the workbench, to enable the assembled **U**-joint shaft to be lowered vertically down into the housing.

Slowly lower the double bearing assembly on the U-joint shaft squarely down into the gearcase. Because there is so little clearance between the bearing assembly and the housing, the assembly can only go in if it is squarely aligned as it is lowered. Tighten the large locknut **ONLY** by hand at this time.

Install two new **O**-rings into the grooves of the U-joint shaft.

Setting the U-Joint Rolling Torque

13- Position the upper gearcase close to the edge of the workbench to allow the U-joint shaft to hang down freely **WITHOUT** making contact with the edge of the workbench.

Obtain a socket the same size as the nut, a 6" extension and an inch-pound torque wrench.

The driveshaft preload is set by tightening the nut 1/16th turn and observing the reading on the torque wrench. At first, a

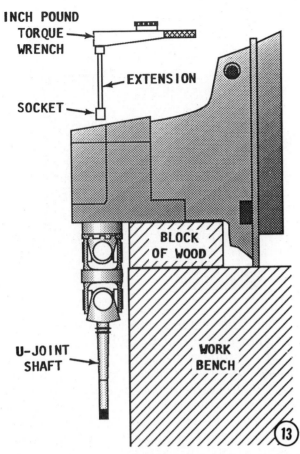

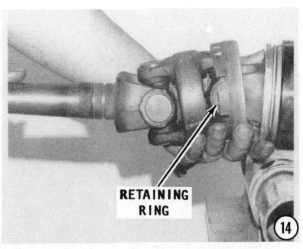

zero reading will be obtained -- because the spacer between the two bearings is free to move.

Continue tightening the nut through 1/16th turn each time until the torque wrench indicates 6-10 in lbs.

If the preload is exceeded by accident, back off the nut a few turns. Tap the end of the shaft through the housing with a piece of wood and start again at the begining of this step until the correct driveshaft rolling torque is obtained.

14- Loosen the **U**-joint shaft large retaining ring and remove the shaft from the housing.

CRITICAL THRUST WASHER WORDS

Two large thrust washers are used in the upper gearcase. During initial assembling at the factory, the thickness of these two washers was determined from dimensions of the upper gearcase housing. The thickness of these thrust washers was embossed on the upper gearcase housing-to-rear cover mating surface. The top number represents the thickness of the upper thrust washer in thousands of an inch and the bottom number represents the thickness of the lower thrust washer in thousands of an inch.

Only three washer sizes were used and each was patriotically color coded as follows:

91 = 0.091" Red
94 = 0.094" White
97 = 0.097" Blue

There are also two small thrust washers in the upper gearcase. The above description and dimensions do not apply to these smaller washers.

As the two large thrust washers were removed during disassembling they should have been identified, as instructed in the procedures. If the washers were not identified or accidentally mixed -- no problem -- simply measure their thickness with a micrometer.

If, after measuring, the thicknesses do not agree with the embossed numbers on the gearcase, obtain the correct thrust washers from the local dealer.

End Thrust Washer Words

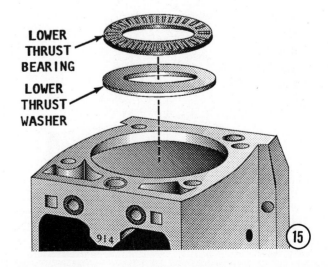

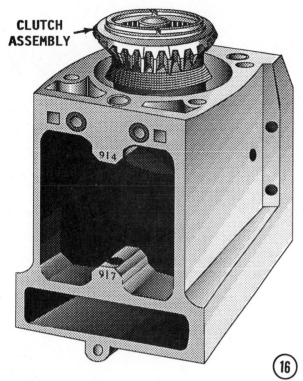

CLUTCH ASSEMBLY

914

917

(16)

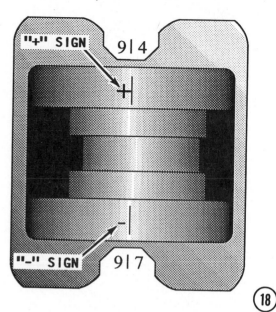

"+" SIGN   914

+

−

"-" SIGN   917

(18)

On with the work:

15- Place the lower thrust washer into the housing with the same face contacting the housing as before removal. Apply a coat of Quicksilver 2-4-C Marine Lubricant to the lower face of the clutch assembly and press the lower thrust bearing onto the lubricated face matching the original contact surfaces.

16- Lower the thrust bearing and clutch assembly into the housing.

Apply a coating of Quicksilver 2-4-C Marine Lubricant to the upper face of the installed gear assembly.

17- Place the upper thrust bearing onto the lubricated top face, matching original contact surfaces. Position the upper thrust

washer over the thrust bearing, again matching contact surfaces.

18- Rotate and observe the brass collars around the two gears. At this point the gears will rotate independently of each other. As they rotate identify a "+" and "-" sign embossed on each of the collars. Position the "+" on the top collar directly above the "-" on the bottom collar.

The "+" and "-" signs **MUST** align in this manner to time the unit correctly. Two "+" signs or two "-" signs must **NEVER** align.

19- Carefully insert the drive gear and **U**-joint shaft into the housing. Mesh the driven gear with the two timed gears **WITHOUT** disturbing the alignment of the timing marks. Tighten the large retaining nut using wrench P/N 91-36235. Install a torque wrench into the square hole in the handle of the tool. The torque wrench **MUST** be in line with the tightening wrench, or a false reading will be obtained. Tighten the nut to a torque value of 150 ft lb (203Nm.)

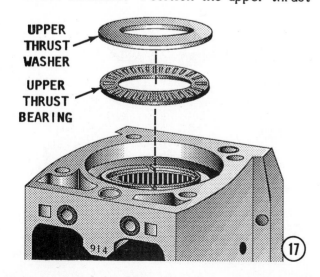

UPPER THRUST WASHER

UPPER THRUST BEARING

914

(17)

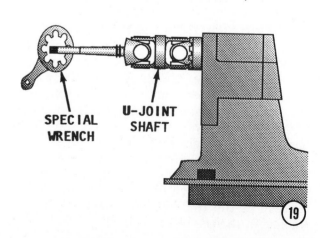

SPECIAL WRENCH   U-JOINT SHAFT

(19)

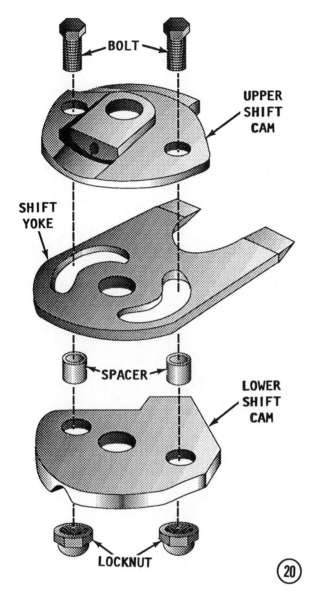

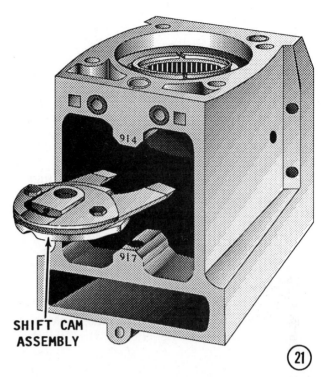

Now, rotate the **U**-joint shaft through a few turns and check to be sure the timing marks on the two gears are still aligned.

If the "+" and "-" signs do not align, the gears were disturbed during assembling, and the gears are not timed correctly. Depending on the degree of miss-timing, such a condition could cause damage to the three meshing gears. At the very least, difficulty in shifting would be experienced and excessive wear on shift linkage components would result. Therefore, remove the **U**-joint shaft and repeat this step until all conditions are met.

20- Perform this step only if the top shift cam was removed as instructed in Step 16 of disassembling.

Slide the bolts through the upper shift cam half. Place a spacer over each bolt. Position the shift yoke over the two bolts. Place the lower shift cam half over the

yoke. Install and tighten two **NEW** locknuts to a torque value of 100 in lb (11.5Nm).

21- Slide the shift cam assembly into the clutch spool. The cam assembly may be installed at 180° to its original position, but it is preferable to install the cam with the bolt heads facing upward. If the nuts were to loosen, the assembly stands a slightly better chance of staying together than if it were installed in the opposite direction with the bolt heads facing downward.

SPECIAL WORDS

Perform the following three steps only if the shift linkage was removed and disassembled in Steps 13 and 14 of disassembling.

22- Hook the curved end of the latch under the pivot on the bar link.

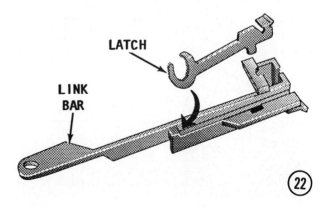

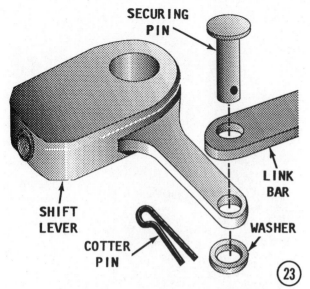

SECURING PIN

SHIFT LEVER

COTTER PIN

LINK BAR

WASHER

②③

23- Begin this step by studying the illustration carefully. It is possible to install the shift lever backwards onto the link bar. Insert the post at the end of the link bar into the hole in the end of the shift lever. Install the washer, and then the cotter pin. Spread the ends of the pin around the post. If a new stainless steel cotter pin is not available, reuse the old one.

24- Slide the assembled shift linkage into the housing with the cotter pin facing toward the left (port side). Wiggle the linkage a little, if necessary, to allow the link bar to enter the housing.

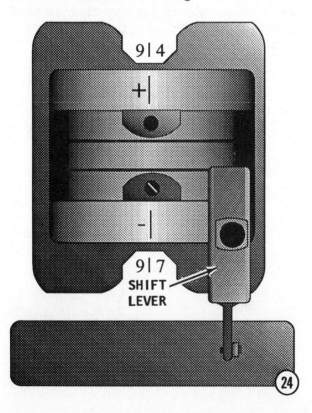

9|4

+

-

9|7
SHIFT LEVER

②④

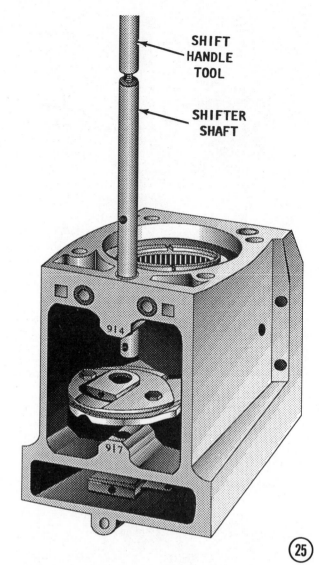

SHIFT HANDLE TOOL

SHIFTER SHAFT

9|4

9|7

②⑤

Rotate the shift lever through 1/4 turn **COUNTERCLOCKWISE** and position the hole in the lever directly under the shift shaft oil seal.

25- Thread shift handle tool P/N 91-17302 (or use the long bolt from Step 11 of disassembling) into the internal threads at the top of the shifter shaft. Push the shifter shaft through the two top bushings in the housing, through the shifter cam, through the lower bushing, through the oil seal, and finally into the shift lever where it will seat. Align the lower threaded hole in the shifter shaft with the hole in the shift lever. Remove the tool or bolt.

26- Install the shift handle tool or bolt through the hole just aligned in the shift lever and thread the tool or bolt into the lower shifter shaft hole. Using the tool or bolt, rotate the shift lever until the upper threaded hole in the shifer shaft aligns with the hole in the lower shift cam. Install the

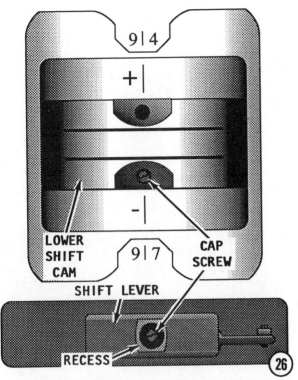

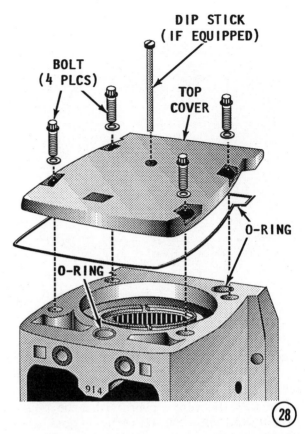

shift cam cap screw and tighten the screw to a torque value of 110 in lb (12.5Nm). Remove the tool or bolt.

Install the shift linkage cap screw and tighten the screw to the same torque value as the other screw.

Before removing the Allen socket used to tighten the cap screw, move the linkage to position the recess above the lower cap

screw centrally in the housing as an assist in the installation of the neutral detent ball in the next step. This places the linkage in the **NEUTRAL** position.

27- Coat the coils of the compression spring with Special Lubricant 101 (bright green in color). Place the spring inside the recess of the cover and install the neutral detent ball cannister over the spring.

Coat the rear cover sealing ring with 3-M Adhesive P/N 92-25234 and install the ring into the groove around the cover.

ALIGNMENT WORDS

The neutral detent ball can only be aligned with the recess above the shift linkage cap screw if the shift lever is in the **NEUTRAL** position and they must align to permit the rear cover to seat. If a problem is encountered, repeat the last paragraph of Step 26 and try again.

28- Coat the new top cover **O**-ring, shifter shaft **O**-ring and oil passage **O**-ring with 3-M Adhesive P/N 92-25234. Install the three **O**-rings. Position the top cover over the housing. Install and tighten the four bolts to a torque value of 20 ft lb (27Nm). Install the dipstick, if equipped.

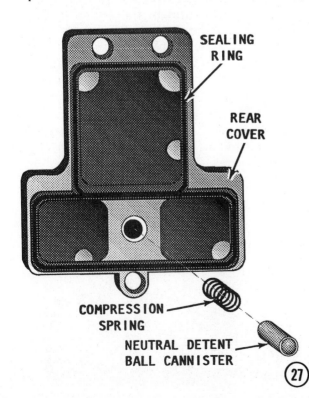

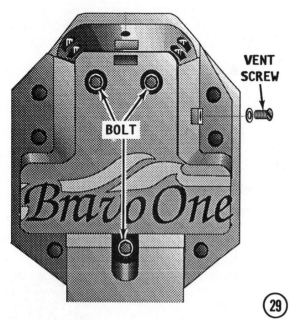

29- Install the vent screw. Coat a new shift linkage **O**-ring and water passage sealing ring with 3-M Adhesive P/N 92-25234 and install them in the front surface of the housing.

Install the rear cover with the neutral detent ball aligned with the recess above the shift linkage cap screw in the shift lever.

Install and tighten the three cover screws to a torque value of 20 ft lb (27Nm).

If the only work to be performed was in the upper gearcase housing, the unit is now ready to be mated with the lower unit. Proceed directly to Section 10-33, on Page 10-106.

If the lower unit is also to be serviced, then proceed with the instructions listed in the following section.

10-32 LOWER UNIT SERVICE

DISASSEMBLING

Preparation for Disassembling

1- If the retainer and driveshaft coupler are still in place on top of the lower driveshaft, remove these two items. (They may have been removed in Step 3 of Section 10-30, when the lower unit was separated from the upper gearcase housing.)

Remove the large water passage O-ring from the forward cavity. Use a suitable tool to pry out the tubular seal from the speedometer water tube bore. Discard the O-ring and the seal.

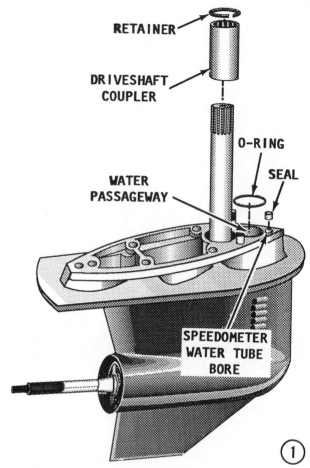

SPECIAL WORDS

The following step involves the use of a special retaining plate clamped over the driveshaft. Stresses on the propeller shaft are transmitted through the pinion gear and drive gears to the driveshaft while the bearing carrier is being removed. The use of this plate is **STRONGLY** recommended by the manufacturer and the authors to prevent driveshaft distortion while the bearing carrier is removed.

2- Obtain Clamp Plate tool P/N 91-43559 and two of the nuts removed when the

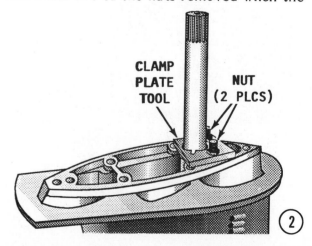

lower unit was separated from the upper gearcase housing (Step 2). Slide the plate over the driveshaft and the two forward studs. Secure the plate to the upper surface of the lower unit with two nuts. Tighten the nuts securely.

Bearing Carrier and Propeller Shaft Removal

3- Use a slotted screwdriver and bend back the tab over the large ring nut.

4- Obtain Bearing Carrier Retainer tool P/N 91-61069 for Bravo "One" or P/N 91-17257 for Bravo "Two". Install the special tool over the end of the propeller shaft with the tool indexing into the slot around the ring nut. Using the appropriate socket and breaker bar, loosen, and then remove the ring nut. Remove the tab washer.

SPECIAL BEARING CARRIER WORDS

The Bravo bearing carrier has **NO** indexing key. However, a small slot may be observed in the top of the carrier. This slot identifies the "top" of the carrier during installation.

5- Two methods may be used to remove the bearing carrier. The first method is recommended by the manufacturer and the authors and involves the use of a special tool. The second method is used by seasoned "salty" mechanics without using the special tool.

First Method

Obtain Bearing Carrier Puller tool P/N 91-90338A1. Hook the end of the special

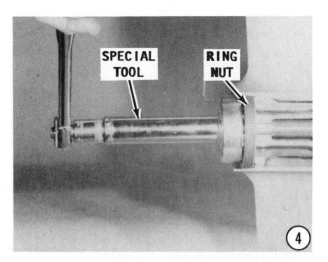

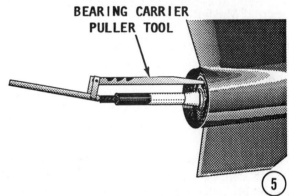

tool behind one of the bearing carrier ribs and position the pivot end of the tool against the end of the propeller shaft, as shown in the accompanying illustration. Pry just a bit, and then move the hook end behind another rib, pry a bit more, and then shift the hook end. By moving the hook end of the tool constantly, the bearing carrier

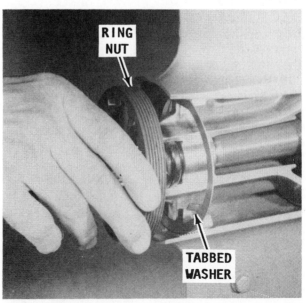

Removing the ring nut and tabbed washer from the lower unit.

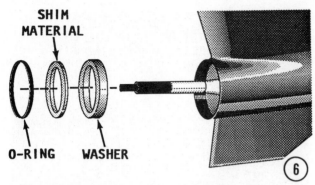

SHIM MATERIAL

O-RING   WASHER

6

will not become cross wedged in the lower unit. Continue working the bearing carrier out until it is free, and then slide it off the propeller shaft.

Second Method

Obtain a slide hammer with a jaw expander attachment. Hook the jaws behind opposite ribs of the bearing carrier and pull the bearing carrier free of the lower unit.

6- Reach inside the lower unit cavity and remove the **O**-ring, shim material, and washer.

SPECIAL SHIM MATERIAL WORDS

Be sure to **SAVE** and identify the shim material removed. This material is critical in obtaining the correct backlash during assembling. Using the old shim material will save considerable time, especially if it should become necessary to begin with no shim material.

7- Pull the propeller shaft out and free of the lower unit.

Special Procedure for "Frozen" Shaft

If the propeller shaft cannot be pulled free of the lower unit, the splines on the forward end of the shaft are "frozen" with the internal splines of the drive gear. To pull the shaft free, obtain a universal bearing separator tool, a pry bar, and approximately six inches of steel tubing with an inside diameter of approximately four inches.

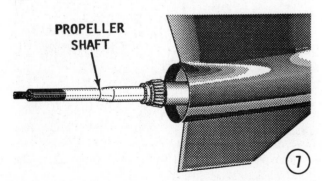

PROPELLER SHAFT

7

Place the bearing separator tool over the propeller shaft followed by the piece of steel tubing. Thread the propeller nut onto the end of the propeller shaft. Insert the end of the pry bar under the propeller nut. **TAKE CARE** not to damage the threads on the end of the shaft. Pry on the nut against the end of the steel tubing to move the propeller shaft. The shaft should move out of the drive gear.

If the attempt to remove the shaft fails, prop the lower unit on the bench with the propeller shaft extending upward. Spray a liberal amount of "liquid wrench" or equivalent into the lower unit and let the unit soak. After a couple hours make another attempt to "pull" the propeller shaft.

Driveshaft Removal

FIRST, THESE WORDS

Temporarily install the ring nut into the lower unit to protect the lower unit threads while performing the following step.

8- Obtain Driveshaft Adaptor tool P/N 91-61077, a suitable size wrench to rotate the tool, a "breaker" bar, and a socket the same size as the pinion nut. Hold the pinion nut steady with the socket and at the same time install the driveshaft tool on the upper end of the driveshaft. With both tools in place, one holding the pinion nut and the other on the driveshaft tool, use "muscle power" and rotate the driveshaft tool **COUNTERCLOCKWISE** to "break" the pinion nut free. Remove the pinion nut and washer from the lower end of the driveshaft. Remove the special tool and nut from the upper end of the driveshaft. Remove the ring nut from the lower unit.

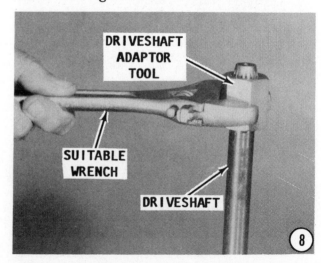

DRIVESHAFT ADAPTOR TOOL

SUITABLE WRENCH

DRIVESHAFT

8

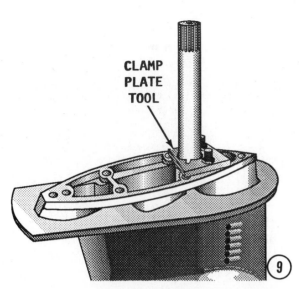

⑨

9- Remove the clamp plate tool retaining nuts, and then slide the tool up and free of the driveshaft.

10- Remove the O-ring, driveshaft spacer, shim material, and the tab washer.

CRITICAL O-RING WORDS

The O-ring used at this location is perhaps the most crucial item in the entire stern drive. Lubricant is shared between the upper gear housing and the lower unit. The lubricant passes freely through the two holes drilled in the driveshaft spacer. The O-ring provides the only seal in the oil passageway between the upper gear housing and the lower unit. No gaskets are used in

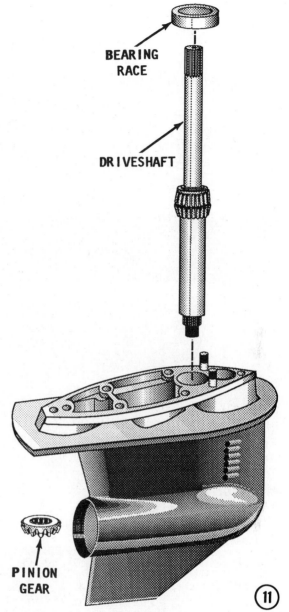

⑪

the Bravo stern drive. A distorted O-ring will fail to provide an adequate seal between the upper gear housing and the lower unit and may allow water to enter or lubricant to be lost.

11- Pull up on the driveshaft and at the same time catch the pinion gear as it is released from the splines at the lower end of the driveshaft. Slide the upper bearing race from the tapered bearing on the driveshaft.

12- Obtain a slide hammer with a jaw expander attachment. Lower the slide hammer down into the driveshaft cavity and hook the expanding jaws under the shim material. Work the slide hammer and remove the lower bearing race and shim material simultaneously.

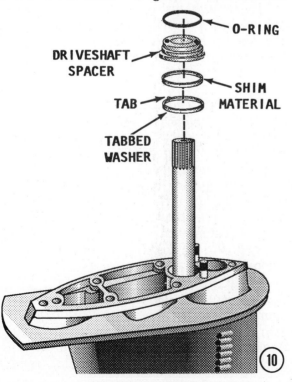

⑩

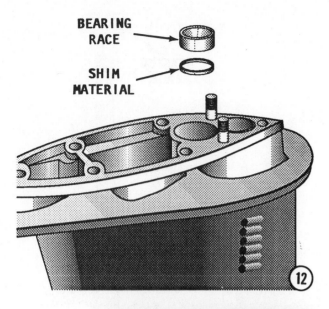

BEARING RACE

SHIM MATERIAL

⑫

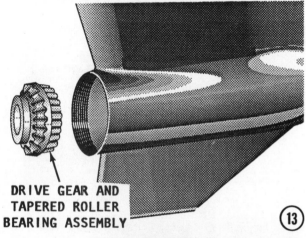

DRIVE GEAR AND TAPERED ROLLER BEARING ASSEMBLY

⑬

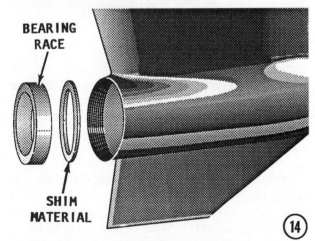

BEARING RACE

SHIM MATERIAL

⑭

CRITICAL SHIM MATERIAL WORDS

Be sure to **SAVE** and identify the shim material removed. This material is critical in obtaining the correct backlash during assembling. Using the old shim material will save considerable time, especially if it should become necessary to begin with no shim material.

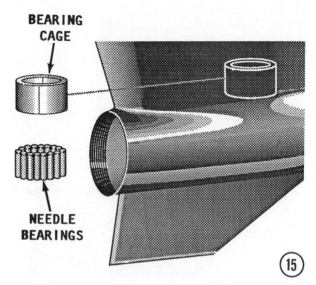

BEARING CAGE

NEEDLE BEARINGS

⑮

13– Remove the drive gear and tapered roller bearing assembly.

14– Use a slide hammer with a sliding jaw attachment to remove the bearing race and shim from the forward end of the lower unit housing.

CRITICAL WORDS AGAIN

Be sure to **SAVE** and identify the shim material removed. This material is critical in obtaining the correct backlash during assembling. Using the old shim material will save considerable time, especially if it should become necessary to begin with no shim material.

15– Reach inside the lower unit housing and up into the driveshaft cavity. Dislodge the needle bearings from the bearing cage just above the area for the pinion gear and catch them as they fall.

SPECIAL NEEDLE BEARING WORDS

If the driveshaft needle bearings are in satisfactory condition, they may be removed, new lubricant applied, and the bearing cage left undisturbed.

If the bearing cage is unsatisfactory for further service, the needle bearings **MUST** be left inside the cage to prevent the cage from breaking or becoming distorted, or wedged in the cavity, when the cage is removed.

15– (Continued) To remove the bearing cage, first check to be sure all the needle bearings are in place inside the cage. Obtain Bearing Remover tool P/N 91-90337, Bearing Driver P/N 91-89868 and Drive Rod P/N 91-37323. Lower the bearing remover

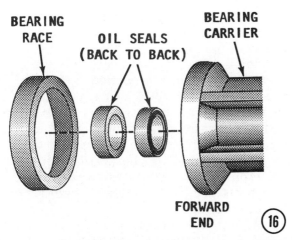

down in the lower unit on top of the cage to be removed. Be sure the tabs on the remover tool align with the slots in the lower unit housing. Place the bearing driver down over the bearing remover. The bearing driver acts as a pilot during the removal operation. Place the driver rod on top of the bearing remover, and then tap lightly on the rod to drive the bearing cage free. Reach in and remove the cage and the needle bearings.

Bearing Carrier Service

FIRST, THESE WORDS

The propeller shaft tapered bearing race can be removed from the bearing carrier without disturbing the two oil seals. Also, the two oil seals may be removed without disturbing the bearing race. Therefore, it is only necessary to remove the item/s unfit for further service, as described in the following step.

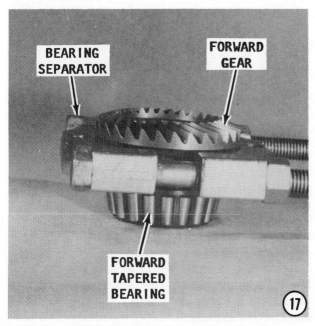

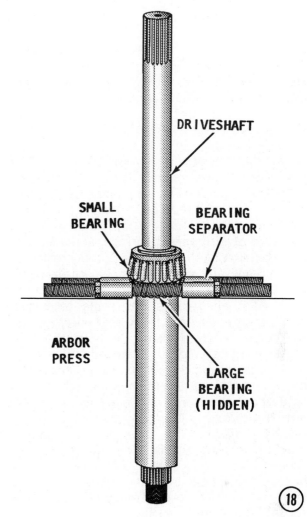

One of the bearing carrier ribs can be clamped in a vise equipped with soft jaws to secure the carrier while pulling the bearing race or the oil seals.

16- Obtain a slide hammer with an expanding jaw attachment. Using the slide hammer, pull the bearing race from the end of the carrier with the larger internal diameter, (the forward end). The two oil seals may be tapped down with a punch from the smaller (aft) end of the carrier. Once the seals have been dislodged, the seals will pass through the bearing race if the race is not disturbed.

Drive Gear Service

17- Position a universal bearing separator tool between the drive gear and the tapered roller bearing. Using an arbor press, separate the gear from the bearing.

Propeller Shaft Bearing Removal

18- Position a universal bearing separator tool between the two bearings on the driveshaft. Place the driveshaft and tool in an arbor press, as shown. Press the bearing free of the shaft. **DO NOT** allow the shaft to fall, once the bearing is free. Turn the shaft end for end and remove the second bearing in the same manner.

CLEANING AND INSPECTING

Wash all parts in solvent and blow them dry with compressed air. Blow all water, oil passageways, and screw holes clean with compressed air. After cleaning, apply a light coating of engine oil to the bright surfaces of all gears, bearings, and shafts as a prevention against rusting and corrosion.

Replace any worn, corroded, or damaged parts. Use a fine file to remove burrs. **REPLACE** all O-rings and seals to ensure satisfactory service from the unit. Clean the corrosion from inside the housing where the bearing carrier was removed.

Check to be sure the water intake is clean and free of any foreign material.

Inspect the lower unit housing inside and out for cracks. The manufacturer recommends replacing the lower unit housing if more than 50% of the skeg has been lost, broken away. A broken piece of skeg cannot be welded back in place because of the tremendous heat generated which will distort the lower unit housing.

Check carefully around screw and shaft holes. Check for burrs around machined faces and holes and check the threads in screw holes.

Check O-ring seal grooves for sharp edges, which could cut a new seal. Clean out the two oil holes in the driveshaft spacer. These two holes are the only oil passageways between the intermediate housing and the lower unit.

Inspect the bearing surfaces of the shafts and splines for wear and burrs. Look for evidence of an inner bearing race turning on the shaft. Check for damaged threads. Measure the runout on all shafts to detect any bent condition. If possible, check the shafts in a lathe for out-of-roundness.

Inspect the gear teeth and shaft holes for wear and burrs. Hold the center race of each bearing and turn the outer race. The bearing must turn freely without binding or evidence of rough spots. Inspect the outside diameter of the outer races and the inside diameter of the inner races for evidence of

turning in the housing or on a shaft. Deep discoloration and scores are evidence of overheating.

Notice an area of scoring on the driveshaft just above the pinion splines. This scoring on a Bravo driveshaft is absolutely **NORMAL**. The driveshaft is made of two different materials, stainless steel and carbon steel. The pinion splines are made of carbon steel, while the remainder of the driveshaft is made of stainless steel. The two parts are "spun welded" together causing the scoring appearance.

Inspect the pinion gear and the drive gear for distortion, burrs, and cracks. Check for any sign of discoloration, which means the gears are running too hot.

Replace the seal and O-rings to ensure maximum service after the work is completed.

ASSEMBLING LOWER UNIT

Driveshaft

WORDS ON BACK TO BACK BEARINGS

The large and small tapered roller bearings are installed onto the driveshaft "back to back". The larger side (the larger diameter), of each taper will face each other, as shown in the accompanying illustration for Step 1.

1- Set up a universal bearing separator tool in an arbor press as a support.

Smaller Bearing: Position the smaller of the new tapered roller bearings on top of the bearing separator, with the taper (the smaller diameter), facing **DOWNWARD**. Press the coupler end of the driveshaft down through the bearing, until the bearing is in the same position as the original bearing. Raise the arbor press, but leave the driveshaft in place with the smaller bearing still on top of the separator tool.

Larger Bearing: With the driveshaft still in the position as described for the smaller bearing, place the larger bearing on the pinion gear end of the driveshaft with the taper (the smaller diameter), facing upward. Now, obtain a suitable mandrel capable of sliding down the driveshaft and with the end against **ONLY** the inner race. Use the arbor press against the mandrel and press the larger bearing down the driveshaft until it barely makes contact with the smaller bearing already installed.

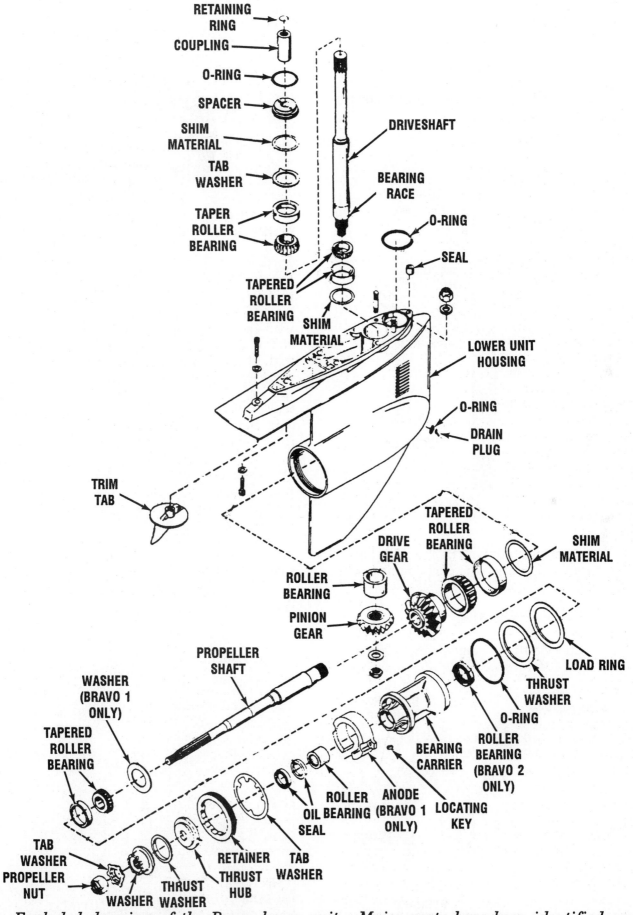

Exploded drawing of the Bravo lower unit. Major parts have been identified and components peculiar to the Bravo 1 or to the Bravo 2 are clearly labeled.

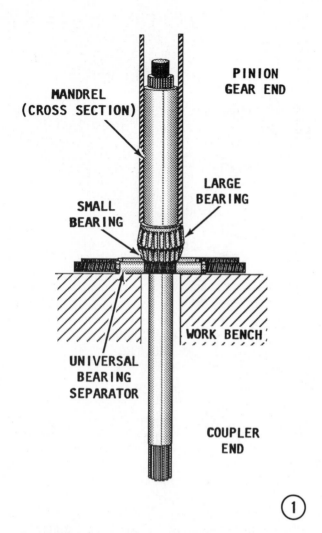

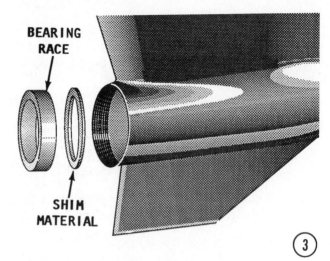

Both bearings will now be "back to back" (the larger diameter of each facing one another).

Bearing Carrier Assembling
2- Pack both oil seal lips with 2-4-C Marine Lubricant, or equivalent. Coat the outside circumference of both seals with

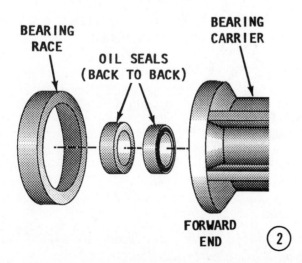

Loctite Type "A". Obtain cup and driver tool P/N 91-89865. Press the first oil seal into the carrier from the wide (forward) end with the lip facing downward toward the aft end of the carrier. Press in the second seal with the lip facing upward toward the forward end of the carrier. In this manner the seals are installed "back to back". The one seal prevents water from entering the lower unit and the other seal prevents lubricant from escaping.

The same driver tool is used to install the tapered bearing race. Place the bearing race squarely into the carrier with the tapered end (the smaller diameter), facing downward (going in first). Using a suitable mandrel and an arbor press, press the race into place until the race is against the shoulder in the carrier.

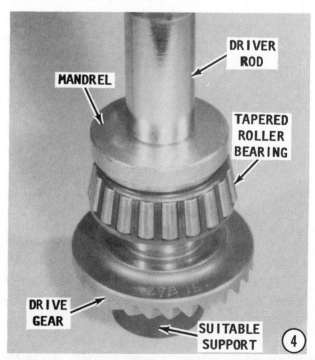

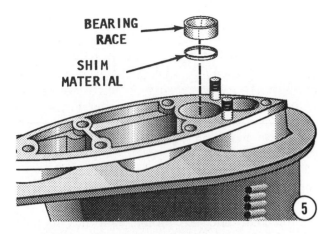

BEARING RACE

SHIM MATERIAL

5

Drive Gear Bearing Race Installation

3- Obtain a suitable driver rod and Driver P/N 91-31106. Install the original amount of shim material into place at the forward end of the lower unit. Position the bearing race squarely in the lower unit and drive it into place until it is fully seated. Once the race is in place the sound of the hammer striking the rod will change quite noticeably.

Drive Gear and Bearing Assembling

4- Position the drive gear tapered roller bearing over the drive gear, with the large side of the taper toward the gear, as shown. Use a suitable support under the drive gear and a suitable mandrel on top of the bearing and press the bearing until it is flush against the shoulder of the gear. **ALWAYS** press on the inner race, **NEVER** on the cage or the roller.

Lower Driveshaft Tapered Roller Bearing Race

5- Install the original amount of shim material into the top of the driveshaft cavity. Position the lower driveshaft tapered

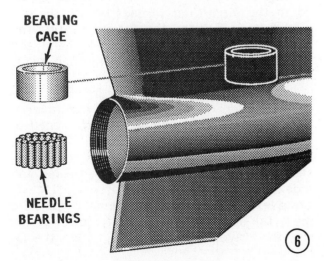

BEARING CAGE

NEEDLE BEARINGS

6

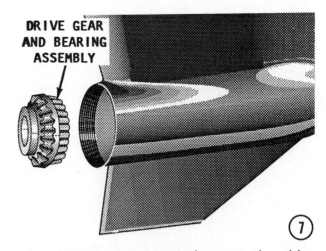

DRIVE GEAR AND BEARING ASSEMBLY

7

roller bearing race squarely over the shim material. Obtain Driver Rod and Driver P/N 91-67443. Drive the bearing race into place until it seats against the shoulder in the lower unit. Once the race is seated, the sound of the hammer striking the rod will change noticeably.

Driveshaft Needle Bearing Race

6- If only the needle bearings were removed, apply a light coating of 2-4-C Marine Lubricant, or equivalent, to the installed bearing cage. Install the needle bearings into the cage one-by-one until they are all in place.

If the entire bearing, including the bearing cage, was removed, proceed as follows: Obtain Bearing Installer tool P/N 91-89867, Pilot tool P/N 91-89868 and Rod P/N 91-31229. Install the new needle bearings into the new cage, using 2-4-C Marine Lubricant on the needles to hold them in place. After all the needles are in place, position the assembled bearing over the bearing installer, with the embossed numbers on the bearing cage facing upward. Set these pieces aside.

Place the pilot tool over the top of the driveshaft cavity. Now, with one hand, hold the assembled bearing and installer tool in the lower unit under the driveshaft cavity. With the other hand, lower the long threaded rod through the pilot and into the driveshaft cavity. Continue lowering the rod down through the bearing, and then thread the end into the installer tool a couple of full turns. The one hand can then be withdrawn from the lower unit. Continue holding the rod with one hand and with the other hand thread a nut onto the upper end of the rod until it is seated against the surface of the pilot tool.

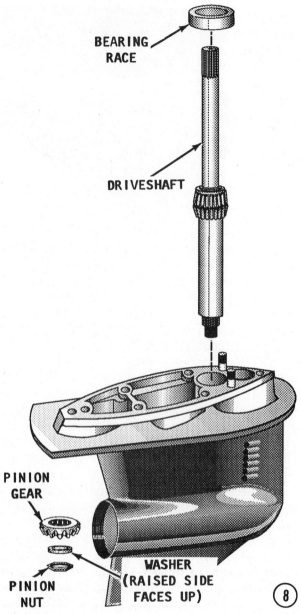

BEARING RACE

DRIVESHAFT

PINION GEAR

PINION NUT

WASHER (RAISED SIDE FACES UP)

(8)

The bearing cage is now ready to be "pulled" up into place. Using a wrench, tighten the nut. The bearing cage will be drawn up into the driveshaft cavity. Continue tightening the nut until the cage seats against the shoulder in the lower unit. Once the cage is seated, movement of the wrench will be restricted almost instantly. Remove the special tools.

SPECIAL WORDS

TAKE CARE not to jar the lower unit from this point on until the driveshaft is installed. The needle bearings may dislodge without warning.

Drive Gear Installation

7- Insert the assembled drive gear and bearing into the lower unit.

Driveshaft Installation

8- Carefully lower the driveshaft into the lower unit. **TAKE CARE** not to dislodge the needle bearings at the lower end of the driveshaft cavity. Reach in with the pinion gear and slide the gear onto the lower end of the driveshaft with the internal splines of the gear indexing with the external splines on the driveshaft.

NOW, SPECIAL WASHER WORDS

A special washer is used between the pinion gear and the pinion gear nut to help distribute loads on the gear and nut. When this washer is new, both sides are flat. However, after use, one side becomes slightly raised and the washer is concave. When a used washer is installed, the raised edge of the washer **MUST** face toward the pinion gear. Stated in another term, the concave side of the washer faces up -- to bear against the lower side of the pinion gear.

8- (Continued) Hold the washer and the nut between your finger tips and thumb. If a used washer is used the concave side **MUST** face upward to bear against the lower edge of the pinion gear. Now, reach in and slide the washer onto the lower end of the driveshaft, and then start the nut.

CRITICAL PINION NUT WORDS

DO NOT use any type of "locking" agent at this time. Backlash and shimming may require the nut to be removed and installed a second time. If no backlash or shimming procedures are to be performed, then a coating of Loctite Type "A" should be applied to the pinion nut threads before the nut is installed. Tighten the nut by rotating the driveshaft **CLOCKWISE** with one hand and holding the nut with the other through just a couple of complete turns.

Slide the upper tapered bearing race down the driveshaft and over the tapered bearing. The pinion nut will be tightened to a given torque value later.

9- Slide the tab washer down the driveshaft. Check to be sure the washer tab indexes with the slot in the driveshaft cavity. Next, slide the original amount of shim material down the driveshaft, followed by the the driveshaft spacer, and then the O-ring.

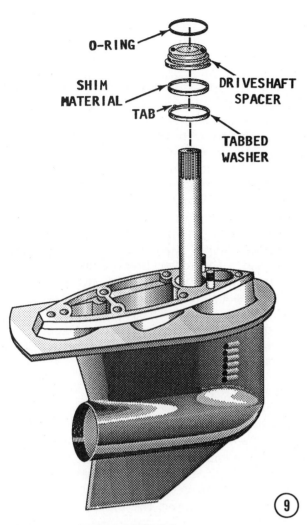

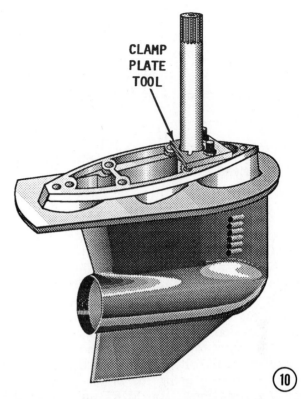

driveshaft distortion while the bearing carrier is being installed.

10- Obtain Clamp Plate tool P/N 91-43559. Install the tool over the driveshaft and secure it in place in the same manner as during disassembling. Tighten the nuts securely.

11- Temporarily install the bearing carrier ring nut to protect the threads in the lower unit opening. Obtain Driveshaft Adaptor tool P/N 91-61077 and a socket the same size as the pinion gear nut. Place the special tool over the driveshaft splines at the upper end of the driveshaft.

Now, using a wrench on the special tool at the upper end of the driveshaft and a socket and torque wrench on the pinion gear nut, rotate the wrench **CLOCKWISE** until the torque wrench indicates a torque value of 113 ft lbs (153 Nm). Remove the tools.

SPECIAL O-RING WORDS

The **O**-ring used at this location is perhaps the most crucial item in the entire stern drive. Lubricant is shared between the upper gear housing and lower unit. The lubricant passes freely through the two holes drilled in the driveshaft spacer. The **O**-ring provides the only seal in the oil passageway between the upper gear housing and the lower unit. No gaskets are used in the Bravo stern drive. A distorted **O**-ring will fail to provide an adequate seal between the upper gear housing and the lower unit and may allow water to enter or lubricant to be lost.

GOOD WORDS

The following step involves the use of a special retaining plate clamped over the driveshaft. Stresses on the propeller shaft are transmitted through the pinion gear and drive gears to the driveshaft while the bearing carrier is being installed. The use of this plate is **STRONGLY** recommended by the manufacturer and the authors to prevent

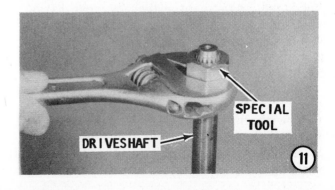

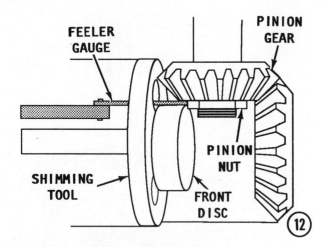

SPECIAL SHIMMING WORDS

Shim material is installed in two locations on the driveshaft. The shim material under the driveshaft spacer is adjusted in thickness to bring the driveshaft preload to specifications.

The shim material installed under the lower tapered roller bearing race is adjusted in thickness to bring the pinion gear height to specifications. Be **SURE** to adjust the shim material at the proper location to achieve the desired results.

The manufacturer stresses the importance of shimming and adjustment procedures for the Bravo stern drive be done in the proper sequence, as follows:

First, pinion gear height.
Second, driveshaft preload.
Next, driven gear backlash.
Finally, propeller shaft rolling torque.

The following steps list the required procedures to obtain the necessary adjustments in the sequence listed.

Checking Pinion Gear Height

12- Obtain Shimming Tool P/N 91-42840 for Bravo "One" and P/N 91-96512 for Bravo "Two" and a feeler gauge. Observe the series of holes in the shimming tool. Each hole has letters or numbers embossed adjacent to the hole. Select the correct hole to make the adjustment either by gear ratio, which for a Bravo is 1.5:1, or the letters B-R-A-V-O embossed by the hole. Insert the shimming tool into the lower unit with the correct hole at the top.

Push the tool into the lower unit housing until the front disc of the tool makes contact with the pinion gear. Insert a feeler gauge into the hole in the tool between the lower edge of the pinion gear and the top surface of the front disc, as indicated in the

accompanying drawing. Determine the clearance between the front disc and the pinion gear. Back out the feeler gauge and shimming tool enough to allow the pinion gear to rotate. Rotate the driveshaft through 120° (approximately 1/3 revolution).

Make another clearance measurement, rotate the driveshaft another 1/3 turn, and then make a third measurement. Determine the average clearance measurement by adding the three measurements and dividing the answer by 3. The specified clearance given by the manufacturer is 0.025" (0.625mm). If the clearance is more than specified, remove the excess thickness of shim material from under the lower tapered bearing race. If the clearance is less than specified, add the necessary thickness of shim material to the shim material already installed under the lower tapered roller bearing race. Repeat the pinion gear height check after making the shim material adjustment.

Driveshaft Preload

13- Remove the clamp plate tool, the O-ring, driveshaft spacer, and shim material from the upper end of the driveshaft. **LEAVE** the tab washer in place.

14- Obtain a depth micrometer capable of reading from zero to one inch. Measure the distance between the top surface of the

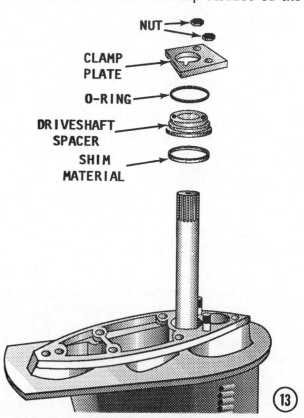

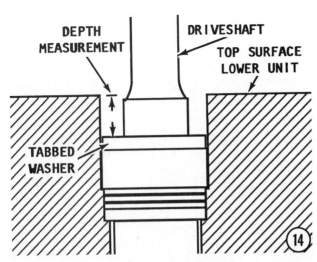

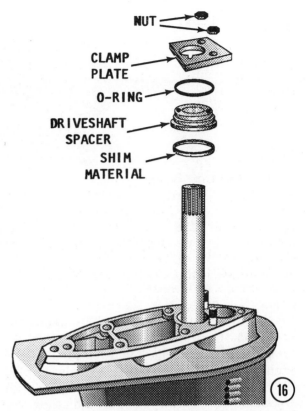

lower unit housing and the tab washer. Make a note of this depth.

15- Using a micrometer, measure the thickness of the driveshaft spacer from the machined shoulder surface to the bottom surface of the spacer, as indicated in the drawing. Make a note of this measurement.

Calculate the required shim material thickness as follows:

Depth measurement (Step 14) minus spacer thickness measurement (this step), plus 0.001" equals shim material thickness required.

Example:

Depth measurement	-- 0.302"
Driveshaft spacer	-- 0.255"
Difference	-- 0.047"
Add	-- 0.001"
Required shim material	-- 0.048"

Therefore, the thickness of the shim material under the driveshaft spacer should be 0.048". Add or subtract shim material, as required to meet specifications. If the original shim material was lost, start with 0.050" (1.27mm)

CRITICAL SHIM MATERIAL WORDS

Correct total shim thickness can only be determined by measuring the thickness of each individual piece and then adding for a total. **NEVER** take a short cut and attempt to place the pieces on top of each other and

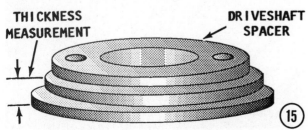

then make one measurement for a total thickness. This practice will never give an accurate thickness total.

16- Install the correct amount of shim material, as determined by the previous step, the driveshaft spacer, O-ring, and then install the clamp plate tool again.

17- Install the propeller shaft, with the forward end of the shaft indexed into the driven gear.

18- Slide the shim material, washer, and O-ring onto the propeller shaft and into the lower unit housing against the housing shoulder.

SPECIAL BACKLASH AND ROLLING TORQUE WORDS

If the drive gear backlash and propeller shaft rolling torque procedure will not be performed, proceed directly to Step 21.

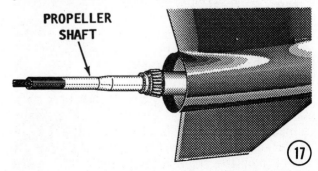

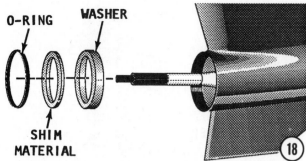

O-RING   WASHER

SHIM
MATERIAL

18

19- Obtain Bearing Carrier Retainer Wrench, P/N 91-61069. Install the bearing carrier into the lower unit. Secure the carrier with the ring nut, but **DO NOT** install the tab washer at this time. Tighten the ring nut to a torque value of 150 ft lb (203Nm) using the special wrench.

Drive Gear Shimming

20- Obtain Dial Indicator Adaptor P/N 91-83155, Backlash Indicator Rod P/N 91-53459 and Dial Indicator P/N 91-58222A1. Secure the dial indicator to the clamp plate and position the dial indicator rod on the embossed "II" mark on the indicator rod. Slowly "rock" the shaft back and forth through about 25° to 30° arc, but not enough to rotate the propeller shaft. If the propeller shaft moves, the arc is too great.

At the outer limits of the arc, a "click" will be heard. This sound occurs when the pinion gear tooth contacts a face of the driven gear tooth. Another "click" will be heard when the pinion gear swings back and the pinion gear tooth contacts the face of the adjacent driven gear tooth. The arc between the two "clicks" represents the backlash (or "free play") between the two gears. Zero the dial indicator gauge at the first "click". As soon as the second "click" is heard, stop all motion and read the maximum deflection on the dial indicator needle. Acceptable backlash is 0.012-0.015" (0.28-0.38mm).

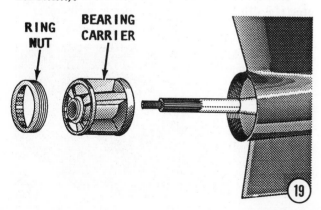

RING
NUT

BEARING
CARRIER

19

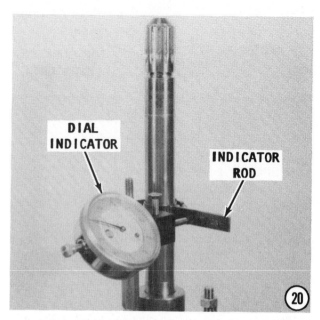

DIAL
INDICATOR

INDICATOR
ROD

20

If the backlash is incorrect, the lower unit must be disassembled, the driven gear bearing race removed and the shim material behind the race adjusted. If the backlash is less than specified, remove shim material from behind the bearing race. If the backlash is more than specified, add shim material behind the bearing race.

If the backlash was within the limits specified, continue with the work by removing the dial indicator setup, but leave the clamp plate in place through Step 24.

21- Remove the ring nut, install the tab washer, with the curled tab indexed in the notch of the bearing carrier and with the outer tab indexed in the slot of the propeller bore. This configuration prevents the bearing carrier from spinning inside the lower unit. There is **NO** key or keyway as found in a standard Mercruiser or Alpha stern drive unit. Using Bearing Carrier Retainer Wrench P/N 91-61069, tighten the ring nut to a torque value of 150 ft lbs (203Nm).

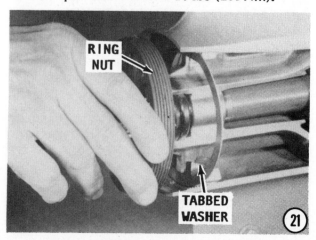

RING
NUT

TABBED
WASHER

21

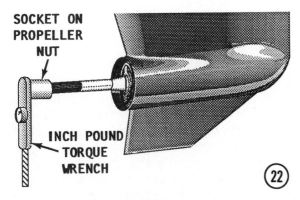

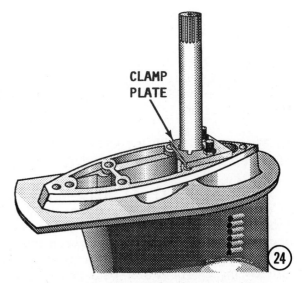

Propeller Shaft Rolling Torque

22- Obtain a socket the same size as the propeller nut and a spare propeller nut. Install the two nuts onto the propeller shaft with the second nut up tight against the first. Now, using the socket and an inch-pound torque wrench with a dial indicator readout, slowly rotate the propeller nut **CLOCKWISE** until a steady reading is observed on the dial. The reading should be as follows:

New bearings -- 8-12 in lbs

Used bearings -- 5-8 in lbs

The manufacturer defines a used bearing as one which has been spun once under load. If the dial reading is lower than specified, remove shim material from in front of the bearing carrier. If the dial reading is higher than specified, add shim material in front of the bearing carrier.

23- Bend one or more of the tabs on the tab washer down over the ring nut.

24- Remove the clamp plate tool from the upper surface of the lower unit.

25- Apply 3-M adhesive to the water passage O-ring, and then position the ring in the forward water passageway. Coat the outside surface of the tubular speedometer seal with 3-M adhesive. Install the seal into the speedometer water tube bore. Press the seal in until the top of the seal is flush with the upper surface of the lower unit. Install

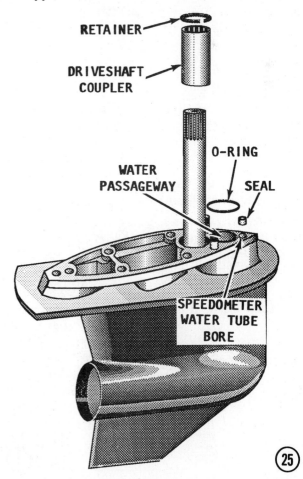

the driveshaft coupler and retainer onto the top of the driveshaft.

EXTRA GOOD WORDS

The lower unit is now ready to be mated with the upper gear housing, as outlined in the next section.

10-33 MATING LOWER UNIT TO UPPER GEARCASE HOUSING

1- Coat a new O-ring with 3-M Adhesive P/N 92-25234 and install it on the upper face of the lower unit at the water passage opening. No gasket is used between the two housings. All sealing is accomplished with two O-rings: The water passageway O-ring and the O-ring installed around the driveshaft spacer. The spacer has two vertical holes drilled through it to allow oil to pass through the spacer and lubricate not only the double roller bearing assembly on the driveshaft but the entire lower unit. Oil passes freely through the double bearing arrangement. Therefore, the oil passageway

O-ring is perhaps the most crucial component in a Bravo stern drive.

CRITICAL MATING WORDS

In the Bravo Stern Drive, no locating pins are utilized in aligning the lower unit with the upper gearcase housing. Therefore, if the bolt holes show serious signs of elongation, or the securing bolts show signs of stress or "necking", the lower unit, the upper gearcase housing and all securing bolts should be replaced. This is not as drastic or expensive as it sounds. All internal components are removed and only the housing castings are replaced.

Install the coupler and retainer onto the upper end of the driveshaft. Raise the lower unit to mate with the upper gearcase housing. The lower end of the short clutch shaft must slide into the coupler at the upper end of the lower driveshaft. It may be necessary to slowly rotate the propeller shaft **COUNTERCLOCKWISE** until the splines index.

2- Place the two washers over the studs above the anti-cavitation plate, one on each side of the unit. Start the two locknuts and tighten them **ONLY** hand tight at this time.

WORDS FROM EXPERIENCE

If the holding fixture allows, carefully tilt the unit to the horizontal position to simplify the task of installing the remaining nuts and washers.

If this is not possible, apply a dab of water resistant lubricant on the **UPPER/-OUTER** face of each of the four washers installed below the anti-cavitation plate. This bit of lubricant will keep them in place

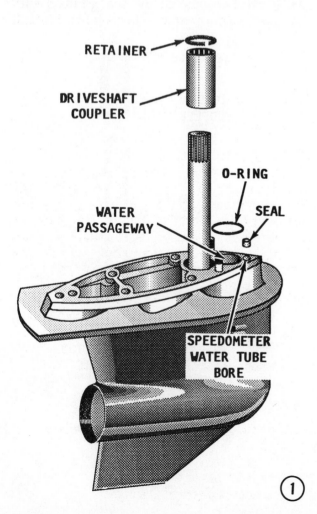

RETAINER

DRIVESHAFT COUPLER

O-RING

WATER PASSAGEWAY

SEAL

SPEEDOMETER WATER TUBE BORE

①

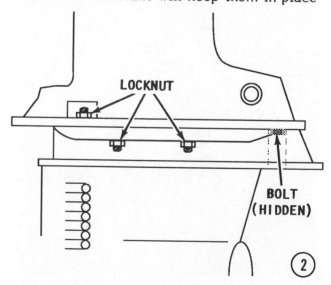

LOCKNUT

BOLT (HIDDEN)

②

against the force of gravity, while the nuts are being installed and tightened. **TAKE CARE** not to smear the lubricant onto the threads of the studs. Any lubricant on the stud threads will result in a false torque wrench reading when the nuts are tightened.

With the washers in place, start the six **NEW** locknuts and tighten them **ONLY** hand tight. **REMEMBER**, one bolt is installed up from the trim tab cavity.

After all nuts and bolts are in place, tighten them alternately and evenly to a torque value of 35 ft lb (47Nm).

IMPORTANT TRIM TAB WORDS

The trim tab performs two very important functions. First, the trim tab compensates for steering torque if the unit is **NOT** equipped with power steering. The trim tab also prevents electrolysis from damaging expensive parts. It should be replaced before half of the material has corroded away. If the trim tab shows no signs of deterioration after the boat has been in use over a period of time, the grounding of the tab and the continuity of the lead wires should be checked to ensure more expensive parts are not being effected by electrolysis.

Units Without Power Steering

For units without power steering, should the boat continually veer to port or starboard while the helmsperson attempts to

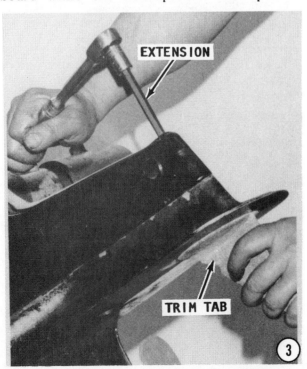

hold a straight course, the trim tab can be adjusted to the side of the "pull" to compensate for steering torque.

Units With Power Steering

If the unit is equipped with power steering, the trim tab should **ALWAYS** be set dead ahead. If the helmsperson experiences a "pull" from a straight course, the cause is not steering torque and should not be corrected by repositioning the trim tab. The problem is in the power steering control valve.

3- Clean off the contact area between the trim tab and the housing to ensure good metal-to-metal contact. **NEVER** paint the contact area or the trim tab. Paint will defeat the purpose of the tab. Install the trim tab bolt into the hole in the upper gearcase housing. Hold the trim tab up against the housing. On units not equipped with power steering, align the tab with the marks made during removal to position the tab in its original setting. If the unit being serviced is equipped with power steering, the trim tab should be set dead ahead. Tighten the bolt to a torque value of 35 ft lb (47Nm).

Install the plastic plug over the trim tab bolt.

10-34 MATING STERN DRIVE TO BELL HOUSING

FIRST, THESE WORDS

If the only work being performed was service of the stern drive, proceed with the following steps. If the powerplant was also removed from the boat for service, then the engine must be aligned and shift cable adjustments made prior to installation of the stern drive. See Chapter 13.

Preliminary Tasks

Check to be sure the remote control unit is in the **NEUTRAL** position.

Apply Quicksilver 2-4-C Marine Lubricant to the following parts:

All six mounting studs.
The driveshaft housing pilot.
The two **O**-rings around the **U**-joint shaft.
The splines of the **U**-joint shaft.
The shift linkage.
The water passage **O**-rings.

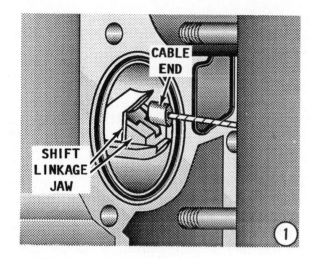

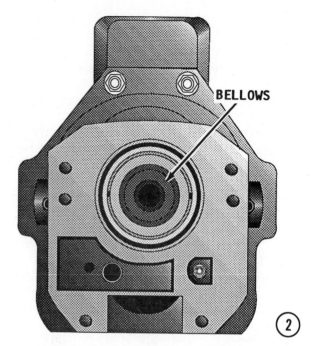

Good Words

No gasket is used between the upper gearcase housing and the bell housing. All sealing is accomplished with O-rings. Therefore, the condition and correct installation of these crucial sealing components is vital to the longevity of the unit.

1- Move the assembled stern drive to within a few inches of the bell housing. Use a pair of needle nose pliers to grasp the lower jaw of the shift linkage and gently pull it out of the housing as far as it will go. As the linkage emerges, the jaws will open.

Guide the end of the shift cable into the open jaws of the linkage. Set the cable end down into the slot in the lower jaw.

Continue to bring the housings together. As the shift cable pushes further into the lower jaw and the linkage slides back into the upper gear housing, the jaws will clamp down over the cable end.

2- Guide the splined end of the U-joint shaft through the bellows and into the engine coupler. Do not allow the splined end of the U-joint shaft to ram into the bellows. The U-joint bellows between the bell housing and the drive shaft housing forms an air tight seal, easily damaged by a heavy blow from the shaft splines.

If the U-joint splines do not index with the splines of the engine coupler, slide the propeller onto the propeller shaft, shift the unit into gear, and **SLOWLY** rotate the propeller **COUNTERCLOCKWISE** until the stern drive can be pushed up against the bell housing.

HELPFUL WORDS

If difficulty is experienced in installing the stern drive to the bell housing, the engine alignment must be checked, see Section 13-4. **NEVER** attempt to force the U-joint shaft into the engine coupler. The engine **MUST** be properly aligned or the female splines in the engine coupler and the male splines of the U-joint shaft will be destroyed in a short time.

Check to be sure the unit is still in **Neutral** gear by rotating the propeller shaft.

3- Secure the driveshaft housing to the bell housing with six **NEW** nylon locknuts on

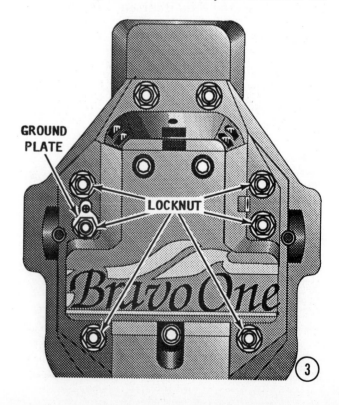

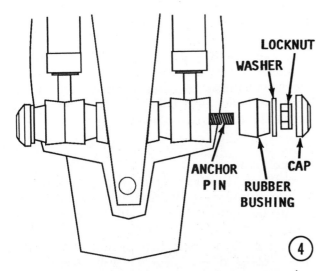

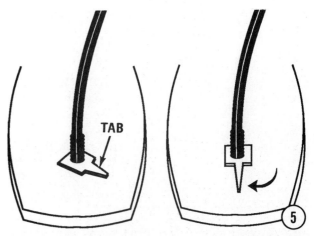

top of five washers and one ground plate (on the portside center stud). Tighten the nuts alternately and eveny to a torque value of 50 ft lb (69Nm).

4- On each side of the lower unit: Loosen the nut securing the hydraulic cylinder to the forward anchor pin.

MORE HELPFUL WORDS

If difficulty is encountered in installing the two sets of rubber bushings onto the aft anchor pin, apply a film of soapy water to the ends of the pin. The bushings will slide on "slicker than a whistle".

Slide the aft anchor pin through the driveshaft housing. Center the pin about the housing. Place one large flat washer over each end of the pin, followed by a rubber bushing over each end, with the taper (the small diameter), facing outward.

Install the cylinders onto the aft anchor pins. Pick up the rubber bushings and observe that one end has a smaller diameter than the other. The small diameter end **MUST FACE** the driveshaft housing. Push a rubber bushing onto the aft anchor pin and into the mounting hole of the cylinder with the small diameter end facing the housing. Slide one small flat washer onto each aft anchor pin. Install a **NEW** nylon locknut onto each anchor pin.

Tighten all four anchor pin locknuts until the washers seat against the anchor pin shoulder.

Install the decorative/protective caps over each locknut.

5- Raise the stern drive to the full up position. Insert the speedometer hose fitting into the hole in the forward end of the anti-cavitation plate. Push the fitting firm-

ly into the hole and rotate the tab **CLOCK-WISE** until the fitting is held snug in the hole.

10-35 PROPELLER INSTALLATION

1- Check to be sure the splines and the threaded end of the propeller shaft are clean and in good condition. Install the propeller thrust hub. Coat the propeller shaft splines with Special Lubricant 101, Quicksilver 2-4-C Marine Lubricant, Perfect or Seal No. 4, and the rest of the shaft with a good grade of anti-corrosion lubricant.

Install the propeller, and then the continuity washer, splined washer, tab washer, and propeller nut.

Position a block of wood between the propeller and the anti-cavitation tab to keep the propeller from turning. Tighten the propeller nut to a torque value of at least 55 ft lb (75Nm). Adjust the nut to fit the tab lock space by tightening, **NEVER** loosening the nut.

2- Bend three of the tab washer tabs into the spline washer using a punch and hammer. The tabs will prevent the nut from backing out.

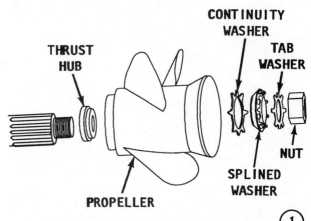

10-36 STERN DRIVE LUBRICATION

Filling Stern Drive Unit

Remove the vent screw in the upper gear housing. Place the stern drive in an approximate vertical position without a list to port or starboard. Insert a lubricant tube into the **DRAIN/FILL** plug opening in the lower unit. Inject lubricant until the excess starts to flow out the **VENT** hole in the upper gear housing. The capacity of the Bravo stern drive is 2 quarts (1.9 Litres).

CRITICAL LUBRICATION WORDS

On the Bravo stern drive, the upper and lower lubricant reservoirs are connected by

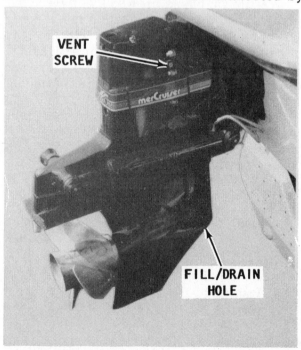

The same gearcase lubricant is circulated through the lower unit and the upper gearcase housing. Therefore, only one opening is used to fill both housings. The vent screw MUST be opened when draining or filling.

an oil passageway. Therefore, the same gearcase lubricant is shared between the upper gearcase housing and the lower unit. Because of this arrangment, both upper and lower housings can be drained and filled from the lower unit **DRAIN/FILL** hole located at the base of the lower unit housing.

Trapped Air

NEVER fill the unit at the vent hole in the upper gearcase housing, because trapped air inside both housings will give a false reading on the dipstick (if equipped), or cause the oil to overflow at the vent hole. These trapped air pockets will eventually work their way upward into the upper gearcase. When this happens all the oil will find its way down into the lower unit, leaving the upper gearcase short of lubricant. Such a condition can and will cause extensive and expensive damage to moving parts in the upper gear case.

Inadequate Lubricant

An inadequate amount of lubricant in the stern drive will cause the oil to whip and foam, greatly reducing its lubricating qualities. Such a condition could lead to disaster -- serious and expensive damage to moving parts.

Adequate Lubricant

When the stern drive is properly filled from the **DRAIN/FILL** hole, any air pockets will eventually work their way upward to the top of the upper gear case housing, and then escape through the vent hole.

Keep the pump nozzle in place and install the **VENT** and **DRAIN/FILL** plugs, with **NEW** washers. As the plugs are tightened to a torque value of 17 in lb (2Nm), check to be sure the washers remain in place to prevent water from leaking into the housing.

Start the engine and make a functional check of the completed work.

CAUTION

Water must circulate through the stern drive, also to and from the engine, anytime the engine is run to prevent damage to the water pump located in the lower unit (original and Alpha drives); or on the front of the engine (Bravo drive). Just a few seconds without water will damage the water pump.

11
BELL HOUSING

11-1 INTRODUCTION

The stern drive unit, which consists of the upper gear housing and the lower unit, is attached to the bell housing. The bell housing is secured to the gimbal ring by two roller bearings. These bearings permit up-and-down trim movement. The gimbal ring is mounted to the gimbal housing by two roller bearings. This second set of roller bearings permits steering movement from side-to-side. Instructions for service of the gimbal housing are presented in Chapter 12.

The bell housing has extended flanges to connect the exhaust, universal and shift cable bellows, and water hoses between the stern drive and the gimbal housing.

The bell housing is held in place by the gimbal ring.

11-2 TROUBLESHOOTING

Water leaking into the boat is usually caused by a defective seal around the gimbal housing, the shift cable bellows or the U-joint bellows.

Driveshaft noises are an indication of a defective bearing, defective U-joints, or possibly misalignment of the engine.

11-3 SERVICE PROCEDURES GENERAL INFORMATION

Bell housing or bellows service: Before the bell housing or bellows can be serviced, the stern drive must be removed, see Chapter 10, and the shift cable bellows removed, see Section 11-5. Replacement of the exhaust bellows is also presented in this chapter, Section 11-6. For replacement of the U-joint bellows, see Section 11-7.

Gimbal housing, inner transom plate, or exhaust separator service: Service for these three items are covered in Chapter 12. However, before these units can be serviced, the engine must be removed, see Chapter 13, and the stern drive removed, see Chapter 10. When the stern drive is replaced, the trim system hydraulic hoses must be properly connected, see Chapter 8; the steering and shift cables connected and adjusted, see Chapter 7; and the trim-limit switch, reverse lock, and trim sender checked and adjusted, see Chapter 8.

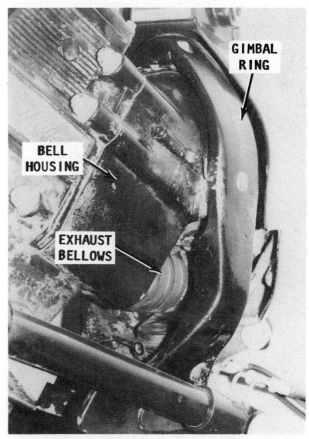

Gimbal ring and bell housing on a late model stern drive unit.

11-4 GIMBAL HOUSING
BALL BEARING SERVICE

REMOVAL

1- Remove the stern drive, see Chapter 10. After the stern drive unit is removed, the gimbal housing bearing can be checked by reaching through the U-joint bellows attached to the bell housing and rotating the inner bearing race. There should be no evidence of binding or rough spots. Check the race for side play by pulling and pushing on the race. If there is any sign of roughness, binding, or excessive side play, the bearing should be replaced. **NEVER** remove the gimbal bearing unless it is to be replaced because it will be damaged during removal. The bearing **MUST** be replaced with the cartridge as an assembly. Therefore, they are packaged and sold as an assembly. A special puller is required to remove the gimbal housing bearing. This puller is designed to establish alignment from the face of the bell housing. Remove the snap ring on the older models, prior to about 1973. The newer models do not have this snap

ring. Assemble the tool C-91-29310 and others identified in the accompanying drawing. Position the plates between the top and middle studs located on the bell housing. Use a 3-jaw puller from the Slide Hammer Puller Set, C-91-34569A1. If the bearing assembly is tight, tap the end of Puller Shaft, C-91-31229, with a mallet while attempting to turn the nut C-11-24156.

Bravo Unit

If servicing a Bravo stern drive, use the same tools to remove the oil seal which contacts the forward end of the bearing carrier in the gimbal housing. The bearing carrier must be removed before the seal can be driven free.

CLEANING AND INSPECTING

Clean all metal parts in solvent and dry them dry with compressed air. **NEVER** spin ball bearings with compressed air, because such action will ruin them.

Inspect the bellows carefully for cracks, cuts, and punctures. Verify the bellows are still flexible. If the condition of the bellows

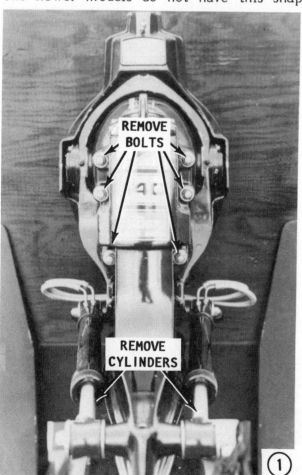

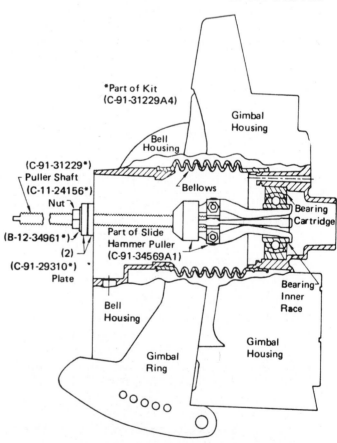

Detailed diagram showing proper installation of the necessary tools required to remove the gimbal bearing.

is doubtful, replace them. In most cases, if the gimbal bearing is damaged due to water, the water has entered through, or around, the bellows.

INSTALLATION

Bravo Units Only

2- Pack the oil seal lip with Multi-Purpose lubricant. Obtain a suitable driver and install the oil seal into the gimbal housing with the lip facing **AFT**. The lip will then contact the bearining carrier after the carrier is installed. The seal is properly installed when it is seated against the shoulder in the housing casting.

All Units
(Including Bravo)

3- If the gimbal housing ball bearing is being replaced, it **MUST** be installed after the bell housing is in place in order to establish an alignment reference. This is accomplished in the following sequence: First, lubricate the outside of a **NEW** gimbal bearing carrier assembly with Multi-Purpose Lubricant, or equivalent. On older units, this bearing arrangement includes the bearing and a retainer ring. The ring fits into the opening of the housing and the bearing snaps into the ring. These two items should **ALWAYS** be replaced as a set. The easy thing to do is merely replace the bearing, but the package contains a new retainer ring, so take the extra time and install it to ensure satisfactory service.

Newer Units Since About 1973
Including Bravo

Newer units have a tolerance band -- a wavy split ring -- around the gimbal bearing carrier, as shown in the accompanying illustration. The opening in the tolerance band

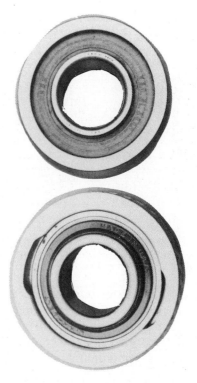

Old style gimbal bearing (top) and the new style bearing (bottom).

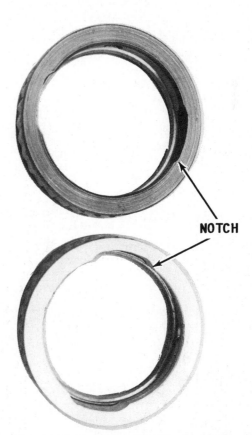

Old style gimbal bearing carrier (top) and the new carrier (bottom). The carriers are most easily identified by their color. The old is black and the new silver. The notches MUST face FORWARD when the carrier is installed.

NOTCH

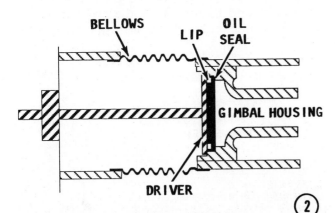

BELLOWS   LIP   OIL SEAL

GIMBAL HOUSING

DRIVER

②

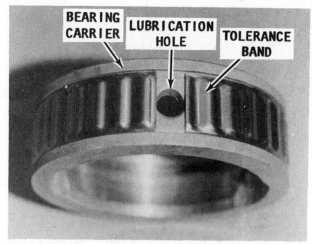

A tolerance band correctly installed around the bearing carrier.

MUST align with the lubrication hole in the bearing carrier **BEFORE** the carrier is installed. The opening must also align with the lubrication hole in the gimbal housing **AFTER** the carrier is installed.

All Units — Old & New

Observe the notches in the bearing carrier, indicated in the accompanying illustration. These notches **MUST** face **FORWARD** when the carrier is installed.

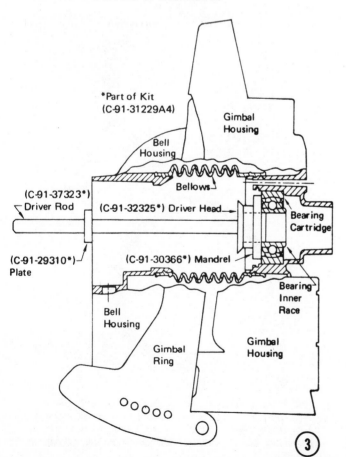

Install the assembled bearing carrier using the tools indicated in the Step 3 illustration. Insert the Driver Head, C-91-32325, through the Mandrel, C-91-30366, and into the inside diameter of the new bearing. Align the bearing with the gimbal housing by positioning Plate, C-91-29310, between the top and middle studs on the bell housing to ensure the driver rod will remain at right angles to the bearing carrier bore. Now, drive the bearing into the gimbal housing cavity with a lead hammer. Lubricate the bearing using only Quicksilver Multi-Purpose Lubricant, C-92-63250 through the grease fitting. To lubricate the bearing properly, pump 40 full strokes, which will deliver about one full ounce of lubricant. If servicing a unit dated prior to about 1973, install the snap ring to secure the bearing in place.

If no other work is to be performed, install the stern drive, see Chapter 10.

CAUTION

Water must circulate through the stern drive, also to and from the engine, anytime the engine is run to prevent damage to the water pump located in the lower unit (original and Alpha drives); or on the front of the engine (Bravo drive). Just a few seconds without water will damage the water pump.

Start the engine and make a functional check of the completed work.

11-5 TRANSOM SHIFT CABLE SERVICE REMOVAL

1- From inside the boat: Disconnect the stern drive unit shift cable from the shift plate mounted on the engine. This is accomplished by removing the nut, washer, and cotter pin.

2- Loosen the set screws from the cable end guide, and then remove the end guide. If only the inner cable is to be replaced, burrs

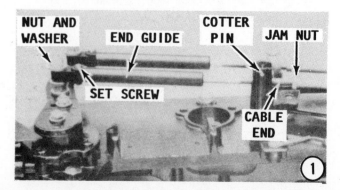

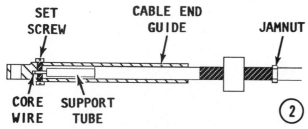

SET SCREW   CABLE END GUIDE   JAMNUT

CORE WIRE   SUPPORT TUBE   ②

made by the set screws must be removed from the core wire, to prevent the inner lining of the shift cable from being damaged when the wire is removed. Loosen the jam nut on the metal cable end, and then turn the metal end out of the cable.

Pull the support tube from the end of the inner core wire.

3- Remove the protective wrapping from the shift cable in the area where the cable passes through the transom. Remove the stern drive, see Chapter 10.

All Units Except Bravo

4- If the inner wire is to replaced, hold the cable slide and pull the inner wire out of the shift cable. Now, remove the safety wire on the end of the cable slide. Remove the Allen-screw and cable slide from the inner shift wire.

CRITICAL INSTRUCTIONS

STOP, and determine if the unit being serviced is an old style or a new style shift cable. Refer to the illustrations on this page to identify the cable being serviced. The locking nut on an old style cable is located on the forward (pivot) end of the bell housing. The locking nut on a new style cable is located on the aft (stud) end of the bell housing. The Bravo cable has two locking nuts, one on each side of the bell housing.

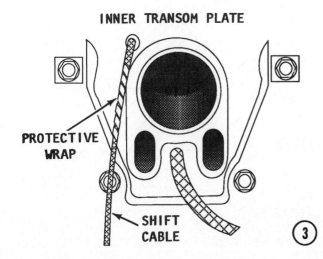

INNER TRANSOM PLATE

PROTECTIVE WRAP

SHIFT CABLE   ③

SAFETY WIRE   ALLEN SCREW   ④

If servicing an **OLD** style shift cable, perform only Steps 5 and 6. If servicing a **NEW** style shift cable, perform only Step 7. If servicing a **Bravo** shift cable, perform only Step 8.

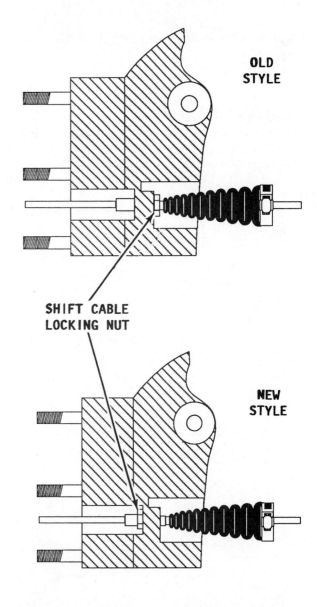

OLD STYLE

SHIFT CABLE LOCKING NUT

NEW STYLE

Identification of old and new style shift cable configurations.

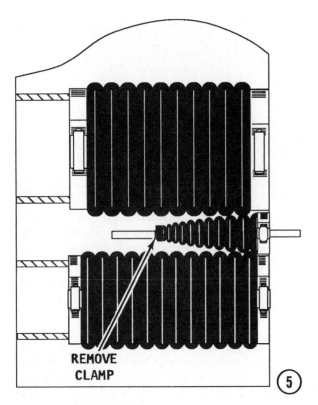

REMOVE
CLAMP

(5)

Old Style Shift Cable

5– If the outer shift cable is to be replaced, the inner core wire must still be removed first. Begin by removing the clamp on the shift bellows at the small end, and then pull the bellows back.

6– Loosen the shift cable locking nut. Now, remove the shift cable from the bell housing, and then pull the shift cable out through the gimbal housing and shift bellows.

New Style Shift Cable

7– Obtain shift cable removal and installation tool C-91-12037. Remove the shift cable bellows clamp at the small end of the

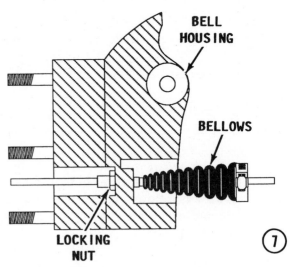

BELL
HOUSING

BELLOWS

LOCKING
NUT

(7)

bellows. Pull the shift cable through the bellows. Using the special tool, completely loosen the locking nut and remove the cable from the bell housing.

Bravo Shift Cable

8– Pull out the shift cable core wire from the aft end of the bell housing. Using two wrenches, hold the larger forward cable locking nut and remove the smaller aft cable locking nut. Pull the shift cable free.

CLEANING AND INSPECTING

Inspect the shift bellows for cracks, cuts, and punctures. Verify the bellows are still flexible. If there is the least doubt about the condition of the bellows, they should be replaced. If the bellows are defective and leak, water will enter the boat.

Inspect the cable locking nut/s threads for any damage such as cross-threading and

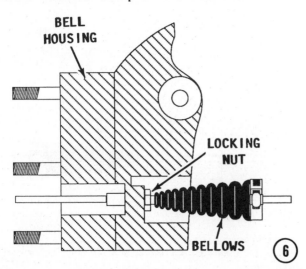

BELL
HOUSING

LOCKING
NUT

BELLOWS

(6)

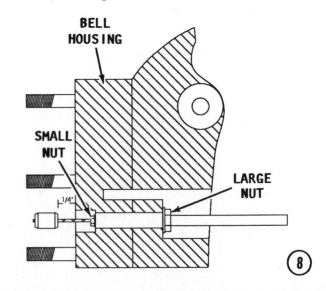

BELL
HOUSING

SMALL
NUT

1/4"

LARGE
NUT

(8)

BROKEN CORE WIRE

Hard shifting may be caused by a frayed inner wire in the shift cable, such as the one shown here from an earlier model. Replacement of the wire will be necessary to correct the problem.

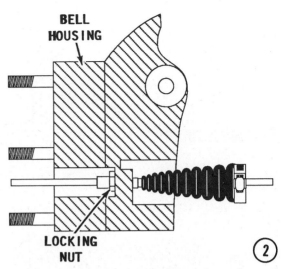

check for any indication of the locking nut separating from the outer casing. Check the length of the shift cable for kinks, cuts or chafing and the inner core wire for signs of unraveling.

INSTALLATION

Old Style Shift Cable

1- From inside the boat, apply a small amount of oil onto the cable, and then feed the cable through the gimbal housing and shift bellows. After the cable has been inserted as far as possible, lift the bell housing and install the shift cable end into the bell housing. Coat the threads of the cable locking nut with Perfect Seal, or equivalent, and then install the nut over the cable end. Tghten the nut securely.

New Style Shift cable

2- Apply a coating of Perfect Seal to the locking nut threads. Secure the shift cable to the bell housing and tighten the nut securely using special cable tool C-91-12037.

Bravo Shift Cable

3- Coat the threads of the large and small locking nuts with Perfect Seal. Install the cable into the bell housing, as shown in the accompanying illustration. Notice how

the smaller nut is installed on the aft end and a seal washer is used between the housing and the larger nut on the forward end. Use a wrench and hold the larger forward locking nut and at the same time tighten the smaller aft nut to a torque value of 65 in lb (7Nm). Install the inner core wire and position the barrel approximately 1/4" (6mm) from the surface of the housing.

FABRICATE SPECIAL PLIERS

The manufacturer strongly recommends the small shift cable bellows clamp be the crimp type **NOT** a worm type clamp **OR** a tie wrap. To crimp such a clamp a special tool is required. However, a pair of pliers may be easily and quickly modified to do the job.

To customize a standard pair of pliers to crimp the bellows clamp, first tack weld a 3/4" nut to the pliers with the gripping surfaces of the pliers contacting two opposite sides of the nut. Clamp the pliers in a vice and drill out most of the threads using

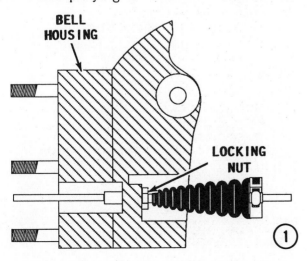

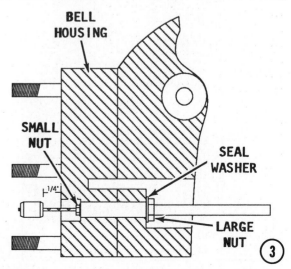

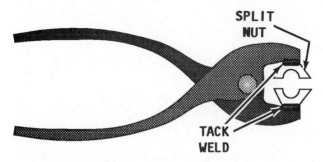

Customized shift bellows crimping pliers, as explained in the text.

a 1/2" drill bit. With the pliers still in the vise cut the nut in half, leaving an equal amount of the nut on each side of the pliers, as shown in the accompanying illustration.

All Units

4— Install the bellow clamp on the small end of the shift bellows with the end of the bellows 2" from the cable locking nut. Use the modified pliers, described in the previous paragraphs, and squeeze the bellows clamp securely around the bellows. After the clamp is properly installed, water will not be able enter the bellows and find its way into the boat.

All Units
Except Bravo

5— Install the inner shift wire through the cable slide from the stern drive side. Install the Allen-screw securely, and then

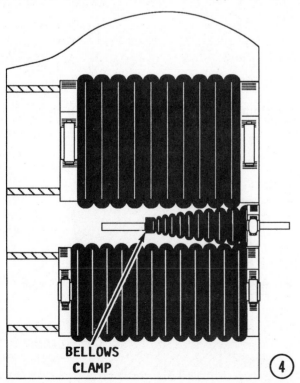

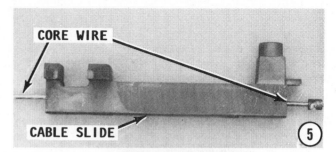

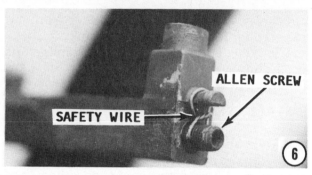

back it off about 1/4 turn. This adjustment will allow the inner wire to rotate freely.

6— Install the safety wire.

7— Coat the inner shift wire with light weight oil, and then feed it into the shift cable until the cable slide enters the bell housing. Now, from inside the boat, carefully pull on the inner wire and be sure it is fully extended.

8— Install the protective wrapping around the cable, as shown.

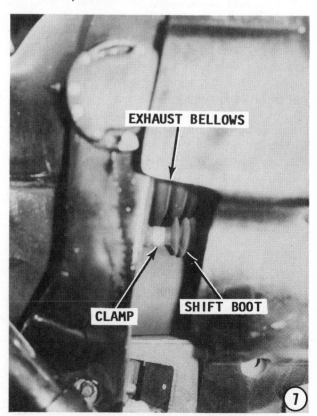

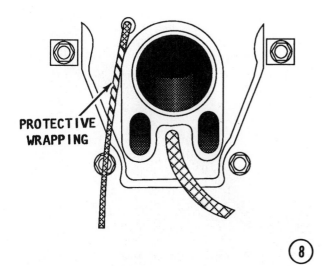

PROTECTIVE WRAPPING

8

9- Before making any measurements of the remote control shift cable, three jobs must be completed in the following sequence. **NEVER** attempt to adjust the shift mechanism **UNLESS** the stern drive is completely installed. First, the guide tube and shift cable end, must be installed, see Chapter 7. Next, the stern drive must be installed, see Chapter 10. Third, the shift mechanism and linkage must be properly adjusted, see Chapter 7.

CAUTION

Water must circulate through the stern drive, also to and from the engine, anytime the engine is run to prevent damage to the water pump located in the lower unit (original and Alpha drives); or on the front of the engine (Bravo drive). Just a few seconds without water will damage the water pump.

Start the engine and make a functional check of the shift mechanism and controls.

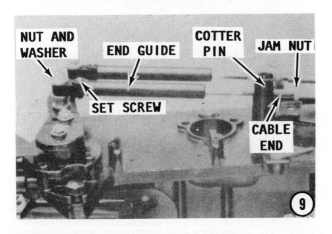

NUT AND WASHER   END GUIDE   COTTER PIN   JAM NUT   SET SCREW   CABLE END

9

11-6 EXHAUST BELLOW SERVICE

REMOVAL

A new exhaust hose, P/N B78458, may be used as a replacement hose to the gimbal housing. When the stern drive is lowered it will fit into the hose. Using this hose will result is a minor amount of exhaust leak, but this loss is actually an advantage because it releases some of the back pressure during engine idle.

The exhaust bellows can be replaced without removing the stern drive or the bell housing. Move the stern drive slightly to the port side, and then remove the two clamps from the underside of the bell housing. Pull the bellows from the flange of the gimbal housing. It may be necessary to exert considerable force to pull the bellows loose because of the adhesive used during installation. Raising the stern drive to the full **UP** position, may be helpful.

CLEANING AND INSPECTING

Observe the white powder on new bellows. The rubber used in the manufacture of bellows is impregnated with this white powder -- a poison to protect the bellows against attack by muskrats. The sharp teeth of these animals can cause extensive damage to the bellows when the boat is left in the water inhabited by muskrats. Damage to the bellows will allow water to enter and cause premature corrosion to bearings, U-joints, and other expensive parts. Because the white powder is a poison **TAKE TIME** to wash thoroughly after handling new bellows.

Clean the bellow mounting flanges of the bell housing with a wire brush or sandpaper, and then wipe the surface clean with lacquer thinner.

Check the flanges and housing for cracks, nicks, or corrosion. Clean the bellow clamps thoroughly to ensure a good ground. New bellows have a clip on each end of the bellows to gound the clamp. Check the clamps for cracks or nicks. Replace the clamps if in doubt as to their condition. **ALWAYS** use stainless steel clamps for satisfactory service.

STORAGE WORDS

When the boat is to be stored for even a short period of time, the stern drive **MUST** be in the full **DOWN** position. The full down position will: prevent the bellows from taking a "set"; prevent undue strain on the underside; and ensure all folds and ribs are in the relaxed position.

INSTALLATION

READ AND OBSERVE THESE WORDS: Bellows Adhesive, C-92-36340A1 **MUST** be used to ensure a satisfactory installation. This adhesive is extremely flammable. Therefore, make every effort to ensure adequate ventilation during its use. Work in the outdoors, if at all possible. Vapors from the adhesive may cause flash fire or ignite explosively. Keep the adhesive away from heat, sparks, and open flame. Observe **NO SMOKING.** Extinguish all flames and pilot lights. Turn off stoves, heaters, electric motors, and all other possible sources of ignition while using the adhesive and until you are convinced all vapors have left the area. Close the container immediately after use. The material contains **TOLUENE** and petroleum distillates. The adhesive is harmful or fatal if swallowed. Avoid pro-

*Bellows Adhesive used to seal bellows. Safety precautions mentioned in the text **MUST** be observed when using this product.*

longed contact with skin or breathing of the vapors. If it is swallowed, do not induce vomiting. Call a physician immediately. Keep the adhesive out-of-reach of children.

After you have read the words in the previous paragraph, apply a coating of Bellow Adhesive, C-92-36340A1, to the inside

Old exhaust bellows (bottom) and the new improved exhaust hose (top). Installation of the new hose is much easier than the bellows. The new hose releases some of the back pressure during engine idle.

Worn and damaged exhaust bellows. Such defective parts should always be replaced to ensure satisfactory service.

Corroded bellows clamp. Electrolysis has caused this damage because the clamp was not properly grounded, or an automotive-type clamp was used.

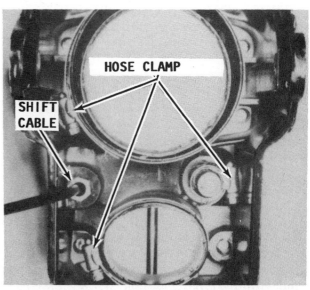

*An old style bell housing without the screwdriver access hole on the port side of the housing. The tightening screws **MUST** be positioned as indicated or they cannot be tightened after the stern drive and bellows are installed.*

of each end of the bellow and around the mounting flanges. Allow the adhesive to dry approximately 10 minutes, or until the material is no longer tacky. Install the clamp on one end of the bellows, and then install it to the gimbal housing flange. Tighten the clamp screw securely onto the bellows.

Start the engine and check for leaks.

CAUTION

Water must circulate through the stern drive, also to and from the engine, anytime the engine is run to prevent damage to the water pump located in the lower unit (original and Alpha drives); or on the front of the engine (Bravo drive). Just a few seconds without water will damage the water pump.

Now, the hard part. Install the bellow clamp on the other end of the bellows, and then pull the bellows over the flange of the stern drive and tighten the clamp.

Start the engine and check for leaks.

CAUTION

Water must circulate through the stern drive, also to and from the engine, anytime the engine is run to prevent damage to the water pump located in the lower unit (original and Alpha drives); or on the front of the engine (Bravo drive). Just a few seconds without water will damage the water pump.

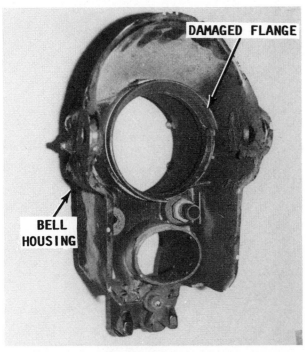

All flanges of the bell housing should be thoroughly cleaned to ensure the adhesive will provide an adequate seal.

New style gimbal housing showing the screwdriver access hole on the port side of the housing. This hole permits release of the clamps securing the U-joint and exhaust bellows.

11-7 U-JOINT BELLOWS SERVICE

REMOVAL

Remove the stern drive, see Chapter 10. **A GOOD WORD:** The bell housing does not have to be removed in order to replace the U-joint bellows. However, this is not an easy job. Therefore, take your time, and work carefully. If the job proves too difficult, you may elect to remove the exhaust bellows and the water intake hose, see Section 11-6.

All Models, Except Bravo

Remove both bellow clamps using a long shank screwdriver. A hole is provided in the side of the bell housing as an assist in removing the clamp. Work the bellows off the gimbal housing and the bell housing flanges. Considerable force may be required to remove the bellows due to the adhesive used during installation.

Bravo Only

Remove the single hose clamp securing the bellows to the gimbal housing using a long shank screwdriver. An access hole is provided in the side of the bell housing to permit loosening the clamp. Work the bellows off the gimbal housing flange. Considerable force may be required to remove the

bellows due to the adhesive used during installation. There is no hose clamp securing the other end of the bellows to the bell housing. Instead, a sleeve fits tightly inside the bellows to hold it against the bell housing flange. The sleeve must be pried free of the bellows. Obtain a can of aerosol cleaner such as "Gunk". Spray the cleaner around the edge of the sleeve to help loosen it. Pry the sleeve free using a thin blade screwdriver. Push the bellows back from from the bell housing flange until it is free. No adhesive is used at this end of the bellows, so just push.

CLEANING AND INSPECTING

Observe the white powder on new bellows. The rubber used in the manufacture of bellows is impregnated with this white powder -- a poison to protect the bellows against attack by muskrats. The sharp teeth of these animals can cause extensive damage to the bellows when the boat is left in the water inhabited by muskrats. Damage to the bellows will allow water to enter and cause premature corrosion to bearings, U-joints, and other expensive parts. Because the white powder is a poison **TAKE TIME** to wash thoroughly after handling new bellows.

Remove all bellow adhesive from the inside diameter of the bellows, if the bel-

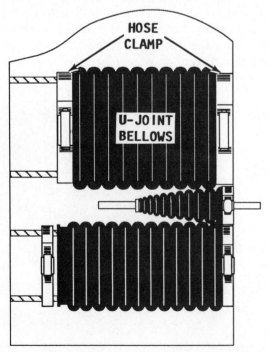

On all models, except Bravo, the U-joint bellows is secured to the bell housing and the gimbal housing with a hose clamp at both ends.

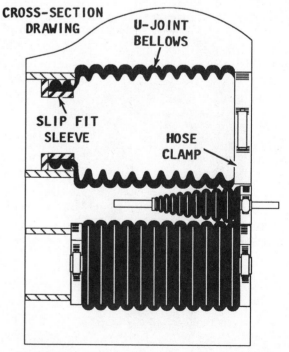

The U-joint bellows on a Bravo stern drive is secured to the bell housing with a slip fit sleeve and to the gimbal housing with a hose clamp.

lows are to be reused. Inspect the bellows for cracks, cuts, punctures and to be sure it is still flexible. If in the least bit of doubt about their condition, install new bellows. Clean the bellow mounting flanges of the bell housing with a wire brush or sandpaper, and then wipe the surface clean with lacquer thinner.

Check the flanges and housing for cracks, nicks, or corrosion. Clean the bellow clamps thoroughly. Check the clamps for cracks or nicks. Replace the clamps if in doubt as to their condition. **ALWAYS** use stainless steel clamps as a replacement.

STORAGE WORDS

When the boat is to be stored for even a short period of time, the stern drive **MUST** be in the full **DOWN** position. The full down position will: prevent the bellows from taking a "set"; prevent undue strain on the underside; and ensure all folds and ribs are in the relaxed position.

INSTALLATION

READ AND OBSERVE THESE WORDS: Bellows Adhesive, C-92-36340A1 **MUST** be used to ensure a satisfactory installation. This adhesive is extremely flammable. Therefore, make every effort to ensure adequate ventilation during its use. Work in the outdoors, if at all possible. Vapors from the adhesive may cause flash fire or ignite explosively. Keep the adhesive away from heat, sparks, and open flame. Observe **NO SMOKING.** Extinguish all flames and pilot lights. Turn off stoves, heaters, electric motors, and all other sources of ignition during use and until you are convinced all vapors have left the area. Close the container immediately after use. The material contains **TOLUENE** and petroleum distillates. The adhesive is harmful or fatal if swallowed. Avoid prolonged contact with skin or breathing of the vapors. If it is swallowed, do not induce vomiting. Call a physician immediately. Keep the adhesive out-of-reach of children.

All Models, Except Bravo

After you have read the words in the previous paragraph, apply a coating of Bellow Adhesive, C-92-36340A1, to the inside of each end of the bellows and flanges. Allow the adhesive to dry approximately 10 minutes, or until the material is no longer tacky. Install the clamp on the small end of the bellows, and then install it to the gimbal housing with the words **AFT-TOP** in the proper position. Install the other clamp on the bellows, and then install it to the bell housing. Tighten the clamp securely.

Bravo Only

Apply a coating of Bellow Adhesive C-92-36340A1 to the inside diameter of the gimbal housing bellow end. **DO NOT** apply the adhesive to the bell housing end. Allow the adhesive to dry for approximately ten minutes, or until the material is no longer tacky. Install a grounding clip over the forward edge of the bellows with the shorter side of the clip on the inside of the bellows. Observe the word **TOP** embossed on the bellows. This word **MUST** face upward after installation. Be sure to position the bead on the inner diameter of the bellows in the groove on the gimbal housing flange. Install the **U**-joint bellows over the flange and position the clamp tightening screw at the 3 o'clock position. Tighten the screw securely.

All Models

If the water hose and the exhaust bellows were removed, install them to the gimbal housing, see Section 11-6.

Replace the stern drive, see Chapter 10.

Start the engine and check the completed work.

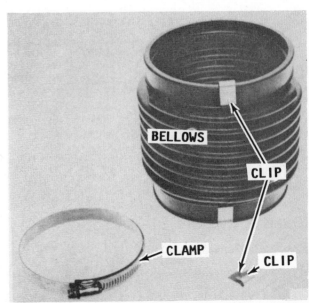

*Clips that **MUST** be used to ground the stainless steel **U**-joint bellow clamps as a prevention against electrolysis.*

CAUTION

Water must circulate through the stern drive, also to and from the engine, anytime the engine is run to prevent damage to the water pump located in the lower unit (original and Alpha drives); or on the front of the engine (Bravo drive). Just a few seconds without water will damage the water pump.

11-8 BELL HOUSING SERVICE

REMOVAL

Several jobs must be done before the bell housing can be removed:

Remove the stern drive, see Chapter 10.

Remove the U-joint bellows, see Section 11-7.

Remove the exhaust bellows, see Section 11-6.

Remove the shift cable, see Section 11-5.

TAKE CARE when working with the bellows not to puncture them. Any damage to the bellows will result in unsatisfactory service by allowing water to leak into the boat. If the bellows are to be replaced, they may be cut and removed from the housings AFTER the bell housing is removed.

Loosen the clamp and remove the bell housing end of the exhaust bellows. Loosen the clamp and remove the gimbal housing end of the U-joint bellows. Remove the clamp from the rear end of the shift cable bellows.

If more detailed information is needed, see Section 11-7 for the U-joint bellows, and Section 11-6 for the exhaust bellows.

Now you are ready to work on the bell housing.

Earlier Models with Pressed-In Hinge Pins

Remove the cotter pins from the hinge pins, one on each side. Use Slide Hammer Puller, C-91-34569A1, and Hinge Pin Puller, C-91-36060, to remove the two hinge pins.

Later Models Including Bravo With Threaded Hinge Pins

Obtain Hinge Pin tool C-91-78310 and remove both hinge pins.

All Models

Loosen the clamp and remove the water hose from the tube in the gimbal housing.

DISASSEMBLING

Check the rubber bumper and if it needs to be replaced, remove it now. Remove the shift lever screw, and then remove the shift lever, washer, and shift shaft. If the shift lever screw is frozen in and refuses to

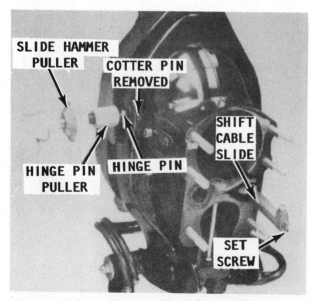

Using a slide hammer puller to remove the hinge pin on the port side. The starboard side pin is removed using the same tool.

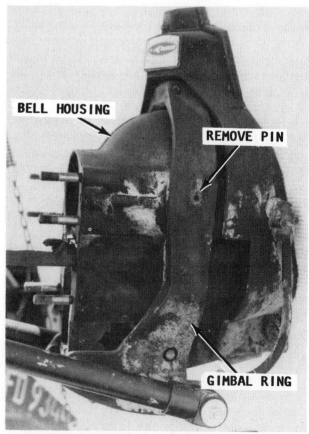

The gimbal ring must be turned to port or starboard to permit removal of the hinge pin on the opposite side.

budge, apply heat to break the Loctite which was probably used during installation. **TAKE CARE** when using heat not to damage the shift shaft oil seal. After the shift lever has been removed, remove the shift shaft oil seal and bushing, but only if it needs to be replaced. Remove and **DISCARD** the O-ring seal in the water passage and the rubber gasket in the universal joint passage. If other service work is to be performed, set the unit to one side. The next assembly to be removed is the gimbal ring, see the next section.

To assemble and install the bell housing, see Section 11-10.

11-9 GIMBAL RING SERVICE

REMOVAL

The following procedures pick-up the work after the bell housing has been removed, as outlined in the previous section.

1- On models prior to about 1973, the hose does not have to be disconnected. Remove the two nuts from the hydraulic hose connection of the trim cylinders, and then lower the connector. Plug the hose connections to prevent losing fluid and to keep contamination out of the hydraulic system. Remove the trim cylinder forward anchor pin, and then place the assembly to one side.

2- Remove the cotter pin, lower swivel pin, and washer.

3- Remove the two bolts (all units except Bravo), or nuts (Bravo) from the top of the gimbal housing.

STEERING LEVER ACCESS HOLE

Observe the plastic plug called out in Illustration No. 3. This plug covers a hole permitting access to the steering lever nut inside the housing. If the unit being servic-

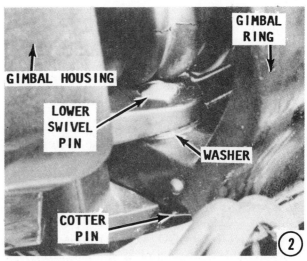

ed is equipped with such a hole and plug, or if the engine and transom assembly has been removed - you are home free - skip to Step 4.

If the unit being serviced has no such plug, or if the engine and transom assembly are still in place - two access holes **MUST** be cut into the gimbal housing, as described in the following paragraphs.

GOOD NEWS

Each hole must be large enough to enable two nuts to be loosened: One - the large nut at the top of the upper swivel shaft. The other, the nut (and bolt) on the steering lever boss. A punch and hammer must be used on the large nut, because there is no room inside the housing for a wrench, with or without an access hole. The small nut can be worked with open end wrench.

Actually, both holes can be any convenient size to permit working on the nuts **AND** which can be successfully plugged at the completion of the work.

The readily available access to these nuts is from the front, requiring the engine and transom assembly to be removed.

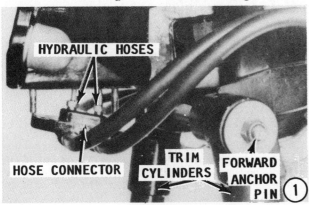

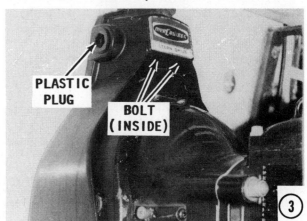

HOWEVER, if the engine and transom assembly are in place **AND** there are no existing access holes drilled in the gimbal housing **AND** the gimbal ring must be removed for service, two access holes must be drilled.

BAD NEWS

Positioning and cutting the holes presents no serious problem. Plugging the holes after the work has been completed -does present a problem.

Two kits are provided by the manufacturer to drill and tap the holes. If these kits are not available, **DO NOT** attempt to make any holes in the gimbal housing, until a satifactory method of plugging the holes is at hand.

Three methods are available to drill and plug the holes. The difference between the first and second method is where the tools are purchased, the first from the dealer, the second from a local hardware store.

TOOLS REQUIRED
First Method

Obtain Tap and Saw Kit C-91-86191A1 and Access Kit C-91-88847A.

Second Method

Purchase two pipe plugs of approximately 1 1/8" diameter and the appropriate hole saw and thread tap to match the threads on the pipe plugs.

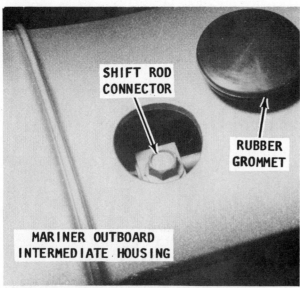

SHIFT ROD CONNECTOR

RUBBER GROMMET

MARINER OUTBOARD INTERMEDIATE HOUSING

A rubber grommet covers an access hole in a Mariner Outboard. Such a plug could easily be used to cover a hole made in the gimbal housing to gain access to the nut and bolt on the steering lever boss, as explained in the text.

Third Method

Purchase a couple of shift rod connector rubber grommets to fit a 3.5hp, 4hp, or 5hp Mariner outboard, 1981 and on, from your local Mariner dealer. The accompanying illustration shows such a grommet being removed from the intermediate housing - an area which is subjected to more water than the top of the gimbal housing!

PROCEDURE
Method One and Two

Pump a liberal amount of 2-4-C lubricant into the lubrication fitting at the top of the gimbal housing to catch metal chips from the cutting process.

All Models, Except Bravo

Place the appropriate template, included in the kit, against the housing and using a punch, make a mark where the pilot hole must be drilled.

Bravo Only

Observe two dimples, one on each side of the gimbal housing. Use a punch to indent the surface in preparation for the holes to be drilled.

All Models

Drill a 1/4" on each side of the gimbal housing, keeping the drill bit perpendicular to the housing surface. Drill the 1 1/8" hole using a hole saw equipped with a 1/4" pilot rod. Make a mark on the tap one inch from the end. Lubricate the grooves of the tap to help catch any metal chips during the threading process. Tap the hole. To avoid damage to the steering lever boss, do not allow the tap to enter the housing past the one inch mark made on the tap.

Make an attempt to remove the metal chips from inside of the housing. Install and check the fit of the threaded plug.

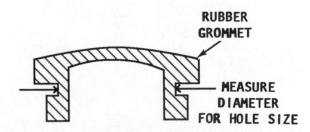

RUBBER GROMMET

MEASURE DIAMETER FOR HOLE SIZE

*A rubber grommet, as described in the text, must first be purchased **BEFORE** an access hole is cut in the gimbal housing. Measure the sealing diameter, as shown, and then obtain a suitable size hole saw and make the cut.*

Method Three

Purchase a couple of shift rod connector rubber grommets to fit a 3.5hp, 4hp, or 5hp Mariner outboard, 1981 and on, from the Mariner dealer. Refer to the accompanying illustration and measure the grommet at the diameter indicated to determine the size of the holes to be drilled. No specification is given here, because the hole size will depend on the grommet purchased. Obtain the exact size hole saw with a pilot rod and a drill bit the same size as the pilot rod. The good news is -- **NO** threads to tap! The bad news is -- for all units except the Bravo, the center of the housing must be determined by estimation without a template. After the first has been drilled, location of the second hole may be more accurately determined.

All Models

Pump a liberal amount of 2-4-C Lubricant into the lubrication fitting at the top of the housing to catch metal chips from the cutting process.

All Models, Except Bravo

Determine the center on one side of the housing and make a mark with a punch identifying location of the first hole. Determine location of the second hole on the other side in a similar manner.

Bravo Only

Observe two dimples, one on each side of the gimbal housing. Use a punch to indent the surface in preparation for the holes to be drilled.

Drill a pilot hole on each side of the gimbal housing, keeping the drill bit perpendicular to the housing surface. Drill the larger hole using a hole saw equipped with a pilot rod.

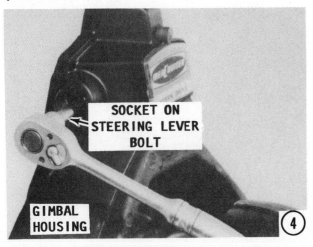

SOCKET ON STEERING LEVER BOLT

GIMBAL HOUSING

④

The Work Continues:

The following procedures may either be performed from the front of the gimbal housing, if the engine and transom assembly has been removed, or through the two access holes just drilled as described in the previous paragraphs.

4- Rotate the wheel to bring the nut and bolt at the steering lever boss into position where they can be loosened. Remove the plug under the upper swivel shaft end.

Use a punch and hammer to rotate the large nut securing the swivel shaft **COUNTERCLOCKWISE** to loosen it. It may be necessary to pull down on the shaft to completely free the nut from the shaft. Remove the nut and small washer. Continue pulling down on the swivel shaft to disengage the shaft from the splines inside the steering lever boss. When the swivel shaft is free, remove the larger washer between the shaft and the steering lever. If the swivel shaft is seized ("frozen"), perform Step 5. If the shaft came free, proceed directly to Step 6.

5- Use a slide hammer puller attached to Puller Head, C-91-38919 (all units except

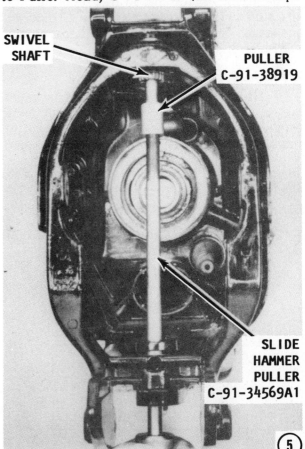

SWIVEL SHAFT

PULLER C-91-38919

SLIDE HAMMER PULLER C-91-34569A1

⑤

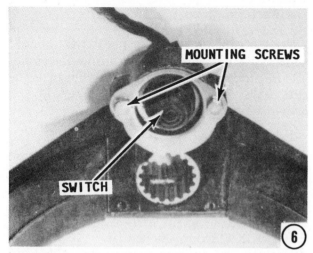

Bravo) or C-91-63616 (Bravo stern drive), to pull the upper swivel shaft. Remove the steering lever and washers. Remove the screws securing the tilt stop switch wires to the gimbal housing. Now, remove the gimbal ring assembly. If other service work is to be performed on other assemblies, such as the gimbal housing and transom plate, set the gimbal ring to one side.

DISASSEMBLING GIMBAL RING

6- Remove the hardware securing the tilt stop switch assembly and the tilt switch actuating lever in place, and then remove the lever and assembly.

All Models, Except Bravo

7- If the lower gimbal ring needle bearing is in satisfactory condition, **DO NOT** remove it. If either or both of the two oil seals, one on each side of the bearing, need

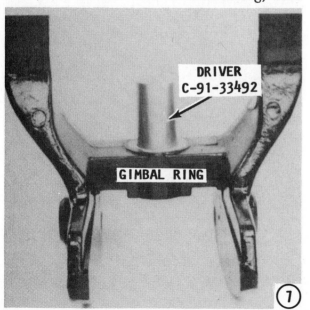

replacement, pry the seals from the housing using a screwdriver. If the bearing also needs replacement, both oil seals **MUST** be removed first. Obtain driver C-91-334942 and drive out the needle bearing.

Bravo Only

Remove the gimbal ring lower swivel shaft bushing using a suitable driver. Remove the gimbal ring upper swivel shaft bushing and single oil seal using a slide hammer with puller jaw attachment.

All Models

8- Remove the gimbal ring hinge pins, using a suitable mandrel if the pins are the press-in type, or obtain Hinge Pin tool C-91-78310 to remove the pins if they are threaded into the housing.

Obtain a suitable driver and drive out both hinge pin bushings.

CLEANING AND INSPECTING

Clean all metal parts in solvent and blow them dry with compressed air. **NEVER** spin ball bearings with compressed air, because such action will ruin them.

Remove all bellow adhesive from the inside diameter of the bellows, if the bel-

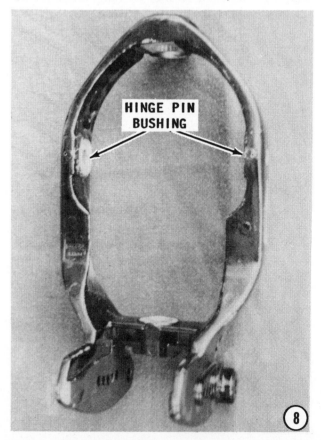

lows are to be reused. Inspect the bellows for cracks, cuts, punctures and to be sure they are still flexible. If in the least bit of doubt about their condition, install new bellows. Clean the bellow mounting flanges of the bell housing with a wire brush or sandpaper, and them wipe the surface clean with lacquer thinner.

Check the shift cables for cuts or damage caused by a cable being pinched or bent too short.

Inspect the water hose for cracks, cuts, punctures, or worn spots. Inspect the shift shaft oil seal for tears, wear, or any roughness. Check the shift shaft and shift shaft bushings for wear.

Inspect the O-ring and rubber gasket for cuts, nicks, hardness, or cracks.

Inspect the surface of the lower swivel pin in the area where the needle bearing rides. Any pitting, grooves, or uneven wear is cause to replace the bearing and swivel pin. Inspect the lip surface of the oil seals for wear, tears, and roughness.

Obtain an ohmmeter and test the tilt-stop switch. **A GOOD WORD:** The switch does not have to be removed in order to be tested. Use the Power-Trim tilting action of the drive unit to actuate the switch contacts. Connect the meter or test lamp to the switch leads. The meter should show continuity, or the lamp should light. Actuate the switch button: The meter should indicate **NO** continuity, or the lamp should **NOT** light.

Inspect the long steering lever retaining bolt. Any grooves found on the bolt are a result of friction against the upper swivel shaft. If grooves are discovered, both the steering lever and bolt must be replaced.

11-10 GIMBAL RING ASSEMBLING AND INSTALLATION

1- Using Bearing Driver tool, C-91-33492, install the lower swivel pin needle bearing. Center the bearing in the bore. Pack the oil seal lips with Multi-Purpose lubricant. Use the same tool to install the two oil seals, one on each side of the bearing. Both oil seals must face **UPWARD** after installation.

Bravo Only

Obtain Resiweld Sealer C-92-65150-1. Apply a coating of this sealer to the outer diameter of the lower bushing. Using a suitable driver, tap the bushing into place. To install the small upper swivel shaft bushing, obtain Bearing and Seal Driver tool C-91-43578. Place the bushing on the tool, and then use a hammer to tap the bushing into place. Install the larger bushing in similar manner. Pack the single oil seal lip with Multi-Purpose lubricant and place the seal on the driver with the lip facing the smaller diameter of the tool. Install the oil seal into the gimbal ring until the seal seats against the larger bushing.

Earlier Models with Tilt Stop Switch

2- Install the tilt stop switch, with the leads positioned as shown in the accompanying illustration. If the leads are not positioned properly, they may become chaffed or pinched.

All Models

3- Apply a coating of Resiweld C-92-65150-1 to the outer diameter of both hinge pin bushings. Using a suitable driver, tap

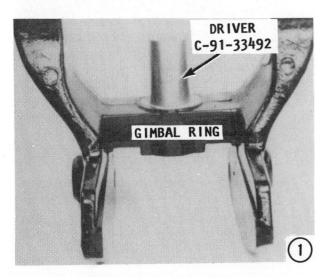

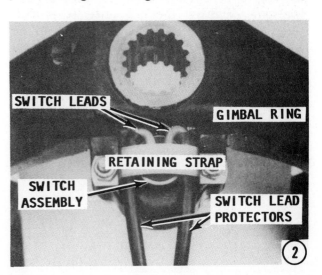

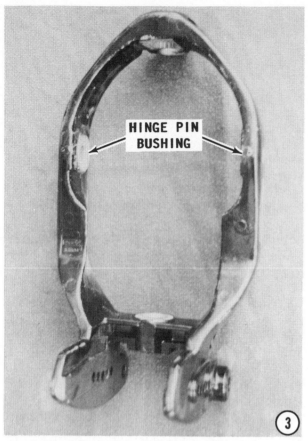

HINGE PIN BUSHING

③

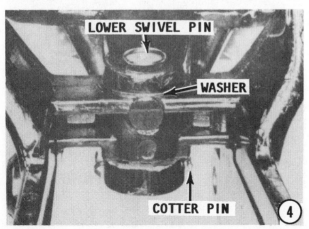

LOWER SWIVEL PIN

WASHER

COTTER PIN

④

PLASTIC PLUG PORT & STARBOARD

UPPER SWIVEL SHAFT

ALLEN SCREW

⑤

the bushings into place on the port and starboard sides of the gimbal ring.

4- Apply a coat of 2-4-C Lubricant onto the surface of the lower swivel pin. Insert the switch harness, if equipped, through the gimbal housing, and then secure it in place with the screws. Place the gimbal ring in position in the gimbal housing. Place a washer between the gimbal ring and the housing. Secure the ring in place with the lower swivel pin and cotter pin.

5- Install the long bolt and nut into the steering lever. A washer is not used at this location. Tighten the nut until it is just "snug". If a new swivel shaft nut is to be used, thread the nut onto the swivel shaft until it is tight, and then remove it. This action cuts the threads into the new nut.

Check to be sure the gimbal ring rests squarely within the gimbal housing. Before installing the swivel shaft, check for a flat area machined onto the shaft splines. If a flat exists, the flat **MUST** face **FORWARD** after installation. Place the larger washer on top of the ring, followed by the steering lever, then the smaller washer and finally, the large nut. Hold these components aligned together while the upper swivel shaft is passed through the gimbal ring and these parts. Start the large nut onto the shaft threads.

6- Tighten the bolt on the steering lever to a torque value of 45 ft lb (61m). Use a punch through one of the side access holes

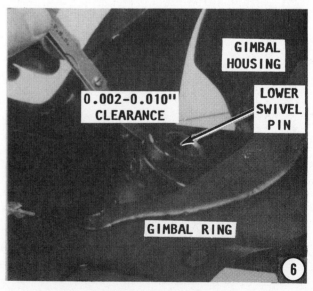

GIMBAL HOUSING

0.002-0.010" CLEARANCE

LOWER SWIVEL PIN

GIMBAL RING

⑥

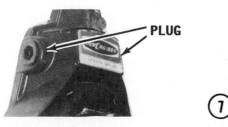

7

to tighten the nut, until a clearance of 0.002-0.010" (0.05-0.25mm) exists between the lower swivel pin washer and the gimbal housing.

Use a raw hide mallet and strike the top of the gimbal ring flanges a couple of times each to seat both pins. Recheck the clearance and adjust the large nut (using the punch and hammer method), as necessary to maintain the specified clearance.

Install the plug over the base of the upper swivel shaft.

7- For all models except Bravo, tighten the two gimbal ring bolts to a torque value of 25 ft lb (34Nm); for the Bravo -- 40 ft lb (47Nm). All models: Apply a coat of Perfect Seal to the threads or the sealing surface of the plugs and close the two large openings drilled earlier on the sides of the gimbal housing for access to the steering lever.

On earlier models: Install the tilt stop actuating lever with the attaching hardware.

8- Install the trim cylinder by first attaching them in place with the forward anchor pin. Next, install the two hydraulic hoses to the connector. Finally, install the connector to the gimbal housing.

9- Connect the steering link rod to the steering lever. Tighten the castellated nut to a torque value of 24 in lbs (3Nm), and then back it off slightly and insert the cotter pin. If the assembly is secured with a Nylok nut instead of the cotter pin, tighten the nut to a torque value of 5 ft lbs (7Nm).

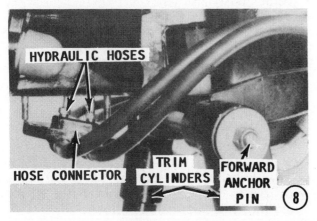

8

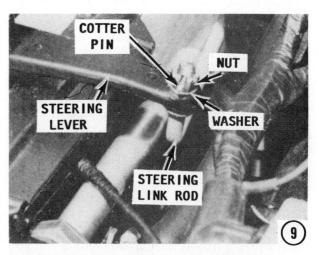

9

11-11 BELL HOUSING ASSEMBLING AND INSTALLATION

All Units Except Bravo

1- Install a **NEW** shift shaft bushing flush with the bottom of the housing. Install a **NEW** oil seal with the lip facing **DOWN**. Install the shift shaft, washer and shift lever. Squeeze just a drop of Loctite "A", or equivalent, to the threads of the locking screw. Install the screw through the shift shaft and into the shaft lever. Coat the gasket and **O**-ring seal with a general adhesive to hold them in place, and then install them around the water passageway. Install the rubber bumper.

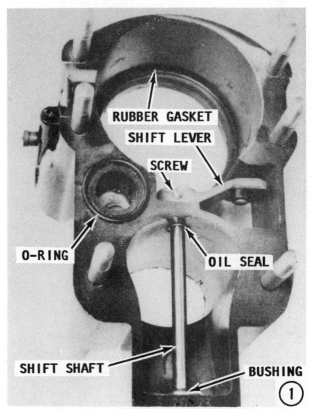

1

All Units

2- READ AND OBSERVE THESE WORDS: Bellows Adhesive, C-92-36340A1 **MUST** be used to ensure a satisfactory installation. This adhesive is extremely flammable. Therefore, make every effort to ensure adequate ventilation during its use. Work in the outdoors, if at all possible. Vapors from the adhesive may cause flash fire or ignite explosively. Keep the adhesive away from heat, sparks, and open flame. Observe **NO SMOKING.** Extinguish all flames and pilot lights. Turn off stoves, heaters, electric motors, and all other sources of ignition during use and until you are convinced all vapors have left the area. Close the container immediately after use. The material contains **TOLUENE** and petroleum distillates. The adhesive is harmful or fatal if swallowed. Avoid prolonged contact with skin or breathing of the vapors. If it is swallowed, do not induce vomiting. Call a physician immediately. Keep the adhesive out-of-reach of children.

After understanding the words in the previous paragraph, apply a coating of Bellow Adhesive, C-92-36340A1, to the inside diameter of **NEW U-**joint bellows. Allow the adhesive to dry approximately 10 minutes, or until the material is no longer tacky.

All Models, Except Bravo

Each of the two ground clips has one side shorter than the other. One clip is hooked onto the end of the **U-**joint bellows marked **"AFT-TOP"**, with the short side of the clip on the inside of the bellows. Position the second bellows clip over the other end of the bellows exactly opposite to the first ground clip. These clips **MUST** be positioned

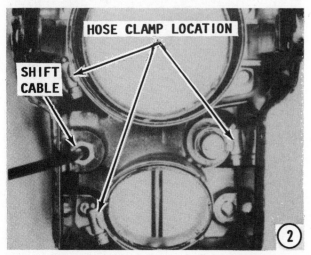

HOSE CLAMP LOCATION

SHIFT CABLE

②

in this manner to prevent the bellows from chaffing and a hole being worn through the rubber.

Install the **U-**joint bellows onto the bell housing with the hose clamp tightening screw positioned at 8 o'clock and the screw head facing down.

Bravo Only

Apply a coating of Bellow Adhesive C-92-36340A1 to the inside diameter of the gimbal housing end of the bellows. **DO NOT** apply the adhesive to the bellow end which will be attached to the bell housing. Allow the adhesive to dry for approximately ten minutes, or until the material is no longer tacky. Install a grounding clip over the forward edge of the bellows with the shorter side of the clip on the inside of the bellows.

Observe the word **"TOP"** embossed on the bellows. This word **MUST** face upward after installation. Be sure to position the bead on the inner diameter of the bellows in the groove on the gimbal housing flange. Install the **U-**joint bellows over the flange and position the hose clamp tightening screw at the 3 o'clock position, with the screw head facing down. Tighten the screw securely.

All Models

3- The accompanying illustration clearly indicates the position of all clamps and clamp screws mentioned in this step.

Check to be sure:

    a- The hose clamp tightening screw on the installed exhaust bellows is in the 12 o'clock position.

    b- The tightening screw is facing starboard.

    c- The hose clamp tightening screw on the installed shift cable bellows is in the 4 o'clock position.

    d- The tightening screw is facing down.

Place the expanded hose clamp which will be installed on the **U-**joint bellows at the gimbal housing flange in the 3 o'clock position. The tightening screw head **MUST** face down. Place the expanded hose clamp, which will be installed on the exhaust bellows at the bell housing flange, in the 9 o'clock position. The screw head **MUST** face down. Lubricate the end of the shift cable with 2-4-C lubricant and insert the shaft cable into the shift cable bellows.

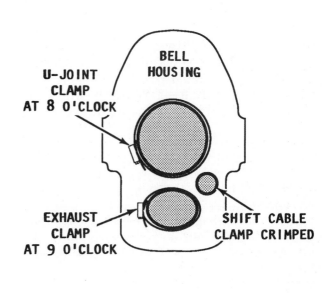

GIMBAL
HOUSING

U-JOINT
CLAMP
AT 3 O'CLOCK

EXHAUST
CLAMP
AT 12 O'CLOCK

SHIFT
CABLE CLAMP
AT 4 O'CLOCK

BELL
HOUSING

U-JOINT
CLAMP
AT 8 O'CLOCK

EXHAUST
CLAMP
AT 9 O'CLOCK

SHIFT CABLE
CLAMP CRIMPED

GIMBAL AND BELL HOUSING CLAMPS MUST BE POSITIONED
AS SHOWN TO PERMIT TIGHTENING THROUGH ACCESS HOLES

③

4- Insert the water tube through the gimbal housing. Route the trim limit switch and trim position sender leads as indicated in the accompanying illustration. Install a sta-strap around both sets of leads approximately 5" (12cm) from the retaining cover.

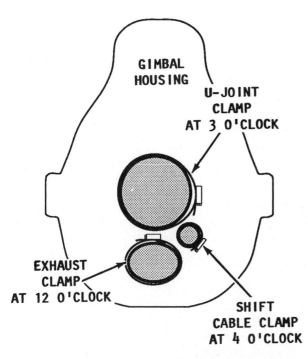

WATER
TUBE

RETAINING
COVER

STA STRAP

ELECTRICAL
LEAD

④

All Models, Except Bravo
5- Bring the bell housing together with the gimbal housing. Push evenly on the perimeter of the bell housing until the **U-**joint bellows slides **OVER** the flange on the gimbal housing.

Bravo Only
Bring the bell housing together with the gimbal housing. Push evenly on the perimeter of the bell housing until the **U-**joint bellows slides **INTO** the flange on the gimbal housing.

COTTER PIN

HINGE PIN

⑤

All Models

Check to be sure the exhaust bellows aligns with the flange on the bell housing.

All Models, Except Bravo

Tighten the hose clamp over the **U**-joint bellows at the gimbal housing flange. The screw should be at the 3 o'clock position and the screw head should be facing down.

Early Models
Equipped with Pressed-In Type Hinge Pins

Coat the two hinge pins with Multi-Purpose lubricant. Insert a pin into each of the two holes. Insert a new cotter pin through the hole in each hinge pin and then bend the cotter pin end over and into the slot provided on the front of the gimbal ring.

Late Models
Equipped with Threaded Type Hinge Pins

Obtain Hinge Pin tool C-91-78310. Coat the hinge pin theads with Locquic Primer "T" and allow to dry. Next, apply a coat of Loctite 35 to the pin threads. Now, position a Synthane washer on each side of the gimbal ring, between the bell housing and the gimbal ring. Install and tighten the hinge pins to a torque value of 60 ft lb (81Nm).

All Models, Except Bravo

6- Use Expander tool C-91-45497A1 to install the exhaust bellows **OVER** the bell housing flange. Tighten the clamp securely at the 9 o'clock position. The screw head

should already be facing down. Install the shift cable inner core, the upper shift shaft and the shift shaft lever. Position the rubber bell housing gasket and water passage **O**-ring against the bell housing surface.

Bravo Only

7- Ease the bellows **INTO** the bell housing flange until the flange beaded edge rests in the second groove from the end of the bellows. Spray engine cleaner around the inside of the flange as an aid to installation. Obtain Sleeve installation tool C-91-43577 and a suitable driver. Position the sleeve into the bell housing flange, and then tap it into place against the bellows, using the installation tool and driver.

Closing Tasks All Models

Install the shift cable, see Section 11-5.
Install the stern drive, see Chapter 10.
Install the trim cylinder, see Chapter 8.
Start the engine and make a thorough check of the completed work.

CAUTION

Water must circulate through the stern drive, also to and from the engine, anytime the engine is run to prevent damage to the water pump located in the lower unit (original and Alpha drives); or on the front of the engine (Bravo drive). Just a few seconds without water will damage the water pump.

EXPANDER TOOL (C91-45497A1)

6

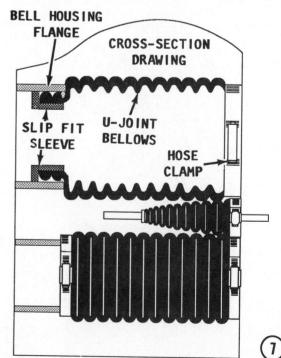

BELL HOUSING FLANGE

CROSS-SECTION DRAWING

SLIP FIT SLEEVE

U-JOINT BELLOWS

HOSE CLAMP

7

12
GIMBAL HOUSING

12-1 INTRODUCTION

The gimbal housing is mounted on the transom of the boat and attached to an inner transom plate. This plate contains the rear engine mounts and attachments for the steering and shift cables. Gimbal housing-transom plate service can generally be performed without removing the gimbal housing. However, in those cases when a part of the housing has been broken, or if the transom plate must be removed, the gimbal housing will have to be removed first.

Before the gimbal housing can be removed, several jobs must be performed in the following order: The stern drive must be removed, see Chapter 10; the engine removed, see Chapter 13; then the bell housing and gimbal ring removed, see Chapter 11.

The exhaust separator or the inner transom plate can be serviced after the engine and stern drive unit have been removed.

If the transom plate and the stern drive unit are removed and installed, the trim system hydraulic hoses must be properly connected; the steering and shift cables connected; and the shift linkage adjusted. The trim-limit switch, reverse lock, and trim sender must also be checked and adjusted.

12-2 GIMBAL HOUSING TRANSOM PLATE REMOVAL

If the gimbal housing is to be removed for service, first remove the stern drive, see

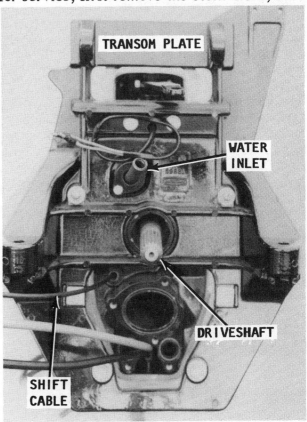

Early model gimbal housing and related parts after the stern drive has been removed.

Transom plate and related parts, after the power plant has been removed.

Chapter 10; remove the engine, see Chapter 13; and then remove the bell housing and gimbal ring, see Chapter 11. Remove the gimbal bearing, see Chapter 11.

The following procedures pick up the work after the tasks listed in the previous paragraph have been completed.

On an In-line engine installation:

Remove the steering link rod. Disconnect and remove the steering cable from the tube. Remove the drive unit and remote control shift cables from the shift plate. Remove the black hydraulic hose from the trim pump and the gray hydraulic hose from the reverse-lock valve. Plug the hose connections and cap the hoses. Remove and cap the black plastic hydraulic hoses from the trim sender. Plug the hose connections.

Remove the hose clamp securing the hydraulic hoses to the inner transom plate. Remove the exhaust elbow. Disconnect the power trim panel wiring harness from the pump. Disconnect the tilt-limit switch from the trim wiring harness and the reverse lock valve cut-out switch from the pump. The cut-out switch needs to be disconnected, only if the pump is mounted separately. Remove the screws and nuts securing the inner and outer transom plates together. Remove the exhaust elbow, then the inner transom plate, and finally the gimbal housing.

On a V8 engine installation:

Remove the steering link rod. Disconnect and remove the steering cable from the tube. Remove the two hydraulic hoses from the trim pump. Plug the hose connections

and cap the hoses. Remove the exhaust separator, with the exhaust elbows and exhaust bellows attached, from the gimbal housing. Disconnect the tilt-limit switch from the trim wiring harness.

If servicing a Bravo stern drive: Obtain Tapered Insert tool C-91-43579, and remove the two tapered inserts from the ends of the water tube, then remove the tube.

All models except the Bravo stern drive: Remove the engine water inlet hose, then remove the water tube cover and grommet and push the tube out of the gimbal housing. Tilt the bell housing up and loosen the hose clamp around the tube at the bell housing flange. Pull the tube away from the flange.

All units

Remove the screws and nuts, securing the inner and outer transom plates together. Remove the exhaust separator, then the inner transom plate, and finally the gimbal housing.

CLEANING AND INSPECTING

Rotate the inside race of the gimbal housing bearing and check for rough spots or any sign of binding. Push and pull on the inner race to check the bearing for side play. The bearing should move only a slight amount. Any excessive movement is just cause to replace the bearing.

Check the upper swivel shaft roller bearing and bushing, by inspecting the area of the swivel shaft, where the bearing and bushing ride. Any pitts, grooves, or uneven wear is cause to replace the bearing, bushing, or swivel shaft.

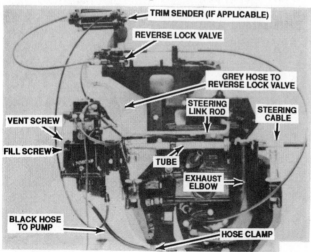

Transom plate area after an in-line engine has been removed.

Transom plate area after a V8 engine has been removed.

Check the transom plate for pitting or worn housing. Clean any old sealer from the back side of the plate. Clean the exhaust outlet. Use a tap, and chase the threads of the screw holes where the exhaust elbow is mounted.

12-3 GIMBAL HOUSING ASSEMBLING

If servicing a Bravo stern drive, Install the water tube through the hole in the gimbal housing. The water tube edge must be flush with the gimbal housing surface. The other end of the water tube must protrude approximately 1/8" (3mm) out from the hole at the bell housing. Use Tapered Insert tool C-91-43579 to install the two tapered inserts into the gimbal housing and the bell housing.

All units except the Bravo stern drive: Secure the water tube to the water hose and tighten the hose clamp securely. Install the water hose and tube through the gimbal housing. Route the trim limit switch and trim position sender leads down over the installed tube. Secure both sets of leads with a sta-strap 5" (12cm) away from their retaining bracket. Tilt the bell housing up and secure the water hose over the flange on the bell housing. Tighten the hose clamp over the tube. Install the grommet and water tube cover and connect the engine water inlet hose to the water tube.

All units
If the gimbal housing ball bearing needs to be replaced, it must be installed after the bell housing is installed, See Chapter 11.

The transom plate is assembled as it is installed.

12-4 GIMBAL HOUSING AND TRANSOM PLATE INSTALLATION

Insert the shift cable, the trim-limit switch lead, and the hydraulic hose through the opening in the transom, and then move the gimbal housing into position. The hydraulic hose **MUST** be positioned on the port side of the exhaust elbow on in-line engines and on the starboard side for V8 engines. **BAD NEWS:** Do not hold onto the trim-limit switch wires to support the gimbal housing while installing the unit. The trim-limit switch or the switch leads will be damaged, if the gimbal housing is supported by the switch leads while fastening the inner transom plate.

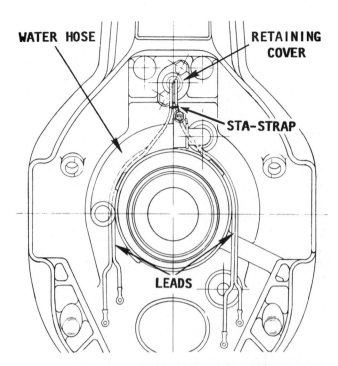

Line drawing to depict the gimbal housing with water tube installed and the trim limit switch and trim position sender leads routed as described in the text.

Hold the gimbal housing in position, and at the same time insert the shift cable and hydraulic hose through the large opening; insert the trim-limit switch leads through the center hole of the three small holes in the inner transom plate; and then set the plate in position.

Thread the two short cap screws, with lockwashers, through the top two holes of the inner transom plate and into the gimbal housing. Insert the two special anode-head bolt assemblies, with rubber seals between the bolt and gimbal housing, through the bottom two holes of the gimbal housing, boat transom, and inner transom plate. Install a flat washer and elastic stop nut on each bolt, but **DO NOT** tighten them at this time. Install a flat washer and elastic stop nut onto each of the two studs protruding through the inner transom plate.

DO NOT attempt to drive the cap screws through the transom, because the threads in the gimbal housing will be damaged. Install the two long cap screws, with a flat washer and lockwasher through the remaining holes in the inner transom plate and into the gimbal housing. The square flat washers **MUST** be against the transom plate.

Now, tighten the transom plate cap screws and elastic stop nuts evenly. Work from the center up, and then down. Tighten the screws and nuts to a torque value of 24 ft lb (33Nm).

On In-line Engines

Check to be sure the mating surfaces on the exhaust elbow and the gimbal housing are clean. Place a **NEW** O-ring seal into the groove in the gimbal housing opening.

Position lockwashers on the hex-head screws, and then install the exhaust elbow to the gimbal housing. Tighten the screws to a torque value of 24 ft lbs (33Nm), and at the same time check to be sure the **O**-ring remains properly seated in the groove. Slide hose clamps onto the rubber exhaust tube, and then install the tube onto the exhaust elbow. Tighten the hose clamps securely.

12-5 HYDRAULIC HOSE CONNECTING — IN-LINE ENGINES

Move quickly while performing this step to prevent spilling more hydraulic fluid than is necessary. Follow the sequence and take care to route the hoses properly, as shown, or the hoses may be damaged and the system become inoperative. Remove the cap from the fitting on the end of the black hydraulic hose, and remove the threaded plug from the pump. Immediately connect the hose to the pump. Remove the cap from the fitting on the end of the gray reverse lock valve hose and remove the threaded plug from the valve. Immediately connect the hose to the valve. Tighten the fittings securely.

Clamp the hydraulic hoses securely to the transom plate. Loosen the reservoir vent screw. This vent screw **MUST** be loosened for proper operation. With the unit in the full **DOWN** position, remove the **FILL** screw and check the fluid level. Fill the system to the bottom of the fill screw threads. **DO NOT OVERFILL.** Install the **FILL** screw. If other service is required on the trim/tilt system, see Chapter 8.

On V8 Engines

Check to be sure the mating surfaces on the exhaust separator and the gimbal housing are clean. Place **NEW** O-ring seals into the grooves in the gimbal housing. Install the exhaust separator, with the exhaust elbows and exhaust bellows attached, to the gimbal housing.

Thread the four hex-head cap screws and the one Allen-head cap screw, with lockwashers, into the gimbal housing. Tighten the screws to a torque value of 24 ft lbs (33Nm), and at the same time check to be sure the **O**-ring seals remain properly seated in the grooves.

12-6 HYDRAULIC HOSE CONNECTING — V8 ENGINES

Route the hydraulic hoses toward the pump, as shown. Move quickly during this step to prevent spilling more hydraulic fluid than necessary. Remove the cap from the hydraulic fitting on the end of the black gimbal housing hose. Remove the threaded plug from the hydraulic pump control valve. Immediately connect the hose to the control valve. Remove the plug from the trim sender and the cap from the black **DOWN** hydraulic hose running from the pump con-

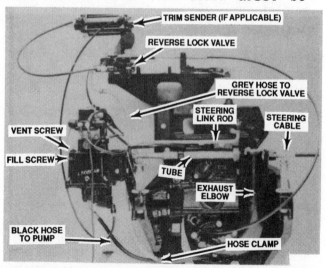

Transom plate area ready for an in-line engine installation.

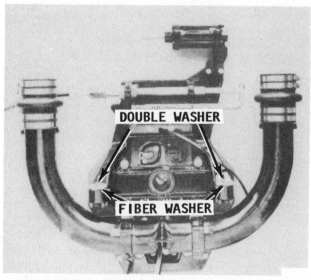

Transom plate area ready for a V8 engine installation.

trol valve. Connect the hose to the trim sender.

Remove the cap from the gray **DOWN** hydraulic hose and remove the plug from the reverse lock valve. Connect the hose to the valve. Remove the cap from the hydraulic hose running from the reverse lock valve to the shaft end of the trim sender. Remove the plug from the sender. Connect the hose to the sender.

Before the tilt system can be properly filled and bled, the stern drive must be installed, see Chapter 10.

Loosen the pump reservoir **VENT** screw. This screw **MUST** be loosened for proper operation of the system. With the drive unit in the full **DOWN** position, remove the **FILL** screw and check the fluid level. Fill the system to the bottom of the fill screw threads. **DO NOT OVERFILL.** Install the **FILL** srew. For service on the tilt/trim system, see Chapter 8.

Bleed the system, see Chapter 8.

Adjust the shift mechanism, if necessary, see Chapter 7.

Start the engine and check the completed work.

CAUTION

Water must circulate through the stern drive, also to and from the engine, anytime the engine is run to prevent damage to the water pump located in the lower unit (original and Alpha drives); or on the front of the engine (Bravo drive). Just a few seconds without water will damage the water pump.

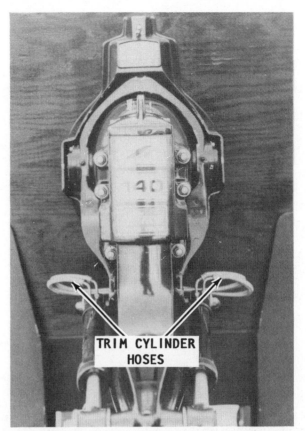

Correct routing of the trim cylinder hoses. Notice how the hoses leave the junction block straight out, then make a uniform bend to the cylinders WITHOUT crossing.

Trim cylinder motor area showing location of the vent screw and the fill screw.

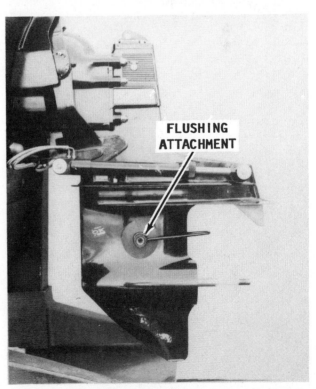

Flush attachment connected to the stern drive. Such a device MUST always be used whenever the engine is started for any reason, while the boat is out of the water.

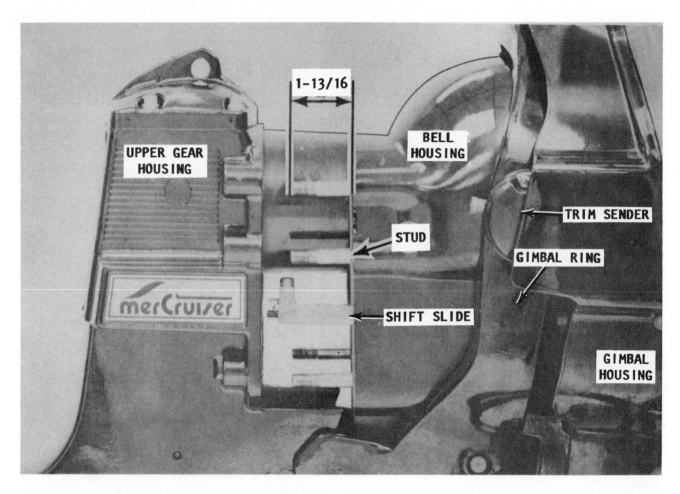

Stern drive installation to the bell housing. The measurement from the bell housing to the end of the mounting studs must be 1-13/16" to ensure the nuts can be properly installed.

13
ENGINE
REMOVAL/INSTALLATION

13-1 INTRODUCTION

In addition to work on the engine block, some service jobs require the powerplant to be removed to gain access to equipment, or to permit removal of specific items. Therefore, detailed procedures are presented in this chapter as an assist in finding the instructions quickly, because they are alone in a separate chapter. Another advantage is because the powerplant must be removed even though no service work on the engine is scheduled.

In some cases, the design of the boat will hinder removing and installing the engine.

Engine hood covers and panels around the engine may have to be removed before the engine will clear.

The stern drive must be removed, before the engines covered in this manual can be removed from the boat, see the detailed instructions outlined in Chapter 10.

Briefly, stern drive removal involves removing the propeller, as a precaution against damaging it; disconnecting the trim cylinders, and tieing them up to prevent damage in case the boat is moved; shifting the unit into **FORWARD** gear; removal of the six elastic stop nuts; and finally, removal of the stern drive unit. All gaskets and **O-rings should** be discarded.

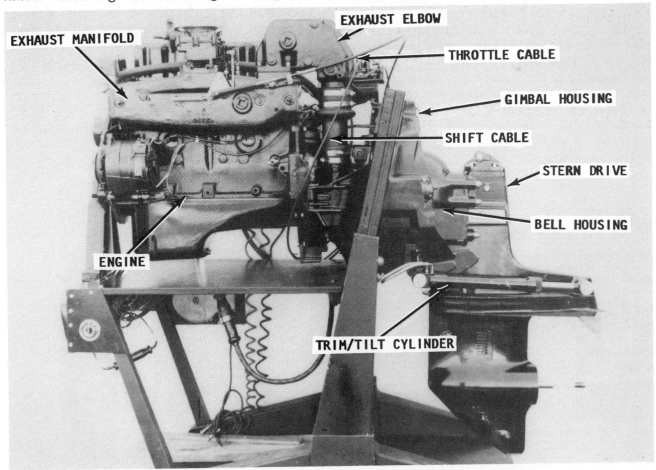

Overall classroom-type view of a complete engine and stern drive installation.

13-2 ENGINE REMOVAL

The following procedures pick up the work after the stern drive has been removed as outlined briefly in the previous section and detailed in Chapter 10.

Disconnect both battery cables at the battery and at the engine. Disconnect the instrument panel harness connector plug from the engine harness receptacle. Disconnect the fuel line from the fuel pump, and **PLUG** the line to prevent fuel from siphoning out of the fuel tank.

Remove the throttle cable from the carburetor and from the anchor plate and remove the remote control shift cable and stern drive shift cable from the shift plate mounted on the engine.

Remove the stern drive shift cable from the J-clamp on the flywheel housing or on the exhaust manifold.

Disconnect the shift cutout switch wires from the terminal block on the shift plate. Remove the trim/tilt pump motor wires from the engine.

Disconnect the reverse lock switch wires from the trim/tilt pump.

Disconnect all accessories connected to the engine and disconnect the lead grounding the flywheel housing stud.

Disconnect the water inlet hose and exhaust elbow bellows. Remove the gray **DOWN** hose from the reverse lock and the black plastic hose from the trim sender. Cap the hoses and plug the hose connections.

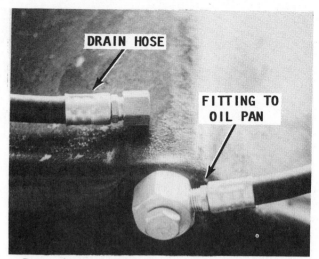

Parts of an oil drain kit that should be installed the first time the engine is removed from the boat.

Disconnect the vacuum hose from the intake manifold. Disconnect the power steering hoses from the power steering unit, again cap the hoses and plug the connections.

Pass a length of chain through the holes in the lifting brackets or eyes on the engine, and then fasten the ends together with a bolt and nut. Attach a suitable lifting device in the center of the chain, and then tie the chains together to prevent the lifting device from riding down the chain as the engine is lifted from the boat.

Support the engine with the lifting device, and remove the rear engine mounting bolts. Next remove the bolts securing the front engine

Location of the stern drive-to-bell housing mounting nuts. The stern drive must be removed before the engine can be removed.

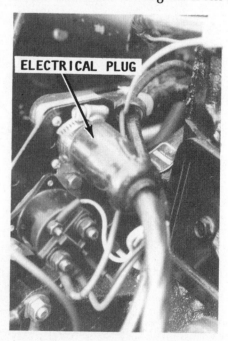

Quick disconnect between the engine and the transom plate.

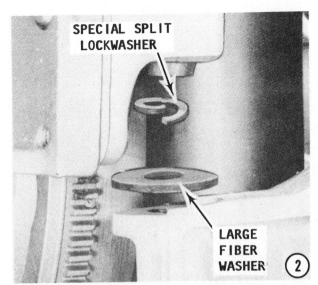

Generation II liquid engine mount.

mounts to the stringer. There is no need to disturb the engine leveling bolts at this time.

GENERATION II MOUNTS

The aft motor mounts for Generation II units utilize the old style mounts. The forward mounts are a liquid filled mount, as shown in the accompanying illustration. Simply remove the one large bolt at each mount.

A GOOD WORD: While the engine is out of the boat, an oil drain kit **SHOULD** be installed. Such a kit may be obtained at very modest cost from any marine dealer.

13-3 INSTALLATION IN-LINE ENGINE

1- The transom plate and exhaust elbow **MUST** be installed before the engine is installed. The Ride Guide steering should also

be in place, see Chapter 7. Run a length of chain through the holes in the lifting brackets or eyes on the engine, and then fasten the ends together with a bolt and nut. Attach a suitable lifting device in the center of the chain, and then tie the chains together to prevent the lifting device from riding down the chain as the engine is lowered into the boat.

Apply a coating of 2-4-C Lubricant to the engine coupling splines. Lower the engine into position.

2- Position a large fiber washer on top of the inner transom plate engine mounting bracket and a special split lockwasher inside the fiber washer.

3- Position the engine over the transom plate mounting brackets. Slide four hose clamps over the rubber exhaust elbow bellows, and then position the bellows over the exhaust manifold outlet. Tighten the hose clamps securely. Place steel washers and

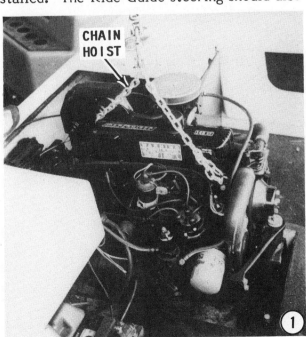

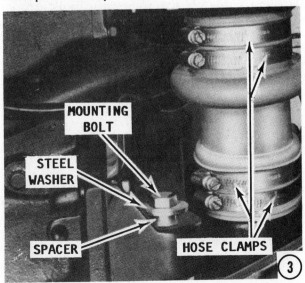

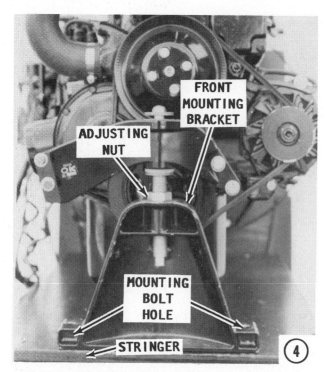

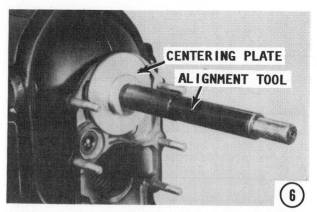

spacers onto the mounting bolts. Insert the bolts through the rear engine mounts, washers, and mounting brackets. Thread elastic stop nuts onto the bolts, and tighten them to a torque value of 60 ft lbs (82Nm).

4- Align the front engine mounting bracket with the holes in the stringer and install the two mounting bolts. If the holes do not align, adjust the front mounting bracket by turning the adjusting nuts until the bracket rests firmly on the mount in the boat to relieve the lifting device tension. Secure the mount to the boat.

5- READ AND BELIEVE: The engine **MUST** be properly aligned or the female splines in the engine coupler and the male splines of the driveshaft will be destroyed in a short time. If the necessary tools are not available, the boat **SHOULD** be taken to a marine dealer for proper alignment. Now,

from outside the boat, insert the small diameter end of Alignment Shaft, C-91-48247, through the U-joint bellows and into the gimbal bearing. If this tool is not available, use the end of Alignmment Tool, C-91-57797A3, which does not have a threaded hole.

6- Move the gimbal bearing with the shaft to index the shaft with the engine coupling splines. The alignment is correct if the shaft enters the coupling freely, with no pressure. **AS AN AID,** apply a small amount of grease onto the end of the alignment tool. Insert the tool through the U-joint bellows, the gimbal bearing, and into the coupler until it is fully seated. Now, carefully remove the tool **WITHOUT** turning it and inpsect the end with the grease. If the grooves are deep on the top side of the tool and light on the bottom, the indication would be that the engine is too high.

7- If the groove marks are reversed, the front of the engine is too low. If the

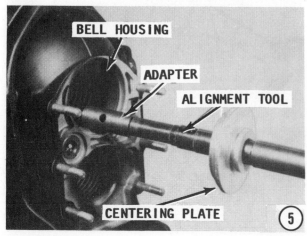

alignment is not correct, and difficulty is encountered in indexing the splines, loosen the front mounting bracket adjusting nuts and rotate them to raise or lower the front of the engine until the alignment shaft slides into the engine coupling splines with no resistance. Once the correct alignment shaft fit is obtained, tighten the lower adjusting nut securely. Again, check the alignment shaft fit. Secure the mounting bracket to the boat.

8- Slide a hose clamp over the water inlet hose, and then install the hose over the copper inlet tube. Tighten the hose clamp securely.

9- Connect the instrument harness to the engine. Connect the ground lead to the stud on the flywheel housing. Connect the shift cutout switch wires to the terminal block on the shift plate and the trim/tilt sender wires to the terminal block mounted on the engine. Connect the vacuum hose to the intake manifold. Install the fuel line to the carburetor and check for leaks. Position the drive unit shift cable in the J-clamp on the flywheel housing, and then bend the clamp closed. Install the remote control shift cable and the stern drive shift cable to the shift plate mounted on the engine, refer to Chapter 7 for cable adjustments.

THROTTLE CABLE INSTALLATION AND ADJUSTMENT

To install the throttle cable: Check to be sure the remote control lever is in the **NEUTRAL** position. Be sure the spacer is in place on the throttle lever stud. Slide the

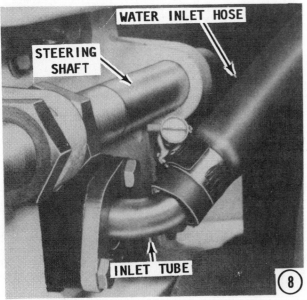

cable end guide over the throttle lever stud and secure the cable to the stud with a flat washer and elastic stop nut. Push the brass barrel gently toward the throttle lever and rotate the brass barrel along the threaded portion of the cable until the hole in the barrel aligns with the anchor stud. Install the barrel onto the stud with the cable **ABOVE** the stud. Secure the barrel to the stud with a flat washer and elastic stop nut.

Place the remote control in the full throttle position, and check all throttle valves are fully open. In this position the throttle lever tang should contact the stop on the carburetor body. Shift the remote control lever to the **NEUTRAL** position. The throttle lever tang should now make contact with the idle rpm adjustment screw.

If all these conditions are met, the throttle cable is correctly adjusted. If not, adjust the position of the brass barrel along the threaded portion of the throttle cable until the tang rests against the stop.

Connect the power steering hoses to the power steering unit and tighten the large hose fitting to a torque valve of 23 ft lb (30Nm) and the small hose fitting to a torque valve of 8.5 ft lb (11.5Nm).

Install the stern drive, see Chapter 10.

CAUTION
Water must circulate through the stern drive, also to and from the engine, anytime the engine is run to prevent damage to the water pump located in the lower unit (original and Alpha drives); or on the front of the engine (Bravo drive). Just a few seconds without water will damage the water pump.

Start the engine and check the completed work.

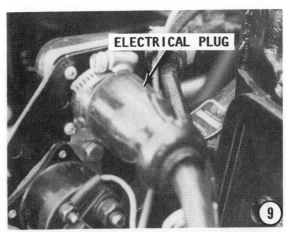

13-4 INSTALLATION
V8 ENGINE

1- The transom plate and exhaust elbow **MUST** be installed before the engine is installed. The Ride Guide steering should also be in place, see Chapter 7. Run a length of chain through the holes in the lifting brackets or eyes on the engine, and then fasten the ends together with a bolt and nut. Attach a suitable lifting device in the center of the chain, and then tie the chains together to prevent the lifting device from riding down the chain as the engine is lowered into the boat.

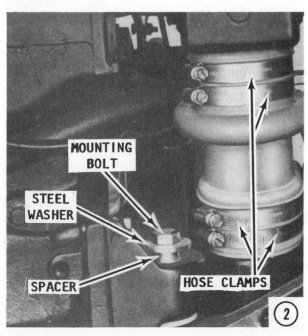

Generation II liquid engine mount.

2- Place one double-wound lockwasher and one fiber washer on top of each inner transom plate engine support. Use a suitable lifting device and suspend the rear end of the engine slightly lower than the front, and with the front mount level. Apply a coating of 2-4-C Lubricant to the engine coupling splines. Lower the engine into position over the transom plate. As the engine is lowered, connect the exhaust bellows to the exhaust elbows. Permit the rear engine mounts to rest on the transom plate engine supports. **DO NOT** relieve the tension on the lifting device. Place one steel washer and one spacer on each rear mount bolt. Insert the bolts down through the engine mounts, washers, and mounting brackets. Thread elastic stop nuts onto the bolts, and then tighten them to a torque value of 60 ft lbs (81Nm). Tighten the exhaust bellows clamps securely.

GENERATION II MOUNTS
The forward engine mounts are a liquid filled mount. Secure the engine to the mount with the single large bolt at each location.

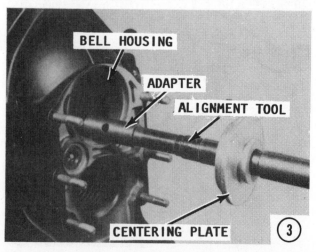

3- READ and **BELIEVE:** The engine **MUST** be properly aligned or the female splines in the engine coupler and the male splines of the driveshaft will be destroyed in a short time. If the necessary tools are not available, the boat **SHOULD** be taken to a marine dealer for proper alignment. Now, from outside the boat, insert the end without the threaded hole of Alignment Tool, C-91-57797A3, through U-joint bellows and into the gimbal bearing. Move the gimbal bearing, with the shaft, to index the shaft with the engine coupling splines. The alignment is correct if the shaft enters the coupling freely, with no pressure. **AS AN AID,** apply a small amount of grease onto the end of the alignment tool. Insert the tool through the U-joint bellows, the gimbal bearing, and into the coupler until it is fully seated. Now, carefully remove the tool **WITHOUT** turning it and inspect the end with the grease. If the grooves are deep on the top side of the tool and light on the bottom, the indication would be that the engine is too high.

4- If the groove marks are reversed, the front of the engine is too low. If the alignment is not correct and difficulty is encountered in indexing the splines, use the lifting device to raise or lower the engine until the alignment shaft slides into the engine coupling splines with no resistance.

5- If the engine mount bases contact the stringers, when the front of the engine is lowered, loosen locking nut "A" and turn the mount adjusting nut "B" counterclockwise to raise the mounts away from the stringers. The first alignment **MUST** be made with the front of the engine supported evenly with the lifting device and the mount bases not contacting the stringers. After the first alignment has been attained, remove the alignment shaft and turn the adjusting nuts "B" on both front mounts clockwise, to lower the mounts until the mount contacts the top of the stringers. Now, stop turning the nuts and make a reference mark on each adjusting nut "B". Turn each nut clockwise **ONE MORE FULL TURN.** The top surface of the stringers **MUST** be positioned with a minimum of 1/4" up-or-down adjustment left after the mount bases are resting on the stringers. This remaining adjustment capability is necessary in order to perform a final engine alignment now and also later after the boat has been in service for a short time. Release the tension on the

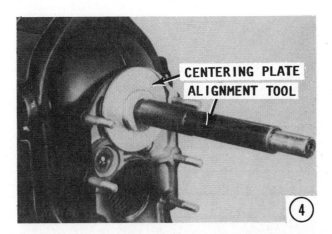

CENTERING PLATE
ALIGNMENT TOOL

4

lifting device and fasten the front mounts, with the slotted mounting hole toward the front of the engine, to the engine stringers. Use 3/8" diameter lag bolts long enough to secure the engine. Again, check the alignment shaft fit. The shaft must enter the engine coupling freely. If it does not, turn both front mount adjusting nuts "B" an equal amount, by alternating from one side to the other, until the shaft indexes with the coupling splines, and bottoms out, without resistance.

6- Slide a hose clamp onto the water inlet hose and connect the hose to the water inlet tube. Tighten the hose clamp securely.

If the engine is equipped with a closed water system, fill the system with a solution of fresh water and rust inhibitor or antifreeze. Connect the cables to the battery terminals and the engine.

Connect the instrument harness to the engine. Connect the ground lead to the stud on the flywheel housing. Connect the shift cutout switch wires to the terminal block on the shift plate and the trim/tilt sender wires to the terminal block mounted on the engine. Connect the vacuum hose to the

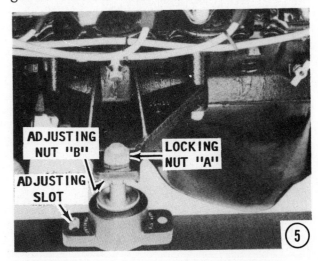

ADJUSTING
NUT "B"

LOCKING
NUT "A"

ADJUSTING
SLOT

5

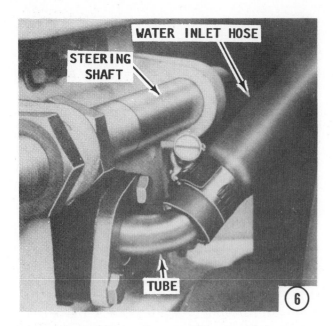

intake manifold. Install the fuel line to the carburetor and check for leaks. Install the remote control shift cable and the stern drive shift cable to the shift plate mounted on the engine, refer to Chapter 7 for cable adjustments.

7- Connect both cables to the battery.

THROTTLE CABLE INSTALLATION AND ADJUSTMENT

To install the throttle cable: check to be sure the remote control lever is in the NEUTRAL position.

All Models Except with EFI

For all models except those equipped with EFI: be sure the spacer is in place on the throttle lever stud.

All Models Including EFI

Slide the cable end guide over the throttle lever stud and secure the cable to the stud with a flat washer and elastic stop nut. Push the brass barrel gently toward the throttle lever and rotate the brass barrel along the threaded portion of the cable until the hole in the barrel aligns with the anchor stud. Install the barrel onto the stud with the cable ABOVE the stud. Secure the barrel to the stud with a flat washer and elastic stop nut.

Models with Carburetors

Place the remote control in the full throttle position -- all throttle valves should be fully open. In this position the throttle

lever tang should against the stop on the carburetor body. Shift the remote control lever to the NEUTRAL position. The throttle lever tang should now make contact with the idle rpm adjustment screw.

If all these conditions are met, the throttle cable is correctly adjusted. If not, adjust the position of the brass barrel along the threaded portion of the throttle cable until the tang rests against the stop.

Models with EFI

Place the remote control lever in the NEUTRAL position. The idle speed adjustment screw must contact the idle speed stop. Adjust the position of the brass barrel on the thottle cable until this condition is met.

All Models

Connect the power steering hoses to the power steering unit and tighten the large hose fitting to a torque valve of 23 ft lb (30Nm) and the small hose fitting to a torque valve of 8.5 ft lb (11.5Nm).

Install the stern drive, see Chapter 10.

CAUTION

Water must circulate through the stern drive, also to and from the engine, anytime the engine is run to prevent damage to the water pump located in the lower unit (original and Alpha drives); or on the front of the engine (Bravo drive). Just a few seconds without water will damage the water pump.

Start the engine and check the completed work.

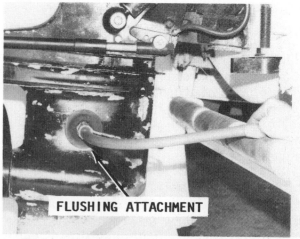

Flush attachment connected to the stern drive. Such a device MUST always be used whenever the engine is started for any reason when the boat is out of the water. Never operate the engine above a fast idle with a flush attachment connected.

14
MAINTENANCE

14-1 INTRODUCTION

The material presented in this chapter is divided into five general areas.

1- General information every boat owner should know.

2- Maintenance tasks that should be performed periodically to keep the boat operating at minimum cost.

3- Care necessary to maintain the appearance of the boat and to give the owner that "Pride of Ownership" look.

4- Winter storage practices to minimize damage during the off-season when the boat is not in use.

In nautical terms, the front of the boat is the **bow;** the rear is the **stern;** the right side, when facing forward, is the **starboard** side; and the left side is the **port** side. All directional references in this manual use this terminology. Therefore, the direction from which an item is viewed is of no consequence, because **starboard** and **port** **NEVER** change no matter where the individual is located.

14-2 ENGINE SERIAL NUMBERS

The engine serial numbers are the manufacturer's key to engine changes. These numbers identify the year of manufacture, the qualified horsepower rating, and the parts book identification. If any correspondence or parts are required, the engine serial number **MUST** be used or proper identification is not possible. The accompanying illustrations will be very helpful in locating the engine identification tag for the various models. **ONE MORE WORD:** The serial

BOW - FORWARD
(FRONT)

STARBOARD
(RIGHT SIDE)

PORT
(LEFT SIDE)

STERN - AFT
(REAR)

Common terminology used throughout the world for reference designation on boats. These are the terms used in this book.

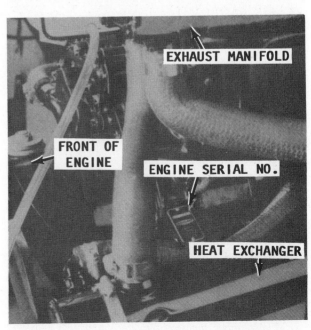

Serial number location for the Model 470 engine.

EXHAUST MANIFOLD

FRONT OF ENGINE

ENGINE SERIAL NO.

HEAT EXCHANGER

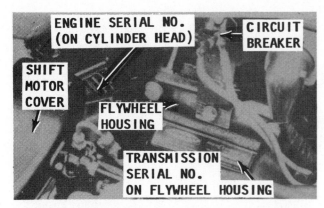

Serial number location for the Model 233 engine.

number establishes the year in which the engine was produced and not necessarily the year of first installation. Look for the identification plate above the cranking motor, or on the starboard side between the flywheel bell housing and the engine block.

14-3 STERN DRIVE IDENTIFICATION AND SERIAL NUMBERS

The accompanying illustrations in this section will be most helpful in identifying your unit by appearance. If your unit appears to be similar to one shown, then identification will be possible through the

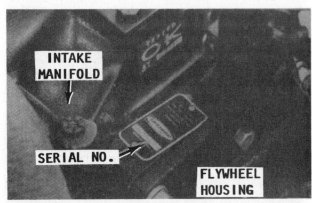

Serial number location for the Model 225S, 888, and the 898 engines.

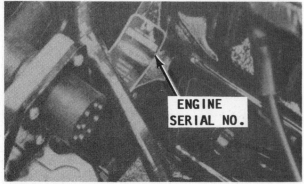

Serial number location for the Model 120, 140, and 165 engines.

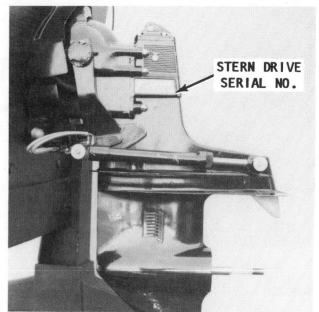

On all units, except the Bravo, stern drive serial numbers are stamped into the casting on the upper gear housing under the decal on the port or starboard side of the housing.

caption under the picture. The captions associated with each illustration will identify the unit if the number cannot be clearly read.

The stern drive serial numbers are the manufacturer's key to model changes over the years. These numbers identify the model year, the gear ratio, and the parts book identification. The stern drive serial number **MUST** be used on any correspondence and parts request, or proper identifi-

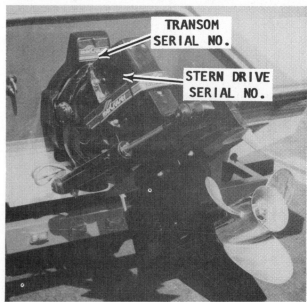

*The Bravo stern drive serial number is found on a decal affixed to the upper gear housing, as shown. The transom serial number is on a decal affixed to the gimbal housing. The two serial numbers **MUST** not be confused.*

cation is not possible. The accompanying illustrations will be very helpful in locating the identification number decal. The number is also stamped into the casting under the decal either on the port or starboard side. If the decal is missing or defaced, the identification number may still be read. **ONE MORE WORD:** As with the engine serial numbers, the stern drive numbers identify the year the unit was produced, and not necessarily the year of first installation.

In 1985, the Alpha/MR units were introduced with the following beginning serial numbers for the **OVERALL** drive gear ratio listed.

```
1.98:1 --   6854393 and above
1.84:1 --   6862702 and above
1.65:1 --   6810538 to 6811037
            6864366 and above
1.50:1 --   6869699 and above
```

Spanner-type universal joint cover nut used on the following stern drive units: Model 1-ABC, 1775192 and below; Model 1-ABC-EZ, 2062140 and below; Model 120, 140, 160, 2495185 and below; Model 120, 140, 160, 2495186 thru 2763441; Model 120, 140, 165, 2763442 thru 3780850 and above; Model 888, 3784374 and below; Model 898, 225S, 233, 3784375 and above; **AND** all Bravo stern drives.

In 1988 the Bravo stern drive was introduced. The transom serial number on a Bravo stern drive is found on a decal affixed to the top of the gimbal housing. The stern drive gear ratio and serial number is on a decal on the port side of the upper gear

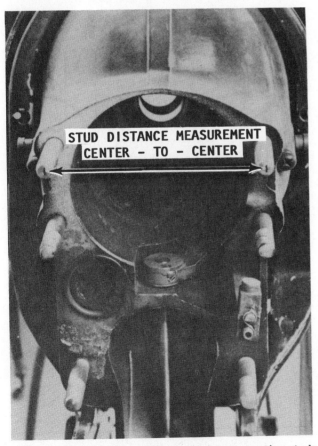

The distance between the stern drive mounting studs of the bell housing is critical. Stern drive models and the distance between the mounting studs are as follows: Measurement - 4-15/16" - Model 1, 1535211 thru 1684188; Model 1-ABC, 1775192 and below; Model 1-ABC-E-Z, 2062140 and below. Measurement - 5-1/16" - Model 120, 140, & 160, 2495185 and below; Model 120, 140, 160, 2495186 thru 3780850 and above; Model 888, 3784374 and below; Model 888, 225S, 233, 3784375 and above.

Single tilt hose installation to the port side of the gimbal housing. This arrangement was used on very early models.

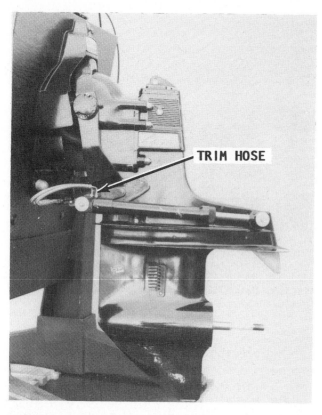

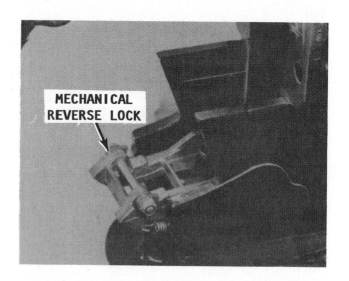

Mechanical reverse lock used on the following stern drive units: Model 1, 1535211 and below; Model 1, 1535212 thru 1684188; Model 1-ABC, 1775192 and below; Model !-ABC-EZ, 2062140 and below; Model 120, 140, 160, 2495185 and below.

Double trim and tilt hose installation connected to the bottom of the gimbal housing. Stern drives with this hose arrangement are: Model 120, 140, 165, 2763442 thru 3780850 and above; Model 888, 3784374 and below; Model 888, 225S, 233, 3784375 and above.

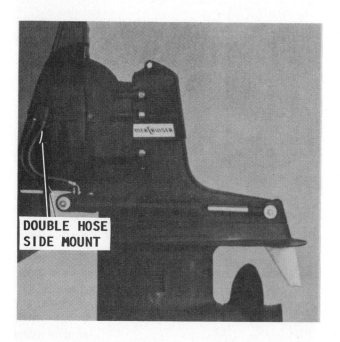

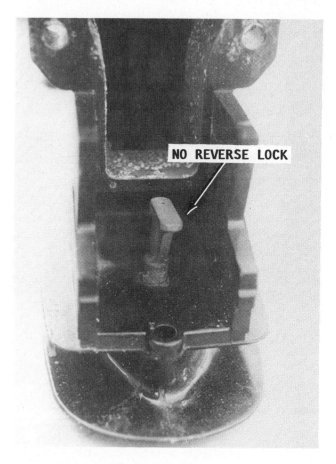

Double trim and tilt hose installation connected to the port side of the gimbal housing. Stern drives with this hose arrangement are: Model 1-ABS-EZ, 2062140 and below; Model 120, 140, 160, 2495185 and below; Model 120, 140, 160, 2495186 thru 2763441.

Stern drive units without the reverse lock. On these models the reverse lock is incorporated in the trim/tilt system: Model 120, 140, 160, 165, 2495186 thru 3780850 and above; Model 888, 3784374 and below; Model 888, 225S, 233, 3784375 and above.

housing. The standard gear ratio for all Bravo units is 1.5:1. No gear ratio change is available (at press time) for the Bravo when operating at high altitudes. However, the manufacturer recommends a propeller change for higher elevations. Consult the local dealer for his propeller suggestions.

A table in the Appendix lists the Overall gear ratios associated with the various stern drives.

14-4 SHIFT IDENTIFICATION ALL UNITS EXCEPT BRAVO

The following paragraphs will be helpful in identifiying the type of shift mechanism installed through the stern drive serial number.

Long adjusting slot used on the following stern drive units: Model 1, 1535211 and below; Model 1, 1535212 thru 1684188; Model 1-ABC, 1775192 and below; Model 1-ABC-EZ, 2062140 and below; Model 120, 140, 160, 2495185 and below; Model 120, 140, 160, 165, 2495186 thru 3780849.

Hand-operated reverse lock used on very early model stern drive units. This arrangement requires the operator to extend himself over the transom to engage the reverse lock.

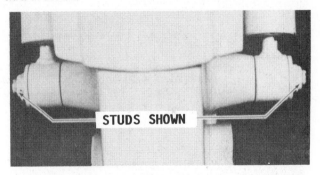

Bolt arrangement used to secure the shock absorber to the upper gear housing on very early model stern drive units. Models using nuts on both ends as shown: Model 1, 1535211 and below, Model 1, 1535212 thru 1563353. Models using bolt with single nut: Model 1 1563354 thru 1684188.

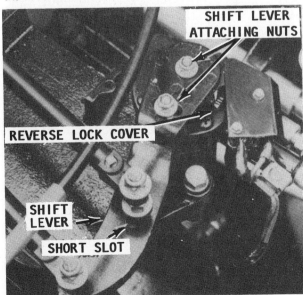

Short adjusting slot shift lever used on the following stern drive units: Model 120, 140, 165, 3780850 and above; Model 888, 225S, 233, 3784375 and above.

Hexagon-type universal joint cover nut used on the following stern drive units: Model 1, 1535211 and below, Model 1, 1535212 thru 1684188.

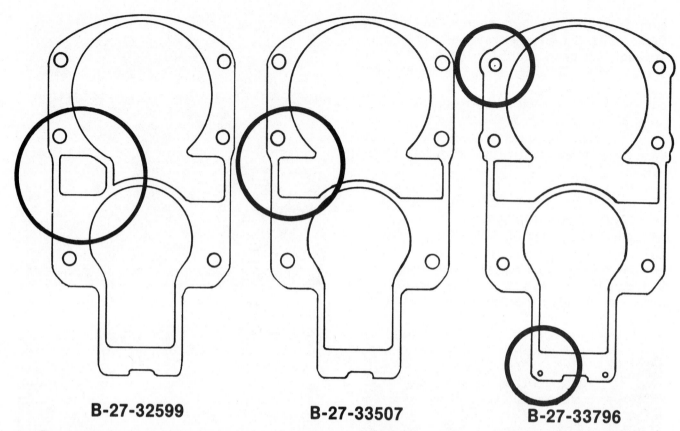

B-27-32599   **B-27-33507**   **B-27-33796**

Shape and hole pattern of the stern drive-to-bell housing gasket. The part number is given under each gasket and MUST be used with the stern drive units as follows: Gasket (left): Model 1, 1535211 and below; Gasket (center): Model 1, 1535212 thru 1536617. Gasket (right): Model 1, 1536618 thru 1684188.

Cam Shift

MerCruiser 1   1535211 and below
                    1535212 thru 1536617
                    1536618 thru 1563353
                    1563354 thru 1684188

MerCruiser 1-ABC 1775192 and below

E-Z SHIFT

MerCruiser 1-ABC 2062140 and below
MerCruiser 120, 140, 160
                    2495185 and below
MerCruiser 120, 140, 160
                    2495186 thru 2763441
MerCruiser 120, 140, 165
                    2763442 thru 3780849
MerCruiser 120, 140, 165
                    3780850 and above
MerCruiser 888 3784374 and above
MerCruiser 888-225S-233
                    3784375 and above

14-5 FIBERGLASS HULLS

Fiberglass-reinforced plastic hulls are tough, durable, and highly resistant to impact. However, like any other material they can be damaged. One of the advantages of this type of construction is the relative ease with which it may be repaired. Because of its break characteristics, and the simple techniques used in restoration, these hulls have gained popularity throughout the world. From the most congested urban marina, to isolated lakes in wilderness areas, to the severe cold of far off northern seas, and in sunny tropic remote rivers of primative islands or continents, fiberglass boats can be found performing their daily task with a minimum of maintenance.

A fiberglass hull has almost no internal stresses. Therefore, when the hull is broken or stove-in, it retains its true form. It will not dent to take an out-of-shape set. When the hull sustains a severe blow, the impact will be either absorbed by deflection of the laminated panel or the blow will result in a definite, localized break. In addition to hull damage, bulkheads, stringers, and other stiffening structures attached to the hull may also be affected and therefore, should be checked. Repairs are usually confined to the general area of the rupture.

14-6 BELOW WATERLINE SERVICE

A foul bottom can seriously affect boat performance. This is one reason why racers, large and small, both powerboat and sail, are constantly giving attention to the condition of the hull below the waterline.

In areas where marine growth is prevalent, a coating of vinyl, anti-fouling bottom paint should be applied. If growth has developed on the bottom, it can be removed with a solution of muriatic acid applied with a brush or swab and then rinsed with clear water. **ALWAYS** use rubber gloves when working with muriatic acid and **TAKE EXTRA CARE** to keep it away from your face and hands. The **FUMES ARE TOXIC.** Therefore, work in a well-ventilated area, or if outside, keep your face on the windward side of the work.

Barnacles have a nasty habit of making their home on the bottom of boats which have not been treated with anti-fouling paint. Actually they will not harm the fiberglass hull, but can develop into a major nuisance.

If barnacles or other crustaceans have attached themselves to the hull, extra work will be required to bring the bottom back to a satisfactory condition. First, if practical, put the boat into a body of fresh water and allow it to remain for a few days. A large percentage of the growth can be removed in this manner. If this remedy is not possible, wash the bottom thoroughly with a high-pressure fresh water source and use a scraper. Small particles of hard shell may still hold fast. These can be removed with sandpaper.

Stern drive with considerable marine growth from extended use in salt water, without a haul-out.

14-7 ENGINE FLUIDS

Engine Oils

The manufacturer recommends the use of Quicksilver 4-cycle Marine Engine Oil for all ambient temperatures. If this oil is not available, a good grade of straight-weight detergent automotive oil may be substituted.

The following weight oils are recommended for the temperatures indicated:

Above 70°F	SE40
32°F to 70°F	SE30
Below 32°F	SE20

The manufacturer discourages the use of multi-viscosity oils, such as 20W-40 or 20W-50 and non-detergent oils. These can be used in an emergency if a straight-weight oil is not available. **NEVER** mix straight-weight oil with multi-viscosity oil. If mixing is necessary in an emergency, then drain the crankcase at the first available opportunity, and replenish with the recommended straight-weight oil.

Oil Changes and Crankcase Capacities

ALWAYS install a new oil filter at the same time the oil is changed. Coat the contact surface of the oil filter sealing ring with clean oil. Install and tighten the new oil filter **BY HAND. NEVER** use an oil filter wrench to tighten the filter.

The following table provides an approximate guide to crankcase oil capacities when installing a new empty oil filter:

*A **NEW** oil filter must always be installed each time the engine oil is changed, to prevent contaminated oil from circulating through the engine.*

4-Cylinder	4 to 5 quarts
6-Cylinder In-line	5 to 6 quarts
V6	4 to 5 quarts
V8	5 to 7 quarts

Add the suggested amount of oil; wait a few minutes to allow the oil to drain down into the crankcase, and then check the level on the dipstick. **ALWAYS** fill the crankcase using the dipstick to determine when the level is correct. Some dipsticks are marked with two lines, meaning **ADD** and **FULL**, some actually have these two words embossed on the stick. Other dipsticks have the words **ADD 1 QT** and **OPERATING RANGE** embossed. **ALWAYS** keep the oil level "a whisker" above the **ADD** or **OPERATING RANGE** line.

Oil Change Intervals

If ambient temperature is above 50°F, the manufacturer recommends engine oil and oil filter be changed every 100 hours of engine operation or every 60 days, whichever occurs first.

If the temperature is lower than 50°F, then the time interval should be cut in half to 50 hours of operation or 30 days, again, whichever occurs first.

Power Steering Fluid

Use only Dexron, or Dexron II automatic transmission fluid in MerCruiser power steering systems. The fluid level **MUST** be checked with the engine **NOT** running. The power steering dipstick is marked **FULL HOT** and **FULL COLD** depending on engine temperature.

Closed Cooling System

The level of coolant in the heat exchanger should be within one inch from the bottom of the filler neck. Allow the engine to cool down before removing the pressure cap. The same safety precautions must be practiced as are followed when opening an automobile radiator cap! The level in the coolant recovery reservoir should be between the **ADD** and **FULL** marks when the engine is at operating temperature. It is perfectly safe to open the reservoir cap when the engine is operating, because the reservoir is not pressurized.

Add rust inhibitor if the anti-freeze has been used for more than one season. If the solution is contaminated, flush and replace with a new mixture of 50/50 ethylene glycol and water.

14-8 OFF-SEASON STORAGE

1- Start the engine and allow it to warm to normal operating temperature.

CAUTION

Water must circulate through the stern drive, also to and from the engine, anytime the engine is run to prevent damage to the water pump located in the lower unit (original and Alpha drives); or on the front of the engine (Bravo drive). Just a few seconds without water will damage the water pump.

If the engine is equipped with a closed-circuit cooling system, check the anti-freeze level at the reservoir. Correct level is between the **ADD** and **FULL** marks with the engine operating at idle speed. Add rust inhibitor if the anti-freeze has been used for more than one season. If the solution is contaminated, flush and replace with a new mixture of 50/50 ethylene glycol and water. Stop the engine, drain the oil in the crankcase, and remove the oil filter. Install a **NEW** filter element, and then fill the crankcase with the prescribed weight and amount of oil. Start the engine and allow it to run at a high idle for a few minutes.

CAUTION

Water must circulate through the stern drive, also to and from the engine, anytime the engine is run to prevent damage to the

Powerplant equipped with a closed cooling system.

water pump located in the lower unit (original and Alpha drives); or on the front of the engine (Bravo drive). Just a few seconds without water will damage the water pump.

Check the oil level as indicated on the dip stick. Some amount of oil may remain in the engine without draining down into the crankcase. A slightly lower reading may, therefore, be indicated on the dipstick. Add only enough oil to bring the reading into the safe running range above the **ADD** or **OPERATING RANGE** mark to allow for the amount of oil held in various areas of the engine.

2- Shut off the gasoline supply at the fuel tank. Disconnect the fuel line between the valve and the fuel pump. Drain the fuel from the line. Insert the end of the line disconnected from the fuel pump into a can containing several ounces of fuel mixed with a rust inhibitor. Start the engine and run it at a fast idle until it stalls from lack of fuel.

CAUTION

Water must circulate through the stern drive, also to and from the engine, anytime the engine is run to prevent damage to the water pump located in the lower unit (original and Alpha drives); or on the front of the engine (Bravo drive). Just a few seconds without water will damage the water pump.

3- Remove the flame arrestor and **SLOWLY** pour about a pint of rust-preventive oil into the carburetor air intakes while running the engine at a fast idle. This can be done at the same time as Step 2, while using up the fuel in the fuel lines, fuel pump, and in the carburetor. Clean the fuel filter and sediment bowl. Install the bowl with a **NEW** gasket and reconnect the fuel line. Clean the flame arrestor in solvent, and then blow it dry with compressed air. Lubricate the alternator, starter, distributor, and control linkage.

4- Remove all of the spark plugs and squirt about a teaspoonful of rust-preventive oil into each cylinder. Crank the engine over several times to allow the oil to coat the cylinder walls. Remove any excess oil from around the spark plug holes, and then install the spark plugs.

5- Remove the rocker arm covers and inspect the valve train mechanism for worn or damaged parts. Clean the inside of the covers. Apply a liberal coating of crankcase oil to the valve mechanism and onto the inside of the covers. Install the covers with **NEW** gaskets. Using new gaskets will ensure a good seal with the valve covers. Clean the outside of the engine, and then wipe it down with an oily rag. Cover the engine with a protective cover, but **ALLOW** for air circulation.

6- Flush the cooling system with fresh water. Allow the water to circulate for at

Injecting just a "squirt" of rust inhibitor through the spark plug openings will provide a protective coating to the cylinder walls during off season storage.

A good maintenance program provides for the inspection of the electrical system, including wiring, connections, and battery.

Water separator unit which should be installed in the fuel line at the first opportunity. Such a kit is not expensive, and contains a throw-a-way filter similar to an oil filter element.

least 5 minutes. If the engine is equipped with a raw-water cooling system, open all of the engine and manifold water jacket drains. Allow the water to drain completely. **ALWAYS** have the stern drive in a horizontal position when draining the system to ensure all of the water is able to leave the system. If the stern drive is not horizontal, water will be trapped inside. If this water should freeze, the water pump will be ruined.

Flame arrestor sealed for winter storage.

Leave all of the drains open. Disconnect the water hoses at the low end and allow them to drain. Remove the drain plugs from the water pumps and allow them to drain.

7- When the boat is to be stored for even a short period, the stern drive **MUST** be in the full **DOWN** position. The full down position will: prevent the bellows from taking a "set"; prevent undue strain on the underside; and ensure all folds and ribs are in the relaxed position.

14-9 STERN DRIVE LUBRICATION

Clean the exterior surface of the unit thoroughly. Inspect the finish for damage or corrosion. Clean any damaged or corroded areas, and then apply primer and matching paint.

Check the entire unit for loose, damaged, or missing parts.

The propeller should be checked regularly to be sure all the blades are in good condition. If any of the blades become bent or nicked, this condition will set up vibrations in the drive unit and the motor. Remove and inspect the propeller. Use a file to trim nicks and burrs. **TAKE CARE** not

Effects of electrolysis due to improper grounding of the various stern drive units. Owner's negligence can be blamed for this condition.

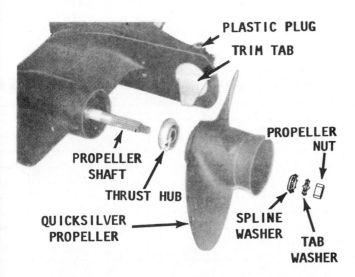

Principle parts of the propeller installation as discussed in the text.

Corroded trim tab which has performed its function. When the trim tab is eaten away by corrosion, more expensive parts are protected.

Hydraulic pump and associated wiring.

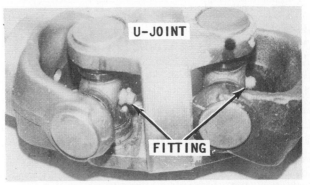

The stern drive should be removed once a year and the universal joint properly lubricated.

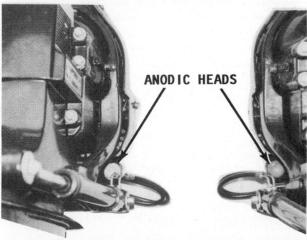

Anodic heads on a dual engine installation. Another set of heads is installed on the outboard side of each unit.

Lubrication fittings on the port side of the stern drive.

Locations	After 1st 20 Hrs. of Operation	Every 50 Hours of Operation	Every 100 Hours of Operation	Once Each Year
Starter Motor and Alternator	None Required			
Change Engine Oil	1		1	
Replace Oil Filter	●		●	
Clean Oil Filler Cap			●	
Clean Flame Arrestor	●		●	
Change Fuel Filter				●
Change Fuel Filter Element (In Carburetor)	●□			●□
Check Fuel System Lines and Connections for Leaks	●	●		
Check Battery Electrolyte Level	●	●		
Check All Electrical Connections	●			●
Check Water Pump and Alternator Belts for Tension	●	●		
Check Cooling System Hoses and Connections for Leaks	●		●	
Check Power Steering Fluid Level		2		
Check for Loose, Damaged or Missing Parts		●		●
Lubricate Throttle and Shift Linkage Pivot Points			1+	1+
Lubricate Distributor Cam				6*□
Inspect Breaker Points		●*		
Check Condition of Spark Plugs				●*
Tighten Engine Mount, Drive, Steering Cable and Trim Cylinder Fasteners	●		●	
Check Engine Alignment	●□			●□
Check Transmission Fluid Level	2	2		
Clean Transmission Oil Strainer Screen				●
Check Condition of Oil Cooler Electrodes	Every 6 Months			
Change Transmission Fluid				2
Check Power Trim Pump Oil Level	1		1	
Check Stern Drive Oil Level	3	3		
Lubricate Drive Unit Upper and Lower Swivel Pins	4+		4+	
Lubricate Steering Cable and Steering Lever	4⊕		4⊕	
Change Stern Drive Unit Oil			3	3
Check Condition of Trim Tab and Anodic Plate	Every 60 Days +			
Inspect Propeller for Possible Damage		●		●
Lubricate Propeller Shaft Splines			5§+	5§+
Inspect Bellows and Clamps	●+			●+
Inspect and Clean Exterior of Drive Unit			●+	●+
Lubricate U-Joint Coupling Splines	●			●□
Lubricate Gimbal Bearing			4	4
Lubricate U-Joint Cross Bearings			7□	7□
Check Stern Drive Unit Water Pump and Impeller				●□
Check Alternator Rear Screen				●+
Clean Crankcase Ventilating System				●
Torque Circulating Water Pump Bolts (Ford Engines)	Every 6 Months or 500 Hours □			
Lubricate Output Housing Bearing			4□	4□
Lubricate Hinge Pins			8+□	8+□

1- *Use Formula 4 Quicksilver Oil*
2- *Use Automatic Transmission Fluid*
3- *Use Quicksilver Super-Duty Lubricant*
4- *Use Multipurpose Lubricant*
* *Does not apply to "Thunderbolt Ignition"*
+ *More frequently if unit is operated in salt water*
  Lubricate during every installation

5- *Use Perfect Seal*
6- *Use suitable high-melting-point, non-bleeding grease*
7- *Use Universal Joint Lubricant*
8- *Use Anti-Corrosion Grease*
□ *By an authorized MerCruiser dealer*
⊕ *Every 60 days in fresh water*
§ *Every 30 days in salt water*

Lubrication and maintenance chart. Recommendations are based on average operating conditions. If the unit is operated at continuous high speed, or under heavy duty service, the inspection and maintenance intervals should be shortened. All tasks do not apply to all models.

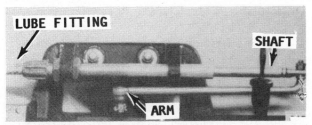

Lubrication fittings inside the boat for the steering mechanism.

to remove any more material than is absolutely necessary. For a complete check, take the propeller to your marine dealer where the proper equipment and knowledgeable mechanics are availble to perform a proper job at modest cost.

Inspect the propeller shaft to be sure it is still true and not bent. If the shaft is not perfectly true, it should be replaced.

Install the thrust hub. Coat the propeller shaft splines with Perfect Seal No. 4, and the rest of the shaft with a good grade of anti-corrosion lubricant. Install the propeller, and then the splined washer, tab washer, and propeller nut.

Position a block of wood between the propeller and the anti-cavitation tab to keep the propeller from turning. Tighten the propeller nut to a torque value of 40 ft lbs (53Nm). Adjust the nut to fit the tab lock space. Bend three of the tab washer tabs into the spline washer using a punch and hammer. The tabs will prevent the nut from backing out. Inspect the U-joint bellows, shift cable bellows, water intake hose, and exhaust tube for any sign of deterioration and damage. Service these units as required. Check to be sure all clamps are secure.

Check the trim system for proper operation. Check the oil level in the pump reservoir and be sure the vent screw is left **OPEN**. Check to be sure all connections are secure. Check and adjust engine alignment. The stern drive must be removed in order to check engine alignment, see Chapter 10. Check and adjust the shift cables. Lubricate all external lubrication points with Multi-Purpose Lubricant. Check and clean the water intake opening. Replace the inserts, if they are cracked or damaged. Check the trim tab and the anodic heads. Replace them, if necessary. The trim tab must make a good ground inside the lower unit. Therefore, the trim tab and the cavity **MUST NOT** be painted. In addition to trimming the boat, the trim tab acts as a zinc to

prevent electrolysis on more expensive parts. It is normal for the tab to show signs of erosion. The tabs are very inexpensive and should be replaced frequently.

Lubrication Points

After every 50 hours of operation, or at least once every season, the following grease fittings should be lubricated:

Outside the Boat

Gimbal housing upper and lower pivot pins and gimbal bearings, with Quicksilver Multi-Purpose Lubricant.

Universal joint bearings with Universal Joint Lubricant once a year or every 100 hours. (This service task requires removal and installation of the stern drive, see Chapter 10.)

Lubricate the hinge pin on each side with Anti-Corosion Grease

Inside the Boat

These points may be lubricated with Quicksilver Multi-Purpose Lubricant:

Ride-Guide steering cable end next to the hand nut. **DO NOT** over-lubricate the cable.

The steering arm pivot socket.

Exposed shaft of the cable passing through the cable guide tube.

Steering link rod to the steering cable.

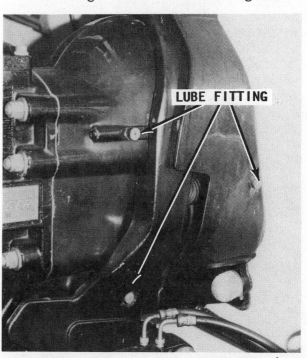

Lubrication fittings on the starboard side of the stern drive.

14-10 DRAINING AND FILLING THE STERN DRIVE

Two types of stern drive fluid reservoirs are used by MerCruiser. Early models, before approximately 1973, have separate fluid reservoirs. Later model units have two reservoirs with interconnected fluid passageways. On the early models, each reservoir is drained and filled independently. On later models, includung the Bravo, both reservoirs can be drained from the lower gear housing, and then filled from the same opening.

NEVER remove the vent or filler plugs when the drive unit is hot. Expanded lubricant would be released through the plug hole. Check the lubricant level after the unit has been allowed to cool. Add only Super-Duty Gear Lubricant. **NEVER** use regular automotive-type grease in the stern drive because it expands and foams too much. Stern drive units do not have provisions to accommodate such expansion.

If the lubricant appears milky brown, or if large amounts of lubricant must be added to bring the lubricant up to the full mark, a thorough check should be made to determine the cause of the loss.

Drain and Fill MerCruiser 1, 1A, 1B, & 1C

1- Draining upper gear housing chamber: Remove the drive unit top cover. Use a suction pump to remove the lubricant.

2- Filling upper gear housing chamber: Install the top cover with a new O-ring seal and tighten the four screws to a torque value of 20 ft-lbs. Remove the **FILL** and **VENT** plugs and gaskets from both sides of the upper gear housing. Position the drive unit approximately vertical and without a list to either port or starboard. Insert a lubricant tube into the **FILL** plug hole and inject lubricant until the excess starts to flow out the **VENT** hole. Install the **VENT** and **FILL** plugs with **NEW** gaskets. Check to be sure the gaskets are properly positioned to prevent water from entering the housing.

3- Draining lower unit: Remove the **FILL** plug from the lower end of the gear housing on the port side and the **VENT** plug just above the anti-cavitation plate.

4- Filling lower unit: Position the drive unit approximately vertical and without a list to either port or starboard. Insert the lubricant tube into the **FILL/DRAIN** hole at

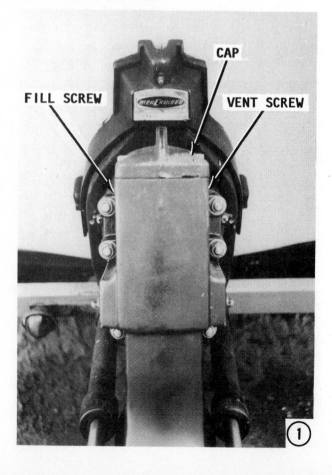

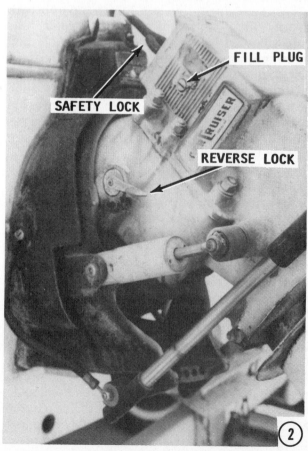

the bottom plug hole, and inject lubricant until the excess begins to come out the **VENT** hole. Install the **VENT** and **FILL** plugs with **NEW** gaskets. Check to be sure the gaskets are properly positioned to prevent water from entering the housing.

Drain and Fill All Other MerCruiser Stern Drives Including Bravo

5- A visual oil reservoir kit is available as an optional accessory from the local MerCruiser marine dealer. This kit includes a see-through transparent lubricant reservoir and the necessary hose and fittings for installation inside the boat in any convenient location.

6- After the kit is installed, the reservoir is filled to the **FULL** mark with regular Quicksilver Super Duty Lower Unit Lubricant. The reservoir allows the lubricant to "breathe" as the running components create channels in the lubricant. Any loss of lubricant can be detected by a change in the level of lubricant in the transparent reservoir after the unit has cooled. Lubricant can be added at any time. A consistent loss of lubricant indicates the stern drive unit is in need of close inspection to determine the cause. Normal maintenance calls for changing the lower unit lubricant every 100 hours of operation, but with the reservoir kit installed, this period recommendation is extended to 200 hours, provided there has not been a consistant loss or discoloration of lubricant.

Draining stern drive unit:

7- Remove the **FILL/DRAIN** plug from the lower end of the lower unit on the port

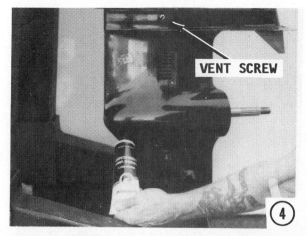

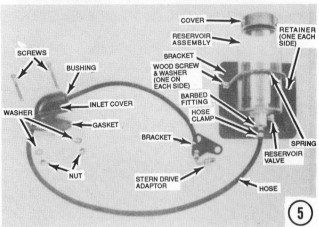

side. Remove the **VENT** plug from the upper gear housing. On these models, the upper and lower gear housing chambers are interconnected. Allow at least **ONE HOUR** for the gear oil to completely drain from the upper gear housing.

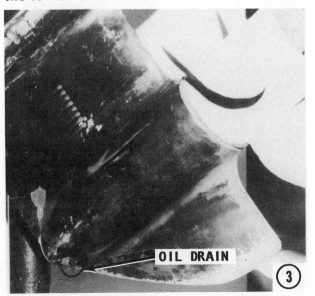

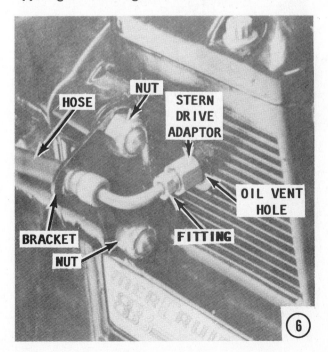

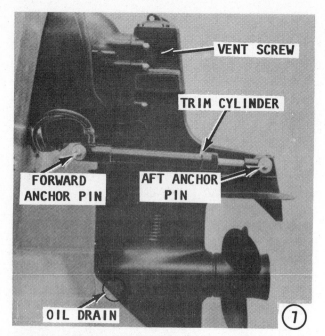

VENT SCREW

TRIM CYLINDER

FORWARD ANCHOR PIN

AFT ANCHOR PIN

OIL DRAIN

⑦

Filling stern drive unit:

8- Place the stern drive in an approximate vertical position without a list to port or starboard. Insert a lubricant tube into the **FILL/DRAIN** plug opening in the lower unit. Inject lubricant until the excess starts to flow out the **VENT** hole in the upper driveshaft housing. After the unit is full of

lubricant, install the **VENT** and **FILL** plugs, with **NEW** washers. As the plugs are tightened, check to be sure the washers remain in place to prevent water from leaking into the housing.

Wipe the outside of the drive unit and the gear case with an oily rag. Check all of the steering connections. Lubricate all joints and pulleys.

GEAR CASE OIL CAPACITIES

Models	Unit	Ounces
1, 1A, 1B, 1C	Upper driveshaft	8
1	Lower gear	18
1A, 1B, 1C	Lower	23
120, 140, 160, 165, 470, 225S, 228, 233, 250, 888 & 898	Upper and lower	28
Bravo	Upper and lower	64

9- Seal off all openings to the carburetor and exhaust system to prevent dust, water, and insects from entering the engine.

10- Remove the batteries from the boat and keep them charged during the storage period. Clean the batteries thoroughly of any dirt or corrosion, and then charge them to full specific gravity reading. After they are fully charged, store them in a clean cool dry place where they will not be damaged or knocked over.

NEVER store the battery with anything on top of it or cover the battery in such a manner as to prevent air from circulating

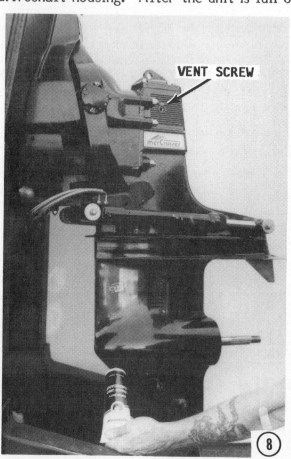

VENT SCREW

⑧

CARBURETOR SEALED

⑨

around the fillercaps. All batteries, both new and old, will discharge during periods of storage, more so if they are hot than if they remain cool. Therefore, the electrolyte level and the specific gravity should be checked at regular intervals. A drop in the specific gravity reading is cause to charge them back to a full reading.

In cold climates, **EXERCISE CARE** in selecting the battery storage area. A fully-charged battery will freeze at about 60 degrees below zero. A discharged battery, almost dead, will have ice forming at about 19 degrees above zero.

ALWAYS remove the drain plug and position the boat with the bow higher than the stern. This will allow any rain water and melted snow to drain from the boat and prevent "trailer sinking". This term is used to describe a boat that has filled with rain water and ruined the interior including the engine because the plug was not removed or the bow was not high enough to allow the water to drain properly.

14-11 PRE-SEASON PREPARATION

1- Lubricate the engine according to the manufacturer's recommendations. Remove, clean, inspect, adjust, and install the spark plugs with new gaskets if they require gaskets. Make a thorough check of the ignition system. This check should include: the points, coil, condenser, condition of the wiring, and the battery electrolyte level and charge.

2- Take time to check the gasoline tank and all of the fuel lines, fittings, couplings, valves, flexible tank fill, vent, and fuel lines. Turn on the gasoline supply valve at the tank. If the gas was not drained at the end of the previous season, make a careful inspection for gum formation. When gasoline is allowed to stand for long periods of time, particularly in the presence of copper, gummy deposits form. This gum can clog the filters, lines, and passageways in the carburetor. See Chapter 4, Fuel System Service.

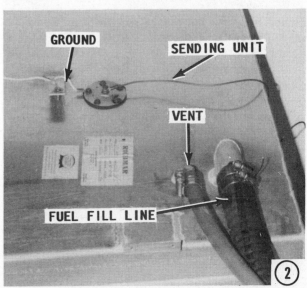

FLAME ARRESTOR

③

3- All marine engines **MUST** be equipped with an effective means of backfire flame control. This can be accomplished by one of two methods. The first and most popular is through installation of a Coast Guard approved flame arrestor on the carburetor. The second method is by ducting the air intakes outside the engine compartment to the atmosphere.

Clean and inspect the flame arrestors. Check and adjust the alternator belt tension, and replace them if they are worn or frayed. Check the oil level in the crankcase. The oil should have been changed prior to storage after the previous season. If the oil was not changed, do so and install a **NEW** oil filter.

4- Close all of the water drains. Check and replace any defective water hoses. Connect and check that the connections do not leak. Replace any spring-type hose clamps with band-type clamps, if they have lost their tension or if they have distorted the water hose. Check the sea cocks of the cooling system. Check to be sure the through-hull fittings are in good condition. If the engine is equipped with a closed-circuit cooling system, check the level of water. See Chapter 9, Cooling System.

5- The engine can be run with the lower unit of the stern drive submerged in water to flush it. If this is not practical, a Flush-it attachment may be used. This unit is attached to the water pickup of the lower unit. Attach a garden hose, turn on the water, allow the water to flow into the engine for awhile, and then run the engine.

CAUTION

Water must circulate through the stern drive, also to and from the engine, anytime the engine is run to prevent damage to the water pump located in the lower unit (original and Alpha drives); or on the front of the engine (Bravo drive). Just a few seconds without water will damage the water pump.

Check the exhaust outlet for water discharge. Check for leaks. Check operation of the thermostat. After the engine has reached operating temperature, tighten the

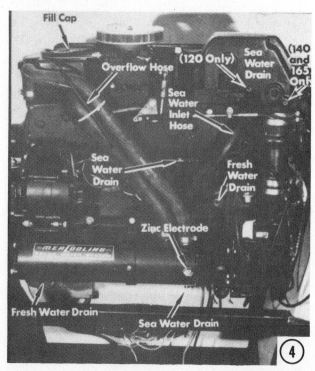

Fill Cap
Overflow Hose
(120 Only)
Sea Water Drain
(140 and 165 Only)
Sea Water Inlet Hose
Sea Water Drain
Fresh Water Drain
Zinc Electrode
Fresh Water Drain
Sea Water Drain

④

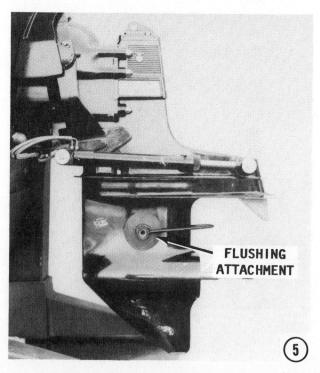

FLUSHING ATTACHMENT

⑤

cylinder head bolts to the torque value given in the Specifications in the Appendix.

6- Check the electrolyte level in the batteries and the voltage for a full charge. Clean and inspect the battery terminals and cable connections. **TAKE TIME** to check the polarity, if a new battery is being installed. Cover the cable connections with grease or special protective compound as a prevention to corrosion formation. Check all electrical wiring and grounding circuits.

7- Check the engine compartment for proper ventilation. **THERE MUST** be adequate means for removing combustible vapors from the boat. Coat Guard standards include specially designed hardware to do the best job of preventing any vapors from accumulating in the engine compartment or in the bilge.

8- Check the tension of the engine drive belt to ensure proper operation of the engine water pump and alternator. If the belt can be depressed more than 1/4" midway between the water pump and alternator, loosen the alternator and make the proper adjustment to the belt.

9- Check all electrical parts in the engine compartment and lower portions of the hull to be sure they are not of a type that could cause ignition of an explosive atmosphere. Rubber caps help keep spark insulators clean and reduce the possibility of arcing. Starters, generators, distributors, alternators, electric fuel pumps, voltage

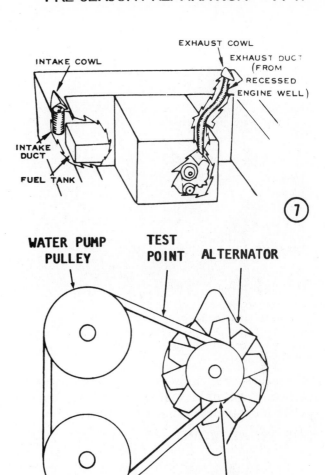

BATTERY SWITCH

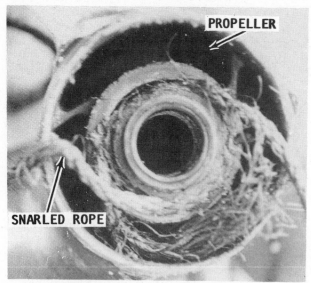

Section of rope entangled behind the propeller. This condition pulled the bearing carrier seal. With the seal gone, the lubricant in the lower unit was lost, and the gears were destroyed.

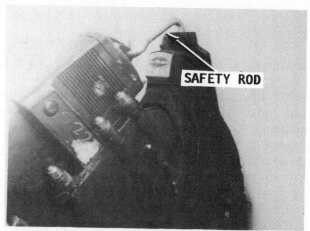

An alternate safety device used when trailering the boat. This safety rod attaches to the gimbal housing and the upper gear housing cap.

regulators, and high-tension wiring harnesses should be of a marine type that cannot cause an explosive mixture to ignite.

ONE FINAL WORD

Before putting the boat in the water, **TAKE TIME** to check to be sure the drain plugs are installed. Countless number of boating excursions have had a very sad beginning because the boat was eased into the water only to have water begin filling the inside.

Keep your gas tank full, the fuel pump pumping, the spark plugs sparking, the lifters lifting, and the pistons, well -- keep them working too.

Joan & Clarence Coles
Seloc Publications

Multipurpose lubricant which may be used as a general all-around lubricant. The one place it should not be used is in the U-joints.

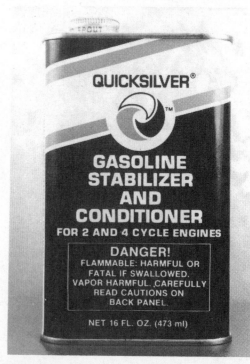

Quicksilver Gasoline Stabilizer and Conditioner may be used to keep the gasoline in the tank fresh. Such an additive will prevent the fuel from "souring" for up to twelve months.

APPENDIX

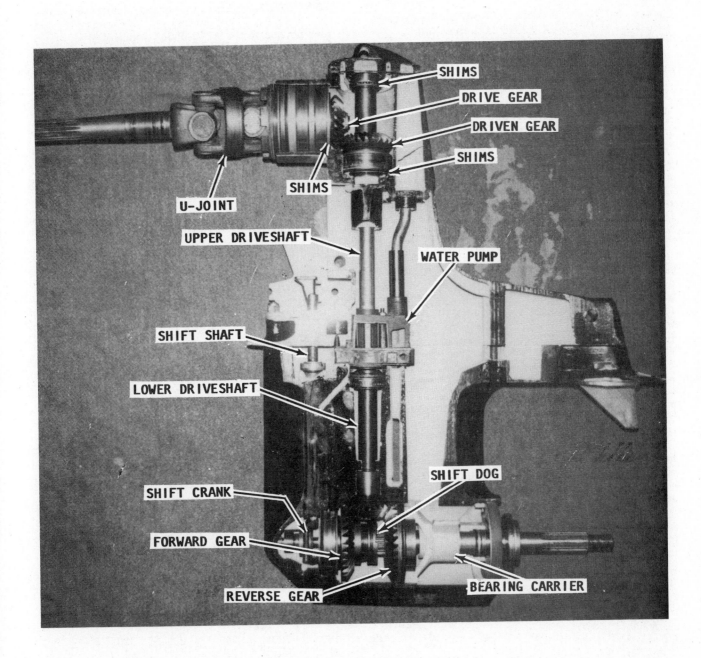

An excellent cutaway view of the complete stern drive unit -- Model R, MR, and Alpha -- exposing the major working parts. Such a view should be most valuable during troubleshooting, disassembly, assembly, and adjustment work.

APPENDIX

METRIC CONVERSION CHART

LINEAR

inches	X 25.4	= millimetres (mm)
feet	X 0.3048	= metres (m)
yards	X 0.9144	= metres (m)
miles	X 1.6093	= kilometres (km)
inches	X 2.54	= centimetres (cm)

AREA

$inches^2$	X 645.16	= $millimetres^2$ (mm^2)
$inches^2$	X 6.452	= $centimetres^2$ (cm^2)
$feet^2$	X 0.0929	= $metres^2$ (m^2)
$yards^2$	X 0.8361	= $metres^2$ (m^2)
acres	X 0.4047	= hectares $(10^4\ m^2)$ (ha)
$miles^2$	X 2.590	= $kilometres^2$ (km^2)

VOLUME

$inches^3$	X 16387	= $millimetres^3$ (mm^3)
$inches^3$	X 16.387	= $centimetres^3$ (cm^3)
$inches^3$	X 0.01639	= litres (l)
quarts	X 0.94635	= litres (l)
gallons	X 3.7854	= litres (l)
$feet^3$	X 28.317	= litres (l)
$feet^3$	X 0.02832	= $metres^3$ (m^3)
fluid oz	X 29.60	= millilitres (ml)
$yards^3$	X 0.7646	= $metres^3$ (m^3)

MASS

ounces (av)	X 28.35	= grams (g)
pounds (av)	X 0.4536	= kilograms (kg)
tons (2000 lb)	X 907.18	= kilograms (kg)
tons (2000 lb)	X 0.90718	= metric tons (t)

FORCE

ounces - f (av)	X 0.278	= newtons (N)
pounds - f (av)	X 4.448	= newtons (N)
kilograms - f	X 9.807	= newtons (N)

ACCELERATION

$feet/sec^2$	X 0.3048	= $metres/sec^2$ (m/S^2)
$inches/sec^2$	X 0.0254	= $metres/sec^2$ (m/s^2)

ENERGY OR WORK (watt-second - joule - newton-metre)

foot-pounds	X 1.3558	= joules (j)
calories	X 4.187	= joules (j)
Btu	X 1055	= joules (j)
watt-hours	X 3500	= joules (j)
kilowatt - hrs	X 3.600	= megajoules (MJ)

FUEL ECONOMY AND FUEL CONSUMPTION

miles/gal	X 0.42514	= kilometres/litre (km/l)

Note:
235.2/(mi/gal) = litres/100km
235.2/(litres/100 km) = mi/gal

LIGHT

footcandles	X 10.76	= $lumens/metre^2$ (lm/m^2)

PRESSURE OR STRESS (newton/sq metre - pascal)

inches HG (60 F)	X 3.377	= kilopascals (kPa)
pounds/sq in	X 6.895	= kilopascals (kPa)
inches H2O (60° F)	X 0.2488	= kilopascals (kPa)
bars	X 100	= kilopascals (kPa)
pounds/sq ft	X 47.88	= pascals (Pa)

POWER

horsepower	X 0.746	= kilowatts (kW)
ft-lbf/min	X 0.0226	= watts (W)

TORQUE

pound-inches	X 0.11299	= newton-metres (N·m)
pound-feet	X 1.3558	= newton-metres (N·m)

VELOCITY

miles/hour	X 1.6093	= kilometres/hour (km/h)
feet/sec	X 0.3048	= metres/sec (m/s)
kilometres/hr	X 0.27778	= metres/sec (m/s)
miles/hour	X 0.4470	= metres/sec (m/s)

TEMPERATURE

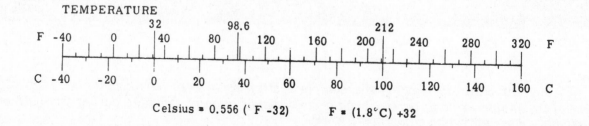

Celsius = 0.556 (°F -32)   F = (1.8°C) +32

DRILL SIZE CONVERSION CHART

SHOWING MILLIMETER SIZES, FRACTIONAL AND DECIMAL INCH SIZES AND NUMBER DRILL SIZES

Milli-Meter	Dec. Equiv.	Frac-tional	Num-ber
.1	.0039		
.15	.0059		
.2	.0079		
.25	.0098		
.3	.0118		
....	.0135		80
.35	.0138		
....	.0415		79
.39	.0156	1/64	....
.4	.0157		
....	.0160		78
.45	.0177		
....	.0180		77
.5	.0197		
....	.0200		76
....	.0210		75
.55	.0217		
....	.0225		74
.6	.0236		
....	.0240		73
....	.0250		72
.65	.0256		
....	.0260		71
....	.0280		70
.7	.0276		
....	.0292		69
.75	.0295		
....	.0310		68
.79	.0312	1/32	....
.8	.0315		
....	.0320		67
....	.0330		66
.85	.0335		
....	.0350		65
.9	.0354		
....	.0360		64
....	.0370		63
.95	.0374		
....	.0380		62
....	.0390		61
1.0	.0394		
....	.0400		60
....	.0410		59
1.05	.0413		
....	.0420		58
....	.0430		57
1.1	.0433		
1.15	.0452		
....	.0465		56
1.19	.0469	3/64	...
1.2	.0472		
1.25	.0492		
1.3	.0512		
....	.0520		55
1.35	.0513		
....	.0550		54
1.4	.0551		
1.45	.0570		
1.5	.0591		
....	.0595		53
1.55	.0610		
1.59	.0625	1/16	....
1.6	.0629		
....	.0635		52
1.65	.0649		
1.7	.0669		
....	.0670		51
1.75	.0689		
....	.0700		50
1.8	.0709		
1.85	.0728		
....	.0730		49
1.9	.0748		
....	.0760		48
1.95	.0767		
1.98	.0781	5/64	....
....	.0785		47
2.0	.0787		
2.05	.0807		
....	.0810		46
....	.0820		45
2.1	.0827		
2.15	.0846		
....	.0860		44
2.2	.0866		
2.25	.0855		
....	.0890		43
2.3	.0905		
2.35	.0925		
....	.0935		42
2.38	.0937	3/32	....
2.4	.0945		
....	.0960		41
2.45	.0964		
....	.0980		40
2.5	.0984		
....	.0995		39
....	.1015		38
2.6	.1024		
....	.1040		37
2.7	.1063		
....	.1065		36
2.75	.1082		
2.78	.1094	7/64	....
....	.1100		35
2.8	.1102		
....	.1110		34
....	.1130		33
2.9	.1141		
....	.1160		32
3.0	.1181		
....	.1200		31
3.1	.1220		
3.18	.1250	1/8	....
3.2	.1260		
3.25	.1279		
....	.1285		30
3.3	.1299		
3.4	.1338		
....	.1360		29
3.5	.1378		
....	.1405		28
3.57	.1406	9/64	....
3.6	.1417		
....	.1440		27
3.7	.1457		
....	.1470		26
3.75	.1476		
....	.1495		25
3.8	.1496		
....	.1520		24
8.9	.1535		
....	.1540		23
3.97	.1562	5/32	....
....	.1570		22
4.0	.1575		
....	.1590		21
....	.1610		20
4.1	.1614		
4.2	.1654		
....	.1660		19
4.25	.1673		
4.3	.1693		
....	.1695		18
4.37	.1719	11/64	....
....	.1730		17
4.4	.1732		
....	.1770		16
4.5	.1771		
....	.1800		15
4.6	.1811		
....	.1820		14
4.7	.1850		13
4.75	.1870		
4.76	.1875	3/16	....
4.8	.1890		12
....	.1910		11
4.9	.1929		
....	.1935		10
....	.1960		9
5.0	.1968		
....	.1990		8
5.1	.2008		
....	.2010		7
5.16	.2031	13/64	....
....	.2040		6
5.2	.2047		
....	.2055		5
5.25	.2067		
5.3	.2086		
....	.2090		4
5.4	.2126		
....	.2130		3
5.5	.2165		
5.56	.2187	7/32	....
5.6	.2205		
....	.2210		2
5.7	.2244		
5.75	.2263		
....	.2280		1
5.8	.2283		
5.9	.2323		
....	.2340		A
5.95	.2344	15/64	....
6.0	.2362		
....	.2380		B
6.1	.2401		
....	.2420		C
6.2	.2441		
6.25	.2460		D
6.3	.2480		
6.35	.2500	1/4	E
6.4	.2520		
6.5	.2559		
....	.2570		F
6.6	.2598		
....	.2610		G
6.7	.2638		
6.75	.2657	16/64	....
6.75	.2657		
....	.2660		H
6.8	.2677		
6.9	.2716		
....	.2720		I
7.0	.2756		
....	.2770		J
7.1	.2795		
....	.2811		K
7.14	.2812	9/32	....
7.2	.2835		
7.25	.2854		
7.3	.2874		
....	.2900		L
7.4	.2913		
....	.2950		M
7.5	.2953		
7.54	.2968	19/64	....
7.6	.2992		
....	.3020		N
7.7	.3031		
7.75	.3051		
7.8	.3071		
7.9	.3110		
7.94	.3125	5/16	....
8.0	.3150		
....	.3160		O
8.1	.3189		
8.2	.3228		
....	.3230		P
8.25	.3248		
8.3	.3268		
8.33	.3281	21/64	....
8.4	.3307		
....	.3320		Q
8.5	.3346		
8.6	.3386		
....	.3390		R
8.7	.3425		
8.73	.3437	11/32	....
8.75	.3445		
8.8	.3465		
....	.3480		S
8.9	.3504		
9.0	.3543		
....	.3580		T
9.1	.3583		
9.13	.3594	23/64	....
9.2	.3622		
9.25	.3641		
9.3	.3661		
....	.3680		U
9.4	.3701		
9.5	.3740		
9.53	.3750	3/8	....
....	.3770		V
9.6	.3780		
9.7	.3819		
9.75	.3838		
9.8	.3858		
....	.3860		W
9.9	.3839		
9.92	.3906	25/64	....
10.0	.3937		
....	.3970		X
....	.4040		Y
10.32	.4062	13/32	....
....	.4130		Z
10.5	.4134		
10.72	.4219	27/64	
11.0	.4330		
11.11	.4375	7/16	
11.5	.4528		
11.51	.4531	29/64	
11.91	.4687	15/32	
12.0	.4724		
12.30	.4843	31/64	
12.5	.4921		
12.7	.5000	1/2	
13.0	.5118		
13.10	.5156	33/64	
13.49	.5312	17/32	
13.5	.5315		
13.89	.5469	35/64	
14.0	.5512		
14.29	.5624	9/16	
14.5	.5709		
14.68	.5781	37/64	
15.0	.5906		
15.08	.5937	19/32	
15.48	.6094	39/64	
15.5	.6102		
15.88	.6250	5/8	
16.0	.6299		
16.27	.6406	41/64	
16.5	.6496		
16.67	.6562	21/32	
17.0	.6693		
17.06	.6719	43/64	
17.46	.6875	11/16	
17.5	.6890		
17.86	.7031	45/64	
18.0	.7087		
18.26	.7187	23/32	
18.5	.7283		
18.65	.7344	47/64	
19.0	.7480		
19.05	.7500	3/4	
19.45	.7656	49/64	
19.5	.7677		
19.84	.7812	25/32	
20.0	.7874		
20.24	.7969	51/64	
20.5	.8071		
20.64	.8125	13/16	
21.0	.8268		
21.04	.8218	53/64	
21.43	.8437	27/32	
21.5	.8465		
21.83	.8594	55/64	
22.0	.8661		
22.23	.8750	7/8	
22.5	.8858		
22.62	.8906	57/64	
23.0	.9055		
23.02	.9062	29/32	
23.42	.9219	59/64	
23.5	.9252		
23.81	.9375	15/16	
24.0	.9449		
24.21	.9531	61/64	
24.5	.9646		
24.61	.9687	31/32	
25.0	.9843		
25.03	.9844	63/64	
25.4	1.0000	1	

TORQUE SPECIFICATIONS

Engine Model	110/120/140hp 2.5/3	470/485	MCM170/190MR 3.7L	185/205MR CID 262	MCM185 229 CID 4.3L	150/160hp 165hp	185/250/260hp MCM260MR 5.0/5.7/7.4L
Cylinder Head	90 - 100 ft lbs. 122 - 136 Nm	130 ft lbs. 176 Nm	130 ft lbs. 176 Nm	75 ft lbs. 102 Nm	65 ft lbs. 88 Nm	95 ft lbs. 128 Nm	①
Connecting Rod	30 - 35 ft lbs. 40 - 47 Nm	40 ft lbs. 54 Nm	40 ft lbs. 54 Nm	45 ft lbs. 61 Nm	45 ft lbs. 61 Nm	35 ft lbs. 47 Nm	②
Camshaft Thrust Plate	72 - 90 in lbs. 8 - 10 Nm	135 in lbs. 15Nm	115 in lbs. 13 Nm	- - -	- - -	80 in lbs. 9 Nm	- - -
Exhaust Manifold	20 - 25 ft lbs. 27 - 33 Nm	25 ft lbs. 33 Nm	25 ft lbs. 33 Nm	20 ft lbs. 27 Nm	20 ft lbs. 27 Nm	25 ft lbs. 33 Nm	20 ft lbs. 27 Nm
Intake Manifold	- - -	35 ft lbs. 47 Nm	25 ft lbs. 33 Nm	40 ft lbs. 54 Nm	30 ft lbs. 40 Nm	30 ft lbs. 40 Nm	25 - 31 ft lbs. 33 - 47 Nm
Flywheel/Coupler	60 - 65 ft lbs. 81 - 88 Nm	60 ft lbs. 81 Nm	30 ft lbs. 40 Nm	40 ft lbs. 54 Nm	40 ft lbs. 54 Nm	- - -	60 - 65 ft lbs. 81 - 88 Nm
Main Bearing Cap	60 - 70 ft lbs. 81 - 94 Nm	55 ft lbs. 74 Nm	55 ft lbs. 74 Nm	85 ft lbs. 115 Nm	70 ft lbs. 94 Nm	87 ft lbs. 115 Nm	③
Oil Filter	Hand Tight	Hand Tight	Hand Tight	Hand Tight	Hand Tight	Hand Tight	Hand Tight
Oil Pan 1/4"	72 - 90 in lbs. 8 - 10 Nm	- - -	- - -	80 in lbs. 9 Nm	80 in lbs. 9 Nm	- - -	72 - 96 in lbs. 8 - 10 Nm
Oil Pan 5/16"	120 -130 in lbs. 13 - 14 Nm	150 in lbs. 17 Nm	130 in lbs. 14 Nm	165 in lbs. 18 Nm	165 in lbs. 18 Nm	75 in lbs. 8 Nm	165 in lbs. 19 Nm
Oil Pan Drain	20 ft lbs. 27 Nm	20 ft lbs. 27 Nm	20 ft lbs. 27 Nm	20 ft lbs. 27 Nm	20 ft lbs. 27 Nm	20 ft lbs. 27 Nm	20 ft lbs. 27 Nm
Oil Pump to Block	115 in lbs. 13 Nm	35ft lbs. 47 Nm	25ft lbs. 34 Nm	65 ft lbs. 88 Nm	65 ft lbs. 88 Nm	115 in lbs. 13 Nm	65 ft lbs. 88 Nm
Oil Pump Cover	70 in lbs. 8 Nm	120 in lbs. 13 Nm	120 in lbs. 13 Nm	80 in lbs. 9 Nm	80 in lbs. 9 Nm	70 in lbs. 8 Nm	72 - 96 in lbs. 8 - 10 Nm
Rocker Arm Cover	55 in lbs. 6 Nm	90 in lbs. 10 Nm	90 in lbs. 10 Nm	60 in lbs. 6 Nm	45 in lbs. 5 Nm	45 in lbs. 5 Nm	50 in lbs. 5.5 Nm
Spark Plug	35 ft lbs. 47 Nm	15 ft lbs. 20 Nm	15 ft lbs. 20 Nm	15 ft lbs. 20 Nm	15 ft lbs. 20 Nm	15 ft lbs. 20 Nm	15 ft lbs. 20 Nm
Torsional Damper	- - -	- - -	- - -	70 ft lbs. 95 Nm	60 ft lbs. 81 Nm	- - -	85 ft lbs. 115 Nm

① 5.0/5.7L—65 ft. lbs. 88 Nm; 7.4L—80 ft. lbs. 108 Nm

② 5.0/5.7L—45 ft. lbs. 61 Nm; 7.4L—(standard) 50 ft. lbs. 68 Nm, (high performance) 65 ft. lbs. 88 Nm

③ 5.0/5.7L—75-85 ft. lbs. 102-115 Nm; 7.4L—110 ft. lbs. 110 Nm

TORQUE SPECIFICATIONS

ALL UNITS EXCEPT BRAVO
(Bravo Torque Values Given In Text)

LOCATION	THREAD	TORQUE VALUE American	TORQUE VALUE Metric
Upper Gear Housing			
Driveshaft housing/upper gear housing, nut (5)	7/16-20	35 ft-lbs	47 N.m
Bell housing/driveshaft housing, nut (6)	7/16-20	50 ft-lbs	68 N.m
Driveshaft housing top cover, screw (4)	3/8-16	20 ft-lbs	27 N.m
Universal joint shaft pinion nut (1)	5/8-16	85 ft-lbs	115 N.m
Universal joint cover retainer (1)	3-7/8-16	200 ft-lbs	271 N.m
Upper driveshaft preload	--	2-6 in.-lbs	3 - 8 N.m
Water pocket cover screw (4)	1/4-20	30-40 in.-lbs	40 - 54 N.m
Trim tab bolt (1)	--	180 in.-lbs	20 N.m
Lower Unit			
Gear housing/driveshaft housing screw (1)	3/8-16	28 ft-lbs	38 N.m
Gear case cover, retainer spool (w/o tab)	4-3/8-16	250 ft-lbs	339 N.m
Trim tab/gear housing screw (1)	7/16-14	180 in.-lbs	20 N.m
Gear housing pinion nut (1)	5/18-18	60-80 ft-lbs	81 - 108 N.m
Water pump housing stud, nut (plastic) (2)	1/4-28	25-30 in.lbs	3 - 3.5 N.m
Water pump housing stud, nut (plastic) (1)	5/16-24	35-40 in.-lbs	4 - 4.5 N.m
Water pump housing screw (plastic) (1)	1/4-20	15-20 in.-lbs	1.7 - 2.3 N.m
Propeller shaft nut (1)	3/4-16	35-45 ft-lbs	47 - 61 N.m

OTHER TORQUE VALUES
Following torque values are for screw sizes not listed elsewhere.

SCREW SIZE	TORQUE VALUE Ft. Lbs	TORQUE VALUE In. Lbs	TORQUE VALUE Metric
#6	---	7 - 10	0.8 N.m
#10	2 - 3	25 - 35	2.8 - 4.0 N.m
#12	3 - 4	35 - 45	4.0 - 5.1 N.m
1/4"	5 - 7	60 - 80	6.8 - 9.5 N.m
5/16"	10 - 12	120 - 140	13.6 - 18.8
3/8"	18 - 20	220 - 240	25 - 27 N.m

TUNE-UP ADJUSTMENTS
See Pages A-12 & A-13 for Special Tune-up Notes

MCM Model	Serial No.	Cu. In. Displ.	Spark Plug AC	Spark Plug Champ.	Spark Plug Auto.	Plug Gap	Point Gap
110	All	153	CR44N Note 1	RN6	404	0.035" 0.89mm	0.022" 0.56mm
120	3825578 and below	153	CR44N Note 1	RN6	404	0.035" 0.89mm	0.022" 0.56mm
120	3825579 and above	153	MR43T	RBL8	144	0.035" 0.89mm	0.022" 0.56mm
120 Alpha/MR	All	153	MR43T	RBL8	144	0.035" 0.89mm	0.022" 0.56mm
2.5L Alpha/MR	All	153	MR43T	RBL8	144	0.035" 0.89mm	0.022" 0.56mm
140 4-Cyl	3826282 and below	181	CR44N Note 1	RN6	404	0.035" 0.89mm	0.022" 0.56mm
140 4-Cyl	3826283 and above	181	MR43T	RBL8	144	0.035" 0.89 mm	0.022" 0.56 mm
3.0L Alpha/MR	All	181	MR43T	RV8C	---	0.035" 0.89mm	0.022" 0.56mm
3.0LX Alpha	All	181	MR43LTS	RS12YC	---	0.035" 0.89mm	Note 9
140 6-Cyl	All	194	CR44N Note 1	RN6	404	0.035" 0.89mm	.016 - .019" .41 - 48mm
150	All	230	CR44N Note 1	RN6	404	0.035" 0.89mm	.016 - .019" .41 - .48mm
160	All	250	CR44N Note 1	RN6	404	0.035" 0.89mm	0.016" 0.41mm
165 4-Cyl	All	224	MR42TS	RV9YC	---	0.035" 0.89mm	0.022" 0.56mm
165 6-Cyl	2771483 and below	250	CR44N Note 1	RN6	404	0.035" 0.89mm	.016 - .019" .41 - .48mm
165 6-Cyl	2771484 and above	250	MR43T	RBL8	144	0.035" 0.89mm	.0.16 - .019" .41 - .48mm
170	All	224	R42TS	RBL9Y	13	0.035" 0.89mm	0.022" 0.56mm

TUNE-UP ADJUSTMENTS

See Pages A-12 & A-13 for Special Tune-up Notes

Point Dwell	Timing	Fuel Pressure	Oil Pressure @ 2000 RPM	Idle RPM In Gear	WOT RPM
28 – 34°	8° BTDC	3.5 – 4.5 psi 24 – 31 kPa	30 – 60 psi 207 – 414 kPa	500 – 600	3900 – 4300
28 – 34°	6° BTDC	3.5 – 4.5 psi 24 – 31 kPa	30 – 60 psi 207 – 414 kPa	500 – 600	3900 – 4300
28 – 34°	8° BTDC	3 – 6 psi 21 – 41 kPa	30 – 60 psi 207 – 414 kPa	650 – 700	3900 – 4300
28 – 34°	8° BTDC	3 – 6 psi 21 – 41 kPa	30 – 60 psi 207 – 414 kPa	650 – 700	4200 – 4600
28 – 34°	8° BTDC	3 – 6 psi 21 – 41 kPa	30 – 60 psi 207 – 414 kPa	650 – 700	4200 – 4600
28 – 34°	6° BTDC	3 – 6 psi 21 – 41 kPa	30 – 60 psi 207 – 414 kPa	500 – 600	4200 – 4600
28 – 34°	6° BTDC	3 – 6 psi 21 – 41 kPa	30 – 60 psi 207 – 414 kPa	650 – 700	4200 – 4600
28 – 34°	6° BTDC	3 – 6 psi 21 – 41 kPa	30 – 60 psi 207 – 414 kPa	650 – 700	4200 – 4600
Note 9	8° BTDC	3 – 6 psi 21 – 41 kPa	30 – 60 psi 207 – 414 kPa	650 – 700	4400 – 4800
28 – 34°	10° BTDC	3.5 – 4.5 psi 24 – 31 kPa	35 psi 241 kPa	500 – 600	3700 – 4100
28 – 34°	6° BTDC	3.5 – 4.5 psi 24 – 31 kPa	35 psi 241 kPa	500 – 600	3900 – 4300
28 – 34°	6° BTDC	3.5 – 4.5 psi 24 – 31 kPa	30 – 55 psi 207 – 379 kPa	500 – 600	3900 – 4300
28 – 34°	4° BTDC	3 – 6 psi 21 – 41 kPa	30 – 60 psi 207 – 414 Kpa	650 – 700	3900 – 4300
28 – 34°	6° BTDC	3.5 – 4.5 psi 24 – 31 kPa	30 – 60 psi 207 – 414 kPa	500 – 600	3900 – 4300
28 – 34°	6° BTDC	3 – 6 psi 21 – 41 kPa	30 – 60 psi 207 – 414 kPa	650 – 700	3900 – 4300
28 – 34°	8° BTDC	3 – 6 psi 21 – 41 kPa	30 – 60 psi 207 – 414 kPa	650 – 700	3800 – 4200

TUNE-UP ADJUSTMENTS
See Pages A-12 & A-13 for Special Tune-up Notes

MCM Model	Serial No.	Cu. In. Displ.	AC	Spark Plug Champ.	Auto.	Plug Gap	Point Gap
170 Alpha/MR	All	224	R42TS	RBL9Y	13	0.035" 0.89mm	0.022" 0.56mm
3.7L Alpha/MR	All	224	R42TS	RBL9Y	13	0.035" 0.89mm	0.022" 0.56mm
470	All	224	R42TS	RBL9Y	13	0.035" 0.89mm	0.022" 0.56mm
175 Alpha/MR	All	262	MR43T	RV8C	144	0.035" 0.89mm	None
180 Alpha/MR	All	262	MR43T	RV8C	144	0.035" 0.89mm	None
185	All	229	MR43T	RBL8	144	0.035" 0.89mm	0.020" 0.51mm
185 Alpha/MR	All	262	MR43T	RV8C	144	0.035" 0.89mm	None
190 Alpha/MR	All	262	MR43T	RV8C	144	0.035" 0.89mm	None
205 Alpha/MR	All	262	MR43T	RV8C	144	0.035" 0.89mm	None
4.3L Alpha/MR	All	262	MR43T	RV8C	144	0.035" 0.89mm	None
485	All	224	R42TS	RBL9Y	13	0.035" 0.89mm	0.022" 0.56mm
488	All	224	R42TS	RBL9Y	13	0.035" 0.89mm	0.022" 0.56mm
888	4169596 and below	302	C83T	F10	115	0.030" 0.76mm	0.017" 0.43mm
888	4169597 and above	302	Note 2	Note 2	Note 2	0.030" 0.76mm	.016 -.019" .41 - .48mm
190 Alpha/MR	All	283	CR43K	RJ6	302	0.035" 0.89mm	0.016" 0.41mm
898	6203178 and below	305	MR44T	RBL8	144	0.035" 0.89mm	.016 - .019" .41 - .48mm

TUNE-UP ADJUSTMENTS
See Pages A-12 & A-13 for Special Tune-up Notes

Point Dwell	Timing	Fuel Pressure	Oil Pressure @ 2000 RPM	Idle RPM In Gear	WOT RPM
28 - 34°	8° BTDC	3 - 6 psi 21 - 41 Kpa	30 - 60 psi 207 - 414 kPa	650 - 700	4200 - 4600
28 - 34°	8° BTDC	3 - 6 psi 21 - 41 Kpa	30 - 60 psi 207 - 414 kPa	650 - 700	4200 - 4600
28 - 34°	8° BTDC	3 - 6 psi 21 - 41 Kpa	30 - 60 psi 207 - 414 kPa	650 - 700	4200 - 4600
None	8° BTDC	3 - 7 psi 21 - 48 kPa	30 - 55 psi 207 - 379 kPa	650 - 700	3800 - 4200
None	8° BTDC	3 - 7 psi 21 - 48 kPa	30 - 55 psi 207 - 379 kPa	650 - 700	4400 - 4800
36 - 41°	8° BTDC	3 - 7 psi 21 - 48 kPa	30 - 55 psi 207 - 379 kPa	650 - 700	4400 - 4800
None	8° BTDC	3 - 7 psi 21 - 48 kPa	30 - 55 psi 207 - 379 kPa	650 - 700	4400 - 4800
None	8° BTDC	3 - 7 psi 21 - 48 kPa	30 - 55 psi 207 - 379 kPa	650 - 700	4400 - 4800
None	8° BTDC	3 - 7 psi 21 - 48 kPa	30 - 55 psi 207 - 379 kPa	650 - 700	4400 - 4800
None	8° BTDC	3 - 7 psi 21 - 48 kPa	30 - 55 psi 207 - 379 kPa	650 - 700	4400 - 4800
28 - 34°	8° BTDC	3 - 6 psi 21 - 41 kPa	30 - 60 psi 207 - 414 kPa	650 - 700	4400 - 4800
28 - 34°	8° BTDC	3 - 6 psi 21 - 41 kPa	30 - 60 psi 207 - 414 kPa	650 - 700	4400 - 4800
26 - 31°	10° BTDC	3 - 7 psi 21 - 48 kPa	40 - 70 psi 276 - 483 kPa	550 - 600	3800 - 4200
28 - 31°	10° BTDC	3 - 7 psi 21 - 48 kPa	40 - 70 psi 276 - 483 kPa	650 - 700	3800 - 4200
28 - 32°	Note 3	5.2 - 6.5 psi 36 - 45 kPa	35 psi 241 kPa	550 - 600	3700 - 4100
26 - 31°	8° BTDC	3 - 7 psi 21 - 48 kPa	30 - 55 psi 207 - 379 kPa	650 - 700	3800 - 4200

TUNE-UP ADJUSTMENTS

See Pages A-12 & A-13 for Special Tune-up Notes

MCM Model	Serial No.	Cu. In. Displ.	AC	Spark Plug Champ.	Auto.	Plug Gap	Point Gap
898	6218462 and above	305	MR44T	RBL8	144	0.035" 0.89mm	None
200	All	292	CR44N Note 1	RN6	404	0.035" 0.89mm	0.016" 0.41mm
200 Alpha/MR	All	305	MR44T	RV8C	144	0.035" 0.89mm	None
215	All	302	C83T	F10	115	0.030" 0.76mm	0.017" 0.43mm
225	2278646 and below	327	C42-1	J4	N/A	0.035" 0.89mm	.016 - .019" .41 - .48mm
225	2278647 thru 3385720	327	CR43K	RJ6	302	0.035" 0.89mm	.016 - .019" .41 - .48mm
225	3385721 and above	302	C83T	F10	115	0.030" 0.76mm	.016 - .019" .41 - .48mm
228	6203671 and below	305	MR44T	RBL8	144	0.035" 0.89mm	.016 - .019" .41 - .48mm
228	6225267 and above	305	MR44T	RBL8	144	0.035" 0.89mm	None
228TR	All	305	MR44T	RBL8	144	0.035" 0.89mm	.016 - .019" .41 - 48mm
230 Alpha/MR	All	305	MR44T	RV8C	144	0.035" 0.89mm	None
5.0L Alpha/MR	All	305	MR44T	RV8C	144	0.035" 0.89mm	None
233	All	351	Note 2	Note 2	Note 2	0.030" 0.76mm	.016 - .019" .41 - .48mm
250	4707999 and below	327	V40K	J2J	2812	None	None
250	4708000 and above	350	MR43T	RBL8	144	0.035" 0.89mm	.016 - .019" .41 - .48mm
255	4175499 and below	351	C83T	F10	115	0.030" 0.77mm	.016 - .019" .41 - .48mm

TUNE-UP ADJUSTMENTS

See Pages A-12 & A-13 for Special Tune-up Notes

Point Dwell	Timing	Fuel Pressure	Oil Pressure @ 2000 RPM	Idle RPM In Gear	WOT RPM
None	8° BTDC	3 - 7 psi 21 - 48 kPa	30 - 55 psi 207 - 379 kPa	650 - 700	3800 - 4200
28 - 34°	8° BTDC	3 - 5 psi 21 - 34 kPa	30 - 55 psi 207 - 379 kPa	500 - 600	3900 - 4300
None	8° BTDC	3 - 7 psi 21 - 48 kPa	30 - 50 psi 207 - 379 kPa	650 - 700	4200 - 4600
26 - 31°	Note 4	3 - 7 psi 21 - 48 kPa	40 - 70 psi 276 - 483 kPa	550 - 600	3800 - 4200
28 - 31°	12° BTDC	3 - 7 psi 21 - 48 kPa	30 - 55 psi 207 - 379 kPa	550 - 600	3800 - 4200
28 - 31°	Note 3	3 - 7 psi 21 - 48 kPa	30 - 55 psi 207 - 379 kPa	550 - 600	3800 - 4200
28 - 31°	10° BTDC	3 - 7 psi 21 - 48 kPa	40 - 70 psi 276 - 483 kPa	550 - 600	3800 - 4200
26 - 31°	8° BTDC	3 - 7 psi 21 - 48 kPa	30 - 55 psi 207 - 379 kPa	650 - 700	4200 - 4600
None	8° BTDC	3 - 7 psi 21 - 48 kPa	30 - 55 psi 207 - 379 kPa	650 - 700	4200 - 4600
26 - 31°	8° BTDC	3 - 7 psi 21 - 48 kPa	30 - 55 psi 207 - 379 kPa	650 - 700	3800 - 4200
None	8° BTDC	3 - 7 psi 21 - 48 kPa	30 - 55 psi 207 - 379 kPa	650 - 700	4200 - 4600
None	8° BTDC	5.5 - 7 psi 38 - 48 kPa	30 - 55 psi 207 - 379 kPa	650 - 700	3800 - 4200
28 - 31°	10° BTDC	3 -7 psi 21 - 48 kPa	40 - 70 psi 276 - 483 kPa	Note 5	3800 - 4200
None	10° BTDC	3 - 7 psi 21 - 48 kPa	30 - 55 psi 207 - 379 kPa	550 - 600	3800 - 4200
28 - 31°	8° BTDC	3 - 7 psi 21 - 48 kPa	30 - 55 psi 207 - 379 kPa	650 - 700	4200 - 4600
28 - 31°	10° BTDC	3 - 7 psi 21 - 48 kPa	40 -70 psi 276 - 483 kPa	550 - 600	3800 - 4200

TUNE-UP ADJUSTMENTS

MCM Model	Serial No.	Cu. In. Displ.	AC	Spark Plug Champ.	Auto.	Plug Gap	Point Gap
255	4175500 and above	350	MR43T	RBL8	144	0.035" 0.89mm	.016 - .019" .41 - .48mm
260	6208787 and below	350	MR43T	RBL8	144	0.035" 0.89mm	.016 - .019" .41 - .48mm
260	6227757 and above	350	MR43T	RBL8	144	0.035" 0.89mm	None
260 Alpha/MR	All	350	MR43T	RBL8	144	0.035" 0.89mm	None
270	All	350	V40K Note 6	J2J Note 6	2812 Note 6	None	None
280	All	350	MR43T	RBL11Y	144	0.035" 0.89mm	.016 - .019" .41 - .48mm
320 EFI Bravo	All	350	MR43T	RV8C	144	0.032" 0.81mm	None
5.7L Bravo	All	350	MR43T	RV8C	144	0.035" 0.89mm	None
310	All	409	43N	N4	393	0.035" 0.89mm	.016 - .019" .41 - .48mm
325	2761141 and below	427	SV4XL	N19V	2812	None	None
325	3043030 and above	427	MR41T	BL3	143	0.035" 0.89mm	None
330	6083181 and below	454	MR43T	RBL8	144	0.035" 0.89mm	.016 - .019" .41 - .48mm
330	6083182 and above	454	MR43T	RBL8	144	0.035" 0.89mm	None
7.4L Bravo	All	454	MR43T	RV8C	144	0.035" 0.89mm	None

NOTE 1 For heavy-duty, high-speed applications, use AC-CR43N or Champion RN5.

NOTE 2 For 5/8", 14mm spark plug; use AC-MR43T, Champion RBL11Y, or Autolite 203, except on Model 888 with Serial Nos. 4615130 thru 4616217 and 4616267 and above. On these engines, use AC-R42TS, Champion RBL9Y, or Autolite 203.
For 13/16", 18mm plug; use AC-C83T, Champion RF10 or F10, or Autolite 115.

NOTE 3 Engines with distributors Nos. 1111076 and 1111249 are timed at 8° BTDC, all others at 12° BTDC.

NOTE 4 Engines with distributor N. C9FJ-12127-A or -B are timed at 12^oBTDC; with distributor No. D1JF-12127-JA or -KA at 10^oBTDC.

TUNE-UP ADJUSTMENTS

Point Dwell	Timing	Fuel Pressure	Oil Pressure @ 2000 RPM	Idle RPM In Gear	WOT RPM
26 – 31°	8° BTDC	3 – 7 psi 21 – 48 kPa	30 – 55 psi 207 – 379 kPa	650 – 700	3800 – 4200
26 – 31°	8° BTDC	3 – 7 psi 21 – 48 kPa	30 – 55 psi 207 – 379 kPa	650 – 700	4200 – 4600
None	8° BTDC	3 – 7 psi 21 – 48 kPa	30 –55 psi 207 – 379 kPa	650 – 700	4200 – 4600
None	8° BTDC	3 – 7 psi 21 – 48 kPa	30 –55 psi 207 – 379 kPa	650 – 700	4200 – 4600
None	10° BTDC	3 – 7 psi 21 – 48 kPa	30 – 55 psi 207 – 379 kPa	550 – 600	3800 – 4200
26 – 31°	6° BTDC	3 – 7 psi 21 – 48 kPa	30 – 55 psi 207 – 379 kPa	850 – 900	4600 – 5000
None	12° BTDC Note 7	Note 8	30 – 60 psi 207 – 414 kPa	750 – 800	4800 – 5200
None	8° BTDC	5.5 – 7.0 psi 38 – 48 kPa	30 – 55 psi 207 – 379 kPa	650 – 700	4200 – 4600
28 – 31°	12° BTDC	5.2 – 6.5 psi 36 – 45 kPa	35 psi 241 kPa	550 – 600	4500
None	10° BTDC	3 – 7 psi 21 –48 kPa	30 –70 psi 207 – 483 kPa	550 – 600	3800 – 4200
None	10° BTDC	3 – 7 psi 21 – 48 kPa	30 – 70 psi 207 – 483 kPa	550 – 600	3800 – 4200
26 – 31°	8° BTDC	3 – 7 psi 21 – 48 kPa	30 – 70 psi 207 – 483 kPa	650 – 700	4200 – 4600
None	8° BTDC	3 – 7 psi 21 – 48 kPa	30 – 70 psi 207 – 483 kPa	650 – 700	4200 – 4600
None	8° BTDC	5.7 – 7.0 psi 38 – 48 kPa	35 – 70 psi 241 – 483 kPa	350 – 700	4200 – 4600

NOTE 5 Serial No. 4173767 and below: 550 - 600 rpm; Serial No. 4173768 and above; 650 - 700 rpm.

NOTE 6 Engines equipped with service replacement heads with tapered spark plug seats, require spark plug AC-MR41T, Champion BL3, or Autolite 143.

NOTE 7 Identify and **GROUND** Red/Yellow lead at ECU. Timing adjustment will not be accurate if this lead is not grounded. Disconnect and protect end of lead from accidental grounding.

NOTE 8 Pressure must be 29 psi (269 kPa) above intake manifold pressure.

NOTE 9 DDIS (Digital Distributorless Ignition System) was considered standard on some 1990 4-cylinder models and as optional on others.

CARBURETOR SPECIFICATIONS

Engine Model	120/140	120/140 2.5/3.0L	470/898	165	470/485
Make & Model	Rochester 2GC	MerCarb 32mm	MerCarb 35mm	Rochester 2GC	Rochester 2GV
Float Level Note 1	21/32" 16.7mm	Note 2 ---	11/16" 17.5mm	21/32" 16.7mm	11/16" 17.5mm
Float Drop Note 3	1-3/4" 44.5mm	1-3/4" 44.5mm	1-3/4" 44.5mm	1-3/4" 44.5mm	1-3/4" 44.5mm Note 2
Pump Rod	7/8" 22.2mm	1-5/32" 29mm	1-5/32" 29mm	15/16" 23.8mm	1-5/32" 29mm
Pump Rod Hole Location	---	---	---	---	---
Accelerator Pump Note 4	---	---	---	---	---
Vacuum Break	---	---	---	---	---
Air Valve Spring Wind up	---	---	---	---	---
Choke Setting	Index Marks Aligned	Index Marks Aligned	Index Marks Aligned	Index Marks Aligned	Index Marks Aligned
Choke Unloader	5/64" 2.0mm	5/64" 2.0mm	5/64" 2.0mm	5/64" 2.0mm	5/64" 2.0mm
Main Jet	---	0.057" 1.45mm	0.065" 1.65mm	---	---
Preliminary Idle Mixture Setting	1-1/4 Turn	1-1/4 Turn	1-1/4 Turn	1-1/4 Turn	1-1/4 Turn

Note 1   Fuel inlet needle is spring loaded. Before checking float level, raise float and allow it to fall by its own weight. Do not force float downward by hand.

Note 2   With spring loaded needle, float level is 5/8" (15.8mm). With solid needle, float level is 1/4" (6.3mm).

Note 3   Float drop measured from air horn (with gasket in place) to bottommost part of float.

Note 4   Accelerator pump measurement taken from flame arrestor mounting surface to pump stem with throttle plate fully closed.

CARBURETOR SPECIFICATIONS

898 200MR	488	165/170MR 185/200MR 3.7/5.0L	185	205/230/260MR 4.3/5.0L 5.7/7.4L	180MR 190MR
Rochester 2GV	Rochester 4MV	MerCarb 35mm	Rochester 4MV	Rochester 4MV	Rochester 4MC
19/32" 15.1mm	1/4" 6.4mm	5/8" 15.8mm	15/64" 5.9mm	15/64" 5.9mm	Note 6
1-3/4" 44.5mm	---	1-3/4" 44.5mm	---	---	----
1-5/32" 29mm	---	1-5/32" 29mm	---	---	----
---	Inner	---	Inner	Inner	Inner
---	23/64" 9.1mm	---	23/64" 9.1mm	23/64" 9.1mm	23/64" 9.1mm
---	13/16" 4.8mm	---	13/16" 4.8mm	5/64" 2.0mm	13/16" 4.8mm
---	1/4 Turn	---	1/4 Turn	1/4 Turn	1/4 Turn
Index Marks Aligned	Index Marks Aligned	Index Marks Aligned	--- Aligned	Note 5 Aligned	Note 7
5/64" 2.0mm	---	5/64" 2.0mm	---	---	----
---	0.066" 1.68mm	0.064" 1.65mm	0.066" 1.68mm	0.066" 1.68mm	0.066" 1.68mm
1-1/4 Turn	2-3 Turns	1-1/4 Turn	2-3 Turns	2-3 Turns	2-3 Turns

Note 5   Choke coil rod adjustment performed with choke valve completely closed, choke rod in bottom of choke lever slot, and choke coil rod pushed down to end of travel.

Note 6   With solid fuel inlet needle: float level is 3/8" (9mm). With spring loaded inlet needle: Float lever must just make contact with needle ball.

Note 7   Position index mark on cover 1/4" (6.4mm) to the right of leanest mark on choke housing.

ENGINE SPECIFICATIONS

EARLY MODEL LATE MODEL	MCM888/225S V8	MCM223 V8	MCM250/260 MCM260/320MR V8
ENGINE BLOCK			
Displacement	302 CID 4.95 Litre	351 CID 5.75 Litre	350 CID 5.74 Litre
Rated H.P.	888-188/225S-225	233	250-250/260-260
Bore & Stroke	4.00 x 3.00" 101.6 x 76.2mm	4.00 x 3.50" 101.6 x 88.9mm	4.00 x 3.48" 101.6 x 88.39mm
Compression Ratio	8:1	8:1	8.7:1
Numbering System (front to rear)			
Port Bank	5-6-7-8	5-6-7-8	1-3-5-7
Starboard Bank	1-2-3-4	1-2-3-4	2-4-6-8
Firing Order	1-3-7-2-6-5-4-8	1-3-7-2-6-5-4-8	1-8-4-3-6-5-7-2
Cylinder Bore			
Diameter	4.0004 - 4.0040" 101.610 - 101.702mm	4.0004 - 4.0036" 101.610 - 101.691mm	3.9995 - 4.0025" 101.587 - 101.663mm
Out-of-Round			
Desired	0.001" Max 0.0254mm Max	0.001" Max 0.0254mm Max	0.001" Max 0.0254mm Max
Allowable	0.005" Max 0.127mm Max	0.005" Max 0.127mm Max	0.002" Max 0.051mm Max
Taper			
Desired			
Thrust Side	0.0005" Max 0.0127mm	0.0005" Max 0.0127mm	0.0005" Max 0.0127mm
Relief Side	0.001" Max 0.0254mm	0.001" Max 0.0254mm	0.001" Max 0.0254mm
Allowable	0.005" Max 0.127mm	0.005" Max 0.127mm	0.005" Max 0.127mm
Piston			
Clearance			
Desired	0.0018 - 0.0026" 0.0457 - 0.0660mm	0.0018 - 0.0026" 0.0457 - 0.0660mm	0.0007 - 0.0017" 0.0178 - 0.0432mm
Allowable	0.0018 - 0.0026" 0.0457 - 0.0660mm	0.0018 - 0.0026" 0.0457 - 0.0660mm	0.0027" Max 0.0686mm Max
Piston Ring - Compression			
Groove Clearance			
Desired Top & 2nd	.002 - .004" 0.0508 - 0.1016mm	.002 - .004" 0.0508 - 0.1016mm	0.0012 - 0.0032" 0.0305 - 0.0813mm
Allowable	0.002 - 0.004" 0.0508 - 0.1016mm	0.002 - 0.004" 0.0508 - 0.1016mm	0.004" Max 0.1016mm Max
Gap			
Desired - Top	0.010 - 0.020" 0.254 - 0.508mm	0.010 - 0.020" 0.254 - 0.508mm	0.010 - 0.023" 0.254 - 0.584
Desired - 2nd	0.010-0.020" 0.254 - 0.508mm	0.010 - 0.020" 0.254 - 0.508	0.010 - 0.023mm 0.254 - 0.584mm
Allowable	0.010 - 0.020" 0.254 - 0.508mm	0.010 - 0.020" 0.254 - 0.508mm	0.010 - 0.024" 0.254 - 0.610mm

ENGINE SPECIFICATIONS

EARLY MODEL LATE MODEL	MCM888/225S V8	MCM223 V8	MCM250/260 MCM260/320MR V8
ENGINE BLOCK (Cont.)			
Piston Ring - Oil			
Groove Clearance			
Desired	Snug	Snug	0.002 - 0.007" 0.051 - 0.178mm
Allowable	Snug	Snug	0.008" Max 0.203mm Max
Gap			
Desired	0.015 - 0.069" 0.381 - 1.753mm	0.015 - 0.055" 0.381 - 1.397mm	0.015 - 0.055" 0.381 - 1.397mm
Allowable	0.015 - 0.069" 0.381 - 1.753mm	0.015 - 0.055 0.381 - 1.397mm	0.015 - 0.065" 0.381 - 1.651mm
Piston Pin			
Diameter	0.9119 - 0.9124" 23.162 - 23.175mm	0.9119 - 0.9124" 23.162 - 23.175mm	0.9270 - 0.9273" 23.546 - 23.553mm
Clearance			
Desired	0.0002 - 0.0004" 0.0051 - 0.0102mm	0.0003 - 0.0005" 0.0076 - 0.0.0127MM	0.0002 - 0.0004" 0.0051 - 0.0102mm
Allowable	0.0008" Max 0.0203mm Max	0.0008" Max 0.0203mm Max	0.0010" 0.0254mm Max
Fit in Rod			
Interference	0.0008-0.0016" 0.0203-0.0406mm	0.0008 - 0.0016" 0.0203 - 0.0406mm	0.0008 - 0.0016" 0.0203 - 0.0406mm
Crankshaft			
End Play	0.004 - 0.008" 0.1016 - 0.2032mm	0.004 - 0.008" 0.1016 - 0.2032mm	0.002 - 0.006" 0.0051 - 0.1524mm
Main Journal			
Diameter	2.2482 - 2.2249" 57.104 - 56.512mm	2.9994 - 3.0002" 76.185 - 76.205mm	See Note 1 Below
Taper			
Desired	0.0003" Max 0.008mm	0.0003" Max 0.008mm	0.0002" Max 0.005mm
Allowable	0.0003" 0.008mm	0.0003" 0.008mm	0.0010" 0.025mm
Out-of-Round			
Desired	0.0004" 0.0102mm	0.0004" 0.0102mm	0.0002" 0.0051mm
Allowable	0.0004" 0.0102mm	0.0004" 0.0102mm	0.0010" Max 0.0254mm Max
Main Bearing Clearance			
Desired			
No. 1	0.0001 - 0.0018"	0.0008 - 0.0026"	0.0008 - 0.0020"
No. 2, 3, 4	0.0005 - 0.0024"	0.0008 - 0.0026"	0.0011 - 0.0020"
No. 5	0.0029 - 0.0045"	0.0008 - 0.0026"	0.0017 - 0.0032"
Allowable	Same as desired	Same as desired	Same as desired

Continued

Note 1 No.1 2.4484-2.4493"; No.2, 3, & 4 2.4481-2.4490"; No.5 2.4479-2.4488"

ENGINE SPECIFICATIONS

EARLY MODEL LATE MODEL	MCM888/225S V8	MCM223 V8	MCM250/260 MCM260/320MR V8
ENGINE BLOCK (Cont.)			
Crankpin			
Diameter	2.1228 - 2.1236" 53.919 - 53.939mm	2.3103 - 2.3111" 58.682 - 58.702mm	2.199 - 2.200 55.855 - 55.880mm
Taper			
Desired	0.0004" 0.0102mm	0.0004" 0.0102mm	0.0003" 0.0076mm
Allowable	.0004" 0.0102mm	0.0004" 0.0102mm	0.0010" 0.0254mm
Out-of-Round			
Desired	0.0004" 0.0102mm	0.0004" 0.0102mm	0.0002" 0.0051mm
Allowable	0.0004" 0.0102mm	0.0004" 0.0102mm	0.0010" 0.0254mm
Rod Bearing Clearance			
Desired	0.0010 - 0.0015" 0.0254 - 0.0381mm	0.0010 - 0.0015" 0.0254 - 0.0381mm	0.0013 - 0.0035" 0.0330 - 0.0889mm
Allowable	0.0008 - 0.0026" 0.0203 - 0.0660mm	0.0008 - 0.0026" 0.0203 - 0.0660mm	0.0035" Max 0.0889mm Max
Connecting Rod Journal			
Diameter	---	---	---
Taper			
Desired	0.0005" Max 0.0127mm Max	0.0005" Max 0.0127mm Max	0.0005" Max 0.0127mm Max
Allowable	0.001" Max 0.0254mm Max	0.001" Max 0.0254mm Max	0.001" Max 0.0254mm Max
Out-of-Round			
Desired	0.0005" Max 0.0127mm Max	0.0005" Max 0.0127mm Max	0.0005" Max 0.0127mm Max
Allowable	0.001" Max 0.0254mm Max	0.001" Max 0.0254mm Max	0.001" Max 0.0254mm Max
Rod Side Clearance	0.010 - 0.020" 0.254 - 0.508mm	0.010" - 0.020" 0.254 - 0.508mm	0.008 - 0.014" 0.203 - 0.356mm
Camshaft			
End Play	N/A	N/A	N/A
Lobe Lift			
Intake	0.260" Max 6.604mm Max	0.278" Max 7.061mm Max	0.263" Max 6.680mm Max
Exhaust	0.278" Max 7.061mm Max	0.283" Max 7.188 mm Max	0.269" Max 6.833mm Max
Journal			
Diameter	See Note 1 Below	See Note 1 Below	1.8682 - 1.8692" 47.452 - 47.478mm
Out-of-Round	0.001" Max 0.0254mm Max	0.001" Max 0.0254mm Max	0.001" Max 0.0254mm Max

Continued

Note 1   No.1 2.805-2.815"   No.2 2.0655-2.0665"   No.3 2.0505-2.0515"   No.4 2.0355-2.0365"
No.5 2.0205-2.0215"

ENGINE SPECIFICATIONS

EARLY MODEL LATE MODEL	MCM888/225S V8	MCM223 V8	MCM250/260 MCM260/320MR V8
CYLINDER HEAD			
Gasket Surface Flatness			
Any 6" (15.24cm) area	0.003" 0.076mm	0.003" 0.076mm	0.003" 0.076mm
Overall Max.	0.007" 0.178mm	0.007" 0.178mm	0.007" 0.178mm
Lifter Type	Hydraulic	Hydraulic	Hydraulic
Rocker Arm Ratio	1.6:1	1.6:1	1.50:1
Valve Lash			
Intake and Exhaust	Fixed	Fixed	3/4 turn down from Zero lash
Face Angle			
Intake & Exhaust	44°	44°	45°
Seat Angle			
Intake & Exhaust	45°	45°	46°
Seat Runout - Maximum			
Intake & Exhaust	0.0015" 0.0381mm	0.0015" 0.0381mm	0.002" 0.0508mm
Seat Width			
Intake	0.0625 - 0.0781" 1.588 - 1.984mm	0.0625 - 0.0781" 1.588 - 1.984mm	0.0312 - 0.0625" 0.792 - 1.588mm
Exhaust	0.0625 - 0.0781" 1.588 - 1.984mm	0.0625 - 0.0781" 1.588 - 1.984mm	0.0625 - 0.0937" 1.588 - 2.3800mm
Stem Clearance			
Intake			
Desired	0.0010 - 0.0027" 0.0254 - 0.0686mm	0.0010 - 0.0027" 0.0254 - 0.0686mm	0.0010 - 0.0027" 0.0254 - 0.0686mm
Allowable	0.0010 - 0.0027" 0.0254 - 0.0686mm	0.0010 - 0.0027" 0.0254 - 0.0686mm	0.0037" Max 0.0940mm Max
Exhaust			
Desired	0.0015 - 0.0032" 0.0381 - 0.0813mm	0.0015 - 0.0032" 0.0381 - 0.0813mm	0.0010 - 0.0027" 0.0254 - 0.0686mm
Allowable	0.0015 - 0.0032" 0.0381 - 0.0813mm	0.0015 - 0.0032" 0.0381 - 0.0813mm	0.0047" Max 0.1194mm Max
Valve Spring			
Free Length	2.07" 52.58mm	2.07" 52.58mm	2.03" 51.56mm
Installed Height			
Intake	1-3/4 - 1-13/16" 44.45 - 46.04mm	1-3/4 - 13/16" 44.45 - 46.04mm	1-23/32" 43.65mm
Exhaust	1-3/4 - 1-13/16" 44.45 - 46.04mm	1-3/4 - 1-13/16" 44.45 - 46.04mm	1-19/32" 40.48mm
Damper			
Free Length	1-3/4 - 1-13/16" 44.45 - 46.04mm	1-3/4 - 1-13/16" 44.45 - 46.04mm	1-55/64" 47.24mm
Approx. No. of Coils	4	4	4

ENGINE SPECIFICATIONS

EARLY MODEL   150/160/165
                                6-Cyl. In-Line

ENGINE BLOCK

Displacement	250 CID
	4.1 Litre
Rated H.P.	Per Model 150/160/165
Bore & Stroke	3.87 x 3.53"
	98.30 x 89.66mm
Compression Ratio	8.5:1
Numbering System	1-2-3-4-5-6
Firing Order	1-5-3-6-2-4
Cylinder Bore	
Diameter	3.8745 - 3.8775"
	98.4123 - 98.4885mm
Out-of-Round	
Desired	0.0005" Max
	0.0127mm
Allowable	0.002" Max
	0.0508mm
Taper	
Desired	
Thrust Side	0.0005" Max
	0.0127mm
Relief Side	0.0005" Max
	0.0127mm
Allowable	0.0005" Max
	0.0127
Piston	
Clearance	
Desired	0.0005 - 0.0015"
	0.0127 - 0.0381mm
Allowable	0.0025"
	0.0635mm
Piston Ring - Compression	
Groove Clearance	
Desired Top & 2nd	.0012-.0027"
	0.0305-0.0686mm
Allowable	0.0012-0.0037"
	0.035-0.0940mm
Gap	
Desired - Top	0.010-0.020"
	0.254-0.508mm
Desired - 2nd	0.010-0.020"
	0.254-0.508mm
Allowable	0.010-0.030"
	0.254-0.762mm
Piston Ring - Oil	
Groove Clearance	
Desired	0.000-0.005"
	0.000-0.0127mm
Allowable	0.000-0.006"
	0.000-0.152mm

EARLY MODEL   150/160/165
                                6-Cyl. In-Line

ENGINE BLOCK (Cont.)

Piston Ring - Oil (Cont.)	
Gap	
Desired	0.015-0.055"
	0.381-1.397mm
Allowable	0.015-0.065"
	0.381-1.651mm
Piston Pin	
Diameter	0.9270-0.9273"
	23.5458-23.5534mm
Clearance	
Desired	0.0015-0.0025"
	0.0381-0.0635mm
Allowable	0.001" Max
	0.0254mm Max
Fit in Rod	
Interference	0.0008-0.0016"
	0.0203-0.0406mm
Crankshaft	
End Play	0.002-0.006
	0.0508-0.1524mm
Runout	0.0015" Max
	0.0381mm Max
Main Journal	
Diameter	2.2983-2.2993"
	58.3768-58.4022mm
Taper	
Desired	0.0002" Max
	0.0051mm Max
Allowable	0.0010" Max
	0.0254mm Max
Out-of-Round	
Desired	0.0002" Max
	0.0051mm Max
Allowable	0.0010" Max
	0.0254mm Max
Main Bearing Clearance	
Desired	0.0003 - 0.0029"
	0.0076 - 0.0737mm
Allowable	0.004"
	0.1016mm

Continued

ENGINE SPECIFICATIONS

EARLY MODEL	150/160/165 6-Cyl. In-Line	EARLY MODEL	150/160/165 6-Cyl. In-Line

ENGINE BLOCK (Cont.)

Crankpin			
Diameter	1.999-2.000" 50.7746-50.8000mm		
Taper			
Desired	0.0003" 0.0076mm		
Allowable	0.001" 0.0254mm		
Out-of-Round			
Desired	0.0002" 0.0051mm		
Allowable	0.001" 0.0254mm		
Rod Bearing Clearance			
Desired	0.0007-0.0027" 0.0178-0.0686mm		
Allowable	0.004" 0.1016mm		
Connecting Rod Journal			
Diameter	Not Available		
Taper			
Desired	0.0005"Max 0.0127mmMax		
Allowable	0.001" Max 0.025mmMax		
Out-of-Round			
Desired	0.0005" Max 0.0127mm Max		
Allowable	0.001" Max 0.025mm Max		
Rod Side Clearance	0.0085 - 0.0135" 0.2159 - 0.3429mm		
Camshaft			
End Play	0.002-0.005" 0.06-0.1mm		
Lobe Lift			
Intake - Model 150	0.1914" Max Note 1		
Exhaust	0.1914" Max Note 2		
Journal			
Diameter	1.8682-1.8692" 47.4523-47.4777mm		
Out-of-Round	0.001" Max 0.025mm Max		

CYLINDER HEAD

Gasket Surface Flatness	
Any 6" (15.24cm) area	0.003" 0.076mm
Overall Max.	0.007" 0.178mm
Lifter Type	Hydraulic
Rocker Arm Ratio	1.75:1
Valve Lash	
Intake/Exhaust	3/4 turn down from Zero lash
Face Angle	
Intake & Exhaust	45°
Seat Angle	
Intake & Exhaust	46°
Seat Runout	
Intake & Exhaust	0.002" Max 0.0508mm Max
Seat Width	
Intake	1/32 - 3/32" 0.792 - 2.380mm
Exhaust	1/16 - 3/32" 1.588 - 2.380mm
Stem Clearance	
Intake	
Desired	0.0010-0.0027" 0.0254-0.0686mm
Allowable	0.0037"Max 0.0940mmMax
Exhaust	
Desired	0.0015-0.0032" 0.0381-0.813mm
Allowable	0.0052"Max 0.1321mmMax
Valve Spring	
Free Length	1.90" 48.26mm
Installed Height (Both)	1.66" 42.07mm
Damper	
Free Length	N/A
Approx. No. of Coils	N/A

Note 1 Model 160/165 - 0.2297" Max
Note 2 Model 160/165 - 0.2297" Max

ENGINE SPECIFICATIONS

EARLY MODEL LATE MODEL	MCM470/485/488 165/170/180/190MR 4-Cyl. In-Line	MCM185R V6	MCM185MR/205MR V6
ENGINE BLOCK			
Displacement	224CID 3.7 Litre	229 CID 3.8 Litre	262 CID 4.3 Litre
Rated Horsepower	170/185/188/190	185	185/205
Bore & Stroke	4.36" x 3.75" 110.7mm x 95.3mm	3.736" x 3.480" 95mm x 88.4mm	4.000 x 3.480" 101.6mm x 88.4mm
Compression Ratio	8.8:1	Port Bank 1-3-5	Port Bank 1-3-5
Numbering System	1-2-3-4	Stbd Bank 2-4-6	Stbd Bank 2-4-6
Firing Order	1-3-4-2	1-6-5-4-3-2	1-6-5-4-3-2
Cylinder Bore			
Diameter	4.3602-4.3609" 110.749-110.767mm	3.736" 95mm	4.000" 101.6mm
Out-of-Round			
Desired	0.0005" Max 0.0127mm	0.001" Max 0.025mm	0.001" Max 0.025mm
Allowable	0.0015" Max 0.0381mm	0.002" Max 0.051mm	0.002" Max 0.051mm
Taper			
Desired			
Thrust Side	0.0005" Max 0.0127mm	0.0005" 0.0127mm	0.0005" 0.0127mm
Relief Side	0.0005" Max 0.0127mm	0.001" 0.025mm	0.001" 0.025mm
Allowable	0.003" Max 0.076mm	0.001" Max 0.025mm	0.001" Max 0.025mm
Piston			
Clearance			
Desired	0.0007"-0.0017" 0.0178-0.0432mm	0.0007"-0.0017" 0.0178-0.0432mm	0.0007"-0.0017" 0.0178-0.0432mm
Allowable	0.0027" 0.0686mm	0.0027" 0.0686mm	0.0027" 0.0686mm
Piston Ring - Compression			
Groove Clearance			
Desired Top & 2nd	0.0025-0.004" 0.07-0.1mm	.0012-.0032" 0.0305-0.0813mm	.0012-.0032" 0.0305-0.0813mm
Allowable	0.0025-0.004" 0.07-0.1mm	+0.001" +0.025mm	+0.001" +0.025mm
Gap			
Desired - Top	0.010-0.020" 0.25-0.5mm	0.010-0.020" 0.254-0.508mm	0.010-0.020" 0.254-0.508mm
Desired - 2nd	0.010-0.020" 0.25-0.5mm	0.010-0.025" 0.254-0.635mm	0.010-0.025" 0.254-0.635mm
Allowable	0.010-0.020" 0.25-0.5mm	+0.10" 0.254mm	+0.10" 0.254mm
Piston Ring - Oil			
Groove Clearance			
Desired	0.0011-0.0065" 0.03-0.15mm	0.002-0.007" 0.051-0.178mm	0.002-0.007" 0.051-0.178mm
Allowable	0.0011-0.0065" 0.03-0.15mm	+0.001" 0.025mm	+0.001" 0.025mm

ENGINE SPECIFICATIONS

EARLY MODEL LATE MODEL	MCM470/485/488 165/170/180/190MR 4-Cyl. In-Line	MCM185R V6	MCM185MR/205MR V6
ENGINE BLOCK (Cont.)			
Piston Ring - Oil (Cont.)			
Gap			
Desired	0.0011-0.0065" 0.03-0.15mm	0.015-0.055" 0.381-1.397mm	0.015-0.055" 0.381-1.397mm
Allowable	0.0011-0.0065" 0.03-0.15mm	0.010" 0.254mm	0.010" 0.254mm
Piston Pin			
Diameter	1.0399-1.0402" 26.413-26.421mm	0.9270-0.9273" 23.5458-23.5534mm	0.9270-0.9273" 23.5458-23.5534mm
Clearance			
Desired	0.0004-0.0006" 0.0102-0.0152mm	0.00025-0.00035" 0.00635-0.00889mm	0.00025-0.00035" 0.00635-0.00889mm
Allowable	0.0004-0.0006" 0.0102-0.0152mm	0.001" Max 0.025mm Max	0.001" Max 0.025mm Max
Fit in Rod			
Interference	0.0006-0.0016" 0.0152-0.0406mm	0.0008-0.0016" 0.0203-0.0406mm	0.0008-0.0016" 0.0203-0.0406mm
Crankshaft			
End Play	0.006-0.010" 0.15-0.25mm	0.002-0.006" 0.051-0.152mm	0.002-0.006" 0.051-0.152mm
Runout	0.0015" Max 0.0381mm Max	0.0015" Max 0.0381mm Max	0.0015" Max 0.0381mm Max
Main Journal			
Diameter	2.7472-2.7482		
No.1	2.7472-2.7482" 69.779-69.804mm	2.4484-2.4493" 62.1894-62.2122mm	2.4484-2.4493" 62.1894-62.2122mm
No.2 & 3	Same as No. 1	2.4481-2.4490" 62.1817-62.2046mm	2.4481-2.4490" 62.1817-62.2046mm
No. 4/5	Same as No. 1	2.4479-2.4488" 62.1767-62.1995mm	2.4479-2.4488" 62.1767-62.1995mm
Taper			
Desired	0.0002" Max 0.0051mm Max	0.0002" Max 0.0051mm Max	0.0002" Max 0.0051mm Max
Allowable	0.0005" Max 0.0127mm Max	0.001" Max 0.025mm Max	0.001" Max 0.025mm Max
Out-of-Round			
Desired	0.0002" Max 0.0051mm Max	0.0002" Max 0.0051mm Max	0.0002" Max 0.0051mm Max
Allowable	0.0005" Max 0.0127mm Max	0.001" Max 0.025mm Max	0.001" Max 0.025mm Max
Main Bearing Clearance			
No. 1			
Desired	0.0009-0.0035" 0.0229-0.0889mm	0.0008-0.0020" 0.0203-0.0508mm	0.0008-0.0020" 0.0203-0.0508mm
Allowable	0.001-0.0035" 0.0254-0.0889mm	0.001-0.0015" 0.0254-0.0381mm	0.001-0.0015" 0.0254-0.0381mm

ENGINE SPECIFICATIONS

EARLY MODEL LATE MODEL	MCM470/485/488 165/170/180/190MR 4-Cyl. In-Line	MCM185R V6	MCM185MR/205MR V6
ENGINE BLOCK (Cont.)			
Crankshaft (Cont.)			
Main Bearing Clearance (Cont.)			
No. 2 & 3			
Desired	0.0009-0.0035" 0.0229-0.0889mm	0.0011-0.0023" 0.0279-0.0584mm	0.0011-0.0023" 0.0279-0.0584mm
Allowable	0.001-0.0035" 0.0254-0.0889mm	0.001-0025" 0.0254-0.0635mm	0.001-0025" 0.0254-0.0635mm
No. 4/5			
Desired	0.0009-0.0035" 0.0229-0.0889mm	0.0017-0.0032" 0.0432-0.0813mm	0.0017-0.0032" 0.0432-0.0813mm
Allowable	0.001-0.0035" 0.0254-0.0889mm	0.0025-0.0035" 0.0635-0.0889mm	0.0025-0.0035" 0.0635-0.0889mm
Connecting Rod Journal			
Diameter	Not Available	2.0986-2.0998" 53.3095-53.3349mm	2.0986-2.0998" 53.3095-53.3349mm
Taper			
Desired	0.0005"Max 0.0127mmMax	0.0005" Max 0.0127mm	0.0005" Max 0.0127mm Max
Allowable	0.001" Max 0.025mmMax	0.001" Max 0.025mm Max	0.001" Max 0.025mm Max
Out-of-Round			
Desired	0.0005" Max 0.0127mm Max	0.0005" Max 0.0127mm Max	0.0005" Max 0.0127mm Max
Allowable	0.001" Max 0.025mm Max	0.001" Max 0.025mm Max	0.001" Max 0.025mm Max
Rod Bearing Clearance			
Desired	0.0009-0.0031" 0.0228-0.0787mm	0.0013-0.0035" 0.0330-0.0889mm	0.0013-0.0035" 0.0330-0.0889mm
Allowable	0.001-0.003" 0.03-0.07mm	0.003" Max 0.0762mm Max	0.003" Max 0.0762mm Max
Rod Side Clearance	0.005-0.012" 0.15-0.3mm	0.008-0.014" 0.203-0.356mm	0.006-0.014" 0.152-0.356mm
Camshaft			
End Play	0.002-0.005" 0.06-0.1mm	0.004-0.012" 0.102-0.304mm	0.004-0.012" 0.102-0.304mm
Lobe Lift		+0.002" ±0.051mm	+0.002" ±0.051mm
Intake	0.287" 7.2898mm	0.357" 9.0678mm	0.273" 6.9342mm
Exhaust	485 .310, others .290" 485 7.874, others 7.366mm	0.390" 9.906mm	0.273" 6.9342mm
Journal			
Diameter	2.1238-2.1248" 53.9445-53.9699mm	1.8682-1.8692" 47.452-47.478mm	1.8682-1.8692" 47.452-47.478mm
Out-of-Round	0.001" Max 0.025mm Max	0.001" Max 0.025mm Max	0.001" Max 0.025mm Max

ENGINE SPECIFICATIONS

EARLY MODEL LATE MODEL	MCM470/485/488 165/170/180/190MR 4-Cyl. In-Line	MCM185R V6	MCM185MR/205MR V6
ENGINE BLOCK (Cont.)			
Camshaft (Cont.)			
Timing Chain Deflection From Taut Position		3/8" 9.5mm	3/8" 9.5mm
Total	1.0" 25.4mm	3/4" 19.1mm	3/4" 19.1mm
CYLINDER HEAD			
Gasket Surface Flatness			
Any 6" (15.24cm) area	0.003" 0.076mm	0.003" 0.076mm	0.003" 0.076mm
Overall Max.	0.007" 0.178mm	0.007" 0.178mm	0.007" 0.178mm
Lifter Type	Hydraulic	Hydraulic	Hydraulic
Rocker Arm Ratio	1.73:1	1.50:1	1.50:1
Valve Lash			
Intake & Exhaust	1 turn down from Zero lash	1 turn down from Zero lash	1 turn down from Zero lash
Face Angle			
Intake	485 29.5°, others 44°	45°	45°
Exhaust	All 44.5°	45°	45°
Seat Angle			
Intake	485 30°, others 45°	46°	46°
Exhaust	All 45°	46°	46°
Seat Runout			
Intake & Exhaust	0.002" Max 0.051mm Max	0.002" Max 0.051mm Max	0.002" Max 0.051mm Max
Seat Width			
Intake	0.060-0.080" 1.524-2.032mm	0.0312-0.0625" 0.79-1.59mm	0.0312-0.0625" 0.79-1.59mm
Exhaust	0.060-0.080" 1.524-2.032mm	0.0625-0.0937" 1.59-2.38mm	0.0625-0.0937" 1.59-2.38mm
Stem Clearance			
Intake			
Desired	0.0010-0.0027" 0.0254-0.0686mm	0.0010-0.0027" 0.0254-0.0686mm	0.0010-0.0027" 0.0254-0.0686mm
Allowable	0.0037"Max 0.0940mmMax	0.0037" 0.0940mm	0.0037" 0.0940mm
Exhaust			
Desired	0.0010-0.0027" 0.0254-0.0686mm	0.0010-0.0027" 0.0254-0.0686mm	0.0010-0.0027" 0.0254-0.0686mm
Allowable	0.0052"Max 0.1321mmMax	0.0047" 0.1194mm	0.0047" 0.1194mm
Valve Spring			
Free Length	2.18" (2-3/16") 55mm	2.03" 51.6mm	2.03" 51.6mm
Installed Height (Both)	1.86"" 47mm	1.7187" 43.7mm	1.7187" 43.7mm
Damper	External		
Free Length		1.86" 47.24mm	1.86" 47.24mm
Approx. No. of Coils		4	4

ENGINE SPECIFICATIONS

EARLY MODEL LATE MODEL	MCM110/120 MCM120MR 4-Cyl. In-Line	MCM140 6-Cyl. In-Line	MCM140 MCM140MR 4-Cyl. In-Line
ENGINE BLOCK			
Displacement	153 CID	194 CID	181 CID
	2.5 Litre	3.18 Litre	3.0 Litre
Rated Horsepower	110/120	140	140
Bore & Stroke	3.87 x 3.25"	3.56 x 3.25"	4.00 x 3.60"
	98.298 x 82.550mm	90.424 x 82.550mm	101.60 x 91.44mm
Compression Ration	8.5:1	8.5:1	8.5:1
Numbering System	1-2-3-4	1-2-3-4-5-6	1-2-3-4
Firing Order	1-3-4-2	1-5-3-6-2-4	1-3-4-2
Cylinder Bore			
Diameter	3.8745-3.8775"	3.5630 - 3.5660"	3.9995 - 4.0075"
	98.4123-98.4885mm	90.5002 - 90.5764mm	101.5873 - 101.7905mm
Out-of-Round			
Desired	0.0005" Max	0.0005" Max	0.0005" Max
	0.0127mm	0.0127mm	0.0127mm
Allowable	0.002" Max	0.002" Max	0.002" Max
	0.0508mm	0.0508mm	0.0508mm
Taper			
Desired			
Thrust Side	0.0005" Max	0.0005" Max	0.0005" Max
	0.0127mm	0.0127mm	0.0127mm
Relief Side	0.0005" Max	0.0005" Max	0.0005" Max
	0.0127mm	0.0127mm	0.0127mm
Allowable	0.0005" Max	0.0005" Max	0.0005" Max
	0.0127	0.0127	0.0127
Piston			
Clearance			
Desired	0.0005-0.0015"	0.0006 - 0.0010"	0.0005 - 0.0015"
	0.0127-0.0381mm	0.0024 - 0.0040mm	0.0127 - 0.0381mm
Allowable	0.0025"	0.0025"	0.0025"
	0.0635mm	0.0635mm	0.0635mm
Piston Ring - Compression			
Groove Clearance			
Desired Top & 2nd	.0012-.0027"	.0012-.0027"	.0012-.0027"
	0.0305-0.0686mm	0.0305-0.0686mm	0.0305-0.0686mm
Allowable	0.0012-0.0037"	0.0012-0.0037"	0.0012-0.0037"
	0.035-0.0940mm	0.035-0.0940mm	0.035-0.0940mm
Gap			
Desired - Top	0.010-0.020"	0.010-0.020"	0.010-0.020"
	0.254-0.508mm	0.254-0.508mm	0.254-0.508mm
Desired - 2nd	0.010-0.020"	0.010-0.020"	0.010-0.020"
	0.254-0.508mm	0.254-0.508mm	0.254-0.508mm
Allowable	0.010-0.030"	0.010-0.030"	0.010-0.030"
	0.254-0.762mm	0.254-0.762mm	0.254-0.762mm
Piston Ring - Oil			
Groove Clearance			
Desired	0.000-0.005"	0.000-0.005"	0.000-0.005"
	0.000-0.0127mm	0.000-0.0127mm	0.000-0.0127mm
Allowable	0.000-0.006"	0.000-0.006"	0.000-0.006"
	0.000-0.152mm	0.000-0.152mm	0.000-0.152mm

ENGINE SPECIFICATIONS

EARLY MODEL LATE MODEL	MCM110/120 MCM120MR 4-Cyl. In-Line	MCM140 6-Cyl. In-Line	MCM140 MCM140MR 4-Cyl. In-Line
ENGINE BLOCK (Cont.)			
Piston Ring - Oil (Cont.)			
Gap			
Desired	0.015-0.055" 0.381-1.397mm	0.015-0.055" 0.381-1.397mm	0.015-0.055" 0.381-1.397mm
Allowable	0.015-0.065" 0.381-1.651mm	0.015-0.065" 0.381-1.651mm	0.015-0.065" 0.381-1.651mm
Piston Pin			
Diameter	0.9270-0.9273" 23.5458-23.5534mm	0.9270-0.9273" 23.5458-23.5534mm	0.9270-0.9273" 23.5458-23.5534mm
Clearance			
Desired	0.0015-0.0025" 0.0381-0.0635mm	0.0015-0.0025" 0.0381-0.0635mm	0.0015-0.0025" 0.0381-0.0635mm
Allowable	0.001" Max 0.0254mm Max	0.001" Max 0.0254mm Max	0.001" Max 0.0254mm Max
Fit in Rod			
Interference	0.0008-0.0016" 0.0203-0.0406mm	0.0008-0.0016" 0.0203-0.0406mm	0.0008-0.0016" 0.0203-0.0406mm
Crankshaft			
End Play	0.002-0.006 0.0508-0.1524mm	0.002-0.006 0.0508-0.1524mm	0.002-0.006 0.0508-0.1524mm
Runout	0.0015" Max 0.0381mm Max	0.0015" Max 0.0381mm Max	0.0015" Max 0.0381mm Max
Main Journal			
Diameter	2.2983-2.2993" 58.3768-58.4022mm	2.2983-2.2993" 58.3768-58.4022mm	2.2983-2.2993" 58.3768-58.4022mm
Taper			
Desired	0.0002" Max 0.0051mm Max	0.0002" Max 0.0051mm Max	0.0002" Max 0.0051mm Max
Allowable	0.0010" Max 0.0254mm Max	0.0010" Max 0.0254mm Max	0.0010" Max 0.0254mm Max
Out-of-Round			
Desired	0.0002" Max 0.0051mm Max	0.0002" Max 0.0051mm Max	0.0002" Max 0.0051mm Max
Allowable	0.0010" Max 0.0254mm Max	0.0010" Max 0.0254mm Max	0.0010" Max 0.0254mm Max
Main Bearing Clearance			
Desired	0.0003-0.0029" 0.0076-0.0737mm	0.0008 - 0.0029" 0.0203 - 0.0737mm	0.0003 - 0.0029" 0.0076 - 0.0737mm
Allowable	0.004" 0.1016mm	0.004" 0.1016mm	0.004" 0.1016mm

ENGINE SPECIFICATIONS

EARLY MODEL LATE MODEL	MCM110/120 MCM120MR 4-Cyl. In-Line	MCM140 6-Cyl. In-Line	MCM140 MCM140MR 4-Cyl. In-Line
ENGINE BLOCK (Cont.)			
Crankpin			
Diameter	1.999-2.000" 50.7746-50.8000mm	1.999-2.000" 50.7746-50.8000mm	2.099 - 2.100" 53.315 - 53.340mm
Taper			
Desired	0.0003" 0.0076mm	0.0003" 0.0076mm	0.0003" 0.0076mm
Allowable	0.001" 0.0254mm	0.001" 0.0254mm	0.001" 0.0254mm
Out-of-Round			
Desired	0.0002" 0.0051mm	0.0002" 0.0051mm	0.0002" 0.0051mm
Allowable	0.001" 0.0254mm	0.001" 0.0254mm	0.001" 0.0254mm
Rod Bearing Clearance			
Desired	0.0007-0.0027" 0.0178-0.0686mm	0.0007-0.0027" 0.0178-0.0686mm	0.0007 - 0.0028" 0.0178 - 0.0711mm
Allowable	0.004" 0.1016mm	0.004" 0.1016mm	0.004" 0.1016mm
Connecting Rod Journal			
Diameter	Not Available	Not Available	Not Available
Taper			
Desired	0.0005"Max 0.0127mmMax	0.0005"Max 0.0127mmMax	0.0005"Max 0.0127mmMax
Allowable	0.001" Max 0.025mmMax	0.001" Max 0.025mmMax	0.001" Max 0.025mmMax
Out-of-Round			
Desired	0.0005" Max 0.0127mm Max	0.0005" Max 0.0127mm Max	0.0005" Max 0.0127mm Max
Allowable	0.001" Max 0.025mm Max	0.001" Max 0.025mm Max	0.001" Max 0.025mm Max
Rod Side Clearance	0.0085-0.0135" 0.2159-0.3429mm	0.0008 - 0.0014" 0.0203 - 0.0356mm	0.0009 - 0.0013" 0.0229 - 0.0330mm
Camshaft			
End Play	0.002-0.005" 0.06-0.1mm	0.002-0.005" 0.06-0.1mm	0.002-0.005" 0.06-0.1mm
Lobe Lift			
Intake	Model 110 - 0.1914" Note 1	0.1914" Max 4.8616mm Max	0.2525" Max 6.4135mm Max
Exhaust	Model 110 - 0.1914" Note 2	0.1914" Max 4.8616mm Max	0.2525" Max 6.4135mm Max
Journal			
Diameter	1.8682-1.8692" 47.4523-47.4777mm	1.8682-1.8692" 47.4523-47.4777mm	1.8682-1.8692" 47.4523-47.4777mm
Out-of-Round	0.001" Max 0.025mm Max	0.001" Max 0.025mm Max	0.001" Max 0.025mm Max

Note 1 Model 120 - 0.2325" (5.9055mm)
Note 2 Model 120 - 0.2325" (5.9055mm)

ENGINE SPECIFICATIONS

EARLY MODEL LATE MODEL	MCM110/120 MCM120MR 4-Cyl. In-Line	MCM140 6-Cyl. In-Line	MCM140 MCM140MR 4-Cyl. In-Line
CYLINDER HEAD			
Gasket Surface Flatness			
Any 6" (15.24cm) area	0.003" 0.076mm	0.003" 0.076mm	0.003" 0.076mm
Overall Max.	0.007" 0.178mm	0.007" 0.178mm	0.007" 0.178mm
Lifter Type	Hydraulic	Hydraulic	Hydraulic
Rocker Arm Ratio	1.75:1	1.75:1	1.75:1
Valve Lash			
Intake/Exhaust	3/4 turn down from Zero lash	3/4 turn down from Zero lash	3/4 turn down from Zero lash
Face Angle			
Intake & Exhaust	45°	45°	45°
Seat Angle			
Intake & Exhaust	46°	46°	46°
Seat Runout			
Intake & Exhaust	0.002" Max 0.0508mm Max	0.002" Max 0.0508mm Max	0.002" Max 0.0508mm Max
Seat Width			
Intake	1/32 - 1/16" 0.7925-1.5875mm	1/32 - 1/16" 0.7925-1.5875mm	1/32 - 1/16" 0.7925-1.5875mm
Exhaust	1/16 - 3/32" 1.588 - 2.380mm	1/16 - 3/32" 1.588 - 2.380mm	1/16 - 3/32" 1.588 - 2.380mm
Stem Clearance			
Intake			
Desired	0.0010-0.0027" 0.0254-0.0686mm	0.0010-0.0027" 0.0254-0.0686mm	0.0010-0.0027" 0.0254-0.0686mm
Allowable	0.0037"Max 0.0940mmMax	0.0037"Max 0.0940mmMax	0.0037"Max 0.0940mmMax
Exhaust			
Desired	0.0015-0.0032" 0.0381-0.813mm	0.0015-0.0032" 0.0381-0.813mm	0.0015-0.0032" 0.0381-0.813mm
Allowable	0.0052"Max 0.1321mmMax	0.0052"Max 0.1321mmMax	0.0052"Max 0.1321mmMax
Valve Spring			
Free Length	2.08" 52.832mm	2.03" 51.562mm	N/A
Installed Height (Both)	1.66" 42.07mm	1.66" 42.07mm	N/A
Damper			
Free Length	1.94" 49.28	N/A	N/A
Approx. No. of Coils	4	N/A	N/A

ENGINE SPECIFICATIONS

EARLY MODEL LATE MODEL	MCM898/228 MCM200MR/230MR V8	MCM250/255/260/270 MCM260MR V8	330IITR/IITRS/340 MCM330TR V8
ENGINE BLOCK			
Displacement	305 CID	350 CID	454 CID
	5.0 Litre	5.74 Litre	7.44 Litre
Rated Horsepower	898 - 198	MCM250 - 250	330 - 330
	228 - 228	MCM255 - 255	330/330IITRS - 330
	MCM200MR - 200	MCM260MR - 260	340 - 330
	MCM230MR - 230	MCM270 - 270	MCM330TR - 330
Bore & Stroke	3.74 x 3.48"	4.000" x 3.480"	4.250 x 4.000"
	94.89mm x 88.39mm	101.60mm x 88.39mm	107.95 x 101.60mm
Numbering System	Port Bank 1-3-5-7	Port Bank 1-3-5-7	Port Bank 1-3-5-7
	Stbd Bank 2-4-6-8	Stbd Bank 2-4-6-8	Stbd Bank 2-4-6-8
Firing Order	1-8-4-3-6-5-7-2	1-8-4-3-6-5-7-2	1-8-4-3-6-5-7-2
Cylinder Bore			
Diameter	3.736"	3.9995-4.0025"	4.2495-4.2525"
	94.89mm	101.5873-101.6635mm	107.94 x 108.01mm
Out-of-Round			
Desired	0.001" Max	0.001" Max	0.001" Max
	0.025mm	0.025mm	0.025mm
Allowable	0.002" Max	0.002" Max	0.002" Max
	0.051mm	0.051mm	0.051mm
Taper			
Desired			
Thrust Side	0.0005"	0.0005"	0.0005"
	0.0127mm	0.0127mm	0.0127mm
Relief Side	0.001"	0.001"	0.001"
	0.025mm	0.025mm	0.025mm
Allowable	0.001" Max	0.001" Max	0.001" Max
	0.025mm	0.025mm	0.025mm
Piston			
Clearance			
Desired	0.0007"-0.0017"	0.0007"-0.0017"	0.0014-0.0024"
	0.0178-0.0432mm	0.0178-0.0432mm	0.0356-0.0610mm
Allowable	0.0027"	0.0027"	0.0027"
	0.0686mm	0.0686mm	0.0686mm
Piston Ring - Compression			
Groove Clearance			
Desired Top & 2nd	.0012-.0032"	.0012-.0032"	.0017-.0032"
	0.0305-0.0813mm	0.0305-0.0813mm	0.0432-0.0813mm
Allowable	+0.001"	+0.001"	0.0033" Max
	+0.025mm	+0.025mm	0.0838mm Max
Gap			
Desired - Top	0.010-0.020"	0.010-0.020"	0.010-0.020"
	0.254-0.508mm	0.254-0.508mm	0.254-0.508mm
Desired - 2nd	0.010-0.025"	0.010-0.025"	0.010-0.020"
	0.254-0.635mm	0.254-0.635mm	0.254-0.508mm
Allowable	0.010"	0.010"	0.010"
	0.254mm	0.254mm	0.254mm

ENGINE SPECIFICATIONS

EARLY MODEL LATE MODEL	MCM898/228 MCM200MR/230MR V8	MCM250/255/260/270 MCM260MR V8	330IITR/IITRS/340 MCM330TR V8
ENGINE BLOCK (Cont.)			
Piston Ring - Oil			
Groove Clearance			
Desired	0.002-0.007" 0.051-0.178mm	0.002-0.007" 0.051-0.178mm	0.005-0.0065" 0.127-0.165mm
Allowable	+0.001" 0.025mm	+0.001" 0.025mm	+0.001" 0.025mm
Gap			
Desired	0.015-0.055" 0.381-1.397mm	0.015-0.055" 0.381-1.397mm	0.015-0.055" 0.381-1.397mm
Allowable	0.010" 0.254mm	0.010" 0.254mm	0.010" 0.254mm
Piston Pin			
Diameter	0.9270-0.9273" 23.5458-23.5534mm	0.9270-0.9273" 23.5458-23.5534mm	0.9895-0.9898" 25.1333-25.1409mm
Clearance			
Desired	0.00025-0.00035" 0.00635-0.00889mm	0.00025-0.00035" 0.00635-0.00889mm	0.00025-0.00035" 0.00635-0.00889mm
Allowable	0.001" Max 0.025mm Max	0.001" Max 0.025mm Max	0.001" Max 0.025mm Max
Fit in Rod			
Interference	0.0008-0.0016" 0.0203-0.0406mm	0.0008-0.0016" 0.0203-0.0406mm	0.0008-0.0016" 0.0203-0.0406mm
Crankshaft			
End Play	0.003-0.007" 0.0762-0.1778mm	0.003-0.007" 0.0762-0.1778mm	0.006-0.010" 0.152-0.254mm
Runout	0.0015" Max 0.0381mm Max	0.0015" Max 0.0381mm Max	0.0015" Max 0.0381mm Max
Main Journal			
Diameter			
No. 1	2.4484-2.4493" 62.1894-62.2122mm	2.4484-2.4493" 62.1894-62.2122mm	2.7485-2.7494" 69.8119-69.8348mm
No. 2,3,4	2.4481-2.4490" 62.1817-62.2046mm	2.4481-2.4490" 62.1817-62.2046mm	2.7481-2.7490" 69.8017-69.8246mm
No. 5	2.4479-2.4488" 62.1767-62.1995mm	2.4479-2.4488" 62.1767-62.1995mm	2.7478-2.7488" 69.7941-69.8195mm
Taper			
Desired	0.0002" Max 0.0051mm Max	0.0002" Max 0.0051mm Max	0.0002" Max 0.0051mm Max
Allowable	0.001" Max 0.025mm Max	0.001" Max 0.025mm Max	0.001" Max 0.025mm Max
Out-of-Round			
Desired	0.0002" Max 0.0051mm Max	0.0002" Max 0.0051mm Max	0.0002" Max 0.0051mm Max
Allowable	0.001" Max 0.025mm Max	0.001" Max 0.025mm Max	0.001" Max 0.025mm Max
Main Bearing Clearance			
No. 1			
Desired	0.0008-0.0020" 0.0203-0.0508mm	0.0008-0.0020" 0.0203-0.0508mm	0.0013-0.0025" 0.0330-0.0635mm
Allowable	0.001-0.0015" 0.0254-0.0381mm	0.001-0.0015" 0.0254-0.0381mm	0.001-0.0015" 0.0254-0.0381mm

ENGINE SPECIFICATIONS

EARLY MODEL LATE MODEL	MCM898/228 MCM200MR/230MR V8	MCM250/255/260/270 MCM260MR V8	330IITR/IITRS/340 MCM330TR V8
ENGINE BLOCK (Cont.)			
Crankshaft (Cont.)			
Main Bearing Clearance (Cont.)			
No. 2,3,4			
Desired	0.0011-0.0023" 0.0279-0.0584mm	0.0011-0.0023" 0.0279-0.0584mm	0.0013-0.0025" 0.0330-0.0635mm
Allowable	0.001-0025" 0.0254-0.0635mm	0.001-0025" 0.0254-0.0635mm	0.001-0025" 0.0254-0.0635mm
No. 5			
Desired	0.0017-0.0032" 0.0432-0.0813mm	0.0017-0.0032" 0.0432-0.0813mm	0.0024-0.0040" 0.0610-0.1016mm
Allowable	0.0025-0.0035" 0.0635-0.0889mm	0.0025-0.0035" 0.0635-0.0889mm	0.0025-0.0035" 0.0635-0.0889mm
Connecting Rod Journal			
Diameter	2.0988-2.0998" 53.3095-53.3349mm	2.0988-2.0998" 53.3095-53.3349mm	2.1985-2.1995" 55.8419-55.8673mm
Taper			
Desired	0.0005" Max 0.0127mm Max	0.0005" Max 0.0127mm Max	0.0005" Max 0.0127mm Max
Allowable	0.001" Max 0.025mm Max	0.001" Max 0.025mm Max	0.001" Max 0.025mm Max
Out-of-Round			
Desired	0.0005" Max 0.0127mm Max	0.0005" Max 0.0127mm Max	0.0005" Max 0.0127mm Max
Allowable	0.001" Max 0.025mm Max	0.001" Max 0.025mm Max	0.001" Max 0.025mm Max
Rod Bearing Clearance			
Desired	0.0013-0.0035" 0.0330-0.0889mm	0.0013-0.0035" 0.0330-0.0889mm	0.0009-0.0025" 0.0229-0.0635mm
Allowable	0.003" Max 0.0762mm Max	0.003" Max 0.0762mm Max	0.003" Max 0.0762mm Max
Rod Side Clearance	0.008-0.014" 0.203-0.356mm	0.008-0.014" 0.203-0.356mm	0.013-0.023" 0.330-0.584mm
Camshaft			
End Play	N/A	N/A	0.001-0.005" 0.025-0.127mm
Lobe Lift	+0.002" +0.051mm	+0.002" +0.051mm	+0.002" +0.051mm
Intake	0.263" 6.680mm	0.263" 6.680mm	0.271" 6.883mm
Exhaust	0.269" 6.833mm	0.269" 6.833mm	0.282" 7.163mm
Journal			
Diameter	1.8682-1.8692" 47.452-47.478mm	1.8682-1.8692" 47.452-47.478mm	1.9482-1.9492" 49.484-49.510mm
Out-of-Round	0.001" Max 0.025mm Max	0.001" Max 0.025mm Max	0.001" Max 0.025mm Max
Runout	0.002" 0.051mm	0.002" 0.051mm	0.002" 0.051mm

ENGINE SPECIFICATIONS

EARLY MODEL LATE MODEL	MCM898/228 MCM200MR/230MR V8	MCM250/255/260/270 MCM260MR V8	330IITR/IITRS/340 MCM330TR V8
ENGINE BLOCK (Cont.)			
Camshaft (Cont.)			
Timing Chain Deflection			
L.H. Rotating Engines Only			
From Taut Position	3/8" 9.5mm	3/8" 9.5mm	3/8" 9.5mm
Total	3/4" 19.1mm	3/4" 19.1mm	3/4" 19.1mm
CYLINDER HEAD			
Gasket Surface Flatness			
Any 6" (15.24cm) area	0.003" 0.076mm	0.003" 0.076mm	0.003" 0.076mm
Overall Max.	0.007" 0.178mm	0.007" 0.178mm	0.007" 0.178mm
Lifter Type	Hydraulic	Hydraulic	Hydraulic
Rocker Arm Ratio	1.50:1	1.50:1	1.70:1
Valve Lash			
Intake & Exhaust	1 turn down from Zero lash	1 turn down from Zero lash	1 turn down from Zero lash
Face Angle			
Intake & Exhaust	45°	45°	45°
Seat Angle			
Intake & Exhaust	46°	46°	46°
Seat Runout			
Intake & Exhaust	0.002" Max 0.051mm Max	0.002" Max 0.051mm Max	0.002" Max 0.051mm Max
Seat Width			
Intake	0.0312-0.0625" 0.79-1.59mm	0.0312-0.0625" 0.79-1.59mm	0.0312-0.0625" 0.79-1.59mm
Exhaust	0.0625-0.0937" 1.59-2.38mm	0.0625-0.0937" 1.59-2.38mm	0.0625-0.0937" 1.59-2.38mm
Stem Clearance			
Intake			
Desired	0.0010-0.0027" 0.0254-0.0686mm	0.0010-0.0027" 0.0254-0.0686mm	0.0010-0.0027" 0.0254-0.0686mm
Allowable	0.0037" 0.0940mm	0.0037" 0.0940mm	0.0037" 0.0940mm
Exhaust			
Desired	0.0010-0.0027" 0.0254-0.0686mm	0.0010-0.0027" 0.0254-0.0686mm	0.0012-0.0029" 0.0305-0.0737mm
Allowable	0.0047" 0.1194mm	0.0047" 0.1194mm	0.0049" 0.1245mm
Valve Spring			2.12" (53.8mm)
Free Length			N/A
w/1 lavender stripe	2.03" 51.6mm	2.03" 51.6mm	
w/2 green stripes	1.91" 48.5mm	1.91" 48.5mm	N/A
Damper			
Free Length	1.86" 47.24mm	1.86" 47.24mm	1.86" 47.24mm
Approx. No. of Coils	4	4	4

ENGINE SPECIFICATIONS

All dimensions are in inches, unless stated otherwise.

ENGINE BLOCK	GMC 502 Cu. In.	ENGINE BLOCK	GMC 502 Cu. In.
GENERAL SPECS.		Crank Pin Diameter	2.199 - 2.200
Type - No. Cylinders	90° - V8	Taper	0.001 In. Max.
Valve Arrangement	In Head	Out of Round	0.001 In. Max.
Bore and Stroke -- inches	4.474 x 4.00	Main Bearing Clearance	
Displacement, Cu. In.	502	No. 1	0.002 In. Max.
Displacement, Litres	8.2	No. 2,3,4,5	0.0035 In. Max.
Cylinder Number - - Front to Rear		**CAMSHAFT**	
Starboard Bank	2-4-6-8	End "Play"	0.004-0.012"
Port Bank	1-3-5-7	Bearing Journal Diameter	1.9497 - 1.9487"
Firing Order	1-8-4-3-6-5-7-2	Runout	0.002" Max.
Compression Ratio	8.8:1	Lobe Lift =0.002 in	
Compression Pressure	125 psi*	Intake	0.234 In.
Water Temp. Control	Thermostat	Exhaust	0.253
Thermostat Open at	140° F	Journal Out of Round	0.001 In. Max
Alternater Belt Adjustment			
Give with finger pressure	1/4 - 3/8 Inch	**VALVE SYSTEM**	
Full Throttle	4600 - 5000 rpm	Lifter Type	Hydraulic
		Rocker Arm Ratio	1.7:1
* Cranking speed at W.O.T. (Wide Open Throttle)		Valve Lash Adjustment	
		Intake & Exhaust (Hot)	3/4 Turn Down from "Zero Lash"
PISTONS, RODS, & CRANKSHAFT			
Pistons		Valve Face Angle	45°
Clearance in Bore	0.0035 In Max	Valve Seat Angle	46°
Piston Rings - Compression		Stem to Guide Clearance	
Groove Clearance - Top	0.0017 - 0.0042 In.	Intake	0.0010 - 0.0037 In
Groove Clearance - 2nd	0.0017 - 0.0042 In.	Exhaust	0.0010 - 0.0047 In
Gap - Top	0.26 In.	Valve Seat Width	
Gap - 2nd	0.018 In.	Intake	1/32 - 1/16 In.
Piston Rings - Oil		Exhaust	1/16 - 3/32 In.
Groove Clearance	0.005 - 0.008 In. Max.	Valve Spring Free Length	2.26 In.
Gap	0.015 - 0.065 In.	Valve Spring Installed Length	1-51/64 In.+1/32
Piston Pins		Pounds @ Inches Closed	140-180 @ 1.88 In
Diameter	0.9895 In.	Pounds @ Inches Open	350-380 @ 1.38 In
Clearance	0.001 In. Max.	Assembled Height	
Fit in Rod	0.013 - 0.0021 In.		
Connecting Rods		**ENGINE LUBRICATION**	
Bearing Clearance	0.0035 In. Max.	Oil Pump Type	Gear
Side Clearance	0.015 - 0.021 In.	Crankcase Capacity	Approx. 6 Qts.
Crankshaft		With new filter	Approx. 7 Qts.
End "Play"	0.006 - 0.010 In.	Oil Filter	
Main Journal Diameter		Purolator P/N	PER-40
No. 1	2.7485 - 2.7494 In.	AC P/N	PF-35
No. 2,3,4	2.7481 - 2.7490 In.	Oil Grade Recommended	
No. 5	2.7478 - 2.7488 In.	32°F and Above	SAE 30
Taper	0.001 In. Max.	0°F to 32°F	SAE 20W-20
Out of Round	0.001 In. Max.	Below 0°F	SAE 10W

Note: The manufacturer does not recommend the use of multi-viscosity engine oils.

UPPER DRIVESHAFT GEAR RATIO

MerCruiser Model	Driveshaft Housing Teeth No. Pinion Gear/ Driven Gear	Overall Drive Gear Ratio	Driven Gear Shimming Tool No. and Face	Gear Oil Capacity Ounces
I Early 110/140	21/28	2:1	91-3348832	
IA EZ and Non-EZ Later 110	21/23	1.84:1	91-30384 "A"	32
	20/24	1.98:1	91-45877 "120" or "B"or 91-60526 "Y"	32
IB EZ and Non-EZ 120	20/24	2.01:1	91-36384 "B"	32
		1.98:1	91-45877 "120" or "B" or 91-60526"Y"	
IC EZ and Non-EZ 150	24/24	1.68:1	91-36384 "C"	32
		1.68:1	91-45877 "160" or "C" or 91-60526 "X"	
120/140 S/N 2791956 and below) Note 1	20/24	1.98:1	91-45877 "120" or "B" or 91-60526 "Y"	32
120/120R/ 120MR/140/ 140R/140MR S/N 2791957 and above) Note 1	20/24	1.98:1	91-60526 "Y"	32
160	24/24	1.68:1	91-45877 "160" or "C" or 91-60526 "X"	32
165	24/24 Standard	1.65:1	91-60526 "X" 91-60523 "Y"	32
	17/19 3500 - 6000'	1.84:1	91-60526 "Y"	
	20/24 Above 6000 ft.	1.98:1	91-60526 "Y"	

UPPER DRIVESHAFT GEAR RATIO
Number of Teeth — Pinion Gear/Driven Gear

MerCruiser Model	Driveshaft Housing Teeth No. Pinion Gear/ Driven Gear	Overall Drive Gear Ratio	Driven Gear Shimming Tool No. and Face	Gear Oil Capacity Ounces
470/470R/ 170MR/485/ 488R/190MR/ 185R/185MR	17/19 Standard	1.84:1	91-60526 "Y"	32
	24/24 Optional	1.65:1	91-60526 "X" 91-60523 "Y"	
	20/24 3500 - 6000	1.98:1	91-60526 "Y"	
888/898/ 225-S/228/ 228R/ 230MR	22/20 Standard	1.50:1	91-60526 "Z"	32
	24/24 Standard & 3500 - 6000'	1.65:1	91-60526 "X" 91-60523 "Y"	
	17/19 Above 600'	1.84:1	91-60526 "Y"	
200MR\ 205MR Alpha 1 Gen. 2	24/24 Standard	1.65:1	91-60526 "X" 91-60523 "Y"	32
	22/20 Optional	1.50:1	91-60526 "Z"	
	17/19 3500 - 6000'	1.84:1	91-60526 "Y"	
	20/24 Above 6000'	1.98:1	91-60526 "Y"	
233/250/260/ 260R/260MR/ 300MR Alpha 1	20/16	1.32:1	91-60526 "Z"	32
	22/20	1.50:1	91-60526 "Z"	
	24/24 3500 - 6000'	1.65:1	91-60526 "X" 91-60523 "Y"	
	17/19 3500 - 6000'	1.84:1	91-60526 "Y"	
Bravo I	23/30	1.65:1	91-42840	32
	27/32	1.50:1		
	27/29	1.36:1		
Bravo II	23/30	2.20:1	91-96512	32
	27/32	2.00:1		
	27/29	1.81:1		

NOTE 1 If the serial number is not legible and cannot be determined, use the following method of inspection to determine the gears for the 120/140 drive unit.
Units with serial number **2791956** and below, the driven gear is attached to the upper driveshaft with a nut.
Units with serial number **2791957** and above, the driven gear is secured to the driveshaft with a press fit.

LOWER UNIT BACKLASH SPECIFICATIONS

Prior to Alpha/MR Units (Mid 1985)

Number of Teeth — Pinion Gear/Driven Gear

	14/28	19/32	20/33	17/28
Stern Drives Using these Gears	Mercruiser I	Mercruiser 1A, 1B, 1C (Non EZ and EZ), 120, 140, 160, 165	120, 140, 165, 470, 485, 888 898, 225-S 228, 233, 250, 260	120, 140, 165, 470, 485, 888 898,225-S 228, 233, 250, 260
Backlash Forward Gear	0.006 - 0.008" 0.15 - 0.20mm	0.006 - 0.008" 0.15 - 0.20mm	0.010 - 0.012" 0.25 - 0.30mm	0.020 - 0.023" 0.51 - 0.58mm
Backlash Reverse Gear	0.015 - 0.020" 0.38 - 0.51mm	0.015 - 0.020" 0.38 - 0.51mm	0.040 - 0.060" 1.02 - 1.52mm Note 1	0.040 - 0.060" 1.02 - 1.52mm Note 1

NOTE 1   On stern drive models 120, 140, and 165, Serial No. 3780849 and below, and on stern drive model 888, Serial No. 3784374 and below having the 17/28 or the 20/33 gear set, the reverse gear backlash **MUST** be set at 0.025 - 0.030", 0.64 - 0.76mm. These stern drive units have a short clutch travel and will not function properly with the higher reverse gear backlash listed. To determine if the drive unit being serviced has a short clutch travel, assuming the Serial Number is not legible, remove the propeller shaft and check shift shaft rotation as follows: if the shift shaft can be rotated only about 45° (1/8 of a circle), the stern drive has the short clutch travel. If the shift shaft can be rotated about 270° (3/4 of a full circle), the stern drive has the long clutch travel.

LOWER UNIT BACKLASH SPECIFICATIONS

MR, Alpha, Alpha 1 - Generation 2, Bravo 1 and Bravo 2 (Mid 1985 and On)

Special Note
The following backlash specifications are not dependent on gear ratios. An **UPWARD** force **MUST** be exerted on the driveshaft while checking forward and reverse gear backlash.

MR and Alpha 1   Forward Gear Backlash -- 0.017 - 0.028" (0.43 - 0.71mm)
Reverse Gear Backlash -- 0.028 - 0.052" (0.71 - 1.32mm)

Alpha 1 Generation 2 Forward Gear Backlash -- 0.017 - 0.028" (0.43 - 0.71mm)
Reverse Gear Backlash -- 0.040 - 0.060" (1.02 - 1.52mm)

Bravo 1 --   One gear only -- 0.012 - 0.015" (0.30 - 0.38mm)

Bravo 2 --   One gear only -- 0.009 - 0.015" (0.23 - 0.38mm)

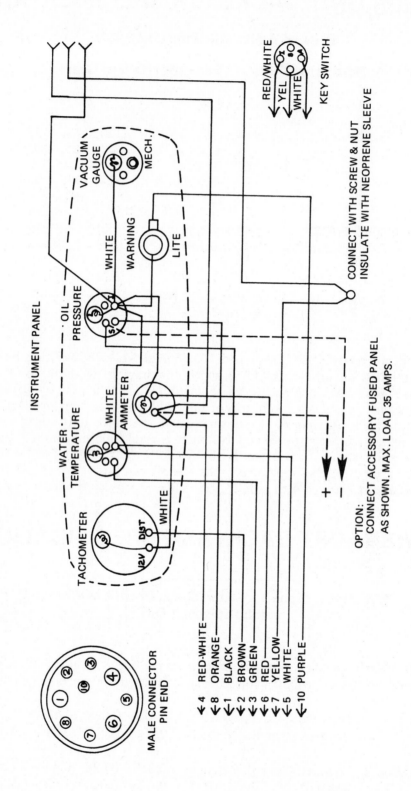

Wire identification and routing -- instrument panel.

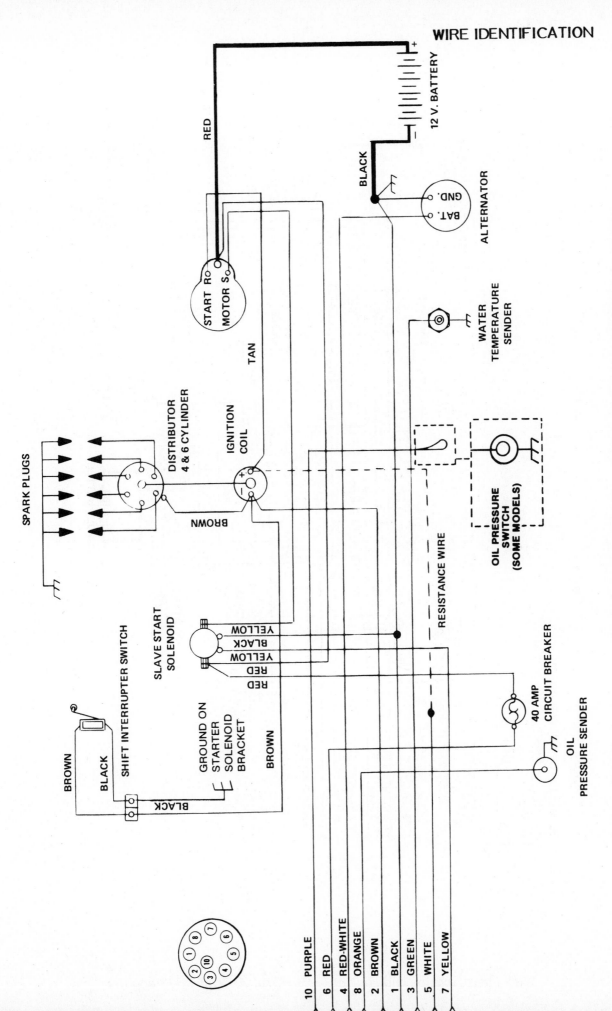

Engine wire identification and routing — Model 120, serial numbers 3770650 and higher; Model 140, serial numbers 3771645 and higher; and Model 165, serial numbers 3774865 and higher.

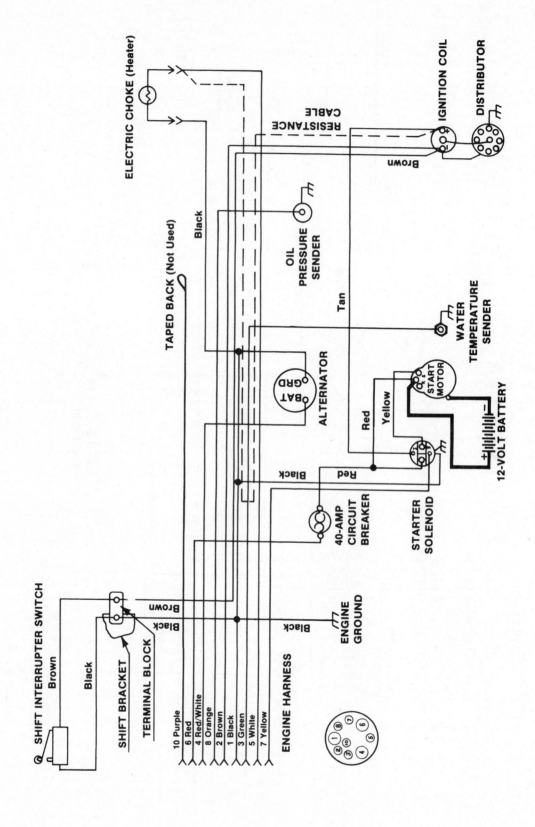

Wire identification and routing -- MerCruiser Model 898 engine.

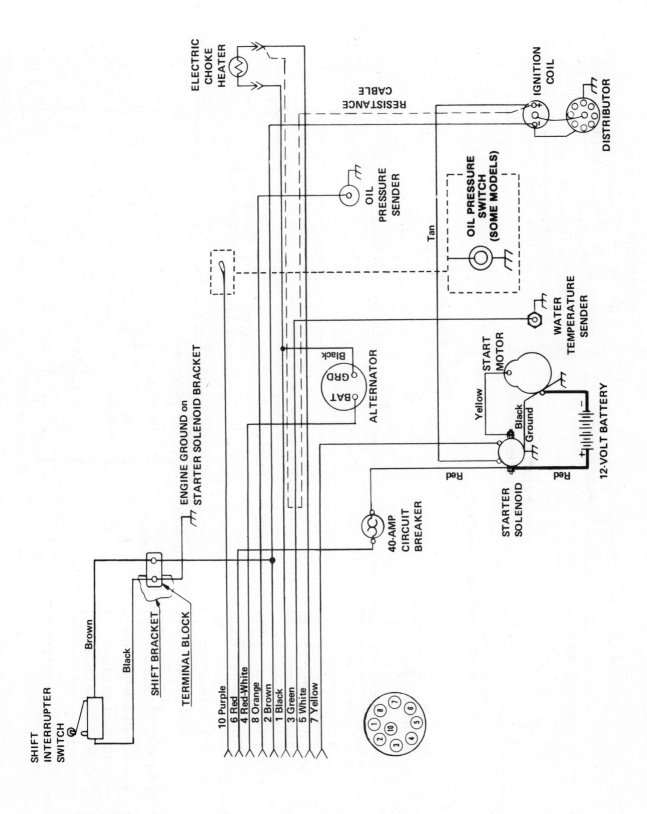

Wire identification and routing — MerCruiser Model 888 engine, serial numbers 3777490 and higher; Model 225-S, serial numbers 3836688 and higher; and all Model 233 engines.

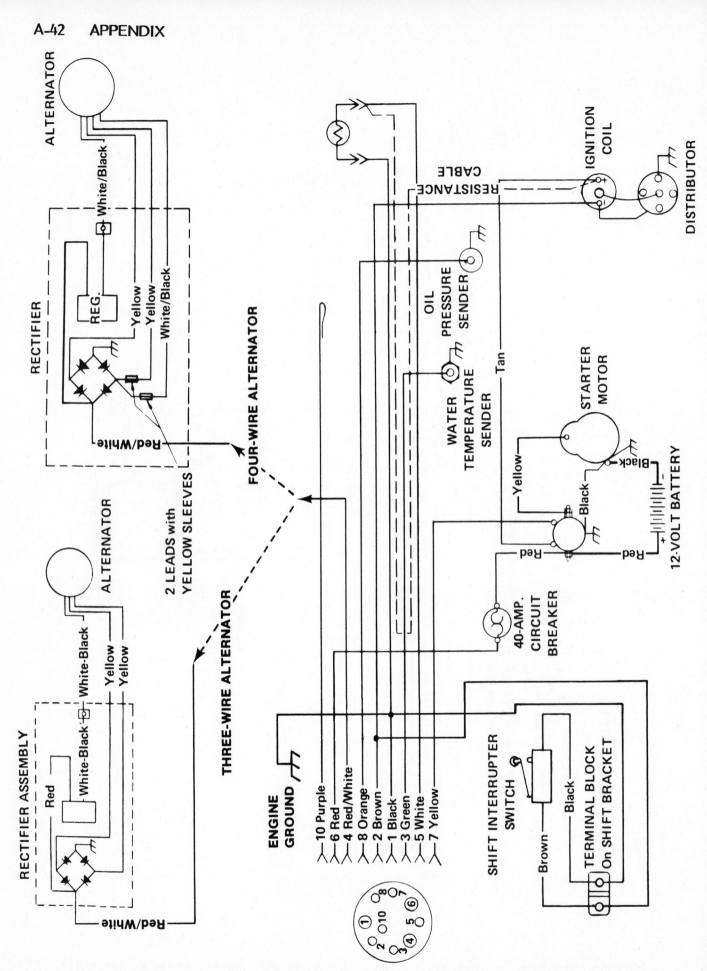

Wire identification and routing — Model 470 and 485 engines with air-cooled voltage regulator.

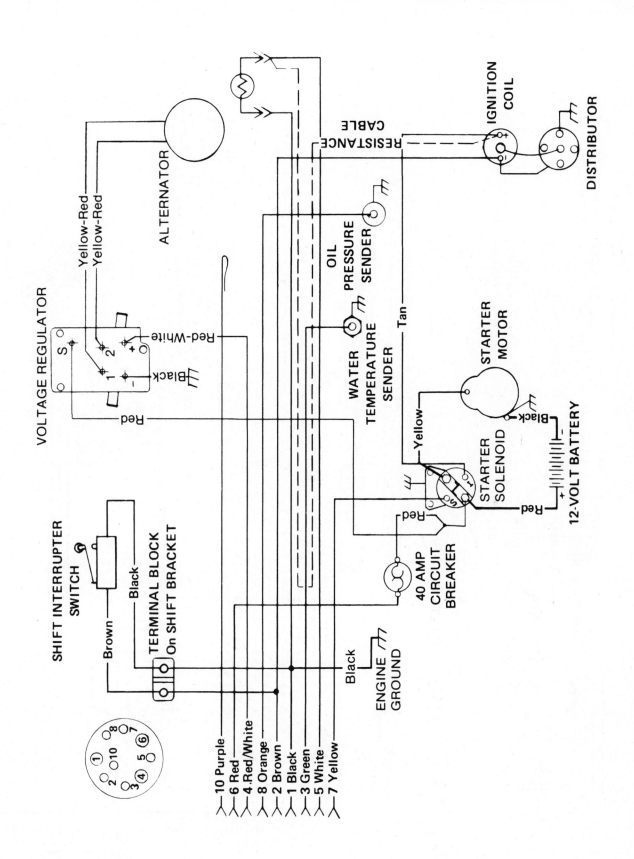

Wire identification and routing -- Model 470 and 485 engines with water-cooled voltage regulator.

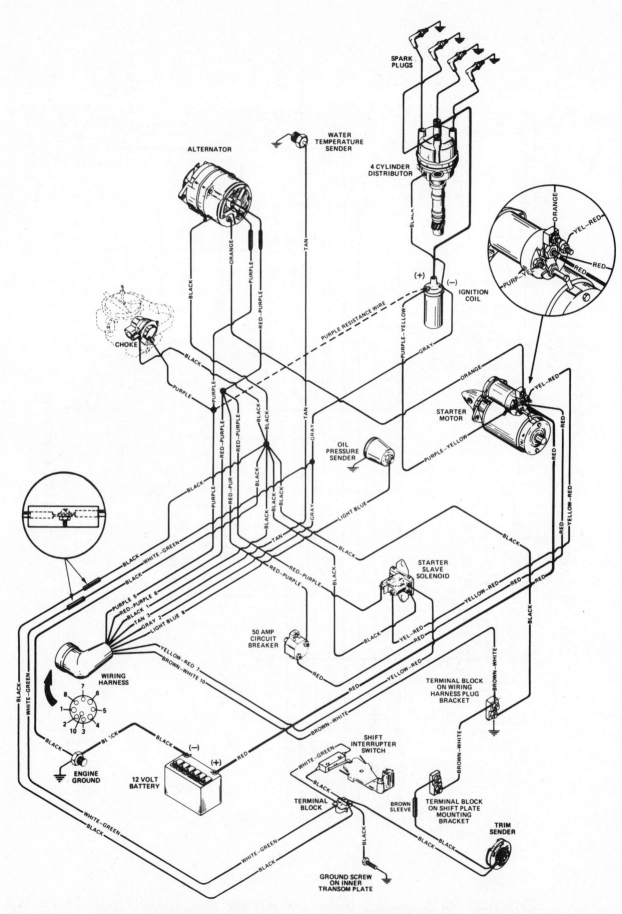

Wire identification and routing -- Model 120 and 140 engines.

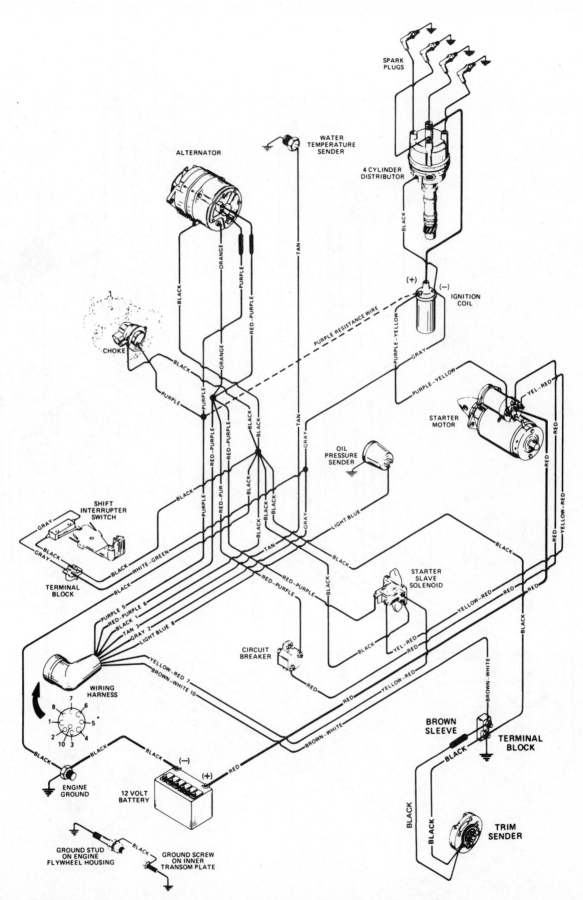

Wire identification and routing -- Model MCM 120R and 140R, also 3.0 Litre engines WITHOUT fuse.

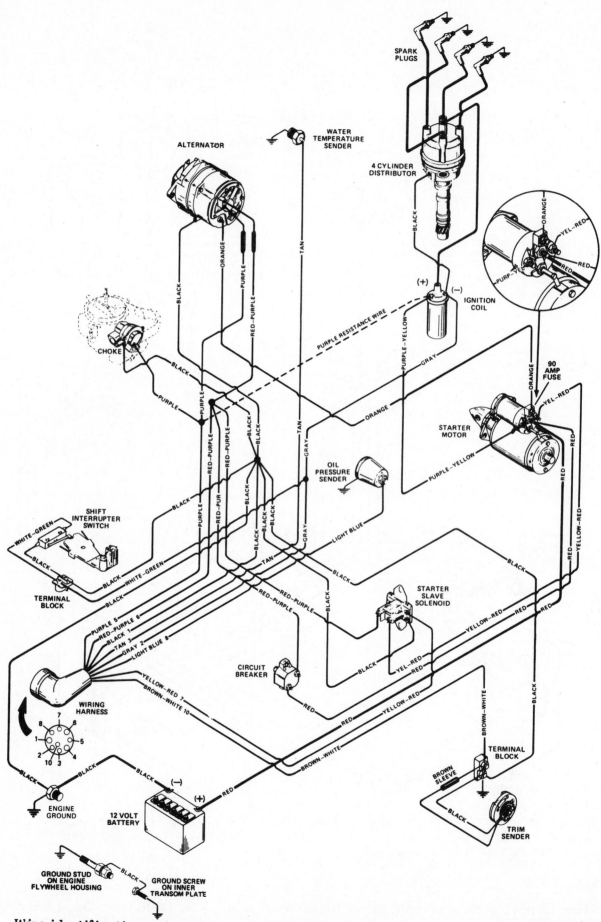

SPARK PLUGS

WATER TEMPERATURE SENDER

4 CYLINDER DISTRIBUTOR

ALTERNATOR

IGNITION COIL

90 AMP FUSE

STARTER MOTOR

CHOKE

OIL PRESSURE SENDER

SHIFT INTERRUPTER SWITCH

STARTER SLAVE SOLENOID

TERMINAL BLOCK

PURPLE 5
RED-PURPLE 6
BLACK 1
TAN 3
LIGHT BLUE 8

WIRING HARNESS

CIRCUIT BREAKER

YELLOW-RED 7
BROWN-WHITE 10

TERMINAL BLOCK

BROWN SLEEVE

ENGINE GROUND

12 VOLT BATTERY

TRIM SENDER

GROUND STUD ON ENGINE FLYWHEEL HOUSING

GROUND SCREW ON INNER TRANSOM PLATE

Wire identification and routing -- Model 120R and 140R, also 3.0 Litre engine WITH fuse.

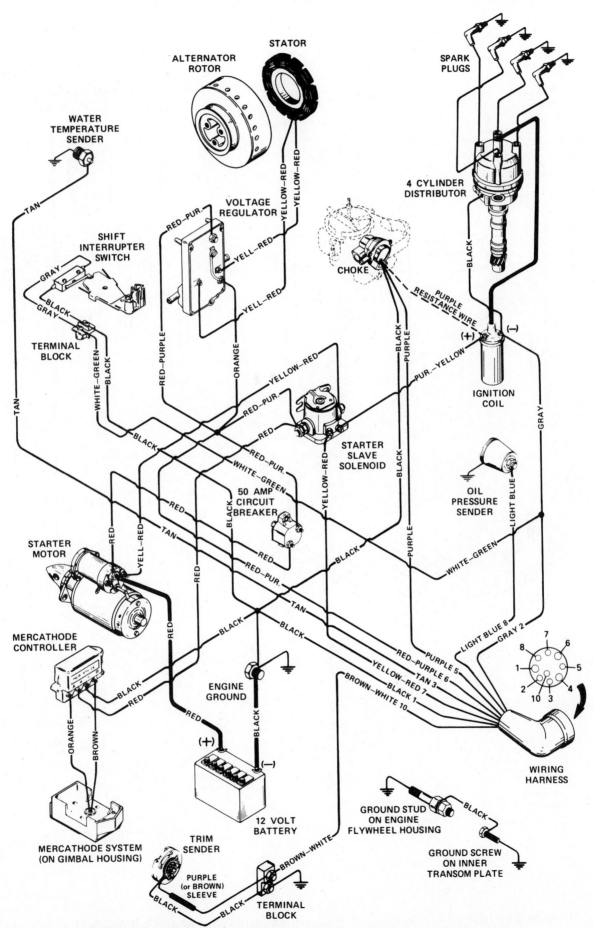

Wire identification and routing — Model MCM 470 and 488 engines.

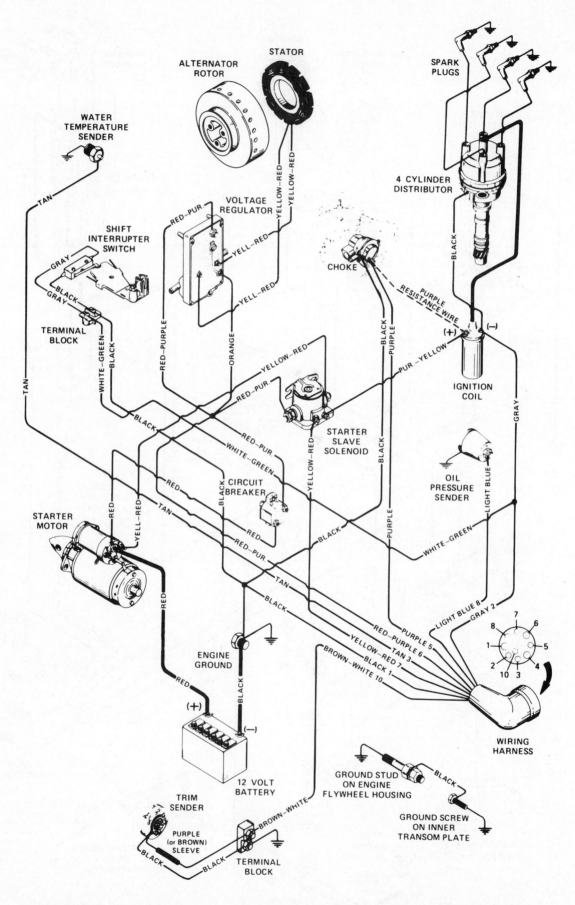

*Wire identification and routing -- Model MCM 470R and 488R **WITHOUT** fuse.*

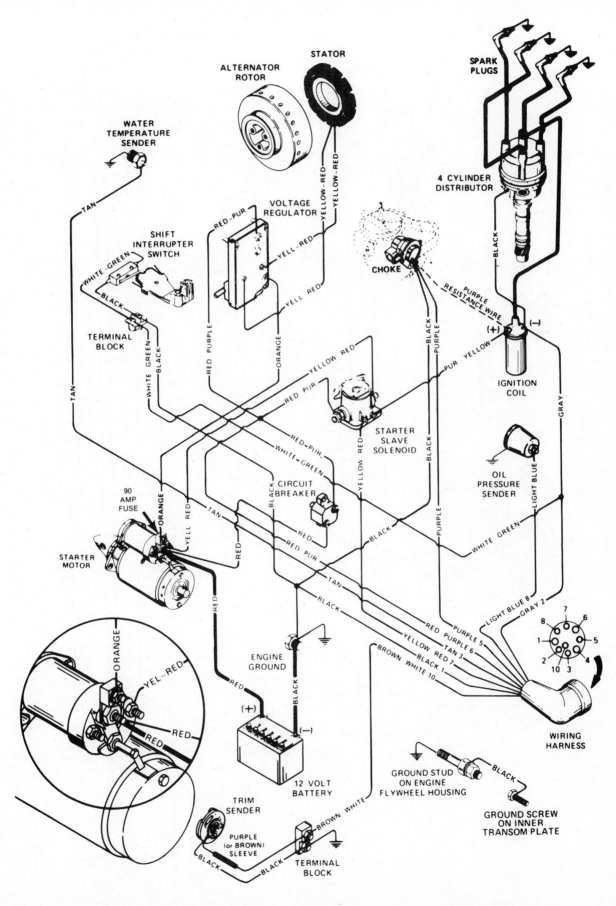

*Wire identification and routing -- Model MCM 470R and 488R **WITH** fuse.*

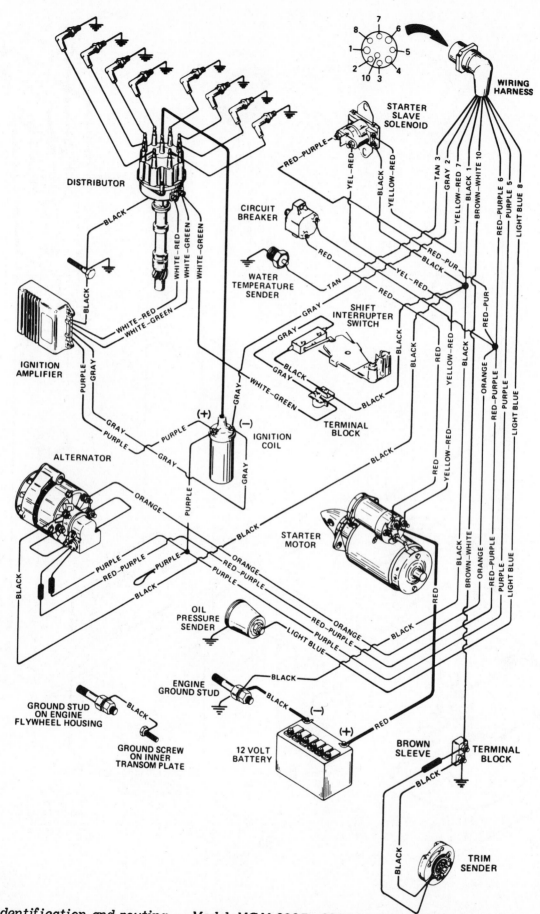

Wire identification and routing — Model MCM 898R, 228R, 260R, 230, also 5.0 Litre and 5.7 Litre engines.

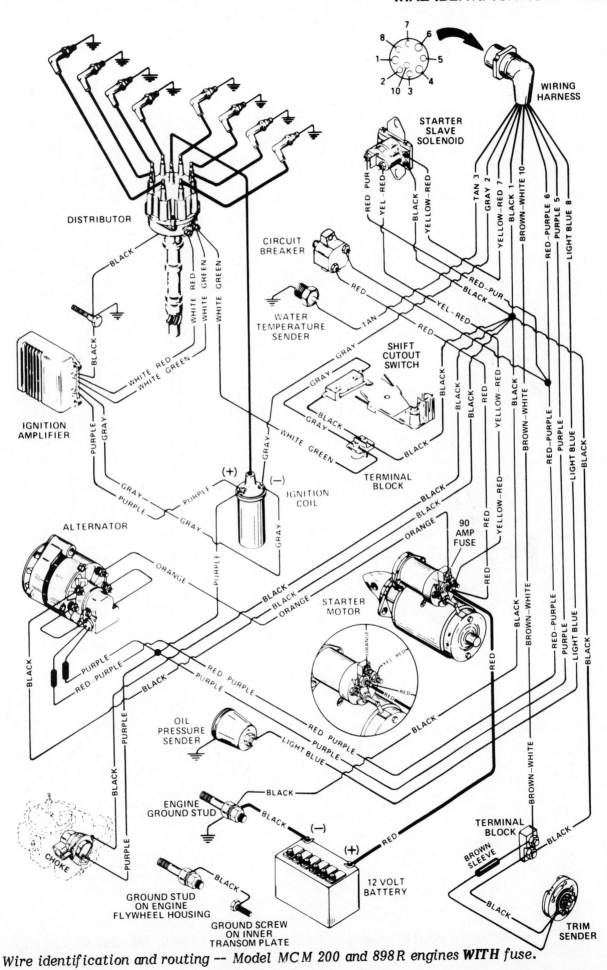

*Wire identification and routing -- Model MCM 200 and 898R engines **WITH** fuse.*

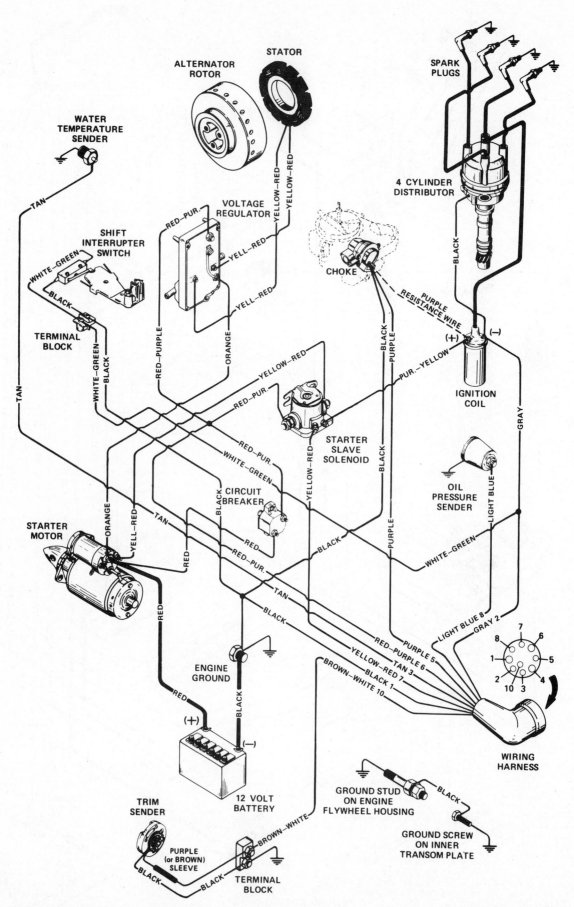

Wire identification and routing — Model MCM 165, 170MR, 180, and 190MR engines.

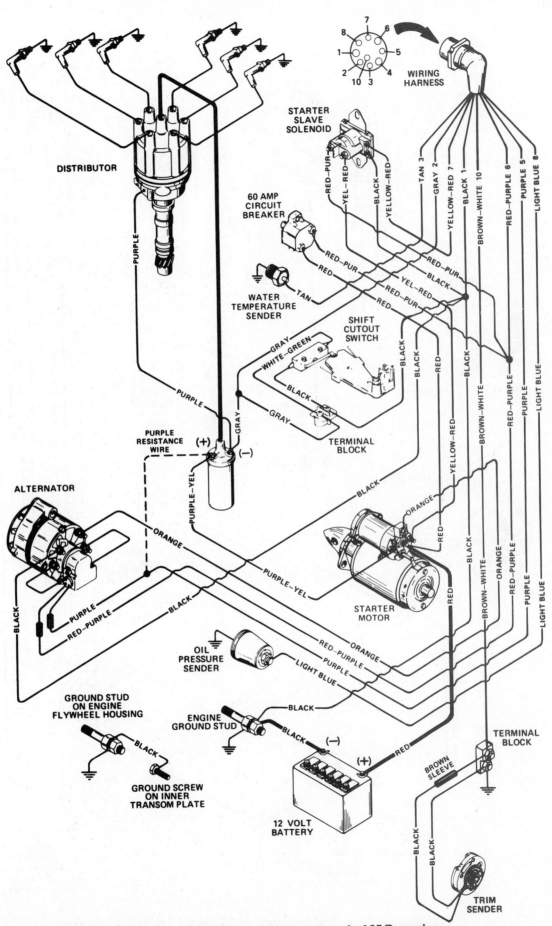

Wire identification and routing -- V6 Model MCM 185 and 185R engines.

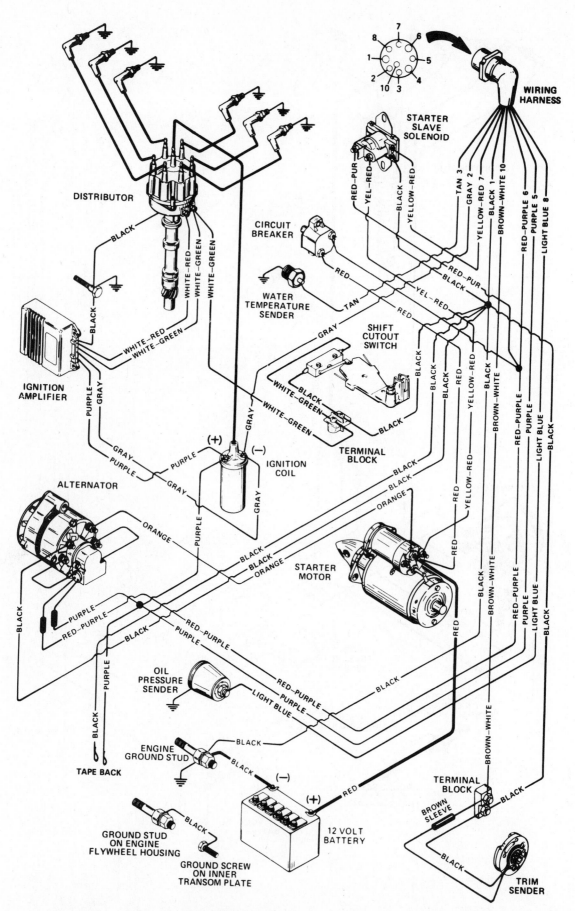

Wire identification and routing -- V6 Model MCM 205 engine.

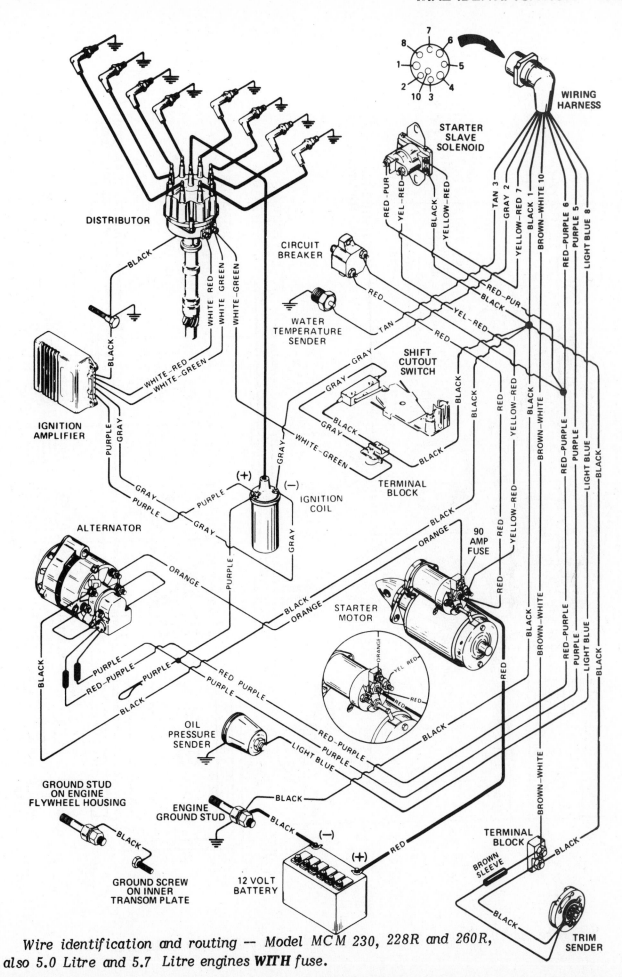

Wire identification and routing -- Model MCM 230, 228R and 260R, also 5.0 Litre and 5.7 Litre engines WITH fuse.

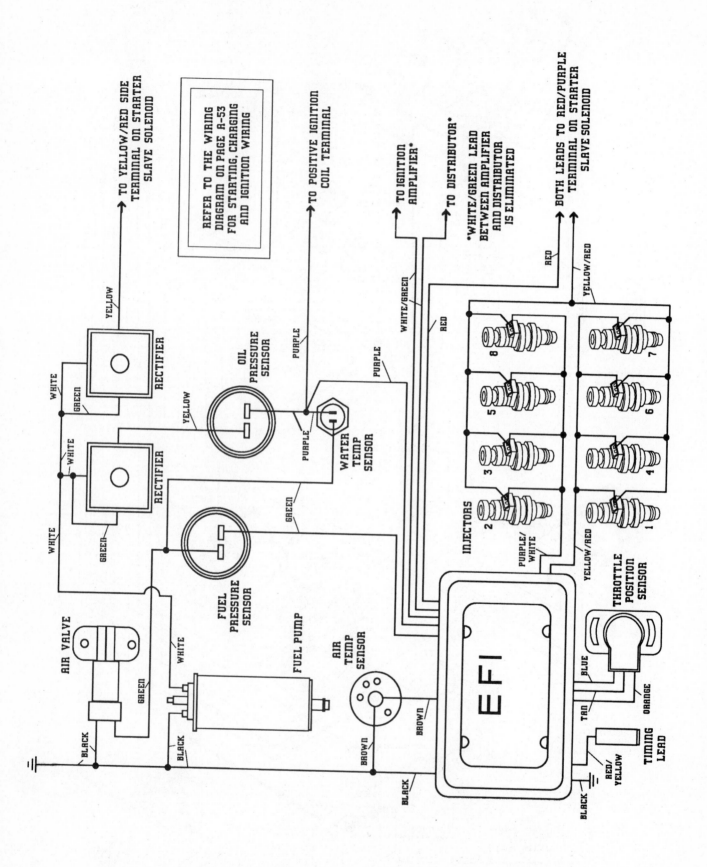

Wire identification and routing — Model MCM 320 engine **WITH** *EFI.*

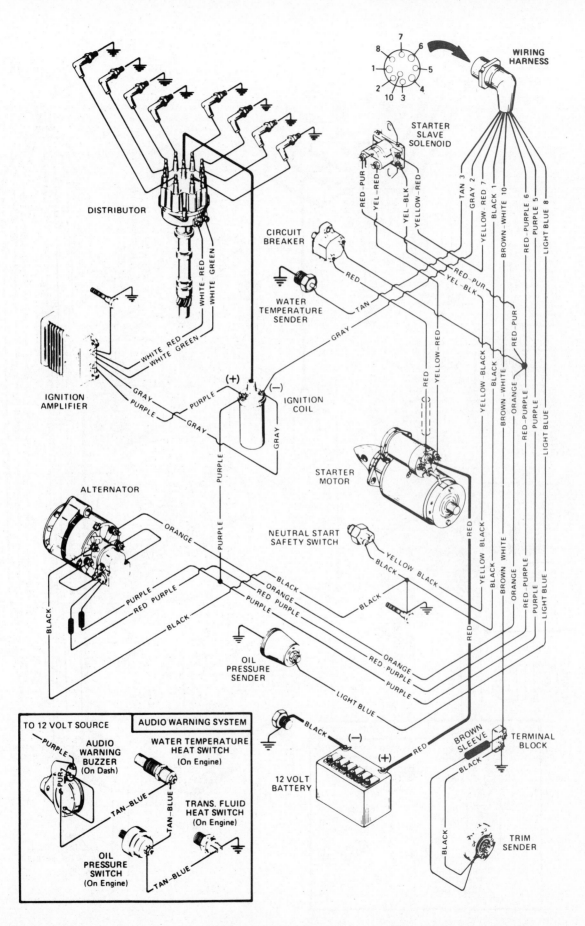

Wire identification and routing -- Model MCM 330 and 7.4 Litre engines WITHOUT fuse.

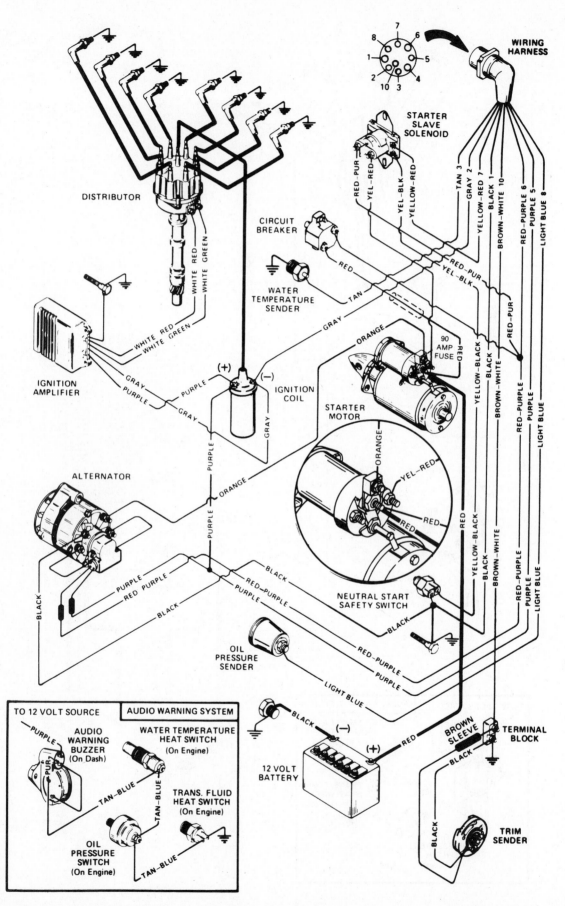

Wire identification and routing -- Model MCM 330 and 7.4 Litre engines WITH fuse.

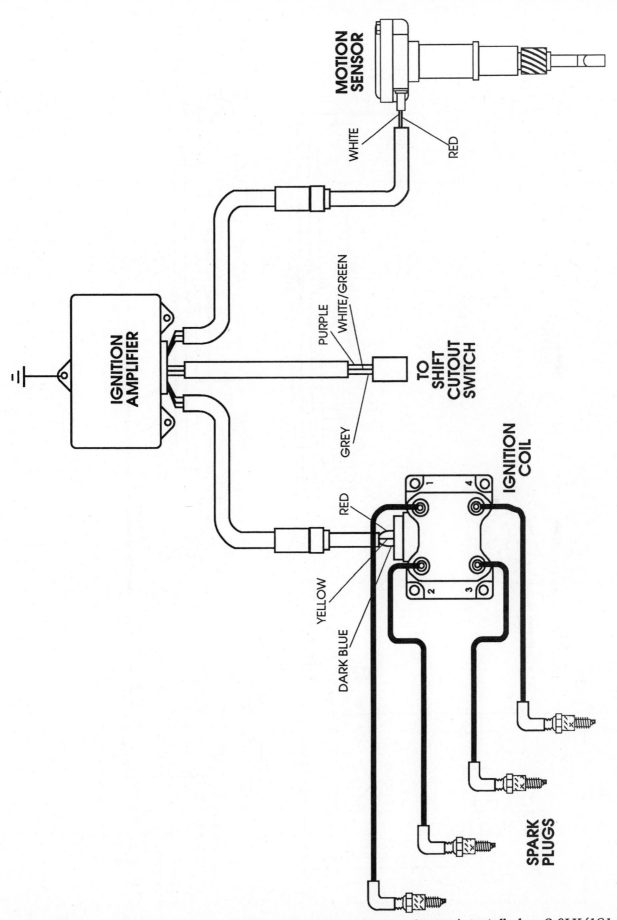

Wiring diagram of a typical DDIS (Digital Distributorless Ignition System), installed on 3.0LX (181 CID) engines -- 1990.

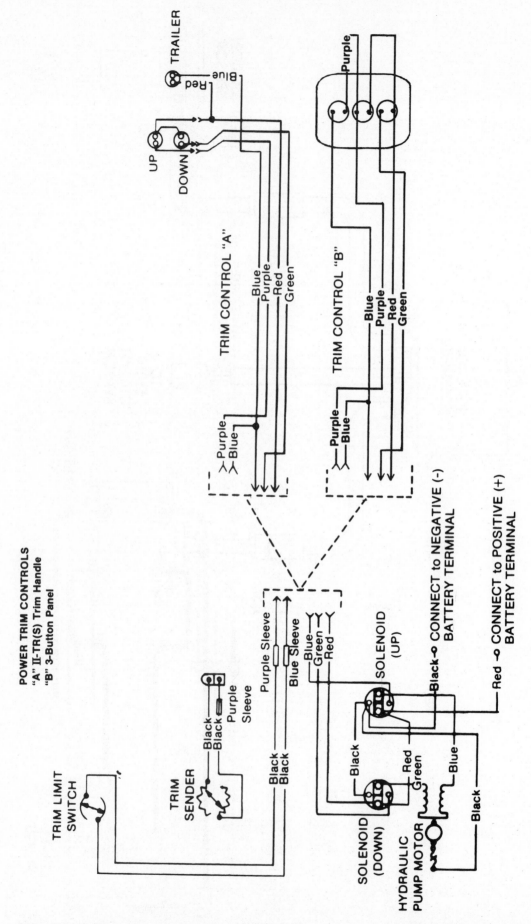

Wire identification and routing -- power trim circuit with double solenoid system.

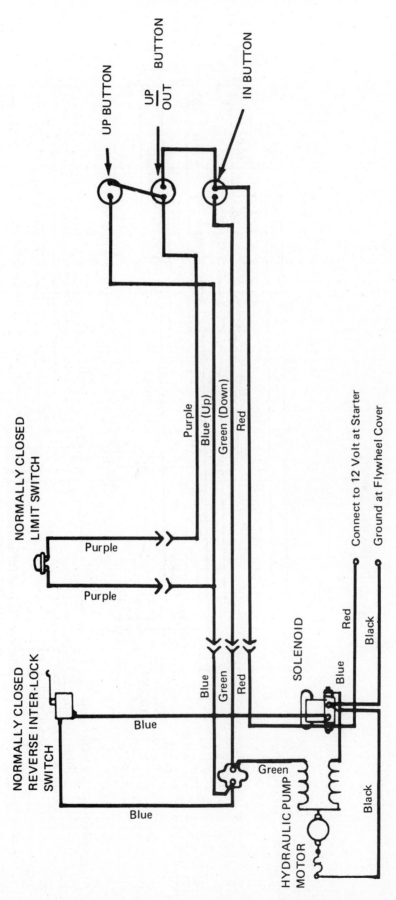

Wire identification and routing -- power trim circuit with late push button and with ONE solenoid.

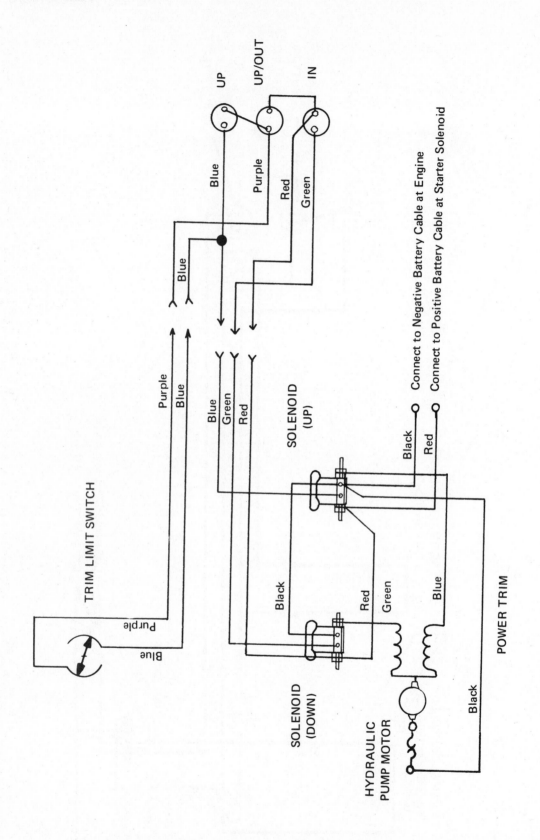

*Wire identification and routing -- power trim with late push button and with **TWO** solenoids.*

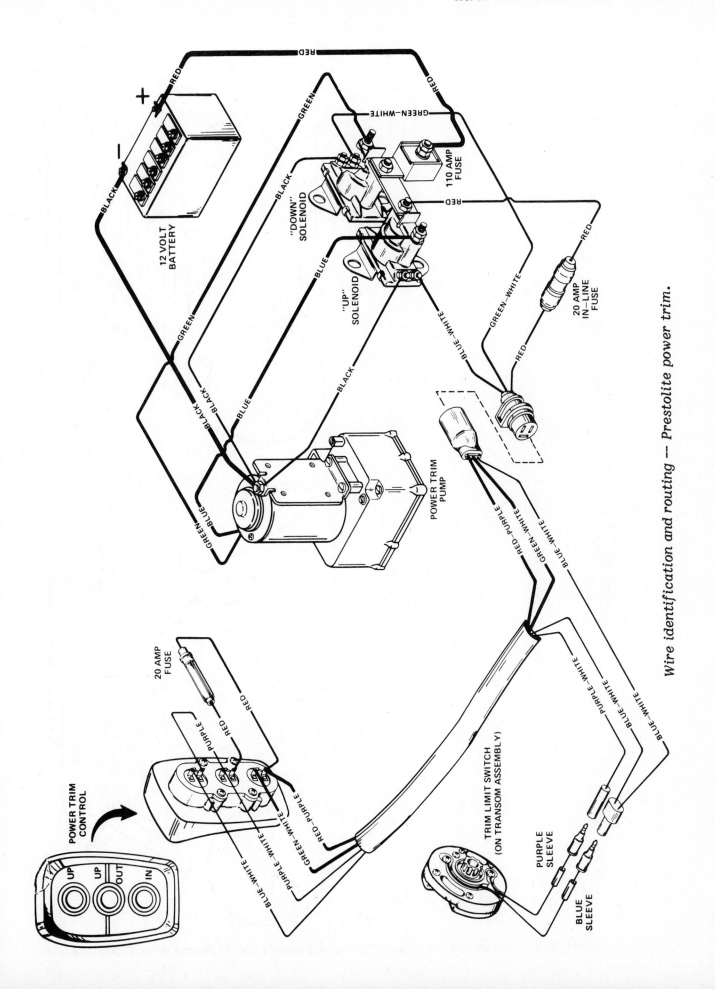

Wire identification and routing — Prestolite power trim.

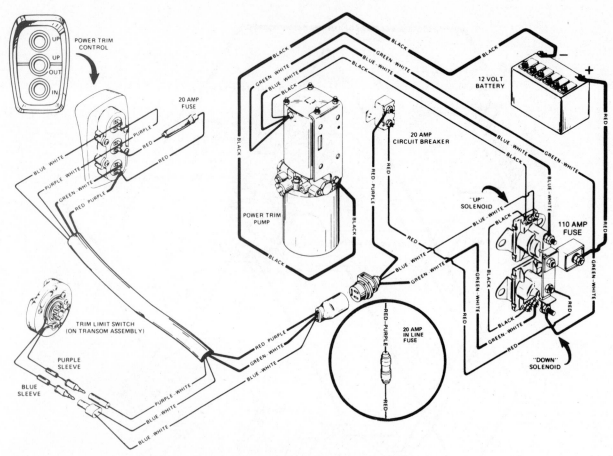

Three-button Panel Trim Control

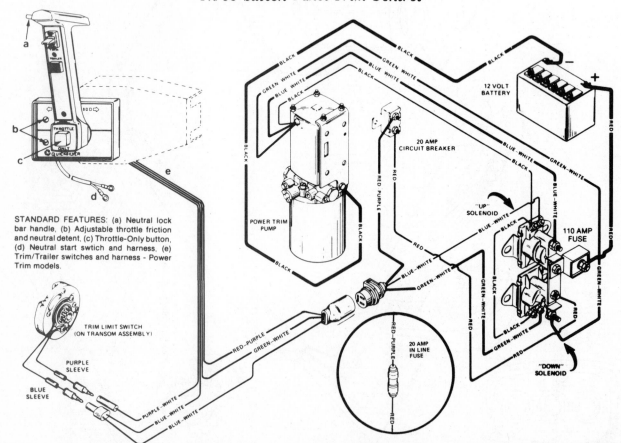

STANDARD FEATURES: (a) Neutral lock bar handle, (b) Adjustable throttle friction and neutral detent, (c) Throttle-Only button, (d) Neutral start swtich and harness, (e) Trim/Trailer switches and harness - Power Trim models.

Wire identification and routing -- Oil Dyne power trim -- Series IV -- Since 1983.

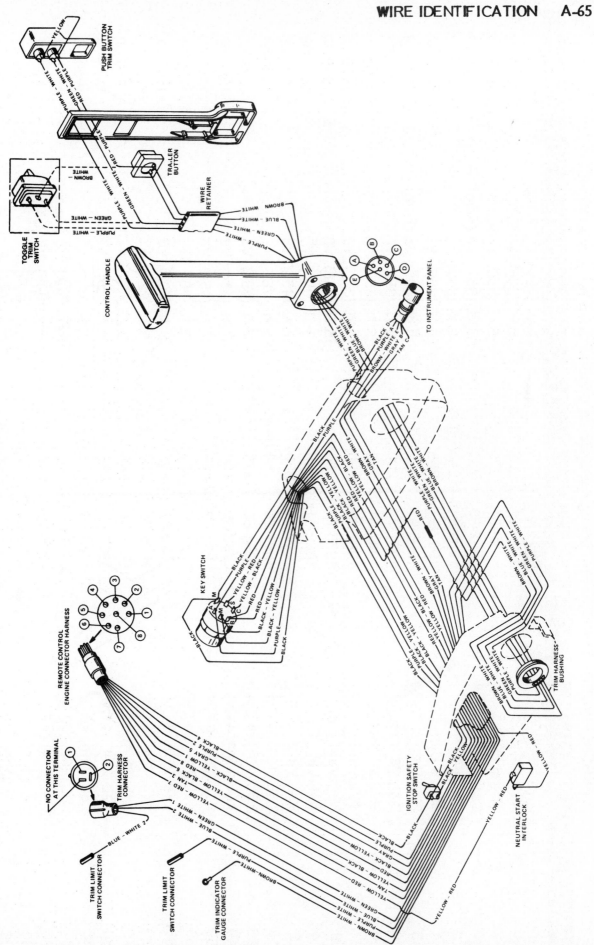

*Wire identification and routing -- "Commander" side mount controls **WITH** power trim/tilt system.*

ENGINE FINDER - The following listings contain all engines covered in this manual

Model/L	Engine/Cylinder	Years	Model/L	Engine/Cylinder	Years
2.5	151, 4 Cyl	1983-89	225	5.0L (302), V8 (Ford)	1973-74
3.0	181, 4 Cyl	1982-91	228	5.0L (305), V8 (GM)	1977-84
3.2	194, 6 Cyl	1964	230	5.0L (305), V8 (GM)	1985-91
3.7	224, 4 Cyl	1987-89	233	5.8L (351), V8 (Ford)	1975-77
3.8	229, V6	1983-84	250	5.7L (350), V8	1977
4.1	250, 6 Cyl	1967-91	255	5.7L (350), V8	1975-76
4.3	262, V6	1985-91	255	5.8L (351), V8 (Ford)	1973-74
5.0	302, V8 (Ford)	1971-74	260	5.7L (350), V8	1978-91
5.0	305, V8 (GM)	1987-91	270	5.7L (350), V8	1969-72
5.7	350, V8	1987-91	280	5.7L (350), V8	1975-76
5.8	351, V8 (Ford)	1973-77	300	5.7L (350), V8	1985-86
7.4	454, V8	1982-91	320	5.7L (350), V8	1987-89
8.2	502, V8	1990-91	330	7.4L (454), V8	1982-87
110	2.5L (151), 4 Cyl	1964-66	350 Mag	5.7L, V8	1986-91
120	2.5L (151), 4 Cyl	1964-89	370	7.4L (454), V8	1978-87
140	3.0L (181), 4 Cyl	1968-89	400	7.4L (454), V8	1980-86
160	4.1L (250), 6 Cyl	1967-69	420	7.4L (454), V8	1987-89
165	3.7L (224), 4 Cyl	1987-89	454 Mag	7.4L, V8	1986-91
165	4.1L (250), 6 Cyl	1970-81	465	8.2L (502), V8	1990-91
170	3.7L (224), 4 Cyl	1985-86	470	3.7L (224), 4 Cyl	1976-84
175	4.3L (262), V6	1987-91	485	3.7L (224), 4 Cyl	1980-82
180	3.7L (224), 4 Cyl	1987-89	488	3.7L (224), 4 Cyl	1983-84
185	3.8L (229), V6	1983-84	502 Mag	8.2L, V8	1990-91
185	4.3L (262), V6	1985-87	888	5.0L (302), V8 (Ford)	1971-77
190	3.7L (224), 4 Cyl	1985-86	898	5.0L (305), V8 (GM)	1977-84
200	5.0L (305), V8 (GM)	1985-91	Alpha	Sterndrive	1985-91
205	4.3L (262), V6	1985-91	Bravo I/II	Sterndrive	1988-91
215	5.0L (302), V8 (Ford)	1970-72	MR	Sterndrive	1985-91
			Type 1	Sterndrive	1964-84